T0335609

Basic Principles of Concrete Structures

Xianglin Gu · Xianyu Jin
Yong Zhou

Basic Principles of Concrete Structures

Xianglin Gu
Tongji University
Shanghai
China

Yong Zhou
Tongji University
Shanghai
China

Xianyu Jin
Zhejiang University
Hangzhou
China

ISBN 978-3-662-48563-7 ISBN 978-3-662-48565-1 (eBook)
DOI 10.1007/978-3-662-48565-1

Jointly published with Tongji University Press, Shanghai, China

Library of Congress Control Number: 2015951754

Springer Heidelberg New York Dordrecht London

Translation from the Chinese language edition: 混凝土结构基本原理, Third edition

Printed on acid-free paper

Springer-Verlag GmbH Berlin Heidelberg is part of Springer Science+Business Media
(www.springer.com)

To engineers who, rather than blindly following the codes of practice, seek to apply the laws of nature.

T.Y. Lin, 1955

Preface

More than ten years ago, when the first version of the textbook *Basic Principles of Concrete Structures* was published in Chinese, I started considering writing a textbook with same contents in English. The primary motivation at the time for this idea was to help Chinese undergraduate students with a language environment so that they could keep learning English continuously, while those foreign undergraduate students in China could be able to study this specialty course more easily when they have a textbook in English. However, this idea turned out to be a tough work for me. As different countries have different codes in civil engineering, those existing textbooks in English from other countries cannot be used directly for students in China. It only made sense that I wrote a new textbook in English to not only introducc basic principles of concrete structures and existing Chinese codes, but also reflect the most recent research results and practical experiences in concrete structures in China. Now I am very happy to see that my dream has come true and the textbook is to be published pretty soon. I hope that this textbook can also serve as a window for those outside China who are interested in China and its developments in civil engineering.

As a basic specialty course for undergraduates majoring in civil engineering, Basic Principles of Concrete Structures is different from either the previously learnt mechanics courses or the designing courses to be learnt. Compared with mechanics courses, quite a number of basic theories of reinforced concrete structures cannot be derived solely by theoretical analysis. And compared with designing courses, this course emphasizes on introductions of basic theories rather than simply being a translation of designing specifications. That means the course of Basic Principles of Concrete Structures should focus on both theoretical derivations and engineering practices. Therefore, based on the latest version of designing codes both for buildings and bridges (GB 50010-2010 and JTG D62-2004), the textbook starts from steel and concrete materials, whose properties are very important to mechanical behavior of reinforced concrete structural members. Step by step, analyses of reinforced concrete members under basic loading types (tension, compression, bending, shearing, and torsion) and environmental actions are introduced. One of the characteristics of the book that I was trying to distinguish it from other textbooks on concrete structures is that more emphasis has been laid on basic theories on concrete structures as well as on applications of the basic theories both in designing new structures and in analyzing existing structures. Examples and problems in each chapter are carefully designed to cover every key knowledge point and practical case.

Professor Xianyu Jin prepared the draft for Chaps. 1, 4, and 5. I prepared the draft for Chaps. 2, 3, and 6–9. Prof. Xiaozu Su prepared the draft for Chap. 10. Dr. Feng Lin prepared the draft for Chap. 11. Prof. Weiping Zhang prepared the draft for Chap. 12. Dr. Qianqian Yu and Ph.D. candidate, Mr. Chao Jiang, drew all of the figures. Dr. Xiaobin Song helped me with the first revision of the draft, while Dr. Yong Zhou helped with the second revision, based on which I finished modifying the final draft. Without help from my colleagues and students, it would be very difficult for me to finish this tough work. I would like to deliver my sincere thanks to all of them. I would also like to thank Mr. Yi Hu from Tongji University Press for his support on the publication of this book. Finally, I would like to thank Publication Foundation of Books at Tongji University for their financial support.

May 2015 Xianglin Gu

Contents

Abstract

Basic Principles of Concrete Structures is one of the key courses for undergraduates majoring in civil engineering. The objective of this book is to help students to completely understand the basic mechanical properties and design methods of structural members made of concrete and reinforcement and to lay the foundation for future study of the design and construction of various types of reinforced concrete structures.

The book consists of 12 chapters, i.e., introduction, materials, bond and anchorage, axially loaded members, flexural members, eccentrically loaded members, shearing, torsion, punching and local bearing, prestressed concrete members, serviceability of members, and durability of reinforced concrete structures.

The book is suitable for teachers and college students majoring in civil engineering and can also be referred by civil engineers.

The textbook is jointly edited by Prof. Xianglin Gu, Prof. Xianyu Jin, and Dr. Yong Zhou.

As a basic specialty course for undergraduates majoring in civil engineering, *Basic Principles of Concrete Structures* is different from either the previously learnt mechanics courses or the design courses to be learnt. Compared with mechanics courses, the basic theories of reinforced concrete structures cannot be solely derived by theoretical analysis. And compared with design courses, this course emphasizes the introduction of basic theories rather than simply being a translation of design specifications. That means the course of *Basic principles of Concrete Structures* should focus on both theoretical derivation and engineering practice. Therefore, based on the latest version of *designing codes both for buildings and bridges* (GB 50010-2010 and JTG D62-2004), the book starts from the steel and concrete materials, whose properties are very important to the mechanical behavior of reinforced concrete members. Step by step, the design and analysis of reinforced concrete members under basic loading types (tension, compression, flexure, shearing, and torsion) and environmental actions are introduced. The characteristic of the book that distinguishes it from other textbooks on reinforced concrete structures is that more emphasis has been laid on the basic theories of reinforced concrete structures and the application of the basic theories in design of new structures and analysis of existing structures. Examples and problems in each chapter are carefully designed to cover every important knowledge point.

Chapter 1
Introduction

1.1 General Concepts of Concrete Structures

This is a textbook for civil engineering students. Understanding the behaviors of concrete structures has always been a major part of civil engineering education. This chapter gives an overview of the most important concepts and theories which make up the basic principles of concrete structures. More importantly, this chapter presents concrete structures as a research field that is constantly expanding to meet the increasing demands for resistance to both natural and man-made disasters.

1.1.1 General Concepts of Reinforced Concrete Structures

Reinforced concrete structures are structures composed of concrete and steel bars. Concrete is strong in compression but weak in tension. Generally, the tensile strength of concrete is only 1/10 of its compressive strength. However, steel is strong in both tension and compression. So it is natural to combine the two materials together so that both of them can play their respective strengths. In other words, concrete mainly resists compression and steel bars sustain all of the tension.

Take beams as an example. For a plain concrete beam subjected to two concentrated loads of the same value P as shown in Fig. 1.1a, the bottom part of the mid-span cross section is subjected to tension while the top part is subjected to compression (Fig. 1.1b). When the external loads are increased to make the stress in the beam's bottom part exceed the tensile strength of concrete, concrete will crack. Furthermore, the beam will rupture after the crack appears, indicating that the load-carrying capacity of plain concrete is very low (just equal to the cracking load P_{cr}) and the failure is brittle (Fig. 1.1c). If an appropriate amount of steel is arranged in the tension zone of the beam, the steel will help the concrete to sustain the tension due to its good tension-resistant property after the concrete cracks

X. Gu et al., *Basic Principles of Concrete Structures*,
DOI 10.1007/978-3-662-48565-1_1

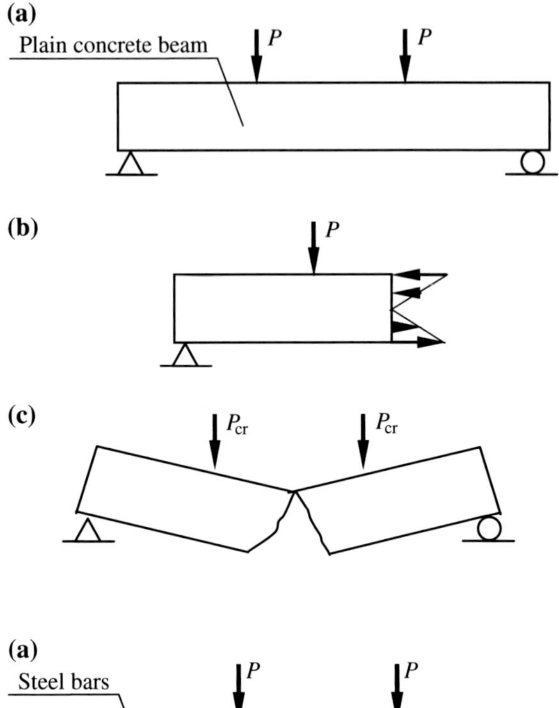

Fig. 1.1 Mechanical performance of a plain concrete beam. **a** A four-point bending plain concrete beam. **b** Stress distribution in a normal section. **c** Rupture of the plain concrete beam

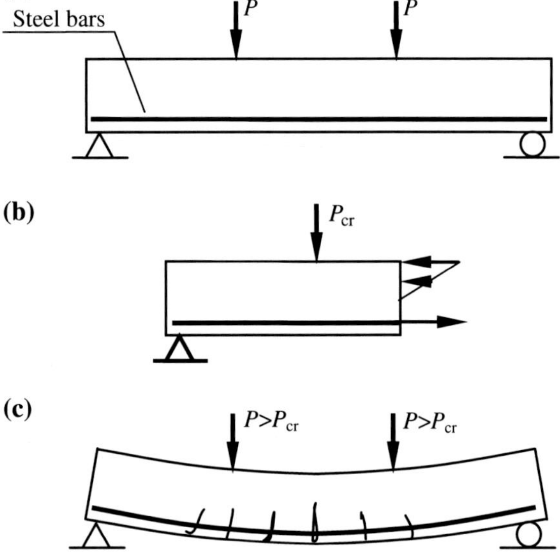

Fig. 1.2 Mechanical performance of a reinforced concrete beam. **a** A four-point bending reinforced concrete beam. **b** Stress distribution in a normal section. **c** Performance of the reinforced concrete beam after cracking

(Fig. 1.2a, b). Thus, the beam can continue to carry loads rather than rupture (Fig. 1.2c). The steel not only increases the load-carrying capacity of the beam, but also improves the beam's deformation ability, thereby presenting obvious warning before the final failure.

In the construction of concrete structures, the primary steps are to (1) fabricate forms according to the dimension of structural members, (2) position reinforcement inside the forms, (3) cast concrete into the forms, and (4) remove the forms after the concrete has hardened to a certain strength value and finish the construction.

1.1.2 Mechanism of Collaboration of Concrete and Steel

The reasons that the two different materials concrete and steel can work together are as follows:

1. Steel bars can be well bonded to concrete; thus, they can jointly resist external loads and deform together.
2. The thermal expansion coefficients of concrete and steel are so close (1.0×10^{-5} – 1.5×10^{-5} for concrete and 1.2×10^{-5} for steel) that the thermal-stress-induced damage to the bond between the two materials can be prevented.
3. Concrete can protect the embedded steel bars from corrosion and high-temperature-induced softening.

1.1.3 General Concepts of Prestressed Concrete Structures

Ducts must be set aside for prestressed tendons in a beam before casting concrete. Tendons can then be passed through the ducts when the strength of concrete has reached a certain value, after which the stretched tendons can be anchored at the beam ends (Fig. 1.3a). In this way, the stretched tendons will generate compressive stress and tensile stress in the concrete at the bottom and top parts of the beam, respectively (Fig. 1.3b). The compressive prestress introduced by the tendons will offset the tensile stress produced by external load P at the beam's bottom (Fig. 1.3c), resulting in no tensile stress or very small tensile stress in the original tension zone (Fig. 1.3d). The crack resistance of concrete will be improved. The beam shown in Fig. 1.3a is called the prestressed concrete beam. Similarly, one must stretch the tendons first, then cast the concrete, and finally release the tendons

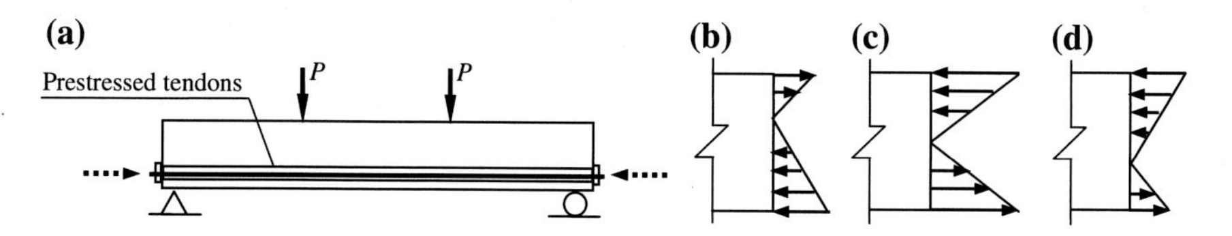

Fig. 1.3 Prestressed concrete beam and stress distribution in the mid-span normal section

after the concrete has reached a certain strength. The compressive prestress can also be established in the concrete through the bond between concrete and tendons.

1.1.4 Members of Concrete Structures

Concrete structures are structural systems composed of different concrete members, including slabs, beams, columns, walls, and foundations. Take a multistory building for example (Fig. 1.4). The main structural members are as follows:

1. Reinforced concrete slabs mainly sustain the self-weight of the slabs and the loads on the slabs' top surface.
2. Reinforced concrete stairs mainly sustain the self-weight of the flight of stairs and the loads on the stair surfaces.
3. Reinforced concrete beams mainly sustain the self-weight of the beams and the loads from the slabs.
4. Reinforced concrete columns mainly sustain the self-weight of the columns and the loads from the beams.

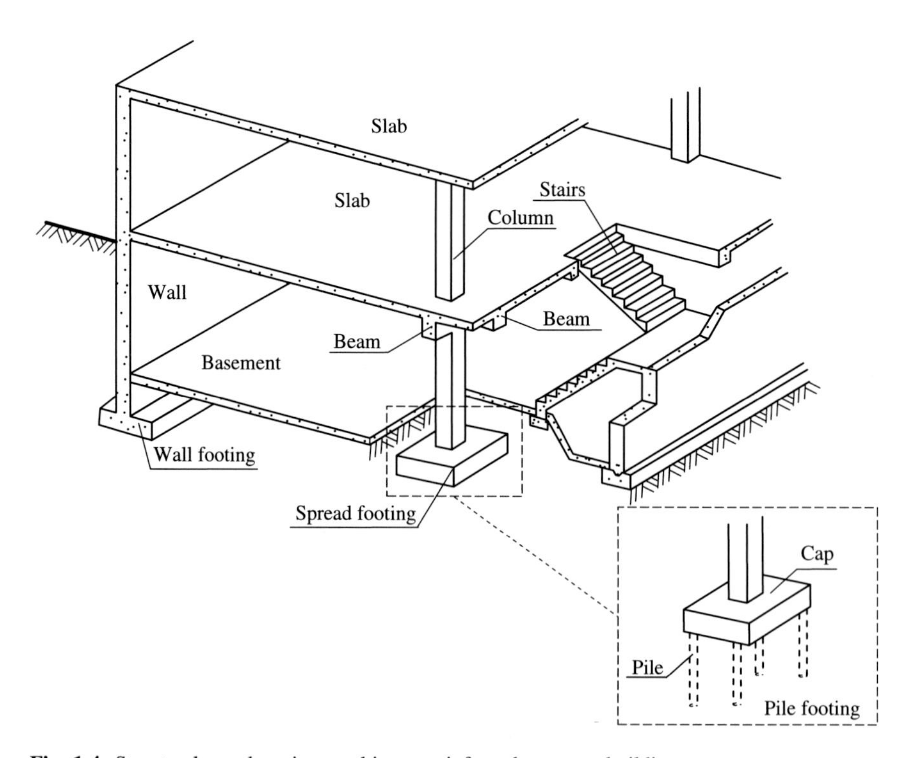

Fig. 1.4 Structural members in a multistory reinforced concrete building

5. Reinforced concrete walls mainly sustain the self-weight of the walls; the lateral earth pressure; and the loads from the slabs, beams, and columns.
6. Reinforced concrete wall footings mainly sustain and pass the loads from the walls to soil.
7. Reinforced concrete spread footings mainly sustain and pass the loads from the columns to soil.

1.1.5 Advantages and Disadvantages of Concrete Structures

1.1.5.1 Advantages

1. Durability
   Concrete strength increases with time. A thick enough concrete cover can protect the embedded steel from corrosion; thus, frequent maintenance and repair of concrete structures are generally unnecessary. Even in severe environment (e.g., exposed to aggressive gas or immersed in seawater), concrete members can generally fulfill their functions if they are rationally designed and specially detailed.
2. Fire resistance
   If no special technical measures are taken, concrete structures can endure 1–3 h of fire without steel-softening-induced structural failure. The fire resistancc of concrete structures is better than that of steel or wood structures.
3. Integrity
   The members of in situ cast concrete structures are firmly connected to effectively resist dynamic loads such as high winds, earthquakes, explosions, and other impacts.
4. Moldability
   Concrete structures can be cast according to designed requirements into various shapes, such as curved beams or arches, curved towers, and spatial shells.
5. Availability of materials
   The materials of concrete used in the largest amount, i.e., sand and gravel, are easy to purchase in local markets, and industry wastes (e.g., blast furnace slag, and fly ash) can be recycled and blended into the concrete during production as artificial aggregates.
6. Economy
   Compared with steel structures, concrete structures mainly use concrete rather than steel to resist compression, which not only gives play to both materials, but also saves a lot of steel.

1.1.5.2 Disadvantages

Concrete structures also have some weaknesses, which to some extent hinder a wider use of them. For example, concrete structures are not suitable for large-span structures, high-rise buildings and seismic-resistant structures because of their large self-weights. The densities of plain concrete and reinforced concrete are 22–24 and 24–25 kN/m^3, respectively. Concrete is prone to crack, and ordinary concrete structures are generally in service with cracks. Special technical measures are needed for concrete structural members when used to build structures with strict requirements for crack control in tight, secure structures, such as pools, underground structures, and containments in nuclear power plants. Cast-in-place concrete structures need a large number of forms. The construction of concrete structures is greatly influenced by the weather. Furthermore, the heat and sound insulations of concrete structures are poor. However, all of these disadvantages will be gradually overcome with the advance of science and technology.

1.2 Historical Development of Concrete Structures

1.2.1 Birth of Concrete Structures

In 1824, a British bricklayer, J. Aspdin, invented Portland cement, laying the foundation for the development of concrete structures. A Frenchman J. L. Lambot patented a rowboat of concrete reinforced with wires and exhibited it in the 1855 Paris World Exposition, symbolizing the birth of concrete structures. In the same year, F. Coigent also applied for a patent for reinforced concrete slabs. From then on, a large amount of reinforced concrete structures (or members) fabricated by experience emerged and were patented. A British textbook published in 1904 listed 43 patents on reinforced concrete, among which 15 were from France, 14 from Germany or the Austro-Hungarian Empire, 8 from the USA, 3 from the UK, and the remaining patents came from 9 other countries.

At the end of the nineteenth century, concrete was introduced into China. Take Shanghai as an example. In 1890, concrete was used for the first time in road paving and the first concrete factory began operation. The factory imported cement from the UK to produce kitchen sinks at the beginning and then expanded its products to concrete beams, slabs, piles, poles, etc. In 1891, Shanghai Municipal Council built the first plain concrete sewer under Wuchang Road. A reinforced concrete floor was used in Municipal City Hall, which was finished in 1896 but does not exist now. The Russo-Chinese Bank building (currently China Foreign Exchange Trading Center, located at East 1st Zhongshan Road 15) adopted steel reinforced concrete with steel columns and beams embedded in concrete. The first reinforced concrete frame structure was the Telephone Company building (currently Shanghai Urban

Telephone Bureau, located at the intersection of Central Jiangxi Road and Hankou Road) constructed in 1908.

Modern prestressed concrete structures were pioneered by a French scholar, E. Freyssinet. In 1928, he suggested using high-strength steel wires as prestressed tendons, invented dedicated anchorage systems, and did some groundbreaking work in the application of prestress techniques to bridges and other structures. His efforts finally brought prestressed concrete structures from the laboratory to real engineering applications.

After prestress is applied, concrete in the beam bottom (Fig. 1.3a) gradually compresses. Simultaneously, the beam also shortens due to the compression of concrete which is a long-term deformation property called *creep*. The shortened length is approximately 1/1000 of the beam span. For ordinary steel bars, the strain will generally not exceed 1.5/1000 during prestress application. Therefore, 2/3 of the prestress is eventually lost due to creep and shrinkage. On the other hand, the strain of high-strength steel bars can reach 7/1000. The prestress loss due to creep and shrinkage is only 1/7 that of applied prestress. Based on the prestress effect in concrete and steel, Freyssinet suggested using high-strength concrete and high-strength steel bars at the same time.

After World War II, prestress technology underwent rapid development, which was further boosted by the foundation of the Federation International de la Precontrainte (FIP). It was reported that as of 1951, 175 prestressed concrete bridges and 50 prestressed concrete frames were built in Europe and 700 prestressed concrete water tanks were constructed in North America.

Prestressed concrete has been used in China since the 1950s and was initially used in steel string prestressed concrete sleepers. Now it is widely used in buildings, bridges, underground structures and special structures (e.g., pools, cooling towers, communication towers, and containments), etc.

1.2.2 Development of Concrete Materials

The development of concrete materials since the birth of concrete has mainly focused on the continuous increase of concrete strength, ceaseless improvement of concrete properties and application of lightweight concrete, no-fines concrete, and fiber reinforced concrete, etc.

The average compressive strength of concrete in the USA was 28 N/mm^2 at the beginning of the 1960s and increased to 42 N/mm^2 in the 1970s. In 1964, using superplasticizer to prepare high-strength concrete originated in Japan and concrete with the strength of 80–90 N/mm^2 could be obtained in construction sites by the end of the 1970s. North American concrete manufacturers have made similar attempts since 1976. After 1990, 60–100 N/mm^2 high-strength concrete could be prepared at construction sites in the USA and Canada and the maximum compressive strength obtained was 120 N/mm^2. In the laboratory, the compressive strength of concrete could even reach 300 N/mm^2.

Before the 1990s, most concrete used in China had a compressive strength of 15–20 N/mm^2 only. With the development of economy and the advance of technology, high-strength concrete gets its use in engineering applications. In railway systems, concrete of 50–60 N/mm^2 is used to fabricate bridges, sleepers, and struts in electrified railway catenaries. The compressive strength of concrete used in highway bridges is as high as 80 N/mm^2. Concrete of 60 N/mm^2 was used in the columns of high-rise buildings in the construction of the Industrial Technology Exchange Center in Liaoning Province, which was completed in 1988. Pump concrete was successfully applied in the construction of Shanghai Hailun Hotel in August 1990, and the Shanghai Xinxin hairdressing salon in September 1990. The use of 60 N/mm^2 concrete in infrastructure, such as concrete water pipes, was also reported. Currently, 60 N/mm^2 concrete has been widely used in civil engineering structures, especially for skyscrapers in China.

Fiber reinforced concrete utilizes various types of fibers (e.g., steel fibers, synthetic fibers, glass fibers and carbon fibers) into ordinary concrete to increase the tensile strength of concrete and to enhance the resistance to cracking, impact, fatigue, and abrasion. Among all kinds of fiber reinforced concrete, steel fiber reinforced concrete is the most mature and has received the widest use. The USA, Japan, and China have all compiled respective design and construction codes or specifications for steel fiber reinforced concrete.

Ongoing research on concrete continues to enhance working properties, reduce bleeding and separation, improve microstructures, and increase corrosion resistance against acid/alkali. In addition, adding smart repair/sensing materials into concrete to give concrete the ability of automatically repairing, recovering and warning of damage has drawn great attention from scholars in many countries. Fiber optic sensing technology has also been used in concrete engineering.

To reduce the self-weight of concrete, lightweight concrete (dry density less than 18 kN/m^3) prepared by mixing binders, porous coarse aggregates, porous or dense fine aggregates and water has been developed by many domestic and overseas researchers. In foreign countries, the compressive strength and density of lightweight concrete in load-carrying members are 30–60 N/mm^2 and 14–18 kN/m^3, respectively, while in China they are 20–40 N/mm^2 and 12–18 kN/m^3, respectively. The floors in the Water Tower Place in Chicago, which was finished in 1976, adopted lightweight concrete with the compressive strength of 35 N/mm^2. The 52-story Shell Plaza in Houston was fully made of lightweight concrete. If the strength of concrete is not a great concern, no-fines concrete with the density of 16–19 kN/m^3 can also be used.

Corrosion of steel in reinforced concrete structures is one of the most important factors that may influence the structure life. Although great efforts have been made by many researchers all around the world, steel corrosion has not been effectively dealt with until recently. In North America, salt is used to thaw snow and deice in winter. Therefore, steel in highway bridges and public garages are seriously corroded. According to a statistical result in 1992, it would take 4–5 billion Canadian dollars and 50 billion US dollars to repair all the public garages in Canada and all the highway bridges in the USA, respectively. In Europe, the economic loss due to steel corrosion is about 10 billion pounds. To replace steel in reinforced concrete by fiber reinforced plastics (FRP) is considered an effective way to solve the problem of steel corrosion.

FRP is a kind of composite material composed of fibers, resin matrix, and some additives. According to fiber type, FRP can be classified as carbon fiber reinforced plastics (CFRP), aramid fiber reinforced plastics (AFRP) and glass fiber reinforced plastics (GFRP). FRP has the advantages of high strength, lightweight, good corrosion resistance, low relaxation, and ease of fabrication. It is an ideal substitute for steel reinforcement, especially when used as a prestressed reinforcement.

As early as in the 1970s, Rehm from University of Stuttgart showed that composite material reinforcement consisting of glass fibers could be used in prestressed concrete structures. In 1992, an American Concrete Institute (ACI) committee drafted the design guidelines for FRP. In 1993, *design guidelines for FRP concrete structures* and *design guidelines for FRP concrete members* were published as a state-level research achievement in Japan. *Canadian highway bridge design code* in 1996 contained the contents of FRP. In the same year, ACI440 published a report about the research state of FRP concrete structures. The American Society of Civil Engineers (ASCE) also organized a special committee to prepare standards on FRP.

In 1980, as an experiment, a short-span FRP concrete pedestrian bridge was built in Muster, Germany. The first GFRP prestressed concrete highway bridge was constructed and put into use in Dusseldorf, Germany in 1986. In 1988, GFRP prestressing system was applied on a two-span bridge in Berlin, Germany. The renovation of the French Mairie d'Ivry substation also utilized a large application of GFRP prestressed reinforcement. CFRP prestressed reinforcement was first adopted in a 7-m-wide and 5.6-m-long bridge in Japan. Twenty-seven GFRP prestressing tendons were arranged in the 1.1-m-thick deck slabs of a three-span highway bridge in Leverkussen, Germany, in 1991. It was Japanese engineers who first introduced FRP prestressing system into building structures. In 1992, the Notsch Bridge in Austria was put into use, which had 41 GFRP prestressing tendons embedded. The first CFRP prestressed concrete highway bridge in Canada was built in Calgary in 1993, and several FRP reinforced concrete and FRP prestressed concrete structures were established in subsequent years. FRP reinforced concrete structures were also researched by Chinese scholars. It is believed that FRP concrete structures will also be widely used in China in the near future.

1.2.3 Development of Structural Systems

Based on different purposes and functions of buildings, various structural systems can be formed by basic concrete members (e.g., beams, slabs, columns, and walls) according to certain rules. Initially, reinforced concrete structures composed of load-carrying reinforced concrete members were dominant. With the development and application of prestress technology, prestressed concrete structures with

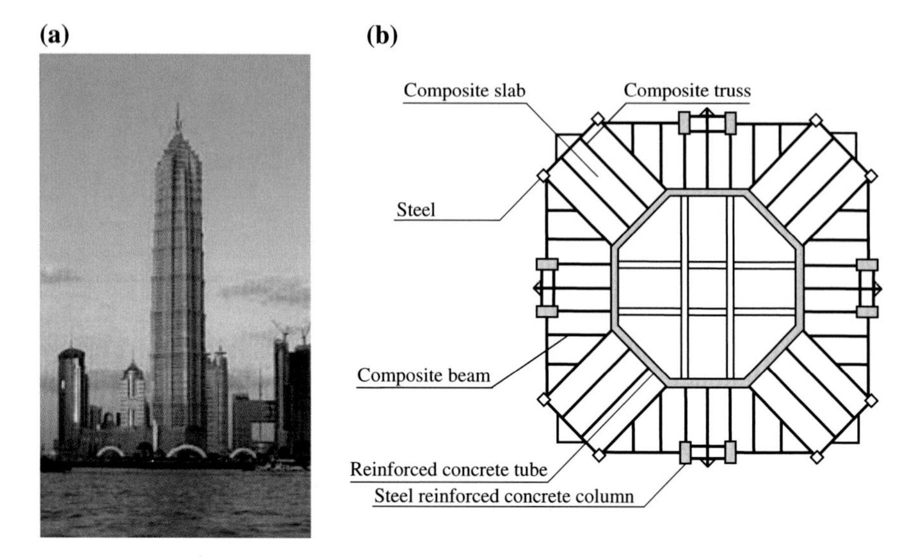

Fig. 1.5 Shanghai Jin Mao Tower. **a** Real photograph. **b** Layout of the structure

prestressed concrete members, as main load-carrying elements, showed apparent superiority in large-span structures and structures requiring high resistance to concrete cracking. In recent years, in order to fulfill the demands for large deformation ability and heavy load-carrying capacity, many strategies have been tried including arranging structural steel in concrete members or combining concrete members with steel members to form steel reinforced concrete structures and filling concrete inside steel tubes to form concrete-filled steel tube structures. These methods have been well developed and put to good use. In addition, steel members, steel reinforced concrete members, and concrete members can be simultaneously used in one structure called a steel-concrete composite structure. As shown in Fig. 1.5, Shanghai Jin Mao Tower has a closed tube formed by reinforced concrete walls, and around the central tube, steel reinforced concrete columns, steel columns, structural steel-concrete composite beams and composite trusses are arranged. The floors are steel-concrete composite slabs. All the members form a steel-concrete composite structural system.

1.2.4 Development in Theoretical Research of Concrete Structures

1.2.4.1 Mechanical Properties of Materials and Structural Members

Research on the mechanical properties of concrete and steel materials as well as on basic concrete structural members is the basis of fundamental theories of concrete structures. Requirements for anyone who wants to pursue a career in researching and understanding the theories and practices of concrete structures are as follows:

1. Understand mechanisms and find laws through experiments to lay the foundation for theoretical analysis;
2. Explain experimental phenomena by theoretical analysis, extrapolate experimental results, establish methods for engineering applications;
3. Accumulate experience, modify theories, perfect theoretical systems, find new problems, and determine the orientations of further research through engineering practices.

The development of statics laid a foundation for the theories of concrete structures. However, the establishment and development of modern concrete structural theories should give the credit to the French gardener J. Monier, who patented reinforced concrete flowerpots, pipes, pools, flat slabs, bridges, and stairs between 1850 and 1875 and obtained several patents issued by German government. Later, the patents were registered by Wayss and Freitag Company. The company subsequently entrusted Mörsch and Bach from University of Stuttgart with the measurement of strength of reinforced concrete structures and simultaneously entrusted Koenen, the chief architect of Prussia, with the computational methods used to measure the strength of reinforced concrete structures. In 1886, Koenen came up with the assumption that the neutral axis of a flexure member was at the center of a cross section. A primitive mechanical model for stress analysis of normal sections of flexure members was thus established. With the application of concrete structures and the progress of relevant research, national and international scholars have carried out broad experimental studies of mechanical properties of materials, performance of structural members under different loading conditions and failure mechanisms. Great progress has been made in developing rules to measure concrete strength and to establish constitutive laws of concrete and steel under uniaxial and multiaxial stress states. Regulations give guidance whereby we can determine the size effect of concrete, define the bond-slip relationship between concrete and steel, and the strength and deformation of confined concrete. Concrete engineers can define the load-deformation relationship of concrete members and use computational methods to determine the load-carrying capacity and deformation ability of concrete members under simple and complex loading conditions. Based on these achievements, complete and sound theoretical models have been tried and proven to establish predictable responses of concrete structures under external loads.

Recently, the durability of concrete structures has drawn the attention of many researchers. Relevant topics such as carbonation of concrete, alkali-aggregate reaction, freeze–thaw damage, corrosion of steel, mechanical properties of reinforced concrete members after the corrosion of steel, and the life cycle of concrete structures have been investigated worldwide and some achievements have been applied to concrete engineering.

1.2.4.2 Design Theories of Concrete Structures and Evaluation of Existing Concrete Structures

Fundamental theories of concrete structures mainly have two aspects of engineering applications. One is in the design of new structures, i.e., designing structural members with known loads. The other is in the evaluation of existing structures, i.e., determining the load-carrying capacity of existing structural members.

In 1894, E. Coignet (the son of F. Coignet) and de Tedeskko developed Koenen's theory and proposed the allowable stress design method of reinforced concrete structures for flexure in their paper to the French Society of Engineers. This elastic mechanics-based method was so simple in mathematics that was instantly accepted by engineers. Although people had known the elastoplastic property of concrete and the ultimate strength theory of reinforced concrete structures for a long time, the allowable stress design method was hardly shaken. Even until 1976, this method was the preferred choice in design codes for building structures in the USA and UK. And ACI 318-95 published in 1995 still listed the allowable stress design method as an alternative in the appendix.

The allowable stress method assumes a linear stress distribution in cross sections, which neglects a characteristic of reinforced concrete, i.e., the stress or internal force redistribution between concrete and steel as well as between different cross sections in a statically indeterminate structure. The ductility necessary for seismic design cannot be fully considered either. On the other hand, the limit states design of reinforced concrete structures is a more general concept. There are limit states other than the ultimate limit state, and serviceability limit states must also be considered. Although the allowable stress design method can be used in limit design in certain conditions, it cannot cover all the contents of the limit design. Moreover, the allowable stress design method can only make a reduction of member strength; furthermore, it is hard to give the reliability of structures by statistics. So it is natural to replace the allowable stress design method by the limit states design method.

In 1932, Полейт from the former Soviet Union proposed the rupture stage design method, which considered the internal forces that cross sections could sustain before failure, and this method was included in the Russian 1939 issued design code. The design standard (temporary) for building structures published by the ministry of industry of the Northeast People's Government in 1952 was also based on this theory. The rupture stage design method was actually an interim version between the allowable stress design method and the limit states design method.

The first design code considering the limit states design was the НИТУ 123 of the former Soviet Union. China first directly referenced НИТУ 123-55 and then issued the *Code for design of reinforced concrete structures* (BIJ 21-66) in 1966 adding part of her own research results. In 1974, BIJ 21-66 was upgraded to TJ 10-74. In 1989, the *Code for design of concrete structures* (GBJ 10-89) was drafted according to the *Unified standard for design of building structures* (GBJ 68-84). The theoretical bases of GBJ 10-89 and current GB 50010 are of no difference.

Both codes assume loads and material strengths as stochastic variables and employ the approximate probability-based limit states design method.

For important concrete structures such as offshore platforms, containments of nuclear power plants, full probability-based limit states design method should be used. Currently, the full life cycle design method becomes a hot topic, which considers the performance degeneration of concrete structures under environmental loading and sets the goal of predicting structure durability. However, this method is still far from being perfected.

The evaluation of existing structures is generally regarded as the inverse process of structural design, and engineers also do it in this way. However, an existing structure as a physical reality has characteristics different from those of a newly designed structure. Firstly, some permanent loads which were treated as random variables in design stage can be considered as determined values in evaluation. Take the commonest permanent load self-weight as an example. To consider its randomness in the design stage is necessary because it will be influenced by uncertainties in materials and construction. But once the structure has been completed, these uncertainties disappear and the self-weight is objectively determined. Secondly, for planned structures, structural parameters (e.g., dimensions and material properties) are all statistical variables. However, for existing structures, these parameters are theoretically determined and most of them can be measured. Thirdly, the service history of an existing structure also provides a lot of useful information, such as the maximum load the structure ever sustained and the corresponding performance under this load. Worldwide researchers have investigated the evaluation of performance of existing structures. Some achievements have been included in codes. For example, ACI 318-95 stipulated that if members dimension and materials strength were both measured, the reduction factors in design and verification formulae could be increased in the evaluation of load-carrying capacity of existing structures. The first author of this book studied the probability models of loads and structural resistances in target periods of usage and proposed an approximate probability-based safety analysis method for existing structural members, which has been adopted by the Standard of Structural Inspection and Assessment for Existing Buildings (DG/TJ 08-804-2005) in Shanghai.

1.2.4.3 Full Life Cycle Maintenance of Concrete Structures

Full life cycle maintenance of concrete structures includes inspection, monitoring, assessment, repair, strengthening, and renovation of existing structures. A great deal of successful research has been carried out in the above fields since the 1990s. A complete theoretical system has been established and used in engineering applications. Currently, corresponding standards or specifications have been issued at different administrative levels (e.g., state and local governments and associations).

1.2.5 *Experiments and Numerical Simulation of Concrete Structures*

Structural experiments play a significant role in the birth and development of theories of concrete structures. The current worldwide design codes for concrete structures are all based on the large quantities of experimental data. For concrete construction of special shapes or complicated structures, the whole structures should be modeled and tested to verify design theories and modify design methods. Along with the progress in developing testing equipment, data acquisition systems and experimental theories, experiments on concrete structures have been developed from simple material property tests to today's series of tests on materials, members, and structures. Loading schemes have been also developed from static tests to static, quasi-static, pseudo-static, and dynamic tests (Fig. 1.6). However, model experiments on structures, especially on large-scale structures, cost huge manpower and money. So the same experiments cannot be repeated too many times. Moreover, scaled model experiments cannot truly reflect the performance of real structures. Therefore, if there is a computational method that can simulate full-scale model tests as the supplement to experimental investigations, the deficiencies mentioned above can be overcome, and the development and application of concrete structural theories will also be greatly pushed forward.

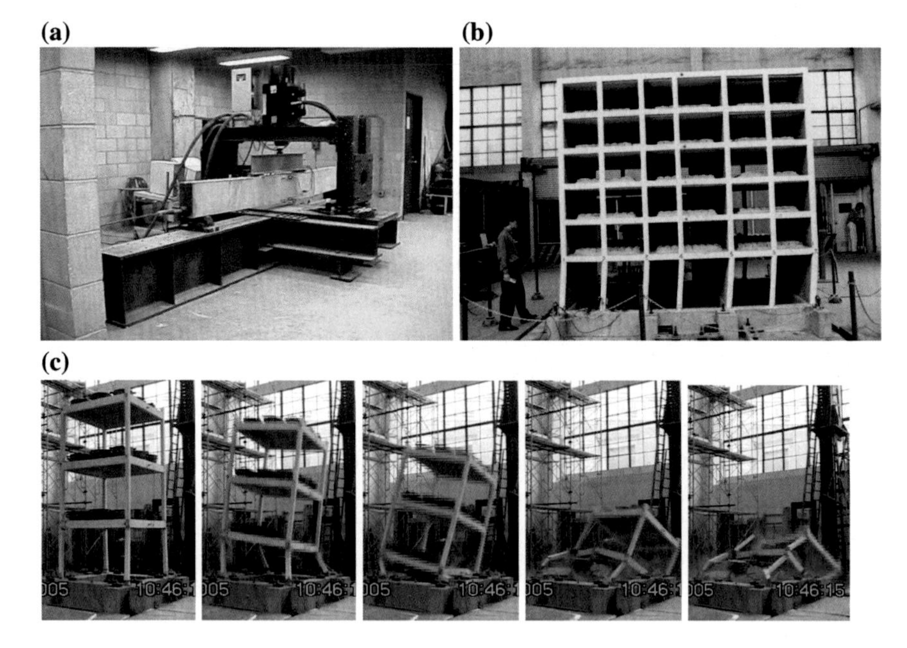

Fig. 1.6 Experiments on concrete structures. **a** Fatigue test on a FRP prestressed concrete beam. **b** Shake table test on a reinforced concrete frame. **c** Collapse test on a reinforced concrete frame

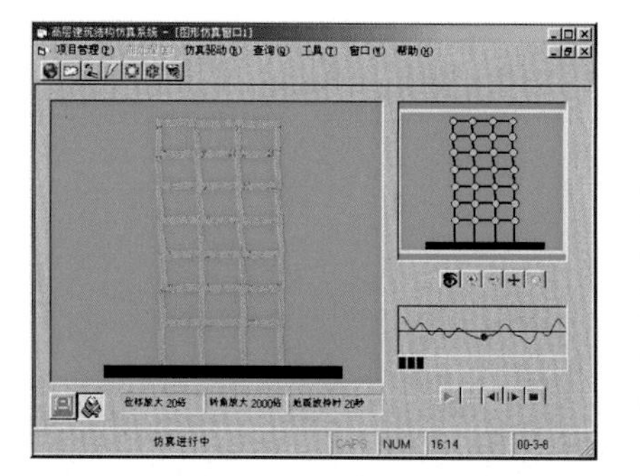

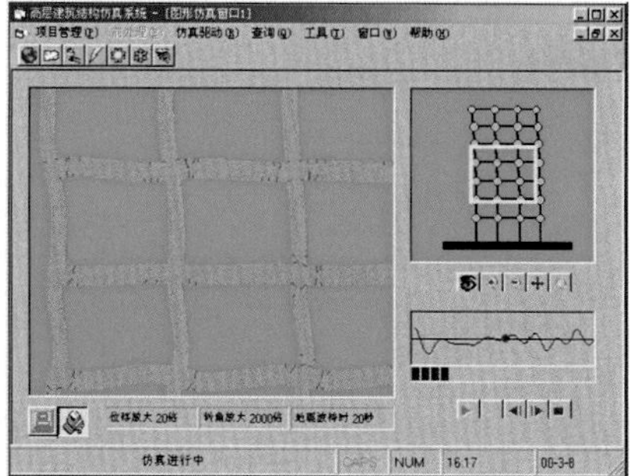

Fig. 1.7 Simulation results of a reinforced concrete frame in seismic test

Since the 1960s, computer simulation technology has been developed from the initial numerical simulation and graphic display of simulation results to an advanced technology closely related to information science, cybernetics, simulation theory, artificial intelligence, multimedia technology, etc. Computer simulation technology can be used when work is difficult or impossible to be accomplished due to many restrictions, such as experiment simulations, disaster prediction, accident reconstruction, scheme optimization, structural performance assessment. This technology has been more and more widely used in concrete structures in recent years owing to many researchers' efforts. For example, scholars from University of Tokyo numerically simulated the collapse process of a reinforced concrete frame under strong earthquake. J. Jiang from Tsinghua University simulated the failure of concrete members. Failure process of basic concrete members and reinforced concrete bar systems under different external disturbances as well as collapse of reinforced concrete frames under monotonic and seismic loadings were investigated at Tongji University with the help of computer simulation technology. Figure 1.7 illustrates the simulation results of a high-rise reinforced concrete frame in a seismic test by implementation of a software developed by Tongji University.

1.3 Applications of Concrete Structures

Concrete structures can be used in every field of civil engineering. In buildings, concrete structures are a major part of the construction. For example, 311 South Wacker Drive in Chicago, USA, is a 65-story, 296-m-high skyscraper completed in 1990. It was the tallest reinforced concrete building in the world at that time. The 105-story, 319.8-m-high Ryugyong Hotel in Pyongyang, DPRK, is also a concrete structure. Montreal Olympic Stadium (Fig. 1.8) and Sydney Opera House (Fig. 1.9) are concrete structures as well. In China, concrete structure buildings are more popular. For instance, there are many concrete structures in the Bund complex built at the beginning of the twentieth century (Fig. 1.10). Although steel structures have

Fig. 1.8 Montreal Olympic Stadium

Fig. 1.9 Sydney Opera House

Fig. 1.10 Shanghai Bund complex

been greatly developed in recent years, most high-rise buildings higher than 100 m adopt concrete structures or steel-concrete composite structures, e.g., 88-story Jin Mao Tower in Shanghai.

Tunnels, bridges, highways, urban viaducts, and subways mostly employ concrete such as the Puxi segment viaduct of the Inner Ring, the towers in Nanpu Bridge and Yangpu Bridge (Fig. 1.11), Metro Line 1–12, and tunnels across Huangpu River in Shanghai, China.

Fig. 1.11 Yang Pu Bridge tower in Shanghai

Concrete structures are also applied in engineering facilities such as dams, gate piers, aqueducts, and harbors. The 285-m-high Grande Dixence Dam in Switzerland established in 1962 is the highest concrete gravity dam. Containments in nuclear power plants (Fig. 1.12), cooling towers in thermal power plants, water tanks, gas tanks, and offshore platforms (Fig. 1.13), etc., are generally concrete structures. Large quantities of high-concrete towers have been built worldwide since the first reinforced concrete TV tower in Stuttgart was designed by Dr. F. Leonharat from University of Stuttgart in 1953. The CN Tower in Toronto is as high as 553.3 m. China has also established some concrete TV towers with six of them higher than 300 m (Fig. 1.14).

It is believed that concrete will get even wider applications.

1.4 Characteristics of the Course and Learning Methods

This course is one of the fundamental specialized courses for undergraduate students majored in civil engineering. From the study of this course, students should know the basic mechanical properties, computational analysis methods and detailing of structural members composed of concrete and reinforcement, understand the distinctions and similarities between this course and previous mechanics courses, acquire the ability of solving real engineering problems in structural design

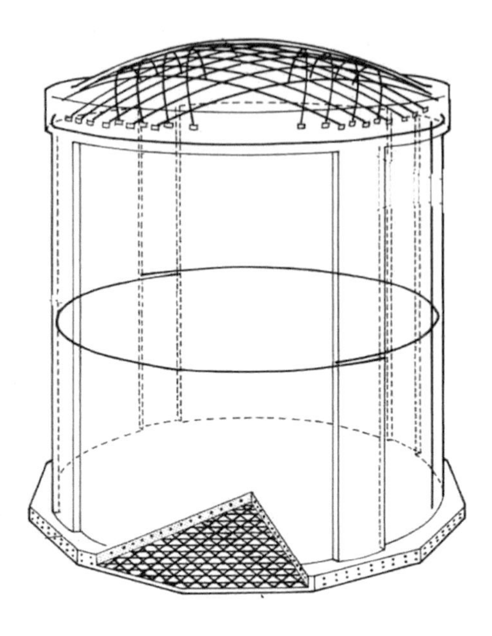

Fig. 1.12 Containment in a nuclear power plant

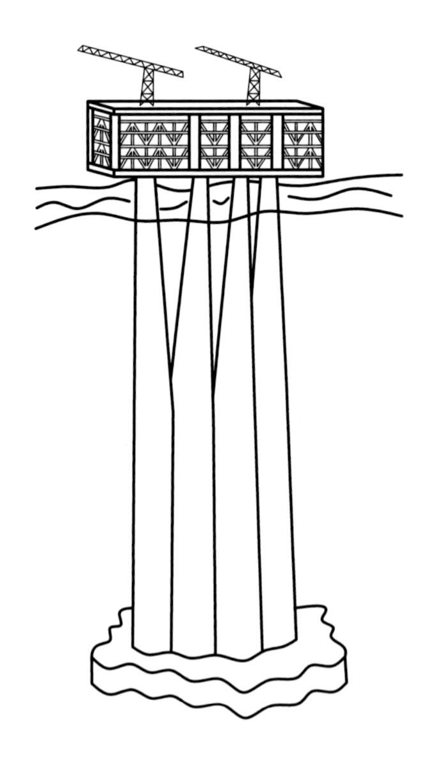

Fig. 1.13 Offshore platform at a depth of 330 m

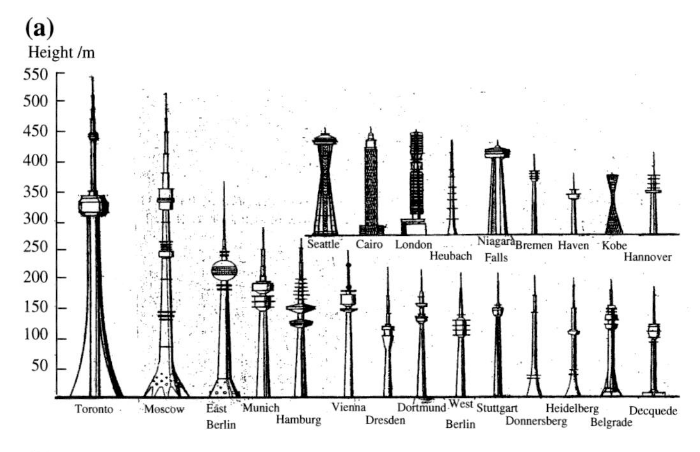

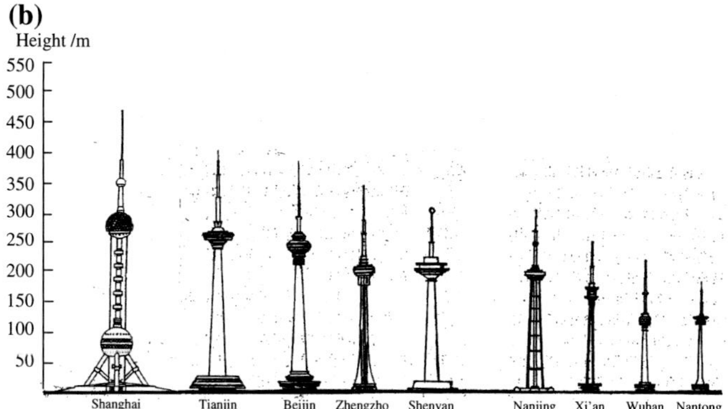

Fig. 1.14 Concrete TV towers. **a** Some of foreign TV towers. **b** Some of TV towers built in China

and assessment, and lay solid foundation for future design courses. To study this course more effectively, students are asked to note the following points:

1. Note the differences and similarities between this course and previous courses, especially as it pertains to the mechanics of materials. You will be expected to know how to use learnt mechanics knowledge to solve problems.
2. Concrete structural theories are mostly based on the experimental research. There has not been a complete or generally accepted theoretical system until now. Many formulae can only be regressed from experimental data rather than derived following strict reasoning. Be careful of the derivation conversion in the study and application. You will be taught to synthesize induction and deduction.
3. To ensure the safety and reliability of structures, quantitative theoretical analysis itself is not enough. Qualitative detailing measures are necessary. These measures are summaries of previous experiences. Although they cannot be explained quantitatively, profound principles are behind them. So in the study,

you should understand the meaning of the detailing measures rather than just memorize them.
4. Study fundamental theories with the goal of applying them in future engineering practice.
5. Relating theories to reality is helpful in the study of this course.

Questions

1.1 What is the basis of the combined action of concrete and steel?
1.2 What are the advantages of reinforced concrete beams compared with plain concrete beams?
1.3 What are the advantages of prestressed concrete beams compared with ordinary concrete beams?
1.4 What are the advantages of concrete structures compared with other structures?

Chapter 2
Mechanical Properties of Concrete and Steel Reinforcement

2.1 Strength and Deformation of Steel Reinforcement

2.1.1 Types and Properties of Steel Reinforcement

2.1.1.1 Stiff Reinforcement and Steel Bars

Both stiff reinforcement and steel bars can be used in reinforced concrete members.

Stiff reinforcement includes shape steels (e.g., angle steel, channel steel, I-shape steel, and pipe) and the skeletons fabricated by welding several pieces of shape steel together. Due to its large stiffness, stiff reinforcement can be used in construction as forms or supports to bear the self-weight of structures and construction loads. This can facilitate shuttering (also known as formwork) and speed construction. Also, structural members reinforced by stiff reinforcement possess higher loading capacity than those reinforced by steel bars.

However, steel bars are more frequently used in ordinary reinforced concrete members. Because steel bars are flexible, they are treated as axially loaded elements, while their own stiffness is meaningless in design.

Most design codes and textbooks (including this book) are referring to steel bars when discussing reinforced concrete structures. Answers to questions concerning concrete members reinforced by stiff reinforcement can be found in special textbooks or standards covering steel-reinforced concrete, concrete-filled steel tubes, etc.

2.1.1.2 Surface Profile of Steel Bars

Steel bars can be classified as plain bars and deformed bars or rebars according to their surface profiles. Deformed bars are the bars with longitudinal and transverse ribs rolled into the surfaces (sometimes without longitudinal ribs). The ribs, which

X. Gu et al., *Basic Principles of Concrete Structures*,
DOI 10.1007/978-3-662-48565-1_2

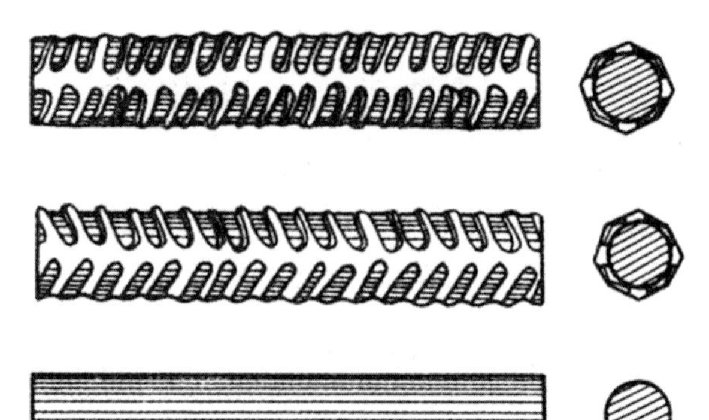

Fig. 2.1 Surface profiles of steel bars

may be in the shape of a spiral, chevron or crescent, etc. (Fig. 2.1), can effectively increase the bonding between steel bars and concrete. The cross-sectional area of a deformed bar varies with its length, so the diameter of the deformed bar is a nominal dimension, i.e., an equivalent diameter is the same as that of a plain bar of identical weight. Generally, the diameters of plain bars are 6, 8, 10, 12, 14, 16, 18, 20 and 22 mm, while the diameters of deformed bars are 6, 8, 10, 12, 14, 16, 18, 20, 22, 25, 28, 32, 36, 40 and 50 mm.

Reinforcement steel bars of small diameter (e.g., <6 mm) are also called steel wires, whose surface is generally smooth. If indentations are rolled into the surface of a steel wire to improve the bond, the steel wire is called indented wire.

2.1.1.3 Reinforcement Cage

Different kinds of reinforcement in structural members can be strapped or welded into reinforcement cages or wire fabrics before being placed in forms. This not only secures the relative position of the reinforcement, but also helps to bring the reinforcement into full play. Figure 2.2 shows part of a reinforcement cage used to support concrete beams.

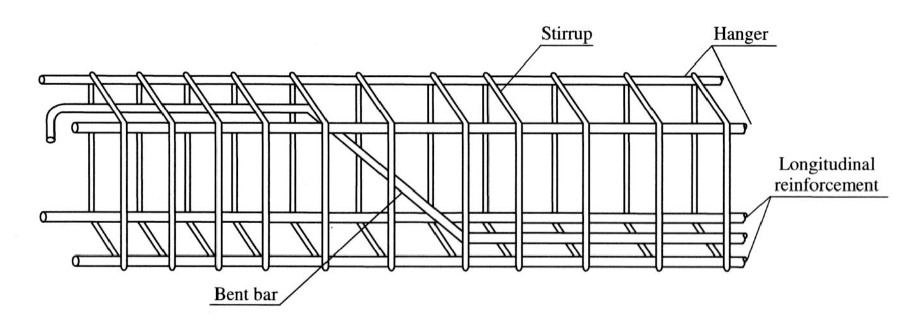

Fig. 2.2 Reinforcement cage

To prevent a plain bar under tension from slipping in concrete, both ends of the bar should be hooked. Sometimes the intermediate segment of reinforcement should be bent due to the design requirement. The detailings of the bent segments and hooks are listed in relevant design codes or acceptance specifications for construction quality. Figure 2.3 illustrates some hooks and bent segments in reinforced concrete structures.

Hook ends are not necessary for plain bars under compression. Because the cross section of the reinforcement under compression tends to expand, the normal pressure caused by surrounding concrete against such deformation can effectively increase the bond between the reinforcement and the concrete.

The ribs of deformed bars allow the bars to form a better bond with concrete, so hooks are unnecessary at the ends. If a hook end is indispensable for a deformed bar to fulfill the requirement of development length, which will be represented in Chap. 3, a right angle hook rather than a half circle hook will be formed for the convenience of machining.

To ensure that the reinforcement will not crack, fracture, or rupture, a cold bending test is commonly used to check the ductility and internal quality of the reinforcement. The cold bending test is performed by bending the reinforcement around a roll shaft. The reinforcement qualifies as acceptable if there is no crack, delamination, or rupture after being bent to stipulated angles. Readers should refer to relevant national standards for detailed specifications of cold bending tests, such as Metallic materials—Bend test (GB/T 232-2010).

Welded steel cages and wire fabrics are well bonded with concrete. Therefore, installing hooks at their ends is not necessary. Moreover, welded steel cages and wire fabrics are suitable for industrial quantity production and are widely used in precast reinforced concrete construction due to the reduction of in situ reinforcement processing. Reinforcement that needs welding should possess good weldability, i.e., no cracks and excessive deformation are allowed after welding under certain technological conditions.

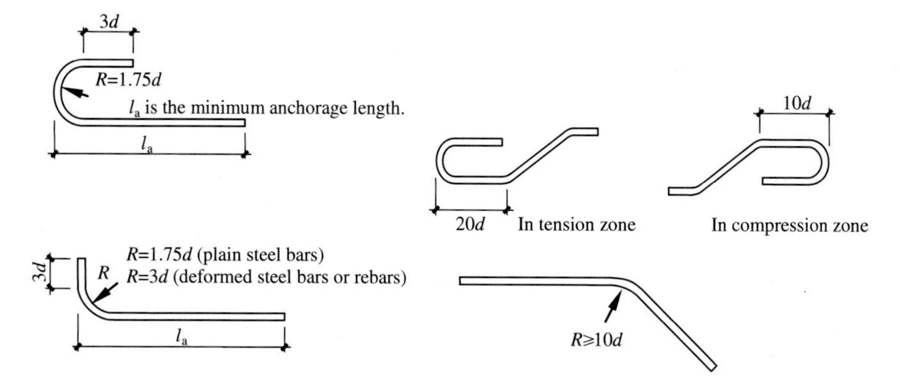

Fig. 2.3 Hooks and bent steel bars

2.1.2 Strength and Deformation of Reinforcement Under Monotonic Loading

2.1.2.1 Stress–Strain Curve

Typical stress–strain curves of steel bars used in reinforced concrete structures are obtained from monotonic tension tests, in which the loads are monotonically applied (without any unloading) until the failure of specimens in a short time.

From monotonic tension tests, researchers can evaluate the strength and deformation of steel bars. Figures 2.4 and 2.5 show two stress–strain curves of steel bars with apparent differences.

For hot-rolled low-carbon steel and hot-rolled low-alloy steel, the stress–strain curve in Fig. 2.4 is recorded. The curve exhibits an initial linear elastic portion (segment *Oa*). The stress corresponding to point *a* is called the proportional limit. In segment *ab*, the strain increases a little bit faster than the stress, although it is not very obvious in the figure. After point *b*, the strain increases a lot with little or no increase in the corresponding stress. The curve extends nearly horizontally to point *c*. Segment *bc* is called the yield plateau. After point *c*, the stress again increases

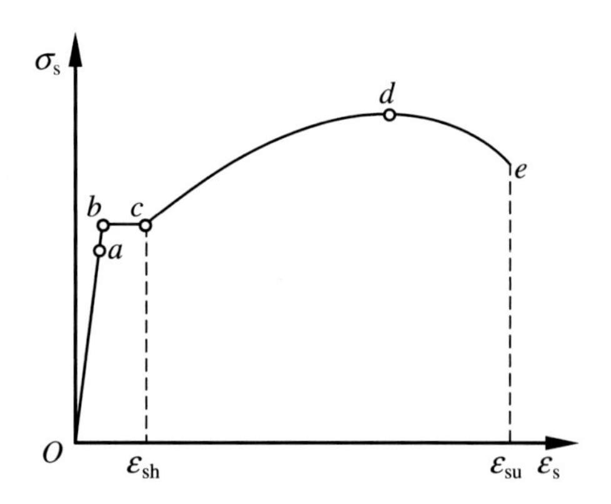

Fig. 2.4 Stress–strain curve for steel bars with a yield plateau

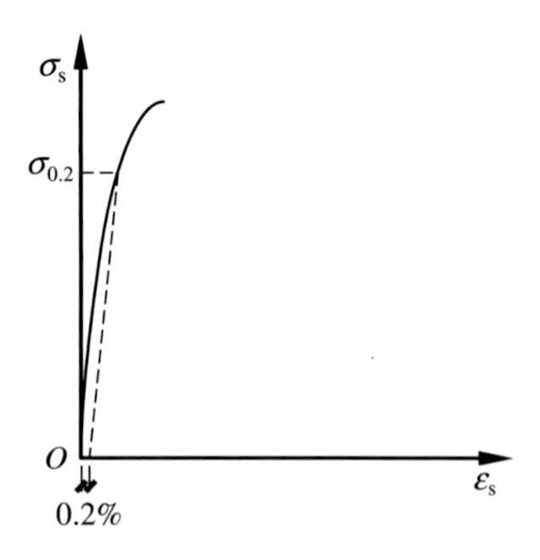

Fig. 2.5 Stress–strain curve for steel bars without a yield plateau

with the strain until point d. The stress corresponding to the highest point d is the ultimate strength of steel bars. Segment cd is the stain-hardening range. After point d, the strain increases rapidly accompanied by the area reduction of the weakest cross section, i.e., necking, and finally fracture occurs at point e.

For high-carbon steel, Fig. 2.5 shows a stress–strain curve, no apparent yield plateau can be observed in the curve. Generally, the stress $\sigma_{0.2}$ corresponding to the residual strain at 0.2 % is taken as the yield strength. It has been stipulated in Chinese metallurgical standards that the yield strength $\sigma_{0.2}$ of reinforcement should not be less than 85 % of the ultimate tensile strength σ_b (0.85 σ_b). Therefore, 0.85 σ_b can be taken as the conditional yield point in real engineering applications.

Steel with a well-defined yield plateau is called mild steel, while steel without a well-defined yield plateau is called hard steel.

In reinforced concrete structures, certain ductility of reinforcement is required. The strain corresponding to the end of a yield plateau ε_{sh} and the ultimate strain ε_{su} are important indices of plasticity, which can ensure apparent warning before either the reinforcement fractures or the members fail.

2.1.2.2 Chemical Composition, Grade, and Type of Reinforcement

According to chemical composition, steel bars can be categorized as carbon steel bars and ordinary low-alloy steel bars.

Carbon steel contains ferrum and other trace elements, such as carbon, silicon, manganese, sulfur, and phosphorus. Experimental results show that the strength of a steel bar increases with carbon content, but at the cost of plasticity and weldability. Generally, carbon steels with a carbon content of less than 0.25 %, 0.25–6 %, and 0.6–1.4 % are called low-carbon steel, medium-carbon steel, and high-carbon steel, respectively. Low- and medium-carbon steels are mild steels, while high-carbon steel is hard steel.

Ordinary low-alloy steel contains added carbon steel alloying elements such as silicon, manganese, vanadium, titanium, and chromium to efficiently increase strength and improve steel properties. Currently in China, ordinary low-alloy steel can be classified into five classes according to the alloying element addition: manganese class (20MnSi, 25MnSi), silicon–vanadium class ($40Si_2MnV$, 45SiMnV), silicon–titanium class ($45Si_2MnTi$), silicon–manganese class ($40Si_2Mn$, $48Si_2Mn$), and silicon–chromium class ($45Si_2Cr$).

According to the processing method, steel bars can be categorized as hot-rolled steel bars, heat treatment steel bars and cold working steel bars. Steel wires include carbon steel wires, indented steel wires, steel strands, and cold stretched low-carbon steel wires. Hot-rolled steel bars can be further classified as hot-rolled plain steel bars HPB300 (also called Grade I reinforcement, denoted by symbol ϕ), hot-rolled ribbed steel bars HRB335, HRB400, and HRB500 (also called Grade II, III, and IV reinforcement, denoted by Φ, Φ, and Φ, respectively), remained heat treatment ribbed steel bars RRB400 (also called Grade III reinforcement, denoted by Φ^{R}), and hot-rolled ribbed fine-grained steel bars HRBF335, HRBF400, and HRBF500 (also

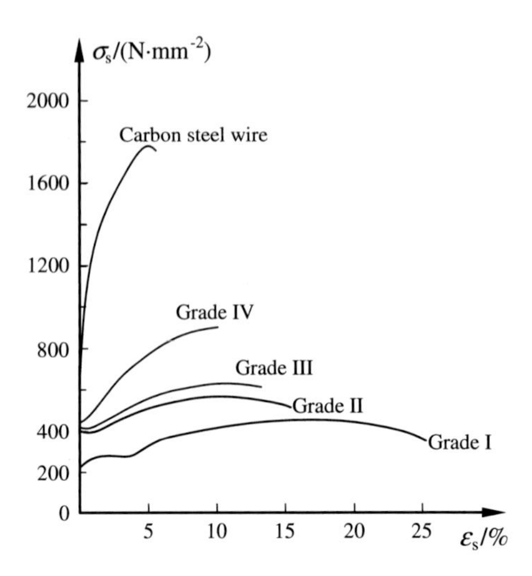

Fig. 2.6 Stress–strain curves of reinforcement for different grades

called Grade II, III, and IV reinforcement, denoted by $\underline{\Phi}^F$, $\underline{\Phi}^F$, and $\underline{\overline{\Phi}}^F$, respectively). Cold stretched steel bars are made by mechanically tensioning hot-rolled steel bars in normal temperature. Heat treatment steel bars are ordinary low-alloy steel bars that have undergone the process of heating, quenching, and tempering. High-carbon-killed steel becomes carbon steel wires after several times of cold stretching, stress relieving, and tempering. Rolling indentations into steel wires make indented steel wires, which enables a more secure bond with surrounding concrete. Steel strands are fabricated by twisting several steel wires of the same diameter. Mechanically, tensioning low-carbon steel wires at room temperature makes cold stretched low-carbon steel wires.

As a general rule, steel bars designated as HPB300, HRB335, HRB400, HRB500, and RRB400 can be used as nonprestressed reinforcement, while carbon steel wires, indented steel wires, steel strands, and cold stretched steel bars provide prestressed reinforcement.

The stress–strain curves of reinforcement for different grades are shown in Fig. 2.6.

2.1.2.3 Strength of Reinforcement

Reinforcement undergoes large plastic strain after yielding. The resulting excessive deformation and crack widths of structural members violate the serviceability requirement. So when the capacity of a reinforced concrete member is calculated, yield point (or conditional yield point) is taken as the upper bound.

Reinforcement strength is obtained by tests. But the strengths of different specimens, even when of the same classification or type, are generally different due to the inherent variability of reinforcement materials. Statistical analysis shows that experimental data of reinforcement strength obeys Gaussian distribution (Fig. 2.7).

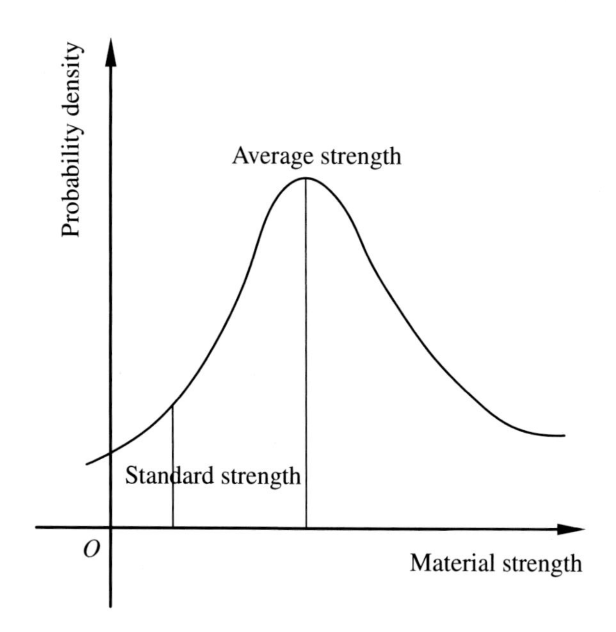

Fig. 2.7 Distribution of tested material strength

Therefore, measured strength with a certain degree of confidence can be taken as the standard strength of reinforcement. As shown in Fig. 2.7, if the degree of confidence is 97.73 %, then the standard strength of reinforcement is equal to the average strength minus twice the standard deviation. Tables 2.2 and 2.3 in the addendum list the standard strength of reinforcement (with a 95 % degree of confidence) stipulated in the *Code for Design of Concrete Structures* (GB 50010). Because GB 50010 adopts the probability limit state design method, the standard strength is further reduced by dividing the coefficient $\gamma_s = 1.1$ to get the design strength of reinforcement. GB 50010 also stipulates that the design strength should be used in the ultimate capacity design of structural members, while the standard strength should be used in the check of deformation and crack width.

Because reinforced concrete design codes in different regions and different engineering professions are based on diverse criteria, such as the *allowable stress design method, ultimate strength design method,* and *limit state design method*, they cannot ensure structural members the same safety margin or reliability. Hence, the same reinforcement may be given different strength values under different specifications or even a different name, which will be introduced in detail in the subsequent structural design course.

2.1.2.4 Theoretical Stress–Strain Models of Reinforcement

In theoretical analyses of reinforced concrete structures, the stress–strain relation curves obtained from experiments are seldom directly employed. Theoretical

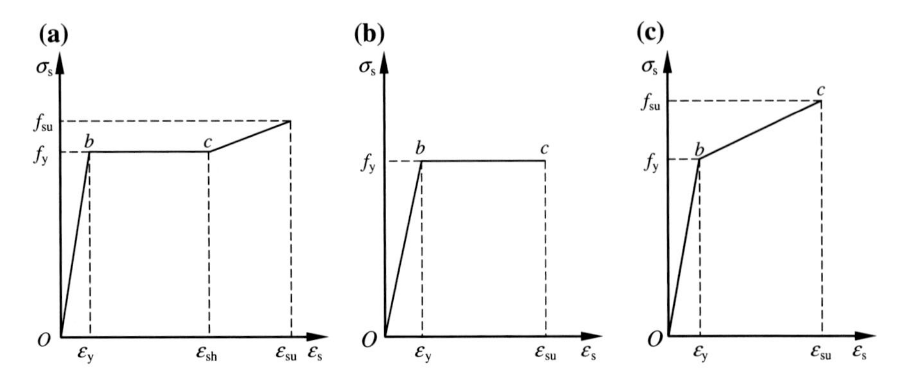

Fig. 2.8 Theoretical stress–strain models of reinforcement. **a** Trilinear model. **b** Bilinear model (I). **c** Bilinear model (II)

models idealized from the experimental curves are generally preferred. Figure 2.8 shows commonly used theoretical stress–strain models of reinforcement.

The trilinear model (Fig. 2.8a), which is suitable for mild steel with a well-defined yield plateau, can depict the strain-hardening stage and correctly evaluate the stress after reinforcement yielding. If the yield plateau is long, the bilinear model (Fig. 2.8b), i.e., the ideal elastic-plastic model, can give adequate analysis results. Note that the ultimate deformation of concrete at the failure of structural members is limited. Even though the corresponding tensile deformation of reinforcement has entered the strain-hardening stage, the extent of its entrance is still limited. Therefore, in practical engineering, the elastic-perfectly plastic model is commonly employed for ordinary steel bars in theoretical analysis, which can be formulated as:

$$\sigma_s = \begin{cases} E_s\varepsilon_s & (\varepsilon_s \leqslant \varepsilon_y) \\ f_y & (\varepsilon_s > \varepsilon_y) \end{cases} \tag{2.1}$$

where

E_s the modulus of elasticity of the steel;
f_y the yield strength of the steel; and
ε_y the yield strain of the steel.

The bilinear model shown in Fig. 2.8c can be used to describe the stress–strain relationship of high-strength steel bars or steel wires, which do not have well-defined yield plateaus.

If the reinforcement does not fail at buckling, the theoretical stress–strain models of steel bars under compression are the same as those under tension.

2.1.3 Cold Working and Heat Treatment of Reinforcement

2.1.3.1 Cold Working of Reinforcement

Cold working such as cold stretching and cold drawing can raise the design strength of hot-rolled steel bars (the yielding stress).

Cold stretching is to stretch reinforcement with a well-defined yield plateau into or beyond its stage of yield. As is illustrated by point *a* in Fig. 2.9, a residual strain *OO′* will remain and cannot be recovered after the stress is released. And if the reinforcement is stretched again right away, the stress–strain curve will follow the path *O′abc* and the yield strength is approximately equal to the stretching stress, which is higher than the yield strength before the cold stretching. But the yield plateau disappears, and the total elongation is reduced from *Oc* to *O′c*, symbolizing worse plasticity. However, if the reinforcement is stretched again after having been placed in natural conditions for a period of time, the yield point can be further increased from point *a* to point *a′*, i.e., a phenomenon called aging hardening. Moreover, an apparent yield plateau can be observed again, and the stress–strain curve will follow the new path *a′b′c′*. The strength increase of steel bars by cold stretching depends on the steel material. The higher the original strength is, the lesser the increase will be. The stretching stress should be rationally selected to keep a certain yield plateau at the same time of increasing the strength. Cold stretching can only increase the tensile strength of reinforcement. When the temperature reaches 700 °C, reinforcement will recover to its original state before cold stretching. So if the reinforcement needs welding, it should be welded first before cold stretching.

Cold drawing is to force reinforcement through a carbide alloy wire drawing die of smaller diameter. The reinforcement will undergo plastic deformation under the

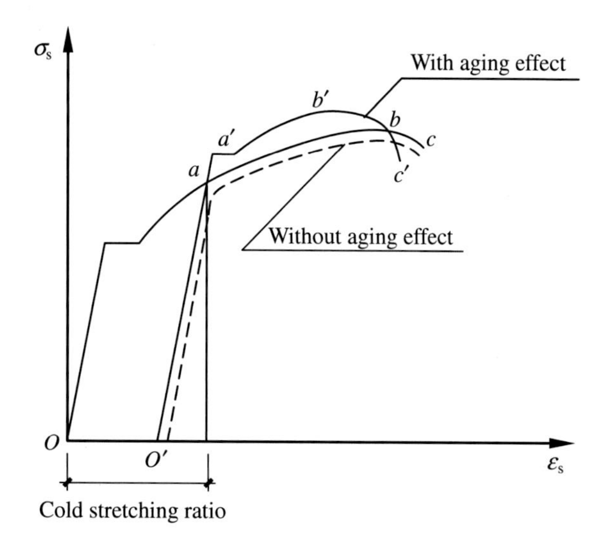

Fig. 2.9 Stress–strain model of a steel bar after cold stretching

simultaneous action of longitudinal tension and transverse pressure. The cross section of the reinforcement will be reduced, while the length increases. The reinforcement strength is apparently raised due to the internal structural change of the steel. After several times of cold drawing, the plasticity of the reinforcement will significantly decrease, which is proven by the disappearance of the stage of yield. Cold drawing can increase the tensile and compressive strength of reinforcement at the same time.

2.1.3.2 Heat Treatment of Reinforcement

Heat treatment of reinforcement is to make low-alloy steel bars of specific strength experience a series of process, i.e., heating, quenching, and tempering. This can greatly increase the strength of the reinforcement with insignificant decrease of plasticity. Heat treatment steel bars include three types: $40Si_2Mn$, $48Si_2Mn$, and $45Si_2Cr$, of which the stress–strain curves have no obvious yield points.

2.1.4 Creep and Relaxation of Reinforcement

The strain of the reinforcement will increase with time if continuously subjected to high stress. This phenomenon is called *creep.*

Relaxation will occur if the length of a steel bar is kept constant, and the stress of the steel bar will decrease with time. Creep and relaxation have the same physical nature. Creep and relaxation increase with time and depend on initial stress, steel material, and temperature. Generally, high initial stress will cause large creep or great relaxation-induced stress loss. The creep and relaxation of cold stretched hot-rolled steel bars are lower than those of cold stretched low-carbon steel wires, carbon steel wires, and steel strands. If the temperature increases, the creep and relaxation will also increase. The relaxation-induced stress loss in prestressed reinforcement must be considered in prestressed concrete structures.

2.1.5 Strength and Deformation of Reinforcement Under Repeated and Reversed Loading

2.1.5.1 Repeated Loading Behavior

Repeated loading is to experience a specimen in one direction loading, unloading, reloading, unloading again, and so on. Figure 2.10 shows a stress–strain curve of a steel specimen under repeated loading. If the load is released before failure, the specimen will recover along the linear stress–strain path *bO′* that is parallel to the

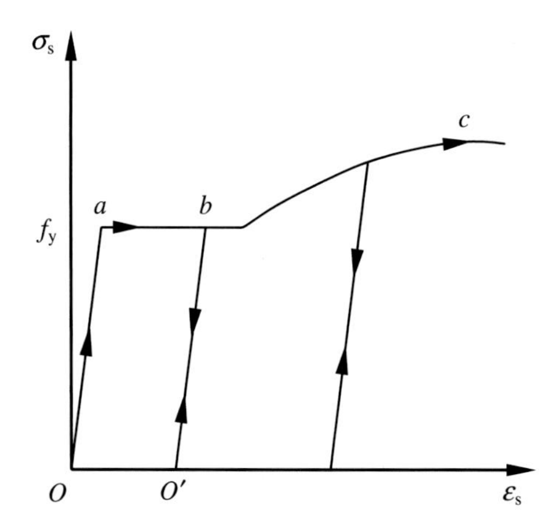

Fig. 2.10 Stress–strain curve of a steel specimen under repeated loading

original elastic portion of the curve Oa. If loaded again, the stress–strain curve will follow the same path $O'b$ up to the original curve. The original curve bc is then closely followed as if unloading had not occurred. Hence, the monotonic stress–strain curve gives a good idealization for the envelope curve of the same specimen being repeatedly loaded.

2.1.5.2 Reversed Loading Behavior

Reversed loading is to load and unload a specimen alternatively in two opposite directions (tension–compression). Figure 2.11 shows a stress–strain curve of a steel specimen under reversed loading. If the stress is released at point b, which lies in the yield plateau, the specimen will recover along the linear stress–strain path bO', that is, parallel to the original elastic portion of the curve Oa. If loaded again in the opposite direction, the plastic deformation will happen at point c, which corresponds to a stress much lower than the initial yield strength. This phenomenon is called the Bauschinger effect.

Reversed loading curves are important when considering the effects of high-intensity seismic loading on members.

2.1.5.3 Fatigue of Reinforcement

When a steel bar is subjected to periodic loading, even though the maximum stress is lower than the strength value under monotonic loading, the steel bar will fail after a certain number of times of loading and unloading between the minimum stress $\sigma^f_{s,\min}$ and the maximum stress $\sigma^f_{s,\max}$. This is called fatigue failure. In engineering applications, fatigue failure may happen to reinforced concrete members (e.g., crane beams, bridge decks, and sleepers) under repeated loading.

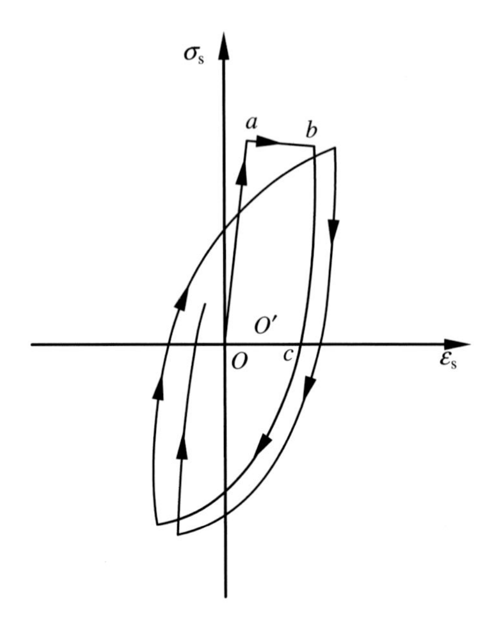

Fig. 2.11 Stress–strain curve of a steel specimen under reversed loading

The fatigue strength of reinforcement is the maximum stress value of a steel specimen before failing by fatigue after a certain number of cyclic loads with the stress restricted in a stipulated range. The fatigue strength depends on the stress amplitude, i.e., the difference between the maximum stress and the minimum stress. The stress amplitudes for ordinary reinforcement and prestressed reinforcement can be calculated, respectively, as follows:

$$\Delta f_{y}^{f} = \sigma_{s,\max}^{f} - \sigma_{s,\min}^{f} \tag{2.2a}$$

$$\Delta f_{py}^{f} = \sigma_{p,\max}^{f} - \sigma_{p,\min}^{f} \tag{2.2b}$$

where

Δf_{y}^{f}, Δf_{py}^{f} the stress amplitudes for ordinary and prestressed reinforcement, respectively;

$\sigma_{s,\max}^{f}$, $\sigma_{s,\min}^{f}$ the maximum stress and the minimum stress of the same layer of ordinary reinforcement at fatigue, respectively; and

$\sigma_{p,\max}^{f}$, $\sigma_{p,\min}^{f}$ the maximum stress and the minimum stress of the same layer of prestressed reinforcement at fatigue, respectively.

In China, the fatigue test is carried out by axially tensioning a steel bar. The times of cyclic loading should be determined in deciding the stress amplitude of reinforced concrete members in service. Two million times of cyclic loading is required in China. In other words, the fatigue strength of reinforcement is usually quantified by the maximum one among all different stress amplitudes that can sustain 2 million times of cyclic loading.

Tables 2.4 and 2.5 in the addendum list the GB 50010 stipulated fatigue stress amplitudes for ordinary reinforcement and prestressed reinforcement, respectively. The fatigue stress ratios in the tables mean the ratios of the minimum stress to the maximum stress of the reinforcement at the same layer.

$$\rho_{s}^{f} = \frac{\sigma_{s,\min}^{f}}{\sigma_{s,\max}^{f}} \tag{2.3a}$$

$$\rho_{p}^{f} = \frac{\sigma_{p,\min}^{f}}{\sigma_{p,\max}^{f}} \tag{2.3b}$$

where ρ_{s}^{f} and ρ_{p}^{f} are fatigue stress ratios of ordinary reinforcement and prestressed reinforcement, respectively.

Excluding stress amplitude, other factors such as surface profile, diameter of reinforcement, the processing method, the operating environment, and the loading rate will also influence the fatigue strength of reinforcement. The stress amplitudes listed in Tables 2.4 and 2.5 have already considered the above-mentioned factors.

2.2 Strength and Deformation of Concrete

Ordinary concrete is a complex multiphase composite material, i.e., a kind of man-made stone created by mixing cement, aggregates and water. The sands, stones and crystals in cement paste and unhydrated cement grains compose the elastic skeleton of concrete to sustain external loads. Under external loads, the C–S–H gel, micro-voids, and micro-cracks at the interface transition zone between mortar and aggregates make concrete deform plastically. Instinct defects such as voids and initial cracks are generally the origin of concrete failure. The propagation of micro-cracks greatly influences the mechanical properties of concrete.

Because it takes several years for cement paste to harden, the strength and deformation of concrete will vary with time. Furthermore, the deformation will gradually increase with time if the concrete is subjected to sustained loads.

2.2.1 Compression of Concrete Cubes

2.2.1.1 Cube Strength of Concrete

The compression test on cubic concrete specimens is very easy and economical. In addition, the measured strength is stable, and cube strength, as an index of evaluation, is deemed as one of the most fundamental indices of concrete strength in China. National *Standard for test method of mechanical properties on ordinary concrete* (GB/T 50081-2002) stipulates that the cube strength (unit: N/mm^2) of concrete is the compressive strength measured according to the standard test method on standard specimens (150 mm cubes), which have been cured after casting for 28 days in a chamber set at 20 °C ± 3 °C with a relative humidity larger than 90 %.

Figure 2.12 shows the setup of the compression experiment on a concrete cube and the failure mode of the specimen. The test method has a great influence on the compressive strength and the failure mode of cubic concrete specimens. After loose concrete is removed, the failed specimen looks like two pyramids connected top against top (Fig. 2.12b). This is because the specimen will shorten vertically and expand laterally when subjected to vertical compression, but the friction between the specimen and the loading plates places both the top and bottom ends of the specimen under a multiaxial loading state, just as if the specimen were restrained by two hoops at the ends. If the top and bottom surfaces of the specimen are greased, the friction between the specimen and the loading plates is significantly reduced. The specimen is almost under a uniaxial compression state. The restriction against lateral expansion of the specimen is approximately constant along the specimen height. Cracks parallel to the loading direction can be observed, and the measured strength is lower than that of the ungreased specimen (Fig. 2.12c). Specimens should not be greased according to the standard test method in China.

The strength of concrete is also related to the strength grade of cement, the water-to-cement ratio, properties of aggregates, methods of forming, age of concrete, environmental conditions during concrete hardening, dimensions and shapes of specimens, and loading rates. Therefore, all nations have their own standard test methods on strength measurement from concrete specimens under uniaxial loading.

For cubic concrete specimens, the faster the loading rate is, the higher the measured strength is. The loading rate is generally specified as 0.3–0.5 N/mm^2 per minute for concrete specimens with a cube strength lower than 30 N/mm^2 and 0.5–0.8 N/mm^2 per minute for concrete specimens with cube strength equal to or greater than 30 N/mm^2.

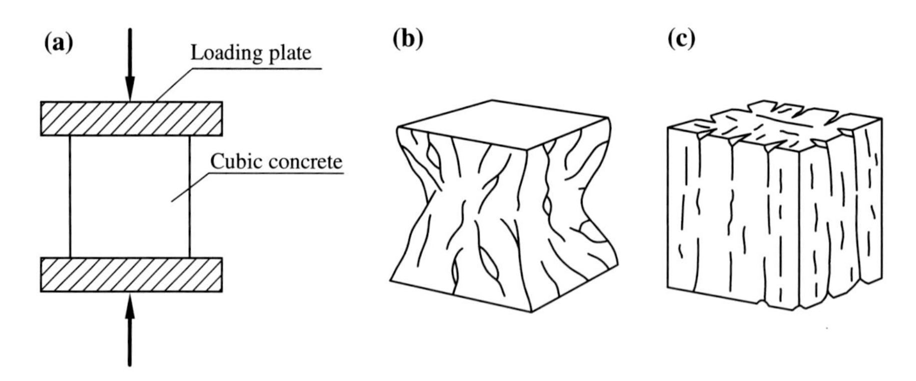

Fig. 2.12 Test setup and failure mode of cubic concrete. **a** Test setup. **b** Failure mode I. **c** Failure mode II

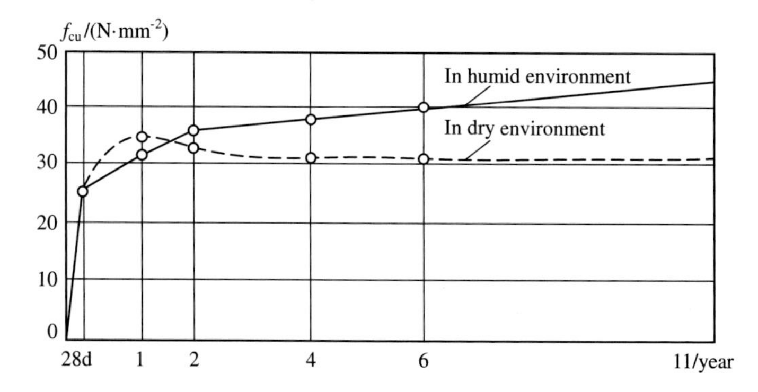

Fig. 2.13 Changes in cube strength of concrete, f_{cu} with aging

The cube strength of concrete will increase with its age. The increase in strength is fast at first, but gradually decreases. This process may take several years or even longer for concrete in a humid environment (Fig. 2.13).

The measured strength of concrete should be rectified if nonstandard cubic specimens are adopted. When the cube strength of concrete is less than 60 MPa, conversion factors of 0.95 and 1.05 should be multiplied to 100 mm × 100 mm × 100 mm and 200 mm × 200 mm × 200 mm specimens, respectively. When the cube strength of concrete is larger than 60 MPa, the conversion factor should be determined by experiments.

2.2.1.2 Strength Grade of Concrete

The cube strength of concrete is the basis to judge the strength grade of concrete. As stipulated in the *Code for Design of Concrete Structures* (GB 50010), concrete strength grade is determined by the characteristic value of the cube strength (denoted by $f_{cu,k}$). In other words, the strength grade of concrete is the cube strength with a 95 % degree of confidence measured according to the standard test method mentioned above. For example, concrete with the strength grade of C30 means its characteristic strength $f_{cu,k} = 30$ N/mm^2. The strength grade of concrete in GB 50010 is within the range of C20–C80. And the concrete with the strength grade equal to or larger than C50 is generally called high-strength concrete.

Statistical analysis shows that the measured cube strength of concrete also obeys Gaussian distribution (Fig. 2.7). If the degree of confidence is taken as 95 %, then the standard strength of concrete is equal to the average strength minus 1.645 times the standard deviation.

GB 50010 specifies that the strength grade of concrete in reinforced concrete structures should not be lower than C20 and should not be lower than C25 for structural members using reinforcement of 400 MPa and structural members under repeated loading. For prestressed concrete structures, C30 is the minimum and C40 or larger is preferred.

2.2.2 *Concrete Under Uniaxial Compression*

2.2.2.1 Experimental Stress–Strain Curve of Concrete Under Axial Compression

The axial compressive strengths measured on prismatic specimens can obviously reflect the real compression capacity of concrete better than the cube strength. *Standard for test method of mechanical properties on ordinary concrete* (GB/T 50081-2002) stipulates that the standard specimen for axial compressive strength should be a prism in the dimension of 150 mm × 150 mm × 300 mm. The fabrication condition for prismatic specimens is the same as that of cubic specimens, and neither prismatic nor cubic specimens are greased. Figure 2.14 shows the setup of the axial compression test on a prismatic specimen and the corresponding failure mode.

Figure 2.15 illustrates a typical measured full stress–strain curve of concrete under axial compression, which can be divided into two parts, i.e., the ascending branch (*Oc*) and the descending branch (*cf*). In the ascending branch, segment *Oa* ($\sigma_c \leqslant 0.3f_c$) is approximately a straight line. The deformation of concrete is mainly due to the elastic deformation of aggregates and cement crystals, while the influence of the viscous flow of hydrated cement paste and the evolution of initial micro-cracks are small. With the increase of stress ($0.3f_c < \sigma_c \leqslant 0.8f_c$), the ascending slope of the curve gradually decreases due to the viscous flow of unhardened gel in concrete and the propagation and growth of micro-cracks. When the stress is increased nearly to the axial compressive strength ($0.8f_c < \sigma_c \leqslant f_c$), large strain energy is stored in the specimen, internal cracks speed their propagation, and the cracks parallel to the axial load link together, which means the specimen is about to fail. Generally, the maximum stress σ_0 corresponding to the peak point c in the stress–strain curve is regarded as the axial compressive strength f_c of concrete, and the strain at point c is called the peak strain ε_0, whose value approximates 0.002.

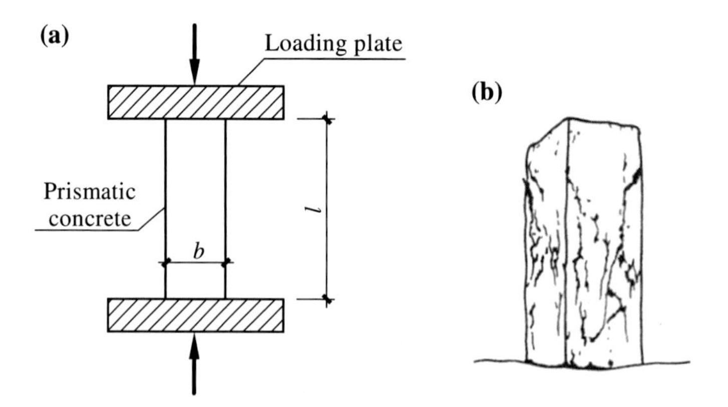

Fig. 2.14 Test setup and failure mode of prismatic concrete. **a** Test setup. **b** Failure mode

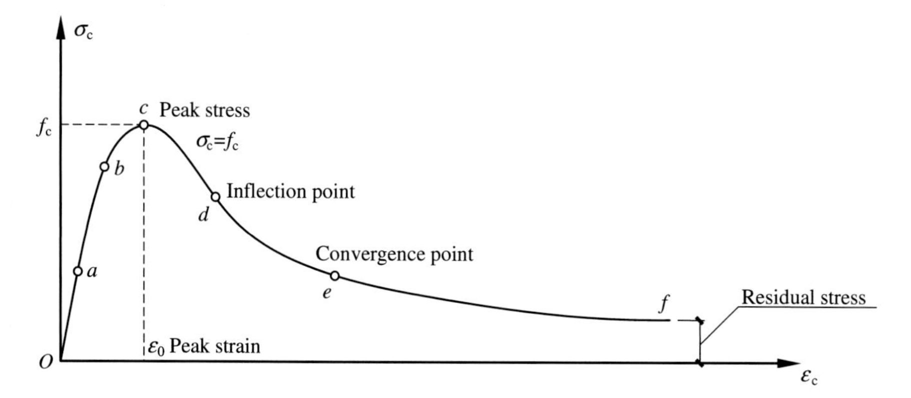

Fig. 2.15 Stress–strain curve of a prismatic concrete under axial compression

After point *c*, further development and connection of continuous cracks damages the prismatic specimen more and more severely, causing the specimen to lose its capacity gradually as shown by the descending branch *cf* in Fig. 2.15. The descending branch is hard to record by ordinary test machines because with the decrease of stress, the strain energy stored in the machine is released and the sudden recovery deformation of the machine will surely crush the already severely damaged specimen. Therefore, machines of large stiffness or with certain auxiliary devices must be adopted, and the strain rate should be strictly controlled so as to obtain the descending branch of the stress–strain curve.

2.2.2.2 Axial Compressive Strength

The higher a prismatic specimen is, the lesser restriction is placed on the transverse deformation at mid-height of the specimen by the friction between loading plates and the specimen. So as the specimen height/width ratio increases, the axial compressive strength decreases (Fig. 2.16). When determining the dimension of a specimen, the specimen height/width ratio should be large enough so as to

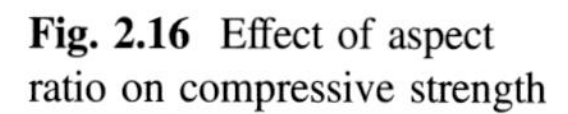
Fig. 2.16 Effect of aspect ratio on compressive strength

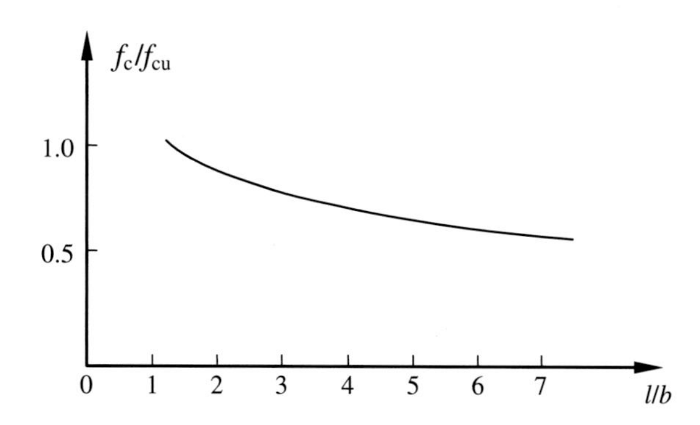

minimize the influence of friction between loading plates and the specimen and to obtain a uniaxial compressive state in the specimen at mid-height. Meanwhile, the height/width ratio cannot be so large as to prevent a large increase in eccentricity, thereby preventing a reduced axial compressive strength from happening before failure. Based on previous research, a height/width ratio of 2–3 is accepted.

GB 50010 stipulates that the measured concrete strength, which is obtained according to the above-mentioned standard test method and has a 95 % degree of confidence, is the characteristic value of the axial compressive strength of concrete, denoted by f_{ck}. The design value of the axial compressive strength is obtained through dividing the characteristic value by a partial safety factor for material $\gamma_c = 1.4$. Table 2.6 in the addendum lists the characteristic values and design values of the axial compressive strength of concrete stipulated in GB 50010.

Similarities in the procedure for measuring the cube strength with nonstandard concrete specimens are evident when measuring the axial compressive strength with nonstandard prismatic specimens, where the measurements must be rectified according to strength grade and dimensions of the specimens. When the strength grade of concrete is less than C60, conversion factors of 0.95 and 1.05 should be multiplied to 100 mm × 100 mm × 300 mm and 200 mm × 200 mm × 400 mm specimens, respectively. When the strength grade of concrete is equal to or larger than C60, the conversion factor should be determined by experiments.

Figure 2.17 shows the relationship between part of the experimental data on axial compressive strength and cube strength obtained in Chinese research institutes. It can be assumed that within a certain range, the axial compressive strength f_c is approximately proportional to the cube strength f_{cu}. Based on experimental

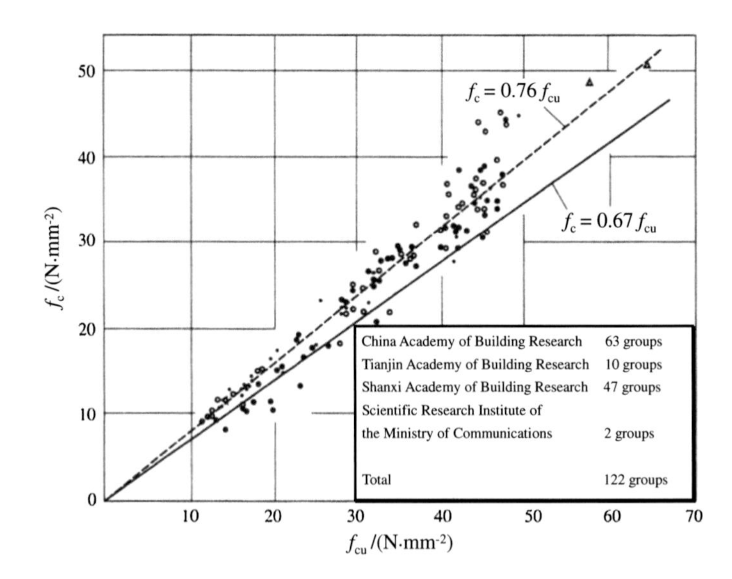

Fig. 2.17 Relationship between axial compressive strength and cubic compressive strength

research, GB 50010 conservatively expresses the relationship between the two strengths as follows:

$$f_{ck} = 0.88\alpha_1\alpha_2 f_{cu,k} \quad (2.4)$$

where

α_1 the ratio of the prism strength to the cube strength. $\alpha_1 = 0.76$ for concrete with the strength grade ⩽C50 and $\alpha_1 = 0.82$ for C80 concrete. The value of α_1 is linearly interpolated between 0.76 and 0.82 for C55–C75 concrete;

α_2 the reduction coefficient considering the brittleness of high-strength concrete. $\alpha_2 = 1.0$ for C40 concrete and $\alpha_2 = 0.87$ for C80 concrete. The value of α_2 is linearly interpolated between 1.0 and 0.87 for intermediate grades; and

0.88 a parameter to consider the strength differences between laboratory specimens and real structural members because they have different fabrication methods, curing conditions, and loading states.

In some countries or regions, cylinder specimens are chosen to determine the axial compressive strength of concrete. For example, concrete cylinders with the diameter of 6 in. (152 mm) and the height of 12 in. are adopted as the standard specimens for the axial compressive strength in America. The cylinder strength is denoted by f'_c. Because of the differences in shape and dimension, cylinder strength is different from prism strength.

Based on foreign research results, the relationship between the cylinder strength f'_c and the cube strength $f_{cu,k}$ is shown in Table 2.1, in which the ratio of $f'_c/f_{cu,k}$ increases with the strength grade of concrete if the strength grade is equal to or larger than C60.

The stress–strain curves for concrete of different strength grades are similar in shape but have substantial distinction. From the experimental curves shown in Fig. 2.18, it can be seen that the peak strains for concrete of different strength grades are nearly the same, but the shape of the descending branches varies greatly. The higher the strength grade is, the steeper the descending slope is. So it is generally accepted that concrete of higher strength has a poorer ductility.

Moreover, axial compressive strength is also influenced by the loading rate. As shown in Fig. 2.19, with the decrease of the loading rate, the peak stress (i.e., the axial compressive strength) also decreases a little bit. However, the strain corresponding to the peak stress increases, and the slope of the descending branch becomes gentle.

Table 2.1 Ratio of f'_c to $f_{cu,k}$

Concrete grade	Under C60	C60	C70	C80
$f'_c/f_{cu,k}$	0.79	0.833	0.857	0.875

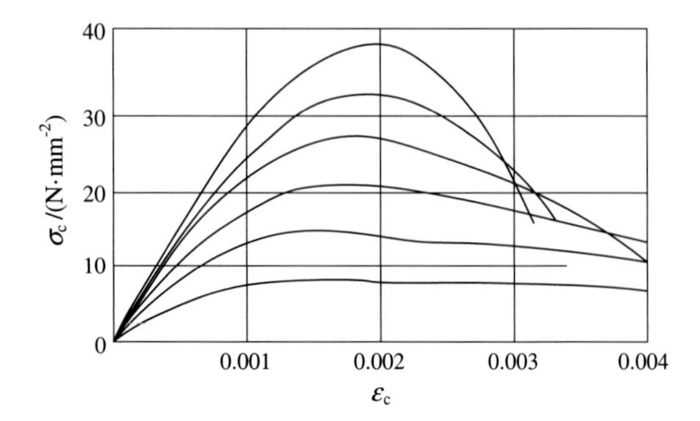

Fig. 2.18 Stress–strain curves for concrete of different strength grades

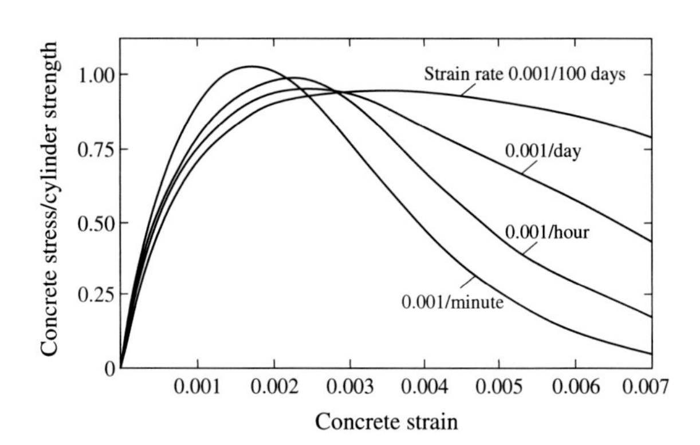

Fig. 2.19 Stress–strain curves for concrete under different strain rates

2.2.2.3 Mathematical Models of the Stress–Strain Relationship of Concrete Under Axial Loading

Researchers at home and abroad have established many mathematical models to describe the stress–strain relationship of concrete under axial loading. Several examples are as follows:

1. Model suggested by Hognestad

The model suggested by Hognestad assumes the ascending branch and the descending branch as a second-order parabola and an oblique straight line (Fig. 2.20), respectively, expressed by Eq. (2.5a and 2.5b).

$$\sigma_c = f_c\left[2\frac{\varepsilon_c}{\varepsilon_0} - \left(\frac{\varepsilon_c}{\varepsilon_0}\right)^2\right] \quad (\varepsilon_c \leqslant \varepsilon_0) \tag{2.5a}$$

$$\sigma_c = f_c\left[1 - 0.15\frac{\varepsilon_c - \varepsilon_0}{\varepsilon_{cu} - \varepsilon_0}\right] \quad (\varepsilon_0 < \varepsilon_c \leqslant \varepsilon_{cu}) \tag{2.5b}$$

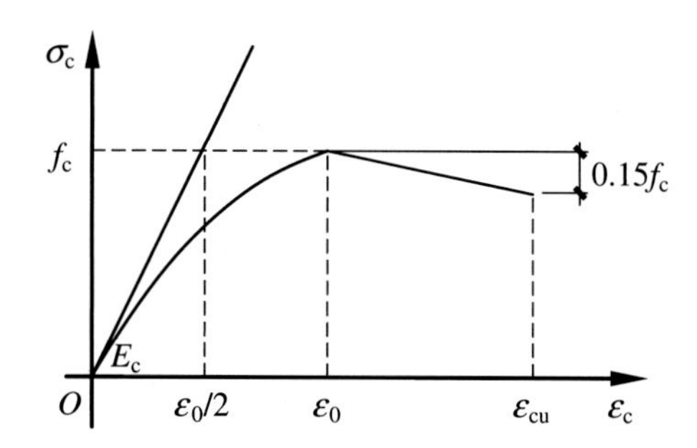

Fig. 2.20 Stress–strain curve suggested by Hognestad

where

f_c the peak stress (the axial compressive strength of concrete);

ε_0 the strain corresponding to the peak stress, taking the value as

$$\varepsilon_0 = 1.8\frac{f_c}{E_c} \tag{2.5c}$$

ε_{cu} the ultimate compressive strain, taking the value as 0.0038; and

E_c the modulus of elasticity of concrete, given by empirical equations (omitted herein).

2. Model suggested by Rüsch

The model suggested by Rüsch also adopts a parabolic ascending branch, but the descending branch is a horizontal straight line (Fig. 2.21), expressed as

$$\sigma_c = f_c\left[2\frac{\varepsilon_c}{\varepsilon_0} - \left(\frac{\varepsilon_c}{\varepsilon_0}\right)^2\right] \quad (\varepsilon_c \leqslant \varepsilon_0) \tag{2.6a}$$

$$\sigma_c = f_c \quad (\varepsilon_0 < \varepsilon_c \leqslant \varepsilon_{cu}) \tag{2.6b}$$

where the strain corresponding to the peak stress is 0.002 and the ultimate strain is 0.0035.

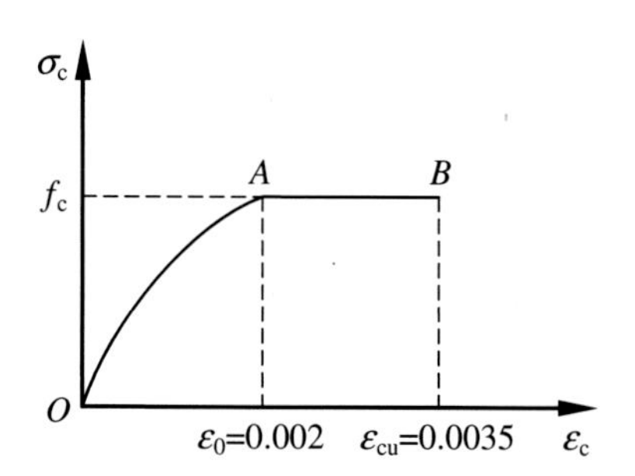

Fig. 2.21 Stress–strain curve suggested by Rüsch

3. Model adopted in GB 50010

Referring to Rüsch's model and research results on high-strength concrete in recent years, GB 50010 proposes the following mathematical expression for the stress–strain relationship of concrete under axial loading:

$$\left.\begin{array}{ll}\sigma_c = f_c\left[1-\left(1-\frac{\varepsilon_c}{\varepsilon_0}\right)^n\right] & \varepsilon_c \leqslant \varepsilon_0 \\ \sigma_c = f_c & \varepsilon_0 < \varepsilon_c \leqslant \varepsilon_{cu}\end{array}\right\} \tag{2.7}$$

$$n = 2 - \frac{1}{60}(f_{cu} - 50) \tag{2.8}$$

$$\varepsilon_0 = 0.002 + 0.5(f_{cu} - 50) \times 10^{-5} \tag{2.9}$$

$$\varepsilon_{cu} = 0.0033 - (f_{cu} - 50) \times 10^{-5} \tag{2.10}$$

where

σ_c the stress in concrete corresponding to the compressive strain ε_c;
f_c the axial compressive strength of concrete;
ε_0 the compressive strain corresponding to f_c. If the calculated $\varepsilon_0 < 0.002$, take $\varepsilon_0 = 0.002$;
ε_{cu} the ultimate compressive strain. If the calculated $\varepsilon_{cu} > 0.0033$, take $\varepsilon_{cu} = 0.0033$;
f_{cu} the cube strength of concrete; and
n a parameter. If the calculated $n > 2.0$, take $n = 2.0$;

When the compressive stress is small (say $\sigma_c \leqslant 0.3f_c$), approximately take

$$\sigma_c = E_c\varepsilon_c \tag{2.11}$$

where E_c is the modulus of elasticity of concrete.

Obviously, when $f_{cu} \leqslant 50$ MPa, Eq. (2.7) becomes Eq. (2.6a, 2.6b).

2.2.2.4 Modulus of Elasticity of Concrete

Modulus of elasticity is the ratio of stress to strain. Because the stress–strain relationship of concrete under axial compression is a curve, the modulus of elasticity is a variable. It can be depicted in three ways, i.e., initial modulus of elasticity, secant modulus, and tangent modulus, whose values are $\tan\alpha_0$, $\tan\alpha_1$, and $\tan\alpha$ as shown in Fig. 2.22, respectively.

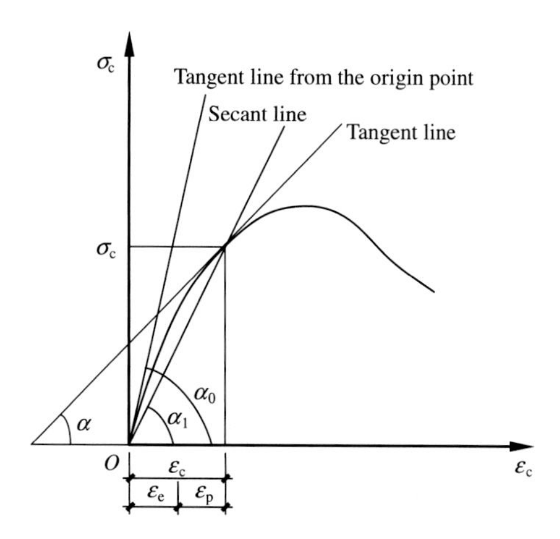

Fig. 2.22 Depiction of concrete elastic modulus

The modulus of elasticity of concrete generally means the initial modulus of elasticity, denoted by E_c. Based on the regression analysis of experimental data, the modulus of elasticity has a relation with cube strength as follows:

$$E_c = \frac{10^5}{2.2 + \frac{34.7}{f_{cu}}} \left(\text{N/mm}^2\right) \tag{2.12}$$

The moduli of elasticity given by GB 50010 are listed in Table 2.6 in the addendum. It is of practical meaning to use secant modulus or tangent modulus in nonlinear analysis of concrete structures, because the two moduli can better reflect the characteristics of stress–strain relationship curves. And the relation between the secant modulus E'_c and the initial modulus of elasticity E_c can be expressed as

$$E'_c = vE_c \tag{2.13}$$

where v is a proportional constant, taking the value as 0.4–1.0 for compression and 1.0 for tension failure.

2.2.2.5 Transverse Deformation of Concrete

Concrete specimens under uniaxial compression will deform not only longitudinally with the compressive strain of ε_v, but also transversely with the transverse strain of ε_h. The transverse deformation coefficient (i.e., the Poisson's ratio) is defined as $v_c = \varepsilon_h/\varepsilon_v$. From the experimental results shown in Fig. 2.23, when the compressive stress is small ($\sigma_c \leqslant 0.5f_c$), v_c is approximately a constant of the value 1/6, which is the Poisson's ratio corresponding to concrete in the elastic stage. When the compressive stress is large ($\sigma_c > 0.5f_c$), v_c increases apparently due to

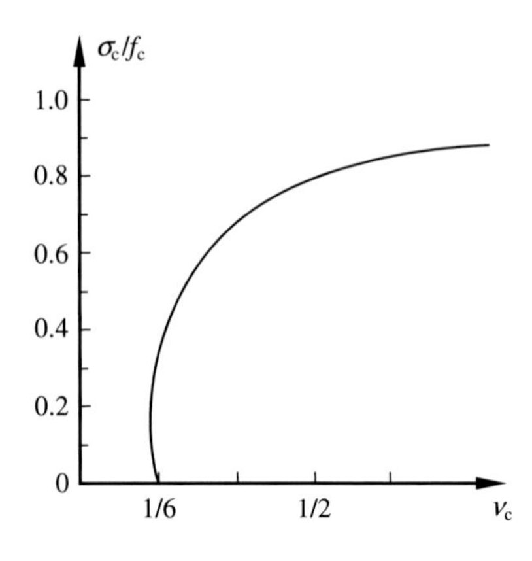

Fig. 2.23 Relationship between compressive stress and Poisson's ratio

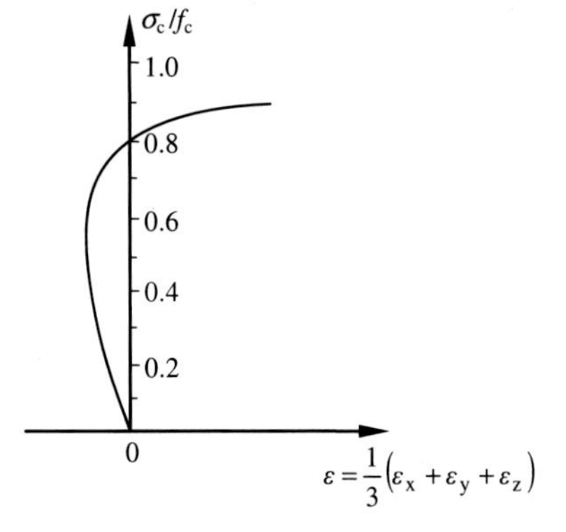

Fig. 2.24 Relationship between stress and averaged strain

internal crack development, and when concrete is nearly failed, ν_c can reach the value as 0.5 or higher.

Figure 2.24 shows the relationship between stress and average strain, which is the average of measured strains in three mutually perpendicular directions. It can be seen that when the stress is small ($\sigma_c \leqslant 0.5f_c$), the specimen volume decreases with the increase of compressive stress. When the stress is large ($\sigma_c > 0.5f_c$), the compressed volume gradually recovers. When the specimen is nearly failed, the volume may even be larger than the original one.

2.2.3 *Concrete Under Uniaxial Tension*

2.2.3.1 Tensile Tests on Concrete

The standard specimens for axial tension test on concrete are prisms with embedded steel bars at both ends (Fig. 2.25).

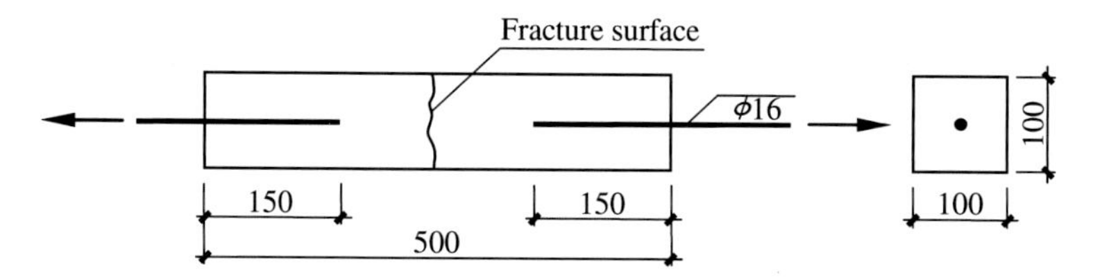

Fig. 2.25 Schematic diagram of the axial tension test

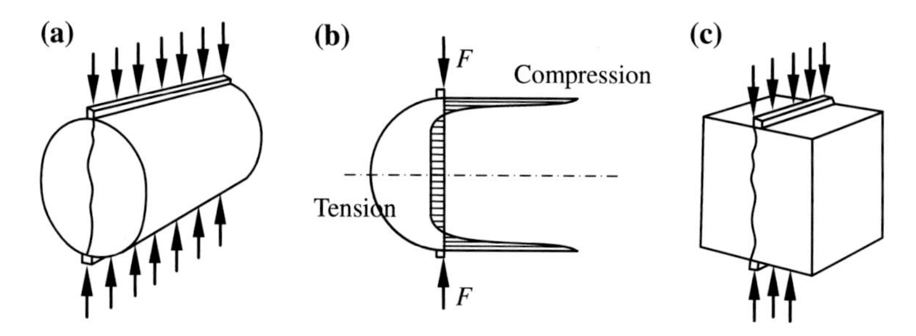

Fig. 2.26 Schematic diagram of the splitting test

However, using specimens shown in Fig. 2.25 cannot ensure the specimens under perfect axial tension, and an eccentric tension will surely influence the accuracy of the tensile strength measurement. So the much simpler splitting test on cylinders or cubes shown in Fig. 2.26 is widely employed at home and abroad to indirectly measure the tensile strength of concrete. From the theory of elasticity, the splitting tensile strength f_{ts} can be calculated according to the following equation

$$f_{ts} = \frac{2F}{\pi d_c l} \tag{2.14}$$

where

F the failure load;
d_c the diameter of cylinders or the dimension of cubes; and
l the length of cylinders or the dimension of cubes.

Experimental results show that splitting tensile strength is only slightly higher than the tensile strength obtained from the direct pulling test.

2.2.3.2 Axial Tensile Strength

Experimental results of ordinary concrete and high-strength concrete under uniaxial tension indicate the following relation between the axial tensile strength and the cube strength

$$f_{\rm t} = 0.395 f_{\rm cu}^{0.55} \tag{2.15}$$

GB 50010 gives the conversion relation between the characteristic value of axial tensile strength and that of cube strength as follows:

$$f_{\rm tk} = 0.88 \times 0.395 f_{\rm cu,k}^{0.55} (1 - 1.645\delta)^{0.45} \times \alpha_2 \tag{2.16}$$

The meanings of 0.88 and the value of α_2 are the same as those in Eq. (2.4). The term of $(1 - 0.645\delta)^{0.45}$ reflects the influence of dispersion degree of experimental data on confidence degree of characteristic strength. δ is the coefficient of variation. Table 2.6 lists the characteristic values and design values of the axial tensile strength of concrete.

The ratio of the axial tensile strength to the cube strength is within 1/17–1/8. The higher the strength grade of concrete is, the smaller the ratio is.

2.2.3.3 Stress–Strain Relationship of Concrete Under Axial Tension

The experimental stress–strain curves of concrete under axial tension are much fewer than those under axial compression. The available experimental results show that the stress–strain curves of concrete under axial tension are similar to those under axial compression in shape and also include the ascending and descending branch. But the slope of the descending branch is steep and may become steeper with the increase of the strength grade of concrete.

Because the axial tensile strength is much lower than the axial compressive strength, the stress–strain relationship of concrete under axial tension can be simulated by a bilinear model and the moduli of elasticity of concrete under tension and compression are assumed to be the same.

2.2.4 Concrete Under Multiaxial Stresses

In real structural members, concrete is usually subjected to multiaxial stresses rather than the ideal uniaxial loading, so investigating the multiaxial behavior of concrete is of great importance for better understanding the properties of concrete members and improving the design and research of concrete structures.

2.2.4.1 Biaxial Behavior of Concrete

The failure curve of concrete under a biaxial stress state (Fig. 2.27) can be obtained by applying normal stresses σ_1 and σ_2 in two mutually perpendicular directions

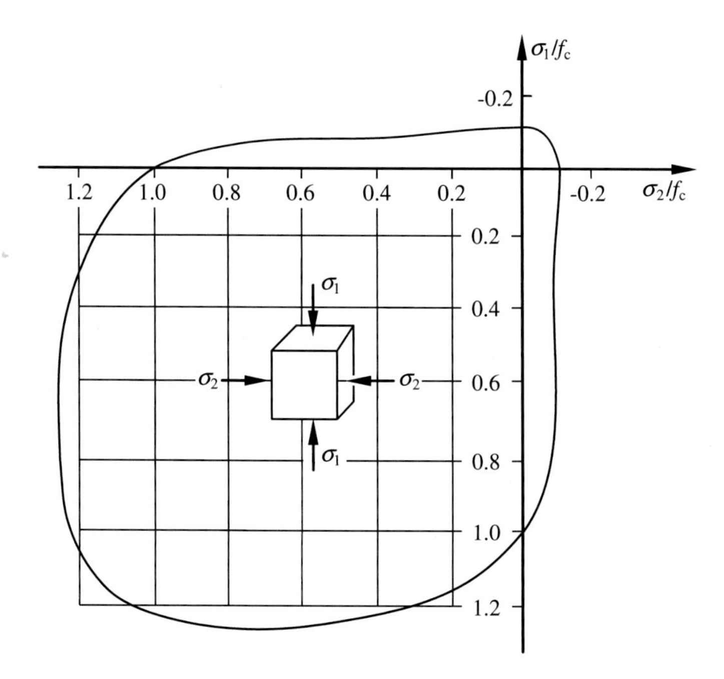

Fig. 2.27 Concrete strength under biaxial stress

while keeping the normal stress of zero in the third direction perpendicular to the aforementioned two directions and recording the strengths of concrete under different stress ratios (σ_1/σ_2).

It can be seen that under biaxial tension (1st quadrant in Fig. 2.27), σ_1 and σ_2 do not greatly influence each other, and the biaxial tension strength of concrete is approximately equal to the uniaxial tensile strength. However, for concrete under biaxial compression, the strength in one direction increases with the buildup of compressive stress in another direction (3rd quadrant in Fig. 2.27). The biaxial compressive strength can be as much as 27 % higher than the uniaxial strength. Combined tension and compression loadings reduce both the tensile and compressive stresses at failure (2nd and 4th quadrant in Fig. 2.27).

2.2.4.2 Strength Under Combined Normal and Shear Stresses

Figure 2.28 shows the failure curve of concrete under combined normal and shear stresses. It is found that the shear strength of concrete will increase with the buildup of compressive stress when the latter is small. But after the compressive stress exceeds (0.5–0.7) f_c, the shear strength will decrease with the increase of compressive stress. On the other hand, the existence of shear stress reduces the compressive strength of concrete. Similarly, the shear strength decreases with the

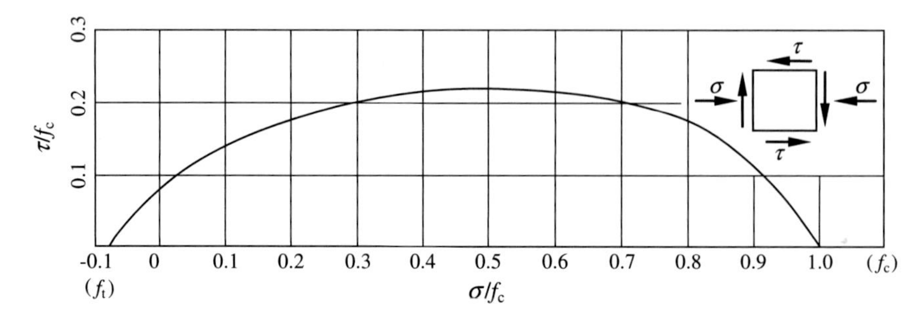

Fig. 2.28 Concrete strength under combined normal and shear stresses

increase of tensile stress. And the tensile strength of concrete reduces in the presence of shear stress.

2.2.4.3 Compressive Strength of Confined Concrete

Overseas experimental research indicates that for axially loaded concrete cylinders, the axial strength will greatly improve if the cylinders are subjected to uniform confining fluid pressure. The increased amplitude is approximately proportional to the confining pressure (Fig. 2.29). When σ_2 is not very large, the ultimate compressive strength f'_{cc} in the direction of σ_1 can be expressed as

$$f'_{cc} = f'_c + 4.1\sigma_2 \tag{2.17}$$

where

f'_c the compressive strength of the unconfined concrete cylinder; and

σ_2 the confining pressure.

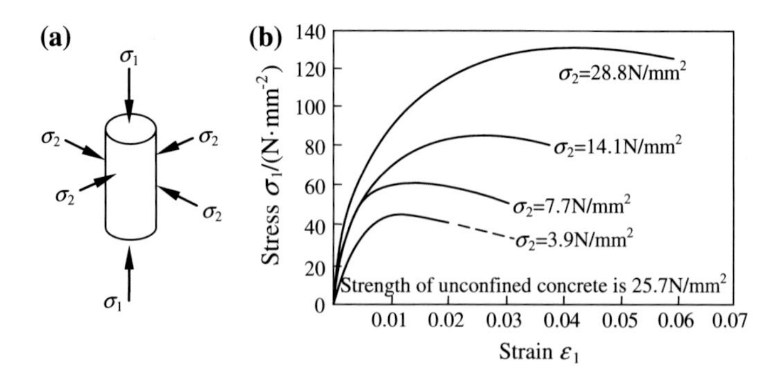

Fig. 2.29 Stress–strain curves of concrete cylinders under three-dimensional compression

The presence of confining pressure restrains the lateral deformation of concrete cylinders and suppresses the initiation and evolution of internal cracks parallel to the direction of σ_1. Therefore, the ductility of concrete is enhanced simultaneously with the increase of axial compressive strength.

Based on the aforementioned mechanism, the load-carrying capacity and deformation property of concrete columns subjected to compression can be improved by the dense arrangement of circular hoops or spirals at the column periphery to restrain the lateral deformation of internal concrete. The calculation based on the arrangement of circular hoops or spirals will be detailed in Chap. 4.

2.2.5 Strength and Deformation of Concrete Under Repeated Loading

The strength and deformation of concrete under repeated loading (several loading and unloading cycles) are greatly different from those under monotonic loading. Fatigue failure can happen to concrete under repeated loading.

Specimens sized 100 mm × 100 mm × 300 mm or 150 mm × 150 mm × 450 mm are usually used in fatigue tests of concrete. And the compressive stress at which the concrete specimen finally fails after 2 million (or even more) times of repeated loading is called the fatigue strength of concrete.

Figure 2.30a shows the stress–strain curves of a concrete prism subjected to one cycle of loading and unloading, in which *Oa* and *ab* are loading and unloading curves, respectively. When the stress is decreased to zero after having reached point *a*, most of the overall strain ε_c corresponding to point *a*, i.e., ε'_e, can be recovered instantaneously during unloading, and a small portion of strain ε''_e can also be recovered after some time, which is referred to as elastic hysteresis. The unrecovered strain ε'_{cr} is called residual strain.

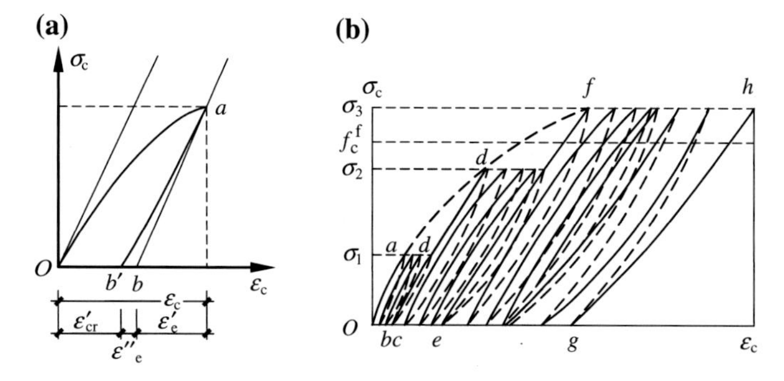

Fig. 2.30 Stress–strain curves of concrete under repeated loading. **a** Stress–strain curves of a concrete prism subjected to one cycle of loading and unloading. **b** Stress–strain curves of a concrete prism subjected to many cycles of loading and unloading

Figure 2.30b shows the stress–strain curves of a concrete prism subjected to many cycles of loading and unloading. When the loading stress is lower than the fatigue strength of concrete f_c^f, e.g., σ_1 or σ_2 in Fig. 2.30b, the stress–strain curve is similar to that in Fig. 2.30a. But the loop formed by loading and unloading curves in one cycle tends to be closed after several repeated times. However, even if the repeated times were as high as several million, the concrete prism would not be failed by fatigue. If the loading stress exceeds the fatigue strength f_c^f, say σ_3 in Fig. 2.30b, the loading curves are initially convex to the stress axis and gradually become convex to the strain axis after many times of repeated loadings. In the final stage, the loading and unloading curves in one cycle cannot form a closed loop and the slope of the stress–strain curve continuously decreases, indicating the imminent fatigue failure of concrete.

Concrete fatigue originates from internal defects such as micro-cracks and micro-voids. The stress concentration in concrete under repeated loading causes defects to develop and form macro-cracks, which finally lead to concrete failure. Fatigue failure is brittle, i.e., giving no apparent warning before failure. Cracks are not wide, but deformation is large.

GB 50010 stipulates that the design value of concrete fatigue strength is determined by multiplying the design strength f_c or f_t by corresponding corrector factors γ_p, which is chosen from Table 2.7 based on the fatigue stress ratio ρ_c^f.

$$\rho_c^f = \frac{\sigma_{c,min}^f}{\sigma_{c,max}^f} \tag{2.18}$$

where $\sigma_{c,\,min}^f$ and $\sigma_{c,\,max}^f$ = the minimum and maximum stresses of concrete in the same fiber, respectively.

2.2.6 *Deformation of Concrete Under Long-Term Loading*

2.2.6.1 Creep

Creep is the increase in strain with time due to a sustained load.

Figure 2.31 illustrates the creep curve of a prismatic concrete specimen. An instantaneous strain ε_c will be recorded when the specimen is loaded to a certain value of stress, (say $0.5f_c$ in Fig. 2.31). If the stress is kept constant, the specimen deformation will continuously increase with time, expressed as the creep strain ε_{cr}. In the first several months of loading, the creep strain increases rapidly and 70–80 % of the total creep strain can be finished in half a year. Then, the increased rate of the creep strain gradually decreases and will stabilize after a long time. The creep strain measured two years later is 1–4 times the instantaneous strain. If unloaded at this time, the specimen will recover part of the deformation (elastic recovery ε'_e), which is smaller than the instantaneous strain at loading. Another part

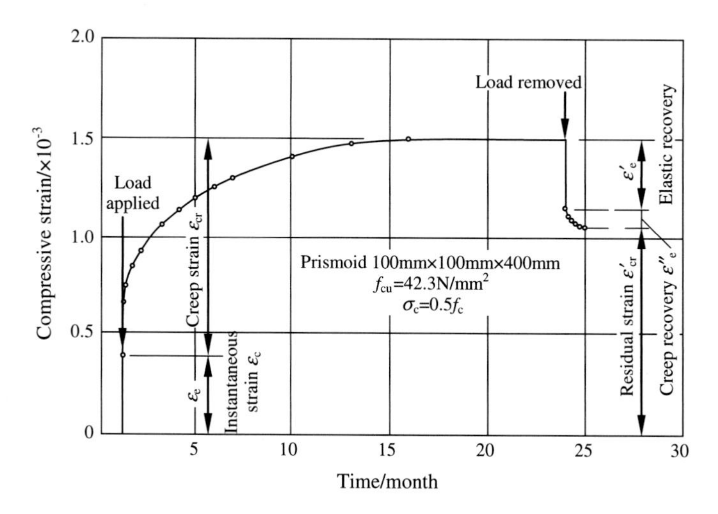

Fig. 2.31 Concrete creep with time

of strain (creep recovery ε''_e) can be recovered in about 15 days after unloading. However, most of the strain is unrecoverable, which is called the residual strain ε'_{cr}.

Creep will greatly influence the properties of concrete structural members. For example, it can enlarge the deformation of structural members, induce stress redistribution in cross sections, and cause prestress loss in prestressed concrete structures.

2.2.6.2 Factors Influencing Creep

Creep of concrete will be influenced by many factors, such as stress magnitude, inherent material characteristics, and environmental conditions.

Experiments show that stress magnitude is one of the most important factors. As shown in Fig. 2.32a, when $\sigma_c \leqslant 0.5f_c$, the spacings between any two adjacent creep curves are nearly equal, indicating that the creep of concrete is approximately proportional to the stress. This is called linear creep. Linear creep accelerates quickly at the beginning of loading and then gradually slows down and can be assumed to be fully stopped after about three years.

Figure 2.32b illustrates the influence of large stresses on the creep of concrete. When the stress is large, say σ_c = (0.5–0.8) f_c, the creep is not proportional to the stress any longer, and the increase rate of creep is greater than that of stress. This is called nonlinear creep. When the stress becomes even larger, say $\sigma_c > 0.8f_c$, the internal cracks in specimens develop in an unstable way, which causes an intensive increase of nonlinear creep and thus leads to the final failure of concrete. Therefore, it is generally agreed that the compressive strength of concrete under sustained loads is just 75–80 % of its short-term counterpart.

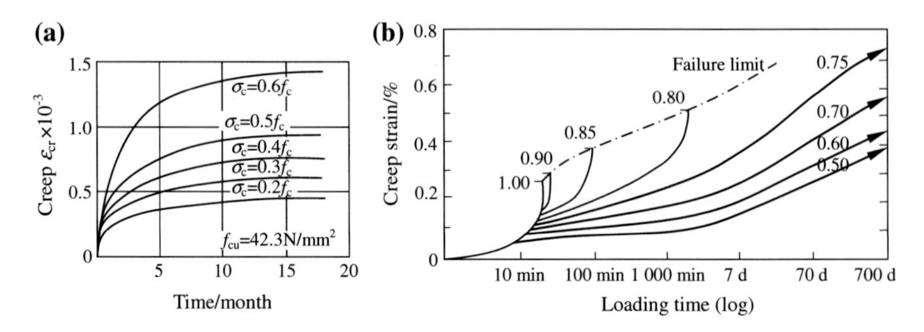

Fig. 2.32 Relationship between creep and compressive stress

The composition of concrete also affects the creep greatly. The larger the water/cement ratio is, the bigger the creep strain is. The increase of cement usage will enlarge the creep strain too. Moreover, the mechanical properties of aggregates have an apparent influence on the creep of concrete. For example, using maximum size solid aggregates allows an increased modulus of elasticity, and an increased volume ratio of aggregates to concrete helps reduce the creep strain.

The creep may be influenced by fabrication methods and curing conditions of concrete as well. Curing concrete in the conditions of high temperature and humidity can promote the hydration of cement so as to reduce the creep strain. On the other hand, if the temperature is high, but the humidity is low during curing, the creep strain will increase. Additionally, the earlier the load is applied, the larger the creep strain is.

Besides, the shape and dimension of members will affect the creep of concrete. Because the moisture in members of large dimension is hard to evaporate, the creep strain is small. The arrangement of reinforcement may also change the creep strain values.

2.2.7 *Shrinkage, Swelling, and Thermal Deformation of Concrete*

Shrinkage refers to the decrease in the volume of a concrete member when it loses moisture by evaporation. The opposite phenomenon, swelling, occurs when the volume increases through water absorption. Shrinkage and swelling represent the volume change of concrete specimens during hardening irrespective of the external load.

Figure 2.33 shows the experimental results of shrinkage and swelling of concrete members. It can be seen that the shrinkage strain increases quickly at the beginning and becomes almost asymptotic after about one year. The test data of shrinkage strain is very scattering, say (2–5) $\times$ 10^{-4}, and the value as 3 $\times$ 10^{-4} is generally

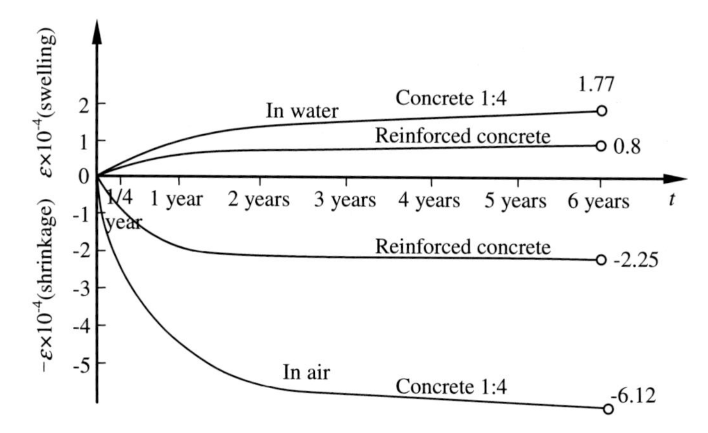

Fig. 2.33 Shrinkage and swelling of concrete with time

taken. In reinforced concrete members, the shrinkage of concrete is restrained. So its value is half that of plain concrete, i.e., 1.5×10^{-4}.

There are several factors that affect the magnitude of drying shrinkage.

2.2.7.1 Type of Cement

The higher the cement grade is, the larger the shrinkage strain is.

2.2.7.2 Amount of Cement

The larger the cement usage or the water/cement ratio is, the bigger the shrinkage strain is.

2.2.7.3 Aggregate Property

The larger the modulus of elasticity based on aggregates is, the smaller the shrinkage strain is.

2.2.7.4 Ambient Conditions

In the hardening and following service stage of concrete, the larger the ambient moisture is, the smaller the shrinkage strain is. The shrinkage strain will decrease with the increase of curing temperature if the relative humidity is large. And an opposite trend will be observed if the ambient condition is dry.

2.2.7.5 Quality of Construction

The denser the concrete is when vibrated, the smaller the shrinkage strain is.

2.2.7.6 Ratio of Volume to Surface Area

The larger the ratio of volume to surface area is, the smaller the shrinkage strain is.

If concrete members are not well cured or the contraction is restrained, shrinkage cracks will appear on the surface or in the interior of the members. Cracks will not only affect the appearance, but also adversely influence the serviceability and durability of concrete.

The swelling of concrete is much smaller than the shrinkage and generally beneficial to the structural members, so it is usually not considered.

The linear expansion coefficient of concrete is related to its mix composition and aggregate property. The value (1.0–1.5) $\times$ 10^{-5} is close to 1.2 $\times$ 10^{-5} of steel. Therefore, the deformation difference between concrete and reinforcement caused by temperature change is small, and no harmful internal stress happens to structural members. However, for mass concrete structures such as pools and chimneys, the influence of thermal stress on structural properties should be taken into account.

Questions:

2.1 How can reinforcement be classified?

2.2 What is the difference between the stress–strain curves of hard steel and mild steel? How is their yield strength determined?

2.3 What mathematical models are used to simulate the stress–strain relationship of reinforcement? How can their applicable conditions be described?

2.4 What is the influence of cold drawing and cold stretching on mechanical properties of steel?

2.5 What are the requirements on properties of steel in reinforced concrete structures?

2.6 How are the cube strength, axial compressive strength (i.e., the prism strength), and the axial tensile strength of concrete determined?

2.7 How is the strength grade of concrete determined? What is the grade range stipulated in GB 50010?

2.8 What is the characteristic of the stress–strain curve of concrete under axial compression? Please provide an example of a frequently used mathematical model used to define the stress–strain relationship.

2.9 How is the modulus of elasticity of concrete determined?

2.10 What is the fatigue strength of concrete? What is the characteristic of the stress–strain curve of concrete under repeated loading?

2.11 What are the shrinkage and creep of concrete? What factors will influence the creep and shrinkage of concrete?

2.12 What are the influences of creep and shrinkage on the mechanical properties of reinforced concrete structural members?

Appendix

See Tables 2.2, 2.3, 2.4, 2.5, 2.6, and 2.7.

Table 2.2 Standard strength, design strength, and elastic modulus of normal steel bars

Type	Symbol	Nominal diameter d/mm	Standard yield strength $f_{yk}/(N \cdot mm^{-2})$	Standard ultimate strength $f_{suk}/(N \cdot mm^{-2})$	Design strength/ $(N \cdot mm^{-2})$		Elastic modulus $E_s/(N \cdot mm^{-2})$	Ultimate strain $\varepsilon_{cu}/\%$
					f_y	f'_y		
HPB 300	ϕ	6–22	300	420	270	270	2.1×10^5	Not less than 10.0
HRB 335	ϕ	6–50	335	455	300	300	2.0×10^5	Not less than 7.5
HRBF 335	ϕ^F							
HRB 400	ϕ	6–50	400	540	360	360		
HRBF 400	ϕ^F							
RRB 400	ϕ^R							
HRB 500	ϕ	6–50	500	630	435	435		
HRBF 500	ϕ^F							

Notes When steel bars with the diameter larger than 40 mm are adopted, experimental studies and reliable engineering experience are essential

Table 2.3 Standard strength, design strength, and elastic modulus of prestressed tendons

Type		Symbol	Diameter d/mm	Standard yield strength f_{pyk}/(N·mm^{-2})	Standard ultimate strength f_{puk}/(N·mm^{-2})	Design strength/(N·mm^{-2})		Elastic modulus E_s/(N·mm^{-2})	Ultimate strain ε_{pu}/%
						f_{py}	f'_{py}		
Prestressed steel wire with medium strength	Smooth	ϕ^{PM}	5, 7, 9	680	800	560	410	2.05 × 10^5	Not less than 3.5
	spiral rib	ϕ^{HM}		820	970	680	410		
				1080	1270	900	410		
stress-relieved steel wire	Smooth spiral rib	ϕ^{P}	5	1330	1570	1110	410		
				1580	1860	1320	410		
		ϕ^{H}	7	1330	1570	1110	410		
			9	1250	1470	1040	410		
				1330	1570	1110	410		
Steel strand	1 × 3 (3-wire twisted)	ϕ^{S}	6.5, 8.6, 10.8, 12.9	1330	1570	1110	390	1.95 × 10^5	
				1580	1860	1320	390		
				1660	1960	1390	390		
	1 × 7 (7-wire twisted)		9.5, 12.7, 15.2	1460	1720	1220	390		
				1580	1860	1320	390		
				1660	1960	1390	390		
			21.6	1460	1720	1220	390		
Prestressed rebar	Spiral rib	ϕ^{T}	18, 25, 32, 40, 50	785	980	650	435	2.00 × 10^5	
				930	1080	770	435		
				1080	1230	900	435		

Notes (1) When the standard strength values of prestressed tendons do not meet the table specifications, they should be converted to Table 2.3 values as shown (2) Compressive strength f'_{py} is not considered for unbonded prestressed tendons

Table 2.4 Limitation of fatigue stress range of steel bars in reinforced concrete structures (N·mm^{-2})

Fatigue stress ratio	Δf_y^f	
	HRB 335	HRB 400
$0 \leqslant \rho_s^f < 0.1$	165	165
$0.1 \leqslant \rho_s^f < 0.2$	155	155
$0.2 \leqslant \rho_s^f < 0.3$	150	150
$0.3 \leqslant \rho_s^f < 0.4$	135	145
$0.4 \leqslant \rho_s^f < 0.5$	125	130
$0.5 \leqslant \rho_s^f < 0.6$	105	115
$0.6 \leqslant \rho_s^f < 0.7$	85	95
$0.7 \leqslant \rho_s^f < 0.8$	65	70
$0.8 \leqslant \rho_s^f < 0.9$	40	45

Notes (1) When flash exposure butt joint is adopted in longitudinal tensile steel bars, the design fatigue strength of welded joints should be multiplied by a factor 0.8 based on the table
(2) Grade RRB 400 steel bars should not be adopted for elements when fatigue check is needed
(3) Steel bars of Grade HRBF 335, HRBF 400, and HRBF 500 are inappropriate for elements requiring a fatigue check. If necessary, an experimental study should be adopted

Table 2.5 Limitation of fatigue stress range of prestressed tendons (N·mm^{-2})

Type		Δf_y^f	
		$0.7 \leqslant \rho_p^f < 0.8$	$0.8 \leqslant \rho_p^f < 0.9$
Stress-relieved steel wire	$f_{puk} = 1770, 1670$	165	165
	$f_{puk} = 1570$	155	155
Steel strand		150	150

Notes (1) When $\rho_s^f \geqslant 0.9$, it is not necessary to check the fatigue strength of prestressed tendons
(2) When well founded, appropriate adjustments can be made to the fatigue stress range in the table

Table 2.6 Standard strength, design strength, elastic modulus, and fatigue modulus of concrete ($N \cdot mm^{-2}$)

Strength grade	Standard strength		Design strength		Elastic modulus E_c	Fatigue modulus E_c^f
	f_{ck}	f_{tk}	f_c	f_t		
C15	10.0	1.27	7.2	0.91	2.20×10^4	
C20	13.4	1.54	9.6	1.10	2.55×10^4	1.10×10^4
C25	16.7	1.78	11.9	1.27	2.80×10^4	1.20×10^4
C30	20.1	2.01	14.3	1.43	3.00×10^4	1.30×10^4
C35	23.4	2.20	16.7	1.57	3.15×10^4	1.40×10^4
C40	26.8	2.39	19.1	1.71	3.25×10^4	1.50×10^4
C45	29.6	2.51	21.1	1.80	3.35×10^4	1.55×10^4
C50	32.4	2.64	23.1	1.89	3.45×10^4	1.60×10^4
C55	35.5	2.74	25.3	1.96	3.55×10^4	1.65×10^4
C60	38.5	2.85	27.5	2.04	3.60×10^4	1.70×10^4
C65	41.5	2.93	29.7	2.09	3.65×10^4	1.75×10^4
C70	44.5	2.99	31.8	2.14	3.70×10^4	1.80×10^4
C75	47.4	3.05	33.8	2.18	3.75×10^4	1.85×10^4
C80	50.2	3.11	35.9	2.22	3.80×10^4	1.90×10^4

Table 2.7 Correction factor γ_p for concrete fatigue compressive strength under different fatigue stress ratios ρ_c^f

ρ_c^f	$0 \leqslant \rho_c^f < 0.1$	$0.1 \leqslant \rho_c^f < 0.2$	$0.2 \leqslant \rho_c^f < 0.3$	$0.3 \leqslant \rho_c^f < 0.4$	$0.4 \leqslant \rho_c^f < 0.5$	$\rho_c^f \geqslant 0.5$
γ_p	0.68	0.74	0.80	0.86	0.93	1.0

Notes If steam curing is adopted, the temperature should not be above 60 °C; If exceeds, the calculated design strength of concrete should be multiplied by a factor 1.2

Chapter 3
Bond and Anchorage

3.1 Bond and Mechanism of Bond Transfer

3.1.1 Bond Before Concrete Cracking

As illustrated in Chap. 1, the prerequisite for combined action of concrete and steel bars requires a strong enough bond between the two materials to sustain the shear stress (called *bond stress*) caused by deformation difference (or relative slip) along the concrete–steel interface. Stress can be transferred between both materials to make them work together with the help of the bond stress. Let us first analyze the bond in an uncracked reinforced concrete beam shown in Fig. 3.1a.

Figure 3.1b shows the moment diagram of the reinforced concrete beam in Fig. 3.1a subjected to two symmetric concentrated loads. Take separated bodies of a beam segment (Fig. 3.1d) and a reinforcement segment (Fig. 3.1c) of the same length Δx from the flexure–shear area (between one support and the adjacent concentrated load) of the beam. From mechanics of materials, the tensions at both ends of the reinforcement segment can be expressed as

$$T_1 = \frac{M_1}{\gamma_s h},\ T_2 = \frac{M_2}{\gamma_s h} = \frac{M_1 + \Delta M}{\gamma_s h} \tag{3.1}$$

where M_1 and M_2 respectively represent moments at the beam segment ends, and $\gamma_s h$ = lever arm of the internal force (γ_s is the lever arm coefficient, and h is the cross-sectional height). So,

$$\Delta T = \frac{\Delta M}{\gamma_s h} \tag{3.2}$$

Then, from the equilibrium of the reinforcement segment, the bond stress at the interface between concrete and reinforcement can be obtained as

X. Gu et al., *Basic Principles of Concrete Structures*,
DOI 10.1007/978-3-662-48565-1_3

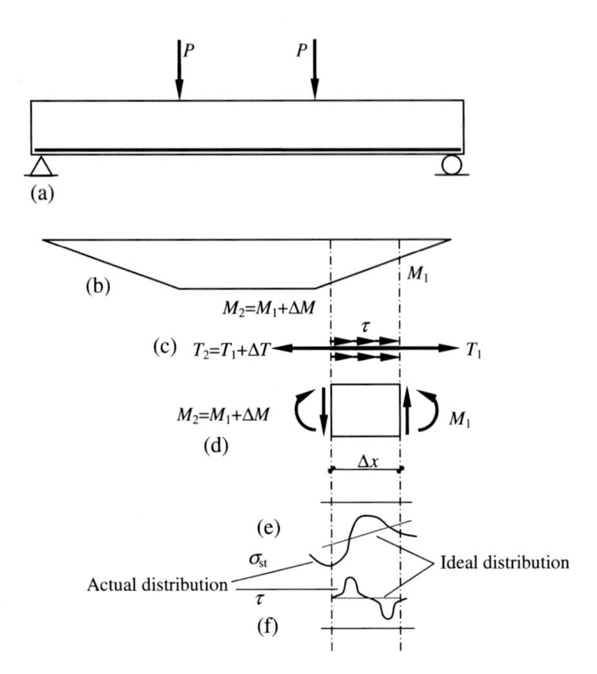

Fig. 3.1 Bond between concrete and steel bars before cracking. **a** Simply supported reinforced concrete beam. **b** Moment diagram. **c** Free-body of reinforcement. **d** Free-body of beam segment. **e** Stress distribution in reinforcement. **f** Bond stress distribution

$$\tau = \frac{\Delta T}{\Delta x \cdot \mu_s} = \frac{\Delta M}{\Delta x} \cdot \frac{1}{\gamma_s h \mu_s} = \frac{V}{\gamma_s h \mu_s} \quad (3.3)$$

where

V shear force in the beam; and

μ_s circumference of the reinforcement.

Equation (3.3) shows that although concrete has not cracked, there is still bond stress between concrete and reinforcement due to the tension difference between any two cross sections in the flexure–shear area (Fig. 3.1e). The bond stress distribution is the same as that of the shear force. In fact, micro-cracks will influence the bond stress distribution (the solid line in Fig. 3.1f, which will be discussed in detail later).

3.1.2 Bond After Concrete Cracking

Cracks will appear as shown in Fig. 3.2 when the concentrated loads in Fig. 3.1a are increased. After cracking, the bond between concrete and reinforcement can be

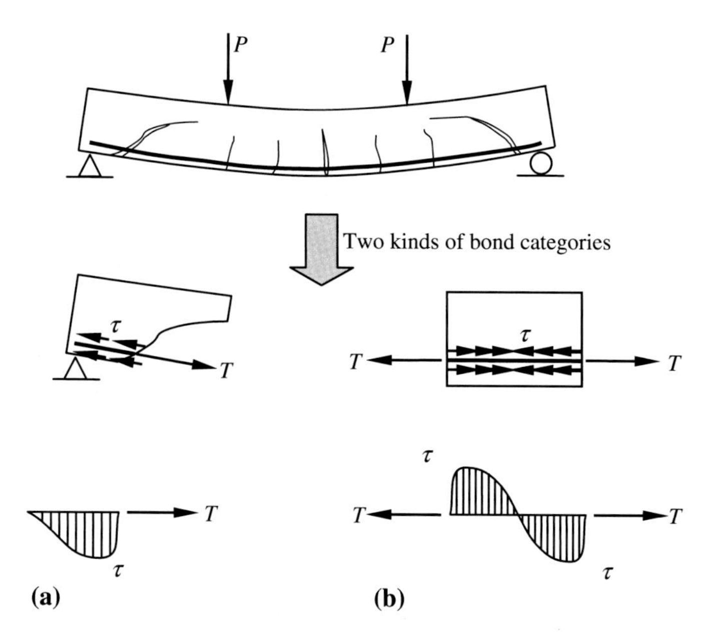

Fig. 3.2 Bond between concrete and reinforcement after concrete cracking. **a** Anchorage bond. **b** In-and-out bond

classified as two categories according to its mechanical properties, i.e., anchorage bond and in-and-out bond.

3.1.2.1 Anchorage Bond

Take the left part of a cracked beam along the diagonal crack (Fig. 3.2a). Cracked concrete cannot carry any tension, so there must be a large tension in the reinforcement intersecting with the diagonal crack. To prevent the reinforcement from being pulled out under tension T, concrete must tightly *grip* the reinforcement to work together. This grip is realized by bond stress accumulation along a certain length of concrete–reinforcement interface. This kind of bond is called anchorage bond.

In fact, the bond between concrete and reinforcement before concrete cracking introduced in Sect. 3.1.1 is also an anchorage bond. In addition, to ensure the performance of reinforcement after being cut for economic reasons, a development length is necessary to gradually accumulate the required tension. If reinforcement is not long enough, or construction joint is indispensable according to the detailing, splicing of bars is needed to transfer the tension from one bar to another. All these factors are relevant to the anchorage bond. An insufficient anchorage bond will lead to premature failure of concrete members.

3.1.2.2 In-and-Out Bond

Take a beam segment between any two adjacent cracks in the pure bending area of the beam (Fig. 3.2b). Right at the cracks, all the tension T is resisted by the reinforcement. But between the cracks, the uncracked concrete is subjected to tension transferred from the reinforcement by the bond between them. Any new crack will similarly produce bond stress at both sides of the crack. This kind of bond is called the in-and-out bond (it transfers the bond stress into the bar and back out again), which can make the uncracked concrete carry part of the tension so as to improve the energy dissipation ability of reinforced concrete beams. The bond between concrete and reinforcement in the axially tensioned specimen in Fig. 4.3 is also an in-and-out bond. Loss and degeneration of the in-and-out bond will reduce the stiffness and increase the crack widths of reinforced concrete members.

3.1.3 Bond Tests

The way to obtain the concrete–steel interface bond is by means of pull-out tests, as shown in Fig. 3.3a. The steps for this procedure are to (1) longitudinally split a steel bar into half, (2) glue strain gauges at certain intervals at the bottom of a longitudinally lathed rectangular groove in the bar center (Fig. 3.3a), (3) after connecting the strain gauges with conductors and taking measures to protect the strain gauges, recover the steel bar by using epoxy to glue the two half steel bars together, (4) embed this steel bar in a concrete specimen, and (5) carry out the pull-out test by increasing the tension at the steel bar end gradually until it is pulled out (Fig. 3.3b). The stress distribution in the steel bar can be calculated from the measured strain distribution as per Hooke's law (Fig. 3.3c). Test results show that after arbitrarily taking a very small element of the length, Δx, from the steel bar, the tensile stresses at the two ends of the element (σ_{s1} and σ_{s2}, respectively) are different from each other (see Fig. 3.3e). If the cross-sectional area of the steel bar is A_s and the circumference is μ_s, then the concrete–steel interface bond stress can be obtained from the force equilibrium of the steel bar element shown in Fig. 3.3e as

$$\tau = \frac{(\sigma_{s1} - \sigma_{s2})A_s}{\Delta x \mu_s} = \frac{\Delta \sigma_s A_s}{\Delta x \mu_s} \tag{3.4}$$

When $\Delta x \to 0$, Eq. (3.4) becomes

$$\tau = \frac{d\sigma_s}{dx} \cdot \frac{A_s}{\mu_s} \tag{3.5}$$

Equation (3.5) gives the bond stress at any position of reinforcement. From Eq. (3.4) or Eq. (3.5), the longitudinal distribution of the bond stress corresponding to any tension T can be calculated as shown in Fig. 3.3d. Two important

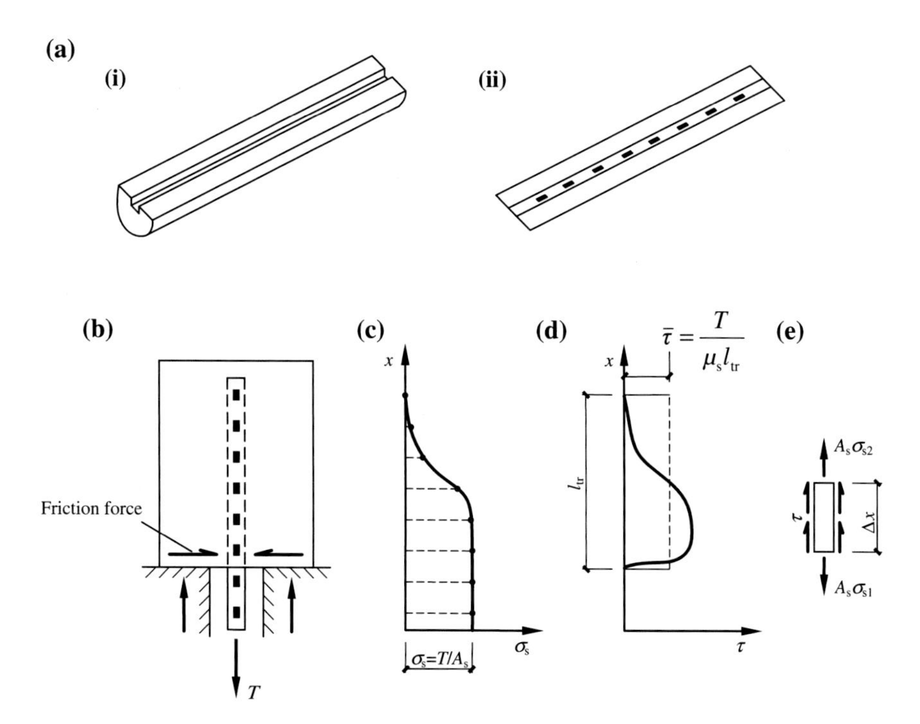

Fig. 3.3 Setup and results of a pull-out test. **a** Preparation of the steel bar. **i** Lathed groove in the split steel bar. **ii** Strain gauge arrangement in the groove. **b** Setup of a pull-out test. **c** Stress in the steel bar. **d** Bond stress. **e** Free body of the steel bar element

conclusions can be drawn from Fig. 3.3d: (1) When the tension T applied on the reinforcement is constant, the bond stress only distributes within a certain length l_{tr} (development length), which is the necessary bond length to transfer tension T from reinforcement to concrete. (2) The bond stress τ varies within the development length l_{tr}, and the average bond stress $\bar{\tau}$ ($\bar{\tau}=\frac{T}{\mu_s l_{tr}}$) is less than the maximum one. It is hard to measure the bond strength of concrete and reinforcement using the experimental setup shown in Fig. 3.3. In addition, reaction at the support during the stretch of the steel bar puts concrete under compression (Fig. 3.3b). There is a friction between the concrete specimen and the support due to the restraint by the support on the lateral expansion of the concrete, which is quite different from the stress state in real structural members. Therefore, researchers have proposed many experimental setups for different purposes such as the pull-out test for local bond research (Fig. 3.4a), half-beam test for anchorage bond (Fig. 3.4b), overhanging-beam test to investigate the splice length (Fig. 3.4c), and the extended length test to determine the cutoff points. With the help of these tests, the bond between concrete and steel can be thoroughly investigated and the research results can be used as the basis for engineering applications.

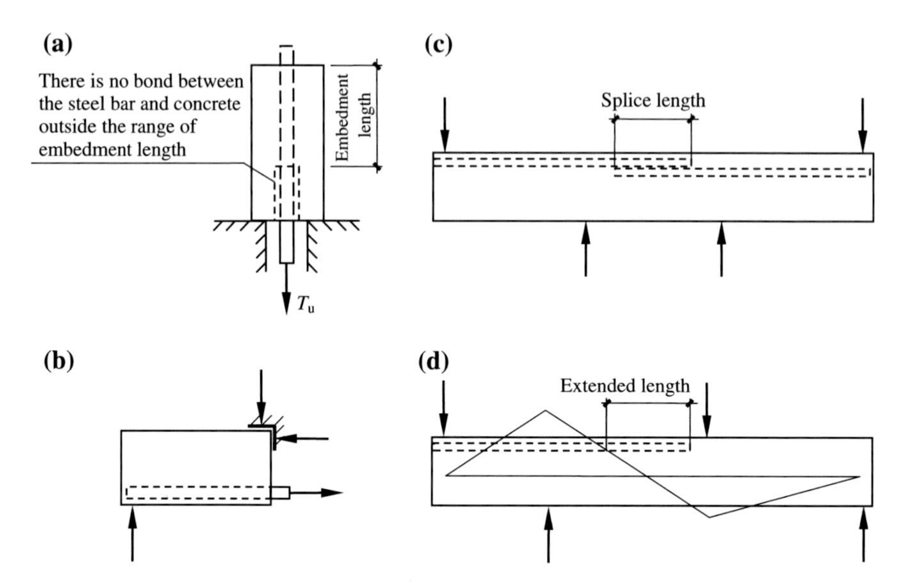

Fig. 3.4 Frequently used experimental setups on bond between concrete and reinforcement. **a** Local bond test. **b** Half-beam test. **c** Overhanging-beam test. **d** Extended length test

Steel bars in the tests mentioned above are all subjected to tension. If steel bars are under compression, there are also corresponding tests to study the concrete–steel bond. For example, squeeze test corresponds to pull-out test.

3.1.4 Mechanism and Failure Mode of Bond

The mechanism of bond transfer using plain round steel bars is different from that of deformed steel bars (steel rebars).

The bond between a plain round bar and concrete consists of three parts: (1) chemical adhesion between mortar paste and the bar surface, (2) friction at the concrete–steel interface, and (3) interlocking of the rough surface with the surrounding concrete. Among these three parts, the role of chemical adhesion can only be credited with a very small portion of the end result. Once slip occurs, bond can be developed only by friction and interlocking. However, when a plain bar is subjected to tension, the friction and interlocking will soon disappear due to the decrease of bar diameter due to the Poisson effect. The failure mode of plain round bars in standard bond tests is the bars pulled out from the encasing concrete with the slip of several millimeters (Fig. 3.5).

Ribs of deformed bars change the action mode between concrete and reinforcement. The bond strength thus significantly increases. The bond between deformed bars and concrete is mainly the interlocking of the ribs with the surrounding concrete

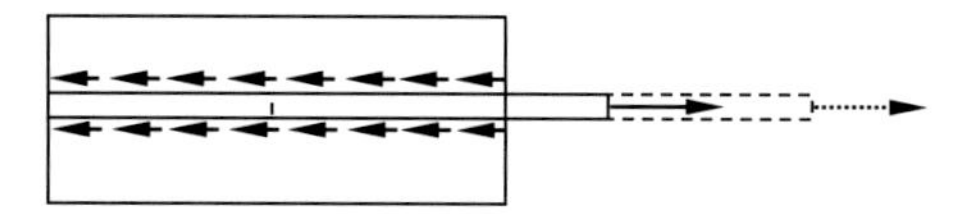

Fig. 3.5 Failure mode of a plain round bar specimen

in addition to the chemical adhesion between mortar paste and bar surface and friction at the concrete–steel interface. The oblique bearing forces by the ribs against concrete resist any slip (Fig. 3.6a). The radial component of the bearing force makes the peripheral concrete under hoop tension, similar to an internally pressured tube (Fig. 3.6b). The longitudinal component (along the reinforcement axis) of the bearing force puts the concrete between two ribs under flexure and makes them shear like a cantilever beam (Fig. 3.6a). So the peripheral concrete is under a very complicated three-dimensional stress state. In real concrete members, when the hoop tension increases to a certain value, concrete will be split at the weakest position along the reinforcement axis, i.e., the bond failure (Fig. 3.6c, d). The root of the cantilever beam between ribs may be torn by the shear stress caused by the longitudinal component of the bearing force (Fig. 3.6e). In addition, the large local compression on the rib–concrete interface may crush the concrete (Fig. 3.6f), and a new slip surface consequently forms along the crushed debris accompanied by a big slip. If the concrete strength is low, the deformed rebar may be pulled out totally (Fig. 3.6g).

The bond mechanism of deformed bars under compression is similar to that under tension, but the bond failure is apparently delayed by the local compression of the bars against the concrete at the bars' ends, and the diameter increase due to the Poisson effect.

3.1.5 Mechanism of Lap Splice

The force carried by the reinforcement at one side can be transferred to the counterpart at the other side in the reinforcement overlapping zone. This is accomplished by the bond between the concrete and reinforcement (Fig. 3.7). Similarly, the lap splice mechanisms of compressed reinforcement and tensioned reinforcement are not identical.

3.2 Bond Strength Between Concrete and Reinforcement

3.2.1 Bond Strength

As shown in Fig. 3.3, the bond stress is unevenly distributed along the axis of reinforcement if its embedded length is long. However, when the embedded length of reinforcement is lower than a certain value, a uniform distribution of the bond stress

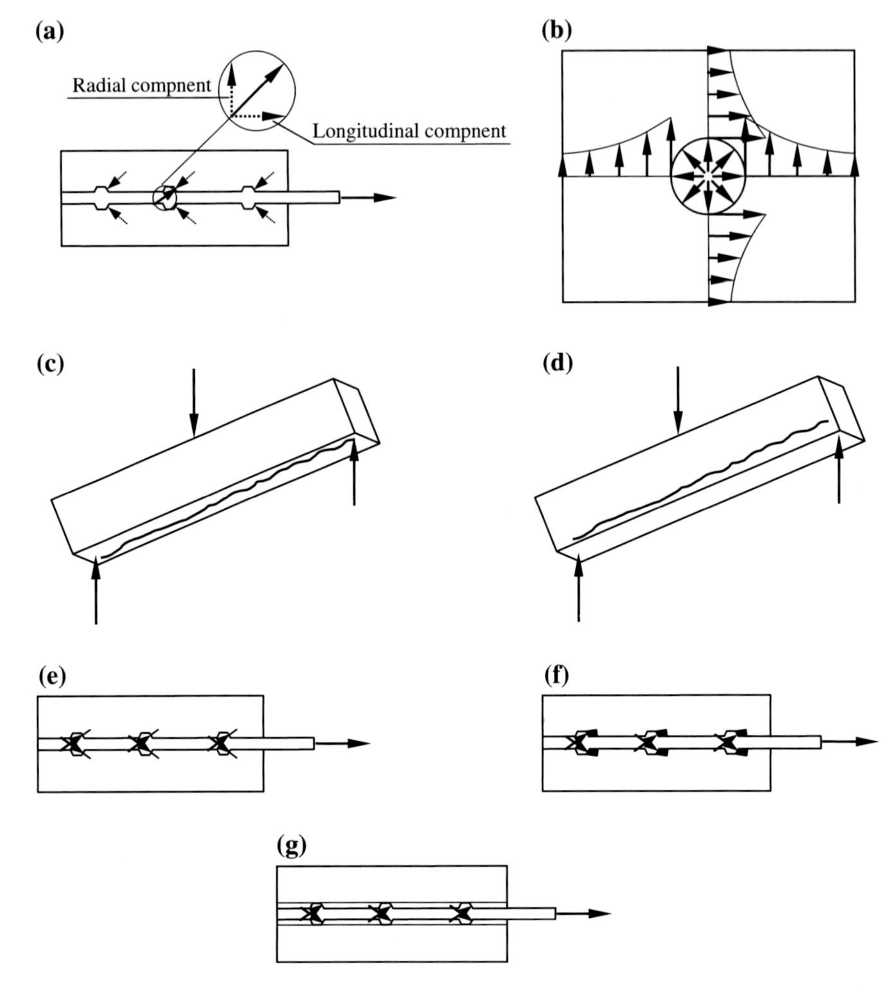

Fig. 3.6 Bond mechanism and bond failure modes of deformed bars. **a** Interlocking at the rib–concrete interface. **b** Tensile stress caused by the radial component of bearing force. **c** Longitudinal cracks at the bottom surface caused by the radial component. **d** Longitudinal cracks at the side surface caused by the radial component. **e** Tearing of concrete caused by the longitudinal component. **f** Local crush of concrete caused by the longitudinal component. **g** Pull-out failure caused by the longitudinal component of bearing force

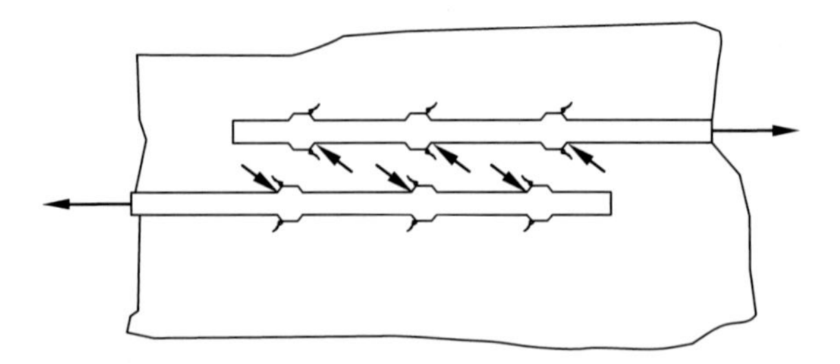

Fig. 3.7 Mechanism of force transfer in lap splice

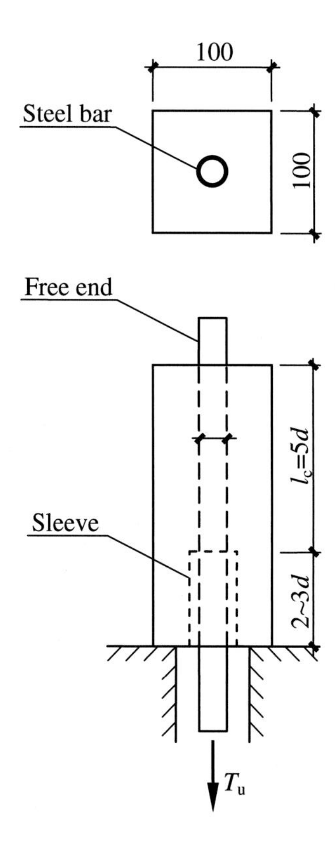

Fig. 3.8 Pull-out test to measure the bond strength

can be assumed. Therefore, the experimental setup in Fig. 3.8 is generally used to measure the bond strength between concrete and reinforcement. The sleeve in the setup is to prevent the concrete from adhering to the reinforcement in this zone. Knowing the pull-out force, T_u, the bond strength can be calculated by Eq. (3.6).

$$\tau_u = \frac{T_u}{\mu_s l_c} \tag{3.6}$$

where

T_u pull-out force;
μ_s circumference of the reinforcement; and
l_c embedded length of the reinforcement (generally, $l_c = 5d$, d is the reinforcement diameter).

3.2.2 *Influential Factors on Bonding Strength*

There are many factors that may influence the bond strength between concrete and reinforcement, such as concrete strength, casting position, concrete cover depth, clear spacing between steel bars, transverse reinforcement, and lateral pressure.

1. The bond strength of both plain round bars and deformed bars will increase with the concrete strength grade and is approximately proportional to the splitting strength of concrete.
2. The bond strength is determined by the positions of steel bars during concrete casting. For horizontal bars at the top of a concrete member with a casting height larger than 300 mm, the sedimentation of concrete due to bleeding and bubble escape leads to a soft and spongy layer of concrete right beneath the bars, which weakens the bond between concrete and reinforcement. The strength of horizontal bars at the top of concrete members is 20–30 % lower than those of vertical steel bars and horizontal bars at the bottom.
3. Deformed bars have greater bond strength than plain round bars. But splitting cracks rather than large slips are more likely to take place in concrete members reinforced by deformed bars. The longitudinal cracks along the reinforcement axis are harmful to concrete members by compromising their safety and durability. Increasing the concrete cover depth and keeping a certain reinforcement spacing can increase the splitting resistance of peripheral concrete so as to ensure bond strength.
4. Transverse reinforcement can delay the propagation of internal cracks and restrict the width of splitting cracks to increase the bond strength. Hence, a certain amount of transverse reinforcement should be arranged in anchorage zone of large diameter reinforcement and reinforcement overlapping zone, e.g., stirrups in beams. When many steel bars are parallel in one row, additional stirrups are necessary to prevent the concrete from peeling off.
5. Lateral pressure on the anchorage zone will restrain the transverse deformation of concrete and increase the friction between steel bars and concrete, so that the bond strength can be increased. Therefore, considering the beneficial effect of bearing stress, the anchorage length can be reasonably reduced at supports (e.g., simply supported ends of beams).
6. Surface characteristics of deformed bars also influence the bond strength but insignificantly.

3.3 Anchorage of Steel Bars in Concrete

3.3.1 Anchorage Length

From the foregoing analysis, it is clear that insufficient anchorage will make concrete members fail immaturely. The concerted action of concrete and steel bars requires reliable anchorage, i.e., the yielding of steel bars happens before anchorage failure, which can be realized by extending the steel bars with a certain length in the concrete. This length is called the anchorage length of steel bars. The minimum anchorage length is actually the development length right at the yielding of steel bars. Hence, the principle to determine the minimum anchorage length is that the yielding of steel bars and the anchorage failure happen at the same time.

Obviously, the pull-out test is the most direct method to determine the anchorage length. However, the anchorage length can also be theoretically derived according to the foregoing bond mechanism between steel bars and concrete. To calculate the minimum anchorage length l_a (Fig. 3.9a, b), a concrete cylinder of $2c'$ in diameter reinforced by a deformed rebar of d in diameter is considered.

Assume: The longitudinal splitting happens first and instantly leads to the anchorage failure; the tensile stress of concrete σ_t caused by internal pressure p is linearly distributed (Fig. 3.9c). Consideration of the equilibrium gives

$$l_a p d = (2c' - d) \cdot \frac{\sigma_t}{2} \cdot l_a \tag{3.7}$$

The pressure p can be solved as:

$$p = \left(\frac{c'}{d} - \frac{1}{2}\right)\sigma_t \tag{3.8}$$

When $\sigma_t = f_t$, longitudinal splits and thus the anchorage failure occurs, and the ultimate internal pressure will be given as

$$p_u = \left(\frac{c'}{d} - \frac{1}{2}\right)f_t \tag{3.9}$$

where f_t = ultimate tensile strength of the concrete.

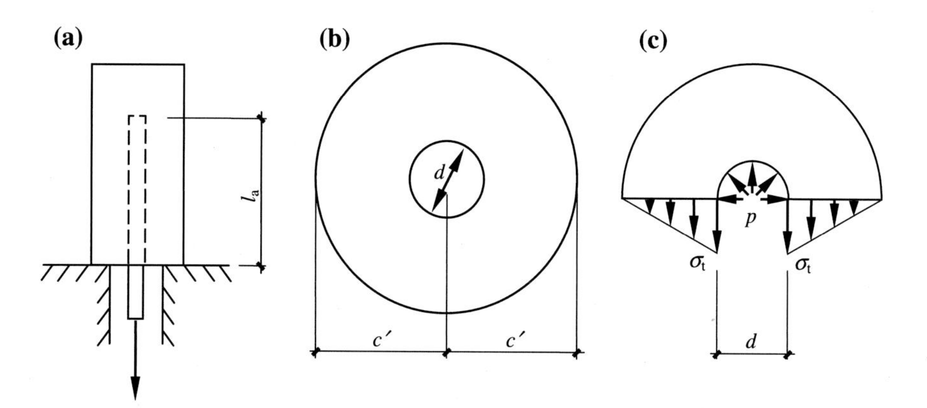

Fig. 3.9 Theoretical model for anchorage length calculation. **a** Cylindrical specimen in the pull-out test. **b** Cross section. **c** Stress distribution

For the deformed rebar, the radial component of the bearing stresses is equal to the longitudinal component when the angle of the ribs is 45°. Therefore,

$$\tau_u = p_u \tag{3.10}$$

which is similar to Eq. (3.6),

$$\tau_u = \frac{T_u}{\mu_s l_a} = \frac{\pi d^2 \cdot f_y/4}{\pi d l_a} = \frac{d f_y}{4 l_a} \tag{3.11}$$

where

T_u pull-out force;
μ_s circumference of the rebar;
τ_u average bond stress; and
f_y yield strength of the rebar.

Substituting Eqs. (3.9) and (3.11) into Eq. (3.10) yields

$$\frac{l_a}{d} = \frac{f_y}{\left(\frac{4c'}{d} - 2\right) f_t} \tag{3.12}$$

Let $c' = 2d$, the minimum anchorage length can be calculated as

$$l_a = \frac{f_y}{6 f_t} \cdot d \tag{3.13}$$

When $c' > 2d$, the anchorage length calculated from Eq. (3.12) is smaller than that calculated from Eq. (3.13). Therefore, a relatively safe design of the anchorage length will be obtained from Eq. (3.13) when concrete cover is thick.

Because concrete always splits longitudinally at the weakest position, the equations derived using cylindrical specimens are applicable to the anchorage length calculation of concrete specimens of any shape (Fig. 3.10).

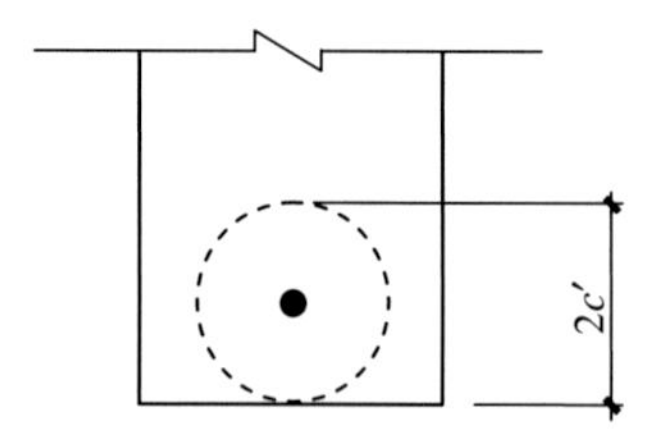

Fig. 3.10 Tension hoop due to reinforcement anchorage

3.3.2 Practical Equation for Anchorage Length Calculation

The theoretical equation (Eq. 3.13) for anchorage length calculation is seldom directly used in engineering applications. Different codes have different calculation methods. In GB 50010, the anchorage length should be calculated according to Eq. (3.14) if the tensile strength of reinforcement is fully utilized.

$$l_a = \alpha \frac{f_y}{f_t} d \tag{3.14}$$

where

l_a minimum anchorage length of a tensioned steel bar;
f_y design value of the tensile strength of a steel bar;
f_t design value of the tensile strength of concrete in the anchorage zone;
d diameter of a steel bar; and
α shape coefficient of a steel bar, taken according to Table 3.1.

When the compressive strength of reinforcement is fully utilized, the anchorage length of compressed reinforcement should not be less than 0.7 times that of tensioned reinforcement.

Equation (3.14) is actually a modification to Eq. (3.13). Based on the minimum anchorage length, many other details can be determined accordingly, such as the splice length for reinforcement overlapping, the extended length for reinforcement cutting off, and the anchorage length at the support. These contents will be discussed in the subsequent chapters.

If the steel bars cannot be extended to the required length due to the space limitation, additional anchorage measures should be taken at the ends of the steel bars, e.g., arranging hooks and welding short steel bars or steel plates.

3.3.3 Hooked Anchorages

90° hooks are widely used as additional anchorage measures in reinforced concrete structures. The behavior of a 90° hook is shown in Fig. 3.11. When the steel bar is tensioned, the tensile force in the bar, T, is resisted by the bond on the surface of the bar and the bearing on the concrete inside the hook (Fig. 3.11a). The varied stresses

Table 3.1 Shape coefficient of the rebar

Rebar type	Hooked plain bar	Deformed bar	Spiral indented steel wire	Steel strand (3-wire twisted)	Steel strand (7-wire twisted)
α	0.16	0.14	0.13	0.16	0.17

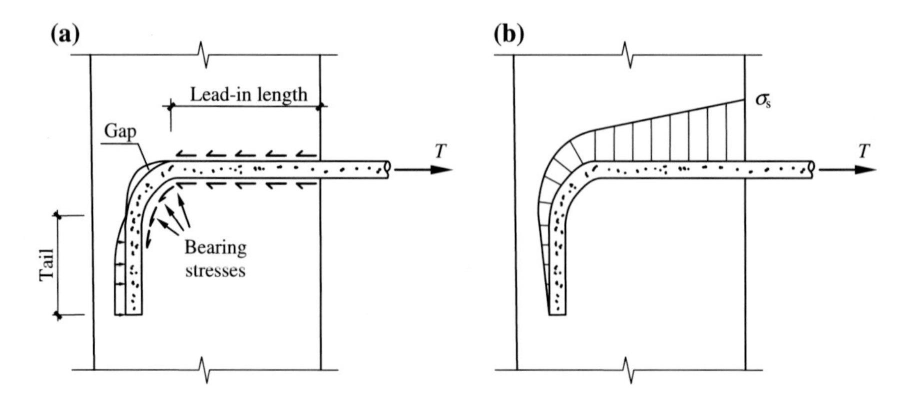

Fig. 3.11 Behavior of a hook. **a** Force acting on bar. **b** Stresses in 90° hook

in the 90° hook is shown in Fig. 3.11b under the tensile force T. The hook moves inward, leaving a gap between it and concrete outside the bend. The compressive force stress inside the bend tends to straighten out the steel bar, producing compressive stress outside the tail. If the lead-in length of a hook is too small, the concrete inside the hook will crush, removing the side cover. If the tail of a hook is close to the outside face, the concrete outside the tail will crack, allowing the tail to straighten. To grantee a hook anchorage is efficient, the lead-in length of the hook is set to be not smaller than $0.4l_a$, and the length of the tail is set to be not smaller than $15d$ in GB 50010 where l_a is calculated by Eq. (3.14) and d is the diameter of the steel bar.

Questions

3.1 What is the bond action between reinforcement and concrete? And how many types of bond actions?
3.2 What components make up the bond between concrete and reinforcement? Which component is dominant?
3.3 How many types of bond failure modes are there for deformed bars?
3.4 What factors mainly affect the bond stresses between reinforcement and concrete?
3.5 What is the principle used for determining the minimum anchorage length? How would you determine the minimum anchorage length?
3.6 For a horizontally cast reinforced concrete beam, what is the difference between the bond strength of the top bar and that of the bottom bar?
3.7 Should the two steel bars be allowed to merge when they are overlapped? Why or why not?
3.8 What is the difference between the development length l_{tr} and the anchorage length l_a?

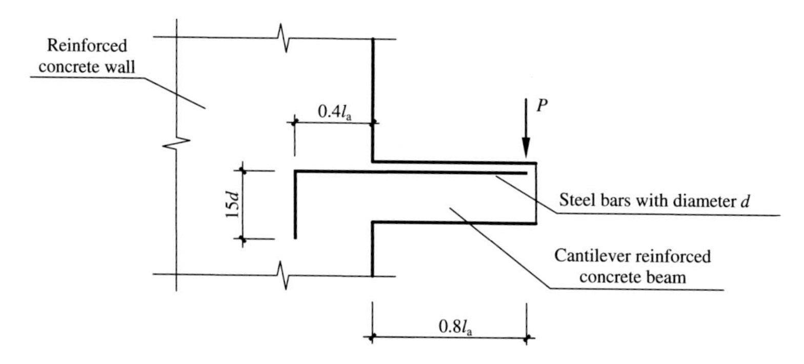

Fig. 3.12 Question 3.9

3.9 A cantilever reinforced concrete beam from a reinforced concrete wall to bear a concentrated load P at the free end of the beam is designed and constructed as shown in Fig. 3.12. The length of the beam is $0.8l_a$ where l_a is determined by Eq. (3.14), based on the strengths of materials and the diameter of the steel bars embedded in the beam. Do you think it is a well designed beam or not? Why?

Chapter 4
Tension and Compression Behavior of Axially Loaded Members

4.1 Engineering Applications and Details of Members

Members subjected to axial tension at the geometric centers of their cross sections are called axially tensioned members. Typical examples in reinforced concrete structures include tensioned web elements and bottom chords in trusses, tie bars in arches, walls of internally pressured tubes, and sidewalls of tanks. Members subjected to axial compression at the geometric center of their cross sections are called axially compressed members. Typical examples are middle columns in multibay and multistory structures mainly sustained by permanent loads, compressed web members, and chords in trusses only loaded at the joints. Figure 4.1 illustrates common engineering applications of axially loaded members.

There are longitudinal steel bars and stirrups in an axially loaded reinforced concrete member (Fig. 4.2a). The stirrups in columns are usually called the ties, and the columns with longitudinal bars and ties are usually called the tied columns, because the longitudinal bars are tied together with smaller bars (ties) at intervals up the columns. Since most of the compression force is resisted by concrete, the function of the longitudinal steel bars includes: (1) helping concrete to resist the compression force so as to reduce the cross-sectional dimension, (2) carrying any possible small moment, and (3) preventing the brittle failure of the member. The stirrups or ties are mainly used to position or tie the longitudinal bars and form a steel skeleton with the longitudinal bars.

For axially compressed members with circular or regular polygonal cross sections, closely spaced continuous spirals or welded circular hoops can be arranged outside the longitudinal bars (Fig. 4.2b). In addition, to prevent the longitudinal bars in a compressive member from buckling and form a steel cage with longitudinal bars, spirals or circular hoops can confine the core concrete so as to increase the load bearing capacity and deformability of the member.

X. Gu et al., *Basic Principles of Concrete Structures*,
DOI 10.1007/978-3-662-48565-1_4

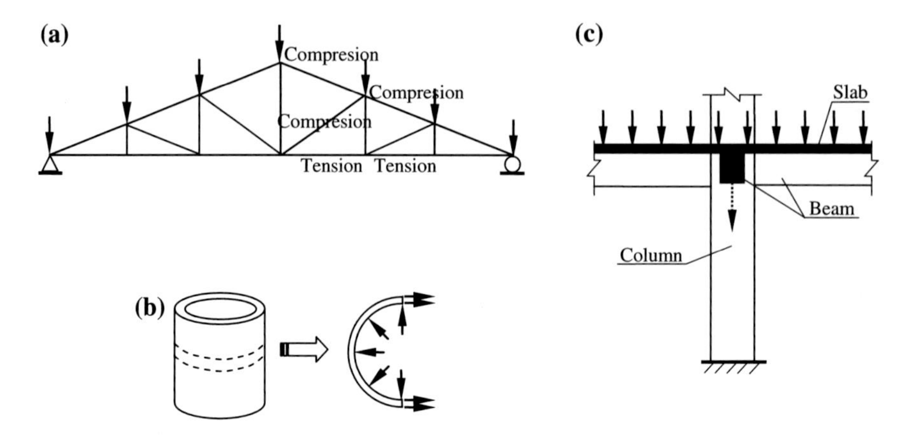

Fig. 4.1 Engineering examples of axially loaded members. **a** Triangle truss. **b** Tank. **c** Column

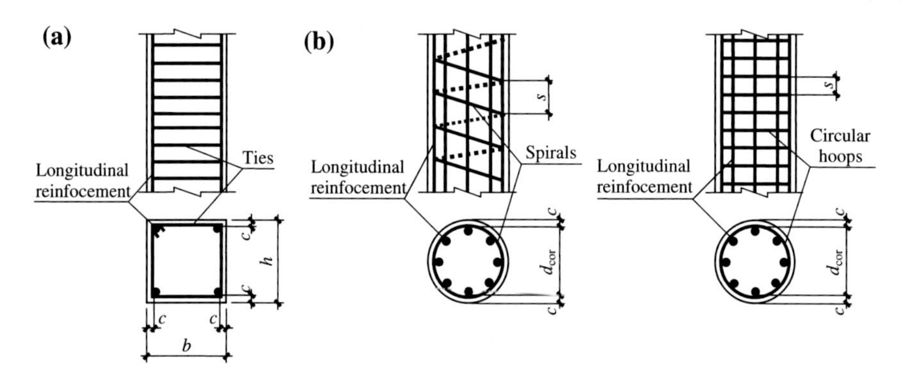

Fig. 4.2 Reinforcement in axially loaded columns. **a** Tied column. **b** Spiral columns

The cross-sectional dimension and reinforcement of an axially loaded reinforced concrete member should satisfy some detailing requirement to facilitate the construction, ensure reliable bond between concrete and steel bars, and provide effective protection to steel bars.

When the dimensions of the cross section of an axially loaded member are not much different in either of the two directions ($h/b \leqslant 4$ in Fig. 4.2a), the member is generally considered to be a tie bar or a column. Otherwise, a slab, a wall, or a shell are more appropriate.

In concrete structures, steel bars are embedded in concrete. The minimum distance from the outer surface of longitudinal steel bars to the nearest surface of the concrete member is called the concrete cover depth (c in Fig. 4.2). The allowed values of the concrete cover depth are distinct in different design codes. For example, in the *Code for design of concrete structures* (GB 50010), the cover depth is the maximum of the largest diameter d of longitudinal steel bars and 15 mm (when the grade of concrete is not larger than C20, the maximum d as 20 mm is suggested) for slabs, walls, or shells in normal indoor conditions. The maximum d as 30 mm is recommended for columns. The concrete cover depth of tie rods can be calculated in the same manner as that of beams, which will be discussed in Chap. 5.

The minimum diameter of longitudinal steel bars in axially loaded structural members is 12 mm. Longitudinal steel bars should be evenly distributed along the circumference in circular columns and with no less than six bars installed. The center-to-center distance between two adjacent longitudinal steel bars should not be larger than 300 mm. For vertically cast members, the clear space between any two longitudinal steel bars should be larger than 50 mm to ensure an easy pass of aggregates through the reinforcement cage. For horizontally cast tie rods, columns, slabs, and shells, the clear space between the steel bars is the same as that for beams, which will be presented in Chap. 5. Usually, the diameter of ties in tied columns should not be less than a quarter of the maximum diameter of longitudinal steel bars, $d/4$ (in which, d is the maximum diameter of longitudinal steel bars), and should be larger than or equal to 6 mm. The pitch should not be larger than 400 mm, the smallest dimension of the cross section of the member, and 15 times the maximum diameter of longitudinal steel bars, $15d$, for the effective restraint to the longitudinal reinforcement, thus reducing the danger of buckling of longitudinal bars as the bar stress approaches yield. Detailing of spirals or circle hoops in spiral columns will be discussed in Sect. 4.7.

4.2 Analysis of Axially Tensioned Structural Members

4.2.1 Experimental Study on Axially Tensioned Structural Members

Figure 4.3 illustrates an axial tension test on a reinforced concrete member. Prior to concrete cracking, the tensile force N_t is carried by both the concrete and the steel bar. After cracking, the concrete at the cracks cannot take any load and the tensile force N_t is completely carried by the steel bar. When the steel yields, the specimen reaches its ultimate tension capacity. The whole process of the tensile test can be divided into three stages.

4.2.1.1 Stage I

Stage I starts from the onset of loading and ends just before the concrete cracking. In this stage, the steel bar transfers the tensile force to concrete through the bond between them so that they can share the load. The shaded area in Fig. 4.3b shows

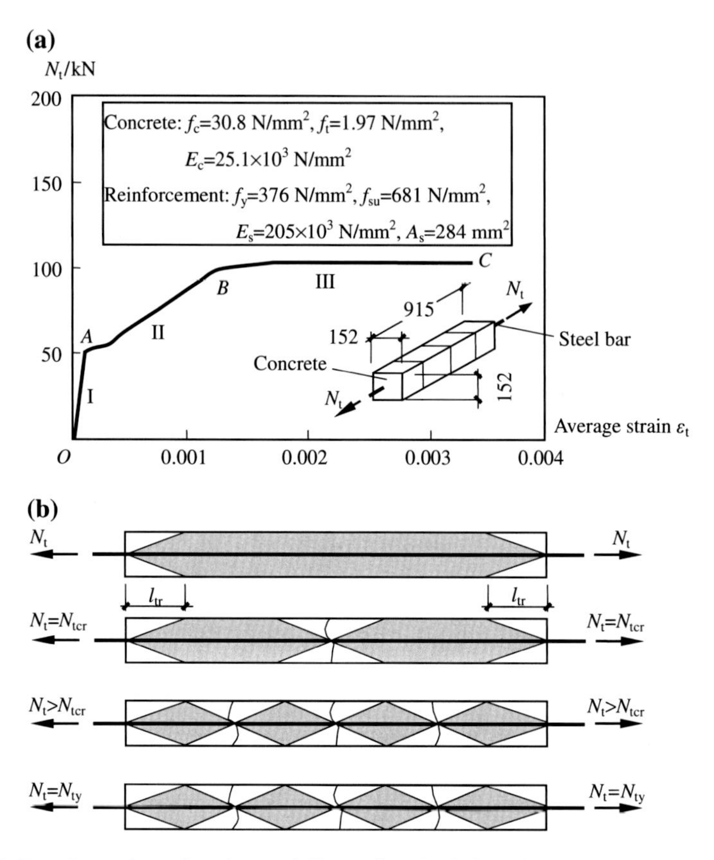

Fig. 4.3 Experimental results of an axially tensioned reinforced concrete member. **a** Load–average strain curve. **b** Propagation of cracks

the tensile force carried by the concrete, in which l_{tr} is the development length and the bond stress is assumed to be a uniform distribution. A linear relationship between N_t and the average tensile strain ε_t is observed in this stage (*OA* in Fig. 4.3a). The end of this stage is generally considered as the critical point for cracking resistance calculation.

4.2.1.2 Stage II

Stage II is from the concrete cracking to the yielding of the steel bar. The first crack appears at the weakest cross section of the specimen. At the same time, the concrete at the crack quits working. The steel bar carries the applied load by itself and transfers the load to surrounding concrete through bonding. With the increase of the applied load, more and more cracks appear. Similarly, the concrete at the cracks cannot take any load, and the steel bar sustains and transfers the load to surrounding concrete. When the distance between two adjacent cracks is shorter than the development length of the bond, the tensile force built up in the concrete is not big enough to crack the concrete. However, the width of the cracks increases with further increase of the load (Fig. 4.3b). And the increase of the average tensile strain of the cracked specimen is larger than that of the uncracked specimen even if the increment of the tensile force is the same, which is shown in Fig. 4.3a as the smaller slope of line *AB* compared to that of line *OA*. The calculations of the crack width and the deformation of axially tensioned members are based on this stage.

4.2.1.3 Stage III

When tensile force reaches the yield load N_{ty}, the steel bar begins to yield and Stage III begins. For an ideal axially loaded member reinforced with more than one steel bar, all the steel bars should yield simultaneously. But in fact, they yield at different times due to nonuniformity of the steel material, misplacement of the steel bars, and so on. During the yielding process of the steel bars, the crack width of the specimen increases rapidly with a slight increase of the tensile load. And when all the steel bars have yielded (as point *C* in Fig. 4.3a), the specimen can be assumed to have failed. Stage III is used to calculate the tensile capacities of members.

If the cross-sectional area of the steel bar in an axially tensioned member (Fig. 4.3a) reduces to a certain extent, two damage modes can be observed: (1) If the tensile load is applied to the steel bar directly, the concrete may not crack even though the steel bar has been broken; (2) if the tensile load is applied to the concrete directly, the steel bar will be broken immediately at the first primary crack because it cannot resist the load transferred from the concrete. In the second mode, the mechanical behavior of the reinforced concrete member is almost the same as that of a plain concrete member. Both members are characterized as linear elastic (*OA* in Fig. 4.3a) and will fail in a brittle way.

From the experiment, it is found that the concrete and the steel have been working together before the specimen cracks. Both materials have the same tensile strain, but because the stresses of the two materials are proportional to their moduli of elasticity (or secant modulus) in the linear elastic stage, the stress in the steel is much higher than that in the concrete. After the concrete has cracked, the stress once sustained by the concrete is transferred to the steel at the crack, which leads to a rapid increase of the stress in the steel, while the stress in the concrete decreases to zero. This kind of stress adjustment between the concrete and the steel over the specimen cross section is called stress redistribution, which is a very important concept in reinforced concrete structures. The specimen will continuously crack until the number of cracks reaches a certain value. The ultimate tensile capacity of the specimen depends on the amount and strength of the steel bars. And the minimum amount of the longitudinal steel bars should be met to ensure that the ultimate tensile capacity is larger than the cracking load of the specimen.

4.2.2 *Relationship Between Tensile Force and Deformation*

4.2.2.1 Stress–Strain Relationship for the Steel and the Concrete

The longitudinal bars in reinforced concrete members subjected to tension are generally made of *mild steel*. From the discussion of the mechanical properties of reinforcement in Chap. 2, it is found that the stress–strain relationship of the longitudinal steel bars can be represented by a bilinear model as shown in Fig. 4.4a. That is to say, when the stress in the steel is lower than the yield strength, the stress–strain curve is an oblique straight line, while after the stress in the steel has reached the yield strength, it will keep constant with the increase of the strain. This stress–strain relationship can be formulated as Eq. (2.1).

In real engineering applications, tensioned concrete members generally cannot unload, so the descending branch in the stress–strain curve of concrete does not exist. From the corresponding discussion in Chap. 2, the stress–strain relationship of the concrete under tension can be simplified as the oblique straight line in Fig. 4.4b, whose function is Eq. (4.1). The slope of the line is the modulus of elasticity of the concrete.

$$\sigma_t = E_c \varepsilon_t \tag{4.1}$$

where

σ_t the stress of the concrete;
ε_t the strain of the concrete; and
E_c the modulus of elasticity of the concrete.

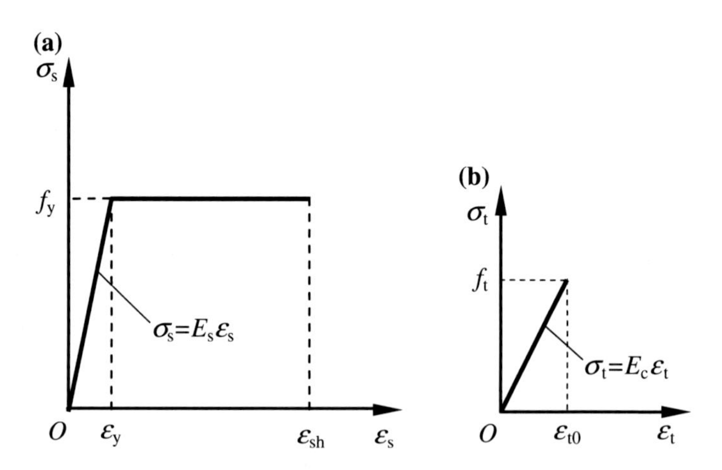

Fig. 4.4 Stress–strain relationships for reinforcement and concrete under tension. **a** Reinforcement. **b** Concrete

When the tensile strain reaches ε_{t0}, the stress of the concrete will get to the tensile strength f_t and the concrete will crack.

4.2.2.2 Tensile Load Versus Deformation Before Concrete Cracking

Prior to concrete cracking, the tensile load is carried by both the concrete and the steel. The two materials also deform compatibly with the same tensile strain (Fig. 4.5a):

$$\varepsilon = \varepsilon_t = \varepsilon_s = \frac{\Delta l}{l} \tag{4.2}$$

According to the equilibrium condition shown in Fig. 4.5b, tensile force can be obtained as

$$N_t = \sigma_t A + \sigma_s A_s \tag{4.3}$$

where

- σ_t stress of concrete;
- σ_s stress of steel bars;
- A sectional area of the member, when $A_s/A \leqslant 3\ \%$, $A = bh$; otherwise, $A = bh - A_s$; and
- A_s sectional area of steel bars.

Substituting the constitutive laws of the two materials [Eqs. (2.1) and (4.1)] and the kinematic equation (4.2) into Eq. (4.3) gives

$$N_t = (E_c A + E_s A_s)\varepsilon = E_c A(1 + \alpha_E \rho)\varepsilon = E_c A_0 \varepsilon \tag{4.4}$$

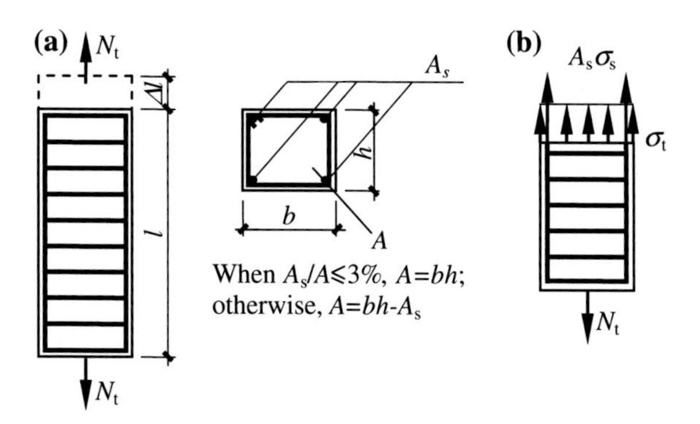

Fig. 4.5 An axially tensioned reinforced concrete member

where

A_0 equivalent sectional area of axially tensioned reinforced concrete member, $A_0 = (1 + \alpha_E \rho)A$;

α_E ratio of the moduli of elasticity of steel and concrete, $\alpha_E = E_s/E_c$; and

ρ reinforcement ratio of longitudinal steel bars, $\rho = A_s/A$.

4.2.2.3 Cracking Load

When $\varepsilon_t = \varepsilon_0$, the concrete will crack. Substituting the cracking strain into Eq. (4.4) gives

$$N_{tcr} = E_c A_0 \varepsilon_{t0} = E_c A(1 + \alpha_E \rho)\varepsilon_{t0} \tag{4.5}$$

4.2.2.4 Tensile Load Versus Deformation After Concrete Cracking

The concrete cannot resist any tension ($\sigma_c = 0$) after cracking. The tensile load will be carried only by the reinforcement. The relationship between the tensile load and the deformation can then be expressed as:

$$N_t = \sigma_s A_s = E_s A_s \varepsilon \tag{4.6}$$

4.2.2.5 Ultimate Tensile Capacity

When the reinforcement yields, the tensioned member will deform further even though the tensile load holds constant. This load is therefore assumed as the ultimate tensile capacity of the member and can be expressed as:

$$N_{tu} = f_y A_s \tag{4.7}$$

where N_{tu} = ultimate tensile capacity of the axially tensioned member and f_y = yield strength of the steel.

Consider the following problem and solution:

Example 4.1 An axially tensioned member is shown in Fig. 4.3a. The following needs to be calculated:

(1) The applied tensile load when the member is elongated $\Delta l = 0.06$ mm, and the stresses of the concrete and the steel at this moment;
(2) The cracking load of the member; and
(3) The ultimate load of the member.

Solution From Fig. 4.3a, it is found that $f_t = 1.97$ N/mm^2, $E_c = 25.1 \times 10^3$ N/mm^2, $f_y = 376$ N/mm^2, $E_s = 205 \times 10^3$ N/mm^2, $A_s = 284$ mm^2, and $A = 23{,}104$ mm^2. So,

$$\rho = \frac{A_s}{A} = \frac{284}{23{,}104} = 1.23\,\% < 3\,\%, \quad \alpha_E = \frac{205 \times 10^3}{25.1 \times 10^3} = 8.17$$

$$\varepsilon_{t0} = \frac{f_t}{E_c} = \frac{1.97}{25.1 \times 10^3} = 0.78 \times 10^4$$

(1) The tensile load and the stresses

When $\Delta l = 0.06$ mm, the strain of the member equals

$$\varepsilon = \varepsilon_t = \varepsilon_s = \frac{\Delta l}{l} = \frac{0.06}{915} = 0.66 \times 10^{-4} < \varepsilon_{t0} = 0.78 \times 10^{-4}$$

The member is not cracked and is still in the elastic stage. So, the tensile load can be calculated by:

$$\begin{aligned} N_t &= E_c A(1 + \alpha_E \rho)\varepsilon = 25.1 \times 10^3 \times 23{,}104 \times (1 + 8.17 \times 0.0123) \times 0.66 \times 10^{-4} \\ &= 42{,}120\,\text{N} = 42.12\,\text{kN} \end{aligned}$$

The stress of the concrete is:

$$\sigma_t = E_c \varepsilon_t = 25.1 \times 10^3 \times 0.66 \times 10^{-4} = 1.66\,\text{N/mm}^2$$

The stress of the steel bar is:

$$\sigma_s = E_s \varepsilon_s = 205 \times 10^3 \times 0.66 \times 10^{-4} = 13.53\,\text{N/mm}^2$$

(2) The cracking load is calculated as:

$$\begin{aligned} N_{tcr} &= E_c A(1 + \alpha_E \rho)\varepsilon_{t0} = f_t A(1 + \alpha_E \rho) = 1.97 \times 23{,}104 \times (1 + 8.17 \times 0.0123) \\ &= 50{,}089\,\text{N} = 50.09\,\text{kN} \end{aligned}$$

(3) The ultimate load is shown by:

$$N_{tu} = f_y A_s = 376 \times 284 = 106{,}784\,\text{N} = 106.78\,\text{kN}$$

Compared with the test results, as shown in Fig. 4.3a, it is shown that the calculation results are correct.

4.3 Applications of the Bearing Capacity Equations for Axially Tensioned Members

The bearing capacity equations can be used in two aspects: (1) bearing capacity calculation of existing structural members; and (2) cross-sectional design of new structural members, which will be discussed separately in the following sections.

4.3.1 Bearing Capacity Calculation of Existing Structural Members

For an existing structural member, the dimension of the cross section (b and h), the amount of the reinforcement (A_s), and the strength of the materials (f_t and f_y) are known. So, the ultimate tensile capacity of the member can be taken as the maximum of N_{tcr} and N_{tu}, which can be calculated by Eqs. (4.5) and (4.7), respectively. The calculation process has been shown in Example 4.1.

4.3.2 Cross-Sectional Design of New Structural Members

This kind of application is to determine the reinforcement amount A_s with the cross section (b and h), the strengths of the materials (f_t and f_y), and the known applied axial tension N_t. To ensure the safety of the cross section under given axial tension, it is required that the ultimate tensile capacity of the member should not be smaller than the applied tensile load, that is, $N_{tu} \geqslant N_t$. So, the reinforcement amount A_s can be calculated according to the following equation:

$$N_t = N_{tu} = f_y A_s \tag{4.8}$$

The ultimate tensile capacity of the cross section must be larger than the cracking load to avoid brittle failure. This can be achieved by checking the longitudinal reinforcement ratio: $\rho = \frac{A_s}{A} \geqslant \rho_{min}$, where, ρ_{min} is the minimum longitudinal reinforcement ratio and can be calculated by equaling the ultimate tensile capacity to the cracking load. From Eqs. (4.7) and (4.5), it can be obtained when

$$f_y A_{smin} = E_c A(1 + \alpha_E \rho_{min})\varepsilon_{t0} \approx A f_t \tag{4.9}$$

$$\rho_{min} = \frac{A_{smin}}{A} = \frac{f_t}{f_y} \tag{4.10}$$

The values of ρ_{min} in design codes may slightly differ from those calculated by Eq. (4.10). Table 4.2 at the end of this chapter lists the allowable values of ρ_{min} for reinforced concrete members extracted from GB 50010.

Example 4.2 A structural member with the cross-sectional dimension $b \times h = 300$ mm × 300 mm is subjected to an axial tension of 640 kN. The tensile strength of the steel and the concrete are f_y = 342 N/mm^2 and f_t = 1.5 N/mm^2, respectively. Calculate the required reinforcement amount based on the ultimate tensile capacity.

Solution $A = 300 \times 300 = 90{,}000\,\text{mm}^2, N_t = 640\,\text{kN}, f_y = 342\,\text{N/mm}^2, f_t = 1.5\,\text{N/mm}^2.$ So,

$$A_s = \frac{N_t}{f_y} = \frac{640{,}000}{342} = 1871\,\text{mm}^2$$

$$\frac{A_s}{A} = \frac{1871}{300 \times 300} = 0.021 > \frac{f_t}{f_y} = \frac{1.5}{342} = 0.004$$

6ϕ20 steel bars are chosen where area A_s = 1884 mm^2 > 1871 mm^2. OK!

4.4 Analysis of Axially Compressed Short Columns

4.4.1 Experimental Study on a Short Column

The specimen shown in Fig. 4.6a has been experimentally studied with an axial compression applied on its top. Under the short-term axial compression, normal strain is uniformly distributed over the cross section of the specimen. Because the steel bars and the concrete are well bonded, their strains are equal. When the applied load is small, it is proportional to the compressive deformation. However, when the load is large, it cannot keep a linear relation with the deformation due to the nonlinearity of the concrete. The deformation increases faster than the load does. And the longitudinal steel bars start to yield at a certain load value. When the load approaches about 90 % of the ultimate load, longitudinal cracks and local crushes of concrete appear on the surface of the specimen. If the load further increases, the concrete cover will spall off and the longitudinal steel bars will buckle outward. The specimen is finally failed by the crushing of concrete under the ultimate load 409.1 kN. The load–deformation curve and the failure mode of the specimen are shown in Fig. 4.6b, c, respectively.

From the test, it can be seen that the loading process of a short column can be divided into two stages: (1) Stage I is from the onset of loading to the yielding of the longitudinal steel bars; and (2) Stage II is from the yielding of the longitudinal steel bars to the crushing of concrete. In both stages, the steel bars and the concrete can work together well and deform compatibly.

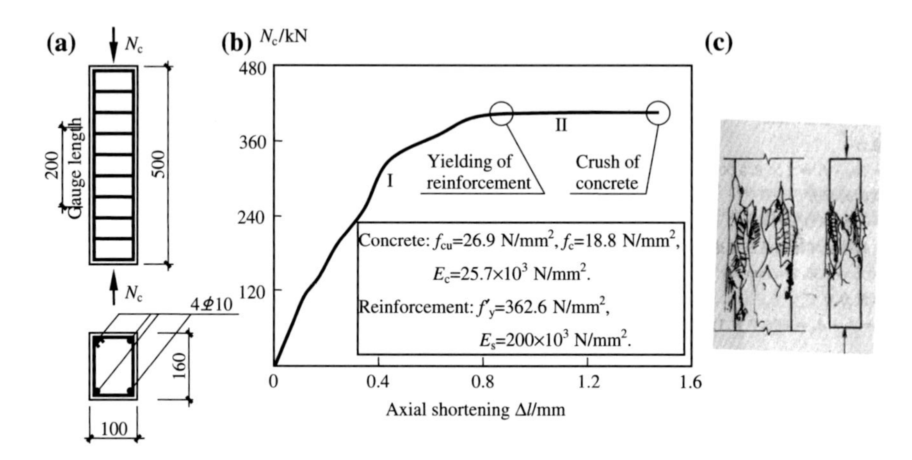

Fig. 4.6 Testing results for an axially compressed short column. **a** Specimen. **b** Load–deformation curve. **c** Failure mode

4.4.2 Load Versus Deformation of Short Columns

4.4.2.1 Stress–Strain Relationships for Reinforcement and Concrete

From Chap. 2, it can be known that the stress–strain curve under compression is the same as that under tension for reinforcement with apparent yielding plateau, provided that no buckling happens during the loading. So, the bilinear stress–strain curve shown in Fig. 4.7a can be used.

$$\sigma'_s = E_s \varepsilon'_s \quad (0 < \varepsilon'_s \leqslant \varepsilon'_y) \tag{4.11a}$$

$$\sigma'_s = f'_y \quad (\varepsilon'_s > \varepsilon'_y) \tag{4.11b}$$

Similar to tensioned structural members, compressed members in real engineering generally cannot unload. So the descending branch of the stress–strain curve of concrete under compression measured in laboratories does not exist in practice. From the discussion in Chap. 2, the stress–strain curve of concrete under compression can be assumed as that shown in Fig. 4.7b when $f_{cu} \leqslant 50$ N/mm^2 (if not otherwise specified, f_{cu} is always smaller than or equal to 50 N/mm^2 in this book), and the mathematic equation of the curve is expressed in Eq. (2.6a).

4.4.2.2 Compressive Load Versus Deformation

In an axially compressed reinforced concrete member, the load is carried by both the concrete and the steel. The two materials deform compatibly with the same strains (Fig. 4.8a):

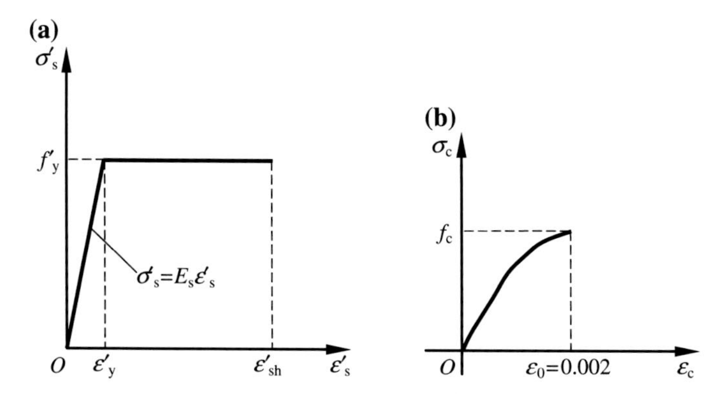

Fig. 4.7 Stress–strain relationships for reinforcement and concrete under compression. **a** Reinforcement. **b** Concrete

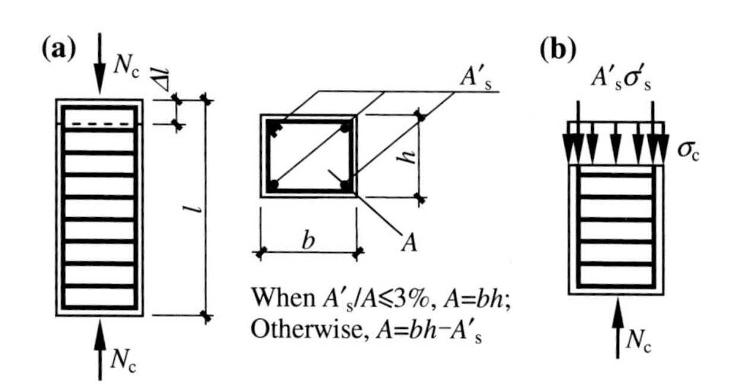

Fig. 4.8 An axially compressed reinforced concrete member

$$\varepsilon = \varepsilon_c = \varepsilon_s' = \frac{\Delta l}{l} \tag{4.12}$$

From the force equilibrium (Fig. 4.8b), it can be obtained:

$$N_c = \sigma_c A + \sigma_s' A_s' \tag{4.13}$$

where

N_c axial compression applied on the member;

σ_c, σ_s' compressive stresses of the concrete and the steel, respectively; and

A, A_s' sectional areas of the structural member and the longitudinal steel bars, respectively.

By substituting Eqs. (4.11a, 4.11b) and (2.6a) into Eq. (4.13) and considering Eq. (4.12), the relationship of axial compression and the deformation can be established as:

$$N_c = 1000\varepsilon(1 - 250\varepsilon) f_c A + E_s \varepsilon A_s' \quad (4.14)$$

where f_c = prism compressive strength of concrete.

Introducing the secant modulus of concrete $E_c' = \nu E_c$, where ν is the elastic coefficient taking into account the reduction of the secant modulus during loading, gives

$$N_c = \nu E_c \varepsilon A + E_s \varepsilon A_s' = \nu E_c \varepsilon A \left(1 + \frac{\alpha_E}{\nu} \rho'\right) \quad (4.15)$$

where $\rho' = A_s'/A$, the longitudinal compression reinforcement ratio.

So, the stress of the concrete is

$$\sigma_c = \frac{N_c}{A\left(1 + \frac{\alpha_E}{\nu} \rho'\right)} \quad (4.16)$$

and the stress of the steel is

$$\sigma_s' = E_s \varepsilon = E_s \frac{\sigma_c}{\nu E_c} = \frac{N_c}{\left(1 + \frac{\nu}{\alpha_E \rho'}\right) A_s'} \quad (4.17)$$

The stress and load relationships of the concrete and the steel are illustrated in Fig. 4.9.

If the axially applied load is small, say smaller than 30 % of the ultimate load, the nonlinear behavior of the concrete can be ignored for simplification. Accordingly, the relationship between the load and the deformation can be expressed as:

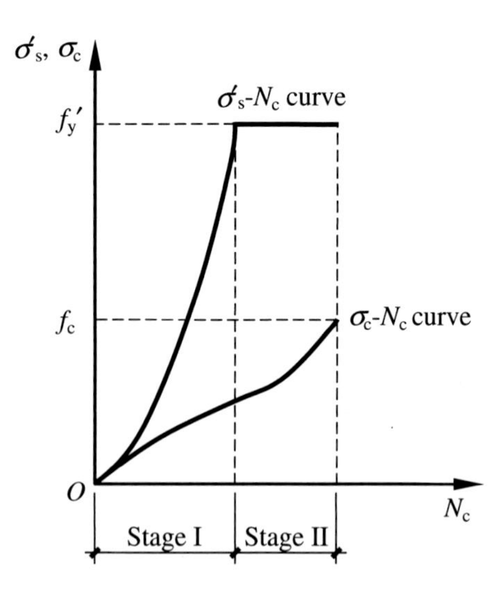

Fig. 4.9 Stress–load curves

$$N_c = (E_c A + E_s A'_s)\varepsilon = E_c A(1 + \alpha_E \rho')\varepsilon \tag{4.18}$$

When $\varepsilon = \varepsilon'_t$, the steel is yielding, symbolizing the start of Stage II. The stress of the steel remains constant, while the concrete stress increases rapidly. The $\sigma_c - N_c$ curve changes from concaving downward to concaving upward. Equation (4.15) can be written as:

$$N_c = 1000\varepsilon(1 - 250\varepsilon)f_c A + f'_y A'_s \tag{4.19}$$

When $\varepsilon = \varepsilon_0 = 0.002$, the concrete is crushed and the column reaches its ultimate compression capacity.

$$N_{cu} = f_c A + f'_y A'_s \tag{4.20}$$

Note that because $\varepsilon = \varepsilon_0 = \varepsilon'_s = 0.002$, the corresponding stress in the longitudinal steel bars is $\sigma'_s = E_s \varepsilon'_s \approx 200 \times 10^3 \times 0.002 = 400\,\text{N/mm}^2$. That is to say, if the strength of the steel bars in an axially compressed column is larger than 400 N/mm^2, the steel will not be fully utilized. So, in the calculation of axially compressed reinforced concrete columns, the maximum steel strength f'_y can only be taken as 400 N/mm^2.

4.4.3 Mechanical Behavior of Short Columns with Sustained Loading

Due to the creep of concrete, the compressive deformation of short columns subjected to constant sustained axial compression will increase with time. And because the concrete and steel bars work together, concrete creep will increase the compressive deformation in the steel bars and thus the load shared by the steel bars. As shown in Fig. 4.10a, b, right at the moment when axial compression N_c is applied, the instantaneous compressive strain of the column is ε_i. From Eqs. (4.16) and (4.17), the stresses in the concrete and steel can be calculated by:

$$\sigma_{c1} = \frac{N_c}{A\left(1 + \frac{\alpha_E}{\nu}\rho'\right)} \tag{4.21}$$

$$\sigma'_{s1} = \frac{N_c}{\left(1 + \frac{\nu}{\alpha_E \rho'}\right)A'_s} \tag{4.22}$$

Concrete creep will happen with the increase of the loading time. The creep strain can be calculated by:

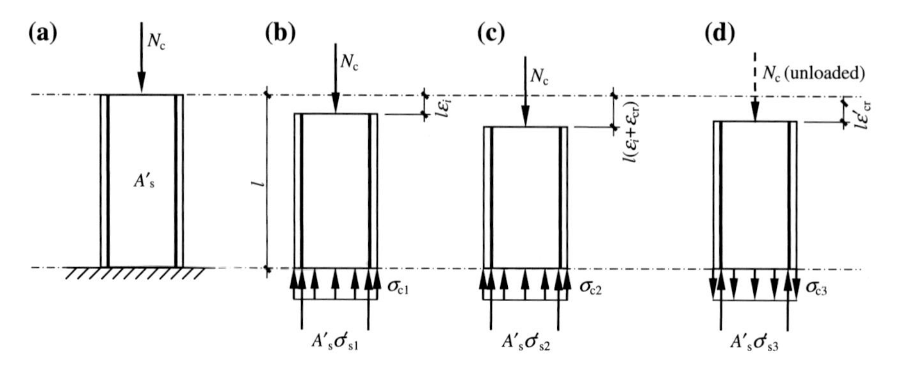

Fig. 4.10 Effect of creep on the mechanical behavior of a short column

$$\varepsilon_{\mathrm{cr}} = C_{\mathrm{t}}\varepsilon_{\mathrm{i}} \tag{4.23}$$

where C_t = creep coefficient.

If the influence of the steel on concrete creep is ignored, the total strain of the member after the creep strain ε_{cr} is (Fig. 4.10c)

$$\varepsilon = \varepsilon_{\mathrm{i}} + \varepsilon_{\mathrm{cr}} = (1 + C_{\mathrm{t}})\varepsilon_{\mathrm{i}} \tag{4.24}$$

The stress in the steel is

$$\sigma'_{\mathrm{s2}} = E_{\mathrm{s}}(1 + C_{\mathrm{t}})\varepsilon_{\mathrm{i}} = (1 + C_{\mathrm{t}})\sigma'_{\mathrm{s1}} = \frac{N_{\mathrm{c}}(1 + C_{\mathrm{t}})}{\left(1 + \frac{v}{\alpha_{\mathrm{E}}\rho'}\right)A'_{\mathrm{s}}} \tag{4.25}$$

For equilibrium, $N_c = A\sigma_{c2} + A'_s\sigma'_{s2}$; the stress of the concrete can be calculated by:

$$\sigma_{\mathrm{c2}} = \left[1 - \frac{\alpha_{\mathrm{E}}(1 + C_{\mathrm{t}})A'_{\mathrm{s}}}{vA\left(1 + \frac{\alpha_{\mathrm{E}}}{v}\rho'\right)}\right]\frac{N_{\mathrm{c}}}{A} = \left(1 - \frac{\alpha_{\mathrm{E}}}{v}\rho' C_{\mathrm{t}}\right)\sigma_{\mathrm{c1}} \tag{4.26}$$

Comparison of Eqs. (4.25) and (4.26) with Eqs. (4.21) and (4.22) shows that $\sigma_{c2} < \sigma_{c1}$ and $\sigma'_{s2} > \sigma'_{s1}$, which is caused by the stress redistribution between the concrete and the steel due to creep.

If the load N_c is removed after being applied for some time, the member cannot recover to the initial configuration as shown in Fig. 4.10d. On assuming the residual strain ε'_{cr} in concrete, the stress in the steel can be determined as:

$$\sigma'_{\mathrm{s3}} = E_{\mathrm{s}}\varepsilon'_{\mathrm{cr}} \tag{4.27}$$

From the equilibrium, it can be concluded that the concrete is tensioned, and the tensile stress in the concrete is

$$\sigma_{c3} = \sigma'_{s3}\frac{A'_s}{A} = E_s\varepsilon'_{cr}\rho' \tag{4.28}$$

Therefore, removing sustained load from short columns will result in tension in the concrete. And, the more reinforcement is, the larger the tensile stress will be in the concrete. More seriously, the concrete may even crack horizontally.

Figure 4.11 shows the variation of the stresses in the steel and the concrete of short columns subjected to sustained loads. It can be seen that, initially, the stresses vary quickly, but gradually, they become stable after some time (say 150 days). The amplitude of stress variation is small for concrete and large for steel. If being unloaded suddenly during sustained loading, the members will recover. But because most of the creep strain is unrecoverable, the steel and concrete are in compression and tension, respectively, when the applied load is zero. If the members are reloaded to the previous load value, the stresses in steel and concrete will vary following the original curves.

Example 4.3 For the model short column shown in Fig. 4.6, calculate

(1) When N_c = 120 kN, the compressive deformation Δl of the member and the respective loads carried by steel bars and concrete if the nonlinear stress–strain relationship for concrete is used;
(2) When N_c = 120 kN, the compressive deformation Δl of the member if the linear stress–strain relationship for concrete is used;
(3) The ultimate compressive capacity of the column; and
(4) The compressive forces carried by concrete and steel bars, respectively, if the sustained compressive load is N_c = 120 kN, and the creep strain of the concrete is ε_{cr} = 0.001.

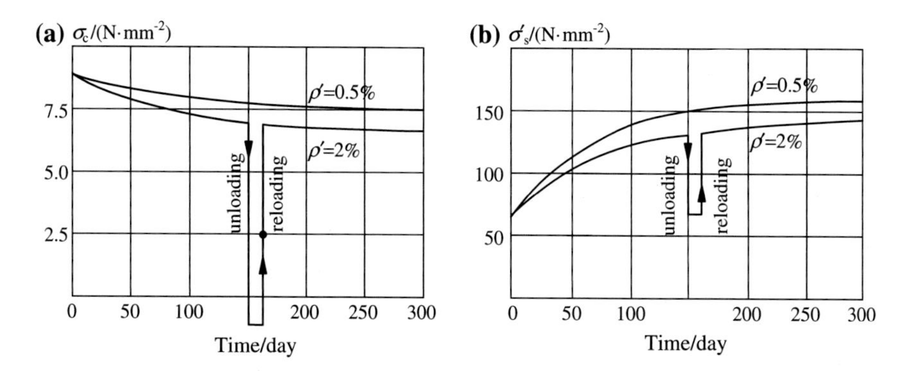

Fig. 4.11 Variation of stresses in the concrete and the steel of short columns subjected to sustained loads. **a** Concrete. **b** Steel

Solution Knowing $f_c = 18.8\ \text{N/mm}^2$, $E_c = 25.7 \times 10^3\ \text{N/mm}^2$, $f'_y = 362.6\ \text{N/mm}^2$, $E_s = 200 \times 10^3\ \text{N/mm}^2$, $A'_s = 314\ \text{mm}^2$, and $bh = 16{,}000\ \text{mm}^2$

Because $A'_s/(bh) = 314/16{,}000 = 1.96\,\% < 3\,\%$, $A = 16{,}000\ \text{mm}^2$

(1) If using the nonlinear stress–strain relationship, it can be obtained from Eq. (4.14)

$$120{,}000 = 1000\varepsilon(1 - 250\varepsilon) \times 18.8 \times 16{,}000 + 2.0 \times 10^5 \times 314\varepsilon$$

The strain can be solved as $\varepsilon = 0.356 \times 10^{-3}$

The compressive deformation is $\Delta l = 0.356 \times 10^{-3} \times 500\ \text{mm} = 0.178\ \text{mm}$

The stress in the steel is $\sigma'_s = E_s\varepsilon = 200 \times 10^3 \times 0.356 \times 10^{-3} = 71.20\ \text{N/mm}^2$

The stress in the concrete is

$$\begin{aligned}\sigma_c &= 1000\varepsilon(1 - 250\varepsilon)f_c \\ &= 1000 \times 0.356 \times 10^{-3} \times (1 - 250 \times 0.356 \times 10^{-3}) \times 18.8 = 6.10\ \text{N/mm}^2\end{aligned}$$

The compressive force carried by the steel is

$$N'_s = \sigma'_s A'_s = 71.20 \times 314 = 22{,}357\ \text{N} = 22.357\ \text{kN}$$

The compressive force carried by the concrete is

$$N'_c = \sigma_c A = 6.10 \times 16{,}000 = 97{,}600\ \text{N} = 97.600\ \text{kN}.$$

(2) $\alpha_E = E_s/E_c = 200 \times 10^3/(25.7 \times 10^3) = 7.78$

If using the linear stress–strain relationship, one has $N_c = E_c A\ (1 + \alpha_E \rho')\varepsilon$

$$20{,}000 = 2.57 \times 10^3 \times 16{,}000 \times (1 + 7.78 \times 0.0196)\varepsilon$$

So, $\varepsilon = \frac{120{,}000}{2.57 \times 10^4 \times 16{,}000 \times (1 + 7.78 \times 0.0196)} = 0.253 \times 10^{-3}$

The compressive deformation of the member is $\Delta l = 0.253 \times 10^{-3} \times 500\ \text{mm} = 0.127\ \text{mm}$

The stress in the steel is $\sigma'_s = E_s\varepsilon = 200 \times 10^3 \times 0.253 \times 10^{-3} = 50.60\ \text{N/mm}^2$

The stress in the concrete is $\sigma_c = E_c\varepsilon = 25.7 \times 10^3 \times 0.253 \times 10^{-3} = 6.50\ \text{N/mm}^2$

The compressive load carried by the steel is

$$N'_s = \sigma'_s A'_s = 50.60 \times 314 = 15{,}888\ \text{N} = 15.888\ \text{kN}$$

The compressive load carried by the concrete is

$$N_c' = \sigma_c A = 6.50 \times 16{,}000 = 104{,}000\,\text{N} = 104.000\,\text{kN}$$

(3) Ultimate compressive capacity of the column

$$N_{cu} = f_c A + f_y' A_s' = 18.8 \times 16{,}000 + 362.6 \times 314 = 414{,}656\,\text{N} = 414.656\,\text{kN}$$

(4) $\varepsilon_{cr} = 0.001$, $\varepsilon_i = 0.356 \times 10^{-3}$, so, $C_t = \varepsilon_{cr}/\varepsilon_i = 0.001/0.356 \times 10^{-3} = 2.809$, $\sigma_{s1}' = 71.20\,\text{N/mm}^2$

The stress in the steel is $\sigma_{s2}' = (1 + C_t)\sigma_{s1}' = (1 + 2.809) \times 71.20 = 271.201\,\text{N/mm}^2$

The compressive force carried by the steel is

$$N_{s2}' = \sigma_{s2}' A_s' = 271.201 \times 314 = 85{,}157\,\text{N} = 85.157\,\text{kN}$$

The compressive force carried by the concrete is

$$N_{c2}' = N_c - N_{s2}' = 120 - 85.157 = 34.843\,\text{kN}.$$

4.5 Analysis of Axially Compressed Slender Columns

4.5.1 Experimental Study on a Slender Column

A 2000-mm-long specimen with all other parameters the same as those of the short column shown in Fig. 4.6 was experimentally studied. The mid-height lateral deflection of the column was monitored by a displacement gauge in addition to the measurement of strains in the concrete and the steel bars. The maximum load (at failure) of the column is 336.9 kN. The load–lateral deflection curve and failure mode of the column are shown in Figs. 4.12 and 4.13, respectively. Based on the results, it can be concluded that the ultimate capacity of a slender column is smaller than that of a short column if all other conditions are the same, such as the materials, reinforcement, and sectional dimension. This is due to the effect of the secondary moment resulting from the lateral deflection of the column. Many factors may cause slender columns bending: inaccurate sectional dimension, inhomogeneous materials, disturbed reinforcement positions, and misalignment of the loading axis and the column axis.

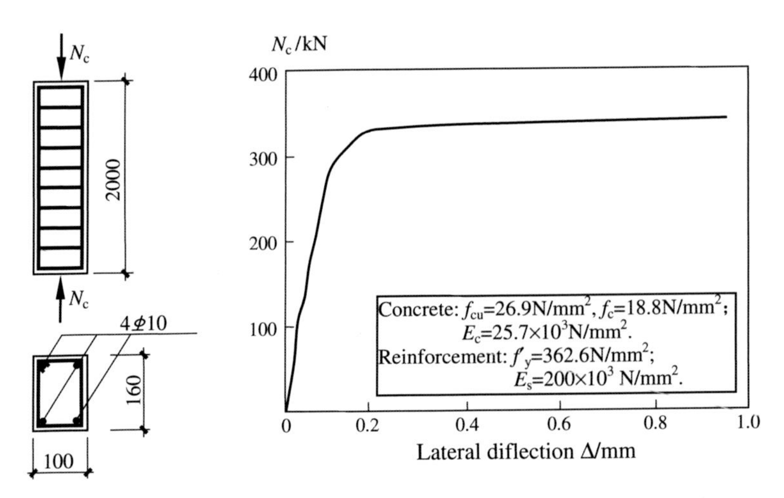

Fig. 4.12 Load–lateral deflection curve of a slender column under axial compressive load

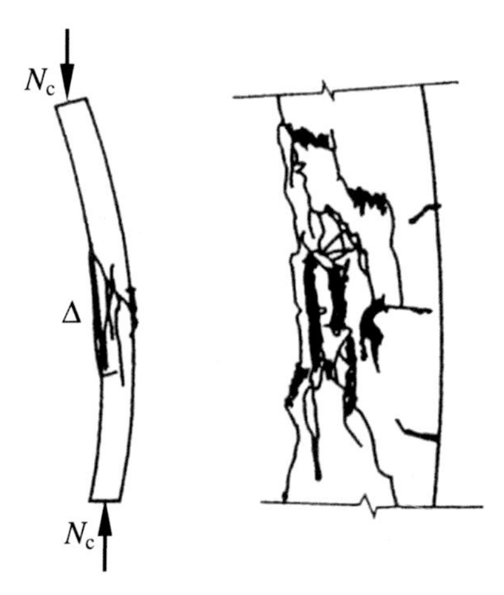

Fig. 4.13 Failure mode of an axially compressed slender column

When the load is small, the full section of a slender column is subjected to compression. However, due to the secondary moment, the compressive stress at one side of the cross section is larger than that at the other side. And, with the increase of the applied load, the stress difference becomes bigger and bigger. In addition, the lateral deflection increases even faster. The concrete at the larger compressive stress side is first crushed, vertical cracks appear on the surface of the column, and the reinforcement buckles outward. The other side of concrete may change from bearing compression to resisting tension and crack horizontally as shown in Fig. 4.13. On the one hand, initial eccentricity leads to the secondary moment. On the other hand, the secondary moment increases lateral deflection. The interaction finally leads to the slender column failure by the combination of flexure and axial loads. If the slenderness ratio is too large, the column may fail by losing its stability.

4.5.2 Stability Coefficient

Stability coefficient φ is defined as the ratio of the ultimate compressive capacities of a slender column and a corresponding short column, which have the same sectional dimension, material, and reinforcement. Therefore, the ultimate compressive capacities of slender columns can be calculated from φ and the capacities of corresponding short columns.

Test results from the Chinese Academy of Building Research and foreign countries show that the coefficient φ mainly depends on the slenderness ratio of columns. For a column with a rectangular section, the slenderness ratio is defined as l_0/b, where l_0 is the effective height of the column and b is the smaller dimension of the rectangular section. Test results of the coefficient φ versus the slenderness ratio are given in Fig. 4.14, from which it can be concluded that the larger the slenderness l_0/b is, the smaller the stability coefficient φ is. Generally, φ can be taken as 1.0 when $l_0/b < 8$. For columns of the same slenderness ratio, φ may be slightly different due to the difference in concrete grade, reinforcement type and reinforcement ratio. The following empirical equations are regressed from experimental data.

$$\varphi = 1.177 - 0.021 l_0/b \quad (l_0/b = 8 \sim 34) \tag{4.29}$$

$$\varphi = 0.87 - 0.012 l_0/b \quad (l_0/b = 35 \sim 50) \tag{4.30}$$

For safety, GB 50010 gives more conservative values of φ than the above equations for columns with a large slenderness ratio, considering the large adverse effects of initial eccentricity and sustained loading on the ultimate compressive

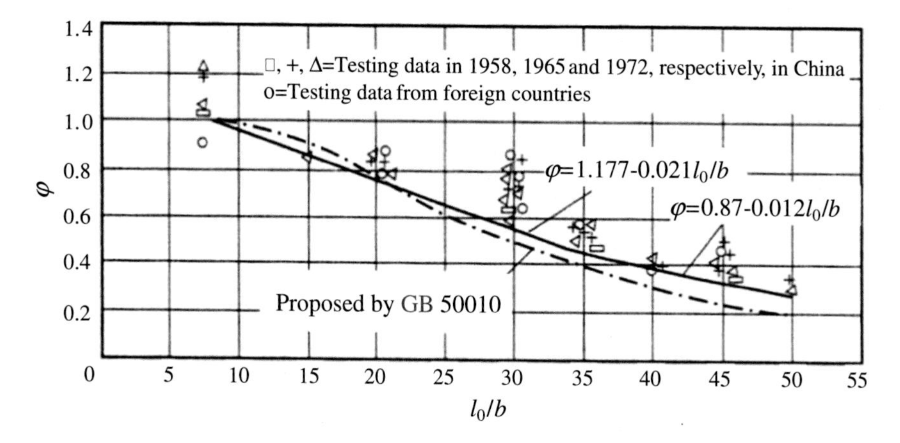

Fig. 4.14 Stability coefficient versus slenderness ratio of columns with rectangular sections

Table 4.1 Stability coefficient of axially compressed reinforced columns

l_0/b	$\leqslant 8$	10	12	14	16	18	20	22	24	26	28
l_0/d_c	$\leqslant 7$	8.5	10.5	12	14	15.5	17	19	21	22.5	24
l_0/i	$\leqslant 28$	35	42	48	55	62	69	76	83	90	97
φ	1.00	0.98	0.95	0.92	0.87	0.81	0.75	0.70	0.65	0.60	0.56
l_0/b	30	32	34	36	38	40	42	44	46	48	50
l_0/d_c	26	28	29.5	31	33	34.5	36.5	38	40	41.5	43
l_0/i	104	111	118	125	132	139	146	153	160	167	174
φ	0.52	0.48	0.44	0.40	0.36	0.32	0.29	0.26	0.23	0.21	0.19

Notes l_0 is the effective height, b is the short dimension of a rectangular section, d_c is the diameter of a circular section, and $i = \sqrt{I/A}$ is the minimum radius of gyration of the section, where I and A are the moment of inertia and area of the section

capacity of columns. Bigger values of φ are assigned to columns of small slenderness ratio according to previous experience. Table 4.1 lists modified φ values. One can obtain φ value from the table by linear interpolation after knowing the slenderness ratio of a column.

4.5.3 *Equation for Ultimate Capacity of Axially Compressed Columns*

For axially compressed members reinforced by moderate strength steel bars, when the compressive stress of concrete reaches its maximum and the steel has yielded, the members are assumed to reach their ultimate compressive capacities. The equation to calculate the ultimate capacity is

$$N_{cu} = \varphi(Af_c + f'_y A'_s) \tag{4.31}$$

where

N_{cu} ultimate compressive capacity of a column;
φ stability coefficient;
f_c axial compressive strength (peak stress) of concrete;
f'_y yield strength of steel;
A sectional area of the column; and
A'_s sectional area of the compression reinforcement.

In engineering applications, different codes may modify Eq. (4.31) a little bit. For example, to account for the effect of accidental moments, GB 50010 multiplies a reduction coefficient 0.9 to the right-hand side of Eq. (4.31) to ensure approximate safety levels for axially and eccentrically compressed structural members.

4.6 Applications of the Bearing Capacity Equation for Axially Compressed Members

4.6.1 *Bearing Capacity Calculation of Existing Structural Members*

For existing structural members, the dimension of the cross section (b and h), effective height l_0, the amount of the reinforcement (A'_s), and the respective strength of the materials (f_c and f'_y) are known. So the ultimate compressive capacity N_{cu} can be calculated as follows:

(1) Calculate φ from l_0/b with the help of Table 4.1;
(2) Check $f'_y \leqslant 400\,\mathrm{N/mm^2}$. If the cube strength of concrete $f_{cu} > 50$ N/mm^2, f'_y should be adjusted according to corresponding concrete strain ε_0;
(3) If $A'_s/(bh) \leqslant 3\,\%$, then $A = bh$. Otherwise, $A = bh - A'_s$;
(4) Calculate N_{cu} using Eq. (4.31).

Example 4.4 Calculate the ultimate compressive capacity of the slender column shown in Fig. 4.12.

Solution Knowing $f_c = 18.8\,\mathrm{N/mm^2}, f'_y = 362.6\,\mathrm{N/mm^2} < 400\,\mathrm{N/mm^2}, A'_s = 314\,\mathrm{mm^2}, b \times h = 16{,}000\,\mathrm{mm^2}, l_0 = 2000\,\mathrm{mm}$

$l_0/b = 2000/100 = 20, \text{so}, \varphi = 0.75.$ from Table 4.1

Since $A'_s/(bh) = 314/16{,}000 = 1.96\,\% < 3\,\%$, $A = 16{,}000$ mm^2

$$N_{cu} = \varphi(Af_c + f'_y A'_s) = 0.75 \times (16{,}000 \times 18.8 + 362.6 \times 314) = 310{,}992\,\mathrm{N} = 310.992\,\mathrm{kN}$$

Compared with the test results shown in Fig. 4.12, it can be seen that the calculation method in Example 4.4 is conservative.

4.6.2 *Cross-Sectional Design of New Structural Members*

This kind of application is to determine the compression reinforcement A'_s with the cross section (b and h), effective height l_0, the strengths of the materials (f_c and f'_y), and the known applied axial compression N_c. To ensure the safety of the cross section under given axial compression, it is required that the ultimate compressive capacity of the member should not be smaller than the applied compressive load, that is, $N_{cu} \geqslant N_c$. So, the reinforcement amount A'_s can be calculated as follows:

(1) Determine φ from Table 4.1 according to l_0/b;
(2) Check $f_y' \leqslant 400\,\text{N/mm}^2$;
(3) Compute A_s' from $N_c = N_{cu} = \varphi\left(Af_c + A_s'f_y'\right)$;
(4) If $A_s'/(bh) \leqslant 3\,\%$, then $A = bh$. Otherwise, recalculate A_s' taking $A = bh - A_s'$
(5) Check $\rho' \geqslant \rho'_{min}$. This is to prevent the brittle failure of the members. Similar to tension steel bars, the allowable minimum reinforcement ratio may be different in different design codes. The minimum reinforcement ratios stipulated by GB 50010 are shown in Table 4.2. If $\rho' < \rho'_{min}$, it is suggested to take $\rho' = \rho'_{min}$.

Theoretically, the larger the reinforcement ratio is, the better an axially compressed column behaves. However, it is generally very difficult to place a large amount of steel bars in a column, especially if lapped splices are used. Usually, the maximum reinforcement ratios for various column sizes range from roughly 3 to 5 %. In addition, the most economical tied-column section generally involves ρ' of 1–2 %. If the calculated $\rho' > 5\,\%$, it is better to redesign the column section.

Example 4.5 The effective length of an axially compressed column is $l_0 = 2000$ mm. The dimension of the section is $bh = 100\text{ mm} \times 160\text{ mm}$. The strength of the concrete and steel is $f_c = 18.8$ MPa and $f_y' = 362.6\,\text{MPa}$, respectively. Consider an axial compression $N_c = 360$ kN to determine the longitudinal reinforcement A_s'.

Solution $f_c = 18.8\,\text{N/mm}^2, f_y' = 362.6\,\text{N/mm}^2 < 400\,\text{N/mm}^2, bh = 16{,}000\,\text{mm}^2$, $l_0 = 2000\,\text{mm}$

$$l_0/b = 2000/100 = 20$$

From Table 4.1, $\varphi = 0.75$. So,

$$A_s' = \frac{N - \varphi A f_c}{\varphi f_y'} = \frac{360{,}000 - 0.75 \times 16{,}000 \times 18.8}{0.75 \times 362.6} = 494\,\text{mm}^2$$

$$A_s'/(bh) = 494/16{,}000 = 3.09\,\% > 3\,\%$$

Taking $A = bh - A_s'$, then

$$A_s' = \frac{N - \varphi bhf_c}{\phi(f_y' - f_c)} = \frac{360{,}000 - 0.75 \times 16{,}000 \times 18.8}{0.75 \times (362.8 - 18.8)} = 521\,\text{mm}^2$$

From Table 4.2, $\rho'_{min} = 0.6\,\%$

$$\frac{A_s'}{bh} = \frac{521}{16{,}000} = 3.26\,\% > \rho'_{min} = 0.6\,\%$$

Arranging 4ϕ14 bars with $A_s' = 615\,\text{mm}^2 > 521\,\text{mm}^2$, OK!

4.7 Analysis of Spiral Columns

4.7.1 Experiment Study on Spiral Columns

The axial compression tests on spiral columns can be done with a test setup similar to that for tied columns, which is shown in Fig. 4.6a. Figure 4.15 presents the axial compression–strain ($N_c - \varepsilon$) curves for different kinds of columns. Figure 4.15 also shows that *A* is a plain concrete column, *B* is a tied column, and *C* series are spiral columns. C_1 has the smallest spiral pitch, while C_3 has the largest one. Experimental results show that the deformations of the tied and spiral columns are barely different when the applied loads are small. After the steel has yielded, the concrete shell outside the ties or spirals spalls off, which reduces the load-carrying area of the concrete and thereby reduces the bearing capacities of the columns. However, closely spaced spirals can prevent the buckling of longitudinal bars so that the latter can still resist loads. With increases of the longitudinal deformation, the transverse expansion of core concrete induces larger hoop tension in spirals. But the tensioned spirals in turn tightly confine the core concrete, which restricts its transverse expansion and places it under a triaxial compression state. The compressive strength and deformation capacity of core concrete are correspondingly enhanced. When the spirals have yielded, their confinement on core concrete will not increase, and the columns start to fail. So, although concrete cover spalls off, the strength increase of core concrete due to spirals confinement compensates for the sectional area reduction of the columns. Columns with appropriately spaced spirals have larger ultimate capacities and better deformation abilities than ordinary tied columns of the same sectional dimensions.

The same behavior can be observed from the test on a column with densely spaced welded circular hoops.

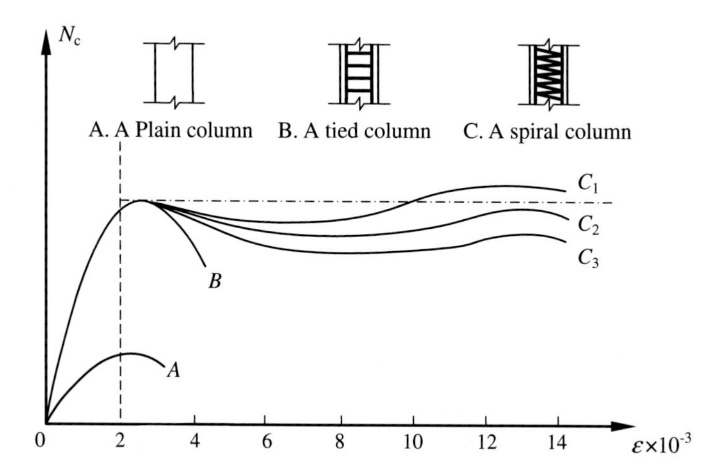

Fig. 4.15 $N_c - \varepsilon$ curves for different kinds of columns under axial compression

From the above discussion, it is found that using spirals or circular hoops as transverse reinforcement can stress core concrete in three directions perpendicular to one another and thus enhance its strength, which will further indirectly increase the ultimate compressive capacities of columns. Therefore, spirals or circular hoops are also called lateral reinforcement.

4.7.2 Ultimate Compressive Capacities of Spiral Columns

From Chap. 2, the axial compressive strength of confined concrete can be approximately taken as

$$f_{cc} = f_c + 4\sigma_r \tag{4.32}$$

where
f_{cc} axial compressive strength of confined concrete; and
σ_r adial pressure acting on core concrete.

When spirals or circular hoops are yielding, σ_r reaches its maximum. From equilibrium of the free body shown in Fig. 4.16, it can be obtained

$$\sigma_r = \frac{2f_y A_{ss1}}{s d_{cor}} = \frac{2f_y A_{ss1} d_{cor} \pi}{4 \cdot \frac{\pi d_{cor}^2}{4} \cdot s} = \frac{f_y A_{ss0}}{2A_{cor}} \tag{4.33}$$

where
A_{ss1} section area of a single spiral or circular hoop leg;
f_y tensile strength of spirals or circular hoops;
s pitch of spirals or hoops;

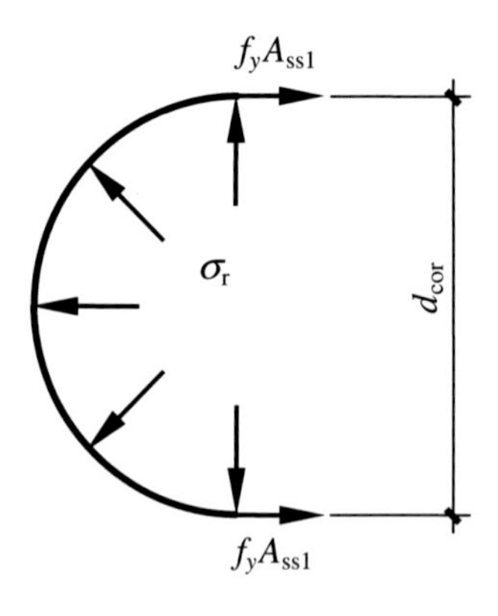

Fig. 4.16 Forces on a spiral

d_{cor} diameter of core concrete, generally $d_{cor} = d_c - 2c$, where d_c stands for the diameter of a column and c is the thickness of concrete cover;

A_{ss0} equivalent area of lateral reinforcement, $A_{ss0} = \frac{\pi d_{cor} A_{ss1}}{s}$; and

A_{cor} section area of core concrete.

From equilibrium in longitudinal direction, the ultimate capacity of a spiral column or a circular-hoop column can be calculated as:

$$N_{cu} = f_{cc}A_{cor} + f'_y A'_s = (f_c + 4\sigma_r)A_{cor} + f'_y A'_s = f_c A_{cor} + f'_y A'_s + 2f_y A_{ss0} \quad (4.34)$$

Equation (4.34) has one more term than Eq. (4.20), i.e., $2f_y A_{ss0}$, which is the capacity increase due to spiral or circular hoops. Experimental data shows that this increase is smaller for high-strength concrete columns. So, a reduction factor α is introduced, with which Eq. (4.34) will be

$$N_{cu} = f_c A_{cor} + f'_y A'_s + 2\alpha f_y A_{ss0} \quad (4.35)$$

When the cube strength of concrete is not larger than 50 N/mm^2, $\alpha = 1.0$. When the cube strength of concrete is 80 N/mm^2, $\alpha = 0.85$. α can be determined by a linear interpolation for concrete with cube strength between 50 and 80 MPa.

The following items should be noted when using Eq. (4.35):

(1) Similar to Eq. (4.31), Eq. (4.35) may be modified in design codes. For example, GB 50010 stipulates that a reduction factor 0.9 should be multiplied to the right-hand side of Eq. (4.35) to ensure approximate safety level between axially and eccentrically compressed structural members.
(2) To prevent the concrete shell from spalling off in the service stage, the ultimate capacity of spiral column calculated by Eq. (4.35) should not be larger than 1.5 times that of an ordinary tied column by Eq. (4.31).
(3) When $l_0/d_c > 12$, the beneficial effect of lateral reinforcement will be neglected and Eq. (4.31) should be used to calculate the ultimate compressive capacity.
(4) If the capacity reduction due to the spalling of concrete cover is larger than the capacity increase due to lateral reinforcement, the beneficial effect of lateral reinforcement will be neglected and Eq. (4.31) should be used to calculate the ultimate compressive capacity.
(5) When the equivalent area of lateral reinforcement A_{ss0} is smaller than 25 % of the total area of longitudinal steel bars, lateral reinforcement is assumed insufficient and cannot provide enough confinement to core concrete. Equation (4.31) should be used to calculate the ultimate compressive capacity.
(6) The pitch of lateral reinforcement should not be larger than 80 mm and $d_{cor}/5$ and not be smaller than 40 mm.

Example 4.6 The effective length of an axially compressed circular column is 2.9 m. The diameter of the cross section is d_c = 420 mm. 8ϕ22 longitudinal steel bars are arranged in the column. The concrete cover is 30 mm thick. The strength of concrete and the steel are f_c = 14.5 N/mm^2 and $f'_y = 240\,\text{N/mm}^2$, respectively. Calculate

(1) The ultimate compressive capacity of the column if calculated as an ordinary tied column;
(2) The ultimate compressive capacity of the column if reinforced by circular hoops of ϕ8@50 mm;
(3) The ultimate compressive capacity if reinforced by circular hoops of ϕ8@80 mm.

Solution l_0/d_c = 2,900/420 = 6.90 < 7, so φ = 1.0.

(1) As a tied column

$$A'_s = 3040\,\text{mm}^2, A = \pi d_c^2/4 = 138474\,\text{mm}^2, 3040/138474 = 2.20\,\% < 3\,\%$$
$$N_{cu} = \varphi(f_c A + f'_y A'_s) = 1.0 \times (14.5 \times 138474 + 240 \times 3040) = 2737 \times 10^3\text{N} = 2737\,\text{kN}$$

(2) Circular hoops of ϕ8@50 mm

$$d_{cor} = 420 - 2 \times 30 = 360\,\text{mm}$$
$$A_{cor} = \frac{\pi \cdot d_{cor}^2}{4} = 101{,}736\,\text{mm}^2$$

$$A_{ss0} = \frac{\pi \cdot d_{cor} \cdot A_{ss1}}{s} = \frac{\pi \times 360 \times 50.3}{50} = 1137\,\text{mm}^2$$

Because the slenderness ratio of the column is smaller than 12, Eq. (4.35) is used to calculate the ultimate capacity. And since f_{cu} < 50 N/mm^2, α = 1.0.

$$\begin{aligned} N_{cu} &= f_c A_{cor} + f'_y A'_s + 2 f'_y A_{ss0} \\ &= 14.5 \times 101736 + 240 \times 3040 + 2 \times 240 \times 1137 = 2751 \times 10^3\,\text{N} = 2751\,\text{kN} \end{aligned}$$

It is larger than 2737 kN calculated as a tied column, which shows that the ultimate compressive capacity of the column is enhanced by the confinement of circular hoops.

(3) Circular hoops of $\phi 8@80$ mm

$$A_{ss0} = \frac{\pi \times 360 \times 50.3}{80} = 711\,\text{mm}^2$$

$$\begin{aligned} N_{cu} &= f_c A_{cor} + f_y' A_s' + 2 f_y' A_{ss0} \\ &= 14.5 \times 101736 + 240 \times 3040 + 2 \times 240 \times 711 = 2546 \times 10^3\,\text{N} = 2546\,\text{kN} \end{aligned}$$

It is smaller than 2737 kN calculated as a tied column, which shows that the capacity increase provided by the circular hoops cannot offset the capacity reduction due to the spalling of the concrete shell. Therefore, the actual ultimate compressive capacity can be taken as the capacity of the tied column 2737 kN.

Questions

4.1 Why don't new cracks appear in an axially tensioned concrete member when the number of cracks has reached a certain value?

4.2 How do you determine the cracking load and the ultimate load of an axially tensioned structural member?

4.3 In the test of a column subjected to short-term axial compression, which one of the stresses in the steel and the concrete increases faster with the applied load? Why?

4.4 How do you determine the ultimate compressive capacity of an axially compressed short column? Why is it not suitable to use high-strength longitudinal steel bars in axially compressed concrete columns?

4.5 Why should the minimum reinforcement ratio be controlled in the design of axially loaded members? How would you determine the minimum reinforcement ratio in axially tensioned or compressed members?

4.6 What is the function of ties in a tied column?

4.7 How do the stresses of steel and concrete change with time when a sustained compression is applied on a reinforced concrete column? Does the creep affect the ultimate compressive capacity of a short column?

4.8 Why should the stability coefficient φ be taken into account in the ultimate compressive capacity calculation of compressed reinforced concrete members? What is the relation between φ and restraint conditions at the both ends of a compressed member?

4.9 Why not consider the contribution of spirals to the ultimate compressive capacity of a spiral column with a slenderness ratio larger than 12?

4.10 What will happen to the ultimate compressive capacity of a compressed short column if the ties can restrain the transverse deformation of the concrete and why?

Problems

4.1 The sectional dimension of an axially tensioned reinforced concrete member is $b \times h = 300$ mm $\times$ 400 mm. 8ϕ20 longitudinal steel bars are in the member. The mechanical properties of the materials are $f_t = 2.0$ N/mm^2, $E_c = 2.1 \times 10^4$ N/mm^2, $f_y = 270$ N/mm^2, and $E_s = 2.1 \times 10^5$ N/mm^2. Calculate the cracking load and the ultimate load of the member.

4.2 An axially tensioned reinforced concrete member has the rectangular section $b \times h = 200$ mm $\times$ 300 mm. The member is 2000 mm long. The tensile strength and the modulus of elasticity of the concrete are $f_t = 2.95$ N/mm^2 and $E_c = 2.55 \times 10^4$ N/mm^2, respectively. The total sectional area of the longitudinal steel bars is $A_s = 615$ mm^2. The steel has the yield strength of $f_y = 270$ N/mm^2 and the modulus of elasticity of $E_s = 2.1 \times 10^5$ N/mm^2. Calculate:

(1) The applied load and the loads shared by the concrete and the steel, when the member is elongated 0.2 mm.
(2) The applied load and the stresses of the concrete and the steel, when the member is elongated 0.5 mm.
(3) The cracking load and the deformation of the member at the moment of cracking.
(4) The ultimate tensile capacity of the member.

4.3 The sectional dimension of an axially tensioned reinforced concrete member is $b \times h = 300$ mm $\times$ 300 mm. There are 8$\underline{\phi}$22 longitudinal steel bars in the member. The mechanical properties of the materials are $f_t = 2.3$ N/mm^2, $E_c = 2.4 \times 10^4$ N/mm^2, $f_y = 345$ N/mm^2, and $E_s = 1.96 \times 10^5$ N/mm^2.

(1) If the member is allowed to crack, calculate the ultimate tensile capacity of the member.
(2) If the member is not allowed to crack, calculate the ultimate tensile capacity of the member.
(3) Compare the two results.

4.4 A reinforced concrete bottom chord is subjected to an axial tension of $N_t = 150$ kN. If the chord is allowed to crack and the yield strength of the steel is $f_y = 270$ MPa, determine the longitudinal reinforcement in the member.

4.5 The sectional dimension of a reinforced concrete short column is $b \times h = 400$ mm $\times$ 400 mm. The length of the column is 2 m. Four $\underline{\phi}$25 longitudinal steel bars are arranged in the column. If the mechanical properties of the materials are $f_c = 19$ N/mm^2, $E_c = 2.55 \times 10^4$ N/mm^2, $f_y' = 357\,\text{N/mm}^2$, and $E_s = 1.96 \times 10^5$ N/mm^2, determine:

(1) The ultimate compressive capacity of the column.

(2) The compressive deformation and the compressive forces resisted by the concrete and the steel, respectively, when an axial compression $N_c = 1200$ kN is applied on the column.
(3) The compressive forces resisted by the concrete and the steel, respectively, after the column has been in service for several years under the constant compression of $N_c = 1200$ kN with a creep strain of $\varepsilon_{cr} = 0.001$.

4.6 A 2-m-long reinforced concrete short column with the sectional dimension of $b \times h = 350$ mm $\times$ 350 mm is axially compressed. There are four ϕ25 longitudinal steel bars in the column. $f_c = 15$ N/mm^2, $f_y' = 270\,\mathrm{N/mm^2}$, $E_s = 1.96 \times 10^5$ N/mm^2, and $\varepsilon_0 = 0.002$. Calculate:

(1) The compression that yields the longitudinal steel bars and the compressive deformation of the column at that time.
(2) The ultimate compressive capacity of the column.

4.7 The effective height of a middle column in a multistorey building is 4.2 m. $f_y' = 310\,\mathrm{N/mm^2}$, $f_c = 10$ N/mm^2. The axial compression on this column is 700 kN. The sectional dimension of the column is $b \times h = 250$ mm $\times$ 250 mm. Determine the longitudinal reinforcement in the column.

4.8 The effective height of an axially compressed column is 4.7 m. The axial compressive strength of the concrete is $f_c = 10$ N/mm^2. Four $\underline{\phi}$20 longitudinal steel bars with the yield strength of $f_y' = 310\,\mathrm{N/mm^2}$ are arranged in the column. Determine:

(1) The ultimate compressive capacity of the column, whose sectional dimension is 300 mm $\times$ 300 mm.
(2) The ultimate compressive capacity of the column, whose sectional dimension is 250 mm $\times$ 250 mm.

4.9 A spiral column with the diameter of 450 mm is subjected to an axial compression 2500 kN. The effective height of the column is $l_0 = 3.5$ m. The axial compressive strength of the concrete is $f_c = 10$ N/mm^2. Eight ϕ22 longitudinal steel bars with the yield strength of $f_y' = 270\,\mathrm{N/mm^2}$ are arranged in the column. If the yield strength of the spiral is also $f_y = 270$ N/mm^2, determine the spiral amount.

4.10 A spiral column with the diameter of 500 mm has the effective height of $l_0 = 4.0$ m. The axial compressive strength of the concrete is $f_c = 10$ N/mm^2. The longitudinal steel bars are six $\underline{\phi}$22 with the yield strength of $f_y' = 380\,\mathrm{N/mm^2}$. The spirals are ϕ10@50 with the yield strength of $f_y = 270$ N/mm^2. If the concrete cover is 30 mm thick, determine the ultimate compressive capacity of the column.

4.11 If all other conditions are the same as those in Problem 4.10 except the spirals are ϕ8@80, calculate the ultimate compressive capacity of the column.

Appendix

See Table 4.2.

Table 4.2 Minimum longitudinal reinforcement ratio in reinforced concrete members (%)

Loading type			Minimum reinforcement ratio
Compressed members	All longitudinal steel bars	Strength grade 500 MPa	0.50
		Strength grade 400 MPa	0.55
		Strength grade 300, 335 MPa	0.60
	Longitudinal steel bars on one side of the member		0.20
Tensile longitudinal steel bars on one side of the flexural member, eccentrically tensioned member, or axially tensioned member			The larger value between 0.20 and $45f_t/f_y$

Notes

(1) When the strength grade of concrete is C60 or larger, the values in the table should be plus 0.1

(2) The compression reinforcement in eccentrically tensioned members should be treated as longitudinal steel bars on one side of compressed member

(3) The reinforcement ratios for compressed members, axially tensioned members, and small eccentrically tensioned members are calculated using the full sectional areas of the members, while the reinforcement ratios for flexural members and large eccentrically tensioned members are calculated using the full sectional areas of the members minus the areas of the compressive flanges

(4) When steel bars are arranged along the perimeter of a section, "Longitudinal steel bars on one side of the member" in the table mean the longitudinal steel bars are on one of the two opposite sides along loading direction

(5) For slabs reinforced by longitudinal steel bars of the strength grades of 400 MPa and 500 MPa, the minimum reinforcement ratio can take the larger value between 0.15 and 45 f_t/f_y

Chapter 5
Bending Behavior of Flexural Members

5.1 Engineering Applications

Flexural members are widely used in civil engineering shown here as reinforced concrete slabs, beams, stairs, and foundations in Fig. 1.4, retaining walls in Fig. 5.1, and girders, bent caps, and crash barriers in beam bridges in Figs. 5.2 and 5.3. Although flexural members have various section shapes, e.g., rectangular section, T-shaped section, box section, I-shaped section, and channel section, they can generally be categorized into two types according to their mechanical properties: rectangular and T-shaped sections (Figs. 5.4 and 5.5). Circular or ring sections are seldom used in practice.

5.2 Mechanical Characteristics and Reinforcement Type of Flexural Members

For a reinforced concrete beam subjected to two symmetric concentrated loads *P* (Fig. 5.6a), the middle segment (between the two concentrated loads) is under pure bending and the two side segments (between the support and the nearby concentrated load) are under the combination of flexure and shear. The distribution of internal forces along the beam is shown in Fig. 5.6b, c. Due to the bending moment, cracks perpendicular to the longitudinal axis of the beam appear in the middle segment (generally called the vertical cracks). And oblique cracks happen to the side segments (generally called the diagonal cracks). Both kinds of cracks are illustrated in Fig. 5.6a. To prevent the flexure failure caused by vertical cracks and shear failure caused by the diagonal cracks, longitudinal steel bars and stirrups and/or bent bars are arranged in the bottom of the beam and in the flexure–shear segment, respectively. In addition, there are auxiliary bars (also called unstressed bars) at the section corners in unstressed part (Fig. 5.6d). Longitudinal bars, bent

X. Gu et al., *Basic Principles of Concrete Structures*,
DOI 10.1007/978-3-662-48565-1_5

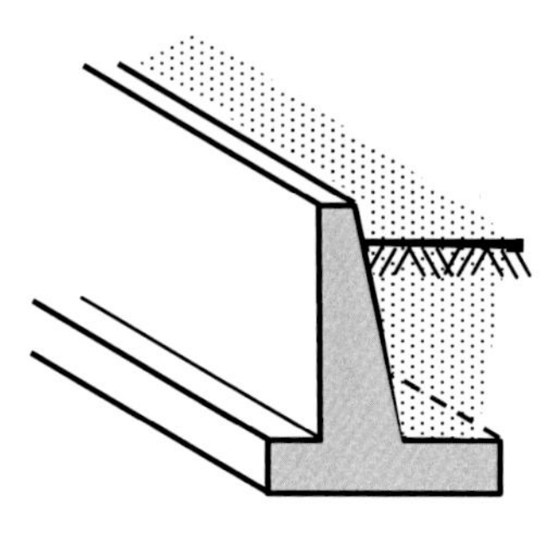

Fig. 5.1 Reinforced concrete retaining wall

Fig. 5.2 Reinforced concrete bridge

Fig. 5.3 Inner ring viaduct in Shanghai, China

Fig. 5.4 Rectangular section

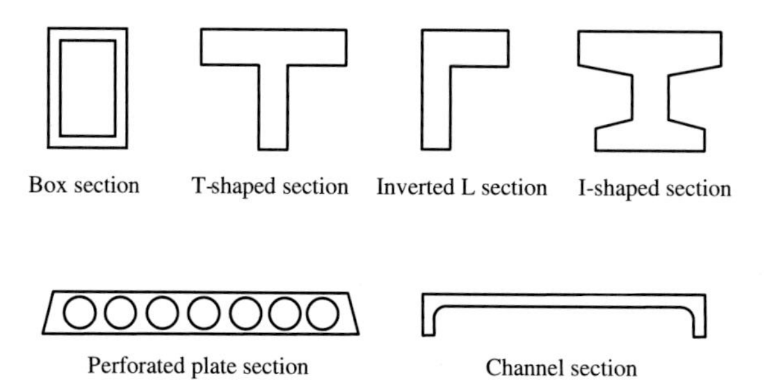

Fig. 5.5 Section shapes classified as T-shaped section

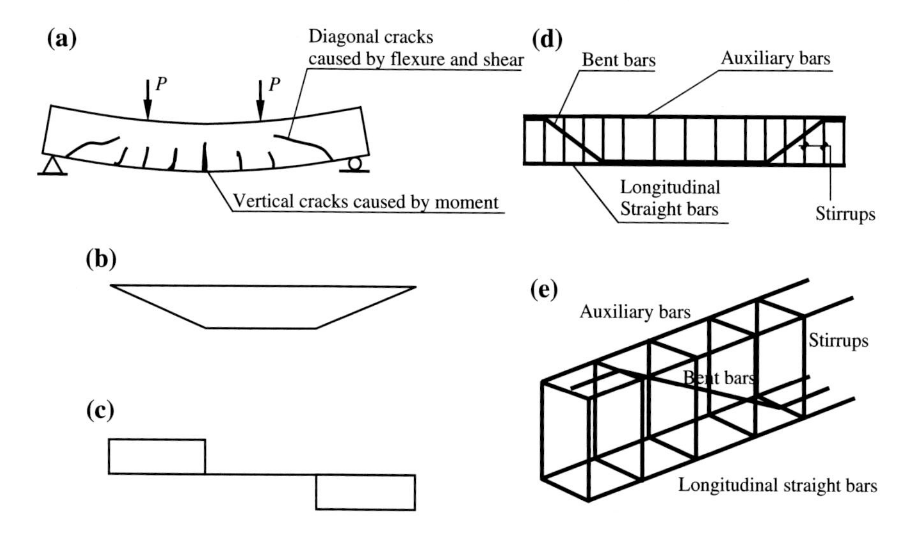

Fig. 5.6 Mechanical characteristics and reinforcement type of beams. **a** Cracks in a beam. **b** Diagram of moment. **c** Diagram of shear. **d** Steel bars in a beam. **e** Cage of steel bars

bars, stirrups, and auxiliary bars are assembled or welded together to form a cage of steel bars (Fig. 5.6e). During construction, reinforced concrete beams are fabricated by setting up the forms first, then putting cages of steel bars into the forms, then casting concrete, and finally vibrating and curing. Compared with beams, reinforced concrete slabs are smaller in thickness but larger in sectional width. Flexure failure rather than shear failure is more commonly observed in slabs. Therefore, only longitudinal steel bars and distribution steel bars used to position the longitudinal steel bars are arranged in reinforced concrete slabs.

5.3 Sectional Dimension and Reinforcement Detailing of Flexural Members

Similar to axially loaded members, flexural members must be constructed in accordance with appropriate detailing requirements to ease construction and achieve a good bond between steel and concrete. Appropriate concrete cover also helps to protect steel bars from corrosion and fire disasters.

The beam width b is usually 150, 180, 200, 220 and 250 mm and thereafter increases by a 50-mm increment. The beam height h increases by a 50-mm increment above 200 mm and a 100-mm increment above 800 mm. The height-to-width (web width for T beams) ratio h/b is 2–3.5 for a rectangular section and 2.5–4 for a T section.

The diameter of longitudinal steel bars in beams is usually 10–28 mm (14–40 mm for bridges). The diameter of auxiliary steel bars is usually 8 mm when the beam span is less than 4 m, 10 mm when the span is between 4 and 6 m, and 12 mm when the span is greater than 6 m. Stirrups generally have the diameter of 6–12 mm. The minimum number of longitudinal bars is 2 (1 is acceptable if $b < 100$ mm). In order for coarse aggregates to pass through a cage of steel bars freely so that the concrete can be appropriately compacted, to ensure a good bond between steel bars and concrete, and to provide enough protection to reinforcement, the clear spacing between longitudinal bars must satisfy the requirements illustrated in Fig. 5.7a.

The thickness of solid reinforced concrete slabs often increases by a 10-mm increment. The minimum thickness is 60 mm for facade panels and residential floor slabs, 70 mm for industrial floor slabs, 120 mm for trough plates in railway bridges, 100 mm for traffic lane panels, and 80 mm for sidewalk slabs. The diameter of longitudinal bars is normally 8–12 mm in solid slabs, while larger diameters can be used in foundation slabs and bridge decks. The distribution bars are usually 6 mm in diameter. The bar spacing is shown in Fig. 5.7b.

Different requirements are specified in various design codes for minimum concrete cover c, which is generally defined as the shortest distance between the outer surfaces of longitudinal bars and the nearby surface of a member. Figure 5.7 illustrates the cover thickness for beams and slabs in a normal indoor environment, as specified by GB 50010. An extra 5 mm should be added to the minimum cover thickness if the concrete grade is lower than C20.

5.4 Experimental Study on Flexural Members

5.4.1 *Test Setup*

Reinforced concrete flexural members are actually composites consisting of two completely different kinds of materials, i.e., steel bars and concrete. The differences in mechanical properties of steel bars and concrete make the behavior of flexural

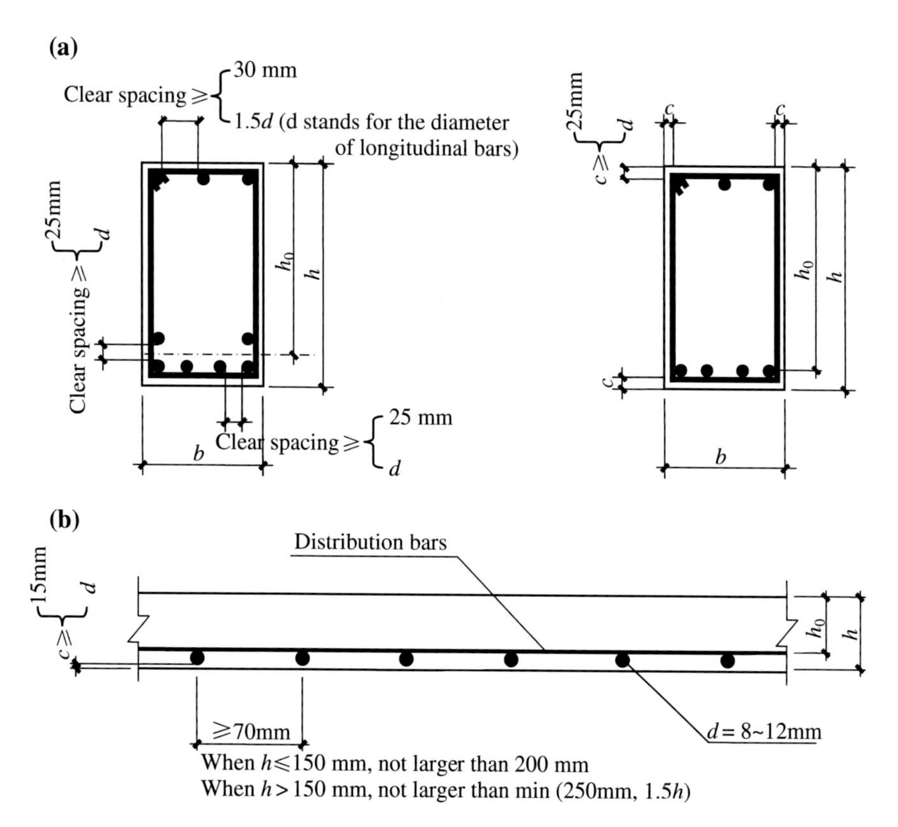

Fig. 5.7 Reinforcement details of reinforced concrete flexural members. **a** A rectangular section beam. **b** A rectangular section slab

reinforced concrete members differ apparently from that of bending members made from single and homogeneous material. To understand the characteristics of flexural reinforced concrete members for accurate analysis and design calculation, experimental study should be carried out first. Figure 5.8 shows a typical test setup for the investigation of mechanical behavior of a simply supported and singly reinforced concrete beam. The specimen is loaded symmetrically with two equal concentrated loads. The beam segment between these two loads is under pure bending as shown by the bending moment and shear diagrams in Fig. 5.8. The failure mechanism of the flexural specimen can be analyzed from the cracking and crushing of concrete in the pure bending segment. The strain distribution along the cross section can be measured by strain gauges of large gauge length, and the deflection of the specimen during loading process can be monitored by a displacement sensor installed at the bottom of the pure bending segment.

Define $\rho = \frac{A_s}{bh_0}$ as the reinforcement ratio of longitudinal bars in a beam, where h_0 (effective depth) is the distance from the extreme compression fiber to the

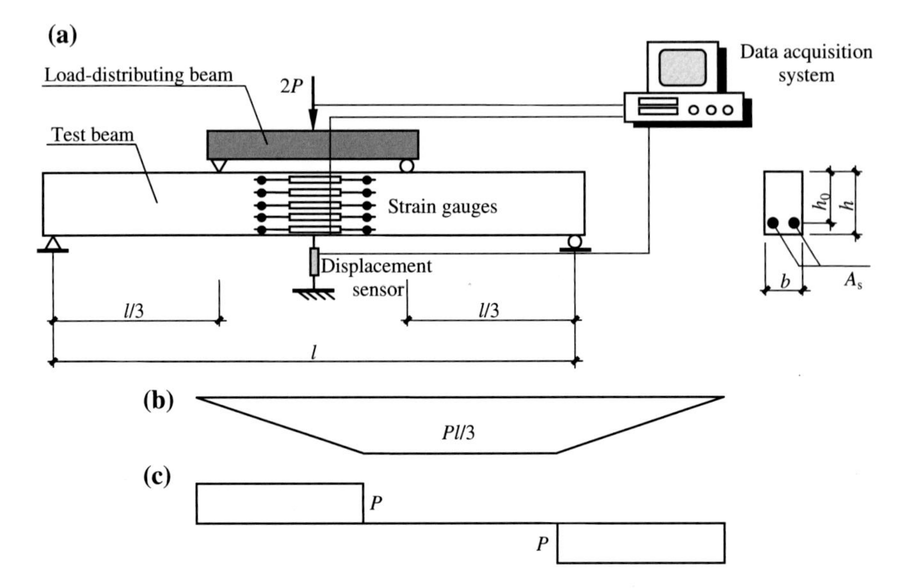

Fig. 5.8 Test setup for a simply supported and singly reinforced concrete beam under flexure. **a** Test setup. **b** Diagram of moment. **c** Diagram of shear

centroid of the steel bars (Figs. 5.7 and 5.8), b is the section width, and A_s is the area of longitudinal bars. The bending behaviors of beams with different reinforcement ratios are studied accordingly.

5.4.2 Experimental Results

5.4.2.1 Failure Process

For beams with appropriate reinforcement ratios (under-reinforced beams), three typical stages can be observed during the loading process.

1. Stage I: Elastic stage.
   When the load is small, the stress along the cross section is linearly distributed and residual deformation barely exists after unloading (Fig. 5.9a). When the maximum tensile stress σ_t^b of the concrete in the beam tension zone reaches the tensile strength f_t and the maximum tensile strain ε_t^b reaches the ultimate tensile strain ε_{tu}, the first vertical crack appears at the weakest section of the pure bending segment, which symbolizes the end of Stage I ($\mathrm{I_a}$). The bending moment M_{cr} at this time is called the cracking moment of the beam (Fig. 5.9b).

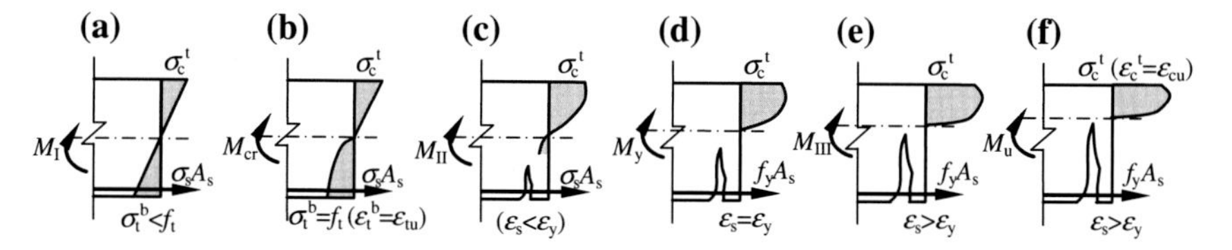

Fig. 5.9 Stress and strain distribution of an under-reinforced beam at different loading stages. **a** I; **b** $\mathrm{I_a}$; **c** II; **d** $\mathrm{II_a}$; **e** III; **f** $\mathrm{III_a}$

2. Stage II: Service stage.
   After cracking, concrete at the crack cannot bear the tensile force. The tensile stress in the reinforcement steel bars across the crack sharply increases and then is transferred to the nearby uncracked concrete through the bond between concrete and steel bars, which will cause new tensile cracks. The compressive stress of the concrete in the compression zone changes from linear distribution to nonlinear distribution with the increasing load increments (Fig. 5.9c). The yielding of tensile steel bars means the end of Stage II ($\mathrm{II_a}$). The moment M_{y} at this time is called the yield moment of the beam (Fig. 5.9d).
3. Stage III: Failure stage.
   After the yielding of tensile steel bars, a small load increment will produce large deformation of the beam. Cracks propagate further and become wider, which pushes the neutral axis upward, and the nonlinear distribution of compressive stress of concrete becomes more apparent (Fig. 5.9e). When the maximum compressive strain $\varepsilon_{\mathrm{c}}^{\mathrm{t}}$ reaches the ultimate value $\varepsilon_{\mathrm{cu}}$, the concrete in the compression zone will crush, symbolizing the flexural failure of the beam ($\mathrm{III_a}$). The moment M_{u} at this time is called the ultimate moment of the beam (Fig. 5.9f).

Figure 5.10a gives the failure mode of an under-reinforced beam and the variation of average strain along the section depth measured by the strain gauges. Although the beam cracks, the plane section assumption is still valid from average strain sense during the whole loading process. The failure of the under-reinforced beam starts from the yielding of longitudinal steel bars and ends at the crushing of concrete. Very large deformation during the failure process gives apparent warning. This kind of failure is called tension failure and is associated with ductile failure.

When the reinforcement ratio ρ of longitudinal steel bars in a beam is very large (an over-reinforced beam), only two stages appear during the loading process. First, the beam is in the linear elastic state. The stress is linearly distributed across the section and the residual deformation barely exists after unloading. When the maximum tensile stress $\sigma_{\mathrm{t}}^{\mathrm{b}}$ of the concrete in the beam tension zone reaches the tensile strength f_{t} and the maximum tensile strain $\varepsilon_{\mathrm{t}}^{\mathrm{b}}$ reaches the ultimate tensile strain $\varepsilon_{\mathrm{tu}}$, a crack initiates in the beam. Cracks continuously increase with the load. But the stress in the steel bars shows very little increase because of the large steel content. Many cracks are closely spaced on the beam surfaces. When the maximum compressive strain $\varepsilon_{\mathrm{c}}^{\mathrm{t}}$ reaches the ultimate value $\varepsilon_{\mathrm{cu}}$, the concrete in the compression

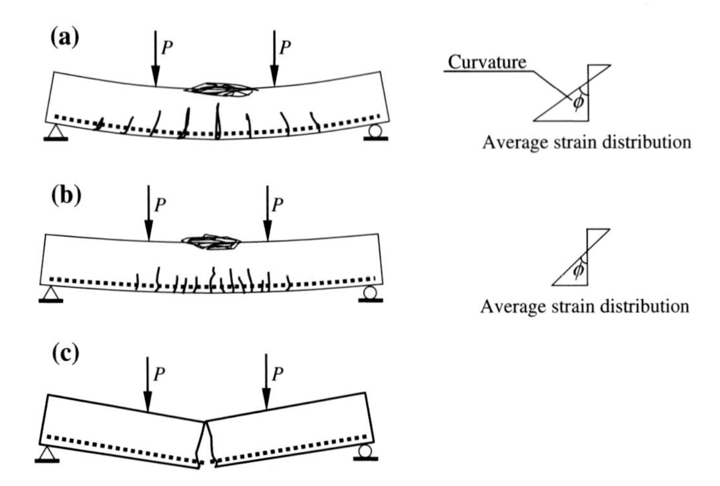

Fig. 5.10 Failure modes of reinforced concrete beams. **a** Tension failure (an under-reinforced beam). **b** Compression failure (an over-reinforced beam). **c** Tension failure (a lightly reinforced beam)

zone will be crushed, symbolizing the flexural failure of the beam. However, the steel bars have not yielded at this moment.

Figure 5.10b gives the failure mode of an over-reinforced beam and the variation of average strain along the section depth measured by the strain gauges. From average strain sense, the plane section assumption is also valid for over-reinforced beams. The steel bars have not yielded even after the concrete has been crushed. Although many cracks appear, the crack width and the beam deformation are small. The failure happens suddenly. This kind of failure is called compression failure, which is also called brittle failure.

When the reinforcement ratio of a beam is too small (a lightly reinforced beam), which is smaller than the minimum required reinforcement (as it is called as in this textbook), only the elastic stage (Stage I) can be observed. After the beam has cracked, the tensile forces, once resisted by the concrete, have been transferred totally to the longitudinal steel bars across the crack. The sudden big stress increase exceeds the ultimate strength of steel bars and they are fractured immediately because little reinforcement was used. The crack rapidly propagates to the beam top, and the beam fails in a brittle way. The compressive strength of concrete has not been fully utilized.

Figure 5.10c gives the failure mode of a lightly reinforced beam. Once the reinforcement has been fractured, the beam is split into two halves. Before failure, there is not any crack on the beam, which deforms elastically. This kind of failure is brittle and dangerous.

5.4.2.2 Conclusion

Figures 5.11 and 5.12 show the moment–curvature curves and load–deflection curves of different reinforced concrete beams, respectively. From these figures, it can be seen that both the flexural bearing capacities and ductility of lightly reinforced concrete beams are poor, and over-reinforced beams have large flexural bearing capacities but poor ductility, both of which do not make good structural members. On the other hand, both flexural bearing capacities and ductility of under-reinforced beams are not only very good, but also suitable for engineering applications.

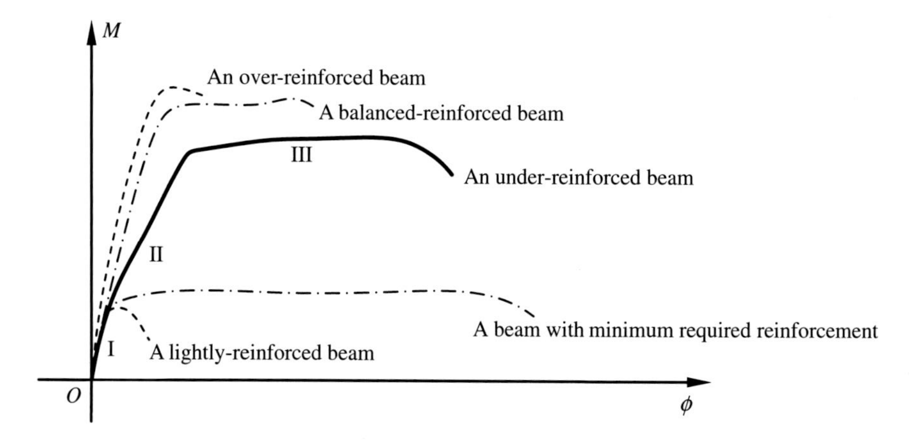

Fig. 5.11 Moment–curvature curves of different reinforced concrete beams

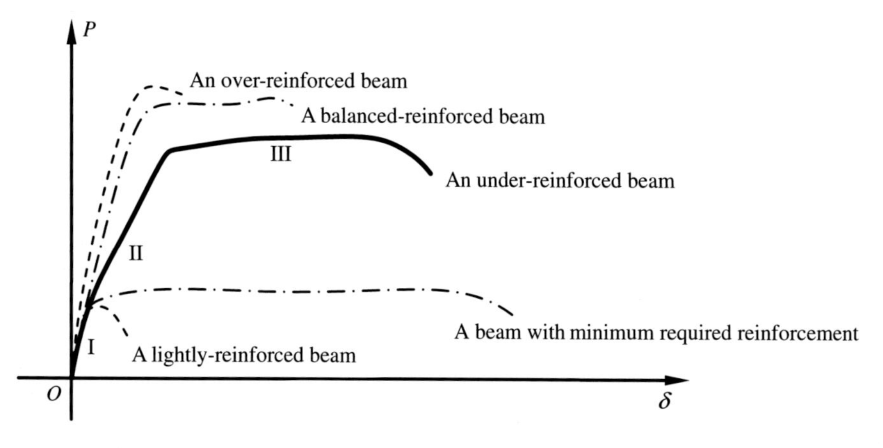

Fig. 5.12 Load–deflection curves of different reinforced concrete beams

Between tension failure and compression failure, there is a balanced failure state (balanced failure). Steel yields and concrete crushes simultaneously. The longitudinal reinforcement ratio of beams under balanced failure is called balanced reinforcement ratio (ρ_b), which is the quantitative index used to distinguish tension failure and compression failure and also the maximum reinforcement ratio of under-reinforced beams.

Similarly, between the under-reinforced beam and the lightly reinforced beam, there is also a balanced state, characterized by equal yield moment and cracking moment. The reinforcement ratio of this kind of member is actually the minimum required reinforcement ratio for under-reinforced beams (ρ_{min}). The beams with the minimum reinforcement ratio have the largest ductility.

5.5 Analysis of Singly Reinforced Rectangular Sections

5.5.1 Basic Assumptions

5.5.1.1 Plane section Assumption

Plane sections before bending remain plane after bending. This assumption, known as the Bernoulli's principle, implies that the longitudinal strain in concrete or steel bars at any point across a section is proportional to the distance from the neutral axis. Experiments have indicated that this assumption is true for uncracked reinforced concrete members and still valid for cracked ones in an average strain sense. Therefore, plane section assumption is nearly correct at all stages up to flexural failure. It is convenient to derive the compatibility (geometrical) relation of a section from the assumption. As shown in Fig. 5.13, if the slip between the reinforcement and the surrounding concrete after section cracking is not considered, the geometrical relation between the strains and curvature is

$$\phi = \frac{\varepsilon_c^t}{\xi_n h_0} = \frac{\varepsilon_c}{y} = \frac{\varepsilon_s'}{\xi_n h_0 - a_s'} = \frac{\varepsilon_s}{(1 - \xi_n) h_0} \tag{5.1}$$

where

ϕ curvature of the section;
ε_c strain in the fiber with a distance of y from the neutral axis;
ε_c^t compression strain in the extreme compression fiber;
$\varepsilon_s, \varepsilon_s'$ strains in longitudinal tension and compression steel bars, respectively;
h_0 effective depth of the section; and
ξ_n relative compression depth, which is the ratio of the compression depth of the concrete to the effective depth of the section; and
a_s' distance from the centroid of the compression steel bars to the extreme compression fiber.

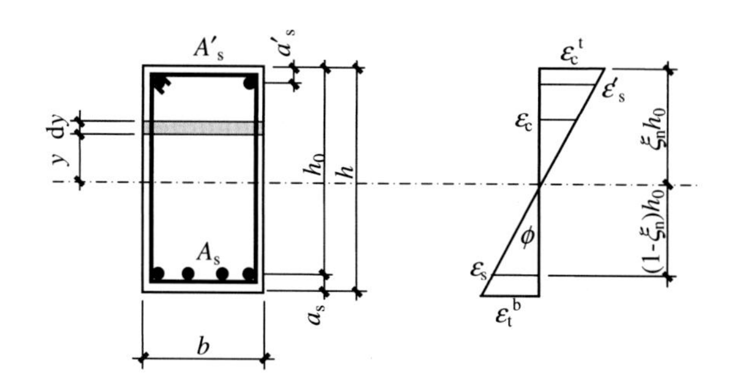

Fig. 5.13 Strain distribution of a rectangular section

5.5.1.2 Perfect Bond Between Steel Bars and Concrete

This assumption ensures that the strains of the steel bars and the concrete are equal at the same point and both materials can work together well.

5.5.1.3 Compressive Stress–Strain Relationship of Concrete

It can be observed from Fig. 5.13 that the strains in the concrete at various points are different across the section height. Fibers of large strains will be restrained by those of small strains. So the stress–strain curve in Fig. 5.14 can be used. GB 50010 suggests the stress–strain relationship as shown in Eq. (2.7).

5.5.1.4 Tensile Stress–Strain Relationship of Concrete

The stress–strain relationship of concrete under tension (Fig. 5.15) can be formulated as follows:

$$\left.\begin{array}{ll} \sigma_t = E_c \varepsilon_t & 0 \leqslant \varepsilon_t \leqslant \varepsilon_{t0} \\ \sigma_t = f_t & \varepsilon_{t0} < \varepsilon_t \leqslant \varepsilon_{tu} \end{array}\right\} \tag{5.2}$$

where

σ_t concrete tensile stress corresponding to the strain of ε_t;
f_t tensile strength of concrete;
ε_{t0} tensile strain corresponding to f_t; and
ε_{tu} ultimate tensile strain of concrete, approximately, $\varepsilon_{tu} = 2\varepsilon_{t0}$.

After cracking ($\varepsilon_t \geqslant \varepsilon_{tu}$), concrete is assumed unable to sustain any tension.

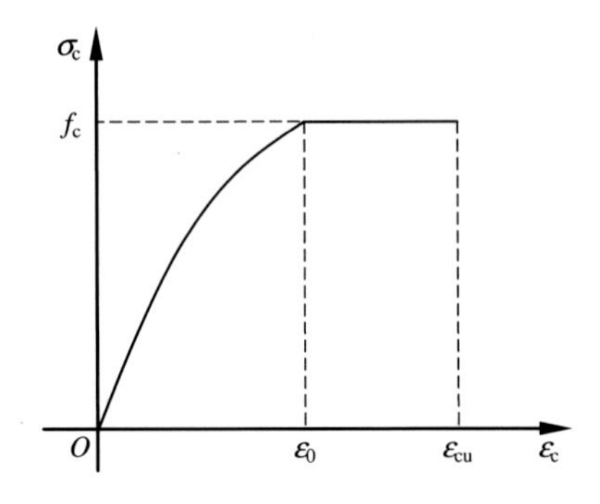

Fig. 5.14 Stress–strain curve of concrete under compression

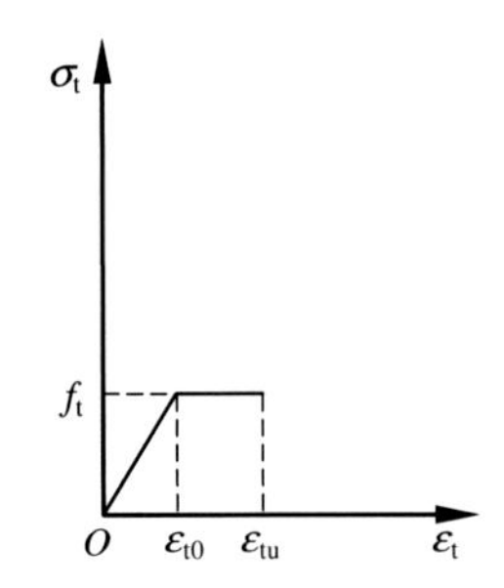

Fig. 5.15 Stress–strain curve of concrete under tension

5.5.1.5 Stress–Strain Relationship of Steel Bars

For the steel bars with apparent yielding plateau, the ideal elastic–plastic stress–strain curve, as shown in Fig. 5.16, is utilized. Since the stress increase of the steel after its yielding in flexural reinforced concrete members is relatively small, the hardening stage will not be considered in Fig. 5.16. Equation (2.1) has already given the mathematic expression of the stress–strain relationship of steel bars.

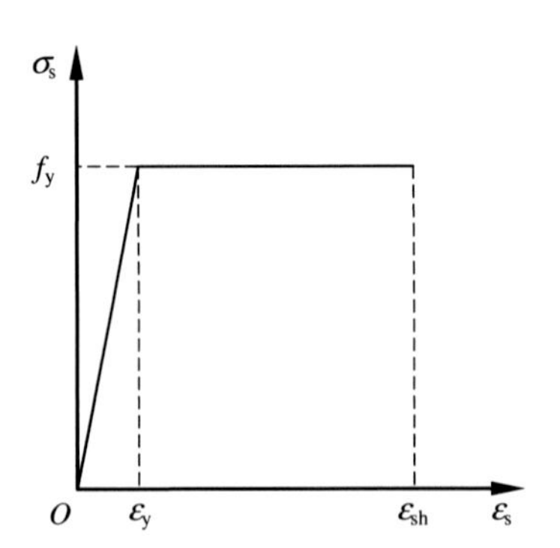

Fig. 5.16 Stress–strain curve of steel bars under tension

5.5.2 Analysis Before Cracking

Figure 5.17 shows the stress and strain distributions across a rectangular section before cracking. From previous analyses, the following linear elastic relation can be used:

$$\varepsilon_c = \frac{\sigma_c}{E_c}, \quad \varepsilon_t = \frac{\sigma_t}{E_c}, \quad \varepsilon_s = \frac{\sigma_s}{E_s} \tag{5.3}$$

Assuming a perfect bond between the concrete and the steel, at the centroid of the tension steels, we have

$$\varepsilon_s = \varepsilon_t \tag{5.4}$$

Substitute Eq. (5.4) into Eq. (5.3), the steel stress can be obtained as

$$\sigma_s = E_s \varepsilon_s = \frac{E_s}{E_c} \sigma_t = \alpha_E \sigma_t \tag{5.5}$$

Then, the total tension force of the steel bars is

$$T = \sigma_s A_s = \alpha_E A_s \sigma_t \tag{5.6}$$

where $\alpha_E = E_s/E_c$ is the ratio of the modulus of elasticity of the steel to that of the concrete.

Equation (5.6) indicates that a reinforced concrete beam can be transformed into a plain concrete beam by replacing the steel bars with concrete of equivalent area $\alpha_E A_s$ and keeping the geometric center unchanged. Then, the formulae in *mechanics of materials* can be used to analyze a reinforced concrete section.

The concrete stress in an arbitrary fiber of the section and the tensile stress in the longitudinal steel bars can be calculated, respectively, as

$$\sigma_{ci} = \frac{M y_0}{I_0} \tag{5.7}$$

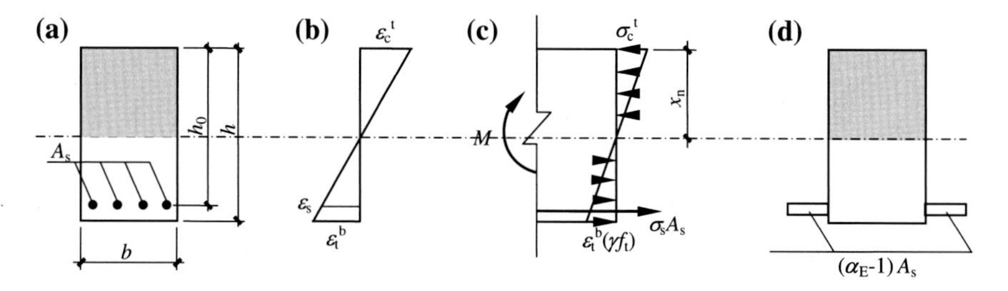

Fig. 5.17 Stress and strain distributions across a section before cracking. **a** Cross section. **b** Strain distribution. **c** Stress distribution. **d** Transformed section

$$\sigma_{\mathrm{s}} = \alpha_{\mathrm{E}} \frac{M(h_0 - x_{\mathrm{n}})}{I_0} \tag{5.8}$$

where

x_{n} compression zone depth of the transformed section;
I_0 inertia moment of the transformed section;
y_0 distance between the fiber and the neutral axis of transformed section; and
h_0 distance between the centroid of the tension reinforcement and the extreme fiber of the compression zone, also called the *effective depth* of the section.

5.5.3 Analysis at Cracking

When $\varepsilon_{\mathrm{t}}^{\mathrm{b}} = \varepsilon_{\mathrm{tu}}$, concrete in the tension zone cracks and quits working. Figure 5.18 shows the stress and strain distributions across the section. From the similar triangles (see Fig. 5.18b), we have

$$\phi_{\mathrm{cr}} = \frac{\varepsilon_{\mathrm{tu}}}{h - x_{\mathrm{cr}}} = \frac{\varepsilon_{\mathrm{c}}^{\mathrm{t}}}{x_{\mathrm{cr}}} = \frac{\varepsilon_{\mathrm{s}}}{h_0 - x_{\mathrm{cr}}} \tag{5.9}$$

where ϕ_{cr} is the curvature corresponding to the cracking moment.

When the section is going to crack, the stress of concrete in the compression zone is still relatively low and a linear distribution can be assumed. According to the stress–strain curve of concrete under tension (see Fig. 5.15) and the strain distribution of concrete in the tension zone (Fig. 5.18b), the corresponding tensile stress distribution of concrete is shown by the dashed line in Fig. 5.18c, which in analysis can be replaced by the rectangular stress distribution (the solid line in Fig. 5.18c), because the tensile stress of the concrete is still very small at that moment. This simplification will cause negligible error but is very convenient for the analysis. According to the stress–strain curves of the concrete and the steel, the following equations can be established:

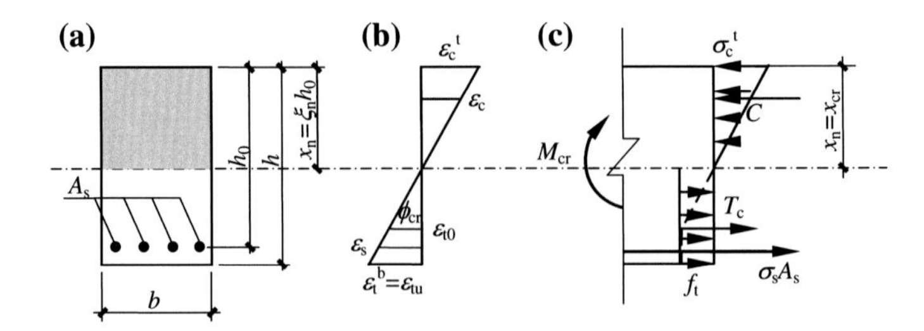

Fig. 5.18 Stress and strain distributions across a cracking section. **a** Cross section. **b** Strain distribution. **c** Stress distribution

$$f_t = E_c\varepsilon_{t0} = 0.5E_c\varepsilon_{tu}, \quad \sigma_c = E_c\varepsilon_c, \quad \sigma_s = E_s\varepsilon_s \tag{5.10}$$

From the force equilibrium in the longitudinal direction, i.e., $\Sigma X = 0$, we have

$$0.5\sigma_c^t bx_{cr} = f_t b(h - x_{cr}) + \sigma_s A_s \tag{5.11}$$

$\varepsilon_s = \varepsilon_{tu}$ can be approximately assumed. Substitute Eq. (5.9) into Eq. (5.10) and then Eq. (5.10) into Eq. (5.11), we have

$$x_{cr} = \frac{1 + \frac{2\alpha_E A_s}{bh}}{1 + \frac{\alpha_E A_s}{bh}} \cdot \frac{h}{2} \tag{5.12}$$

For general reinforced concrete beams, $A_s/(bh)$ = (0.5–2)%, α_E = 6.7–8.0. Substituting them into Eq. (5.12) obtains $x_{cr} \approx 0.5h$. Then, the curvature of the cracked section can be expressed as follows:

$$\phi_{cr} = \frac{2\varepsilon_{tu}}{h} \tag{5.13}$$

For equilibrium, $\Sigma M = 0$, where M is taken about the acting point of the resultant compressive force of the compression zone, the cracking moment of the section can be calculated by

$$M_{cr} = f_t b(h - x_{cr})\left(\frac{h - x_{cr}}{2} + \frac{2x_{cr}}{3}\right) + 2\alpha_E f_t A_s\left(h_0 - \frac{x_{cr}}{3}\right) \tag{5.14}$$

By assuming $h_0 = 0.92h$, letting $\alpha_A = 2\alpha_E \cdot \frac{A_s}{bh}$ and considering $x_{cr} = 0.5h$, we have

$$M_{cr} = 0.292(1 + 2.5\alpha_A)f_t bh^2 \tag{5.15}$$

Similarly, the cracking moment can also be calculated by elastic theory. Introducing a coefficient γ to consider the plasticity of the concrete, a linear stress distribution before the cracking can be assumed. When the stress in the extreme tension fiber reaches γf_t, the section cracks and the cracking moment is

$$M_{cr} = \gamma f_t \frac{I_0}{y_0} = \gamma f_t W_0 \tag{5.16}$$

where

W_0 modulus of the transformed section; and

γ a coefficient considering the plasticity of concrete, whose value will be discussed in Chap. 10.

Example 5.1 A rectangular section with the width b = 250 mm and the depth h = 600 mm is shown in Fig. 5.19. The thickness of the concrete cover is expressed as c = 25 mm. The material properties of the steel and the concrete are as follows: f_c = 23 N/mm^2, f_t = 2.6 N/mm^2, E_c = 2.51 × 10^4 N/mm^2, f_y = 357 N/mm^2, E_s = 1.97 × 10^5 N/mm^2. Calculate:

1. the cracking moment M_{cr} and the corresponding σ_s, σ_c^t, and ϕ_{cr} when the section is reinforced with two ⌀22 tension steel bars (A_s = 760 mm^2);
2. the cracking moment M_{cr} and the corresponding σ_s, σ_c^t, and ϕ_{cr} when the section is reinforced with four ⌀22 tension steel bars (A_s = 1520 mm^2).

Solution

1. When A_s = 760 mm^2, calculate x_{cr} and M_{cr} according to Eqs. (5.12) and (5.14), respectively.

$$\alpha_E = \frac{E_s}{E_c} = \frac{1.97 \times 10^5}{2.51 \times 10^4} = 7.849$$

$$\frac{\alpha_E A_s}{bh} = \frac{7.849 \times 760}{250 \times 600} = 0.04, \quad \alpha_A = \frac{2\alpha_E A_s}{bh} = 0.08$$

$$x_{cr} = \frac{1 + \frac{2\alpha_E A_s}{bh}}{1 + \frac{\alpha_E A_s}{bh}} \cdot \frac{h}{2} = \frac{1 + 0.08}{1 + 0.04} \times \frac{600}{2} = 312\,\text{mm}$$

$$h_0 = 600 - 25 - \frac{22}{2} = 564\,\text{mm}$$

$$\begin{aligned} M_{cr} &= f_t b(h - x_{cr})\left(\frac{h - x_{cr}}{2} + \frac{2x_{cr}}{3}\right) + 2\alpha_E f_t A_s\left(h_0 - \frac{x_{cr}}{3}\right) \\ &= 2.6 \times 250 \times (600 - 312)\left(\frac{600 - 312}{2} + \frac{2 \times 312}{3}\right) \\ &\quad + 2 \times 7.849 \times 2.6 \times 760 \times \left(564 - \frac{312}{3}\right) = 80.16\,\text{kN}\cdot\text{m} \end{aligned}$$

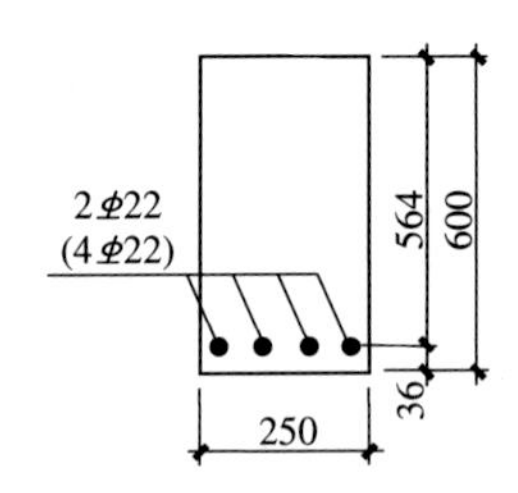

Fig. 5.19 Dimension and reinforcement of a beam section

And from Eq. (5.15), we have

$$
\begin{aligned}
M_{\mathrm{cr}} &= 0.292(1+2.5\alpha_{\mathrm{A}})f_{\mathrm{t}}bh^2 \\
&= 0.292 \times (1+2.5 \times 0.080) \times 2.6 \times 250 \times 600^2 = 81.99\,\mathrm{kN{\cdot}m}
\end{aligned}
$$

The error between the results by Eqs. (5.14) and (5.15) is approximately 2 %.

$$\sigma_{\mathrm{s}} = 2\alpha_{\mathrm{E}}f_{\mathrm{t}} = 2 \times 7.849 \times 2.6 = 40.81\,\mathrm{N/mm^2}$$

From Eq. (5.11), we get

$$
\begin{aligned}
\sigma_{\mathrm{c}}^{\mathrm{t}} &= \frac{f_{\mathrm{t}}b(h-x_{\mathrm{cr}})+\sigma_{\mathrm{s}}A_{\mathrm{s}}}{0.5bx_{\mathrm{cr}}} = \frac{2.6 \times 250 \times (600-312)+40.81 \times 760}{0.5 \times 250 \times 312} = 5.60\,\mathrm{N/mm^2} \\
\Phi_{\mathrm{cr}} &= \frac{f_{\mathrm{t}}}{0.5E_{\mathrm{c}}(h-x_{\mathrm{cr}})} = \frac{2.6}{0.5 \times 2.51 \times 10^4 \times (600-312)} = 7.2 \times 10^{-7}\,\mathrm{mm^{-1}}
\end{aligned}
$$

2. When A_{s} = 1520 mm^2, we have

$$
\begin{aligned}
\frac{\alpha_{\mathrm{E}}A_{\mathrm{s}}}{bh} &= \frac{7.849 \times 1520}{250 \times 600} = 0.080, \quad \alpha_{\mathrm{A}} = \frac{2\alpha_{\mathrm{E}}A_{\mathrm{s}}}{bh} = 0.160 \\
x_{\mathrm{cr}} &= \frac{1+\frac{2\alpha_{\mathrm{E}}A_{\mathrm{s}}}{bh}}{1+\frac{\alpha_{\mathrm{E}}A_{\mathrm{s}}}{bh}} \cdot \frac{h}{2} = \frac{1+0.16}{1+0.08} \times \frac{600}{2} = 322\,\mathrm{mm}, \quad h_0 = 564\,\mathrm{mm}
\end{aligned}
$$

$$
\begin{aligned}
M_{\mathrm{cr}} &= f_{\mathrm{t}}b(h-x_{\mathrm{cr}})\left(\frac{h-x_{\mathrm{cr}}}{2}+\frac{2x_{\mathrm{cr}}}{3}\right)+2\alpha_{\mathrm{E}}f_{\mathrm{t}}A_{\mathrm{s}}\left(h_0-\frac{x_{\mathrm{cr}}}{3}\right) \\
&= 2.6 \times 250 \times (600-322)\left(\frac{600-322}{2}+\frac{2 \times 322}{3}\right) \\
&\quad + 2 \times 7.849 \times 2.6 \times 1520 \times \left(564-\frac{322}{3}\right) = 92.24\,\mathrm{kN{\cdot}m}
\end{aligned}
$$

And from Eq. (5.15), M_{cr} can be calculated as

$$
\begin{aligned}
M_{\mathrm{cr}} &= 0.292(1+2.5\alpha_{\mathrm{A}})f_{\mathrm{t}}bh^2 \\
&= 0.292 \times (1+2.5 \times 0.16) \times 2.6 \times 250 \times 600^2 = 95.66\,\mathrm{kN{\cdot}m}
\end{aligned}
$$

The error between the results by Eqs. (5.14) and (5.15) is approximately 4 %.

$$\sigma_{\mathrm{s}} = 2\alpha_{\mathrm{E}}f_{\mathrm{t}} = 2 \times 7.849 \times 2.6 = 40.81\,\mathrm{N/mm^2}$$

From Eq. (5.11), we get

$$\sigma_c^t = \frac{f_t b(h - x_{cr}) + \sigma_s A_s}{0.5 b x_{cr}} = \frac{2.6 \times 250 \times (600 - 322) + 40.81 \times 1520}{0.5 \times 250 \times 322} = 6.03\,\text{N/mm}^2$$

$$\phi_{cr} = \frac{f_t}{0.5 E_c (h - x_{cr})} = \frac{2.6}{0.5 \times 2.51 \times 10^4 \times (600 - 322)} = 7.45 \times 10^{-7}\,\text{mm}^{-1}$$

It can be seen from the example that the cracking moment can only be increased by 15 % when the reinforcement content is doubled. Therefore, it is not an effective way to improve the cracking resistance of a section by increasing the amount of reinforcement.

5.5.4 Analysis After Cracking

5.5.4.1 Concrete in the Compression Zone Behaves Elastically

After the section has cracked, the concrete in the tension zone is considered failed. When M is relatively low, a linear distribution of σ_c in compression zone can be assumed as shown in Fig. 5.20.

According to the plane section assumption, the following equation holds:

$$\phi = \frac{\varepsilon_c^t}{\xi_n h_0} = \frac{\varepsilon_s}{(1 - \xi_n) h_0} = \frac{\varepsilon_c}{y} \tag{5.17}$$

Combined with Eq. (2.11), it can be obtained that

$$\sigma_c = E_c \varepsilon_c = E_c \varepsilon_c^t \frac{y}{\xi_n h_0} = \sigma_c^t \frac{y}{\xi_n h_0} \tag{5.18}$$

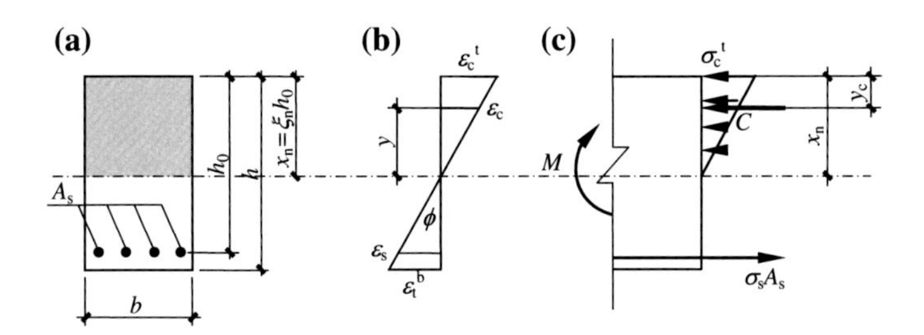

Fig. 5.20 Stress and strain distributions across a cracked section when the concrete in the compression zone behaves elastically. **a** Cross section. **b** Strain distribution. **c** Stress distribution

For equilibrium, $\Sigma X = 0$. We have

$$0.5\sigma_c^t b\xi_n h_0 = \sigma_s A_s = E_s \varepsilon_s A_s = E_s \frac{(1-\xi_n)h_0}{\xi_n h_0}\varepsilon_c^t A_s = \alpha_E \frac{1-\xi_n}{\xi_n}\sigma_c^t A_s \tag{5.19}$$

Further simplification of Eq. (5.19) yields

$$\xi_n^2 + 2\alpha_E \rho \xi_n - 2\alpha_E \rho = 0 \tag{5.20}$$

And

$$\xi_n = \sqrt{(\alpha_E \rho)^2 + 2\alpha_E \rho} - \alpha_E \rho \tag{5.21}$$

The moment corresponding to this stress distribution is ($\Sigma M = 0$):

$$M = 0.5\sigma_c^t b\xi_n h_0^2\left(1 - \frac{1}{3}\xi_n\right) = \sigma_s A_s h_0\left(1 - \frac{1}{3}\xi_n\right) \tag{5.22}$$

5.5.4.2 Concrete in the Compression Zone is in the Elastoplastic Stage and $\varepsilon_c^t < \varepsilon_0$

Figure 5.21 gives the calculation diagram in this stage. Take flexural reinforced concrete members with a concrete grade of no more than C50 for instance. By neglecting the contribution of the concrete in the tension zone and applying the stress–strain relationship of the concrete under compression Eq. (2.7) and the geometrical relationship Eq. (5.17), the internal compressive force C on the concrete in the compression zone and the distance y_c from the acting point of the force to the extreme compression fiber can be obtained as follows:

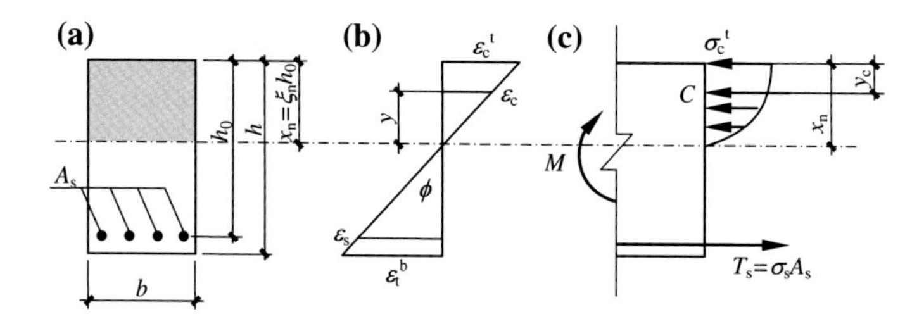

Fig. 5.21 Stress and strain distributions across a cracked section when the concrete in the compression zone is in elastoplastic stage and $\varepsilon_c^t < \varepsilon_0$. **a** Cross section. **b** Strain distribution. **c** Stress distribution

$$
\begin{aligned}
C &= f_c b \int_0^{\xi_n h_0} \left(2\frac{\varepsilon_c}{\varepsilon_0} - \frac{\varepsilon_c^2}{\varepsilon_0^2}\right) dy \\
&= f_c b \int_0^{\xi_n h_0} \left(2\frac{\varepsilon_c^t}{\xi_n h_0 \varepsilon_0} y - \frac{\varepsilon_c^{t2}}{\xi_n^2 h_0^2 \varepsilon_0^2} y^2\right) dy \\
&= f_c b \xi_n h_0 \left(\frac{\varepsilon_c^t}{\varepsilon_0} - \frac{\varepsilon_c^{t2}}{3\varepsilon_0^2}\right)
\end{aligned}
\tag{5.23}
$$

$$
y_c = \xi_n h_0 - \frac{f_c b \int_0^{\xi_n h_0} \left(2\frac{\varepsilon_c}{\varepsilon_0} - \frac{\varepsilon_c^2}{\varepsilon_0^2}\right) y dy}{f_c b \int_0^{\xi_n h_0} \left(2\frac{\varepsilon_c}{\varepsilon_0} - \frac{\varepsilon_c^2}{\varepsilon_0^2}\right) dy} = \xi_n h_0 \cdot \frac{\frac{1}{3} - \frac{\varepsilon_c^t}{12\varepsilon_0}}{1 - \frac{\varepsilon_c^t}{3\varepsilon_0}}
\tag{5.24}
$$

Similarly, from the stress–strain relationship of steel under tension in Eq. (2.1) and the geometrical relationship in Eq. (5.17), the tensile force of the steel bars can be calculated as follows:

$$
T_s = \sigma_s A_s = E_s \varepsilon_s A_s = E_s \frac{1 - \xi_n}{\xi_n} \varepsilon_c^t A_s
\tag{5.25}
$$

Because $\Sigma X = 0$,

$$
f_c b \xi_n h_0 \left(\frac{\varepsilon_c^t}{\varepsilon_0} - \frac{\varepsilon_c^{t2}}{3\varepsilon_0^2}\right) = E_s \frac{1 - \xi_n}{\xi_n} \varepsilon_c^t A_s
\tag{5.26}
$$

Simplifying Eq. (5.26) yields

$$
f_c \xi_n^2 \left(\frac{1}{\varepsilon_0} - \frac{\varepsilon_c^t}{3\varepsilon_0^2}\right) = E_s (1 - \xi_n) \rho
\tag{5.27}
$$

Solving this quadratic equation gives ξ_n. And further considering the moment equilibrium $\Sigma M = 0$ yields

$$
\begin{aligned}
M &= f_c b \xi_n h_0^2 \left(\frac{\varepsilon_c^t}{\varepsilon_0} - \frac{\varepsilon_c^{t2}}{3\varepsilon_0^2}\right) \left(1 - \xi_n \frac{\frac{1}{3} - \frac{\varepsilon_c^t}{12\varepsilon_0}}{1 - \frac{\varepsilon_c^t}{3\varepsilon_0}}\right) \\
&= \sigma_s A_s h_0 \left(1 - \xi_n \frac{\frac{1}{3} - \frac{\varepsilon_c^t}{12\varepsilon_0}}{1 - \frac{\varepsilon_c^t}{3\varepsilon_0}}\right) \quad (\sigma_s \leqslant f_y).
\end{aligned}
\tag{5.28}
$$

5.5.4.3 Concrete in the Compression Zone is in the Elastoplastic Stage and $\varepsilon_0 \leqslant \varepsilon_c^t \leqslant \varepsilon_{cu}$

Figure 5.22 shows the calculation diagram in this stage. Still take flexural reinforced concrete members with the concrete grade of no more than C50 for instance. Similar analysis gives

$$C = f_c b \xi_n h_0 \left(1 - \frac{1}{3}\frac{\varepsilon_0}{\varepsilon_c^t}\right) \tag{5.29}$$

$$y_c = \xi_n h_0 \left[1 - \frac{\frac{1}{2} - \frac{1}{12}\left(\frac{\varepsilon_0}{\varepsilon_c^t}\right)^2}{1 - \frac{1}{3}\frac{\varepsilon_0}{\varepsilon_c^t}}\right] \tag{5.30}$$

The tensile force in the steel is calculated by Eq. (5.25). And $\Sigma X = 0$ gives

$$f_c b \xi_n h_0 \left(1 - \frac{\varepsilon_0}{3\varepsilon_c^t}\right) = E_s \frac{1 - \xi_n}{\xi_n} \varepsilon_c^t A_s \tag{5.31}$$

Simplifying Eq. (5.31) yields

$$f_c \xi_n^2 \left(1 - \frac{\varepsilon_0}{3\varepsilon_c^t}\right) = E_s (1 - \xi_n) \varepsilon_c^t \rho \tag{5.32}$$

Solving this quadratic equation yields ξ_n. And further considering the moment equilibrium $\Sigma M = 0$ yields

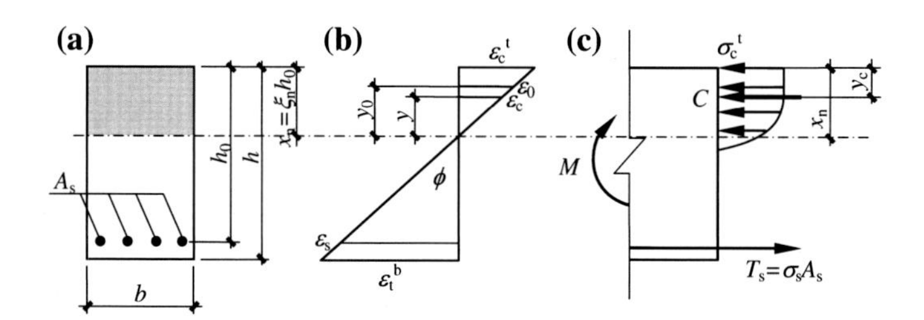

Fig. 5.22 Stress and strain distributions across a cracked section when the concrete in the compression zone is in elastoplastic stage and $\varepsilon_0 \leqslant \varepsilon_c^t \leqslant \varepsilon_{cu}$. **a** Cross section. **b** Strain distribution. **c** Stress distribution

$$
\begin{aligned}
M &= f_c b \xi_n h_0^2 \left(1 - \frac{1}{3}\frac{\varepsilon_0}{\varepsilon_c^t}\right)\left\{1 - \xi_n \left[1 - \frac{\frac{1}{2} - \frac{1}{12}\left(\frac{\varepsilon_0}{\varepsilon_c^t}\right)^2}{1 - \frac{1}{3}\frac{\varepsilon_0}{\varepsilon_c^t}}\right]\right\} \\
&= \sigma_s A_s h_0 \left\{1 - \xi_n \left[1 - \frac{\frac{1}{2} - \frac{1}{12}\left(\frac{\varepsilon_0}{\varepsilon_c^t}\right)^2}{1 - \frac{1}{3}\frac{\varepsilon_0}{\varepsilon_c^t}}\right]\right\} \quad (\sigma_s \leqslant f_y)
\end{aligned}
\tag{5.33}
$$

5.5.5 *Analysis at Ultimate State*

When $\varepsilon_c^t = \varepsilon_{cu}$, the concrete in the compression zone is crushed and the section fails. As for the concrete with the strength grade of no more than C50, $\varepsilon_c^t = \varepsilon_{cu} = 0.0033$ and $\varepsilon_0 = 0.002$. Substituting the values into Eqs. (5.32) and (5.33) gives

$$f_c \xi_n^2 - 0.00414 E_s (1 - \xi_n)\rho = 0 \tag{5.34}$$

$$M_u = 0.798 f_c \xi_n b h_0^2 (1 - 0.412\xi_n) = \sigma_s A_s h_0 (1 - 0.412\xi_n) \quad (\sigma_s \leqslant f_y) \tag{5.35}$$

where M_u is the ultimate moment of the section or the flexural bearing capacity of the section.

For under-reinforced beams, the tension steels have yielded before the concrete is crushed (Stage III_a). Substituting $\sigma_s = f_y$ yields

$$\xi_n = 1.253\rho \frac{f_y}{f_c} \tag{5.36}$$

$$M_u = 0.798 f_c \xi_n b h_0^2 (1 - 0.412\xi_n) = f_y A_s h_0 (1 - 0.412\xi_n) \tag{5.37}$$

Example 5.2 All conditions are the same as those in Example 5.1. When ε_c^t = 0.0008, 0.0010, 0.0020, 0.0025 and 0.0033, respectively,

1. calculate the section moment M and the corresponding curvature ϕ and steel stress σ_s when the section is reinforced with two ϕ22 ($A_s = 760$ mm^2) bars in the tension zone;
2. calculate the section moment M and the corresponding curvature ϕ and steel stress σ_s when the section is reinforced with four ϕ22 ($A_s = 1520$ mm^2) bars in the tension zone;
3. plot the M-ϕ curves of the cross section according to the results of Examples 5.1 and 5.2.

Solution

1. When A_s = 760 mm^2,

$$\rho = \frac{A_s}{bh_0} = \frac{760}{250 \times 564} = 5.39 \times 10^{-3}$$

(1) When $\varepsilon_c^t = 0.0008$, $\varepsilon_c^t < \varepsilon_0$. From Eq. (5.27), it can be obtained

$$23\xi_n^2\left(\frac{1}{0.002} - \frac{0.0008}{3 \times 0.002^2}\right) = 1.97 \times 10^5 \times (1 - \xi_n) \times 5.39 \times 10^{-3}$$
$$9.386\xi_n^2 + \xi_n - 1 = 0$$

Solution of the quadratic equation yields

$$\xi_n = 0.227$$
$$\sigma_s = E_s\frac{1 - \xi_n}{\xi_n}\varepsilon_c^t = 1.97 \times 10^5 \times \frac{1 - 0.277}{0.277} \times 0.0008$$
$$= 411.35\,\text{N/mm}^2 > f_y = 357\,\text{N/mm}^2$$

Since the steel bars have already yielded, ξ_n needs to be recalculated with $\sigma_s = f_y$.

$$23\xi_n^2\left(\frac{0.008}{0.002} - \frac{0.0008}{3 \times 0.002^2}\right) = 5.39 \times 10^{-3} \times 357$$
$$7.973\xi_n = 1.924, \quad \xi_n = 0.241$$

From Eq. (5.28),

$$M = \sigma_s A_s h_0\left[1 - \xi_n\frac{\frac{1}{3} - \frac{\varepsilon_c^t}{12\varepsilon_0}}{1 - \frac{\varepsilon_c^t}{3\varepsilon_0}}\right] = 357 \times 760$$
$$\times 564\left[1 - 0.241 \times \frac{\frac{1}{3} - \frac{0.0008}{12 \times 0.002}}{1 - \frac{0.0008}{3 \times 0.002}}\right] = 140.26\,\text{kN}\cdot\text{m}$$

$$\phi = \frac{\varepsilon_c^t}{\xi_n h_0} = \frac{0.0008}{0.241 \times 564} = 5.89 \times 10^{-6}\,\text{mm}^{-1}$$

(2) When $\varepsilon_c^t = 0.0010$, the steel bars have already yielded, so $\sigma_s = f_y$ and $\varepsilon_c^t < \varepsilon_0$. From Eq. (5.26), it is clear that

$$23\xi_n\left(\frac{0.0010}{0.002} - \frac{0.0010^2}{3 \times 0.002^2}\right) = 5.39 \times 10^{-3} \times 357$$
$$9.583\xi_n = 1.924, \quad \xi_n = 0.201$$

From Eq. (5.28),

$$M = 357 \times 760 \times 564 \times \left[1 - 0.201 \times \frac{\frac{1}{3} - \frac{0.0010}{12\times0.002}}{1 - \frac{0.0010}{3\times0.002}}\right] = 142.26\,\text{kN}\cdot\text{m}$$
$$\phi = \frac{\varepsilon_c^t}{\xi_n h_0} = \frac{0.0010}{0.201 \times 564} = 8.82 \times 10^{-6}\,\text{mm}^{-1}$$

(3) When $\varepsilon_c^t = 0.0015$, the steel bars have already yielded, so $\sigma_s = f_y$ and $\varepsilon_c^t < \varepsilon_0$. From Eq. (5.26), we have

$$23\xi_n\left(\frac{0.0015}{0.002} - \frac{0.0015^2}{3 \times 0.002^2}\right) = 5.39 \times 10^{-3} \times 357$$
$$12.938\xi_n = 1.924, \quad \xi_n = 0.149$$

From Eq. (5.28),

$$M = 357 \times 760 \times 564 \times \left[1 - 0.149 \times \frac{\frac{1}{3} - \frac{0.0015}{12\times0.002}}{1 - \frac{0.0015}{3\times0.002}}\right] = 144.79\,\text{kN}\cdot\text{m}$$
$$\phi = \frac{\varepsilon_c^t}{\xi_n h_0} = \frac{0.0015}{0.149 \times 564} = 17.85 \times 10^{-6}\,\text{mm}^{-1}$$

(4) When $\varepsilon_c^t = 0.0020$, the steel bars have already yielded, so $\sigma_s = f_y$ and $\varepsilon_0 \leqslant \varepsilon_c^t \leqslant \varepsilon_{cu}$. From Eq. (5.31), we have

$$23\xi_n\left(1 - \frac{0.002}{3 \times 0.0020}\right) = 5.39 \times 10^{-3} \times 357$$
$$15.333\xi_n = 1.924, \quad \xi_n = 0.125$$

From Eq. (5.33),

$$M = \sigma_s A_s h_0 \left\{ 1 - \xi_n \left[1 - \frac{\frac{1}{2} - \frac{1}{12}\left(\frac{\varepsilon_0}{\varepsilon_c^t}\right)^2}{1 - \frac{\varepsilon_0}{3\varepsilon_c^t}} \right] \right\}$$

$$= 357 \times 760 \times 564 \times \left\{ 1 - 0.125 \times \left[1 - \frac{\frac{1}{2} - \frac{1}{12}\left(\frac{0.002}{0.0020}\right)^2}{1 - \frac{0.002}{3 \times 0.0020}} \right] \right\} = 145.85\,\text{kN·m}$$

$$\phi = \frac{\varepsilon_c^t}{\xi_n h_0} = \frac{0.0020}{0.125 \times 564} = 28.37 \times 10^{-6}\,\text{mm}^{-1}$$

(5) When $\varepsilon_c^t = 0.0025$, the steel bars have already yielded, so $\sigma_s = f_y$ and $\varepsilon_0 \leqslant \varepsilon_c^t \leqslant \varepsilon_{cu}$. From Eq. (5.31), it is obtained that

$$23\xi_n \left(1 - \frac{0.002}{3 \times 0.0025} \right) = 5.39 \times 10^{-3} \times 357$$

$$16.867\xi_n = 1.924, \quad \xi_n = 0.114$$

From Eq. (5.33),

$$M = 357 \times 760 \times 564 \times \left\{ 1 - 0.114 \times \left[1 - \frac{\frac{1}{2} - \frac{1}{12}\left(\frac{0.002}{0.0025}\right)^2}{1 - \frac{0.002}{3 \times 0.0025}} \right] \right\} = 146.21\,\text{kN·m}$$

$$\phi = \frac{\varepsilon_c^t}{\xi_n h_0} = \frac{0.0025}{0.114 \times 564} = 38.88 \times 10^{-6}\,\text{mm}^{-1}$$

(6) When $\varepsilon_c^t = 0.0033$, the steel bars have already yielded, so $\sigma_s = f_y$ and $\varepsilon_c^t = \varepsilon_{cu}$. From Eq. (5.36), it is obtained that

$$\xi_n = 1.253\rho\frac{f_y}{f_c} = 1.253 \times 5.39 \times 10^{-3} \times \frac{357}{23} = 0.105$$

From Eq. (5.37),

$$M_u = f_y A_s h_0 (1 - 0.412\xi_n) = 357 \times 760 \times 564 \times (1 - 0.412 \times 0.105) = 146.40\,\text{kN·m}$$

$$\phi_u = \frac{\varepsilon_c^t}{\xi_n h_0} = \frac{0.0033}{0.105 \times 564} = 55.72 \times 10^{-6}\,\text{mm}^{-1}$$

2. When $A_s = 1520\ \text{mm}^2$,

$$\rho = \frac{A_s}{bh_0} = \frac{1520}{250 \times 564} = 10.78 \times 10^{-3}$$

(1) When $\varepsilon_c^t = 0.0008$, $\varepsilon_c^t < \varepsilon_0$. From Eq. (5.27), it is obtained that

$$23\xi_n^2\left(\frac{1}{0.002} - \frac{0.0008}{3 \times 0.002^2}\right) = 1.97 \times 10^5 \times (1 - \xi_n) \times 10.78 \times 10^{-3}$$
$$4.693\xi_n^2 + \xi_n - 1 = 0$$

Solution of the quadratic equation yields

$$\xi_n = 0.367$$
$$\sigma_s = E_s \frac{1 - \xi_n}{\xi_n}\varepsilon_c^t = 1.97 \times 10^5 \times \frac{1 - 0.367}{0.367} \times 0.0008 = 271.83\,\text{N/mm}^2$$

From Eq. (5.28),

$$M = \sigma_s A_s h_0 \left[1 - \xi_n \frac{\frac{1}{3} - \frac{\varepsilon_c^t}{12\varepsilon_0}}{1 - \frac{\varepsilon_c^t}{3\varepsilon_0}}\right]$$
$$= 271.83 \times 1520 \times 564 \times \left[1 - 0.367 \times \frac{\frac{1}{3} - \frac{0.0008}{12\times0.002}}{1 - \frac{0.0008}{3\times0.002}}\right] = 203.43\,\text{kN}\cdot\text{m}$$
$$\phi_u = \frac{\varepsilon_c^t}{\xi_n h_0} = \frac{0.0008}{0.367 \times 564} = 3.86 \times 10^{-6}\,\text{mm}^{-1}$$

(2) When $\varepsilon_c^t = 0.0010$, $\varepsilon_c^t < \varepsilon_0$. From Eq. (5.27), it is obtained that

$$23\xi_n^2\left(\frac{1}{0.002} - \frac{0.0010}{3 \times 0.002^2}\right) = 1.97 \times 10^5 \times (1 - \xi_n) \times 10.78 \times 10^{-3}$$
$$4.513\xi_n^2 + \xi_n - 1 = 0$$

Solution of the quadratic equation yields

$$\xi_n = 0.373$$
$$\sigma_s = E_s \frac{1 - \xi_n}{\xi_n}\varepsilon_c^t = 1.97 \times 10^5 \times \frac{1 - 0.373}{0.373} \times 0.0010$$
$$= 331.15\,\text{N/mm}^2 < f_y = 357\,\text{N/mm}^2$$

From Eq. (5.28),

$$M = 331.15 \times 1520 \times 564\left[1 - 0.373 \times \frac{\frac{1}{3} - \frac{0.0010}{12\times0.002}}{1 - \frac{0.0010}{3\times0.002}}\right] = 246.84\,\text{kN}\cdot\text{m}$$
$$\phi = \frac{\varepsilon_c^t}{\xi_n h_0} = \frac{0.0010}{0.373 \times 564} = 4.75 \times 10^{-6}\,\text{mm}^{-1}$$

(3) When $\varepsilon_c^t = 0.0015$, $\varepsilon_c^t < \varepsilon_0$. From Eq. (5.27), it is obtained that

$$23\xi_n^2\left(\frac{1}{0.002} - \frac{0.0015}{3 \times 0.002^2}\right) = 1.97 \times 10^5 \times (1 - \xi_n) \times 10.78 \times 10^{-3}$$

$$4.061\xi_n^2 + \xi_n - 1 = 0$$

Solution of the quadratic equation yields

$$\xi_n = 0.388$$

$$\sigma_s = E_s \frac{1 - \xi_n}{\xi_n} \varepsilon_c^t = 1.97 \times 10^5 \times \frac{1 - 0.388}{0.388} \times 0.0015$$

$$= 466.10\,\text{N/mm}^2 > f_y = 357\,\text{N/mm}^2$$

The steel bars have already yielded. Recalculate ξ_n with $\sigma_s = f_y$. From Eq. (5.26),

$$23\xi_n^2\left(\frac{0.0015}{0.002} - \frac{0.0015^2}{3 \times 0.002^2}\right) = 10.78 \times 10^{-3} \times 357$$

$$12.938\xi_n = 3.849, \quad \xi_n = 0.297$$

From Eq. (5.28),

$$M = 357 \times 1520 \times 564 \times \left[1 - 0.297 \times \frac{\frac{1}{3} - \frac{0.0015}{12 \times 0.002}}{1 - \frac{0.0015}{3 \times 0.002}}\right] = 273.23\,\text{kN}\cdot\text{m}$$

$$\phi = \frac{\varepsilon_c^t}{\xi_n h_0} = \frac{0.0015}{0.297 \times 564} = 8.95 \times 10^{-6}\,\text{mm}^{-1}$$

(4) When $\varepsilon_c^t = 0.0020$, the steel bars have already yielded, so $\sigma_s = f_y$ and $\varepsilon_0 \leqslant \varepsilon_c^t \leqslant \varepsilon_{cu}$. From Eq. (5.31), it is obtained that

$$23\xi_n\left(1 - \frac{0.002}{3 \times 0.0020}\right) = 10.78 \times 10^{-3} \times 357$$

$$15.333\xi_n = 3.849, \quad \xi_n = 0.251$$

From Eq. (5.33),

$$
\begin{aligned}
M &= \sigma_s A_s h_0 \left\{ 1 - \xi_n \left[1 - \frac{\frac{1}{2} - \frac{1}{12}\left(\frac{\varepsilon_0}{\varepsilon_c^t}\right)^2}{1 - \frac{\varepsilon_0}{3\varepsilon_c^t}} \right] \right\} \\
&= 357 \times 1520 \times 564 \times \left(1 - 0.251 \times \left(1 - \frac{\frac{1}{2} - \frac{1}{12}\left(\frac{0.002}{0.0020}\right)^2}{1 - \frac{0.002}{3 \times 0.0020}}\right)\right) \\
&= 277.24\,\text{kN}\cdot\text{m}
\end{aligned}
$$

$$
\phi = \frac{\varepsilon_c^t}{\xi_n h_0} = \frac{0.0020}{0.251 \times 564} = 14.13 \times 10^{-6}\,\text{mm}^{-1}
$$

(5) When $\varepsilon_c^t = 0.0025$, the steel bars have already yielded, so $\sigma_s = f_y$ and $\varepsilon_0 \leqslant \varepsilon_c^t \leqslant \varepsilon_{cu}$. From Eq. (5.31), it is obtained that

$$
\begin{aligned}
23\xi_n\left(1 - \frac{0.002}{3 \times 0.0025}\right) &= 10.78 \times 10^{-3} \times 357 \\
16.867\xi_n = 3.849, &\quad \xi_n = 0.228
\end{aligned}
$$

From Eq. (5.32),

$$
M = 357 \times 1520 \times 564 \times \left\{ 1 - 0.228 \times \left[1 - \frac{\frac{1}{2} - \frac{1}{12}\left(\frac{0.002}{0.0025}\right)^2}{1 - \frac{0.002}{3 \times 0.0025}} \right] \right\} = 278.77\,\text{kN}\cdot\text{m}
$$

$$
\phi = \frac{\varepsilon_c^t}{\xi_n h_0} = \frac{0.0025}{0.228 \times 564} = 19.44 \times 10^{-6}\,\text{mm}^{-1}
$$

(6) When $\varepsilon_c^t = 0.0033$, the steel bars have already yielded, so $\sigma_s = f_y$ and $\varepsilon_c^t = \varepsilon_{cu}$. From Eq. (5.36), it is obtained that

$$
\xi_n = 1.253\rho\frac{f_y}{f_c} = 1.253 \times 10.78 \times 10^{-3} \times \frac{357}{23} = 0.210
$$

From Eq. (5.37),

$$
\begin{aligned}
M_u &= f_y A_s h_0 (1 - 0.412\xi_n) = 357 \times 1520 \times 564 \times (1 - 0.412 \times 0.210) \\
&= 279.57\,\text{kN}\cdot\text{m}
\end{aligned}
$$

$$
\phi_u = \frac{\varepsilon_c^t}{\xi_n h_0} = \frac{0.0033}{0.210 \times 564} = 27.86 \times 10^{-6}\,\text{mm}^{-1}
$$

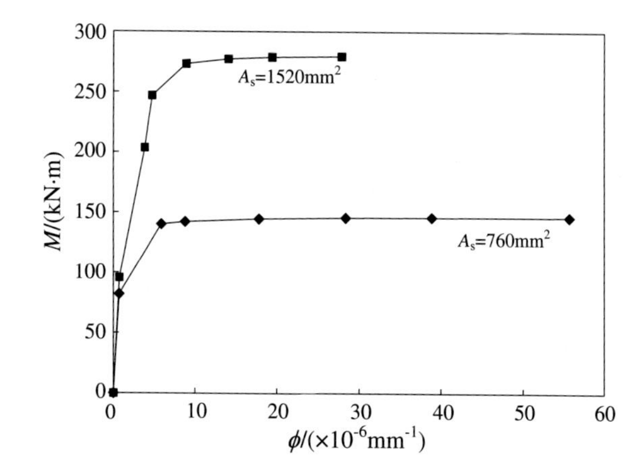

Fig. 5.23 *M*-ϕ curves of the section reinforced by different amount of reinforcements

3. *M*-ϕ curves

   According to the results of Examples 5.1 and 5.2, Fig. 5.23 shows the *M*-ϕ curves of a rectangular section singly reinforced by different amount of reinforcement. It can be seen that the flexural bearing capacity of the section increases with the reinforcement at the sacrifice of the ductility.

5.6 Simplified Analysis of Singly Reinforced Rectangular Sections

The mechanical behavior of flexural members of singly reinforced rectangular sections in a whole loading process was analyzed in Sect. 5.5. However, to ensure the members' safety in real engineering applications, we are more concerned about their ultimate flexural bearing capacities. For a singly reinforced rectangular section, its ultimate flexural bearing capacity is right at the maximum moment, which has been formulated in Eqs. (5.35) and (5.37). But the equations are too complex to use, so a simplification of these equations is necessary.

5.6.1 Equivalent Rectangular Stress Block

The main reason for the complexity of Eqs. (5.35) and (5.37) is the curved stress distribution of the concrete in the compression zone (Figs. 5.22c and 5.24c). If the curved stress distribution is replaced by a rectangular stress block, the analysis will be much easier to understand without loss of accuracy. The equivalent principle is that the resultant forces (*C*) from the curved stress distribution and the rectangular

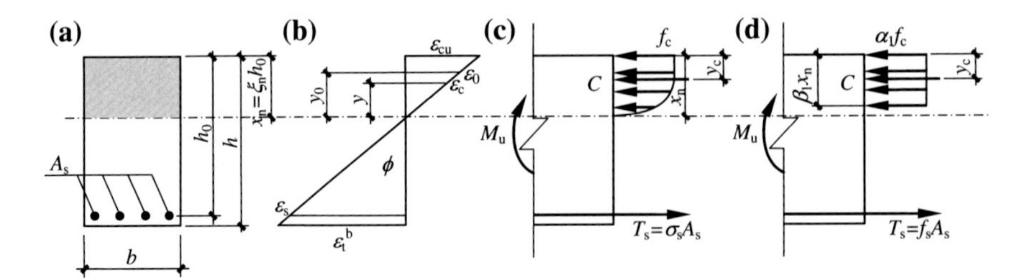

Fig. 5.24 Calculation diagram of the ultimate flexural bearing capacity of a singly reinforced rectangular section. **a** Cross section. **b** Strain distribution. **c** Curved stress distribution. **d** Equivalent rectangular stress distribution

stress block should be equal, and the acting points (y_c) of the resultant forces should be the same. If the equivalent compressive stress of the concrete and the depth of rectangular stress block are assumed as $\alpha_1 f_c$ and $\beta_1 x_n$, respectively, as shown in Fig. 5.24d, it can be ascertained according to the equivalent principle that

$$C = f_c b \xi_n h_0 \left(1 - \frac{1}{3}\frac{\varepsilon_0}{\varepsilon_{cu}}\right) = \alpha_1 f_c \beta_1 \xi_n b h_0 \tag{5.38}$$

$$y_c = \xi_n h_0 \left[1 - \frac{\frac{1}{2} - \frac{1}{12}\left(\frac{\varepsilon_0}{\varepsilon_{cu}}\right)^2}{1 - \frac{1}{3}\frac{\varepsilon_0}{\varepsilon_{cu}}}\right] = 0.5\beta_1 \xi_n h_0 \tag{5.39}$$

Simplifying the two equations yields

$$\alpha_1 = \frac{1}{\beta_1}\left(1 - \frac{1}{3}\frac{\varepsilon_0}{\varepsilon_{cu}}\right) \tag{5.40}$$

$$\beta_1 = \frac{1 - \frac{2}{3}\frac{\varepsilon_0}{\varepsilon_{cu}} + \frac{1}{6}\left(\frac{\varepsilon_0}{\varepsilon_{cu}}\right)^2}{1 - \frac{1}{3}\frac{\varepsilon_0}{\varepsilon_{cu}}} \tag{5.41}$$

For the concrete with the strength grade of no more than C50 (f_{cu} = 50 MPa), α_1 = 0.969 and β_1 = 0.824 can be obtained after substituting ε_{cu} = 0.0033 and ε_0 = 0.002 into Eqs. (5.40) and (5.41). For simplicity, α_1 and β_1 are rounded to 1.00 and 0.80, respectively. When f_{cu} = 80 MPa (C80), similar analysis gives α_1 = 0.94 and β_1 = 0.74. For the concrete with the strength grade between C50 and C80, α_1 and β_1 can be obtained using the linear interpolation method.

5.6.2 *Compression Zone Depth of a Balanced-Reinforced Section*

A balanced-reinforced section is defined as one that fails with the yielding of the tension steel bars and the crushing of the compressed concrete simultaneously. For a balanced-reinforced section, the depth and the relative depth of the compression zone are assumed as x_{nb} and ξ_{nb}, respectively. According to the strain distribution of a balanced-reinforced section (see Fig. 5.25), it can be obtained as follows:

$$\xi_{nb} = \frac{x_{nb}}{h_0} = \frac{\varepsilon_{cu}}{\varepsilon_{cu} + \varepsilon_y} \tag{5.42}$$

If x_b and ξ_b are used to denote the depth and the relative depth of the rectangular stress block of the balanced-reinforced section, Eq. (5.42) can be transformed as

$$\xi_b = \frac{x_b}{h_0} = \frac{\beta_1 x_{nb}}{h_0} = \frac{\beta_1 \varepsilon_{cu}}{\varepsilon_{cu} + \varepsilon_y} = \frac{\beta_1}{1 + \frac{\varepsilon_y}{\varepsilon_{cu}}} = \frac{\beta_1}{1 + \frac{f_y}{E_s \varepsilon_{cu}}} \tag{5.43}$$

For the concrete with $f_{cu} \leqslant 50$ MPa, Eq. (5.43) can be further simplified as

$$\xi_b = \frac{0.8}{1 + \frac{f_y}{0.0033 E_s}} \tag{5.44}$$

As shown in Fig. 5.25, from the comparison of the relative compression zone depth between a section and a balanced one, we can determine the type of a flexural member, to which the section belongs, as follows:

When $\xi < \xi_b$, i.e., $\xi_n < \xi_{nb}$, it is an under-reinforced member;
When $\xi > \xi_b$, i.e., $\xi_n > \xi_{nb}$, it is an over-reinforced member;
When $\xi = \xi_b$, i.e., $\xi_n = \xi_{nb}$, it is a balanced-reinforced member.

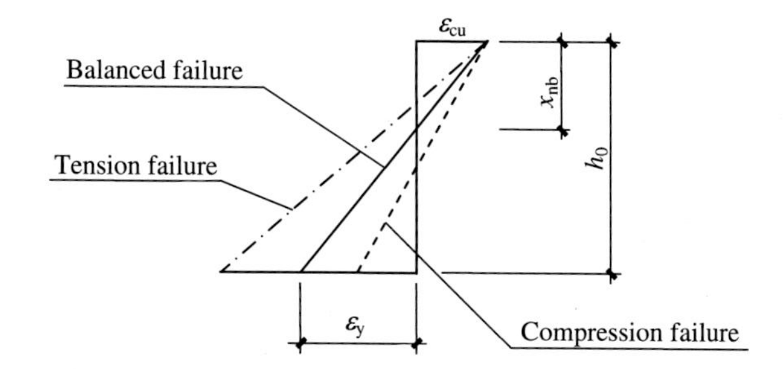

Fig. 5.25 Strain distribution of a balanced section

5.6.3 *Calculation of the Flexural Bearing Capacity of a Singly Reinforced Rectangular Section*

5.6.3.1 Basic Formulae

Referring to the calculation diagram shown in Fig. 5.24d and considering the axial force equilibrium $\Sigma X = 0$ and the bending moment equilibrium $\Sigma M = 0$, we can get the basic formula for the simplified calculation of the flexural bearing capacity of a singly reinforced rectangular section

$$\alpha_1 f_c bx = \sigma_s A_s \tag{5.45}$$

$$M_u = \alpha_1 f_c bx\left(h_0 - \frac{x}{2}\right) = \sigma_s A_s\left(h_0 - \frac{x}{2}\right) \tag{5.46}$$

5.6.3.2 The Flexural Bearing Capacities of Under-Reinforced Members

For under-reinforced members (including balanced-reinforced members), i.e., $\xi \leqslant \xi_b$, when their sections fail, the steels yield before (or right at the moment of) the crushing of the concrete. By substituting $\sigma_s = f_y$ into Eqs. (5.45) and (5.46), we have

$$\alpha_1 f_c bx = f_y A_s \tag{5.47}$$

$$M_u = \alpha_1 f_c bx\left(h_0 - \frac{x}{2}\right) = f_y A_s\left(h_0 - \frac{x}{2}\right) \tag{5.48}$$

Further simplification of the above two formulae gives

$$\xi = \frac{x}{h_0} = \frac{f_y A_s}{\alpha_1 f_c bh_0} = \rho\frac{f_y}{\alpha_1 f_c} \tag{5.49}$$

$$\begin{aligned} M_u &= \alpha_1 f_c bh_0^2 \xi(1 - 0.5\xi) = \alpha_s \alpha_1 f_c bh_0^2 \\ &= A_s f_y h_0 (1 - 0.5\xi) = A_s f_y \gamma_s h_0 \end{aligned} \tag{5.50}$$

Equation (5.49) indicates that the relative depth of the rectangular stress block ξ not only reflects the longitudinal reinforcement ratio, but also relates to the yield strength of the steel and the strength of the concrete. In Eq. (5.50), α_s indicates the magnitude of the section modulus and is called the section modulus coefficient; γ_s reflects the ratio of the internal lever arm to the effective depth of the section and is called the internal lever arm coefficient.

$$\alpha_s = \xi(1 - 0.5\xi) \tag{5.51}$$

$$\gamma_s = 1 - 0.5\xi \tag{5.52}$$

5.6.3.3 The Maximum Reinforcement Ratios of Under-Reinforced Members

The reinforcement ratio of a balanced-reinforced member is the maximum reinforcement ratio of a corresponding under-reinforced member. By substituting $\xi = \xi_b$ into Eq. (5.49), the maximum reinforcement ratios of under-reinforced members can be obtained by the following equation:

$$\rho_{max} = \rho_b = \xi_b \frac{\alpha_1 f_c}{f_y} \tag{5.53}$$

Similarly, by substituting $\xi = \xi_b$ into Eq. (5.50), the maximum flexural bearing capacity and the maximum section modulus coefficient for under-reinforced sections can be expressed as

$$M_{u,max} = \alpha_1 f_c b h_0^2 \xi_b (1 - 0.5\xi_b) = \alpha_{s,max} \alpha_1 f_c b h_0^2 \tag{5.54}$$

$$\alpha_{s,max} = \xi_b (1 - 0.5\xi_b) \tag{5.55}$$

To avoid the compression failure in the design of flexural members, we should ensure that $\xi \leqslant \xi_b$, or $\rho \leqslant \rho_{max}$, or $M \leqslant M_{u,max}$, or $\alpha_s \leqslant \alpha_{s,max}$.

5.6.3.4 The Minimum Reinforcement Ratios of Under-Reinforced Members

To avoid the brittle failure of a flexural member due to insufficient reinforcement content, the minimum reinforcement ratio must be met. The experimental results in Sect. 5.4 show that the failure characteristic of a flexural member with the minimum reinforcement ratio is the same magnitude of the yield moment and the cracking moment. Since the reinforcement content is small, the cracking moment of the flexural member is almost the same as that of a plain concrete member of the same materials and dimensions. Using Eq. (5.15) and neglecting the contribution of the steel bars, the cracking moment of the plain concrete flexural member can be calculated as

$$M_{cr} = 0.292 f_t b h^2 = 0.292 f_t b (1.05 h_0)^2 = 0.322 f_t b h_0^2 \tag{5.56}$$

Also because the reinforcement content is small, the stress of the compressed concrete is relatively low when the steel yields. Therefore, it can be assumed that

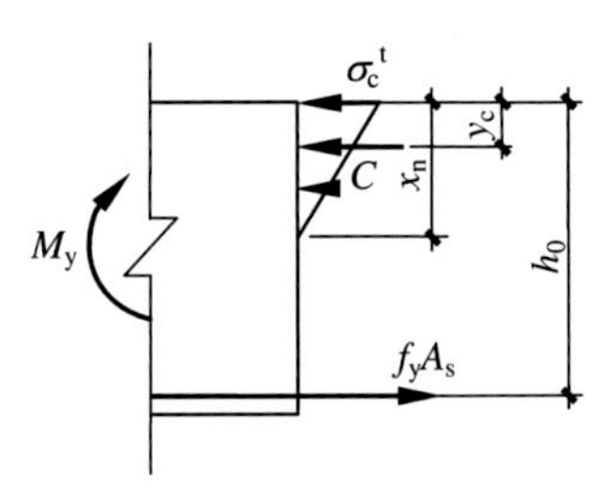

Fig. 5.26 Stress distribution when the steel yields ($\rho_{\min}$)

the compressive stress is linearly distributed as shown in Fig. 5.26. From the bending moment equilibrium $\Sigma M = 0$, it can be shown that

$$M_y = f_y A_s \left(h_0 - \frac{x_n}{3}\right) \approx f_y A_s \cdot 0.9 h_0 \tag{5.57}$$

Taking $M_{cr} = M_y$ yields

$$\rho_{\min} = \frac{A_s}{bh_0} = 0.36 \frac{f_t}{f_y} \tag{5.58}$$

In practice, different codes modify Eq. (5.58) accordingly. For example, to prevent the instantaneous fracture of the steel at the moment of the concrete cracking, GB 50010 enlarges the minimum reinforcement ratio as

$$\rho_{\min} = 0.45 \frac{f_t}{f_y} \tag{5.59}$$

In addition, the following equation is used in checking the minimum reinforcement ratio.

$$\rho = \frac{A_s}{bh} \geqslant \rho_{\min} = 0.45 \frac{f_t}{f_y} \tag{5.60}$$

In Eq. (5.60), replacing h_0 by h is equivalent to increasing the minimum reinforcement ratio by about 10 %.

5.6.3.5 Flexural Bearing Capacities of Over-Reinforced Members

When an over-reinforced member fails, the tension reinforcement does not yield. Therefore, the tensile stress of the steel should be determined before calculating the flexural bearing capacity. The strain and the stress of the steel at arbitrary positions of a section can be determined from Fig. 5.27.

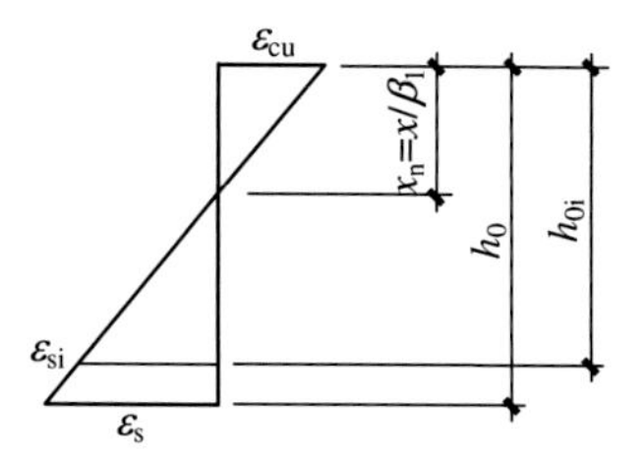

Fig. 5.27 Strain distribution of an over-reinforced section

$$\varepsilon_{\mathrm{si}} = \frac{h_{0\mathrm{i}} - x_{\mathrm{n}}}{x_{\mathrm{n}}}\varepsilon_{\mathrm{cu}} = \varepsilon_{\mathrm{cu}}\left(\frac{h_{0\mathrm{i}}\beta_1}{x} - 1\right) = \varepsilon_{\mathrm{cu}}\left(\frac{h_{0\mathrm{i}}\beta_1}{\xi h_0} - 1\right) \tag{5.61}$$

$$\sigma_{\mathrm{si}} = E_s\varepsilon_{\mathrm{cu}}\left(\frac{h_{0\mathrm{i}}\beta_1}{\xi h_0} - 1\right) \tag{5.62}$$

For sections with only one row of steel bars, it is found that

$$\sigma_{\mathrm{s}} = E_{\mathrm{s}}\varepsilon_{\mathrm{cu}}\left(\frac{\beta_1}{\xi} - 1\right) \tag{5.63}$$

If the concrete with a grade of no more than C50 is considered, Eq. (5.63) becomes

$$\sigma_{\mathrm{s}} = 0.0033E_{\mathrm{s}}\left(\frac{0.8}{\xi} - 1\right) \tag{5.64}$$

Equation (5.64) indicates a hyperbola. Therefore, a high-order equation of ξ needs to be solved when it is substituted into the equilibrium equations. When comparing the experimental and the numerical results shown in Fig. 5.28, a linear relationship between σ_{s} and ξ can be obtained with $\xi = \xi_{\mathrm{b}}$ and $\xi = 0.8$ as boundaries:

$$\sigma_{\mathrm{s}} = f_{\mathrm{y}}\frac{\xi - 0.8}{\xi_{\mathrm{b}} - 0.8} \tag{5.65}$$

Equation (5.65) can greatly simplify the analysis. Solving Eqs. (5.45), (5.46), and (5.65) will give the flexural bearing capacities of over-reinforced members.

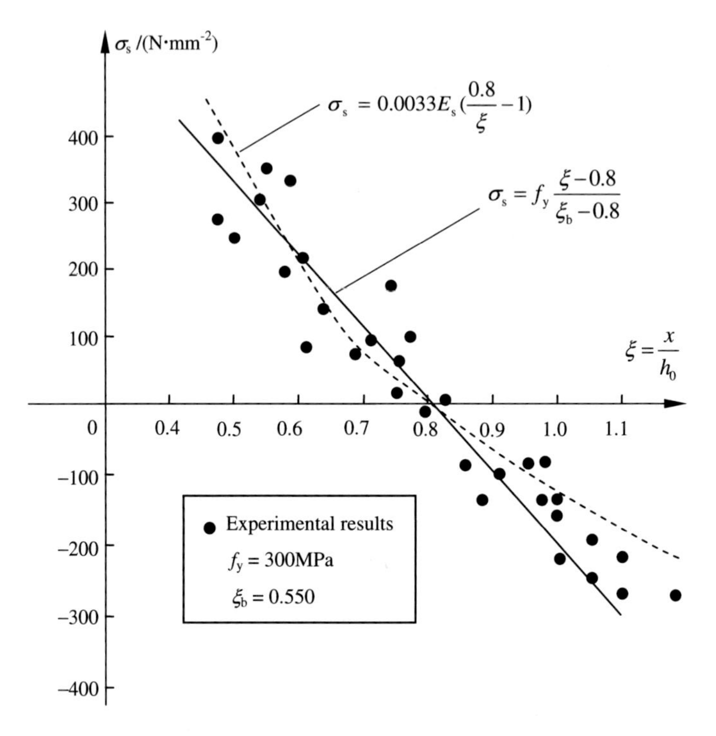

Fig. 5.28 σ_s versus ξ for over-reinforced sections

5.7 Applications of the Equations for Flexural Bearing Capacities of Singly Reinforced Rectangular Sections

5.7.1 Bearing Capacity Calculation of Existing Structural Members

Knowing the cross-sectional dimensions (b, h, and h_0), reinforcement (A_s), and material properties (f_c, f_t, and f_y), the flexural bearing capacity M_u can be evaluated according to the following steps:

1. Calculate the reinforcement ratio: $\rho = \frac{A_s}{bh_0}$ (or $\rho = \frac{A_s}{bh}$);
2. If $\rho < \rho_{min}$, use Eq. (5.15): $M_u = M_{cr}$;
3. If $\rho_{min} \leqslant \rho \leqslant \rho_b$, calculate M_u from the equations for under-reinforced beams, i.e., solve x from Eq. (5.47) and then calculate M_u by Eq. (5.48);
4. If $\rho > \rho_b$, calculate M_u from the equations for over-reinforced beams, i.e., substitute Eq. (5.65) into Eq. (5.45) to get x (or ξ) first, then substitute x (or ξ) and Eq. (5.65) into Eq. (5.46) to calculate M_u.

h_0 can be calculated as follows:

For single-row steel bars, $h_0 = h - c - d/2$;
For double-row steel bars, $h_0 = h - [c + d + \max(25/2, d/2)]$;
where c is the thickness of the concrete cover (Fig. 5.7) and d is the diameter of the steel bar.

Example 5.3 All conditions are the same as those in Example 5.1:

1. When the section is reinforced by 2⌀16, 2⌀22, 4⌀22, and 8⌀28, respectively, calculate the flexural bearing capacity of the section;
2. When $f_c = 32$ N/mm^2 and $f_t = 3.0$ N/mm^2, calculate the flexural bearing capacity of the section that is reinforced by 4⌀22;
3. When $f_y = 460$ N/mm^2, calculate the flexural bearing capacity of the section that is reinforced by 4⌀22;
4. When $h = 700$ mm, calculate the flexural bearing capacity of the section that is reinforced by 4⌀22 (Fig. 5.29).

Solution

1. When the section is reinforced by 2⌀16, 2⌀22, 4⌀22, and 8⌀28, respectively,

$$\rho_{\min} = 0.45\frac{f_t}{f_y} = \frac{0.45 \times 2.6}{357} = 3.277 \times 10^{-3}$$

$$\xi_b = \frac{0.8}{1 + \frac{f_y}{0.0033E_s}} = \frac{0.8}{1 + \frac{357}{0.0033 \times 1.97 \times 10^5}} = 0.516$$

$$\rho_b = \xi_b \frac{\alpha_1 f_c}{f_y} = 0.516 \times \frac{1.0 \times 23}{357} = 3.324 \times 10^{-2}$$

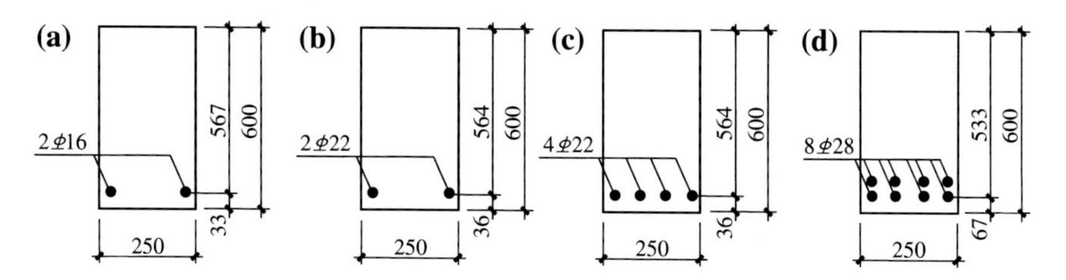

Fig. 5.29 Reinforcement conditions of the section (Example 5.3)

(1) $2\phi16$ (A_s = 402 mm^2)

$$\rho = \frac{A_s}{bh} = \frac{402}{250 \times 600} = 2.680 \times 10^{-3} < \rho_{min} = 3.277 \times 10^{-3}$$
$$\alpha_E = 7.849, \quad \alpha_A = 2\alpha_E \frac{A_s}{bh} = 2 \times 7.849 \times 2.680 \times 10^{-3} = 0.042$$
$$M_u = M_{cr} = 0.292 \times (1 + 2.5\alpha_A) f_t bh^2$$
$$= 0.292 \times (1 + 2.5 \times 0.042) \times 2.6 \times 250 \times 600^2 = 75.50\,\text{kN}\cdot\text{m}$$

(2) $2\phi22$ (A_s = 760 mm^2)

$$\rho = \frac{A_s}{bh} = \frac{760}{250 \times 600} = 5.067 \times 10^{-3} > \rho_{min} = 3.277 \times 10^{-3}$$
$$h_0 = 600 - 25 - \frac{22}{2} = 564\,\text{mm}$$

$\rho = \frac{A_s}{bh_0} = \frac{760}{250 \times 564} = 5.390 \times 10^{-3} < \rho_b = 3.324 \times 10^{-2}$, an under-reinforced section.

From Eq. (5.47), $x = \frac{f_y A_s}{\alpha_1 f_c b} = \frac{357 \times 760}{1.0 \times 23 \times 250} = 47\,\text{mm}$

From Eq. (5.48), $M_u = f_y A_s \left(h_0 - \frac{x}{2}\right) = 357 \times 760 \times \left(564 - \frac{47}{2}\right) = 146.65\,\text{kN}\cdot\text{m}$

It is very similar to the result obtained in Example 5.2 (M_u = 146.40 kN·m), which implies the satisfying accuracy of the simplified method.

(3) $4\phi22$ (A_s = 1520 mm^2)

$$h_0 = 564\,\text{mm}$$

$\rho = \frac{A_s}{bh_0} = \frac{1520}{250 \times 564} = 10.780 \times 10^{-3} < \rho_b = 3.324 \times 10^{-2}$, an under-reinforced section.

From Eq. (5.47), $x = \frac{f_y A_s}{\alpha_1 f_c b} = \frac{357 \times 1520}{1.0 \times 23 \times 250} = 94\,\text{mm}$

From Eq. (5.48), $M_u = f_y A_s \left(h_0 - \frac{x}{2}\right) = 357 \times 1520 \times \left(564 - \frac{94}{2}\right) = 280.54\,\text{kN}\cdot\text{m}$

It is also quite close to the result of Example 5.2 (M_u = 279.57 kN·m).

(4) $8\phi28$ (A_s = 4924 mm^2)

$$h_0 = 600 - 25 - 28 - 14 = 533\,\text{mm}$$

$\rho = \frac{A_s}{bh_0} = \frac{4924}{250 \times 533} = 3.695 \times 10^{-3} > \rho_b = 3.324 \times 10^{-2}$, an over-reinforced section.

Substituting Eq. (5.65) into Eq. (5.45), we have

$$\xi = \frac{0.8}{1 + \frac{\alpha_1 f_c b h_0}{f_y A_s}(0.8 - \xi_b)} = \frac{0.8}{1 + \frac{1.0 \times 23 \times 250 \times 533}{357 \times 4924}(0.8 - 0.156)} = 0.535$$

$$x = \xi h_0 = 0.535 \times 533 = 285\,\text{mm}$$

$$\sigma_s = f_y \frac{\xi - 0.8}{\xi_b - 0.8} = 357 \times \frac{0.535 - 0.8}{0.516 - 0.8} = 333.12\,\text{N/mm}^2$$

$$M_u = \sigma_s A_s \left(h_0 - \frac{x}{2}\right) = 333.12 \times 4924 \times \left(533 - \frac{285}{2}\right) = 640.53\,\text{kN}\cdot\text{m}$$

2. When f_c = 32 N/mm^2 and 4ϕ22 are used,

$$\rho_{\min} = 0.45 \frac{f_t}{f_y} = 0.45 \times \frac{3.0}{357} = 3.782 \times 10^{-3}$$

$$\rho_b = \xi_b \frac{\alpha_1 f_c}{f_y} = 0.516 \times \frac{1.0 \times 32}{357} = 4.625 \times 10^{-2}$$

$$\rho = \frac{A_s}{bh} = \frac{1520}{250 \times 600} = 10.133 \times 10^{-3} > \rho_{\min} = 3.782 \times 10^{-3}$$

$$\rho = \frac{A_s}{bh_0} = \frac{1520}{250 \times 564} = 10.780 \times 10^{-3} < \rho_b = 4.625 \times 10^{-2}$$

From Eq. (5.47), $x = \frac{f_y A_s}{\alpha_1 f_c b} = \frac{357 \times 1520}{1.0 \times 32 \times 250} = 68\,\text{mm}$

From Eq. (5.48), $M_u = f_y A_s \left(h_0 - \frac{x}{2}\right) = 357 \times 1520 \times \left(564 - \frac{68}{2}\right) = 287.60\,\text{kN}\cdot\text{m}$

3. When f_y = 460 N/mm^2 and 4ϕ22 are used,

$$\xi_b = \frac{0.8}{1 + \frac{460}{0.0033 \times 1.97 \times 10^5}} = 0.468$$

$$\rho_b = \xi_b \frac{\alpha_1 f_c}{f_y} = 0.468 \times \frac{1.0 \times 23}{460} = 2.340 \times 10^{-2}$$

$$\rho = \frac{A_s}{bh_0} = 10.780 \times 10^{-3} < \rho_b = 2.340 \times 10^{-2}$$

From Eq. (5.47), $x = \frac{f_y A_s}{\alpha_1 f_c b} = \frac{460 \times 1520}{1.0 \times 23 \times 250} = 122\,\text{mm}$

From Eq. (5.48), $M_u = f_y A_s \left(h_0 - \frac{x}{2}\right) = 460 \times 1520 \times \left(564 - \frac{122}{2}\right) = 351.70\,\text{kN}\cdot\text{m}$

4. When h = 700 mm and 4ϕ22 are used,

$$\rho = \frac{A_s}{bh} = \frac{1520}{250 \times 700} = 8.686 \times 10^{-3} > \rho_{\min} = 3.277 \times 10^{-3}, \quad h_0 = 664\,\text{mm}$$

$$\rho = \frac{A_s}{bh_0} = \frac{1520}{250 \times 664} = 9.157 \times 10^{-3} < \rho_b = 3.324 \times 10^{-2}$$

From Eq. (5.47), $x = \frac{f_y A_s}{\alpha_1 f_c b} = \frac{357 \times 1520}{1.0 \times 23 \times 250} = 94\,\mathrm{mm}$

From Eq. (5.48), $M_u = f_y A_s \left(h_0 - \frac{x}{2}\right) = 357 \times 1520 \times \left(664 - \frac{94}{2}\right) = 334.81\,\mathrm{kN \cdot m}$

From the results of Example 5.3, the main factors that influence the flexural bearing capacities of singly reinforced rectangular sections can be summarized as follows:

1. The flexural bearing capacity of a section increases with the longitudinal reinforcement, but cannot increase infinitely. As shown by Example 5.3(1), the flexural bearing capacity is nearly doubled from 146.65 to 280.54 kN·m when the reinforcement increases from 2ϕ22 to 4ϕ22 (doubled). However, when the reinforcement increases from 4ϕ22 to 8ϕ28 (more than tripled), the flexural bearing capacity is only slightly doubled from 280.54 to 640.53 kN·m. The previously increasing speed is obviously slowed down.
2. Increasing the strength of the steel can enhance the flexural bearing capacity of a section. From the comparison of the results of Example 5.3(1) and (2), it can be seen that the flexural bearing capacity is increased by 2.5 % with a 39 % increase of the concrete strength.
3. The flexural bearing capacity of a section increases insignificantly with the concrete strength. From the comparison of the results of Example 5.3(1) and (2), it can be seen that the flexural bearing capacity is increased by 2.5 % with a 39 % increase of the concrete strength.
4. Increasing the sectional dimensions will significantly increase the flexural bearing capacity. From the comparison of the results of Example 5.3(1) and (4), it can be seen that the flexural bearing capacity is increased by 19 % with a 17 % increase of the section height.

5.7.2 Cross-Sectional Design of New Structural Members

For this kind of problem, the cross-sectional dimensions (b, h, and h_0), material properties (f_c, f_t, and f_y), and the external bending moment M are already known and the reinforcement content (A_s) should be calculated. To ensure that the designed section will not fail under the given bending moment, $M_u \geqslant M$ must be satisfied. And the calculation should be carried out as follows:

1. Calculate $M_{u,max}$ from Eq. (5.54). If $M > M_{u,max}$, enlarge the cross section and recalculate; if $M \leqslant M_{u,max}$, go to the next step.
2. Solve Eq. (5.66) for A_s.

$$\begin{cases} \alpha_1 f_c bx = f_y A_s \\ M = M_u = \alpha_1 f_c bx\left(h_0 - \frac{x}{2}\right) = f_y A_s\left(h_0 - \frac{x}{2}\right) \end{cases} \tag{5.66}$$

3. Calculate the reinforcement ratio: $\rho = \frac{A_s}{bh}$.
4. If $\rho \geqslant \rho_{min}$, the calculation is completed.
5. If $\rho < \rho_{min}$, take $A_s = \rho_{min} bh$ and finish the calculation.

In the design, because the diameter of the reinforcement is unknown, the effective depth h_0 of the cross section can be estimated according to the following principles:

For beams of single-row steel bars, $h_0 = h{-}35$(mm);
For beams of double-row steel bars, $h_0 = h{-}60$(mm);
For slabs, $h_0 = h{-}20$(mm).

Example 5.4 The dimensions of the section and the strengths of the materials are the same as those in Example 5.1. Calculate the steel areas required for the bending moments of 70, 600 and 900 kN·m, respectively.

Solution From Example 5.3, it is known that $\xi_b = 0.516$.
From Eq. (5.55), $\alpha_{s,max} = 0.516 \times (1 - 0.5 \times 0.516) = 0.383$.
Assume $h_0 = 600{-}35 = 565$ mm.
From Eq. (5.54), we have

$$\begin{aligned} M_{u,max} &= \alpha_{s,max}\alpha_1 f_c b h_0^2 = 0.383 \times 1.0 \times 23 \times 250 \times 565^2 \\ &= 703.01\,\text{kN}\cdot\text{m} \end{aligned}$$

1. $M = 70$ kN·m $< M_{u,max} = 703.01$ kN·m
   From Eq. (5.66), it is obtained that

$$\begin{cases} 1.0 \times 23 \times 250x = 357A_s \\ 70 \times 10^6 = 1.0 \times 23 \times 250x\left(565 - \frac{x}{2}\right) \end{cases}$$

   Solution of the equation set yields

$$x = 22\text{mm}, \quad A_s = 355\,\text{mm}^2$$

$$\rho_{min} = 0.45\frac{f_t}{f_y} = \frac{0.45 \times 2.6}{357} = 3.277 \times 10^{-3}$$

$$A_{s,min} = \rho_{min} bh = 3.277 \times 10^{-3} \times 250 \times 600 = 491\,\text{mm}^2 > A_s = 355\,\text{mm}^2$$

   Select 2ϕ18 ($A_s = 509$ mm$^2 > A_{s,min} = 491$ mm^2).
2. $M = 600$ kN·m $< M_{u,max} = 703.01$ kN m
   Because M is approaching to $M_{u,max}$, much reinforcement is needed. So it is appropriate to lay the reinforcement in two rows, and hence, $h_0 = h{-}60 = 600{-}60 = 540$ mm.
   From Eq. (5.66), it can be obtained that

$$\begin{cases} 1.0 \times 23 \times 250x = 357A_s \\ 640 \times 10^6 = 1.0 \times 23 \times 250x\left(540 - \frac{x}{2}\right) \end{cases}$$

Solution of the equation set yields

$$x = 252\,\text{mm}, \quad A_s = 4059\,\text{mm}^2 > A_{s,\min} = 491\,\text{mm}^2$$

Select $6\phi25 + 2\phi28$ ($A_s = 4175\ \text{mm}^2$).

3. $M = 900\ \text{kN·m} > M_{u,max} = 703.01\ \text{kN·m}$
   The cross section must be enlarged. Take $b \times h = 300\ \text{mm} \times 700\ \text{mm}$, and thus, $h_0 = 700{-}60 = 640\ \text{mm}$.
   From Eq. (5.54), it can be obtained that

$$\begin{aligned} M_{u,\max} &= \alpha_{s,\max}\alpha_1 f_c b h_0^2 = 0.383 \times 1.0 \times 23 \times 300 \times 640^2 \\ &= 1082.45\,\text{kN·m} > M = 900\,\text{kN·m} \end{aligned}$$

From Eq. (5.66), it is found that

$$\begin{cases} 1.0 \times 23 \times 300x = 357A_s \\ 900 \times 10^6 = 1.0 \times 23 \times 300x\left(640 - \frac{x}{2}\right) \end{cases}$$

Solution of the equation set yields

$$\begin{aligned} &x = 253\,\text{mm}, \\ &\quad A_s = 4890\,\text{mm}^2 > A_{s,\min} = 3.277 \times 10^{-3} \times 300 \times 700 = 688\,\text{mm}^2 \end{aligned}$$

Select $8\phi28$ ($A_s = 4\ 924\ \text{mm}^2$).

5.8 Analysis of Doubly Reinforced Sections

In actual day-to-day concrete engineering, when the section of a flexural member is subjected to a large bending moment, while its depth is restricted for architectural reasons (e.g., a need for limited headroom in multistory buildings), compression steels A_s' may be arranged in the compression zone of the section to share the moment. This section is called the doubly reinforced section, as shown in Fig. 5.30. If the bending moment on a section can change the sign under different conditions, the doubly reinforced section should also be adopted. The flexural property of a doubly reinforced section is similar to that of a singly reinforced section. But the

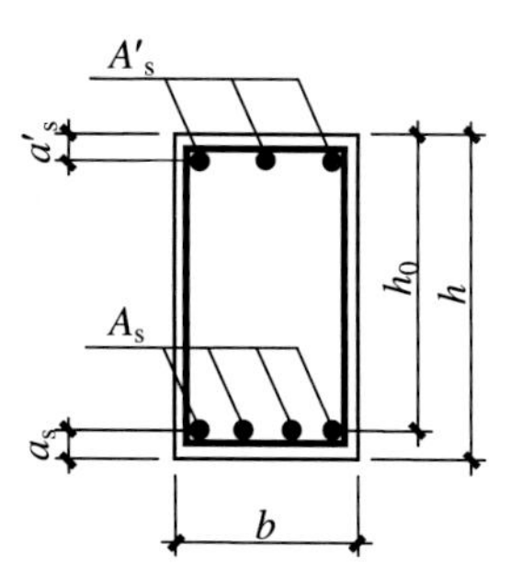

Fig. 5.30 A doubly reinforced section

compression steel in the compression zone will make the doubly reinforced section have its own characteristics, which will be the focus of this section.

5.8.1 Detailing Requirement on Doubly Reinforced Sections

In addition to positioning the longitudinal reinforcement and forming the steel cage, the stirrups in flexural members of doubly reinforced sections can laterally restrain the compression steel bars to prevent the members from failing immaturely by the buckling of the compression steel bars. Actually, the main function of the stirrups in flexural members is to sustain the shear force with the concrete, which will be introduced in Chap. 7. The stirrups should satisfy the following details:

1. The stirrups should be closed, and the spacing should be no more than $15d$ and 400 mm.
2. The stirrup diameter should be at least $d/4$, where d is the maximum diameter of the longitudinal reinforcement.
3. When there are more than 3 (or 4 for $b \leqslant 400$ mm) longitudinal steel bars, additional stirrups should be arranged.
4. When there are more than 5 compression steel bars with the diameter larger than 18 mm, the spacing of the stirrups should be at most $10d$.

5.8.2 Experimental Results

Because doubly reinforced sections generally have greater tension reinforcement, failure due to light reinforcement will not happen.

For under-reinforced doubly reinforced sections, three stages from loading onset to failure, which are similar to those of singly reinforced sections, can be observed, i.e., elastic stage (I), service stage (II), and failure stage (III), and three critical states correspond to the ends of each stage: cracking of the concrete (I_a), yielding of the tension steels (II_a), and crushing of the compressed concrete (III_a).

For over-reinforced doubly reinforced sections, there are only two stages (I and II). The stress in the tension steel bars will increase after the concrete cracks. But the concrete in the compression zone will be crushed before the yielding of the tension steel bars and the sections fail.

If a certain depth is kept for the compression zone, the compression steel bars will generally yield at failure even if sections are doubly under-reinforced or doubly over-reinforced.

5.8.3 Analysis of Doubly Reinforced Sections

5.8.3.1 Elastic Stage

According to the strain and stress distributions across a doubly reinforced rectangular section (Fig. 5.31), a similar analysis to that of a singly reinforced rectangular section can be made, which has shown that relevant formulae in *Mechanics of Materials* may be used in the analysis as long as the compression steels are equivalently transformed to the concrete (Fig. 5.31d).

When $\varepsilon_t^b = \varepsilon_{tu}$, the concrete in the tension zone cracks and is considered failed. Similar to a singly reinforced rectangular section, when a doubly reinforced rectangular section is close to cracking, the tensile stress of the concrete in the tension zone forms an approximate rectangular-shaped distribution, as shown by the dashed line in Fig. 5.31c. Considering the effect of the compressive steel bars, we can get the cracking moment as follows:

$$M_{cr} = 0.292(1 + 2.5\alpha_A)f_t bh^2 + \sigma_s' A_s' \left(\frac{1}{3} x_{cr} - a_s' \right) \tag{5.67}$$

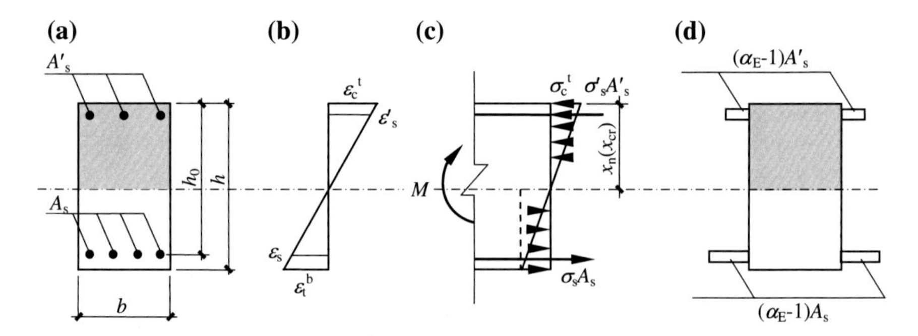

Fig. 5.31 Calculation diagram for a doubly reinforced rectangular section in elastic stage. **a** Cross section. **b** Strain distribution. **c** Stress distribution. **d** Transformed section

where

σ'_s stress in the compression steel bars;

A'_s area of the compression steel bars; and

a'_s depth from the centroid of the compression steel bars to the extreme compression fiber, and other symbols have the same meanings as in Eq. (5.15).

By substituting the strain distribution in Fig. 5.31b and the stress–strain relationship of the compression steel bars (the same as that of the tension steel bars), Eq. (5.67) is transferred to

$$M_{cr} = 0.292(1 + 2.5\alpha_A) f_t bh^2 + 2\frac{E_s}{E_c} f_t A'_s \left(\frac{1}{3}x_{cr} - a'_s\right) \frac{x_{cr} - a'_s}{h - x_{cr}} \quad (5.68)$$

Assuming $\alpha'_A = 2\alpha_E(A'/bh)$, $a'_s/h = 0.08$, $h_0/h = 0.92$, and $x_{cr} = 0.5h$, Eq. (5.68) can be further simplified as follows:

$$M_{cr} = 0.292(1 + 2.5\alpha_A + 0.25\alpha'_A) f_t bh^2 \quad (5.69)$$

5.8.3.2 Service Stage

After the section cracks, the concrete stress distribution in the compression zone, the resultant compressive force C and its acting point can be calculated in the same way as that for singly reinforced rectangular sections, except that the effect of the compression steel bars should be considered in establishing the equilibrium equations, which will not be elaborated herein.

5.8.3.3 Flexural Bearing Capacity

When $\varepsilon_c^t = \varepsilon_{cu}$, the concrete in the compression zone is crushed, symbolizing the section failure. Take the flexural members with the concrete grade of no more than C50 for example. The strain and stress distributions across the section are shown in Figs. 5.32b, c. From the force equilibrium $\Sigma X = 0$, it can be obtained that

$$0.798 f_c \xi_n^2 = E_s(1 - \xi_n)\,\varepsilon_{cu}\rho - E_s\left(\xi_n - \frac{a'_s}{h_0}\right)\varepsilon_{cu}\rho' \quad (5.70)$$

Solution of this quadratic equation yields ξ_n. Further with $\Sigma M = 0$, we have

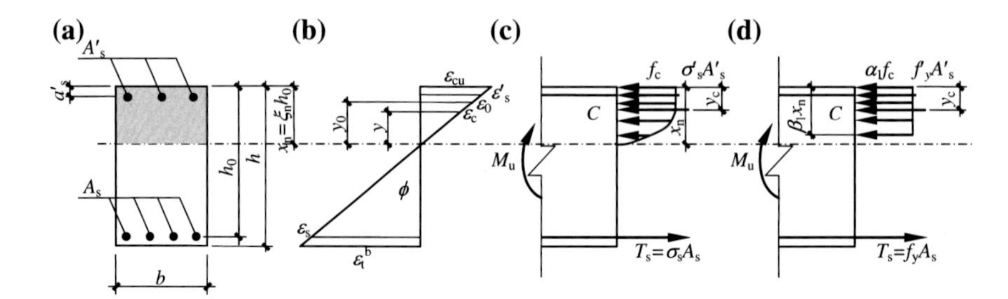

Fig. 5.32 Calculation diagram for the flexural bearing capacities of doubly reinforced rectangular sections. **a** Cross section. **b** Strain distribution. **c** Curved stress distribution. **d** Equivalent rectangular stress distribution

$$\begin{aligned} M_u &= 0.798 f_c b \xi_n h_0^2 (1 - 0.412\xi_n) + \sigma'_s A'_s h_0 \left(1 - \frac{a'_s}{h_0}\right) \\ &= \sigma_s A_s h_0 (1 - 0.412\xi_n) + \sigma'_s A'_s h_0 \left(0.412\xi_n - \frac{a'_s}{h_0}\right) \end{aligned} \tag{5.71}$$

Experimental results indicate that as long as a certain depth of the compression zone can be kept, the compressive steels will generally yield before the section fails. Therefore, if the value of a certain depth can be found, whether or not the compression steels are yielding will be determined. According to the strain distribution across the section shown in Fig. 5.32b, when the concrete grade is no more than C50, there is

$$\sigma'_s = 0.0033 E_s \left(\frac{a'_s}{x_n} - 1\right) \tag{5.72}$$

Substituting $E_s = 2\times10^5$ N/mm^2 and $a'_s = 0.5 \times 0.8x_n$ into Eq. (5.72), we have $\sigma'_s = -396\,\text{N/mm}^2$. Then, it can be concluded that when $x_n \geqslant 2.5a'_s$, HPB300, HRB335, HRBF335, HRB400, HRBF400, and RRB400 have yielded when the section fails. It is also found from the strain distribution across the section shown in Fig. 5.32b that if $x_n \leqslant \xi_{nb} h_0$, the tension steel bars have yielded at the section failure. ξ_{nb} can be calculated in the same way as that for singly reinforced rectangular sections. When both the tension and the compression steel bars can yield, Eqs. (5.70) and (5.71) can be further simplified as

$$\xi_n = 1.253\left(\rho\frac{f_y}{f_c} - \rho'\frac{f_y}{f_c}\right) \tag{5.73}$$

$$\begin{aligned} M_u &= f_y A_s h_0 (1 - 0.412\xi_n) + f'_y A'_s h_0 \left(0.412\xi_n - \frac{a'_s}{h_0}\right) \\ &= f_c b h_0^2 \xi_n (0.798 - 0.329\xi_n) + f'_y A'_s h_0 \left(1 - \frac{a'_s}{h_0}\right) \end{aligned} \tag{5.74}$$

Example 5.5 The dimensions of the section and the strengths of the materials of a reinforced concrete beam are the same as those in Example 5.1. When the section is reinforced with 4ϕ22 tension steel bars and 2ϕ16 compression steel bars, calculate:

1. The cracking moment M_{cr} and the corresponding curvature ϕ_{cr};
2. The flexure bearing capacity M_u and the corresponding curvature ϕ_{cr}.

Solution

$$A'_s = 402\,\text{mm}^2, \quad A_s = 1520\,\text{mm}^2, \quad \alpha_E = \frac{E_s}{E_c} = 7.849$$

From Example 5.1, $\alpha_A = 0.160$, $\alpha'_A = \frac{2\alpha_E A'_s}{bh} = \frac{2\times 7.849\times 402}{250\times 600} = 0.042$

1. From Eq. (5.69), it can be obtained that

$$\begin{aligned} M_{cr} &= 0.292(1 + 2.5\alpha_A + 0.25\alpha'_A) f_t b h^2 \\ &= 0.292 \times (1 + 2.5 \times 0.160 + 0.25 \times 0.042) \times 2.6 \times 250 \times 600^2 = 96.38\,\text{kN·m} \end{aligned}$$

$$\phi_{cr} = \frac{\varepsilon_{tu}}{h/2} = \frac{4f_t}{E_c h} = \frac{4 \times 2.6}{2.51 \times 10^4 \times 600} = 6.91 \times 10^{-7}\,\text{mm}^{-1}$$

2. $\alpha'_s = 25 + 16/2 = 33\,\text{mm}$, $h_0 = 564\,\text{mm}$

$$\xi_{nb} = \frac{1}{1 + \frac{f_y}{0.0033E_s}} = \frac{1}{1 + \frac{357}{0.0033\times 1.97\times 10^5}} = 0.646$$

$$\rho = \frac{A_s}{bh_0} = \frac{1520}{250 \times 564} = 1.078 \times 10^{-2}, \quad \rho' = \frac{A'_s}{bh_0} = \frac{402}{250 \times 564} = 2.851 \times 10^{-3}$$

$$\begin{aligned} \xi_n &= 1.253\left(\rho\frac{f_y}{f_c} - \rho'\frac{f'_y}{f_c}\right) = 1.253 \times \left(1.078 \times 10^{-2} \times \frac{357}{23} - 2.851 \times 10^{-3} \times \frac{357}{23}\right) \\ &= 0.154 \end{aligned}$$

$\xi_n < \xi_{nb}$, the tension steel can yield.
$x_n = \xi_n h_0 = 0.154 \times 564 = 87$ mm $> 2.5a'_s = 83$ mm, the compression steel can also yield.

$$\begin{aligned} M_u &= f_y A_s h_0 (1 - 0.412\xi_n) + f'_y A'_s h_0 \left(0.412\xi_n - \frac{a'_s}{h_0}\right) \\ &= 357 \times 1520 \times 564 \times (1 - 0.412 \times 0.154) + 357 \times 402 \times 564 \\ &\quad \times \left(0.412 \times 0.154 - \frac{33}{564}\right) = 287.20\,\text{kN}\cdot\text{m} \end{aligned}$$

$$\phi_u = \frac{\varepsilon_{cu}}{\xi_n h_0} = \frac{0.0033}{0.154 \times 564} = 38.0 \times 10^{-6}\,\text{mm}^{-1}$$

By comparing the results of Example 5.5 with those of Examples 5.1 and 5.2, it can be seen that the compression steel can not only increase the cracking moment and the flexure bearing capacity of the section, but also improve the ductility, which is proved by the 36 % increase of the ultimate curvature of the section in Example 5.5 compared with that of the singly reinforced rectangular section (4ϕ22) in Example 5.2.

5.8.4 Simplified Calculation of the Flexural Bearing Capacities of Doubly Reinforced Sections

Similar to singly reinforced rectangular sections, the curved stress distribution in the compression zone of a doubly reinforced section can be regarded as an equivalent rectangular stress block with the resultant compressive force and its acting point unchanged, as shown in Fig. 5.32d. α_1 and β_1 are determined in the same way as that for singly reinforced sections. If both the compression steel and the tension steel yield before the section fails, the formulae of the simplified calculation can be derived from $\Sigma X = 0$ and $\Sigma M = 0$ as

$$\begin{cases} \alpha_1 f_c bx + f'_y A'_s = f_y A_s \\ M_u = \alpha_1 f_c bx\left(h_0 - \frac{x}{2}\right) + f'_y A'_s \left(h_0 - a'_s\right) \end{cases} \tag{5.75}$$

The tension steel A_s can be decomposed into two parts:

$$A_s = A_{s1} + A_{s2}$$

The tensile force sustained by A_{s1} is equal to the compressive force in the concrete, and the two forces form the resisting moment M_{u1} (equivalent to a singly reinforced rectangular section). A_{s2} is the steel used to balance the compression steel A'_s. The forces in A_{s2} and A'_s make up the resisting moment M'_u. The calculating diagram for the flexural bearing capacity with the tension steel bars decomposed is shown in Fig. 5.33, from which Eq. (5.75) can also be calculated as

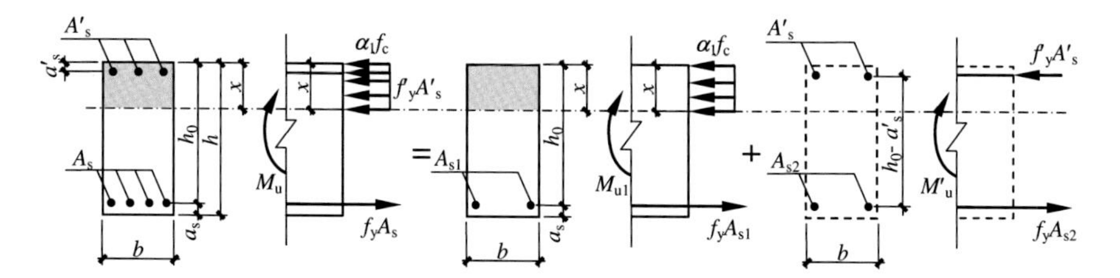

Fig. 5.33 Calculation diagram for the flexural bearing capacity of a doubly reinforced rectangular section with the tension steel bars decomposed

$$\begin{cases} \alpha_1 f_c bx + f_y' A_s' = f_y A_{s1} + f_y A_{s2} \\ M_u = M_{u1} + M_u' \\ M_{u1} = \alpha_1 f_c bx\left(h_0 - \dfrac{x}{2}\right) = f_y A_{s1}\left(h_0 - \dfrac{x}{2}\right) \\ M_u' = f_y' A_s'(h_0 - a_s') \end{cases} \tag{5.76}$$

Since there are generally sufficient tension steel bars in doubly reinforced rectangular sections, $\rho \geqslant \rho_{\min}$ can be automatically satisfied. To ensure the yielding of the tension steels, one of the following conditions must be satisfied:

$$\xi = \frac{x}{h_0} \leqslant \xi_b \tag{5.77a}$$

or

$$\rho_1 = \frac{A_{s1}}{bh_0} \leqslant \rho_{\max} = \xi_b \frac{\alpha_1 f_c}{f_y} \tag{5.77b}$$

or

$$M_1 \leqslant M_{u1,\max} = \alpha_{s,\max} \alpha_1 f_c b h_0^2 \tag{5.77c}$$

where M_1 is the moment resisted by the concrete and the corresponding tension steel bars.

For the yielding of the compression steel bars, $x = 0.8x_n \geqslant 2a_s'$ must be satisfied. Otherwise, the flexural strength of doubly reinforced rectangular sections should be calculated as

$$\begin{cases} \alpha_1 f_c bx - \sigma_s' A_s' = f_y A_s \\ M_u = \alpha_1 f_c bx(h_0 - \frac{x}{2}) + \sigma_s' A_s'(h_0 - a_s') \\ \sigma_s' = E_s \varepsilon_{cu}\left(\frac{\beta_1 a_s'}{\xi h_0} - 1\right) \end{cases} \tag{5.78}$$

or approximately take $x = 2a_s'$ and calculate M_u as

$$M_u = f_y A_s h_0 \left(1 - \frac{a'_s}{h_0}\right) \tag{5.79}$$

5.9 Applications of the Equations for Flexural Bearing Capacities of Doubly Reinforced Rectangular Sections

5.9.1 Bearing Capacity Calculation of Existing Structural Members

Knowing the cross-sectional dimensions (b, h, and h_0), reinforcement (A_s and A'_s), and material properties (f_c, f_t, f_y, and f'_y), the flexural bearing capacity M_u can be evaluated as follows:

1. Decompose the tension steel bars into two parts: $A_{s2} = A'_s f'_y / f_y$ and $A_{s1} = A_s - A_{s2}$;
2. Calculate $M'_u = f'_y A'_s (h_0 - a'_s)$;
3. Calculate x in the same way as that for a singly reinforced rectangular section with the tension steels of A_{s1};
4. If $2a'_s \leqslant x \leqslant \xi_b h_0$, calculate M_{u1} in the same way as that for a singly reinforced rectangular section, and thus, the flexural strength is $M_u = M_{u1} + M'_u$;
5. If $x > \xi_b h_0$, calculate M_{u1} in the same way as that for a singly reinforced rectangular section, and thus, the flexural strength is $M_u = M_{u1} + M'_u$;
6. If $x < 2a'_s$, calculate M_u from Eq. (5.78) or Eq. (5.79).

h_0 can be calculated in the same way as that for a singly reinforced rectangular cross section.

Example 5.6 The dimensions of the section and the strengths of the materials of a reinforced concrete beam are the same as those in Example 5.1. Calculate the flexural bearing capacity of the section when:

1. the beam is reinforced with 4ϕ22 tension steel bars and 2ϕ16 compression steel bars;
2. the beam is reinforced with 8ϕ28 tension steel bars and 2ϕ16 compression steel bars;
3. the beam is reinforced with 4ϕ22 tension steel bars and 4ϕ22 compression steel bars.

Solution

1. $A_s = 1520\,\text{mm}^2$, $A'_s = 402\,\text{mm}^2$, $h_0 = 564\,\text{mm}$, $a'_s = 33\,\text{mm}$

$$A_{s1} = A_s - A'_s = 1520 - 402 = 1118\,\text{mm}^2$$
$$M'_u = f'_y A'_s \left(h_0 - a'_s\right) = 357 \times 402 \times (564 - 33) = 76.21\,\text{kN}\cdot\text{m}$$

From Eq. (5.47), it is obtained that

$$x = \frac{f_y A_{s1}}{\alpha_1 f_c b} = \frac{357 \times 1118}{1.00 \times 23 \times 250} = 69\,\text{mm}$$

$2a'_s = 66\,\text{mm}$. It is known from the Example 5.3 that $\xi_b = 0.516$, and $\xi_b h_0 = 0.516 \times 564 = 291$ mm. $2a'_s \leqslant x \leqslant \xi_b h_0$ indicates that this is an under-reinforced section.

$$M_{u1} = f_y A_{s1}\left(h_0 - \frac{x}{2}\right) = 357 \times 1118 \times \left(564 - \frac{69}{2}\right) = 211.34\,\text{kN}\cdot\text{m}$$
$$M_u = M_{u1} + M'_u = 211.34 + 76.21 = 287.55\,\text{kN}\cdot\text{m}$$

The calculated flexural bearing capacity is very close to that of Example 5.5, which means that the accuracy of the simplified calculation method is satisfactory.

2. $A_s = 4924\,\text{mm}^2$, $A'_s = 402\,\text{mm}^2$, $h_0 = 533\,\text{mm}$, $a'_s = 33\,\text{mm}$, $\xi_b = 0.516$

$$A_{s1} = A_s - A'_s = 4924 - 402 = 4522\,\text{mm}^2$$
$$M'_u = f'_y A'_s \left(h_0 - a'_s\right) = 357 \times 402 \times (533 - 33) = 71.76\,\text{kN}\cdot\text{m}$$

From Eq. (5.47), it is found that

$$x = \frac{f_y A_{s1}}{\alpha_1 f_c b} = \frac{357 \times 4522}{1.00 \times 23 \times 250} = 281\,\text{mm}$$
$$2a'_s = 66\,\text{mm}, \quad \xi_b h_0 = 0.516 \times 533 = 275\,\text{mm}$$

$x \geqslant \xi_b h_0$ indicates that this is an over-reinforced section.
Substituting Eq. (5.65) into Eq. (5.47), we have

$$\xi = \frac{0.8}{1 + \frac{\alpha_1 f_c b h_0}{f_y A_{s1}}(0.8 - \xi_b)} = \frac{0.8}{1 + \frac{1.0 \times 23 \times 250 \times 533}{357 \times 4522}(0.8 - 0.516)} = 0.520$$
$$x = \xi h_0 = 0.520 \times 533 = 277\,\text{mm}$$
$$\sigma_s = f_y \frac{\xi - 0.8}{\xi_b - 0.8} = 357 \times \frac{0.520 - 0.8}{0.516 - 0.8} = 351.97\,\text{N/mm}^2$$
$$M_{u1} = \sigma_s A_{s1}\left(h_0 - \frac{x}{2}\right) = 351.97 \times 4522 \times \left(533 - \frac{277}{2}\right) = 627.89\,\text{kN}\cdot\text{m}$$
$$M_u = M_{u1} + M'_u = 627.89 + 71.76 = 699.65\,\text{kN}\cdot\text{m}$$

3. $A_s = 1520\,\text{mm}^2$, $A'_s = 1520\,\text{mm}^2$, $h_0 = 564\,\text{mm}$, $a'_s = 25 + 22/2 = 36\,\text{mm}$, $\xi_b = 0.516$

$$A_{s1} = A_s - A'_s = 1520 - 1520 = 0\,\text{mm}^2$$
$$M'_u = f'_y A'_s (h_0 - a'_s) = 357 \times 1520 \times (564 - 36) = 286.51\,\text{kN·m}$$

From Eq. (5.47), it is obtained that

$$x = \frac{f_y A_{s1}}{\alpha_1 f_c b} = \frac{357 \times 0}{1.0 \times 23 \times 250} = 0\,\text{mm}$$

$x < 2a'_s = 72\,\text{mm}$ indicates that the compression steel bars cannot yield. From Eq. (5.79), it is obtained that

$$M_u = f_y A_s h_0 \left(1 - \frac{a'_s}{h_0}\right) = 357 \times 1520 \times 564 \times \left(1 - \frac{36}{564}\right) = 286.51\,\text{kN·m}$$

Comparing the solutions in Question (1) and Question (3) in this example, it can be seen that if the compression steel bars exceed a certain content, they will not yield when the concrete in the compression zone is crushed. Therefore, increasing the content of the compression steels cannot increase the flexural bearing capacity of the section.

5.9.2 *Cross-Sectional Design of New Structural Members*

The cross-sectional design based on the flexural bearing capacity can be classified into two cases:

1. Case 1
   Given the section dimensions (b, h, and h_0), material strengths (f_c, f_t, f_y, f'_y), and external bending moment M, the problem here is to find A_s and A'_s. Similar to a singly reinforced rectangular section, it is required that the flexural bearing capacity of the section is no less than the external moment, i.e., $M_u \geqslant M$. Therefore, the calculation can be done as follows:

   (1) In order to fully utilize the concrete in the compression zone, assume $x = \xi_b h_0$;
   (2) Calculate A_{s1} and M_1: $A_{s1} = \alpha_1 f_c bx/f_y$, $M_1 = M_{u1} = A_{s1} f_y (h_0 - 0.5x)$;
   (3) Calculate M' and A_{s2}: $M' = M'_u = M - M_1$, $A_{s2} = M'/(h_0 - a'_s) f_y$;
   (4) Calculate A'_s and A_s: $A'_s = A_{s2} f_y / f'_y$, $A_s = A_{s1} + A_{s2}$.
   Since $x = \xi_b h_0$ has been assumed in the first step, all the applicable conditions are automatically satisfied.

2. Case 2
   Given the section dimensions (b, h, and h_0), material strengths (f_c, f_t, f_y, f'_y), compression steel bars (A'_s), and external bending moment M, the problem here is to find A_s. Calculations can be done as follows:

   (1) Calculate A_{s2} and M': $A_{s2} = A'_s f'_y / f_y$, $M' = A_{s2} f_y (h_0 - a'_s)$;
   (2) Calculate M_1: $M_1 = M - M'$;
   (3) Calculate x in the same way as that for a singly reinforced rectangular section subjected to bending moment M_1;
   (4) If $x < 2a'_s$, calculate A_s in the same way as that for a singly under-reinforced rectangular section; however, the minimum reinforcement ratio must be checked;
   (5) If $2a'_s \leqslant x \leqslant \xi_b h_0$, calculate A_{s1} in the same way as that for a singly under-reinforced rectangular section;
   (6) $A_s = A_{s1} + A_{s2}$;
   (7) If $x > \xi_b h_0$, assume that A'_s is unknown and calculate A'_s and A_s by following the steps in Case 1.
   h_0 can be calculated in the same way as that for a singly reinforced rectangular section.

Example 5.7 The dimensions of the section and the strengths of the materials of a reinforced concrete beam are the same as those in Example 5.1. The compression steel bars of 2ϕ16 have been arranged in the compressive zone. Try to find the required tension steels A_s when the external bending moments M are 650 kN·m and 900 kN·m, respectively.

Solution

$$A'_s = 402\text{mm},\ a'_s = 33\text{mm},\ \xi_b = 0.516$$

Because the bending moment is large, it is reasonable to assume that the tension steel bars will be arranged in two rows. So, $h_0 = 600-60 = 540$ mm.

1. When $M = 650$ kN m,

$$A_{s2} = A'_s f'_y / f_y = 402 \times 357/357 = 402\,\text{mm}^2$$
$$M' = f_y A_{s2}\left(h_0 - a'_s\right) = 357 \times 402 \times (540 - 33) = 72.76\,\text{kN}\cdot\text{m}$$
$$M_1 = M - M' = 650 - 72.76 = 577.24\,\text{kN}\cdot\text{m}$$

From Eq. (5.66), there is

$$\begin{cases} 1.0 \times 23 \times 250x = 357A_{s1} \\ 577.24 \times 10^6 = 1.0 \times 23 \times 250x\left(540 - \frac{x}{2}\right) \end{cases}$$

Solution of the equation set yields

$$x = 239\,\text{mm}$$
$$2a'_s = 66\ \text{mm}, \quad \xi_b h_0 = 0.516 \times 540 = 279\,\text{mm}$$

$2a'_s < x < \xi_b h_0$. Substituting x into the equation set, we have

$$A_{s1} = 3849\,\text{mm}^2$$
$$A_s = A_{s1} + A_{s2} = 3849 + 402 = 4251\,\text{mm}^2$$

Select 6ϕ28 + 2ϕ20 (A_s = 4321 mm^2).

2. When M = 900 kN m,

$$A_{s2} = A'_s f'_y / f_y = 402 \times 357/357 = 402\,\text{mm}^2$$
$$M' = f_y A_{s2}(h_0 - a'_s) = 357 \times 402 \times (540 - 33) = 72.76\,\text{kN·m}$$
$$M_1 = M - M' = 900 - 72.76 = 827.24\,\text{kN·m}$$

From Eq. (5.66), there is

$$\begin{cases} 1.0 \times 23 \times 250x = 357 A_{s1} \\ 827.24 \times 10^6 = 1.0 \times 23 \times 250x\left(540 - \frac{x}{2}\right) \end{cases}$$

Solution of the equation set yields x = 478 mm > $\xi_b h_0$ = 0.516 × 540 = 279 mm, so assume A'_s unknown and redesign the cross section.
Take x = $\xi_b h_0$ = 0.516 × 540 = 279 mm, a'_s = 35 mm

$$A_{s1} = \frac{\alpha_1 f_c bx}{f_y} = \frac{1.0 \times 23 \times 250 \times 279}{357} = 4493\,\text{mm}^2$$
$$M_1 = A_{s1} f_y (h_0 - 0.5x) = 4493 \times 357 \times (540 - 0.5 \times 279) = 642.40\,\text{kN·m}$$
$$M_1 = M - M' = 900 - 642.40 = 257.60\,\text{kN·m}$$
$$A_{s2} = \frac{M}{(h_0 - a'_s) f_y} = \frac{257.60 \times 10^6}{(540 - 35) \times 357} = 1429\,\text{mm}^2$$
$$A'_s = \frac{A_{s2} f_y}{f'_y} = \frac{1429 \times 357}{357} = 1429\,\text{mm}^2$$
$$A_s = A_{s1} + A_{s2} = 5922\,\text{mm}^2$$

Choose 3ϕ25 (A'_s = 1473 mm^2) as compression steel bars and 6ϕ30 + 2ϕ34 (A_s = 6054 mm^2) as tension steel bars.

5.10 Analysis of T Sections

T sections are similar to singly reinforced rectangular sections in material compositions, i.e., without compression steels. The overhang of the concrete flange of a T section is equivalent in mechanical behavior to the compression steels in a doubly reinforced rectangular section. Therefore, we would focus on the simplified calculation method for the flexural bearing capacity of T sections in this section rather than on the detailed analysis of their mechanical properties. Because the concrete in the tension zone makes no contribution when the failure is imminent, the flexural strength calculation method of I and box sections is similar to that of T sections. Therefore, the analysis of T sections is of great importance.

5.10.1 Effective Compressed Flange Width of T Beams

Experimental results show that the longitudinal compressive stress is not uniformly distributed in the compressed flange. The farther the flange concrete from the web is, the smaller the compressive stress is, as shown in Fig. 5.34a. If the flange is considered to be uniformly stressed (Fig. 5.34b), the value of b_f' (the effective compressed flange width) must be specified.

GB 50010 stipulates that the effective width b_f' of the compressed flange of T, I, or inverted L sections should take the minimum values in Table 5.1.

5.10.2 Simplified Calculation Method for the Flexural Bearing Capacities of T Sections

5.10.2.1 Classification of T sections

According to the position of the neutral axis (nominal neutral axis for simplified calculation), T sections can be sorted into two types:

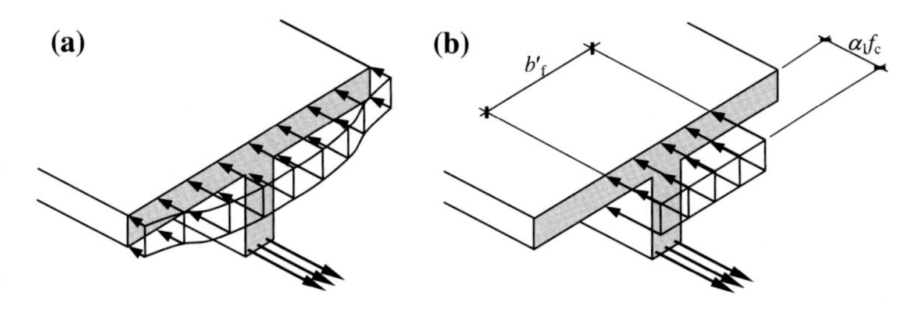

Fig. 5.34 Stress distributions in the compressed flange of a T section. **a** Actual stress distribution. **b** Equivalent distribution

Table 5.1 Effective width b_f' of the compressed flange

Category		T and I sections		Inverted L sections
		Monolithic beam–slab	Independent beam	Monolithic beam–slab
1	Span length l_0	$l_0/3$	$l_0/3$	$l_0/6$
2	Clear web spacing s_n	$b + s_n$	–	$b + s_n/2$
3	Flange thickness h_f'	$b + 12h_f'$	b	$b + 5h_f'$

Notes (1) The in situ cast reinforced concrete slab shown in Fig. 1.4 is one of the monolithic beam–slab systems. If lateral diaphragms with the spacing less than the web spacing are provided within the span length of a monolithic beam–slab system, the effective flange width will not be restricted by Category 3
(2) b is the width of the web
(3) For haunched beams of T, I, or inverted L sections, when the depth of the haunch in the compression zone $h_f \geqslant h_f'$ and the width of the haunch $b_h \geqslant 3h_h$, the effective flange width can be the values in Category 3 plus $2b_h$ for T and I sections, plus b_h for inverted L sections, respectively
(4) If an independent beam is cracked longitudinally at the web–flange junction, the effective flange width can only take the web width b

(1) Type I: The neutral axis lies in the compressed flange (Fig. 5.36);
(2) Type II: The neutral axis lies in the web (Fig. 5.37).

From the simplified calculation diagram for a T section with the whole flange under compression (Fig. 5.35), it is found that a T section can be classified as Type I if either of the following two conditions is satisfied (otherwise, it will be considered as Type II).

$$f_y A_s \leqslant \alpha_1 f_c b_f' h_f' \tag{5.80}$$

$$M \leqslant M_u = \alpha_1 f_c b_f' h_f' \left(h_0 - \frac{h_f'}{2} \right) \tag{5.81}$$

where
M external bending moment on the section;

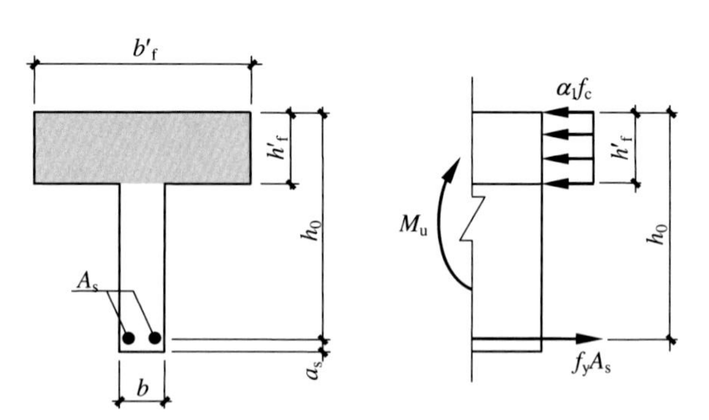

Fig. 5.35 Calculation diagram of a T section with the whole flange under compression

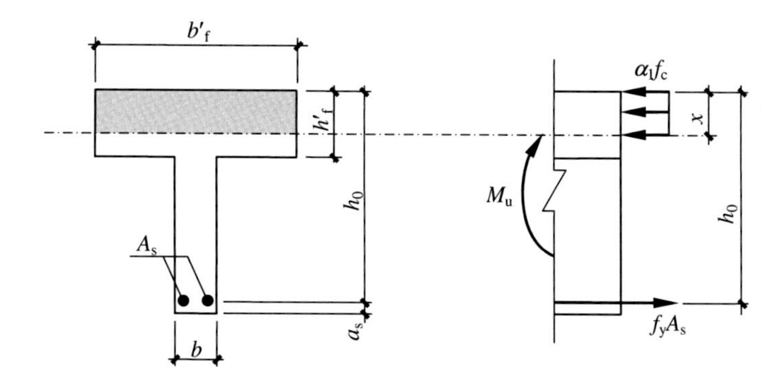

Fig. 5.36 Calculation diagram of a Type I T section

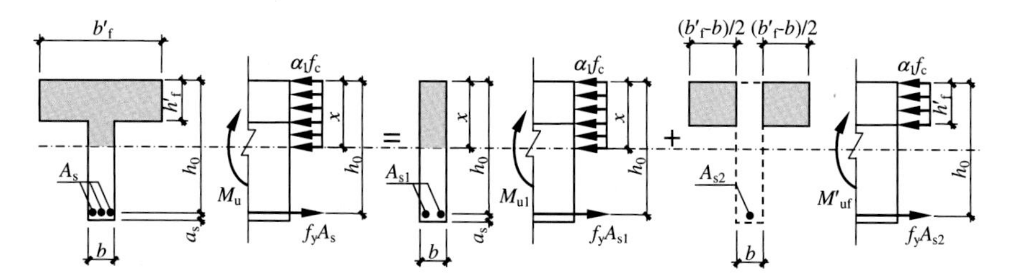

Fig. 5.37 Simplified calculation diagram for a Type II T section

b'_f effective width of the compressed flange; and
h'_f depth of the compressed flange.

5.10.2.2 Type I T Sections

Because the concrete below the neutral axis is assumed ineffective, the flexural bearing capacity of a Type I T section can be calculated in the same way as that for a singly reinforced rectangular section with the dimensions of $b'_f \times h$ as shown in Fig. 5.36. And the simplified calculation equations are

$$\begin{cases} \alpha_1 f_c b'_f x = f_y A_s \\ M_u = \alpha_1 f_c b'_f x\left(h_0 - \frac{x}{2}\right) = f_y A_s\left(h_0 - \frac{x}{2}\right) \end{cases} \tag{5.82}$$

Since for Type I T sections, the neutral axis lies within the flange, and x is small, a tension failure (under-reinforced) generally occurs, i.e., $x \leqslant \xi_b h_0$.

To avoid the failure due to insufficient reinforcement, $A_s \geqslant \rho_{min} bh$ should be ensured. ρ_{min} is calculated from Eq. (5.59). Since the cracking moment of a T section is nearly equal to that of a rectangular section with the same dimension as the web only, the dimensions $b \times h$ rather than $b_{\mathrm{f}}' \times h$ should be used when the minimum reinforcement ratio is checked.

5.10.2.3 Type II T Sections

The mechanical behavior of a Type II T section is similar to that of a doubly reinforced rectangular section. The flexural bearing capacity of the section can be decomposed into two parts: M_{u1} produced by the compressed concrete over the web and the corresponding tension steel bars A_{s1} and M_{uf}' produced by the compressed concrete in the overhang of the flange and the corresponding tension steel A_{s2}. The simplified calculation diagram is shown in Fig. 5.37. By the equilibrium equations of $\Sigma X = 0$ and $\Sigma M = 0$, it can be obtained that

$$\begin{cases} \alpha_1 f_{\mathrm{c}} bx + \alpha_1 f_{\mathrm{c}} (b_{\mathrm{f}}' - b) h_{\mathrm{f}}' = f_{\mathrm{y}} A_{\mathrm{s}} \\ M_{\mathrm{u}} = M_{\mathrm{u1}} + M_{\mathrm{uf}}' = \alpha_1 f_{\mathrm{c}} bx \left(h_0 - \frac{x}{2}\right) + \alpha_1 f_{\mathrm{c}} (b_{\mathrm{f}}' - b) h_{\mathrm{f}}' \left(h_0 - \frac{h_{\mathrm{f}}'}{2}\right) \end{cases} \tag{5.83}$$

The neutral axis of a Type II T section lies in the web, so x is relatively large and the section failure due to insufficient reinforcement hardly occurs, i.e., $A_{\mathrm{s}} \geqslant \rho_{\min} bh$ is generally automatically satisfied. However, this equality still needs to be checked.

To prevent the compression failure from happening, either one of the following conditions must be satisfied:

$$\xi = \frac{x}{h_0} \leqslant \xi_{\mathrm{b}} \tag{5.84a}$$

or

$$\rho_1 = \frac{A_{\mathrm{s1}}}{bh_0} \leqslant \rho_{\max} = \xi_{\mathrm{b}} \frac{\alpha_1 f_{\mathrm{c}}}{f_{\mathrm{y}}} \tag{5.84b}$$

or

$$M_1 \leqslant M_{\mathrm{u1,max}} = \alpha_{\mathrm{s,max}} \alpha_1 f_{\mathrm{c}} bh_0^2 \tag{5.84c}$$

where M_1 is the external moment resisted by the compressed concrete in the web and the corresponding tension steel.

5.11 Applications of the Equations for Flexural Bearing Capacities of T Sections

5.11.1 Bearing Capacity Calculation of Existing Structural Members

Given the section dimensions (b, h, h_0, b_f', and h_f'), reinforcement (A_s), and material strengths (f_c, f_t, f_y), the ultimate moment M_u of a T section can be evaluated as follows:

1. Determine the type of T section: If $f_y A_s \leqslant \alpha_1 f_c\, b_f' h_f'$, it is a Type I T section; otherwise, it is a Type II T section;
2. For a Type I T section, calculate the flexural bearing capacity as that for a singly reinforced rectangular section with the dimension of $b_f' \times h$. If $A_s < \rho_{min}\, bh$, the flexural bearing capacity is equal to the cracking moment of a rectangular section of $b \times h$;
3. For a Type II T section, the next two steps should be followed:
4. Calculate M_{uf}': $M_{uf}' = \alpha_1 f_c \left(b_f' - b\right) h_f' \left(h_0 - \frac{h_f'}{2}\right)$;
5. Calculate M_{u1} as that for a singly reinforced rectangular section with the dimension of $b \times h$ and $M_u = M_{u1} + M_{uf}'$.

The method to determine h_0 is the same as that for a singly reinforced rectangular section.

Example 5.8 The T section of a reinforced concrete beam is shown in Fig. 5.38. The effective span of the beams is l_0 = 8500 mm. The material properties are f_c = 15 N/mm^2, f_t = 1.43 N/mm^2, f_y = 320 N/mm^2, and E_s = 1.97 × 10^5 N/mm^2. Calculate the flexural bearing capacity of the section with the following reinforcement:

1. 4ϕ20 in the tension zone;
2. 8ϕ25 in the tension zone.

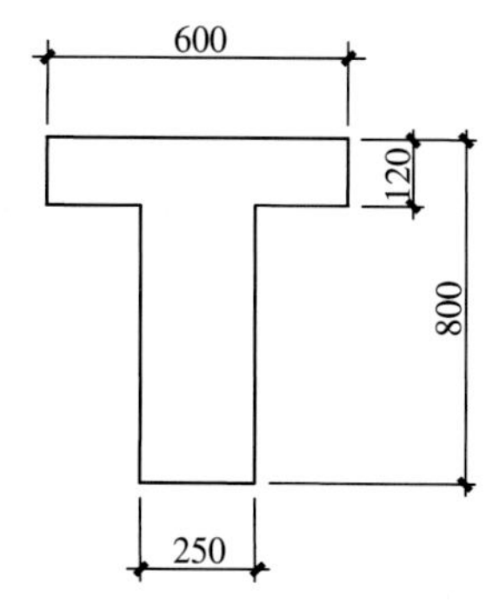

Fig. 5.38 The T section of a beam

Solution $b'_f = 600$ mm, $b'_f < l_0/3 = 8500/3 = 2833$ mm, $b'_f < b + 12\,h'_f = 1690$ mm. It is satisfactory.

1.
$$\begin{aligned} A_s &= 1256\text{mm}^2 > \rho_{\min} bh = 0.45 \cdot \frac{f_t}{f_y} \cdot bh \\ &= 0.45 \times \frac{1.43}{320} \times 250 \times 800 = 402\,\text{mm} \\ h_0 &= 800 - 25 - 20/2 = 765\,\text{mm} \\ f_y A_s &= 320 \times 1256 = 401.92\,\text{kN} \\ \alpha_1 f_c b'_f h'_f &= 1.0 \times 15 \times 600 \times 120 = 1080\,\text{kN} \end{aligned}$$

$f_y A_s < \alpha_1 f_c b'_f h'_f$, so it is a Type I T section.

From Eq. (5.47), it is obtained that

$$x = \frac{f_y A_s}{\alpha_1 f_c b'_f} = \frac{401.92 \times 10^3}{1.0 \times 15 \times 600} = 45\,\text{mm}$$

From Eq. (5.48), it is obtained that

$$M_u = f_y A_s \left(h_0 - \frac{x}{2}\right) = 320 \times 1256 \times \left(765 - \frac{45}{2}\right) = 298.43\,\text{kN·m}$$

2. $A_s = 3928\text{mm}^2 > \rho_{\min} bh = 0.45 \cdot \frac{f_t}{f_y} \cdot bh = 0.45 \times \frac{1.43}{320} \times 250 \times 800 = 402\,\text{mm}$

$$h_0 = 800 - 25 - 25 - 25/2 = 737.5\,\text{mm}$$
$$f_y A_s = 320 \times 3928 = 1256.96\,\text{kN},\ \alpha_1 f_c b'_f h'_f = 1.0 \times 15 \times 600 \times 120 = 1080\,\text{kN}$$

$f_y A_s > \alpha_1 f_c b'_f h'_f$, so it is a Type II T section.

$$\begin{aligned} M'_{uf} &= \alpha_1 f_c \left(b'_f - b\right) h'_f \left(h_0 - \frac{h'_f}{2}\right) \\ &= 1.0 \times 15 \times (600 - 250) \times 120 \times \left(737.5 - \frac{120}{2}\right) = 426.83\,\text{kN·m} \end{aligned}$$

From the first equation in Eq. (5.83), it is obtained that

$$x = \frac{f_y A_s - \alpha_1 f_c \left(b'_f - b\right) h'_f}{\alpha_1 f_c b} = \frac{1256.96 \times 10^3 - 1.0 \times 15 \times (600 - 250) \times 120}{1.0 \times 15 \times 250}$$
$$= 167\,\text{mm}$$

$$\xi_b = \frac{0.8}{1 + \frac{f_y}{0.0033 E_s}} = \frac{0.8}{1 + \frac{320}{0.0033 \times 1.97 \times 10^3}} = 0.536$$

$$x < \xi_b h_0 = 0.536 \times 737.5 = 395\,\text{mm}$$

$$M_{u1} = \alpha_1 f_c b x \left(h_0 - \frac{x}{2}\right)$$
$$= 1.0 \times 15 \times 250 \times 167 \times \left(737.5 - \frac{167}{2}\right) = 409.57\,\text{kN·m}$$

$$M_u = M_{u1} + M'_{uf} = 409.57 + 426.83 = 836.40\,\text{kN·m}$$

5.11.2 Cross-Sectional Design of New Structural Members

Given the section dimensions (b, h, h_0, b'_f, h'_f), material strengths (f_c, f_t, f_y), and bending moment M, find A_s. Similar to singly and doubly reinforced rectangular sections, T sections should not fail under given external bending moment M, i.e., $M_u \geqslant M$. Therefore, the calculation should be done as the following steps:

1. Determine the section type: If $M \leqslant \alpha_1 f_c b'_f h'_f \cdot \left(h_0 - \frac{h'_f}{2}\right)$, it is a Type I T section; if $M > \alpha_1 f_c b'_f h'_f \cdot \left(h_0 - \frac{h'_f}{2}\right)$, it is a Type II T section;
2. For a Type I T section, it can be designed as a singly reinforced rectangular section of $b'_f \times h$, and $A_s \geqslant \rho_{min} bh$ must be checked;
3. For a Type II T section, the next steps should be followed:
4. Calculate A_{s2} and M': $A_{s2} = \alpha_1 f_c \left(b'_f - b\right) h'_f / f_y$, $M' = M'_{uf} = A_{s2} f_y \left(h_0 - \frac{h'_f}{2}\right)$;
5. Calculate M_1: $M_1 = M - M'$;
6. Based on M_1, calculate x as that for a singly reinforced rectangular section of $b \times h$;
7. If $x \leqslant \xi_b h_0$, calculate A_{s1} as that for a singly under-reinforced rectangular section and $A_s = A_{s1} + A_{s2}$;
8. If $x > \xi_b h_0$, enlarge the section and restart the design from Step 1 or arrange the compression steels A'_f in the flange, then calculate A_s.

The method to determine h_0 is the same as that for a singly reinforced rectangular section.

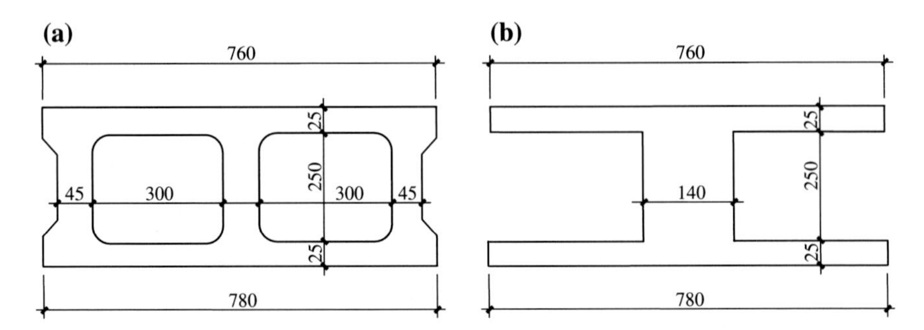

Fig. 5.39 Section dimensions of a precast spectacle plate. **a** Original section. **b** Transformed section

Example 5.9 The section dimensions of a precast spectacle plate are shown in Fig. 5.39a. Given the effective span of the slab l_0 = 4800 mm, f_c = 14 N/mm^2, f_t = 1.3 N/mm^2, f_y = 300 N/mm^2, and $E_s = 1.97 \times 10^5$ N/mm^2, find the required steel areas when the plate is subjected to the external bending moments of M = 26 and 45 kN m, respectively.

Solution Transform the section first (Fig. 5.39b), and then $b'_f = 760\,\text{mm}$, $b'_f < l_0/3 = 4800/3 = 1600\,\text{mm}$, $b'_f > b + 12h'_f = 440\,\text{mm}$

According to Table 5.1, assume b'_f = 440 mm, h_0 = 300−20 = 280 mm

$$\alpha_1 f_c b'_f h'_f \left(h_0 - \frac{h'_f}{2}\right) = 1.0 \times 14 \times 440 \times 25 \times \left(280 - \frac{25}{2}\right) = 41.20\,\text{kN}\cdot\text{m}$$

1. M = 26 kN m,

   M < 41.20 kN m, so it is a Type I T section and can be designed as that for a singly reinforced rectangular section of 440 mm × 300 mm. From Eq. (5.66), it is obtained that

$$\begin{cases} 1.0 \times 14 \times 440x = 300A_s \\ 26 \times 10^6 = 1.0 \times 14 \times 440x\left(280 - \frac{x}{2}\right) \end{cases}$$

   Solution of the equation set yields x = 16 mm and A_s = 329 mm^2. Different from the T section, the tension flange has a significant influence on the cracking moment of an I section, which should be taken into account when checking the minimum tension steel ratio.

$$\rho_{\min}[bh + (b_f - b)h_f] = 0.45\frac{f_t}{f_y}[bh + (b_f - b)h_f]$$
$$= 0.45 \times \frac{1.3}{300} \times (140 \times 300 + 640 \times 25) = 113\,\text{mm}^2$$

   A_s > 113 mm^2, it is OK. Select 3ϕ12 (A_s = 339 mm^2).

2. $M = 45$ kN m,
   $M > 41.20$ kN m, so it is a Type II T section.

$$A_{s2} = \alpha_1 f_c (b'_f - b) h'_f / f_y = 1.0 \times 14 \times (440 - 140) \times 25/300 = 350\,\text{mm}^2$$

$$M' = M'_{uf} = f_y A_{s2}\left(h_0 - \frac{h'_f}{2}\right) = 300 \times 350 \times \left(280 - \frac{25}{2}\right) = 28.09\,\text{kN·m}$$

$$M_1 = M - M' = 45 - 28.09 = 16.91\,\text{kN·m}$$

From Eq. (5.66), it is obtained that

$$\begin{cases} 1.0 \times 14 \times 140x = 300A_{s1} \\ 16.91 \times 10^6 = 1.0 \times 14 \times 140x\left(280 - \frac{x}{2}\right) \end{cases}$$

Solving the equations yields $x = 33$ mm, $\xi_b = \frac{0.8}{1 + \frac{300}{0.0033 \times 197 \times 10^5}} = 0.547$. $x < \xi_b h_0 = 0.547 \times 280 = 153$ mm, it is OK. By substituting x into the first equation of Eq. (5.66), it can be obtained that

$$A_{s1} = 216\,\text{mm}^2$$

$$A_s = A_{s1} + A_{s2} = 216 + 350 = 566\,\text{mm}^2$$

Select $3\phi12 + 3\phi10$ ($A_s = 575$ mm^2).

5.12 Deep Flexural Members

5.12.1 Basic Concepts and Applications

Beams of the span/depth ratio $l_0/h < 5.0$ are generally called deep flexural members (Fig. 5.40), among which simply supported beams of $l_0/h \leqslant 2.0$ and continuous beams with any span of $l_0/h \leqslant 2.5$ are called deep beams. Beams that are neither deep beams nor ordinary beams ($l_0/h \geqslant 5.0$) are called short beams. For deep flexural members, the effect of the shear force becomes more and more apparent with the decrease of the span/depth ratio. So their mechanical behaviors are different from ordinary beams.

Deep flexural members have wide applications in real engineering, such as transfer girders in multistory or high-rise buildings (Fig. 5.41a), raft foundation beams (Fig. 5.41b), and sidewalls of silos (Fig. 5.41c).

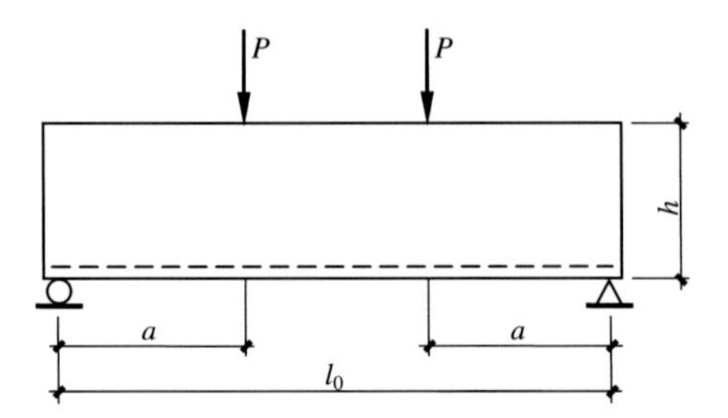

Fig. 5.40 Deep flexural members

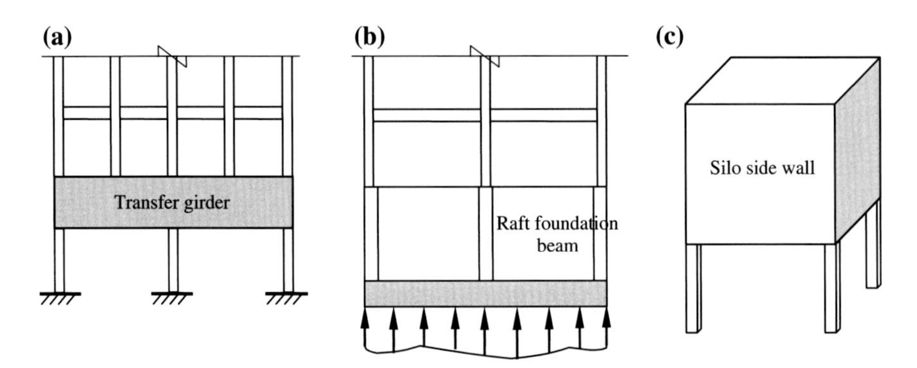

Fig. 5.41 Applications of deep flexural members

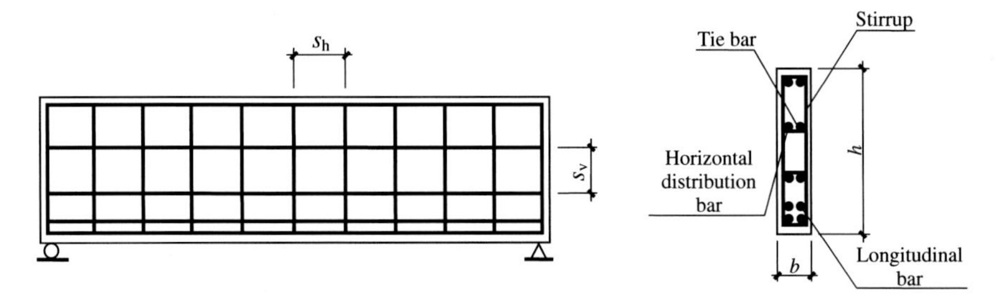

Fig. 5.42 Reinforcement in deep flexural members

The longitudinal reinforcement in deep flexural members is generally of small diameter and distributed within the range $0.2h$ from the extreme tension fiber. Due to the large depth, except for the longitudinal bars and stirrups, there are also horizontal distribution bars tied by lacing wires in deep flexural members, as shown in Fig. 5.42.

5.12.2 Mechanical Properties and Failure Modes of Deep Flexural Members

The loading process of deep flexural members can also be divided into elastic stage, service stage, and failure stage.

From the results of photoelastic tests or numerical analyses based on theories of elastic mechanics, the stress distribution of a deep flexural member in elastic stage can be obtained as shown in Fig. 5.43a, where the dashed and the solid lines are trajectories of principal compressive stresses and tensile stresses, respectively. It can be seen that the trajectory of the maximum principal tensile stress is parallel to the bottom surface of the beam and the trajectory of the maximum principal compressive stress is parallel to the line connecting the support and the acting point of the concentrated load. Figure 5.43b shows the strain distribution across the mid-span section, which apparently does not obey the plane section assumption.

When the maximum tensile stress in the extreme tension fiber exceeds the tensile strength of the concrete, a vertical crack appears in the mid-span of the beam. The load at this moment is about 1/3–1/2 of the ultimate one. With the increase of the load, the width and number of the cracks continuously increase and diagonal cracks also appear near the supports. The mechanical properties of the beam change greatly after the appearance of the diagonal cracks, i.e., the beam mechanism is weakened, while the arch mechanism is enhanced. As shown in Fig. 5.44, the longitudinal bars can be assumed as the bottom chord of an arch, and the concrete between the dashed lines is identified as arch ribs. Two different failure modes may happen depending on the amount of the longitudinal bars. If the steel content is low, with the increase of the crack width, the longitudinal bars are yielding, which leads to a flexure failure. On the other hand, if the steel content is high, the concrete in the arch rib will be crushed before the yielding of the longitudinal bars, indicating that a diagonal compression failure has occurred. In addition, bearing failure or bond failure may also happen near the acting points of concentrated loads or the supports.

The mechanical properties and failure modes of short beams are similar to those of ordinary beams. The failure modes include tension failure, compression failure, and the failure due to insufficient reinforcement.

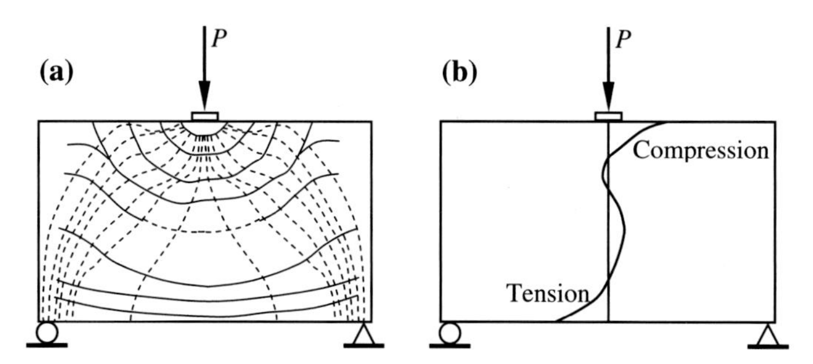

Fig. 5.43 Principal stress trajectories and strain distribution. **a** Principal stress trajectories. **b** Strain distribution across the mid-span section

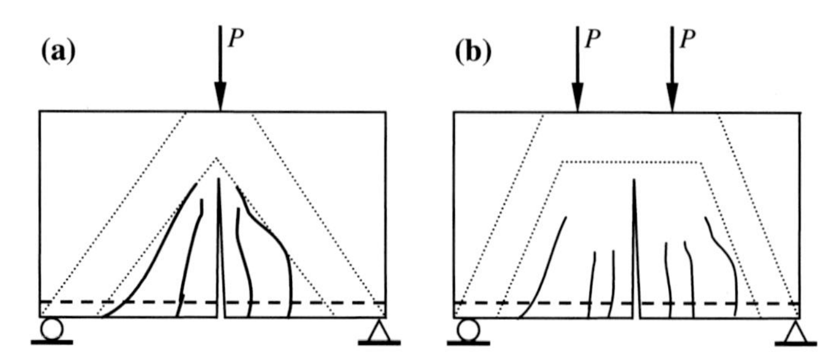

Fig. 5.44 Failure modes of deep beams (*dashed line* means arch mechanism). **a** One concentrated load. **b** Two concentrated loads

5.12.3 Flexural Bearing Capacities of Deep Beams

5.12.3.1 Balanced Reinforcement Ratio of Deep Beams

Between the flexure failure mode and the diagonal compression failure mode, which are distinguished by whether the longitudinal bars are yielding or not, there must be an intermediate failure mode, i.e., the yielding of the longitudinal bars and the crushing of the concrete happen at the same time, which is called the flexure–shear balanced failure mode. The reinforcement ratio corresponding to the flexure–shear balanced failure is called the balanced reinforcement ratio of deep beams. Based on the regression analysis of the experimental data with 271 members, the balanced reinforcement ratio can be determined by

$$\rho_{\mathrm{bm}} = 0.19\lambda \frac{f_{\mathrm{c}}}{f_{\mathrm{y}}} \tag{5.85}$$

where

λ shear-span ratio, $\lambda = a/h$; and

a horizontal distance between the concentrated load and the nearest support, and for uniformly distributed load, $a = l_0/4$.

For more details about λ, one can refer to Chap. 7.

As for the restrained deep beams and the continuous deep beams, the balanced steel ratio is

$$\rho_{\mathrm{bm}} = \frac{0.19\lambda}{1 + 1.48\psi} \frac{f_{\mathrm{c}}}{f_{\mathrm{y}}} \tag{5.86}$$

where ψ is the absolute value of the ratio of the maximum bending moment at the support to that at the mid-span.

5.12.3.2 Formula for the Flexural Strength of Deep Beams

The loads on deep beams are generally very large. And the small deformation before failure due to the large stiffness implies that the failure of deep beams is sudden. From previous analysis, it is known that the plane section assumption is inapplicable to the deep beams, and thus, the formulae for the flexural bearing capacities of ordinary beams cannot be used. For safety reasons, deep beams are assumed to fail if the longitudinal bars (including the horizontal distribution bars) in the zone of $h/3$ height measured from the extreme tension fiber are yielding.

Based on the experimental results of 93 simply supported beams tested in China, the yield moment of deep beams can be obtained as follows:

$$M_y = \left(f_y A_s + 0.33\rho_h f_{yh} bh\right)\gamma_s h_0 \tag{5.87}$$

$$\gamma_s = 1 - \left(1 - 0.1\frac{l_0}{h}\right)\left(\rho + 0.5\rho_h\frac{f_{yh}}{f_y}\right)\frac{f_y}{f_c} \tag{5.88}$$

where

ρ_h reinforcement ratio of horizontal bars, $\rho_h = \frac{A_{sh}}{bs_v}$;
A_{sh} whole areas of horizontal bars within the vertical spacing s_v;
f_{yh} yield strength of horizontal bars; and
γ_s coefficient for the internal lever arm.

5.12.4 Flexural Bearing Capacities of Short Beams

Because the plane section assumption is still valid for short beams and the mechanical properties of short beams and ordinary beams are similar, the flexural bearing capacities of short beams can be obtained as

$$M_y = f_y A_s h_0 (0.9 - 0.33\xi) \tag{5.89}$$

5.12.5 Unified Formulae for the Flexural Bearing Capacities of Deep Flexural Members

GB 50010 stipulates the following unified formulae for the flexural bearing capacities of deep flexural members,

$$\begin{cases} \alpha_1 f_c bx = f_y A_s \\ M_u = f_y A_s z \end{cases} \tag{5.90}$$

$$z = \alpha_d \left(h_0 - \frac{x}{2} \right) \tag{5.91}$$

$$\alpha_d = 0.8 + 0.04 \frac{l_0}{4} \tag{5.92}$$

where

z internal lever arm; when $l_0 < h$, $z = 0.6\ l_0$;
x compression zone depth, when $x < 0.2h_0$, $x = 0.2h_0$;
α_d correction factor of internal lever arm; and
h_0 effective depth of sections, $h_0 = h - a_s$. When $l_0/h \leqslant 2$, $a_s = 0.1h$ for mid-span sections and $a_s = 0.2h$ for sections at supports. When $l_0/h > 2$, h_0 is equal to the distance measured from the centroid of the steel bars to the extreme tension fiber.

The formulae presented above are suitable for both deep beams and ordinary beams, and conservative for the former, because the contribution of the horizontal bars (about 10–30 % of the flexural bearing capacities) is neglected.

The minimum reinforcement ratios of different kinds of steel bars in deep beams are listed in Table 5.2, and the minimum reinforcement ratios in short beams are the same as those in ordinary beams.

Example 5.10 The effective span of a simply supported deep beam shown in Fig. 5.45 is 6000 mm. Given f_c = 13.5 N/mm², f_y = 290 N/mm², F = 1090 kN, q = 40 kN/m (self-weight included). Calculate the required longitudinal steel area in the tension zone.

Table 5.2 Minimum reinforcement ratios as percentage for deep beams (%)

Reinforcement type	Longitudinal bars $\rho = A_s/(bh)$	Horizontal bars $\rho_{sh} = A_{sh}/(bs_v)$	Vertical bars $\rho_{sv} = A_{sv}/(bs_h)$
HPB300	0.25	0.25	0.20
HRB335, HRBF335, HRBF400, HRBF400, RRB400	0.20	0.20	0.15
HRB500, HRBF500	0.15	0.15	0.1

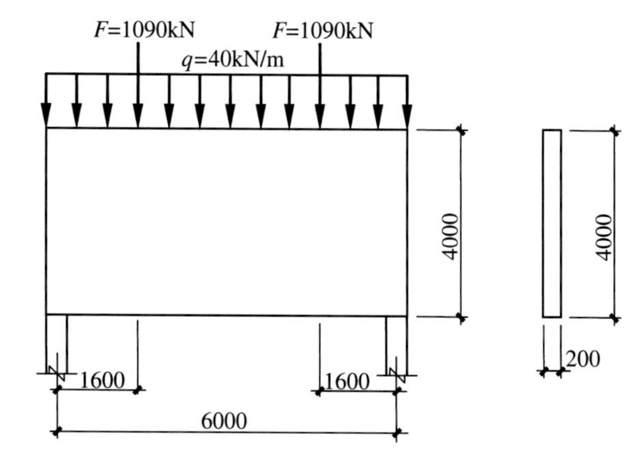

Fig. 5.45 Dimensions of a deep beam and the loads on it

Solution

$$M = 1090 \times 1.6 + \frac{40 \times 6^2}{8} = 1924\,\text{kN·m}$$

$$\frac{l_0}{h} = \frac{6}{4} = 1.5 < 2, \quad a_s = 0.1h = 0.1 \times 4000 = 400\,\text{mm}$$

$$h_0 = h - a_s = 4000 - 400 = 3600\,\text{mm}$$

From Eqs. (5.90), (5.91), and (5.92), it is obtained that

$$\alpha_d = 0.8 + 0.04\frac{l_0}{h} = 0.8 + 0.04 \times \frac{6}{4} = 0.86$$

$$\begin{cases} 1.0 \times 13.5 \times 200x = 290A_s \\ 1924 \times 10^6 = 1.0 \times 13.5 \times 200 \times 0.86x\left(3600 - \frac{x}{2}\right) \end{cases}$$

Solution of the equation set yields x = 238 mm < $0.2h_0$ = 720 mm. Taking x = 720 mm, we have

$$A_s = 6703\,\text{mm}^2 > 0.25\% \times 200 \times 4000 = 2000\,\text{mm}^2$$

Select 14ϕ25 (A_s = 6869 mm^2).

5.13 Ductility of Normal Sections of Flexural Members

Ductility reflects the deformation ability of reinforced concrete members. The ductility coefficient defined in Eq. (5.93) can be used to quantify the ductility of normal sections of flexural members.

$$\mu = \frac{\phi_u}{\phi_y} \tag{5.93}$$

where

ϕ_u ultimate curvature of a section; and

ϕ_y curvature at first yield of the tension steel.

The larger the ductility coefficient is, the better the ductility of a section is. And the better the ductility of a section is, the stronger the energy dissipation ability of a member is (Fig. 5.46).

In addition to material properties, the steel content greatly influences the ductility of a section. When two sections are identical in dimensions and material properties but different in the steel contents, the total area of the longitudinal bars is A_{s1} in one section and A_{s2} in another, and $A_{s1} < A_{s2}$.

As shown in Fig. 5.47a, because $A_{s1} < A_{s2}$, the compression zone depth of the section reinforced by A_{s1} is smaller than that reinforced by A_{s2}. So, $\phi_{y1} < \phi_{y2}$. Similarly, $\phi_{u1} > \phi_{u2}$ can be obtained from Fig. 5.47b. Substituting the two inequalities into Eq. (5.93), one can find that

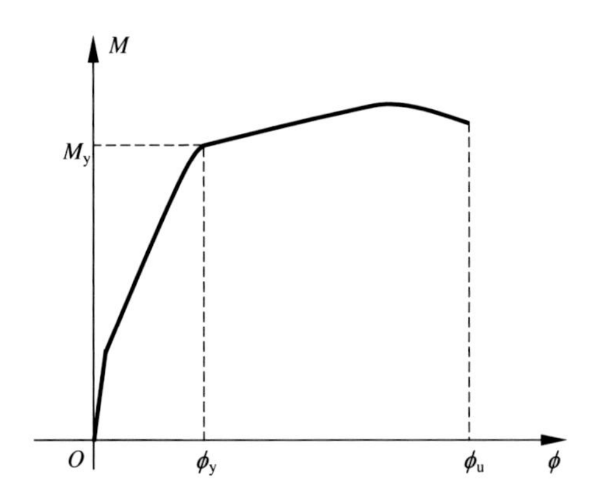

Fig. 5.46 Moment versus curvature diagram of a normal section

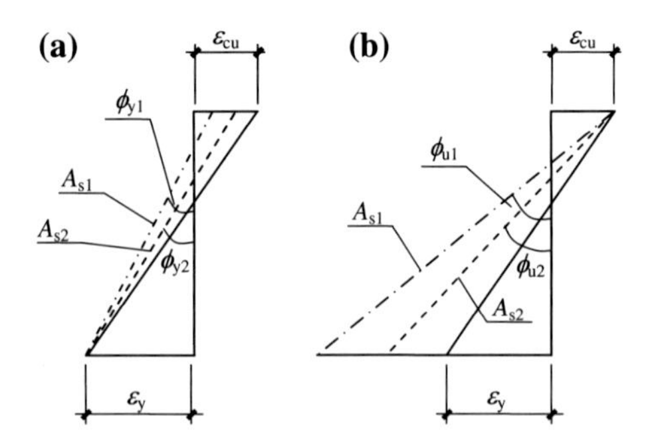

Fig. 5.47 Strain distribution across a normal section. **a** At first yield of the tension steel. **b** At crushing of the concrete

$$\mu_1 = \frac{\phi_{u1}}{\phi_{y1}} > \mu_2 = \frac{\phi_{u2}}{\phi_{y2}} \tag{5.94}$$

Equation (5.94) implies that the fewer the longitudinal bars are, the better the ductility is. This conclusion can also be seen from the moment–curvature diagrams of normal sections of different steel contents in Fig. 5.11. For under-reinforced beams, the minimum reinforcement ratio corresponds to the best ductility, and the balanced reinforcement ratio corresponds to the worst ductility.

Questions

5.1 What are the reinforcement details in reinforced concrete beams?
5.2 What are the reinforcement details in reinforced concrete slabs?
5.3 Why stipulate the minimum spacing and the minimum concrete cover depth for longitudinal bars in reinforced concrete beams and slabs?
5.4 What are the generally used diameters for longitudinal bars?
5.5 What are the flexural failure modes for reinforced concrete beams and how are they characterized?
5.6 What is a balanced failure?
5.7 What principle is used to determine the minimum reinforcement ratio of the longitudinal steel bars in reinforced concrete beams?
5.8 How do the flexural bearing capacity and the ductility of reinforced concrete beams change with the increase of the longitudinal bars?
5.9 What state does the extreme concrete tension fiber reach when the tension zone is assumed to be cracked?
5.10 How does one determine whether an under-reinforced concrete beam is failed by flexure?
5.11 How does one determine whether an over-reinforced concrete beam is failed by flexure?
5.12 Why can reinforced concrete flexural members be thought to conform to the plane section assumption?
5.13 How does one make the actual stress distribution of the concrete in the compression zone equivalent to a rectangular stress block?
5.14 How does one determine the depth of a balanced compression zone?
5.15 What is the contribution of the compression steels in reinforced concrete flexural members?
5.16 During the design of a doubly reinforced rectangular section, how does one ensure that the section fails at the same time as the longitudinal steels yield?
5.17 How does one calculate the flexural bearing capacity of a doubly reinforced rectangular section if $x < 2a'_s$?
5.18 Why should the effective width of the compression flange of a T section be specified?
5.19 How does one check the minimum reinforcement ratio of a Type I T section? Why?

5.20 If the longitudinal steel bars are uniformly distributed across a reinforced concrete rectangular section, is the flexural bearing capacity calculated by $M_u = A_s f_y \left(h_0 - \frac{x}{2}\right)$, and is that calculation consistent with the actual flexural bearing capacity? Why?

5.21 What are the failure modes of deep beams? What is the characteristic of each failure mode?

5.22 What are the reinforcement details in deep beams?

5.23 What is the ductility of a reinforced concrete flexural member?

5.24 How does the reinforcement ratio influence the ductility of a reinforced concrete flexural member?

Problems

5.1 A reinforced concrete beam has the section of $b = 250$ mm and $h = 6000$ mm. The concrete cover is c = 25 mm thick. The material properties are $f_c = 23$ N/mm^2, $f_t = 2.6$ N/mm^2, $E_c = 2.51 \times 10^4$ N/mm^2, $f_y = 357$ N/mm^2, and $E_s = 1.97 \times 10^5$ N/mm^2. Longitudinal steel bars of 3ϕ25 ($A_s = 1472$ mm^2) are placed in the tension zone. Calculate:

(1) σ_s, σ_c^t, and ϕ when M = 50 kN m;
(2) The cracking moment M_{cr} and the corresponding σ_s, σ_c^t, and ϕ_{cr}.

5.2 All the conditions are the same as those in Problem 5.1,

(1) Calculate the ultimate moment M_u and the corresponding ϕ_u;
(2) Calculate ultimate moment M_u by using the simplified method assuming equivalent rectangular stress distribution;
(3) Compare the results above.

5.3 All the conditions are the same as those in Problem 5.1, and calculate the flexural bearing capacity when the cross section is reinforced with 2ϕ12 and 10ϕ28 longitudinal bars, respectively.

5.4 Given a simply supported reinforced concrete slab with the effective span $l_0 = 2.8$ m. The slab is 90 mm thick and reinforced by ϕ10@200 longitudinal bars. The material properties are f_c = 11.9 N/mm^2, f_t = 1.6 N/mm^2, f_y = 280 N/mm^2, and E_s = 2.0 × 10^5 N/mm^2. Calculate the maximum load (per square meter) that the slab can sustain.

5.5 The dimensions of a simply supported reinforced concrete beam are b = 220 mm and h = 500 mm. The effective span l_0 is 6 m. The beam is subjected to a uniformly distributed load of q = 24kN/m (self-weight included) and is reinforced by 2ϕ22 + 2ϕ20 tension steel bars. The material properties are f_c = 13 N/mm^2, f_t = 1.2 N/mm^2, f_y = 365 N/mm^2, and E_s = 1.97 × 10^5 N/mm^2. Try to determine whether the beam is safe.

5.6 Given a rectangular section of width b = 200 mm and depth h = 500 mm. $f_c = 13$ N/mm^2, $f_t = 1.2$ N/mm^2, $f_y = 310$ N/mm^2, and $E_s = 1.97 \times 10^5$N/mm^2. When the bending moment M = 40, 60, 80, 100, 120, 140, 160, 180, 200, and 220 kN m, respectively, calculate the corresponding A_s. Plot the M-A_s diagram and discuss it.

5.7 Given a simply supported reinforced concrete slab of the effective span $l_0 = 1.92$ m and the thickness $h = 80$ mm. The slab is subjected to a uniformly distributed load of $q = 4$ kN/m (self-weight included). $f_c = 13$ N/mm^2, $f_t = 1.2$ N/mm^2, $f_y = 270$ N/mm^2, and $E_s = 1.90 \times 10^5$ N/mm^2. Find the required steel area of this slab.

5.8 A doubly reinforced rectangular section of $b \times h$=400 mm × 1200 mm is reinforced by 4ϕ28 compression steel bars and 12ϕ28 tension steel bars. $f_c = 14.3$ N/mm^2, $f_t = 1.43$ N/mm^2, $E_c = 3.0 \times 10^4$ N/mm^2, $f_y = 310$ N/mm^2, and $E_s = 1.97 \times 10^5$ N/mm^2. Calculate:

(1) the cracking moment M_{cr} and the corresponding ϕ_{cr};
(2) the ultimate moment M_u and the corresponding ϕ_u (general method);
(3) the ultimate moment M_u and the corresponding ϕ_u (simplified method).

5.9 A simply supported beam is shown in Fig. 5.48. Given $f_c = 16$ N/mm^2, $f_t = 1.5$ N/mm^2, $f_y = 365$ N/mm^2, and $E_s = 1.97 \times 10^5$ N/mm^2. Calculate the maximum uniformly distributed load q (self-weight included) that the beam can sustain.

5.10 A two-span continuous beam is shown in Fig. 5.49. Given that $f_c = 13$ N/mm^2, $f_t = 1.2$ N/mm^2, $f_y = 370$ N/mm^2, and $E_s = 1.97 \times 10^5$ N/mm^2, calculate the maximum load P that the beam can sustain (neglect the self-weight of the beam and calculate the moments in an elastic way).

5.11 A doubly reinforced section of $b \times h = 200$ mm × 500 mm is subjected to the bending moment of $M = 250$ kN m. Given that $f_c = 16$ N/mm^2, $f_t = 1.5$ N/mm^2, $f_y = 320$ N/mm^2, and $E_s = 1.97 \times 10^5$ N/mm^2, find A_s and A'_s.

5.12 A doubly reinforced section of $b = 250$ mm and $h = 600$ mm is subjected to $M = 330$ kN m. $f_c = 13$ N/mm^2, $f_t = 1.2$ N/mm^2, $f_y = 310$ N/mm^2, and $E_s = 1.97 \times 10^5$ N/mm^2. If 3ϕ22 longitudinal bars have already been placed in the compression zone, find A_s.

5.13 A doubly reinforced section of $b = 250$ mm and $h = 600$ mm is subjected to $M = 390$ kN m. $f_c = 13$ N/mm^2, $f_t = 1.2$ N/mm^2, $f_y = 370$ N/mm^2, and $E_s = 1.97 \times 10^5$ N/mm^2. If 2ϕ14 longitudinal bars have already been placed in the compression zone, find A_s.

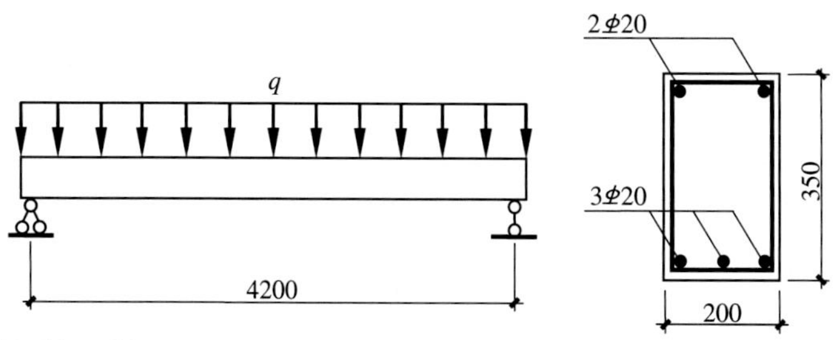

Fig. 5.48 Problem 5.9

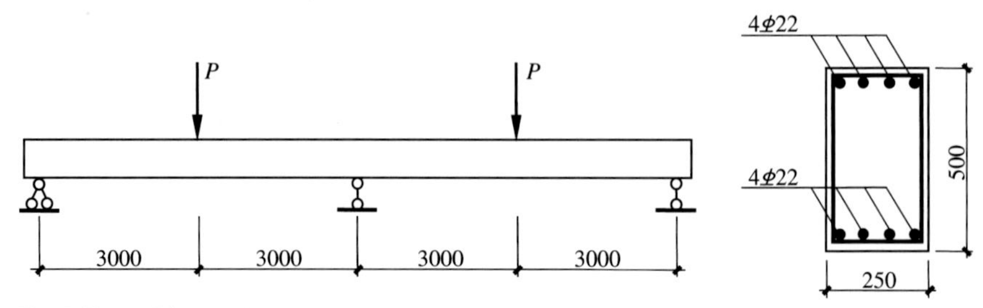

Fig. 5.49 Problem 5.10

5.14 Given a T section beam where $b_f' = 280$ mm, $h_f' = 120$ mm, $b = 180$ mm, $h = 450$ mm, the effective span is $l_0 = 5$ m, and it is reinforced with 3⌀25 steel bars. $f_c = 13$ N/mm^2, $f_t = 1.2$ N/mm^2, $f_y = 310$ N/mm^2, $E_s = 1.97 \times 10^5$ N/mm^2. Try to calculate the flexural bearing capacity.

5.15 The T section of a reinforced concrete beam has the dimensions of $b_f' = 2500$ mm, $h_f' = 180$ mm, $b = 400$ mm, and $h = 1200$ mm and is reinforced by 12⌀25 tension steel bars. The effective span of the beam is $l_0 = 12$ m. $f_c = 14.3$ N/mm^2, $f_t = 1.4$ N/mm^2, $f_y = 370$ N/mm^2, and $E_s = 1.97 \times 10^5$ N/mm^2. Calculate the flexural bearing capacity of the section.

5.16 Given that $f_c = 14.3$ N/mm^2, $f_t = 1.4$ N/mm^2, $f_y = 370$ N/mm^2, and $E_s = 1.97 \times 10^5$ N/mm^2, calculate the flexural bearing capacities of the three sections shown in Fig. 5.50.

5.17 A simply supported beam is shown in Fig. 5.51. Given that $f_c = 13$ N/mm^2, $f_t = 1.2$ N/mm^2, $f_y = 310$ N/mm^2, and $E_s = 1.97 \times 10^5$ N/mm^2, find A_s.

5.18 The section of a precast hollow slab with the effective span of $l_0 = 3.6$ m is shown in Fig. 5.52. Given that $f_c = 14$ N/mm^2, $f_t = 1.4$ N/mm^2, $f_y = 270$ N/mm^2, and $E_s = 2.0 \times 10^5$ N/mm^2, calculate the flexural bearing capacity of the slab (Hint: First transform the circular holes to rectangular holes with the centroid, with area and then with inertia moment unchanged, follow Example 5.9).

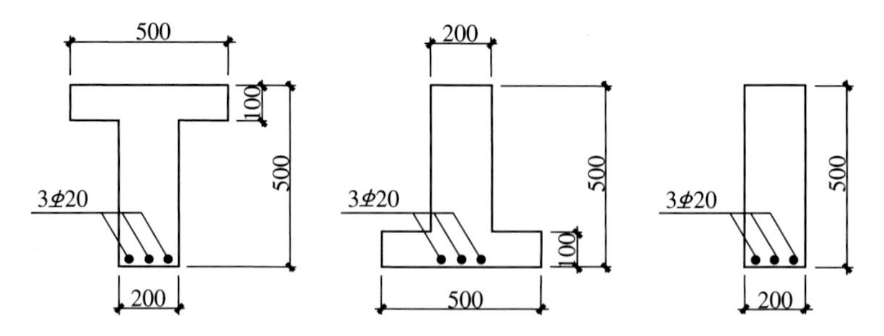

Fig. 5.50 Problem 5.16

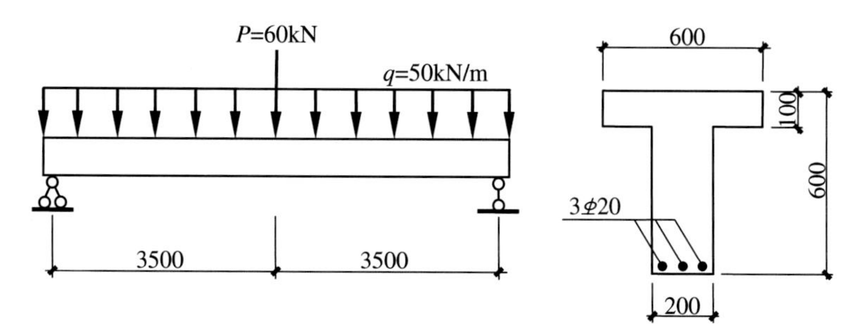

Fig. 5.51 Problem 5.17

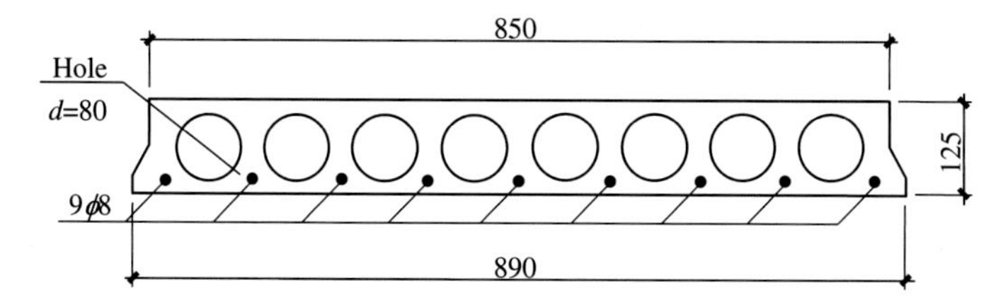

Fig. 5.52 Problem 5.18

5.19 The section of a simply supported deep beam is of $b \times h = 310$ mm × 6200 mm. The beam has an effective span of $l_0 = 6$ m and is uniformly reinforced with 16⌀20 longitudinal tension steel bars. $f_c = 20.1$ N/mm^2, $f_t = 2.1$ N/mm^2, $f_y = 390$ N/mm^2, and $E_s = 1.97 \times 10^5$ N/mm^2. Calculate the flexural bearing capacity of the deep beam.

Chapter 6
Compression and Tension Behavior of Eccentrically Loaded Members

6.1 Engineering Applications and Reinforcement Detailing

In an eccentrically loaded member, the axial compressive load N_c or axial tension load N_t is applied at an eccentricity e_0 on the member (Fig. 6.1a), which is equivalent to the combination of an axial load and bending moment $M = N_c e_0$ or $M = N_t e_0$ (Fig. 6.1b). The eccentrically loaded members include the eccentrically compressed members and the eccentrically tensioned members.

Eccentrically compressed members are commonly used structural members in reinforced concrete structures. Bent columns in single-story industrial buildings (Fig. 6.2a), columns in reinforced concrete frame structures (Fig. 6.2b), upper chords in arched roof trusses (Fig. 6.2a), shear walls in high-rise buildings (Fig. 6.2c), main arches in arch bridges (Fig. 6.2f), and bridge piers are all examples of eccentrically compressed members.

If a load (e.g., hung weight) is directly applied on the lower chord of a truss, the chord will behave as an eccentrically tensioned member, because it is subjected to the combination of axial tension and bending moment. Besides, walls in water pools (Fig. 6.2d) and silos (Fig. 6.2e) are also the examples of eccentrically tensioned members.

Usually, the sections of eccentrically loaded members are of rectangular shape. Circular sections, I sections, T sections, L sections, and cross sections can also be used. Rectangular and circular sections are easy to construct but waste materials compared with I or T sections. If the flanges of I or T sections are too thin, they will prematurely crack, which influences the durability and ultimate bearing capacity of the sections. Further consideration of the stabilities of the flanges and the webs dictate that the minimum thicknesses of flanges and webs should not be lower than 120 and 100 mm, respectively. In seismic zones, the thicknesses of flanges and webs should be increased.

X. Gu et al., *Basic Principles of Concrete Structures*,
DOI 10.1007/978-3-662-48565-1_6

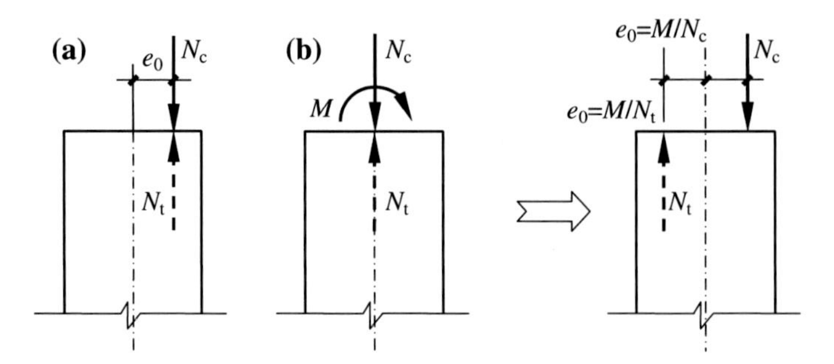

Fig. 6.1 Eccentrically loaded members. **a** An eccentrically compressed (or tensioned) member and **b** equivalent loading

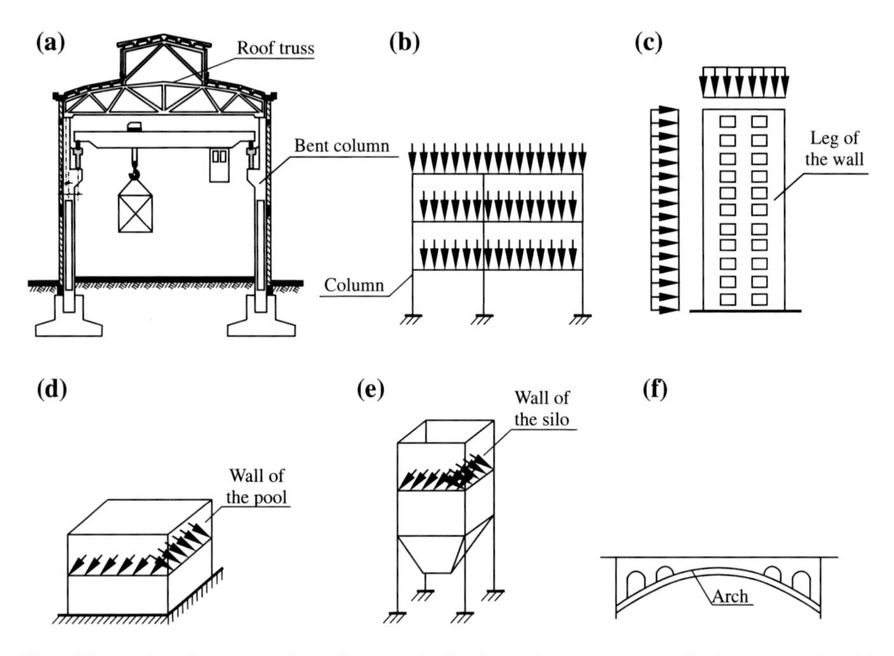

Fig. 6.2 Engineering examples of eccentrically loaded members. **a** Single-story industrial building, **b** frame, **c** reinforced concrete shear wall, **d** water pool, **e** silo, **f** arch bridge

In an eccentrically loaded member, longitudinal steel bars and closed ties are generally arranged (Figs. 6.3, 6.4). The longitudinal steel bars are placed in the moment direction or evenly distributed around the perimeter of a section. The ties are arranged according to the shape of the section and the number and positions of the longitudinal steel bars. Additional ties should be placed if the dimension of a section on the longitudinal bars side is larger than 400 mm and more than three longitudinal bars are used, or if more than four longitudinal bars are used, although the dimension is less than 400 mm. The ties in eccentrically loaded members can

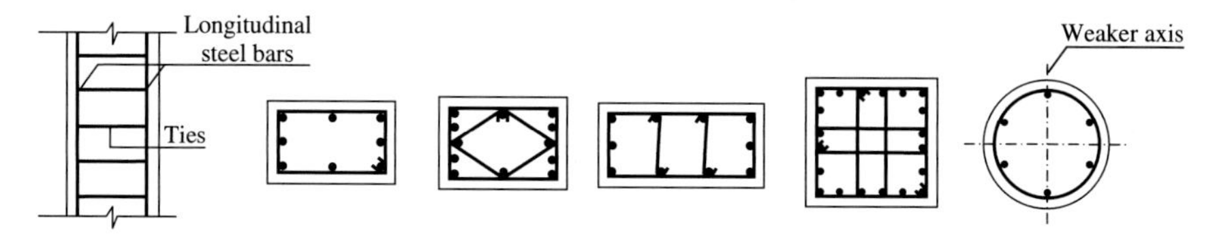

Fig. 6.3 Arrangement of longitudinal reinforcement and ties in rectangular and circular sections

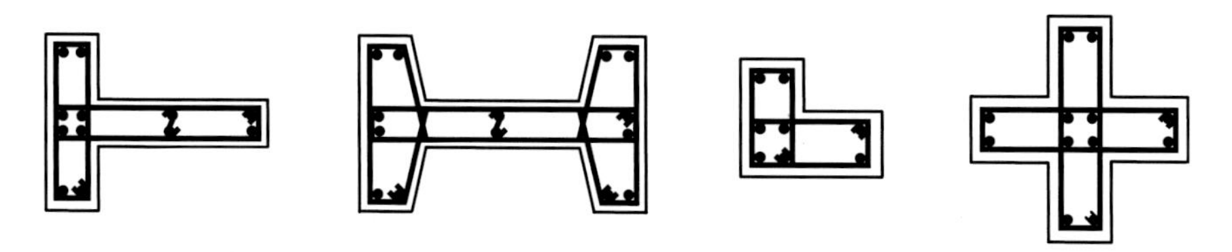

Fig. 6.4 Arrangement of longitudinal reinforcement and ties in complicated sections

also help the concrete to resist lateral shear forces even though these ties have the same functions when attached to axially loaded members. The basic requirements on the reinforcement, the pitch and the concrete cover depth in an eccentrically loaded member are the same as those in an axially loaded member.

6.2 Interaction Diagram

To conceptually illustrate the interaction between moment and axial loads in an eccentrically loaded member, an idealized homogeneous and elastic eccentrically loaded member with a compressive strength f_c and tensile strength f_t will be discussed in this section.

If $f_t = f_c$, for an eccentrically loaded elastic member, as shown in Fig. 6.5, failure would occur in compression when the maximum compressive stress reached f_c, as given by

$$\frac{N_c}{A} + \frac{M}{W} = f_c \tag{6.1}$$

where

A area of the cross section;

W modulus of the cross section, which is defined as the ratio of the moment of inertia of the section I to the distance from the centroidal axis to the most highly compressed surface x;

N_c axial load, positive in compression; and

M moment, positive as shown in Fig. 6.5.

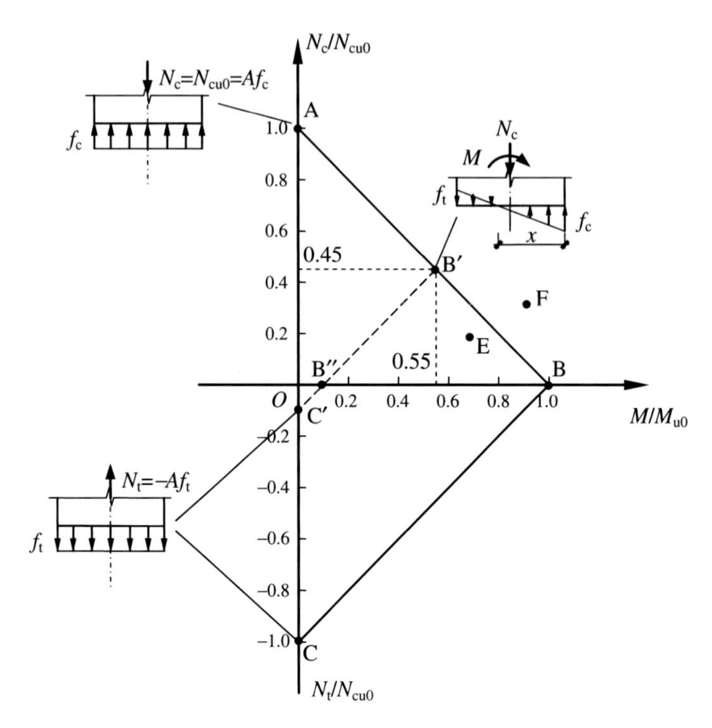

Fig. 6.5 Interaction diagram for an elastic eccentrically loaded member when $f_t = f_c$ or $f_t = 0.1f_c$

Dividing both sides of Eq. (6.1) by f_c gives

$$\frac{N_c}{Af_c} + \frac{M}{Wf_c} = 1 \tag{6.2}$$

The maximum axial load a column can bear occurs when $M = 0$ and $N_{cmax} = N_{cu0} = Af_c$. Similarly, the maximum moment that can be borne occurs when $N_c = 0$ and $M_{max} = M_{u0} = Wf_c$. Substituting N_{cu0} and M_{u0} for Af_c and Wf_c respectively gives

$$\frac{N_c}{N_{cu0}} + \frac{M}{M_{u0}} = 1 \tag{6.3}$$

This equation shows the interaction between axial load and moment at failure. It is plotted as line AB in Fig. 6.5. A similar equation for a tensile load N_t and moment M governed by f_t gives the line of BC in Fig. 6.5. Lines AB and BC in Fig. 6.5 are referred to as an interaction diagram. Points on the lines represent the bearing capacities of the eccentrically loaded member or combinations of N_c (or N_t) and M corresponding to the failure of the section. A point inside the diagram, such as E, in

Fig. 6.5 represents a combination of N_c and M that will not cause failure. Combinations of N_c (or N_t) and M falling on the lines or outside the lines, such as F in Fig. 6.5, will equal or exceed the bearing capacity of the section and cause failure.

If $f_t = 0.1 f_c$ for an eccentrically loaded elastic member, the interaction diagram becomes Lines AB′ and B′C′ as shown in Fig. 6.5. Line AB′ indicates load combinations corresponding to failure initiated by compression (governed by f_c), while Line B′C′ indicates load combinations corresponding to failure initiated by tension (governed by f_t). Point B′ in Fig. 6.5 represents a balanced failure, in which the tensile strength and compressive strength of the material are reached simultaneously on opposite edges of the section.

Concrete is not elastic and has a tensile strength that is much lower than compressive strength. However, the tensile resistance of a reinforced concrete member is enhanced by reinforcing steel bars. For these reasons, the calculation of an interaction diagram for an eccentrically loaded reinforced concrete member is more complex than for an elastic member. However, the general shape resembles Curve AB′C′ shown in Fig. 6.5.

6.3 Experimental Studies on Eccentrically Compressed Members

6.3.1 Experimental Results

Figure 6.6 shows the result of a short column with a large eccentricity subjected to a compressive load (the column is called the large eccentrically compressed column for short). When the applied load is small, the column is in elastic stage. The

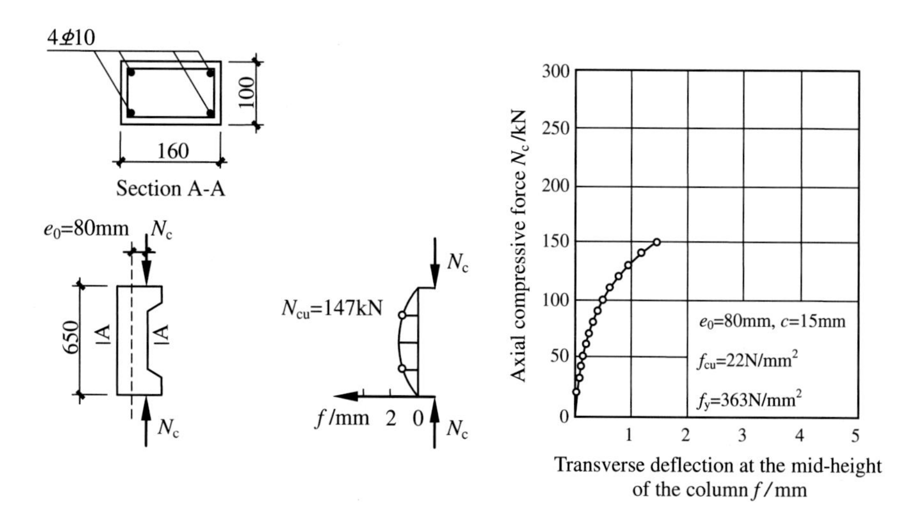

Fig. 6.6 Experimental results of an eccentrically loaded short column with a large eccentricity

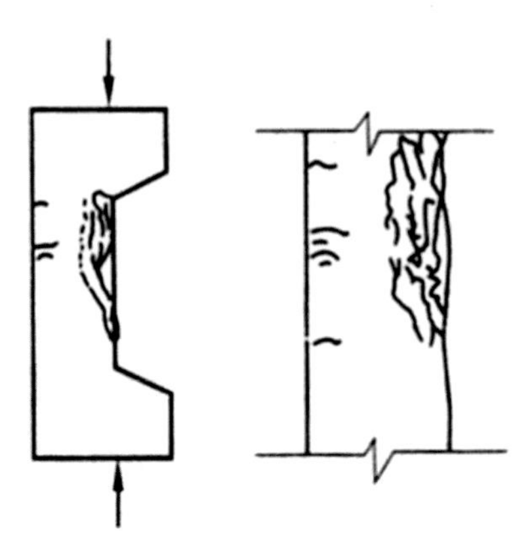

Fig. 6.7 Typical failure mode of a large eccentrically compressed column

stresses in both the concrete and the steel bars are very small, and the transverse deflection at a column's mid-height varies linearly with the load. With the increase of the load, transverse cracks appear on the surface of the concrete in the tension zone and the strain and stress in the longitudinal bars farthest from the load increase at a faster rate. More and more cracks appear in the tension zone and propagate toward the compression zone. The depth of the compression zone gradually decreases, and the compressive stress in the concrete increases. When the strain in the longitudinal steel bars farthest from the load reaches the yield strain, the steels are yielding and a principal crack can be observed. When the strain in the extreme compression fiber has reached the maximum compressive strain of the concrete, longitudinal cracks appear in the compression zone, and the concrete is crushed, which signifies section failure. Meanwhile, the longitudinal steel bars close to the load are also yielding under compression. The failure mode is shown in Fig. 6.7. The crushed concrete can be seen as an approximate triangular shape.

Figure 6.8 presents the result of a short column with a small eccentricity subjected to a compressive load (the column is called the small eccentric compression column for short). With the increase of the load, the compressive stress in the concrete close to the load continuously increases to compressive strength when the concrete is crushed. Meanwhile, the longitudinal bars close to the load are also yielding under compression, while their counterparts are less stressed. At failure, the compression zone depth is large. The cracking load is very near the ultimate one. No apparent warning is given before failure. Compared to those with large eccentrically compressed members, the transverse deflections of small eccentrically compressed members are much smaller. The failure mode is shown in Fig. 6.9.

From the experimental results shown in Figs. 6.6 and 6.8, it can be concluded that:

1. The ultimate bearing capacity of an eccentrically compressed short column decreases with the increase of eccentricity.

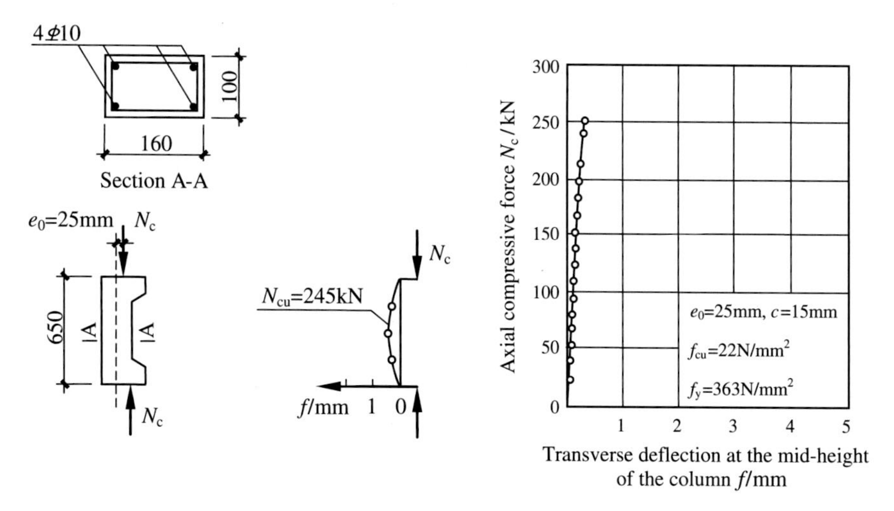

Fig. 6.8 Experimental results of an eccentrically loaded short column with a small eccentricity

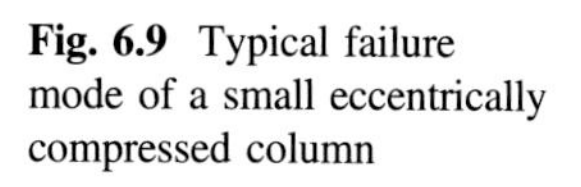

Fig. 6.9 Typical failure mode of a small eccentrically compressed column

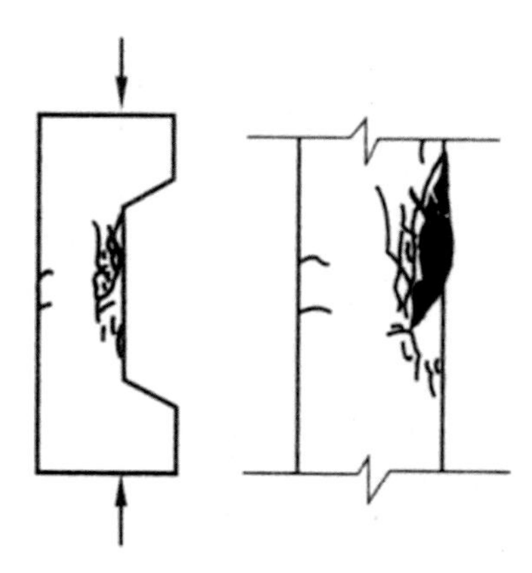

2. With the increase of eccentricity, two different kinds of failure modes can be observed for an eccentrically loaded member:

   (1) *Compression failure* (or *small eccentric compression failure*): The failure is initiated from the crushing of concrete in compression. The longitudinal steel bars farthest from the load are either in tension or in compression at failure. If the longitudinal steel bars are in tension but not yielding, the failure phenomenon of the member is similar to that of an axially loaded member;

   (2) *Tension failure* (or *large eccentric compression failure*): The failure is initiated from the tensioned yielding of the longitudinal steel bars farthest from the load. After the longitudinal steel bars yield, the principal crack propagates continuously, causing the concrete in the compression zone to be

further stressed until finally crushed. The whole failure process of the member is similar to that of a doubly reinforced beam section.

(3) There is obviously a balanced failure between the compression failure and the tension failure, which is characterized by the simultaneous occurrence of the concrete crushing and the longitudinal steel bars' yielding.

(4) The plane section assumption is still valid (in terms of the average strain) in the normal section analysis of eccentrically loaded columns.

6.3.2 Analysis of Failure Modes

The failure mode of an eccentrically loaded member is determined not only by the eccentricity, but also by the steel contents. When e_0 is very small and A_s is appropriate, the full section of the member will be under compression. A_s' can yield at failure, but A_s generally cannot. However, if A_s is much smaller than A_s', although the geometrical centroid lies between N_c and A_s, the physical centroid may lie between N_c and A_s', causing the member to fail by the crushing of the A_s side concrete (Fig. 6.10a). When e_0 is small and A_s is appropriate, A_s may be under tension but will not yield at failure (Fig. 6.10b). When both e_0 and A_s are large, A_s may still not yield at failure because of too many tensioned steel bars (Fig. 6.10c). When e_0 is large and A_s is appropriate, the big tensile stress in A_s will yield first and the concrete in the compression zone is subsequently crushed (Fig. 6.10d). From the failure characteristics of the members shown in Fig. 6.10, it can be concluded that the members in Fig. 6.10a–c are in compression failure, while the member in Fig. 6.10d is in tension failure. The member in Fig. 6.10a more closely resembles an axially loaded member, and the member in Fig. 6.10d more closely resembles a flexural member.

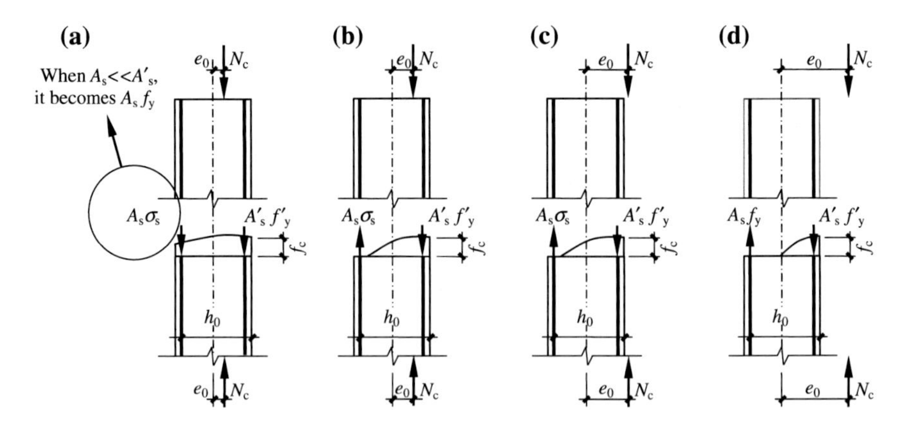

Fig. 6.10 Failure modes of eccentrically compressed members with different eccentricities and steel contents. **a** e_0 is very small, and A_s is appropriate, **b** e_0 is small, **c** e_0 is large, and A_s is large too, **d** e_0 is large, and A_s is appropriate

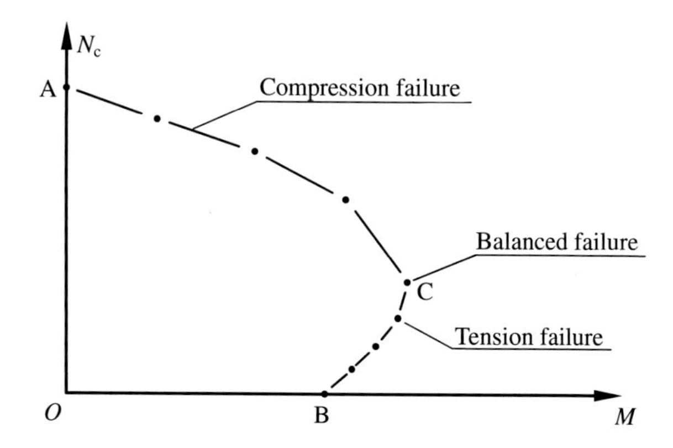

Fig. 6.11 Interaction diagram of N_{cu} and M_u

6.3.3 N_{cu}–M_u Interaction Diagram

By testing a series of specimens, which are identical in cross-sectional dimensions, height, reinforcement, and materials under compression of different eccentricities, the ultimate moments M_u and the ultimate axial forces N_{cu} can be obtained in pairs. The N_{cu}–M_u interaction diagram shown in Fig. 6.11 is just the plot of the results with an ACB curve. Even though the interaction relationship between N_{cu} and M_u is not linear, the shape of the ACB curve is very similar to that of Lines AB′ and B′B″ in Fig. 6.5. The diagram shown in Fig. 6.11 illustrates that in small eccentric compression failure, the flexural bearing capacity of the member decreases with the increase of the axial force, while in large eccentric compression failure, the trend is on the contrary. The latter is because the tensile stress caused by the moment is partially counteracted by the axial force-induced compressive stress and the failure is delayed. In a balanced failure, point C in Fig. 6.11, the flexural bearing capacity of the member reaches the maximum. In Fig. 6.11, point A ($M = 0$) represents the axial compression failure, point B ($N_c = 0$) represents the flexural failure, and point C represents the balanced failure. It can be seen that the flexural bearing capacity of the member in tension failure (curve BC) is larger than that in pure bending failure, while the ultimate axial force in compression failure (curve AC) is smaller than that in axial compression failure.

6.3.4 Slenderness Ratio Influence on Ultimate Capacities of Members

Any eccentrically compressed member undergoes lateral deflection f. Therefore, the bending moment actually sustained by an eccentrically loaded member is $M = N_c (e_0 + f)$ rather than the initial value $M_0 = N_c \cdot e_0$. This kind of deformation-induced internal force increase is called the P-Δ effect or the secondary effect.

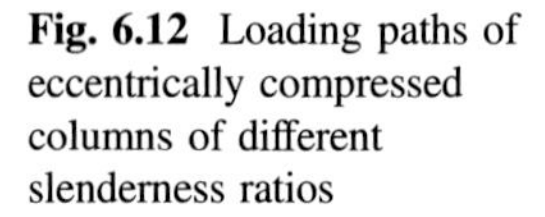

Fig. 6.12 Loading paths of eccentrically compressed columns of different slenderness ratios

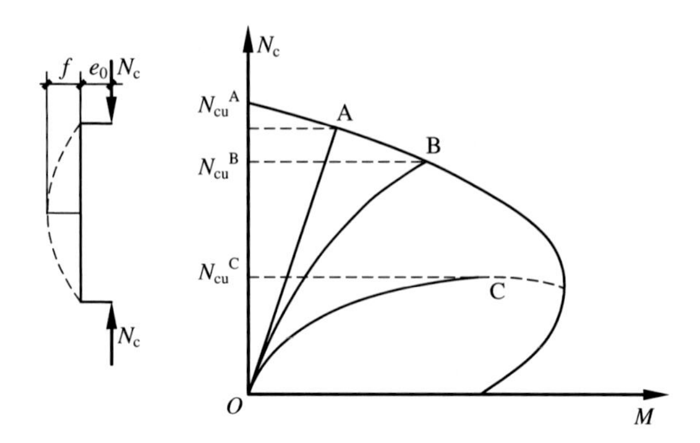

The P-Δ effect can be generally neglected in short columns, but it must be considered in slender columns. Whether a column is a short one or a slender one can be determined by the slenderness ratio, which is defined as the ratio of the effective length l_0 to the sectional dimension (h or d_c) in the direction of the moment. When $l_0/h \leqslant 5$ (for rectangular, T, and I sections) or $l_0/d_c \leqslant 5$ (for circular sections), it is a short column. When $5 < l_0/h$ (or l_0/d_c) $\leqslant 30$, it is a slender column. And when l_0/h (or l_0/d_c) > 30, it is a very slender column.

For members of the same sectional dimensions, materials, and reinforcement, their bearing capacities are different if their effective lengths are not the same. Figure 6.12 presents the loading paths of three eccentrically compressed specimens with different slenderness ratios. The envelope in the figure is the N_{cu}–M_u interaction diagram for short columns.

For the short column, whose slenderness ratio is very small, when the sectional moment increases from $N_c e_0$ to $N_c(e_0 + f)$, the secondary moment ΔM is so small (because $f < e_0$) that it can be ignored. Hence, its loading path from beginning to failure can be well represented by the straight line OA. When the line OA intersects the envelop curve at point A, compression failure happens to the column and the maximum compressive load is N_{cu}^{A}. The materials at the critical section can reach their strengths, so it is a material failure.

For the slender column, large lateral deflection leads to a large secondary moment. The interaction between the deflection and the moment results in the nonlinear relationship between N_c and M as shown by the line OB in Fig. 6.12. It can be seen that the maximum compressive load N_{cu}^{B} apparently reduces due to the secondary moment increase. The materials at the critical section can reach their strengths at the failure, so it is still a material failure in terms of the failure characteristic.

For the very slender column, P-Δ effect is more obvious. When the column is loaded to the maximum compressive force, the failure point C does not intersect the N_{cu}–M_u interaction diagram, as shown by the line OC. The stresses in the materials at the critical section calculated by equilibrium conditions are far less than their strengths at failure, so it is actually a stability failure.

6.4 Two Key Issues Related to Analysis of Eccentrically Compressed Members

6.4.1 Additional Eccentricity e_a

Due to the uncertainty of the loading position, nonuniformity of the concrete material, asymmetry of the reinforcement, and the construction error, the actual eccentricity may not be equal to the ideal eccentricity $e_0 = M/N_c$. Even for axially loaded members, $e_0 = 0$ is impossible. The increase of the eccentricity will enlarge the moment M. To consider this adverse influence, the following equation is used to calculate the initial eccentricity:

$$e_i = e_0 + e_a \tag{6.4}$$

where

e_i initial eccentricity;
e_0 eccentricity of the axial compressive force, $e_0 = M/N_c$; and
e_a additional eccentricity.

Different codes take different values of e_a. GB 50010 stipulates that e_a is equal to the maximum between 20 mm and 1/30 of the maximum sectional dimension in the bending direction.

6.4.2 Moment Magnifying Coefficient

As discussed above, the P-Δ effect must be taken into account for slender and very slender columns. Introducing the magnifying coefficient is an effective way to accomplish this purpose.

Assume $N_c(e_i + f) = \eta_s N_c e_i$, then

$$\eta_s = 1 + \frac{f}{e_i} \tag{6.5}$$

where the moment magnifying coefficient η_s is the ratio of the total moment $N_c(e_i + f)$, which has considered the P-Δ effect, to the initial moment $N_c e_i$. f is the lateral deflection caused by the initial moment $N_c e_i$. It can be seen that the key to determine η_s is the calculation of f. Experimental results show that the deflection curve of an eccentrically compressed simply supported column can be approximately assumed as a sinusoid (Fig. 6.13), which can be expressed as

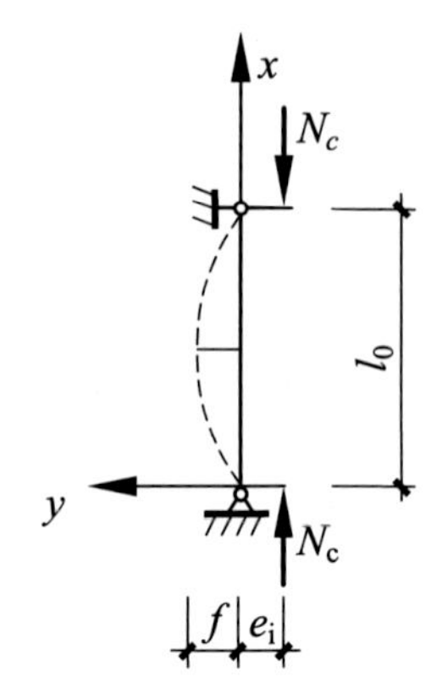

Fig. 6.13 Deflection curve of a column

$$y = f\sin\frac{\pi x}{l_0} \tag{6.6}$$

The curvature of the deflection curve is approximately

$$\phi = \frac{M}{EI} = -\frac{\mathrm{d}^2 y}{\mathrm{d}x^2} \tag{6.7}$$

Differentiating Eq. (6.6) twice, and noting that when $x = \frac{l_0}{2}$, y takes the maximum value f, we have

$$\phi = f \cdot \frac{\pi^2}{l_0^2} \approx 10 \cdot \frac{f}{l_0^2} \tag{6.8}$$

According to the plane section assumption shown in Fig. 6.14, it can be expressed as

$$\phi = \frac{\varepsilon_\mathrm{c} + \varepsilon_\mathrm{s}}{h_0} \tag{6.9}$$

When the balanced failure occurs, the curvature of the critical section is

$$\phi_\mathrm{b} = \frac{K\varepsilon_\mathrm{cu} + \varepsilon_\mathrm{y}}{h_0} \tag{6.10}$$

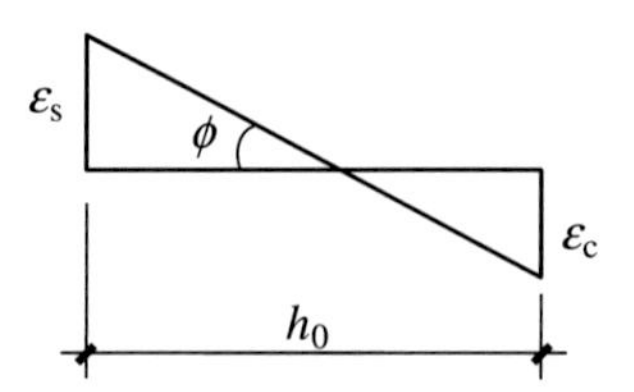

Fig. 6.14 Strain distribution of the critical section

where K is the compressive strain magnifying coefficient considering concrete creep under long-term load, usually K = 1.25. For mild steel, it is acceptable to take $\varepsilon_y = f_y/E_y \approx 0.0020$. When the cube strength of concrete is not larger than 50 N/mm^2, $\varepsilon_{cu} = 0.0033$. Then,

$$\phi_b = \frac{1.25 \times 0.0033 + 0.0020}{h_0} = \frac{1}{163.3h_0} \tag{6.11}$$

Actually, balanced failure cannot happen to all eccentrically compressed members. And the bending moment at the critical section of an eccentrically compressed member under any compression failure or tension failure is always less than the corresponding values under balanced failure. This rule also holds true for the curvature. By modifying ϕ_b with the above consideration, the curvature corresponding to the maximum bending moment can be obtained as

$$\phi = \phi_b \zeta_c = \frac{1}{163.3h_0}\zeta_c \tag{6.12}$$

where ζ_c is the modifying coefficient.

Substituting Eq. (6.12) into Eq. (6.8) and rearranging the equation, we have

$$f = \frac{1}{1633} \cdot \frac{l_0^2}{h_0} \cdot \zeta_c \tag{6.13}$$

By substituting Eq. (6.13) into Eq. (6.5) and assuming $h = 1.1h_0$, it can be obtained that

$$\eta_s \approx 1 + \frac{1}{1300 \cdot \frac{e_i}{h_0}} \cdot \left(\frac{l_0}{h}\right)^2 \cdot \zeta_c \tag{6.14}$$

where l_0 is the effective length of an eccentrically compressed member. How to determine the effective length will be left to the structural design course; h_0 is the effective depth of a section and can be determined similarly as that for flexural structural members.

GB 50010 adopts Eq. (6.14) to calculate the moment magnifying coefficient η_s for eccentrically compressed members, and based on experimental results, it takes

$$\zeta_c = \frac{0.5 f_c A}{N_c} \tag{6.15}$$

where A is the sectional area of the member. $A = bh + 2(b'_f - b)h'_f$ for T section and I section. $\zeta_c = 1.0$ if $\zeta_c > 1.0$.

If $l_0/h \leqslant 5$, $l_0/d_c \leqslant 5$ or $l_0/i \leqslant 17.5$, $\eta_s = 1.0$.

One should refer to relevant codes on how to consider P-Δ effect for concrete structures in highways, railways, and bridges. However, the calculation principle is the same as discussed herein.

6.5 Analysis of Eccentrically Compressed Members of Rectangular Section

When balanced failure happens to an eccentrically compressed member, the steel bars farthest away from the axial load are yielding, and at the same time, the concrete on the opposite side is crushed. The failure phenomenon is nearly the same as that of a large eccentric compression failure, hence, the balance failure can also be included in large eccentric compression failure. Furthermore, eccentrically compressed members can be classified as large eccentric compression members and small eccentric compression members.

In a large eccentric compression member, the steel bars farthest away from the axial load yield first. Then, the concrete in the compression zone is crushed. The stress distribution and the failure mode at the critical section are similar to those of a doubly under-reinforced rectangular section under flexure. Hence, the analysis method for both large eccentric compression sections and under-reinforced rectangular sections is the same.

In a small eccentric compression member, the steel bars farthest away from the axial load are generally subjected to very small tensile stress or even compressive stress if N_c is very large and e_0 is very small. The section under small eccentric compression has a similar failure phenomenon as a doubly over-reinforced rectangular section under flexure. But the loading state of a small eccentric compression section depends not only on the bending moment on the section, but also on the magnitude of the axial force. In other words, the steel bars farthest away from the axial force cannot yield under tension even after the concrete close to the axial force has been crushed, which is the characteristic of compression failure. So limiting the steel ratio cannot prevent compression failure from happening to a small eccentric compression section.

The whole loading process from the onset to the failure of eccentrically compressed members can be analyzed in a way similar to the analysis of flexural members. But in real engineering, the ultimate capacity is the primary concern, so this chapter will focus on the calculation method of the ultimate capacity of the eccentrically compressed method.

6.5.1 Ultimate Bearing Capacities of Large Eccentrically Compressed Sections

From the failure mode of a large eccentrically compressed section, the stress and strain distributions across the section at failure are illustrated in Fig. 6.15. In Sect. 6.3, the moment magnifying coefficient is introduced to consider the P-Δ effect. From Fig. 6.1, it can be seen that amplifying the moment is equivalent to amplifying the eccentricity for eccentrically compressed members. So to make the calculation diagram clear, amplifying the eccentricity e_i to $\eta_s e_i$ is adopted in Fig. 6.15.

The strain and stress in the extreme compression fiber reach the ultimate values ε_{cu} and f_c, respectively. According to the plane section assumption, the strain is linearly distributed across the section, thus, the stress distribution in the compression zone is similar to the stress–strain curve of the concrete with the starting point at the neutral axis.

When a member starts to fail, the concrete far away from the axial force has already cracked. Although a small portion of concrete near the neutral axis is uncracked, it can only sustain few tensile forces. Therefore, the tension resistance of the concrete is often neglected.

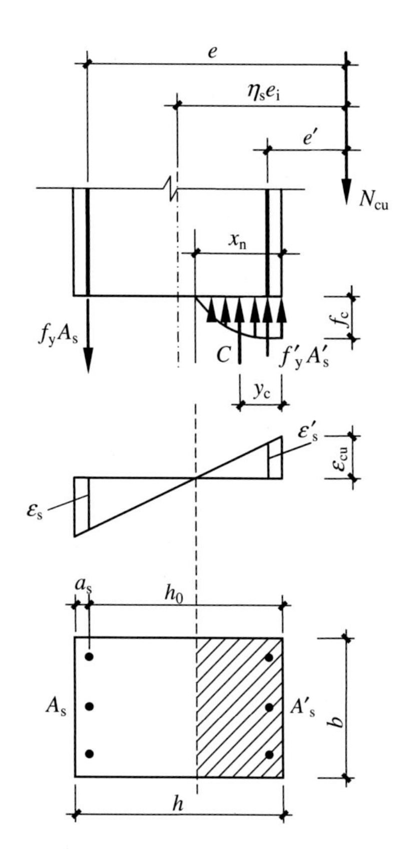

Fig. 6.15 Stress and strain distributions of a large eccentrically compressed section

The steel bars farthest away from the axial force have yielded at failure with the stress equaling to the yield strength f_y. The stress of the steel bars in the compression zone depends on the compression zone depth x_n.

From the linear strain distribution shown in Fig. 6.15, we have

$$\frac{\varepsilon_{cu}}{x_n} = \frac{\varepsilon'_s}{x_n - a'_s} \tag{6.16}$$

$$x_n = \frac{\varepsilon_{cu}}{\varepsilon_{cu} - \varepsilon'_s} \cdot a'_s \tag{6.17}$$

where

ε_{cu} ultimate compressive strain of the concrete;

ε'_s strain of the compression steels; and

a'_s distance from the centroid of the compression steels to the extreme compression concrete fiber.

If the cube strength of the concrete is expressed as $f_{cu} \leqslant 50$ N/mm^2, the ultimate compressive strain of the concrete can be taken as $\varepsilon_{cu} = 0.0033$. If the compression steels are assumed to yield, their yield strain can be determined; hence, $\varepsilon'_s = \varepsilon'_y = 0.0017$ for HRB335. Substituting the above values into Eq. (6.17), we get $x_n = 2.06\ a'_s$, which means that if $x_n \geqslant 2.06\ a'_s$, the compression steels can yield at failure. For other kinds of steels, the conclusion is still applicable. $x_n \geqslant 2.06\ a'_s$ can be generally satisfied in large eccentric compression sections, so it is valid to assume the yielding of the compression steels in the analysis of large eccentrically compressed sections.

From the force equilibrium $\Sigma N = 0$,

$$N_{cu} = \int_0^{x_n} \sigma_c b \mathrm{d}x + f'_y A'_s - f_y A_s \tag{6.18}$$

From the moment equilibrium $\Sigma M = 0$,

$$N_{cu} e = \int_0^{x_n} \sigma_c b \mathrm{d}x (h_0 - x) + f'_y A'_s (h_0 - a'_s) \tag{6.19}$$

where

N_{cu} ultimate axial compressive capacity of the section;

f'_y, f_y yield strengths of tension and compression steels, respectively;

A'_s, A_s sectional areas of compression and tension steels, respectively; and

e distance from the acting point of the axial force N_{cu} to the centroid on the tension steels, $e = \eta_s e_i + \frac{h}{2} - a_s$.

The distance from the acting point of the resultant compressive force C to the extreme compression fiber can be calculated as

$$y_c = \frac{\int_0^{x_n} \sigma_c bx \mathrm{d}x}{\int_0^{x_n} \sigma_c b \mathrm{d}x} \tag{6.20}$$

Similar to a flexural section, the strains of the concrete vary along the compression zone depth of a large eccentrically compressed section. Less strained fibers will restrain more strained ones. So the stress–strain relationship of the concrete shown in Fig. 5.14 and formulated in Eq. (2.7) can be used in the analysis. If the cube strength of the concrete is not larger than 50 MPa, the resultant compressive force C and the corresponding distance y_c can be derived and they are the same as Eqs. (5.29) and (5.30). Substituting $\varepsilon_0 = 0.002$ and $\varepsilon_{cu} = 0.0033$ into Eqs. (5.29) and (5.30), we have $C = 0.798 f_c b x_n$ and $y_c = 0.412 x_n$. Further substituting C and y_c into Eqs. (6.18) and (6.19), we finally obtain

$$N_{cu} = 0.798 f_c b x_n + f_y' A_s' - f_y A_s \tag{6.21}$$

$$N_{cu} e = 0.798 f_c b x_n (h_0 - 0.412 x_n) + f_y' A_s' (h_0 - a_s') \tag{6.22}$$

If the dimensions, material properties, and eccentricity of a section are known, the ultimate compressive capacity N_{cu} can be calculated. It must be noted that if other stress–strain relationships of the concrete are used in the calculation, the results may be slightly different, but the method is the same as shown in Eqs. (6.17–6.22).

Example 6.1 Calculate the ultimate compressive capacity of the column section shown in Fig. 6.6.

Solution We know that $b \times h = 100$ mm $\times$ 160 mm, $a_s = a_s' = 20$ mm, $f_{cu} = 22$ N/mm^2, $f_y = f_y' = 363$ N/mm^2, $A_s = A_s' = 157$ mm^2.

According to the result regressed from experimental data in Fig. 2.17, $f_c = 0.76 f_{cu} = 16.72$ N/mm^2.

Because

$$\frac{h}{30} = \frac{160}{30} = 5.333 \text{ mm}, e_a = \max(20,\ h/30) = 20 \text{ mm},$$

$$e_i = 80 + 20 = 100 \text{ mm}$$

$$l_0/h = 4.06 < 5, \text{so } \eta_s = 1.0$$

$$e = \eta_s e_i + \frac{h}{2} - a_s = 1.0 \times 100 + \frac{160}{2} - 20 = 160 \text{ mm}$$

$$h_0 = h - a_s = 160 - 20 = 140 \text{ mm}$$

From Eq. (6.21),

$$N_{cu} = 0.798 \times 16.72 \times 100x_n + 157 \times 363 - 157 \times 363 = 1334.256x_n$$

From Eq. (6.22),

$$\begin{aligned} N_{cu} \times 160 &= 0.798 \times 16.72 \times 100x_n(140 - 0.412x_n) \\ &\quad + 363 \times 157 \times (140 - 20)x_n^2 + 48.544x_n - 12439.892 = 0 \\ x_n &= 89.877\,\text{mm} \\ N_{cu} &= 1334.256 \times 89.877 = 119{,}919\,\text{N} = 119.919\,\text{kN} \end{aligned}$$

Compared with the test result (147 kN) in Fig. 6.6, it can be seen that the calculation error is −18 %, which means that the calculation result is conservative. The main reason for the big error is that e_a = 20 mm is conservative for small sections.

6.5.2 *Ultimate Bearing Capacities of Small Eccentrically Compressed Sections*

The failure characteristic of small eccentric compression sections is that the steel bars farthest away from the axial force will not yield under tension. When the eccentricity e_0 is large, there is a tension zone on the side farthest away from the axial force. The concrete and the steel bars in this zone are subjected to tension.

If the tensile stress in the concrete exceeds its tensile strength, the concrete will crack. When the eccentricity e_0 is very small, the full section may be under compression. And the compressive stress in the concrete far away from the axial force will be smaller. The stress and strain distributions across a small eccentrically compressed section are illustrated in Fig. 6.16.

For a partially compressed section, the concrete stress distribution across the section is similar to that across a large eccentric compression section (Fig. 6.16a). The stress in the extreme compression fiber can reach the compressive strength of the concrete. If the concrete far away from the axial force cracks, it can be assumed to have quit working. Even if the tensile stress is so small that it cannot crack the concrete, the contribution of the concrete to tension resistance can still be neglected because the total tensile force is so minimal.

For a fully compressed section (Fig. 6.16b), the stress in the extreme compression fiber can reach the compressive strength f_c of the concrete at failure, while the stress on the opposite side is smaller. If the eccentricity is so small that the section can be assumed under axial compression, all of the concrete section may be compressed to f_c.

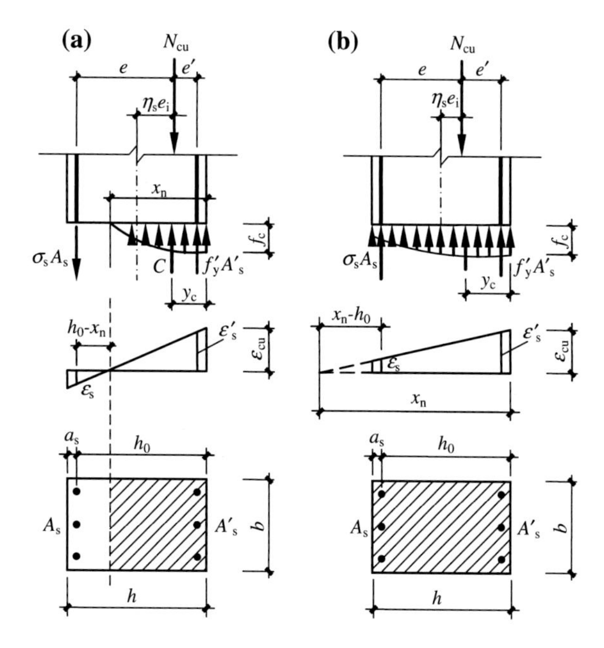

Fig. 6.16 Stress and strain distributions of a small eccentrically compressed section. **a** A partially compressed section, **b** a fully compressed section

The stress in the steel bars farthest away from the axial force can be obtained from the plane section assumption as (Fig. 6.16a)

$$\frac{\varepsilon_s}{h_0 - x_n} = \frac{\varepsilon_{cu}}{x_n} \tag{6.23}$$

$$\varepsilon_s = \frac{h_0 - x_n}{x_n} \cdot \varepsilon_{cu} = \left(\frac{1}{\xi_n} - 1\right)\varepsilon_{cu} \tag{6.24}$$

where

ξ_n relative depth of the compression zone, $\xi_n = \frac{x_n}{h_0}$; and

ε_s real strain of the steel bars far away from the axial force.

From Eq. (2.1), we have

$$\sigma_s = \varepsilon_s E_s = \left(\frac{1}{\xi_n} - 1\right) E_s \varepsilon_{cu} \tag{6.25}$$

where

σ_s stress in the steel bars far away from the axial force; and

E_s modulus of elasticity of the steel bars far away from the axial force.

Obviously, for the case in Fig. 6.16a, σ_s is the tensile stress and $\sigma_s < f_y$. From the equilibrium conditions, we have

$$\Sigma N = 0, \quad N_{cu} = \int_0^{x_n} \sigma_c b dx + f'_y A'_s - \sigma_s A_s \tag{6.26}$$

$$\Sigma M = 0, \quad N_{cu} e = \int_0^{x_n} \sigma_c b dx (h_0 - x) + f'_y A'_s (h_0 - a'_s) \tag{6.27}$$

where $e = \eta_s e_i + \frac{h}{2} - a_s$.

If the stress–strain relationship of the concrete in Fig. 5.14 is used, and the cube strength of the concrete is not larger than 50 MPa, Eqs. (6.26) and (6.27) may be written as

$$N_{cu} = 0.798 f_c b x_n + f'_y A'_s - \sigma_s A_s \tag{6.28}$$

$$N_{cu} e = 0.798 f_c b x_n (h_0 - 0.412 x_n) + f'_y A'_s (h_0 - a'_s) \tag{6.29}$$

For the case in Fig. 6.16b, there are

$$\frac{\varepsilon_s}{x_n - h_0} = \frac{\varepsilon_{cu}}{x_n} \tag{6.30}$$

$$\varepsilon_s = \frac{x_n - h_0}{x_n} \cdot \varepsilon_{cu} = \left(1 - \frac{1}{\xi_n}\right) \varepsilon_{cu} \tag{6.31}$$

$$\sigma_s = E_s \varepsilon_s = E_s \varepsilon_{cu} \left(1 - \frac{1}{\xi_n}\right) \tag{6.32}$$

σ_s is the compressive stress and $\sigma_s \leqslant f'_y$.

Based on the equilibrium conditions, there are

$$N_{cu} = \int_0^{h} \sigma_c b dx + f'_y A'_s + \sigma_s A_s \tag{6.33}$$

$$N_{cu} e = \int_0^{h} \sigma_c b dx (h_0 - x) + f'_y A'_s (h_0 - a'_s) \tag{6.34}$$

where $e = \eta_s e_i + \frac{h}{2} - a_s$.

If the stress–strain relationship of the concrete shown in Fig. 5.14 is used, and the cube strength of the concrete is not larger than 50 MPa, the resultant force in Eqs. (6.33) and (6.34) can be calculated as follows:

$$\int_0^h \sigma_c b dx = 0.798 f_c b x_n - 0.667 f_c b x_n \left(1 - \frac{h}{x_n}\right)^2 \left[3.3 - 2.723\left(1 - \frac{h}{x_n}\right)\right] \quad (6.35)$$

and the distance from the acting point of the resultant compressive force C to the extreme compression fiber y_c is

$$y_c = \frac{0.318 - 0.222\left(1 - \frac{h}{x_n}\right)^2 \left(1 + 2\frac{h}{x_n}\right)\left[3.3 - 2.723\left(1 - \frac{h}{x_n}\right)\right]}{0.798 - 0.667\left(1 - \frac{h}{x_n}\right)^2 \left[3.3 - 2.723\left(1 - \frac{h}{x_n}\right)\right]} x_n \quad (6.36)$$

Substituting $\int_0^h \sigma_c b dx$ and y_c into Eqs. (6.33) and (6.34), we can solve x_n and N_{cu}.

Example 6.2 Calculate the ultimate compressive capacity N_{cu} of the column shown in Fig. 6.8.

Solution We know that $b \times h$ = 100 mm × 160 mm, $a_s = a'_s$ = 20 mm, h_0 = 140 mm, f_{cu} = 22 N/mm^2, $f'_y = f_y$ = 363 N/mm^2, e_0 = 25 mm, η_s = 1.0, $A'_s = A_s$ = 157 mm^2.

According to the result regressed from experimental data in Fig. 2.17, f_c = 0. 76f_{cu} = 16.72 N/mm^2, $E_s = 2 \times 10^5$ N/mm^2.

Because

$$\frac{h}{30} = \frac{160}{30} = 5.333\,\text{mm}, e_a = \max(20, h/30) = 20\,\text{mm}$$

$$e_i = e_0 + e_a = 25 + 20 = 45\,\text{mm}, \ \eta_s = 1.0$$

$$e = \eta_s e_i + \frac{h}{2} - a_s = 1.0 \times 45 + \frac{160}{2} - 20 = 105\,\text{mm}$$

Assuming that the section is partially compressed first, we get

$$\sigma_s = 0.0033 \times 2 \times 10^5 \left(\frac{1}{\xi_n} - 1\right)$$

Substituting σ_s into Eqs. (6.28) and (6.29), we obtain

$$N_{cu} = 0.798 \times 16.72 \times 100 \times 140\xi_n + 363 \times 157 - 660\left(\frac{1}{\xi_n} - 1\right) \times 157$$

$$N_{cu} \times 105 = 0.798 \times 16.72 \times 100 \times 140^2 \left(\xi_n - 0.412\xi_n^2\right) + 363 \times 157 \times (140 - 20)$$

Solution of the two equations yields

$$\xi_n^3 - 0.607\xi_n^2 + 0.930\xi_n - 1.010 = 0$$
$$\xi_n = 0.911, \quad x_n = \xi_n h_0 = 0.911 \times 140 = 127.54\,\text{mm} < 160\,\text{mm}$$

It is true that the section is partially compressed. So,

$$\sigma_s = 660 \times \left(\frac{1}{0.911} - 1\right) = 64.479\,\text{N/mm}^2$$
$$N_{cu} = 0.798 \times 16.72 \times 100 \times 140 \times 0.911 + 363 \times 157 - 64.794 \times 157 = 216.996\,\text{kN}$$

Compared with the test result (247 kN) in Fig. 6.8, it can be seen that the calculation result is smaller (relative error −11 %). The main reason for the big error is that e_a = 20 mm is conservative for small sections.

6.5.3 Balanced Sections

Balanced failure is that the strain in the extreme compression fiber reaches the ultimate compressive strain of the concrete ε_{cu} caused by the yielding of the steel bars farthest away from the axial force. The stress and stain distributions of a balanced section are shown in Fig. 6.17.

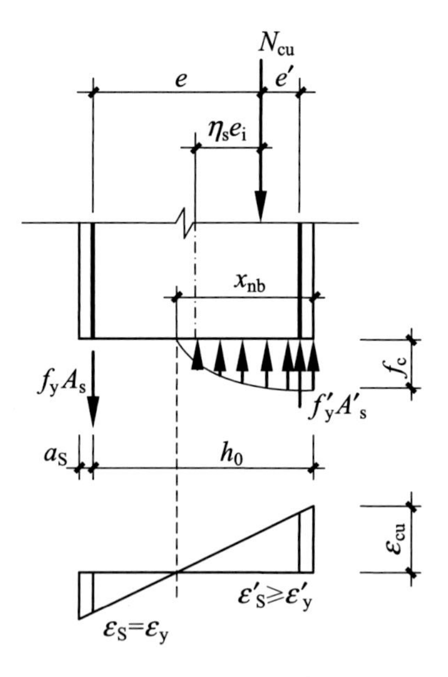

Fig. 6.17 Stress and strain distributions of a balanced section

According to the plane section assumption, we have

$$\frac{\varepsilon_y}{h_0 - x_{nb}} = \frac{\varepsilon_{cu}}{x_{nb}} \tag{6.37}$$

So,

$$\frac{x_{nb}}{h_0} = \frac{\varepsilon_{cu}}{\varepsilon_y + \varepsilon_{cu}} = \frac{1}{1 + \frac{\varepsilon_y}{\varepsilon_{cu}}} \tag{6.38}$$

Assuming $\frac{x_{nb}}{h_0} = \xi_{nb}$ and noting $f_y = \varepsilon_y E_s$, we have

$$\xi_{nb} = \frac{1}{1 + \frac{f_y}{E_s \varepsilon_{cu}}} \tag{6.39}$$

where

x_{nb} compression zone depth of the balanced section;
ε_y yield strain of the steel bars farthest away from the axial force; and
ξ_{nb} relative depth of the compression zone of the balanced section.

Obviously, ξ_{nb} can be used to distinguish whether a section is a large eccentrically compressed section or a small eccentrically compressed section. When $\xi_n \leqslant \xi_{nb}$, the section is a large eccentrically compressed section. On the other hand, when $\xi_n > \xi_{nb}$, the section is a small eccentrically compressed section.

6.5.4 *Simplified Calculation Method to Determine Ultimate Bearing Capacities of Eccentrically Compressed Sections*

6.5.4.1 Basic Principles

Similar to flexural members, the actual stress distribution over the compression zone can be replaced by an equivalent rectangular stress block, which will yield the same resultant compressive force at the same acting point. The width and height of the equivalent rectangular stress block are $x = \beta_1 x_n$ and $\alpha_1 f_c$, respectively, as shown in Fig. 6.18. α_1 and β_1 can be determined by Eqs. (5.40) and (5.41) according to the basic principles. If the cube strength of the concrete is not larger than 50 MPa, α_1 = 1.00 and β_1 = 0.80. If the cube strength of the concrete is 80 MPa, α_1 = 0.94 and β_1 = 0.74. For the concrete with a cube strength larger than 50 MPa and smaller than 80 MPa, α_1 and β_1 can be determined by linear interpolation.

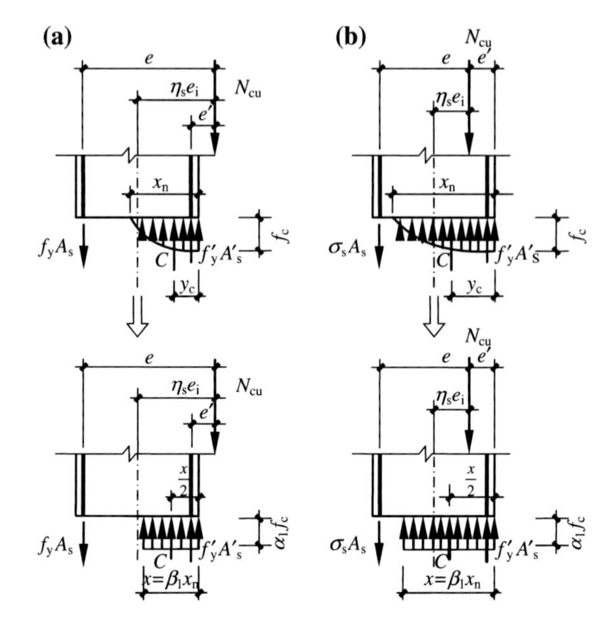

Fig. 6.18 Simplified stress distributions across eccentrically compressed sections. **a** A large eccentrically compressed section, **b** a small eccentrically compressed section

6.5.4.2 Limit Between Small and Large Eccentrically Compressed Sections

By assuming $\xi_b = \beta_1 \xi_{nb}$, Eq. (6.39) can be written as

$$\xi_b = \frac{\beta_1}{1 + \frac{f_y}{E_s \varepsilon_{cu}}} \tag{6.40}$$

where ξ_b = relative depth of the rectangular stress block of the balanced section, $\xi_b = \frac{x_b}{h_0}$; x_b = depth of the rectangular stress block of the balanced section. When the cube strength of the concrete is not larger than 50 MPa, Eq. (6.40) can be rewritten as

$$\xi_b = \frac{0.8}{1 + \frac{f_y}{E_s \varepsilon_{cu}}} \tag{6.41}$$

When $\xi \leqslant \xi_b$, the section is a large eccentrically compressed section; when $\xi > \xi_b$, the section is a small eccentrically compressed section.

6.5.4.3 Simplified Calculation Equations for the Ultimate Compressive Capacity of a Large Eccentrically Compressed Section

By the equilibrium shown in Fig. 6.18a, the simplified equations for the ultimate compressive capacity are

$$N_{\rm cu} = \alpha_1 f_{\rm c} bx + f'_{\rm y} A'_{\rm s} - f_{\rm y} A_{\rm s} \tag{6.42}$$

$$N_{\rm cu} e = \alpha_1 f_{\rm c} bx \left(h_0 - \frac{x}{2} \right) + f'_{\rm y} A'_{\rm s} (h_0 - a'_{\rm s}) \tag{6.43}$$

where $e = \eta_{\rm s} e_{\rm i} + \frac{h}{2} - a_{\rm s}$. The applicable conditions are $x \leqslant x_{\rm b}$ (or $\xi \leqslant \xi_{\rm b}$) and $x \geqslant 2$ $a'_{\rm s}$.

Given the dimensions, material properties, reinforcement, and eccentricity, the ultimate compressive capacity of the section $N_{\rm cu}$ can be obtained by solving Eqs. (6.42) and (6.43).

6.5.4.4 Simplified Calculation Equations for the Ultimate Compressive Capacity of a Small Eccentrically Compressed Section

By the equilibrium shown in Fig. 6.18b, the simplified equations for the ultimate compressive capacity are

$$N_{\rm cu} = \alpha_1 f_{\rm c} bx + f'_{\rm y} A'_{\rm s} - \sigma_{\rm s} A_{\rm s} \tag{6.44}$$

and

$$N_{\rm cu} e = \alpha_1 f_{\rm c} bx \left(h_0 - \frac{x}{2} \right) + f'_{\rm y} A'_{\rm s} (h_0 - a'_{\rm s}) \tag{6.45}$$

Substituting $\xi = \beta_1 \xi_{\rm n}$ into Eq. (6.25), we have

$$\sigma_{\rm s} = E_{\rm s} \varepsilon_{\rm cu} \left(\frac{\beta_1}{\xi} - 1 \right) \tag{6.46}$$

When the cube strength of the concrete is not larger than 50 MPa, Eq. (6.46) becomes

$$\sigma_{\rm s} = E_{\rm s} \varepsilon_{\rm cu} \left(\frac{0.8}{\xi} - 1 \right) \tag{6.47}$$

Similar to over-reinforced sections, to avoid solving high-order equations, Eq. (6.47) can be simplified as

$$\sigma_s = \frac{0.8-\xi}{0.8-\xi_b} f_y \quad (-f_y \leqslant \sigma_s \leqslant f_y) \tag{6.48}$$

Solving Eqs. (6.44), (6.45), and (6.48) simultaneously gives the ultimate compressive capacity of the section N_{cu}.

Example 6.3 Calculate the ultimate compressive capacity of the column shown in Fig. 6.6 using the simplified method.

Solution We know that $b \times h$ = 100 mm × 160 mm, $a_s = a'_s$ = 20 mm, f_{cu} = 22 N/mm^2, $f_y = f'_y$ = 363 N/mm^2, $A_s = A'_s$ = 157 mm^2, f_c = 16.72 N/mm^2, $E_s = 2.0 \times 10^5$ N/mm^2.

From Example 6.1, it can be seen that e = 160 mm and h_0 = 140 mm.

Assume that the section is a large eccentrically compressed section first. Using Eqs. (6.42) and (6.43), we have

$$N_{cu} = 16.72 \times 100x + 363 \times 157 - 363 \times 157 = 1672x$$

$$N_{cu} \times 160 = 16.72 \times 100x\left(140 - \frac{x}{2}\right) + 363 \times 157 \times (140 - 20)$$

Solution of the two equations gives

$$x^2 + 40x - 8180.526 = 0$$

$$x = 72.631\,\text{mm}$$

$$\xi_b = \frac{0.8}{1 + \frac{f_y}{E_s \varepsilon_{cu}}} = \frac{0.8}{1 + \frac{363}{2\times10^5\times0.0033}} = 0.516$$

$$x = 72.631\,\text{mm} \approx \xi_b h_0 = 0.516 \times 140 = 72.24\,\text{mm}$$

The section can be considered as a large eccentricity section. So,

$$N_{cu} = 1672 \times 72.631 = 121.439\ \text{kN}$$

The result is almost the same as that in Example 6.1, which means that the simplified method is applicable.

Example 6.4 Calculate the ultimate compressive capacity of the column in Fig. 6.8 using the simplified equations.

Solution We know that $b \times h$ = 100 mm × 160 mm, $a_s = a'_s$ = 20 mm, h_0 = 140 mm, f_{cu} = 22 N/mm^2, $f'_y = f_y$ = 363 N/mm^2, $A'_s = A_s$ = 157 mm^2, f_c = 16.72 N/mm^2, $E_s = 2\times10^5$ N/mm^2.

From Example 6.2, e = 105 mm.

$$\xi_b = \frac{0.8}{1 + \frac{f_y}{E_s \varepsilon_{cu}}} = \frac{0.8}{1 + \frac{363}{2\times10^5\times0.0033}} = 0.516$$

Assume that the section is a large eccentrically compressed section first. Using Eqs. (6.42) and (6.43), we have

$$N_{cu} = 16.72 \times 100x = 1672x$$

$$N_{cu} \times 105 = 16.72 \times 100x\left(140 - \frac{x}{2}\right) + 363 \times 157 \times (140 - 20)$$

Solving the two equations simultaneously gives

$$x^2 - 70.000x - 8180.526 = 0$$

$$x = 131.982\,\text{mm}$$

Because $\xi = \frac{x}{h_0} = \frac{131.982}{140} = 0.943 > \xi_b = 0.516$, the section is actually a small eccentrically compressed section. And the value of ξ needs to be recalculated.

Using Eqs. (6.44), (6.45), and (6.48), we have

$$N_{cu} = 1672x + 363 \times 157 - 157 \times \frac{0.8 - \xi}{0.8 - \xi_b} \times 363$$

$$N_{cu} \times 105 = 16.72 \times 100x\left(140 - \frac{x}{2}\right) + 363 \times 157 \times (140 - 20)$$

Solving the two equations gives

$$x^2 + 110.030x - 21185.902 = 0$$

x = 100.589 mm > $\xi_b h_0$ = 0.516 × 140 = 72.24 mm. It is true that the section is a small eccentric section. So,

$$\sigma_s = \frac{0.8 - \xi}{0.8 - \xi_b} f_y = \frac{0.8 - \frac{100.589}{140}}{0.8 - 0.516} \times 363 = 104.180\,\text{N/mm}^2$$

$$N_{cu} = 1672 \times 100.589 + 363 \times 157 - 104.189 \times 157 = 208{,}831\,\text{N} = 208.831\,\text{kN}$$

It is little bit smaller than the result in Example 6.2, which means that the simplified method is still applicable.

6.6 Applications of the Ultimate Bearing Capacity Equations for Eccentrically Compressed Members

6.6.1 Design of Asymmetrically Reinforced Sections

When the steel contents at the two opposite faces of a section in the eccentricity direction are not equal, that is $A_s \neq A'_s$, the section is an asymmetrically reinforced section. In the design of an asymmetrically reinforced section, generally the dimensions $b \times h$ of the section, the materials strengths f_c, f_y, and f'_y, the effective length l_0 of the member, the internal force N_c, and the eccentricity e_0 (or N_c and M, $e_0 = M/N_c$) are known, and the reinforcements A_s and A_s' need to be calculated. Sometimes A'_s is also known, and only A_s is to be designed. To make sure that the section can sustain the combined action of the axial force and the moment, $N_c \leqslant N_{cu}$ and $M \leqslant N_{cu}e_0$ must be satisfied. Because large eccentricity sections and small eccentricity sections are obviously different in mechanical property and failure mode, they should be distinguished first; then, these sections should be designed by corresponding methods.

6.6.1.1 Practical Method to Distinguish Large and Small Eccentrically Compressed Sections in Design

As discussed previously, when $\xi \leqslant \xi_b$, the section is a large eccentrically compressed section, and when $\xi > \xi_b$, the section is a small eccentricity section. However, in a section design, ξ is unknown a priori. So the above criterion cannot be used directly. A two-step discriminance is thus adopted: (1) An initial guess is made according to the eccentricity value. From previous engineering practice, if the eccentricity is $\eta_s e_i > 0.3h_0$, the section can be assumed as a large eccentricity; otherwise, the section has a small eccentricity. (2) Calculate ξ by corresponding equations based on the initial guess. Then, the final decision can be made with ξ.

6.6.1.2 Design of Large Eccentrically Compressed Sections

Based on the design principles discussed before and Eqs. (6.42) and (6.43), the basic equations for a large eccentrically compressed section design are

$$N_c = N_{cu} = \alpha_1 f_c b h_0 \xi + f'_y A'_s - f_y A_s \tag{6.49}$$

and

$$N_c e = N_{cu} e = \alpha_1 f_c b h_0^2 \xi \left(1 - \frac{\xi}{2}\right) + f'_y A'_s \left(h_0 - a'_s\right) \tag{6.50}$$

Two cases may be involved in the section design:

Case I Both A_s and A'_s are to be determined;
Case II The compression reinforcement A'_s is given, and A_s is to be determined.

In Case I, there are three unknowns A_s, A'_s, and x in Eqs. (6.49) and (6.50). It is surely impossible to solve three unknowns uniquely from two equations. The designer must determine any one of the three unknowns and solve the other two. Currently, the requirement of minimum steel contents ($A_s + A'_s$) is generally accepted as an addition condition, with which x (or ξ) will be solved first and then A'_s and A_s can be calculated. To ensure that the steel contents are the minimum, the concrete should be fully utilized. Therefore, assuming $x = \xi_b h_0$, the section can be designed as follows:

1. If $l_0/h \leqslant 5$, $\eta_s = 1.0$. Otherwise, calculate η_s by Eq. (6.14).
2. Calculate e_i by Eq. (6.4).
3. If $\eta_s e_i > 0.3\ h_0$, start as a large eccentrically compressed section. Otherwise, start as a small eccentrically compressed section, which will be discussed later.
4. Calculate ξ_b by Eq. (6.40).
5. Take $x = \xi_b h_0$ or $\xi = \xi_b$.
6. Calculate A'_s by Eq. (6.50).
7. Calculate A_s by Eq. (6.49).
8. Check $\rho \geqslant \rho_{min}$ and $\rho' \geqslant \rho'_{min}$. Similar to the design of axially compressed structural members and flexural members, the minimum amount of reinforcement must be checked. Table 4.2 lists the minimum longitudinal reinforcement ratio for eccentrically compressed members stipulated in GB 50010. If $\rho < \rho_{min}$, take $A_s = \rho_{min} bh$. If $\rho' < \rho'_{min}$, which means that the value of $\xi = \xi_b$ is too big, take $A'_s = \rho'_{min} bh$ and recalculate A_s with A'_s known (Case II).
9. Check the compressive capacity of the member (as an axially compressed one) in the out-of-plane direction with the calculated A_s and A'_s. If the compressive capacity is lower than the axial force N_c, more reinforcement, larger sectional dimensions, or stronger concrete should be adopted.

In Case II, because the compression reinforcement A'_s is given, the two unknowns A_s and x can be solved uniquely from the two basic equations. The detailed design steps are as follows:

1. Check $\rho' \geqslant \rho'_{min}$. If $\rho' < \rho'_{min}$, design the section as if A'_s was unknown (Case I).
2. If $l_0/h \leqslant 5$, $\eta_s = 1.0$. Otherwise, calculate η_s by Eq. (6.14).
3. Calculate e_i by Eq. (6.4).
4. If $\eta_s e_i > 0.3h_0$, start as a large eccentrically compressed section. Otherwise, start as a small eccentrically compressed section, which will be discussed later.
5. Calculate ξ by Eq. (6.50).
6. Calculate ξ_b by Eq. (6.40).
7. Check $\xi \leqslant \xi_b$. If it is true, which means that the initial guess in Step 4 is right, i.e., the section is truly a large eccentric section. For a large eccentrically compressed section, continue the calculation. Otherwise, the section must be

designed as a small eccentrically compressed section, which will be discussed later.

8. If $x = \xi h_0 > 2\,a'_s$, calculate A_s by Eq. (6.49).
9. If $x = \xi h_0 < 2\,a'_s$, A'_s cannot yield. Either one of the following two methods can be used in the design: The first one is to replace f'_y in Eq. (6.49) and Eq. (6.50) by $\sigma'_s = E_s \varepsilon_{cu}\left(\frac{\beta_1 a'_s}{\xi h_0} - 1\right)$; then, calculate ξ and A_s. The second method assumes $x = 2\,a'_s$ and then calculates A_s by Eq. (6.50) (taking moment about A'_s in Fig. 6.18a).

$$A_s = \frac{Ne'}{f_y(h_0 - a'_s)} \tag{6.51}$$

where $e' = \eta_s e_i - \frac{h}{2} + a'_s$.

10. If $A_s < \rho_{min} bh$, take $A_s = \rho_{min} bh$.
11. Check the compressive capacity of the member (as an axially compressed one) in the out-of-plane direction with the calculated A_s and A'_s.

6.6.1.3 Design of Small Eccentricity Sections

According to the design principle discussed at the beginning of this section and the ultimate bearing capacity equations for small eccentrically compressed sections, Eqs. (6.44) and (6.45), the basic design equations can be written as

$$N_c = N_{cu} = \alpha_1 f_c b h_0 \xi + f'_y A'_s - \sigma_s A_s \tag{6.52}$$

and

$$N_c e = N_{cu} e = \alpha_1 f_c b h_0^2 \xi \left(1 - \frac{\xi}{2}\right) + f'_y A'_s (h_0 - a'_s) \tag{6.53}$$

where σ_s is given by Eq. (6.48).

In this case, four unknowns ξ, A'_s, A_s, and σ_s are to be determined from three available equations. The requirement of minimum steel contents ($A'_s + A_s$) can still be used herein as an additional condition to calculate ξ. But the calculation involves the troublesome solution of a cubic equation of ξ.

From another point of view, the steels' A_s at the face far away from the axial force does not yield at the ultimate state, so the less the steels' A_s is, the more economical the section is. Therefore, A_s can be determined directly from the minimum reinforcement ratio, i.e., $A_s = \rho_{min} bh$. Then, A'_s and ξ can be uniquely solved from Eqs. (6.52) and (6.53). The detailed steps are as follows:

1. If $l_0/h \leqslant 5$, $\eta_s = 1.0$. Otherwise, calculate η_s by Eq. (6.14).
2. Calculate e_i by Eq. (6.4).

3. If $\eta_s e_i \leqslant 0.3h_0$, start as a small eccentrically compressed section. Otherwise, start as a large eccentrically compressed section.
4. Calculate ξ_b by Eq. (6.40).
5. Take $A_s = \rho_{min}bh$.
6. Substitute Eq. (6.48) into Eq. (6.52), and calculate ξ and A'_s by solving Eqs. (6.52) and (6.53) simultaneously.
7. Check $\xi > \xi_b$. If it is true, which means that the initial guess in Step 3 is right, i.e., the section is really a small eccentrically compressed section, continue the calculation. Otherwise, the section must be designed as a large eccentrically compressed section.
8. Check $A'_s \geqslant \rho'_{min}bh$. If it is not satisfied, take $A'_s = \rho'_{min}bh$.
9. Check the content of A_s. From the failure modes of eccentrically compressed members shown in Fig. 6.10, it is found that if both e_0 and A_s are very small and A'_s is large, the physical centroid of the section may shift to the right-hand side of the axial force (Fig. 6.19a), which causes larger compressive strain at the face far from the axial force. Therefore, the failure will start at the face far from the axial force. The steel bars near the axial force may not yield, and the compressive stress in the surrounding concrete is below its compressive strength.
   To avoid such situation, A_s cannot be too small. In Fig. 6.19b, assuming $e'_i = e_0 - e_a$, $\eta_s = 1.0$, and taking moment about A'_s, we have

$$N_{cu}e' = \alpha_1 f_c bh\left(h'_0 - \frac{h}{2}\right) + f_y A_s (h'_0 - a_s) \tag{6.54}$$

where

h'_0 distance from the center of A'_s to the extreme fiber at A_s face; and
e' distance from the axial force to the center of A'_s.

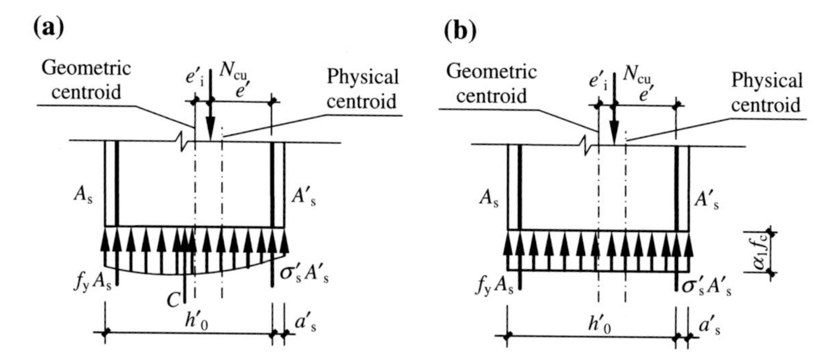

Fig. 6.19 Stress distribution when e_0 is very small. **a** Actual stress distribution, **b** simplified stress distribution

$$e' = \frac{h}{2} - e'_i - a'_s \tag{6.55}$$

To guarantee that A_s does not yield first under N_c, the value of A_s must meet the requirement in Eq. (6.56).

$$A_s \geqslant \frac{N_c e' - \alpha_1 f_c bh\left(h'_0 - \frac{h}{2}\right)}{f_y(h'_0 - a_s)} \tag{6.56}$$

10. Check the compressive capacity of the member (as an axially compressed one) in the out-of-plane direction with the calculated A_s and A'_s. If the compressive capacity is lower than the axial force N_c, more reinforcement, larger sectional dimensions, or stronger concrete should be adopted.

Example 6.5 Consider an eccentrically compressed column of a rectangular section: If $N_c = 800$ kN, $M = 270$ kN·m, $b \times h = 300$ mm × 500 mm, $a_s = a'_s = 40$ mm, $l_0 = 4.2$ m, the properties of the concrete are $f_c = 16.7$ N/mm^2, $\varepsilon_{cu} = 0.0033$, and the properties of the steels are $f'_y = f_y = 300$ N/mm^2, $E_s = 2 \times 10^5$ N/mm^2, determine A'_s and A_s.

Solution

1. Calculate e_i, η_s, and e

$h_0 = h - a_s = 500 - 40 = 460\,\text{mm}$, $\frac{l_0}{h} = \frac{4200}{500} = 8.4 > 5$, η_s must be calculated.

$$e_0 = \frac{M}{N} = \frac{270}{800} = 0.338\,\text{m} = 338\,\text{mm},\ \frac{h}{30} = \frac{500}{30} = 16.7\,\text{m},\ \text{take } e_a = 20\,\text{mm},$$

$$e_i = e_0 + e_a = 338 + 20 = 358\,\text{mm},\ \frac{e_i}{h_0} = \frac{358}{460} = 0.778,$$

$$\zeta_c = \frac{0.5 f_c A}{N_c} = \frac{0.5 \times 16.7 \times 300 \times 500}{800{,}000} = 1.566 > 1.0,\ \text{take } \zeta_c = 1.0$$

$$\eta_s = 1 + \frac{1}{1300 \times 0.778} \times 8.4^2 \times 1.0 = 1.070$$

$$e = \eta_s e_i + \frac{h}{2} - a_s = 1.070 \times 358 + \frac{500}{2} - 40 = 593\,\text{mm}$$

2. Determine whether it is a large eccentrically compressed section or not

$$\eta_s e_i = 1.070 \times 358 = 383\,\text{mm} > 0.3h_0 = 138\,\text{mm}$$

It can be assumed as a large eccentrically compressed section.

3. Determine A_s' and A_s

$$\xi_b = \frac{0.8}{1+\frac{f_y}{E_s\varepsilon_{cu}}} = \frac{0.8}{1+\frac{300}{2\times10^5\times0.0033}} = 0.55$$

Taking $\xi = \xi_b = 0.55$,

$$A_s' = \frac{N_c e - \alpha_1 f_c b h_0^2(\xi - 0.5\xi^2)}{f_y'(h_0 - a_s')}$$
$$= \frac{800\times10^3\times593 - 16.7\times300\times460^2\times(0.55-0.5\times0.55^2)}{300\times(460-40)} = 410\,\text{mm}^2$$

From Table 4.2, $\rho'_{min} = 0.2\ \%$.

$$A_s' = 410\,\text{mm}^2 > \rho'_{min}bh = 0.002\times300\times500 = 300\,\text{mm}^2,\ \text{OK}.$$
$$A_s = \frac{\alpha_1 f_c b h_0 \xi + f_y' A_s' - N_c}{f_y}$$
$$= \frac{16.7\times300\times460\times0.55 + 300\times410 - 800\times10^3}{300} = 1905\,\text{mm}^2$$

From Table 4.2, $\rho_{min} = 0.2\ \%$.

$$A_s = 1905\,\text{mm}^2 > \rho_{min}bh = 0.002\times300\times500 = 300\,\text{mm}^2,\text{OK}.$$

4. The boundary condition in the out-of-plane direction (b direction) is the same as that in the eccentric direction (h direction). So, $l_0 = 4.2$ m.

$\frac{l_0}{b} = \frac{4200}{300} = 14$. From Table 4.1, $\varphi = 0.92$

$$\frac{A_s'}{bh} = \frac{410+1905}{300\times500} = 1.543\,\% < 3\,\%$$

So,

$$N_{cu} = \varphi\left(Af_c + f_y'A_s'\right) = 0.92\times(300\times500\times16.7 + 2315\times300)$$
$$= 2,943,540\,\text{N} = 2943.540 > N_c = 800\,\text{kN},\text{OK}.$$

Example 6.6 All other conditions are the same as those in Example 6.5, except $A_s' = 1140\ \text{mm}^2$ is given; determine A_s.

Solution

1. $A_s' = 1140\,\text{mm}^2 > \rho_{\min}' bh = 0.002 \times 300 \times 500 = 300\,\text{mm}^2$, OK.
2. Calculate e_i, η_s, and e

$$e_i = 358\ \text{mm}, \eta_s = 1.070, e = 593\ \text{mm}$$

3. Determine whether it is a large eccentrically compressed section or not

$$\eta_s e_i = 1.070 \times 358 = 383\,\text{mm} > 0.3h_0 = 138\,\text{mm}$$

   It can be assumed as a large eccentrically compressed section.
4. Determine A_s

$$A_s' = 1140\,\text{mm}^2$$

$$\xi = 1 - \sqrt{1 - 2 \times \frac{N_c e - f_y' A_s' (h_0 - a_s')}{\alpha_1 f_c b h_0^2}}$$

$$= 1 - \sqrt{1 - 2 \times \frac{800 \times 10^3 \times 593 - 300 \times 1140 \times (460 - 40)}{1.0 \times 16.7 \times 300 \times 460^2}} = 0.387 < \xi_b = 0.55$$

   It is truly a large eccentric section.
   Because $x = \xi h_0 = 0.387 \times 460 = 178.02$ mm $> 2\,a_s' = 70$ mm,

$$A_s = \frac{\alpha_1 f_c b h_0 \xi + f_y' A_s' - N_c}{f_y} = \frac{1.0 \times 16.7 \times 300 \times 460 \times 0.387 + 300 \times 1140 - 800 \times 10^3}{300}$$
$$= 1446\,\text{mm}^2 > \rho_{\min} bh = 0.002 \times 300 \times 500 = 300\,\text{mm}^2, \text{OK}.$$

5. Check the compressive capacity of the member in the out-of-plane direction (omitted here for brevity).

The total steel area $A_s + A_s'$ in Example 6.6 is 271 mm^2 (12 %) more than that in Example 6.5, because the requirement of minimum $A_s + A_s'$ is basically met in Example 6.5, while the reinforcement is inappropriately arranged in Example 6.6.

Example 6.7 Consider an eccentrically compressed column of rectangular section, if $N_c = 400$ kN, $M = 280$ kN·m, $b \times h = 300$ mm $\times$ 500 mm, $a_s = a_s' = 40$ mm, $l_0 = 3.9$ m, the strength of the concrete is $f_c = 16.7$ N/mm^2, the ultimate strain of the concrete is $\varepsilon_{cu} = 0.0033$, the strength of the steel is $f_y' = f_y = 300$ N/mm^2, and the modulus of elasticity of the steel is $E_s = 2 \times 10^5$ N/mm^2, determine A_s' and A_s.

Solution

1. Calculate e_i, η_s, and e
   $h_0 = h - a_s = 500 - 40 = 460\,\text{mm}$, $\frac{l_0}{h} = \frac{3900}{500} = 7.8 > 5$, η_s must be calculated.

$$e_0 = \frac{M}{N} = \frac{280}{400} = 0.70\,\text{m} = 700\,\text{mm},\ \frac{h}{30} = \frac{500}{30} = 16.7\,\text{mm. Take}\ e_a = 20\,\text{mm.}$$

$$e_i = e_0 + e_a = 700 + 20 = 720\,\text{mm},\ \frac{e_i}{h_0} = \frac{720}{460} = 1.565$$

$$\zeta_c = \frac{0.5 f_c A}{N_c} = \frac{0.5 \times 16.7 \times 300 \times 500}{400{,}000} > 1.0,\ \text{take}\ \zeta_c = 1.0.$$

$$\eta_s = 1 + \frac{1}{1300 \times 1.565} \times 7.8^2 \times 1.0 = 1.030$$

$$e = \eta_s e_i + \frac{h}{2} - a_s = 1.030 \times 720 + \frac{500}{2} - 40 = 951\,\text{mm}$$

2. Determine whether it is a large eccentricity section or not

$$\eta_s e_i = 1.030 \times 720 = 742\,\text{mm} > 0.3h_0 = 138\,\text{mm}$$

   It can be assumed as a large eccentricity section.
3. Determine A'_s and A_s

$$\xi_b = \frac{0.8}{1 + \frac{f_y}{E_s \varepsilon_{cu}}} = \frac{0.8}{1 + \frac{300}{2 \times 10^5 \times 0.0033}} = 0.55,\ \text{Take}\ \xi = \xi_b = 0.55$$

$$\begin{aligned} A'_s &= \frac{N_c e - \alpha_1 f_c b h_0^2 (\xi - 0.5\xi^2)}{f'_y (h_0 - a'_s)} \\ &= \frac{400 \times 10^3 \times 951 - 16.7 \times 300 \times 460^2 \times (0.55 - 0.5 \times 0.55^2)}{300 \times (460 - 40)} \\ &= -336\,\text{mm}^2 < 0 \end{aligned}$$

Take $A'_s = \rho'_{\min} bh = 0.002 \times 300 \times 500 = 300\,\text{mm}^2$

$$\begin{aligned} \xi &= 1 - \sqrt{1 - 2 \times \frac{N_c e - f'_y A'_s (h_0 - a'_s)}{\alpha_1 f_c b h_0^2}} \\ &= 1 - \sqrt{1 - 2 \times \frac{400 \times 10^3 \times 951 - 300 \times 300 \times (460 - 40)}{16.7 \times 300 \times 460^2}} = 0.405 < \xi_b \end{aligned}$$

It is truly a large eccentrically compressed section.

$$A_s = \frac{\alpha_1 f_c b h_0 \xi + f'_y A'_s - N_c}{f_y} = \frac{16.7 \times 300 \times 460 \times 0.405 + 300 \times 300 - 400 \times 10^3}{300}.$$
$$= 2078\,\text{mm}^2 > \rho_{\min} bh = 0.002 \times 300 \times 500 = 300\,\text{mm}^2, \text{OK}.$$

4. Check the compressive capacity of the member in the out-of-plane direction $\frac{l_0}{b} = \frac{3900}{300} = 13$. From Table 4.1, $\varphi = 0.935$.

$$\frac{A'_s}{bh} = \frac{300 + 2078}{300 \times 500} = 1.585\% < 3\%$$
$$\text{So}\, N_{cu} = \varphi\left(Af_c + f'_y A'_s\right) = 0.935 \times (16.7 \times 300 \times 500 + 2378 \times 300)$$
$$= 3{,}009{,}204\,\text{N} = 3009.204\,\text{kN} > N_c = 280\,\text{kN}, \text{OK}.$$

Example 6.8 All other conditions are the same as those in Example 6.7 except N_c = 580 kN; determine A'_s and A_s.

Solution

1. Calculate e_i, η_s, and e

$h_0 = h - a_s = 500 - 40 = 460\,\text{mm}$, $\frac{l_0}{h} = \frac{3900}{500} = 7.8 > 5$, η_s must be calculated.

$$e_0 = \frac{M}{N} = \frac{280}{580} = 0.483\,\text{m} = 483\,\text{mm}, \quad \frac{h}{30} = \frac{500}{30} = 16.7\text{mm, take}\, e_a = 20\,\text{mm}.$$
$$e_i = e_0 + e_a = 483 + 20 = 503\,\text{mm}, \quad \frac{e_i}{h_0} = \frac{503}{460} = 1.093$$
$$\zeta_c = \frac{0.5 f_c A}{N_c} = \frac{0.5 \times 16.7 \times 300 \times 500}{580{,}000} = 2.159 > 1.0, \text{take}\, \zeta_c = 1.0.$$
$$\eta_s = 1 + \frac{1}{1300 \times 1.093} \times 7.8^2 \times 1.0 = 1.043$$
$$e = \eta_s e_i + \frac{h}{2} - a_s = 1.043 \times 503 + \frac{500}{2} - 40 = 735\,\text{mm}$$

2. Determine whether it is a large eccentricity section or not

$$\eta_s e_i = 1.043 \times 503 = 525\,\text{mm} > 0.3h_0 = 0.3 \times 460 = 138\,\text{mm}$$

It can be assumed as a large eccentricity section.

3. Calculate A'_s, A_s

$$\xi_b = \frac{0.8}{1+\frac{f_y}{E_s\varepsilon_{cu}}} = \frac{0.8}{1+\frac{300}{2\times10^5\times0.0033}} = 0.55. \text{ Take } \xi = \xi_b = 0.55$$

$$A'_s = \frac{N_c e - \alpha_1 f_c b h_0^2(\xi - 0.5\xi^2)}{f'_y(h_0 - a'_s)}$$
$$= \frac{580\times10^3\times735 - 16.7\times300\times460^2\times(0.55-0.5\times0.55^2)}{300\times(460-40)} = 28\,\text{mm}^2$$

$$A'_s < \rho'_{min}bh = 0.002\times300\times500 = 300\,\text{mm}^2, \text{ take } A'_s = 300\,\text{mm}^2.$$

$$\xi = 1-\sqrt{1-2\times\frac{N_c e - f'_y A'_s(h_0 - a'_s)}{\alpha_1 f_c b h_0^2}}$$
$$= 1-\sqrt{1-2\times\frac{580\times10^3\times735-300\times300\times(460-40)}{16.7\times300\times460^2}} = 0.483 < \xi_b$$

It is truly a large eccentricity section.

$$A_s = \frac{\alpha_1 f_c b h_0\xi + f'_y A'_s - N_c}{f_y}$$
$$= \frac{16.7\times300\times460\times0.481+300\times300-580\times10^3}{300}$$
$$= 2062\,\text{mm}^2 > \rho_{min}bh = 0.002\times300\times500 = 300\,\text{mm}^2, \text{ OK.}$$

4. Check the compressive capacity of the member in the out-of-plane direction

$$\frac{l_0}{b} = \frac{3900}{300} = 13. \text{ From Table 4.1}, \varphi = 0.935.$$

$$\frac{A'_s}{bh} = \frac{300+2062}{300\times500} = 1.575\,\% < 3\,\%$$

$$\text{So}, N_{cu} = \varphi\left(Af_c + f'_y A'_s\right) = 0.935\times(16.7\times300\times500+2362\times300)$$
$$= 3{,}004{,}716\,\text{N} = 3004.716\,\text{kN} > N_c = 580\,\text{kN, OK.}$$

Compared with Example 6.7, it can be found that the steel content calculated in Example 6.8 is smaller even though the axial load N_c is increased by 180 kN. This means that axial load in a certain value range can increase the ultimate capacity of a large eccentricity section, which is in accordance with the N_{cu}–M_u interaction diagram.

Example 6.9 Consider an eccentrically compressed column of rectangular section. If N = 2500 kN, M = 180 kN·m, $b \times h$ = 300 mm × 500 mm, $a_s = a'_s$ = 40 mm, l_0 = 3.9 m, the strength of the concrete is f_c = 16.7 N/mm^2, the ultimate strain of the concrete is ε_{cu} = 0.0033, the strength and the modulus of elasticity of the steel are $f'_y = f_y$ = 300 N/mm^2 and $E_s = 2\times10^5$N/mm^2, respectively, determine A'_s and A_s.

Solution

1. Calculate e_i, η_s, and e

$$h_0 = h - a_s = 500 - 40 = 460\,\text{mm},\ \tfrac{l_0}{h} = \tfrac{3900}{500} = 7.8 > 5,\ \eta_s \text{ must be calculated.}$$

$$e_0 = \frac{M}{N} = \frac{180}{2500} = 0.072\,\text{m} = 72\,\text{mm},\ \frac{h}{30} = \frac{500}{30} = 16.7\,\text{mm},\ \text{take}\ e_a = 20\,\text{mm}.$$

$$e_i = e_0 + e_a = 72 + 20 = 92\,\text{mm},\ \frac{e_i}{h_0} = \frac{92}{460} = 0.2$$

$$\zeta_c = \frac{0.5 f_c A}{N_c} = \frac{0.5 \times 16.7 \times 300 \times 500}{2{,}500{,}000} = 0.501$$

$$\eta_s = 1 + \frac{1}{1300 \times 0.2} \times 7.8^2 \times 0.501 = 1.117$$

$$e = \eta_s e_i + \frac{h}{2} - a_s = 1.117 \times 92 + \frac{500}{2} - 40 = 313\,\text{mm}$$

2. Determine whether it is a large eccentricity section or not

$$\eta_s e_i = 1.117 \times 92 = 103\,\text{mm} < 0.3 h_0 = 0.3 \times 460 = 138\,\text{mm}$$

   It can be assumed as a small eccentricity section.

3. Calculate A'_s and A_s

$$\xi_b = \frac{0.8}{1 + \frac{f_y}{E_s \varepsilon_{cu}}} = \frac{0.8}{1 + \frac{300}{2 \times 10^5 \times 0.0033}} = 0.55$$

   Take $A_s = \rho_{\min} bh = 0.002 \times 300 \times 500 = 300\,\text{mm}^2$

$$\xi = -B_1 + \sqrt{B_1^2 - 2C_1}$$

$$B_1 = \frac{f_y A_s (h_0 - a'_s)}{\alpha_1 f_c b h_0^2 (0.8 - \xi_b)} - \frac{a'_s}{h_0} = \frac{300 \times 300 \times (460 - 40)}{16.7 \times 300 \times 460^2 \times (0.8 - 0.55)} - \frac{40}{460} = 0.056$$

$$C_1 = \frac{N_c (e - h_0 + a'_s)(0.8 - \xi_b) - 0.8 f_y A_s (h_0 - a'_s)}{\alpha_1 f_c b h_0^2 (0.8 - \xi_b)}$$

$$= \frac{2500 \times 10^3 \times (313 - 460 + 40) \times (0.8 - 0.55) - 0.8 \times 300 \times 300 \times (460 - 40)}{16.7 \times 300 \times 460^2 \times (0.8 - 0.55)}$$

$$= -0.366$$

$$\xi = -0.056 + \sqrt{0.056^2 - 2 \times (-0.366)} = 0.801 > \xi_b = 0.55$$

   It is truly a small eccentricity section. So,

$$A_s' = \frac{N_c e - \alpha_1 f_c b h_0^2 (\xi - 0.5\xi^2)}{f_y'(h_0 - a_s')}$$
$$= \frac{2500 \times 10^3 \times 313 - 16.7 \times 300 \times 460^2 \times (0.801 - 0.5 \times 0.801^2)}{300 \times (460 - 40)}$$
$$= 2170\,\text{mm}^2 > \rho_{\text{min}}' bh = 0.002 \times 300 \times 500 = 300\,\text{mm}^2,\ \text{OK.}$$

4. Check the content of A_s

$$e_i' = e_0 - e_a = 72 - 20 = 52\,\text{mm},$$
$$e' = \frac{h}{2} - e_i' - a_s' = \frac{500}{2} - 52 - 40 = 158\,\text{mm},$$
$$h_0' = h - a_s' = 500 - 40 = 460\,\text{mm}$$
$$A_s \geqslant \frac{N_c e' - \alpha_1 f_c b h (h_0' - \frac{h}{2})}{f_y (h_0' - a_s)}$$
$$= \frac{2500 \times 10^3 \times 158 - 16.7 \times 300 \times 500 \times (460 - 250)}{300 \times (460 - 40)}$$
$$= -1040\,\text{mm}^2$$

Hence, the content of A_s determined in Step (3) is sufficient. So, take $A_s = 300\ \text{mm}^2$.

5. Check the compressive capacity of the member in the out-of-plane direction

$$\frac{l_0}{b} = \frac{3900}{300} = 13$$

From Table 4.1, $\varphi = 0.935$

$$\frac{A_s'}{bh} = \frac{1000 + 2170}{300 \times 500} = 2.113\,\% < 3\,\%$$
$$\text{So}, N_{cu} = \varphi\left(Af_c + f_y' A_s'\right) = 0.935 \times (16.7 \times 300 \times 500 + 3170 \times 300)$$
$$= 3{,}231{,}360\,\text{N} = 3231.36\,\text{kN} > N_c = 2500\,\text{kN},\ \text{OK.}$$

Example 6.10 All other conditions are the same as those in Example 6.9 except N_c = 3220 kN and M = 48 kN·m; determine A_s' and A_s.

Solution

1. Calculate e_i, η_s, and e

   $h_0 = h - a_s = 500 - 40 = 460\,\text{mm}$, $\frac{l_0}{h} = \frac{3900}{500} = 7.8 > 5$, η_s must be calculated.

$$e_0 = \frac{M}{N} = \frac{48}{3220} = 0.0149\,\text{m} = 15\,\text{mm}, \quad \frac{h}{30} = \frac{500}{30} = 16.7\,\text{mm, take } e_a = 20\,\text{mm}.$$

$$e_i = e_0 + e_a = 15 + 20 = 35\,\text{mm}, \quad \frac{e_i}{h_0} = \frac{35}{460} = 0.076$$

$$\zeta_c = \frac{0.5 f_c A}{N_c} = \frac{0.5 \times 16.7 \times 300 \times 500}{3{,}220{,}000} = 0.389$$

$$\eta_s = 1 + \frac{1}{1300 \times 0.076} \times 7.8^2 \times 0.389 = 1.240$$

$$e = \eta_s e_i + \frac{h}{2} - a_s = 1.240 \times 35 + \frac{500}{2} - 40 = 253\,\text{mm}$$

2. Determine whether it is a large eccentrically compressed section or not

$$\eta_s e_i = 1.240 \times 35 = 43\,\text{mm} < 0.3 h_0 = 0.3 \times 460 = 138\,\text{mm}$$

It can be assumed as a small eccentrically compressed section.

3. Calculate A'_s and A_s

$$\xi_b = \frac{0.8}{1 + \frac{f_y}{E_s \varepsilon_{cu}}} = \frac{0.8}{1 + \frac{300}{2 \times 10^5 \times 0.0033}} = 0.55$$

Take $A_s = \rho_{\min} bh = 0.002 \times 300 \times 500 = 300\,\text{mm}^2$

$$\xi = -B_1 + \sqrt{B_1^2 - 2C_1}$$

$$B_1 = \frac{f_y A_s (h_0 - a'_s)}{\alpha_1 f_c b h_0^2 (0.8 - \xi_b)} - \frac{a'_s}{h_0} = \frac{300 \times 300 \times (460 - 40)}{16.7 \times 300 \times 460^2 \times (0.8 - 0.55)} - \frac{40}{460} = 0.056$$

$$C_1 = \frac{N_c (e - h_0 + a'_s)(0.80 - \xi_b) - 0.8 f_y A_s (h_0 - a'_s)}{\alpha_1 f_c b h_0^2 (0.8 - \xi_b)}$$

$$= \frac{3220 \times 10^3 \times (253 - 460 + 40) \times (0.8 - 0.55) - 0.8 \times 300 \times 300 \times (460 - 40)}{16.7 \times 300 \times 460^2 \times (0.8 - 0.55)}$$

$$= -0.621$$

$$\xi = -0.056 + \sqrt{0.056^2 - 2 \times (-0.621)} = 1.060 > \xi_b$$

It is truly a small eccentrically compressed section. So

$$A'_s = \frac{N_c e - \alpha_1 f_c b h_0^2 (\xi - 0.5\xi^2)}{f'_y (h_0 - a'_s)}$$

$$= \frac{3220 \times 10^3 \times 253 - 16.7 \times 300 \times 460^2 \times (1.060 - 0.5 \times 1.060^2)}{300 \times (460 - 40)}$$

$$= 2273\,\text{mm}^2 > \rho_{\min} bh = 0.002 \times 300 \times 500 = 300\,\text{mm}^2, \text{ OK.}$$

4. Check the content of A_s.

$$e' = \frac{h}{2} - e_i' - a_s',\ e_i' = e_0 - e_a = 15 - 20 = -5\,\text{mm}$$

$$e' = \frac{500}{2} - (-5) - 40 = 215\,\text{mm},\ h_0' = 500 - 40 = 460\,\text{mm}$$

$$A_s \geqslant \frac{N_c e' - \alpha_1 f_c bh(h_0' - \frac{h}{2})}{f_y(h_0' - a_s)}$$

$$= \frac{3220 \times 10^3 \times 215 - 16.7 \times 300 \times 500 \times (460 - 0.5 \times 500)}{300 \times (460 - 40)} = 1319\,\text{mm}^2$$

Obviously, the content of A_s determined in Step (3) is not sufficient. So, take $A_s = 1319\ \text{mm}^2$.

5. Check the compressive capacity of the member in the out-of-plane direction

$$\frac{l_0}{b} = \frac{3900}{300} = 13$$

From Table 4.1, $\varphi = 0.935$

$$A_s' = 2273 + 1319 = 3592\,\text{mm}^2,\ \rho' = \frac{3592}{300 \times 500} = 2.395\,\% < 3\,\%$$

$$N_{cu} = \varphi\left(Af_c + f_y'A_s'\right) = 0.935 \times (16.7 \times 300 \times 500 + 300 \times 3592)$$

$$= 3{,}349{,}731\,\text{N} = 3349.731\,\text{kN} > N_c = 3220\,\text{kN, OK.}$$

6.6.2 *Evaluation of Ultimate Compressive Capacities of Existing Asymmetrically Reinforced Eccentrically Compressed Members*

6.6.2.1 Determine N_{cu} When e_0 is Known

The effective length l_0, sectional dimensions $b \times h$, eccentricity of the axial compressive force e_0, material properties f_c, f_y, f_y', and E_s, and reinforcements A_s and A_s' are generally given in this kind of problem, and N_{cu} can be evaluated according to the following steps:

1. If $l_0/h \leqslant 5$, $\eta_s = 1.0$. Otherwise, calculate η_s using Eq. (6.14).
2. Calculate e_i using Eq. (6.4).
3. Calculate ξ_b using Eq. (6.40).

4. Assume that the section is a large eccentrically compressed section first, and calculate ξ using Eqs. (6.42) and (6.43).
5. If $\xi \leqslant \xi_b$ and $\xi h_0 \geqslant 2\,a'_s$, it is a large eccentrically compressed section and A'_s can yield. N_{cu} can be calculated directly using Eq. (6.42).
6. If $\xi \leqslant \xi_b$ and $\xi h_0 < 2\,a'_s$, it is a large eccentrically compressed section and A'_s cannot yield. N_{cu} can be calculated using either one of the following two methods: The first one is to calculate ξ and N_{cu} by substituting $\sigma'_s = E_s \varepsilon_{cu}\left(\frac{\beta_1 a'_s}{\xi h_0} - 1\right)$ into Eqs. (6.42) and (6.43) and the second one is to calculate N_{cu} directly using the equation $N_{cu} = A_s f_y \frac{(h_0 - a'_s)}{e'}$ by assuming $x = 2\,a'_s$.
7. If $\xi > \xi_b$, the assumption in Step (4) is not valid. The section is a small eccentrically compressed section. Calculate ξ and N_{cu} by solving Eqs. (6.44), (6.45), and (6.48) simultaneously. It should be noted that the sign of σ_s may be positive or negative.
8. Calculate the compressive capacity N_{cu} of the member (as an axially compressed one) in the out-of-plane direction.
9. Take the minimum value between the in-plane and out-of-plane compressive capacities as the final N_{cu}.

The detailed calculation has been shown in Examples 6.3 and 6.4.

6.6.2.2 Determine M_u When N_c is Known

The effective length l_0, sectional dimensions $b \times h$, material properties f_c, f_y, f'_y, and E_s, reinforcements A'_s and A_s, and axial force N_c are generally given in this kind of problem, and M_u needs to be determined.

Rewriting Eqs. (6.42) and (6.43), we have

$$N_c = \alpha_1 f_c bx + f'_y A'_s - f_y A_s \tag{6.57}$$

$$M_u = N_c e_0 = \frac{-N_c\left(\eta_s e_a + \frac{h}{2} - a_s\right) + \alpha_1 f_c bx\left(h_0 - \frac{x}{2}\right) + f'_y A'_s\left(h_0 - a'_s\right)}{\eta_s} \tag{6.58}$$

Rewriting Eqs. (6.44) and (6.45), we have

$$N_c = \alpha_1 f_c bx + f'_y A'_s - \sigma_s A_s \tag{6.59}$$

$$M_u = N_c e_0 = \frac{\alpha_1 f_c bx\left(h_0 - \frac{x}{2}\right) + f'_y A'_s\left(h_0 - a'_s\right) - N_c\left(\eta_s e_a + \frac{h}{2} - a_s\right)}{\eta_s} \tag{6.60}$$

M_u can be evaluated according to the following steps:

1. Check $N_c \leqslant N_{cu}$. If it is true, continue the calculation. Otherwise, $M_u = 0$.
2. If $l_0/h \leqslant 5$, $\eta_s = 1.0$. Otherwise, calculate η_s using Eq. (6.11).
3. Calculate e_a.
4. Calculate ξ_b using Eq. (6.40).
5. Assume that the section is a large eccentric section first, and calculate x (or ξ) using Eq. (6.57).
6. If $\xi \leqslant \xi_b$ and $x = \xi h_0 \geqslant 2\,a'_s$, it is a large eccentrically compressed section and A'_s can yield. M_u can be calculated directly using Eq. (6.58).
7. If $\xi\ \xi_b$ and $X = \xi h_{\leqslant 0} < 2\,a'_s$, it is a large eccentrically compressed section and A'_s cannot yield. N_{cu} can be calculated using either one of the following two methods: (1) calculate x (or ξ) by substituting $\sigma'_s = E_s\varepsilon_{cu}\left(\frac{\beta_1 a'_s}{\xi h_0} - 1\right)$ into Eq. (6.57) first, and then evaluate M_u by substituting σ'_s and x (or ξ) into Eq. (6.58); (2) evaluate M_u directly using Eq. (6.61) by assuming $x = 2\,a'_s$.

$$M_u = N_c e_0 = \frac{f_y A_s\left(h_0 - a'_s\right) - N_c\left(\eta_s e_a - \frac{h}{2} + a'_s\right)}{\eta_s} \tag{6.61}$$

8. If $\xi > \xi_b$, the assumption in Step (5) is not valid, and the section is a small eccentrically compressed section. Substitute Eq. (6.48) into Eq. (6.59) to obtain x (or ξ), and then evaluate M_u using Eq. (6.60).

6.6.3 *Design of Symmetrically Reinforced Sections*

In real engineering practice, sometimes sections of eccentrically compressed members suffer moments in opposite directions. For example, the direction of moments on columns in frames and bent frames and on bridge piers, which are subjected to seismic load or wind load of undetermined directions, will vary with the load direction. Therefore, symmetric reinforcement, i.e., the reinforcement at the two opposite faces of a section is of the same type and area, is a very common practice in eccentrically compressed members. Similar to asymmetrically reinforced sections, whether the section has a large or small eccentricity should be determined first in the design of a symmetrically reinforced section. Then, the calculations will go on according to the characteristics of large or small eccentrically compressed sections.

6.6.3.1 Determine Section Type

Because $A'_s = A_s$ and $f'_y = f_y$, in the basic equations for large eccentricity sections, i.e., Eqs. (6.49) and (6.50), there are only two unknowns, which can be solved uniquely without any additional conditions. And because $f'_y A'_s$ and $f_y A_s$ are equal in the magnitude but opposite in the sign, they will cancel out. ξ can be directly obtained as

$$\xi = \frac{N_c}{\alpha_1 f_c b h_0} \tag{6.62}$$

Obviously, when $\xi \leqslant \xi_b$, the section has a large eccentricity; otherwise, it has a small eccentricity. But if the section has a small eccentricity, the basic equations for small eccentric sections should be used to calculate ξ rather than Eq. (6.62).

In balanced failure, $\xi = \xi_b$; so, Eq. (6.49) becomes

$$N_{cb} = \alpha_1 f_c b h_0 \xi_b \tag{6.63}$$

When $N_c \leqslant N_{cb}$, the section has a large eccentricity; otherwise, it has a small eccentricity.

The section type can be determined by either Eq. (6.62) or Eq. (6.63) for convenience.

6.6.3.2 Design of Large Eccentrically Compressed Sections

Given N_c, M, l_0, b, h, f_c, f_y, f_y' ($f_y' = f_y$) and E_s, A_s' and A_s ($A_s' = A_s$) can be calculated according to the following steps:

1. If $l_0/h \leqslant 5$, $\eta_s = 1.0$. Otherwise, calculate η_s using Eq. (6.14).
2. Calculate e_i using Eq. (6.4).
3. Calculate ξ_b using Eq. (6.40) or calculate N_{cb} by Eq. (6.63).
4. Calculate ξ by Eq. (6.62).
5. If $\xi \leqslant \xi_b$ (or $N_c \leqslant N_{cb}$), i.e., the section has large eccentricity, continue the calculation. Otherwise, the section should be redesigned as having a small eccentricity.
6. If $x = \xi h_0 \geqslant 2\,a_s'$, calculate $A_s' = A_s$ using Eq. (6.50).
7. If $x = \xi h_0 < 2\,a_s'$, A_s' cannot yield. Either one of the following two methods can be used in the design: The first one is to replace f_y' in Eq. (6.49) and Eq. (6.50) by $\sigma_s' = E_s \varepsilon_{cu}\left(\frac{\beta_1 a_s'}{\xi h_0} - 1\right)$, and then calculate ξ and $A_s = A_s'$. The second method assumes $x = 2\,a_s'$; then, calculate $A_s = A_s'$ by Eq. (6.51).
8. If $A_s = A_s' < \rho_{min} bh$, take $A_s = A_s' = \rho_{min} bh$.
9. Check the compressive capacity of the member (as an axially compressed member) in the out-of-plane direction with the calculated A_s and A_s'.

6.6.3.3 Design of Small Eccentrically Compressed Sections

The design of small eccentricity sections is to determine A_s' and A_s ($A_s' = A_s$) when N_c, M, l_0, b, h, f_c, f_y, f_y' ($f_y' = f_y$), and E_s are given.

For small eccentricity sections, because $\sigma_s < f_y$, ξ should be determined by solving Eqs. (6.52) and (6.53) simultaneously. Substituting σ_s calculated by Eq. (6.48) into Eq. (6.52) and considering $f_y'A_s' = f_yA_s$, we have

$$f_y'A_s' = \frac{(N_c - \alpha_1 f_c b h_0 \xi)(\beta_1 - \xi_b)}{\xi - \xi_b} \tag{6.64}$$

Substituting Eq. (6.64) into Eq. (6.53), we get a cubic equation

$$\begin{aligned} &0.5\xi^3 - (1+0.5\xi_b)\xi^2 + \left[\frac{N_c e}{\alpha_1 f_c b h_0^2} + (\beta_1 - \xi_b)\left(1 - \frac{a_s'}{h_0}\right) + \xi_b\right]\xi \\ &- \alpha_1 \frac{N_c}{f_c b h_0}\left[\frac{e}{h_0}\xi_b + (\beta_1 - \xi_b)\left(1 - \frac{a_s'}{h_0}\right)\right] \\ &= 0 \end{aligned} \tag{6.65}$$

Solution of this formula yields the value of ξ.

The calculation of ξ can be simplified to avoid solving the cubic equation. From the foregoing derivation of Eq. (6.65), it can be found that if $\xi - 0.5\xi^2$ is replaced by a linear expression of ξ or a constant, the order of the equation will be reduced. So the relation between $\xi - 0.5\xi^2$ and ξ is investigated for small eccentricity sections ($\xi > \xi_b$) and shown by the quadratic curve (the solid line) in Fig. 6.20. The value of $\xi - 0.5\xi^2$ varies between 0.375 and 0.500 when the value of ξ varies between 0.5 and 1.0. When $\xi = 1.0$, $\xi - 0.5\xi^2$ takes the maximum value. And when $\xi > 1.0$, the curve is descending. Because the variation amplitude of $\xi - 0.5\xi^2$ is not big for small eccentricity sections, $\xi - 0.5\xi^2 = 0.43$ (the dash and dot line in Fig. 6.20) is assumed to simplify the calculation of ξ, which is about the average value of the maximum and minimum of the quadratic curve.

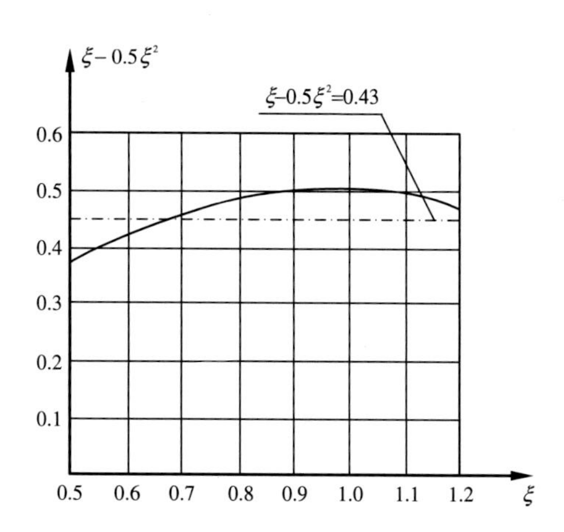

Fig. 6.20 Relationship between $\xi - 0.5\xi^2$ and ξ

Substituting $\xi - 0.5\xi^2 = 0.43$ into Eq. (6.53), we have

$$N_c e = 0.43\alpha_1 f_c b h_0^2 + f_y' A_s' (h_0 - a_s') \tag{6.66}$$

So,

$$f_y' A_s' = \frac{N_c e - 0.43\alpha_1 f_c b h_0^2}{h_0 - a_s'}$$

Substituting Eq. (6.64) into the above equation, we have

$$\frac{(N_c - \alpha_1 f_c b h_0 \xi)(\beta_1 - \xi_b)}{\xi - \xi_b} = \frac{N_c e - 0.43\alpha_1 f_c b h_0^2}{h - a_s'}$$

$$\begin{aligned} \frac{N_c e - 0.43\alpha_1 f_c b h_0^2}{(\beta_1 - \xi_b)(h - a_s')}(\xi - \xi_b) &= N_c - \alpha_1 f_c b h_0 \xi \\ &= N_c - \alpha_1 f_c b h_0 (\xi - \xi_b) - \alpha_1 f_c b h_0 \xi_b \end{aligned} \tag{6.67}$$

Rearranging the equation above, we obtain

$$\xi = \frac{N_c - \alpha_1 f_c b h_0 \xi_b}{\frac{N_c e - 0.43\alpha_1 f_c b h_0^2}{(\beta_1 - \xi_b)(h_0 - a_s')} + \alpha_1 f_c b h_0} + \xi_b \tag{6.68}$$

Using Eq. (6.68), ξ can be determined directly, and thus, it is not necessary to solve the cubic equation (6.65). And from the above analysis, it can be seen that the error of this approximate calculation of ξ is so small that it can be neglected in designs. The detailed steps for the design of a small eccentricity column section are as follows:

1. If $l_0/h \leqslant 5$, $\eta_s = 1.0$. Otherwise, calculate η_s using Eq. (6.14).
2. Calculate e_i using Eq. (6.4).
3. Calculate ξ_b using Eq. (6.40) or calculate N_{cb} by Eq. (6.63).
4. Calculate ξ by Eq. (6.62).
5. If $\xi > \xi_b$ (or $N_c > N_{cb}$), i.e., the section is a small eccentricity one, continue the calculation. Otherwise, the section should be redesigned as a large eccentricity.
6. Recalculate ξ using Eq. (6.68).
7. Determine $A_s' = A_s$ by Eq. (6.53). Due to symmetric reinforcement, A_s will not yield first under compression. So it is not necessary to check the A_s content using Eq. (6.56).
8. If $A_s = A_s' < \rho_{min} bh(\rho'_{min} bh)$,take $A_s = A_s' = \rho_{min} bh$.
9. Check the compressive capacity of the member (as axially compressed) in the out-of-plane direction.

Example 6.11 In Example 6.7, if the reinforcement is symmetrically placed, determine the reinforcement.

Solution

1. Calculate e_i, η_s, and e
   From Example 6.7, we know that h_0 = 460 mm, η_s = 1.030, e_0 = 700 mm, e_a = 20 mm, e_i = 720 mm, and e = 951 mm.
2. Determine whether it is a large eccentrically compressed section or not
   $\xi = \frac{N_c}{f_c b h_0} = \frac{400\times10^3}{16.7\times300\times460} = 0.174 < \xi_b = 0.55$, and it is a large eccentrically compressed section.
3. Determine A'_s and A_s

$$x = \xi h_0 = 0.174 \times 460 = 80\,\text{mm} = 2a'_s = 80\,\text{mm}$$

$$\begin{aligned} A'_s &= \frac{N_c e - \alpha_1 f_c b h_0^2 (\xi - 0.5\xi^2)}{f'_y (h_0 - a'_s)} \\ &= \frac{400 \times 10^3 \times 951 - 16.7 \times 300 \times 460^2 \times (0.174 - 0.5 \times 0.174^2)}{300 \times (460 - 40)} \\ &= 1682\,\text{mm}^2 > \rho'_{\min} bh = 0.002 \times 300 \times 500 = 300\,\text{mm}^2,\ \text{OK.} \\ A_s &= A'_s = 1682\,\text{mm}^2 > \rho_{\min} bh = 0.002 \times 300 \times 500 = 300\,\text{mm}^2,\ \text{OK.} \end{aligned}$$

4. Check the compressive capacity of the member in the out-of-plane direction (omitted for brevity).

Example 6.12 In Example 6.9, if the reinforcement is symmetrically placed, determine the reinforcement.

Solution

1. Calculate e_i, η_s and e
   From Example 6.9, we know that e_i = 92 mm, η_s = 1.117, e = 313 mm.
2. Determine whether it is a large eccentrically compressed section or not

$$N_{cb} = f_c b h_0 \xi_b = 16.7 \times 300 \times 460 \times 0.55 = 1267.53\ \text{kN}$$

   N_c = 2500 kN > N_{cb}, it is a small eccentrically compressed section.
3. Calculate ξ
   Calculate ξ using Eq. (6.65) first,

$$\xi^3 - 2.55\xi^2 + 3.050\xi - 1.318 = 0$$
$$\xi = 0.797$$

   Calculate ξ using Eq. (6.68),

$$\xi = \frac{N_c - \alpha_1 f_c b h_0 \xi_b}{\frac{N_c e - 0.43\alpha_1 f_c b h_0^2}{(\beta_1 - \xi_b)(h_0 - a_s')} + \alpha_1 f_c b h_0} + \xi_b$$

$$= \frac{2500 \times 10^3 - 16.7 \times 300 \times 460 \times 0.55}{\frac{2500\times10^3\times313-0.43\times16.7\times300\times460^2}{(0.8-0.55)(460-40)} + 16.7 \times 300 \times 460} + 0.55$$

$$= 0.776$$

4. Calculate A_s' when $\xi = 0.797$,

$$A_s' = \frac{N_c e - \alpha_1 f_c b h_0^2 (\xi - 0.5\xi^2)}{f_y'(h_0 - a_s')}$$

$$= \frac{2500 \times 10^3 \times 313 - 16.7 \times 300 \times 460^2 \times (0.797 - 0.5 \times 0.797^2)}{300 \times (460 - 40)}$$

$$= 2177\,\text{mm}^2 > \rho'_{min} bh = 0.002 \times 300 \times 500 = 300\,\text{mm}^2$$

When $\xi = 0.776$,

$$A_s' = \frac{N_c e - \alpha_1 f_c b h_0^2 (\xi - 0.5\xi^2)}{f_y'(h_0 - a_s')}$$

$$= \frac{2500 \times 10^3 \times 313 - 16.7 \times 300 \times 460^2 \times (0.778 - 0.5 \times 0.778^2)}{300 \times (460 - 40)}$$

$$= 2211\,\text{mm}^2 > \rho'_{min} bh = 0.002 \times 300 \times 500 = 300\,\text{mm}^2$$

The reinforcement calculated by the two different ξ values differs by only $\frac{2211-2177}{2177} = 1.6\,\%$. And the approximate equation gives a more conservative result.

5. Check the compressive capacity of the member in the out-of-plane direction (omitted for brevity).

After comparing the calculation results of Examples 6.7, 6.9, 6.11 and 6.12, we can conclude that:

1. Symmetrically reinforced sections generally have larger steel contents than asymmetrically reinforced sections. So from the point of view of economy, symmetric reinforcement is not optimum.
2. For small eccentricity sections, the steel content near the axial force does not differ too much between a symmetric reinforcement design and an asymmetric reinforcement design. For instance, the steels contents are nearly the same in Examples 6.9 and 6.12, of which all other conditions are the same except for the reinforcement arrangement. This is mainly because the stresses in the steels far from the axial force are small. Increasing the steel content in the symmetric

reinforcement design cannot significantly enhance the ultimate compressive capacity of the section.

6.6.4 *Evaluation of Ultimate Compressive Capacities of Existing Symmetrically Reinforced Eccentrically Compressed Members*

The calculation method of the ultimate compressive capacity of existing symmetrically reinforced eccentrically compressed members is similar to that of existing asymmetrically reinforced eccentrically compressed members. It is left to the readers to conclude the calculation steps.

6.7 Analysis of Eccentrically Compressed Members of I Section

Eccentrically compressed members of I section are widely used in engineering practice, such as the columns in single-story factories. The mechanical behavior of eccentrically compressed members of I sections is almost the same as that of members of rectangular sections. In this section, the simplified analysis method for eccentrically compressed members of I section will be briefly introduced.

6.7.1 *Basic Equations for Ultimate Compressive Capacities of Large Eccentrically Compressed I Sections*

The neutral axis of an I section may be located in the flange near the axial force (Case 1) or in the web of the section (Case 2). Rectangular stress blocks can be used to simplify the real compressive stress distribution in concrete (Fig. 6.21).

Case 1 If $x \leqslant h'_f$, the neutral axis will be located in the compression flange and the mechanical behavior of the section is equivalent to that of a rectangular section of $b'_f \times h$ (Fig. 6.21a). For equilibrium,

$$N_{cu} = \alpha_1 f_c b'_f x + f'_y A'_s - f_y A_s \quad (6.69)$$

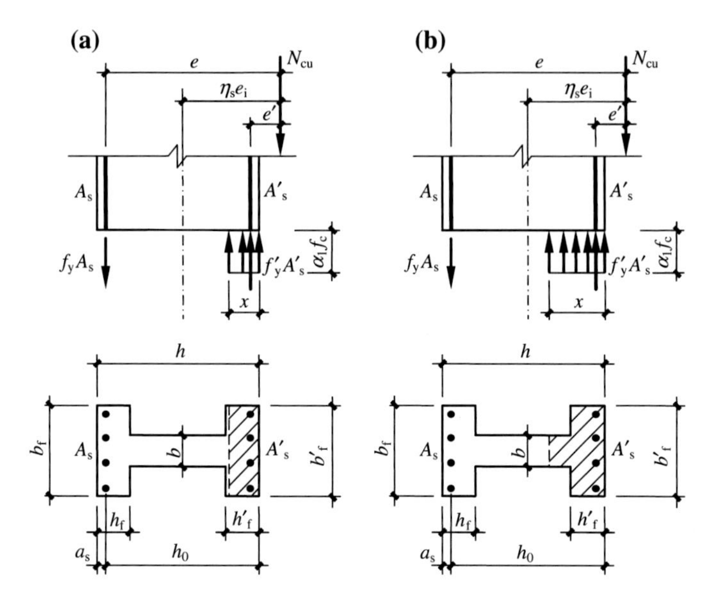

Fig. 6.21 Simplified stress distributions across large eccentrically compressed I sections. **a** Case 1, **b** Case 2

$$N_{cu}e = \alpha_1 f_c b'_f x(h_0 - \frac{x}{2}) + f'_y A'_s (h_0 - a'_s) \quad (6.70)$$

where $e = \eta_s e_i + \frac{h}{2} - a_s$.

Case 2 If $x > h'_f$, the neutral axis will be located in the web of the section and the mechanical behavior of the section is similar to that of a T section under flexure (Fig. 6.21b). For equilibrium,

$$N_{cu} = \alpha_1 f_c bx + \alpha_1 f_c (b'_f - b) h'_f + f'_y A'_s - f_y A_s \quad (6.71)$$

$$N_{cu}e = \alpha_1 f_c bx(h_0 - \frac{x}{2}) + \alpha_1 f_c (b'_f - b) h'_f (h_0 - \frac{h'_f}{2}) f'_y A'_s (h_0 - a'_s) \quad (6.72)$$

where $e = \eta_s e_i + \frac{h}{2} - a_s$.

The applicable conditions of the above equations are $x \leqslant x_b$ (or $\xi \leqslant \xi_b$) and $x \geqslant 2a'_s$. For Case 2, the condition of $x \geqslant 2a'_s$ can be generally met automatically.

6.7.2 Basic Equations for Ultimate Compressive Capacities of Small Eccentrically Compressed I Sections

For small eccentrically compressed I sections, the neutral axis may be in the web, or in the flange far away from the axial force when the eccentricity is very small, or even out of the section (the full section is under compression). Considering the location of the neutral axis, the corresponding stress distributions across small eccentrically compressed I sections are illustrated in Fig. 6.22.

Case 1 If $x < h - h_{\mathrm{f}}$, the neutral axis locates in the web of the section (Fig. 6.22a). The equilibrium equations are

$$N_{\mathrm{cu}} = \alpha_1 f_{\mathrm{c}} bx + \alpha_1 f_{\mathrm{c}}(b'_{\mathrm{f}} - b)h'_{\mathrm{f}} + f'_{\mathrm{y}} A'_{\mathrm{s}} - \sigma_{\mathrm{s}} A_{\mathrm{s}} \tag{6.73}$$

$$N_{\mathrm{cu}} e = \alpha_1 f_{\mathrm{c}} bx(h_0 - \frac{x}{2}) + \alpha_1 f_{\mathrm{c}}(b'_{\mathrm{f}} - b)h'_{\mathrm{f}}\left(h_0 - \frac{h'_{\mathrm{f}}}{2}\right) + f'_{\mathrm{y}} A'_{\mathrm{s}}(h_0 - a'_{\mathrm{s}}) \tag{6.74}$$

where $e = \eta_{\mathrm{s}} e_{\mathrm{i}} + \frac{h}{2} - a_{\mathrm{s}}$. σ_{s} can be calculated by using Eq. (6.46) or Eq. (6.48) (if $f_{\mathrm{cu}} > 50$ MPa, 0.8 in Eq. (6.48) should be replaced by β_1).

Case 2 If $h - h_{\mathrm{f}} < x < h$, the neutral axis locates in the flange far away from the axial force (Fig. 6.22b). The equilibrium equations are

$$N_{\mathrm{cu}} = \alpha_1 f_{\mathrm{c}} bx + \alpha_1 f_{\mathrm{c}}(b'_{\mathrm{f}} - b)h'_{\mathrm{f}} + \alpha_1 f_{\mathrm{c}}(b_{\mathrm{f}} - b)(h_{\mathrm{f}} - h + x) + f'_{\mathrm{y}} A'_{\mathrm{s}} - \sigma_{\mathrm{s}} A_{\mathrm{s}} \tag{6.75}$$

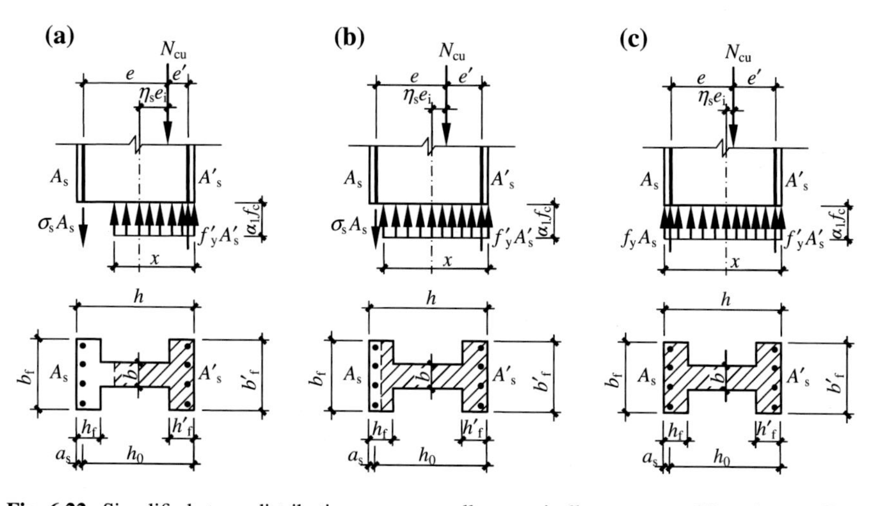

Fig. 6.22 Simplified stress distributions across small eccentrically compressed I sections. **a** Case 1, **b** Case 2, **c** Case 3

$$N_{cu}e = \alpha_1 f_c bx(h_0 - \frac{x}{2}) + \alpha_1 f_c(b'_f - b)h'_f(h_0 - \frac{h'_f}{2}) + \alpha_1 f_c(b_f - b)(h_f - h + x)(\frac{2h_0 + h_f - h - x}{2}) + f'_y A'_s(h_0 - a'_s) \quad (6.76)$$

Case 3 If $x = h$, the full section is under compression (Fig. 6.22c). The equilibrium equations can be expressed as follows:

$$N_{cu} = \alpha_1 f_c bx + \alpha_1 f_c(b'_f - b)h'_f + \alpha_1 f_c(b_f - b)h_f + f'_y A'_s + f_y A_s \quad (6.77)$$

$$N_{cu}e = \alpha_1 f_c bh\left(h_0 - \frac{h}{2}\right) + \alpha_1 f_c(b'_f - b)h'_f\left(h_0 - \frac{h'_f}{2}\right) + \alpha_1 f_c(b_f - b)h_f(\frac{h_f}{2} - a_s) + f'_y A'_s(h_0 - a'_s) \quad (6.78)$$

6.8 Applications of the Ultimate Capacity Equations for Eccentrically Compressed Members of I Section

6.8.1 Design of I Sections

I sections are generally symmetrically reinforced. To guarantee the safety of the designed section, the loads on the section must satisfy $N_c \leqslant N_{cu}$ and $M \leqslant N_{cu}e_0$.

6.8.1.1 Practical Method to Distinguish Large and Small Eccentricity Sections in Design

Because $f_y A_s = f'_y A'_s$, the ultimate compressive capacity of an I section under balanced failure is

$$N_{cb} = \alpha_1 f_c bh_0 \xi_b + \alpha_1 f_c(b'_f - b)h'_f \quad (6.79)$$

When $N_c \leqslant N_{cb}$, it has a large eccentrically compressed section; when $N_c > N_{cu}$, it has a small eccentrically compressed section.

Of course, we can also use ξ to distinguish large and small eccentrically compressed sections.

$$\xi = \frac{N_c - \alpha_1 f_c(b'_f - b)h'_f}{\alpha_1 f_c bh_0} \quad (6.80)$$

When $\xi \leqslant \xi_b$, it is a large eccentrically compressed section; when $\xi > \xi_b$, it is a small eccentrically compressed section.

6.8.1.2 Design of Large Eccentrically Compressed I Sections

I sections with the neutral axes in the flange near the axial force can be designed as rectangular sections of $b'_f \times h$. It should be noted that if $x < 2\,a'_s$, $A_s = A'_s$ can be calculated either by Eq. (6.51) with $x = 2\,a'_s$ assumed or by Eqs. (6.49) and (6.50) with σ'_s obtained first.

For I sections with the neutral axes in the web, it is found that Eqs. (6.71) and (6.72) are different from Eqs. (6.49) and (6.50) in two constant terms, i.e.,

$$\left.\begin{aligned} C_1 &= \alpha_1 f_c (b'_f - b) h'_f \\ M_1 &= \alpha_1 f_c (b'_f - b) h'_f \left(h_0 - \frac{h'_f}{2} \right) \end{aligned}\right\} \tag{6.81}$$

Noting $f_y A_s = f'_y A'_s$, we can rewrite Eqs. (6.71) and (6.72) as

$$N_c = \alpha_1 f_c b x + C_1 \tag{6.82}$$

$$N_c e = \alpha_1 f_c b x \left(h_0 - \frac{x}{2} \right) + f'_y A'_s (h_0 - a'_s) + M_1 \tag{6.83}$$

So the calculation method and steps are similar to those for rectangular sections. Eq. (6.80) can be used to calculate x or ξ, and the steel content can be determined by the following equations;

$$A'_s = \frac{N_c e - \alpha_1 f_c b h_0^2 (\xi - 0.5\xi^2) - M_1}{f'_y (h_0 - a'_s)} \tag{6.84}$$

$$A_s = A'_s \tag{6.85}$$

6.8.1.3 Design of Small Eccentrically Compressed I Sections

For small eccentrically compressed I sections, the stress in the steels far away from the axial force can be calculated by Eq. (6.48) (if $f_{cu} > 50$ MPa, 0.8 in Eq. (6.48) should be replaced by β_1).

Case 1 By replacing the constant terms in Eqs. (6.73) and (6.74) by Eq. (6.81), the section design is similar to that of rectangular sections. In the calculation, ξ can be determined by

$$\xi = \frac{N_c - \alpha_1 f_c b h_0 \xi_b - C_1}{\frac{N_c e - 0.43\alpha_1 f_c b h_0^2 - M_1}{(\beta_1 - \xi_b)(h_0 - a_s')} + \alpha_1 f_c b h_0} + \xi_b \tag{6.86}$$

Case 2 Because the stress and the resultant force in the flange far away from the axial force are very small, the contribution of this flange to the ultimate capacity of the section can be neglected. Using the equations for Case 1 to design sections of Case 2 will not result in big errors, and the results are conservative.

Case 3 $A_s' = A_s$ can be determined by directly using Eq. (6.78).

If the section is asymmetrically reinforced, A_s must meet the requirement of Eq. (6.87) to avoid the steels far away from the axial force yielding first.

$$\begin{aligned} N_c e' = {} & \alpha_1 f_c b h \left(h_0 - \frac{h}{2} \right) + \alpha_1 f_c (b_f' - b) h_f' \left(\frac{h_f'}{2} - a_s' \right) \\ & + \alpha_1 f_c (b_f - b) h_f \left(h_0' - \frac{h_f}{2} \right) + f_y A_s (h_0' - a_s) \end{aligned} \tag{6.87}$$

The detailed design of I sections can refer to that of rectangular sections and will not be listed herein for brevity.

Example 6.13 There is an I section column in a single-story factory. The effective length of the column is l_0 = 7.56 m. The section of the column is shown in Fig. 6.23a. The material properties are f_c = 16.7 N/mm^2, $f_y' = f_y$ = 300 N/mm^2, and E_s = 2 × 10^5 N/mm^2. If the symmetrically reinforced section is subjected to N_c = 900 kN and M = 360 kN·m, determine A_s' and A_s.

Solution Calculate A_s' and A_s based on the simplified section (Fig. 6.23b). Assume $a_s = a_s'$ = 40 mm and ξ_b = 0.55.

1. Calculate e_i, η_s, and e

   $h_0 = h - a_s = 700 - 40 = 660\,\text{mm}$, $\frac{l_0}{h} = \frac{7560}{700} = 10.8 > 5$, η_s should be calculated.

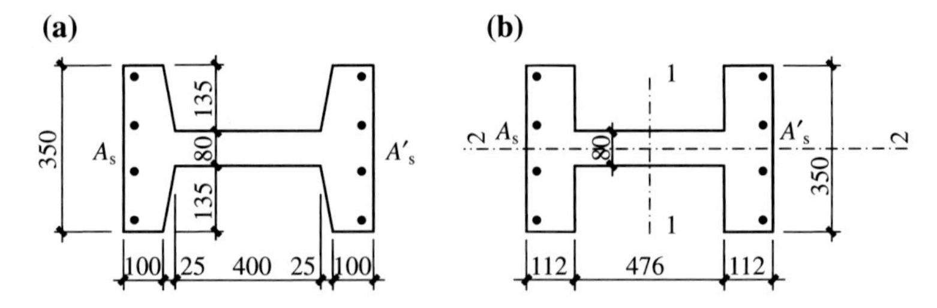

Fig. 6.23 Sectional dimensions and reinforcement. **a** The real section. **b** The simplified section

$$e_0 = \frac{M}{N} = \frac{360}{900} = 0.40\,\text{m} = 400\,\text{mm},\ \frac{h}{30} = \frac{700}{30} = 23.3 > 20,\ \text{take}\ e_a = 23\,\text{mm}.$$

$$e_i = e_0 + e_a = 400 + 23 = 423\,\text{mm},\ \frac{e_i}{h_0} = \frac{423}{660} = 0.641$$

$$\zeta_c = \frac{0.5 f_c A}{N_c} = \frac{0.5 \times 16.7 \times 116{,}480}{900{,}000} = 1.081 > 1.0,\ \text{take}\ \zeta_c = 1.0.$$

$$\eta_s = 1 + \frac{1}{1300 \times 0.641} \times 10.8^2 \times 1.0 = 1.140$$

$$e = \eta_s e_i + \frac{h}{2} - a_s = 1.140 \times 423 + \frac{700}{2} - 40 = 792\,\text{mm}$$

2. Determine whether it is a large eccentrically compressed section or not

$$C_1 = \alpha_1 f_c \left(b'_f - b\right) h'_f = 16.7 \times (350 - 80) \times 112 = 505{,}008\,\text{N}$$

$$\xi = \frac{N_c - C_1}{\alpha_1 f_c b h_0} = \frac{900 \times 10^3 - 505{,}008}{16.7 \times 80 \times 660} = 0.448 < \xi_b = 0.55$$

It is a large eccentrically compressed section.
$x = \xi h_0 = 0.448 \times 660 = 296\ \text{mm} > 112\ \text{mm}$, the neutral axis locates in the web.

3. Determine A'_s and A_s

$$M_1 = \alpha_1 f_c \left(b'_f - b\right) h'_f \left(h_0 - \frac{h'_f}{2}\right) = 16.7 \times (350 - 80) \times 112 \times \left(660 - \frac{112}{2}\right) = 305{,}024{,}832\,\text{N} \cdot \text{mm}$$

$$\begin{aligned} A'_s &= \frac{N_c e - \alpha_1 f_c b h_0^2 (\xi - 0.5\xi^2) - M_1}{f'_y \left(h_0 - a'_s\right)} \\ &= \frac{900 \times 10^3 \times 792 - 16.7 \times 80 \times 660^2 \times (0.448 - 0.5 \times 0.448^2) - 305{,}024{,}832}{300 \times (660 - 40)} \\ &= 1105\,\text{mm}^2 > \rho'_{\min} A = 0.002 \times (80 \times 700 + 2 \times 270 \times 112) = 233\,\text{mm}^2, \text{OK.} \end{aligned}$$

$$A_s = A'_s = 1105\,\text{mm}^2$$

4. Check the compressive capacity of the member in the out-of-plane direction. From Fig. 6.23b, the area and the moment of inertia of the section can be calculated as $A = 116{,}480\ \text{mm}^2$, $I = 8.2064 \times 10^8\ \text{mm}^4$

$$i = \sqrt{\frac{I}{A}} = \sqrt{\frac{8.2064 \times 10^8}{116{,}480}} = 83.94\,\text{mm}$$

$\frac{l_0}{i} = 90.07\,\text{mm}$. According to Table 4.1, $\varphi = 0.60$

$$\frac{A_s'}{A} = \frac{2 \times 1105}{116{,}480} = 1.897\,\% < 3\,\%$$

$$N_{cu} = \varphi\left(f_c A + f_y' A_s'\right) = 0.6 \times (16.7 \times 116{,}480 + 300 \times 2210)$$
$$= 1564.930\,\text{kN} > N_c = 900\,\text{kN, OK.}$$

Example 6.14 In Example 6.13, if the applied loads are N_c = 1500 kN and M = 260 kN·m, determine A_s' and A_s.

Solution Calculate A_s' and A_s based on the simplified section (Fig. 6.23b).
Assume $a_s = a_s'$ = 40 mm and ξ_b = 0.55.

1. Calculate e_i, η_s, and e
$h_0 = h - a_s = 700 - 40 = 660\,\text{mm}$, $\frac{l_0}{h} = \frac{7560}{700} = 10.8 > 5$, η_s should be calculated.

$$e_0 = \frac{M}{N} = \frac{260}{1500} = 0.173\,\text{m} = 173\,\text{mm},\ \frac{h}{30} = \frac{700}{30} = 23.3 > 20,\ \text{take}\ e_a = 23\,\text{mm.}$$
$$e_i = e_0 + e_a = 173 + 23 = 196\,\text{mm},\ \frac{e_i}{h_0} = \frac{196}{660} = 0.298$$
$$\zeta_c = \frac{0.5 f_c A}{N_c} = \frac{0.5 \times 16.7 \times 116{,}480}{1{,}500{,}000} = 0.648$$
$$\eta_s = 1 + \frac{1}{1300 \times 0.298} \times 10.8^2 \times 0.648 = 1.195$$
$$e = \eta_s e_i + \frac{h}{2} - a_s = 1.195 \times 196 + \frac{700}{2} - 40 = 544\,\text{mm}$$

2. Determine whether it is a large eccentrically compressed section or not

$$C_1 = \alpha_1 f_c (b_f' - b) h_f' = 16.7 \times (350 - 80) \times 112 = 505{,}008\,\text{N}$$
$$\xi = \frac{N_c - C_1}{\alpha_1 f_c b h_0} = \frac{1500 \times 10^3 - 505{,}008}{16.7 \times 80 \times 660} = 1.128 > \xi_b = 0.55$$

It is a small eccentrically compressed section.

3. Calculate ξ

$$M_1 = \alpha_1 f_c (b_f' - b) h_f' \left(h_0 - \frac{h_f'}{2}\right) = 16.7 \times (350 - 80) \times 112 \times \left(660 - \frac{112}{2}\right) = 305{,}024{,}832\,\text{N}\cdot\text{mm}$$
$$\xi = \frac{N_c - \alpha_1 f_c b h_0 \xi_b - C_1}{\frac{N_c e - 0.43\alpha_1 f_c b h_0^2 - M_1}{(\beta_1 - \xi_b)(h_0 - a_s')} + \alpha_1 f_c b h_0} + \xi_b$$
$$= \frac{1500 \times 10^3 - 16.7 \times 80 \times 660 \times 0.55 - 505{,}008}{\frac{1500 \times 10^3 \times 544 - 0.43 \times 16.7 \times 80 \times 660 - 305{,}024{,}832}{(0.8 - 0.55) \times (660 - 40)} + 16.7 \times 80 \times 660} + 0.55$$
$$= 0.672$$
$$x = \xi h_0 = 0.672 \times 660 = 444\,\text{mm} < h - h_f = 700 - 112 = 588\,\text{mm}$$

The neutral axis locates in the web.

4. Calculate A_s'

$$A_s' = \frac{N_c e - \alpha_1 f_c b h_0^2 (\xi - 0.5\xi^2) - M_1}{f_y'(h_0 - a_s')}$$
$$= \frac{1500 \times 10^3 \times 554 - 16.7 \times 80 \times 660^2 \times (0.672 - 0.5 \times 0.672^2) - 305{,}024{,}832}{300 \times (660 - 40)}$$
$$= 1351 \text{ mm}^2 > \rho_{\min}' bh = 233 \text{ mm}^2, \text{ OK.}$$
$$A_s = A_s' = 1351 \text{ mm}^2$$

5. Check the compressive capacity of the member in the out-of-plane direction. From Fig. 6.23b, the area and the moment of inertia of the section can be calculated as A = 116,480 mm^2, $I = 8.2064 \times 10^8$ mm^4
Radius of gyration

$$i = \sqrt{\frac{I}{A}} = \sqrt{\frac{8.2064 \times 10^8}{116{,}480}} = 83.94 \text{ mm}$$

$\frac{l_0}{i} = \frac{7560}{83.94} = 90.07$ mm, $\varphi = 0.60$ from Table 4.1.

$$A_s' = 1351 \times 2 = 2702 \text{ mm}^2$$
$$\frac{A_s'}{A} = \frac{2702}{116{,}480} = 2.320\,\% < 3\,\%$$
$$N_{cu} = \varphi\left(f_c A + f_y' A_s'\right) = 0.60 \times (16.7 \times 116{,}480 + 300 \times 2702)$$
$$= 1653.490 \text{ kN} > N_c = 1500 \text{ kN, OK.}$$

6.8.2 *Evaluation of the Ultimate Compressive Capacities of Existing Eccentrically Compressed Members of I Sections*

The evaluation of ultimate compressive capacities of existing eccentrically compressed members of I sections is similar to that of members of rectangular sections and will not be detailed herein.

6.9 Analysis of Eccentrically Compressed Members with Biaxial Bending

Corner columns in reinforced concrete structures are typical examples of eccentrically compressed members with biaxial bending. Due to the randomness of the direction of seismic actions, reinforced concrete columns are usually subjected to the combination of axial force N_c and bending moments M_x and M_y around two principal axes.

Experimental results indicate that the plane section assumption is still valid for normal sections of eccentrically compressed members with biaxial bending in the sense of average strain, and the sections fail when the extreme compressive strain in the concrete reaches the ultimate value. Therefore, the strain at any point in the section can be evaluated when the neutral axis has been positioned. Then, the stresses in the steels and the concrete can be determined by corresponding constitutive laws, which have been introduced in Chap. 2. Finally, the ultimate compressive capacity of the section can be calculated by equilibrium conditions. However, integral operation and iteration operation are necessary in calculating the internal forces from the stresses and locating the neutral axis, respectively. So the calculation is very complicated and hard to perform.

In engineering practice, the principle of stress superposition in elastic theory is used to calculate the ultimate compressive capacity of eccentrically compressed members with biaxial bending approximately.

Assume that N_{cu0} is the axial compressive capacity of a member without considering the stability factor φ. N_{cux} and N_{cuy} are the compressive capacities of the member based on all longitudinal reinforcement when the axial force acts on x-axis and y-axis, respectively, and the additional eccentricity and moment magnifying coefficient ($\eta_{sx}e_{ix}$, $\eta_{sy}e_{iy}$) have been considered; N_{cu} is the compressive capacity of the member when the eccentricities ($\eta_{sx}e_{ix}$, $\eta_{sy}e_{iy}$) happen simultaneously Fig. (6.24); A_0 is the transformed area of the section; W_x and W_y are the transformed section moduli in x and y directions, respectively; the maximum stresses that the section can sustain under N_{cu0}, N_{cux}, N_{cuy}, and N_{cu}, respectively, are all f_c in elastic stage, i.e.,

$$\frac{N_{cu0}}{A_0} = f_c \tag{6.88}$$

$$N_{cux}\left(\frac{1}{A_0} + \frac{\eta_{sx}e_{ix}}{W_x}\right) = f_c \tag{6.89}$$

$$N_{cuy}\left(\frac{1}{A_0} + \frac{\eta_{sy}e_{iy}}{W_y}\right) = f_c \tag{6.90}$$

$$N_{cu}\left(\frac{1}{A_0} + \frac{\eta_{sx}e_{ix}}{W_x} + \frac{\eta_{sy}e_{iy}}{W_y}\right) = f_c \tag{6.91}$$

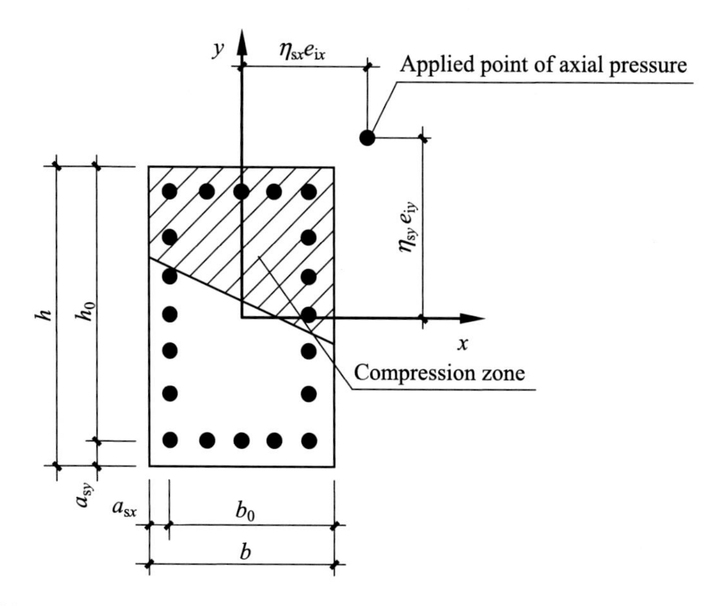

Fig. 6.24 An eccentrically compressed member with biaxial bending

Canceling f_c, A_0, W_x, and W_y in the above equations gives

$$\frac{1}{N_{cu}} = \frac{1}{N_{cux}} + \frac{1}{N_{cuy}} - \frac{1}{N_{cu0}} \tag{6.92}$$

$$N_{cu} = \frac{1}{\frac{1}{N_{cux}} + \frac{1}{N_{cuy}} - \frac{1}{N_{cu0}}} \tag{6.93}$$

Obviously, if N_{cu0}, N_{cux}, and N_{cuy} can be determined, then the ultimate compressive capacity of the member can be calculated directly by Eq. (6.93). N_{cu0} can be easily obtained by Eq. (4.20). And all longitudinal reinforcement should be considered in the calculation of N_{cux} and N_{cuy}. From Fig. 6.24, it is observed that part of the steels cannot yield when the member fails with uniaxial bending. So the stress in each steel bar must be calculated first by Eq. (5.62); then, N_{cux} and N_{cuy} can be determined by using corresponding equations.

Equation (6.93) can be used in design and capacity evaluation. The readers can summarize the detailed calculation steps by following the previous examples.

6.10 Analysis of Eccentrically Compressed Members of Circular Section

Eccentrically compressed members of circular section are widely used in public buildings and bridge structures. Generally, longitudinal reinforcement of circular sections is positioned evenly along the perimeter. In order to avoid failures around the weaker axis, at least six steel bars are required (Fig. 6.3). Different from the eccentrically compressed members of rectangular and I sections, the longitudinal reinforcement of circular sections will yield successively when approaching failure and the section width is not constant, both of which increase difficulties of stress analysis on the normal section. A simplified approach for the calculation of ultimate compressive capacities of circular sections will be introduced in this section.

6.10.1 *Stress and Strain Distributions Across the Section at Failure*

Given a circular section as shown in Fig. 6.25a, r is the radius and r_s is the radius of the circle composed at the center of longitudinal steel bars. When the number of longitudinal steel bars arranged evenly along the perimeter exceeds 6, the reinforcement is transferred to a steel loop with an area of A_s and a radius of r_s.

The strain distribution is obtained based on the plane section assumption and displayed in Fig. 6.25b. Thereafter, the stress distributions of the concrete and steel

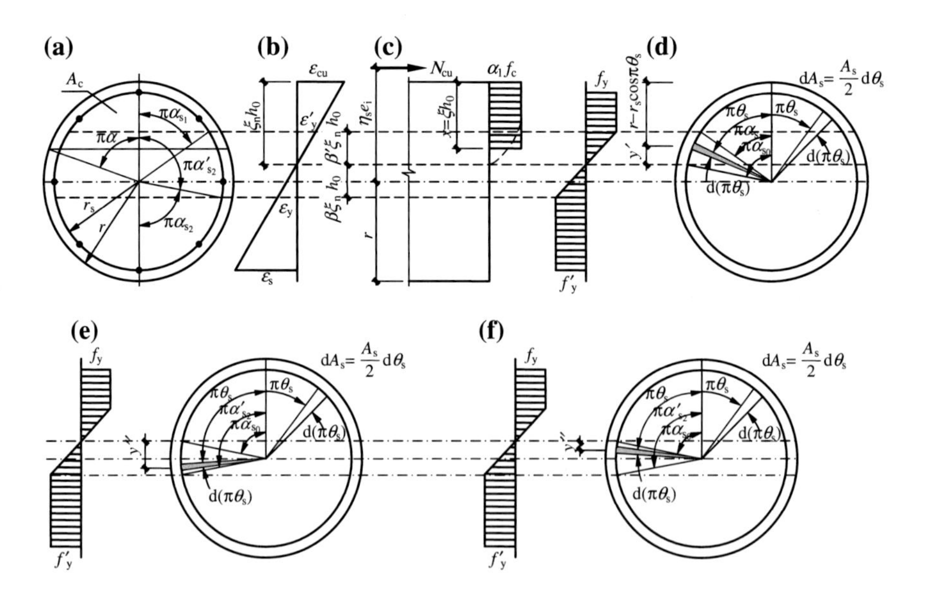

Fig. 6.25 Stress and strain distributions of eccentrically compressed members of circular section

loop are evaluated according to the stress–strain relationships and plotted in Fig. 6.25c and d, respectively. The compressive area of the concrete is an arch. The equivalent compressive area of concrete is denoted as A_c with a central angle of $2\pi\alpha$. The effective depth of the circular section is $h_0 = r + r_s$. Assume that the depth of the concrete in compression is $\xi_n h_0 = \xi_n(r + r_s)$. The central angle of the steel loop in compression is $2\pi\alpha_{s0}$, where α_{s0} is the ratio of the compressive area to the total area of the steel loop. The central angle corresponding to the intersection points between the circle with the radius of r_s and a line located at a distance of y to the extreme compression fiber of concrete is set as $2\pi\theta_s$. We then get

$$y = r - r_s \cos \pi\theta_s \tag{6.94}$$

$$\theta_s = \frac{1}{\pi}\cos^{-1}\left[\frac{r}{r_s} - \frac{y}{r_s}\right] \tag{6.95}$$

The values of θ_s corresponding to the heights of the neutral axis, equivalent compressive zone, and critical equivalent compressive zone, and the distances from the yielding points in compression and tension of the steel loop to the extreme compressive fiber, respectively, are calculated by Eq. (6.95) and listed in Table 6.1. α_{s1} is the relative area of the steel loop in the plastic compression zone, and $\alpha_{s2} = 1 - \alpha'_{s2}$ is the relative area of the steel loop in the plastic tension zone. Obviously, when $\xi_n = \xi_{nb}$ or $\alpha_s = \alpha_{sb}$, i.e., $(1 + \beta)\,\xi_n = 1.0$, $\alpha'_{s2} = 1.0$, and $\alpha_{s2} = 0$. β and β' are the ratios of compressive and tensile yielding strains of steel bars to the ultimate compressive strain of concrete, respectively.

$$\beta = \beta' = \frac{f_y}{\varepsilon_{cu}E_s} = \frac{f'_y}{\varepsilon_{cu}E_s} \tag{6.96}$$

Table 6.1 Key geometric parameters of the circular section

Definition	y	θ_s
Depth of the neutral axis	$\xi_n(r+r_s)$	$\alpha_{s0} = \frac{1}{\pi}\cos^{-1}\left[\frac{r}{r_s} - \xi_n\left(1+\frac{r}{r_s}\right)\right]$
Depth of the equivalent compression zone	$\beta_1\xi_n(r+r_s)$	$\alpha_s = \frac{1}{\pi}\cos^{-1}\left[\frac{r}{r_s} - \beta_1\xi_n\left(1+\frac{r}{r_s}\right)\right]$
Depth of the critical equivalent compression zone	$\beta_1\xi_{nb}(r+r_s)$	$\alpha_{sb} = \frac{1}{\pi}\cos^{-1}\left[\frac{r}{r_s} - \beta_1\xi_{nb}\left(1+\frac{r}{r_s}\right)\right]$
Distance from the compressive yielding point (f'_y) of the steel loop to the extreme compressive fiber	$(1-\beta')\xi_n(r+r_s)$	$\alpha_1 = \frac{1}{\pi}\cos^{-1}\left[\frac{r}{r_s} - (1-\beta')\xi_n\left(1+\frac{r}{r_s}\right)\right]$
Distance from the tensile yielding point (f_y) of the steel loop to the extreme compressive fiber	$(1+\beta)\xi_n(r+r_s)$	$\alpha'_{s2} = \frac{1}{\pi}\cos^{-1}\left[\frac{r}{r_s} - (1+\beta)\xi_n\left(1+\frac{r}{r_s}\right)\right]$

6.10.2 Calculation of Normal Section's Ultimate Bearing Capacities

6.10.2.1 Cross-Sectional Analysis

Assume that compressive forces and anticlockwise moments are positive and tensile forces and clockwise moments are negative.

1. Resultant force C and moment M_c of concrete in the compression zone
   The compressive stress of the equivalent rectangular stress block is $\alpha_1 f_c$. The area of the equivalent compressive area of concrete A_c is

$$A_c = r^2(\pi\alpha - \sin\pi\alpha\cos\pi\alpha) = \alpha\left(1 - \frac{\sin 2\pi\alpha}{2\pi\alpha}\right)A \tag{6.97}$$

   where $A = \pi r^2$ is the cross-sectional area.
   Hence, the resultant force of concrete in the compression zone C is

$$C = \alpha_1 f_c A_c = \alpha_1 f_c \alpha\left(1 - \frac{\sin 2\pi\alpha}{2\pi\alpha}\right)A \tag{6.98}$$

   The corresponding moment M_c about the center of the section can be obtained from Fig. 6.25a, d and expressed as

$$M_c = 2\int_0^{\alpha} \alpha_1 f_c r\cos\pi\theta dA_c = \frac{2}{3}\alpha_1 f_c A r\frac{\sin^3\pi\alpha}{\pi} \tag{6.99}$$

2. Resultant force and moment of the steel loop in the compression zone
   The resultant compressive force C_1 and moment about the center of the cross section M_{c1} of the steel loop in the plastic zone ($\sigma_s = f'_y$) are

$$C_1 = \alpha_{s1} f'_y A_s \tag{6.100}$$

$$M_{c1} = f'_y A_s r_s \frac{\sin\pi\alpha_{s1}}{\pi} \tag{6.101}$$

   Assume that the stress of the steel loop at a distance of y' from the neutral axis is σ'_s, and the corresponding central angle is $2\pi\theta_s$. It can be obtained from Fig. 6.25d that

$$\sigma'_s = \frac{f'_y y'}{\beta'\xi_n h_0} = \frac{f'_y}{\beta'\xi_n h_0}(\xi_n h_0 - r + r_s\cos\pi\theta_s) \tag{6.102}$$

Hence, the resultant compressive force C_2 of the steel loop in the elastic zone where α_s ranges from α_{s1} to α_{s0} is

$$C_2 = 2\int_{\alpha_{s1}}^{\alpha_{s0}} \sigma_s' \mathrm{d}A_s = \int_{\alpha_{s1}}^{\alpha_{s0}} \sigma_s' A_s \mathrm{d}\theta_s = f_y' A_s k_c \tag{6.103}$$

where

$$k_c = \frac{[\xi_n(1+r/r_s) - r/r_s]\pi(\alpha_{s0}-\alpha_{s1}) + \sin\pi\alpha_{s0} - \sin\pi\alpha_{s1}}{\pi\beta'\xi_n(1+r/r_s)} \tag{6.104}$$

The corresponding moment M_{C2} is

$$M_{c2} = \int_{\alpha_{s1}}^{\alpha_{s0}} \sigma_s' A_s r_s \cos\pi\theta_s \mathrm{d}\theta_s = f_y' A_s r_s \frac{m_c}{\pi} \tag{6.105}$$

where

$$m_c = \frac{[\xi_n(1+r/r_s) - r/r_s](\sin\pi\alpha_{s0} - \sin\pi\alpha_{s1}) + \frac{\pi(\alpha_{s0}-\alpha_{s1})}{2} + \frac{\sin 2\pi\alpha_{s0} - \sin 2\pi\alpha_{s1}}{4}}{\beta'\xi_n(1+r/r_s)} \tag{6.106}$$

3. Resultant force and moment of the steel loop in the tension zone
The resultant tensile force T_1 and moment about the center of the cross section M_{t1} of the steel loop in the plastic zone ($\sigma_s = f_y$) are

$$T_1 = -\alpha_2 f_y A_s \tag{6.107}$$

$$M_{t1} = f_y A_s r_s \frac{\sin\pi\alpha_{s2}}{\pi} \tag{6.108}$$

Assume that the stress of the steel loop at a distance of y'' from the neutral axis is σ_s, and the corresponding central angle is $2\pi\theta_s$. It can be obtained from Fig. 6.25e that

$$\begin{aligned} \sigma_s &= \frac{f_y y}{\beta\xi_n h_0} = \frac{f_y}{\beta\xi_n h_0}(r - \xi_n h_0 + r_s\cos(\pi - \pi\theta_s)) \\ &= \frac{f_y}{\beta\xi_n h_0}(r - \xi_n h_0 - r_s\cos\pi\theta_s) \end{aligned} \tag{6.109}$$

Here, $\pi\theta_s$ is an obtuse angle. When $\pi\theta_s$ is an acute angle, as shown in Fig. 6.25f, σ_s can also be expressed as $\sigma_s = \frac{f_y}{\beta\xi_n h_0}(r - \xi_n h_0 - r_s\cos\pi\theta_s)$.

Consequently, the resultant tensile force T_2 of the steel loop in the elastic zone where α_s ranges from α_{s0} to α'_{s2} is

$$T_2 = -\int_{\alpha_{s0}}^{\alpha'_{s2}} \sigma_s A_s \mathrm{d}\theta_s = -f_y A_s k_t \tag{6.110}$$

where

$$k_t = \frac{[\xi_n(1 + r/r_s) - r/r_s]\pi(\alpha_{s0} - \alpha'_{s2}) + \sin\pi\alpha_{s0} - \sin\pi\alpha'_{s2}}{\pi\beta\xi_n(1 + r/r_s)} \tag{6.111}$$

When $\pi\theta_s$ is an obtuse angle (Fig. 6.25e), the moment is positive (anticlockwise) and can be expressed as $M_{t2-1} = \int_{\alpha_{s0}}^{1/2} \sigma_s A_s r_s \cos(\pi - \pi\theta_s)\mathrm{d}\theta_s$; when $\pi\theta_s$ is an acute angle (Fig. 6.25f), the moment is negative (clockwise) and can be expressed as $M_{t2-2} = -\int_{1/2}^{\alpha'_{s2}} \sigma_s A_s r_s \cos\pi\theta_s \mathrm{d}\theta_s$. They are found to be equal and therefore expressed as

$$M_{t2} = \int_{\alpha_{s0}}^{\alpha'_{s2}} \sigma_s A_s r_s \cos(\pi - \pi\theta_s)\mathrm{d}\theta_s = f_y A_s r_s \frac{m_t}{\pi} \tag{6.112}$$

where

$$m_t = \frac{[\xi_n(1 + r/r_s) - r/r_s](\sin\pi\alpha'_{s2} - \sin\pi\alpha_{s0}) + \frac{\pi(\alpha'_{s2} - \alpha_{s0})}{2} + \frac{\sin 2\pi\alpha'_{s2} - \sin 2\pi\alpha_{s0}}{4}}{\beta\xi_n(1 + r/r_s)} \tag{6.113}$$

6.10.2.2 Calculation of Ultimate Bearing Capacities

The ultimate bearing capacities are obtained according to the equilibrium conditions.

$$N_{cu} = \alpha_1 f_c \alpha\left(1 - \frac{\sin 2\pi\alpha}{2\pi\alpha}\right)A + f'_y A_s(\alpha_{s1} + k_c) - f_y A_s(\alpha_{s2} + k_t) \tag{6.114}$$

$$N_{cu}\eta_s e_i = \frac{2}{3}\alpha_1 f_c A r \frac{\sin^3\pi\alpha}{\pi} + f'_y A_s r_s \frac{\sin\pi\alpha_{s1} + m_c}{\pi} + f_y A_s r_s \frac{\sin\pi\alpha_{s2} + m_t}{\pi} \tag{6.115}$$

When r/r_s, β', and β are known, α_{s0}, α_s, α_{s1}, and α_{s2} corresponding to different values of ξ_n can be evaluated by Table 6.1, and also, k_c, m_c, k_t, and m_t can be calculated by Eqs. (6.104), (6.106), (6.111), and (6.113), respectively. Afterward, the ultimate bearing capacities are obtained. However, considering the complicated expressions of Eqs. (6.114) and (6.115), it is necessary to simplify the formulae.

6.10.3 Simplified Calculation of Ultimate Bearing Capacities

For concrete, when $f_c \leqslant 50$ MPa, $\varepsilon_{cu} = 0.0033$; when $f_c = 80$ MPa, $\varepsilon_{cu} = 0.0030$. For steel bars of grades HRB335, HRB400, and HRB500, $E_s = 2.0 \times 10^5$ MPa. According to Eq. (6.96), β' and β range from 0.51 to 0.83. α_1 and β_1 are also calculated by using Eqs. (5.40) and (5.41), respectively. The detailed values are listed in Table 6.2.

Assume $h/h_0 = \frac{2r}{r + r_s} = 1.05 \sim 1.10$, and the average value is 1.075; hence, $r/r_s = 1.16$. The relationship between the central angle of the concrete in compression

Table 6.2 β', β, α_1, and β_1 corresponding to steel bars and concrete with different strengths

Steel bar	Concrete	$\beta = \beta'$	β_1	α_1
HRB335	C30, C40, C50	0.51	0.8	1.0
HRB400	C30, C40, C50	0.61	0.8	1.0
HRB500	C30, C40, C50	0.76	0.8	1.0
HRB335	C60	0.52	0.78	0.98
HRB400	C60	0.63	0.78	0.98
HRB500	C60	0.78	0.78	0.98
HRB335	C70	0.54	0.76	0.96
HRB400	C70	0.65	0.76	0.96
HRB500	C70	0.81	0.76	0.96
HRB335	C80	0.56	0.74	0.94
HRB400	C80	0.67	0.74	0.94
HRB500	C80	0.83	0.74	0.94

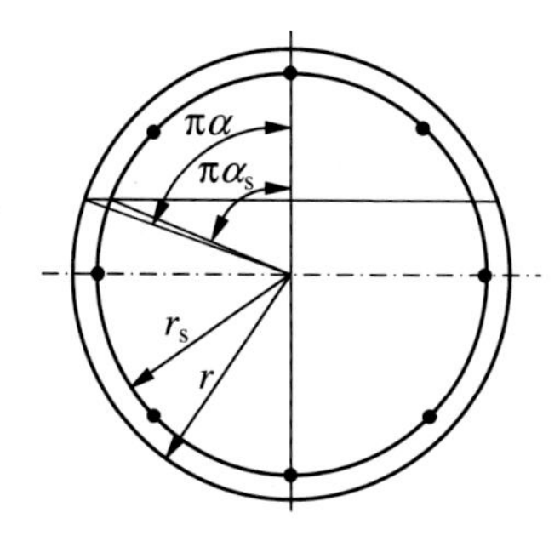

Fig. 6.26 The central angle of the concrete in compression 2α and the central angle of the steel bar in compression $2\alpha_s$

2α and the central angle of the steel bar in compression $2\alpha_s$ is $r\cos(\pi\alpha) = r_s\cos(\pi\alpha_s)$, as shown in Fig. 6.26. When $r/r_s = 1.16$, the values of $\alpha_{s1} + k_c$, $\alpha_{s2} + k_t$, $\sin\pi\alpha_{s1} + m_c$, and $\sin\pi\alpha_{s2} + m_t$ versus α corresponding to different values of β', β, α_1, and β_1 are plotted in Fig. 6.27. It is indicated that the value of β has a significant effect on $\alpha_{s1} + k_c$ and $\sin\pi\alpha_{s1} + m_c$, but only has a very limited effect on $\alpha_{s2} + k_t$ and $\sin\pi\alpha_{s2} + m_t$.

According to the regression analysis, the relationship between $\alpha_{s1} + k_c$ and α could be expressed as

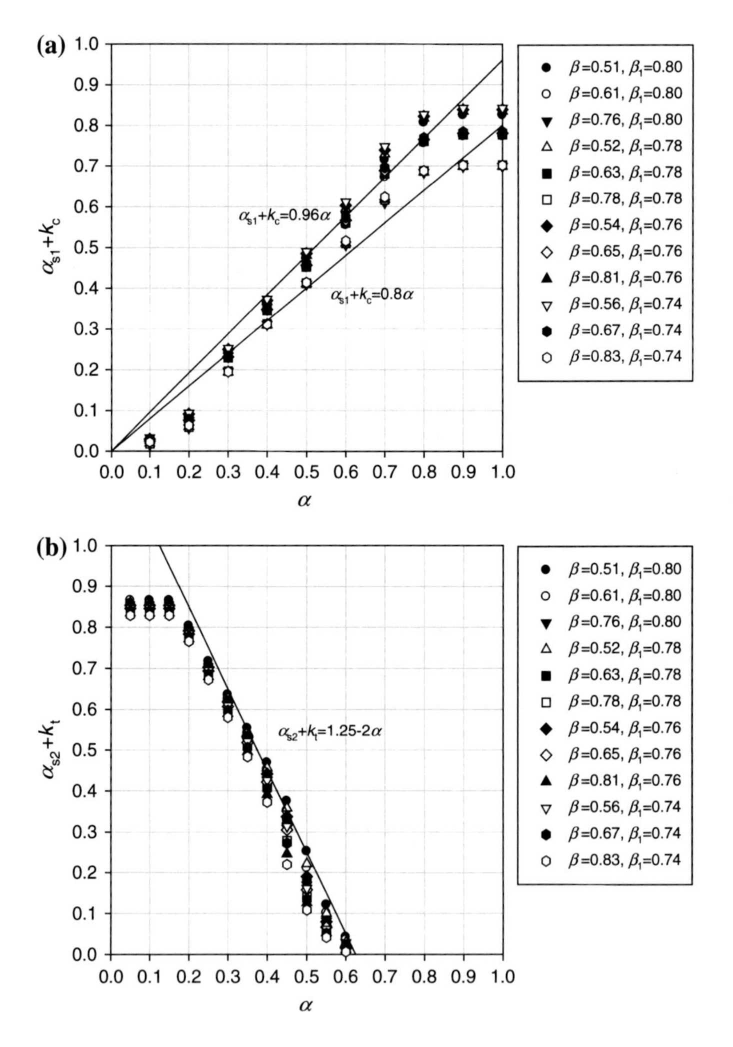

Fig. 6.27 The relationship between $\alpha_{s1} + k_c$, $\alpha_{s2} + k_t$, $\sin\pi\alpha_{s1} + m_c$, $\sin\pi\alpha_{s2} + m_t$, and α. **a** $\alpha_{s1} + k_c$ versus α. **b** $\alpha_{s2} + k_t$ versus α. **c** $\sin\pi\alpha_{s1} + m_c$ versus α. **d** $\sin\pi\alpha_{s2} + m_t$ versus α

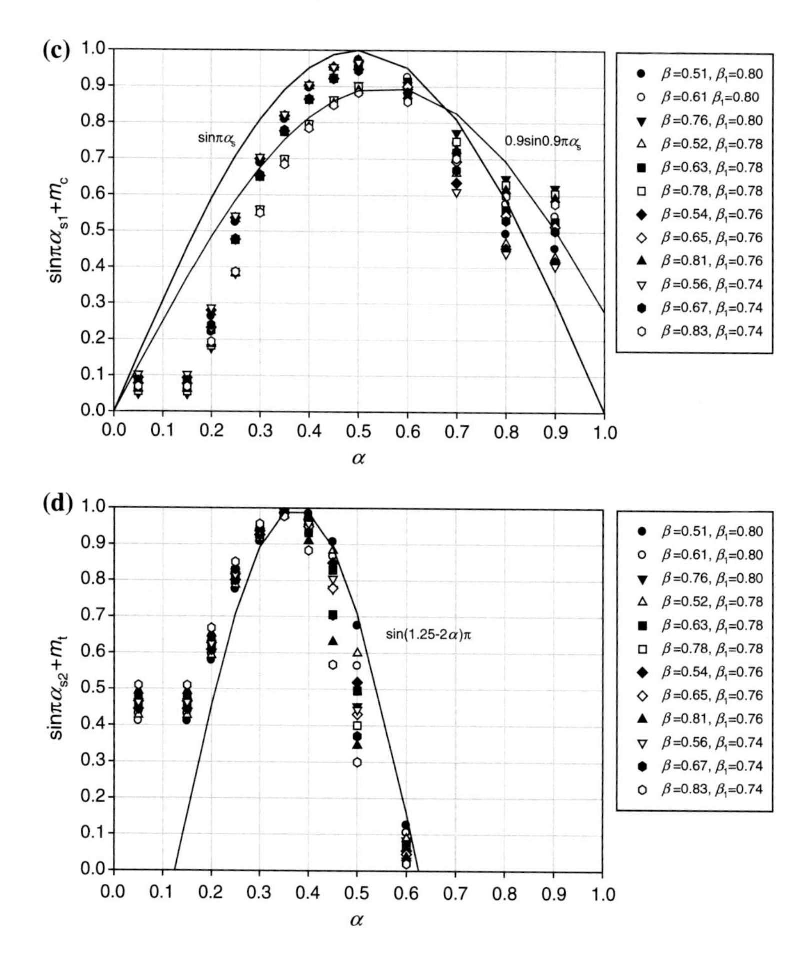

Fig. 6.27 (continued)

$$\alpha_{s1} + k_c = p\alpha \tag{6.116}$$

where when $\beta = 0.51$, $p = 0.96 \approx 1$; when $\beta = 0.83$, $p = 0.8$; when β is between 0.51 and 0.83, p can be obtained by using the linear interpolation method.

The relationship between $\alpha_{s2} + k_t$ and α can be expressed as

$$\alpha_{s2} + k_t = \alpha_t \tag{6.117}$$

where $\alpha_t = 1.25 - 2\alpha$.

The relationship between $\sin\pi\alpha_{s1} + m_c$ and α could be expressed as

$$\sin\pi\alpha_{s1} + m_c \approx q\sin q\pi\alpha \tag{6.118}$$

where when $\beta = 0.51$, $q = 1$, and when $\beta = 0.83$, $q = 0.9$. When β is between 0.51 and 0.83, q can be obtained by using the linear interpolation method.

The relationship between $\sin\pi\alpha_{s2} + m_t$ and α can be expressed as

$$\sin\pi\alpha_{s2} + m_t \approx \sin\alpha_t \tag{6.119}$$

where $\alpha_t = 1.25 - 2\alpha$.

With all the aforementioned simplifications, the calculation of ultimate bearing capacities of the normal section is expressed as

$$N_{cu} = \alpha_1 f_c \alpha \left(1 - \frac{\sin 2\pi\alpha}{2\pi\alpha}\right) A + f_y' A_s p\alpha - f_y A_s \alpha_t \tag{6.120}$$

$$N_{cu}\eta_s e_i = \frac{2}{3}\alpha_1 f_c A r \frac{\sin^3 \pi\alpha}{\pi} + f_y' A_s r_s \frac{q \sin q\pi\alpha}{\pi} + f_y A_s r_s \frac{\sin \pi\alpha_t}{\pi} \tag{6.121}$$

where $\alpha_t = 1.25 - 2\alpha$; when $\alpha > 0.625$, take $\alpha_t = 0$.

This simplification is adopted in the calculation of ultimate bearing capacities of eccentrically compressed members of circular sections in *Code for Design of Concrete Structures* (GB 50010). Readers can summarize the steps and perform practical calculations by using this code.

6.11 Analysis of Eccentrically Tensioned Members

When the axial tensile force lies between the reinforcements at the two opposite faces, it is a small eccentrically tensioned section. When the axial tensile force lies outside the reinforcements, it is a large eccentrically tensioned section.

6.11.1 Ultimate Tension Capacities of Small Eccentrically Tensioned Sections

Small eccentrically tensioned sections can be classified as fully tensioned ones and partially tensioned ones depending on the value of e_0. For fully tensioned sections, e_0 is small. So the concrete across the full section is subjected to tension, and the tensile stress in the concrete near the axial force is larger. For partially tensioned sections, e_0 is big and the concrete far from the axial force may be under compression.

With the increase of the axial force, the stress in the concrete also increases. When the stress in the extreme tension fiber reaches the tensile strength of concrete,

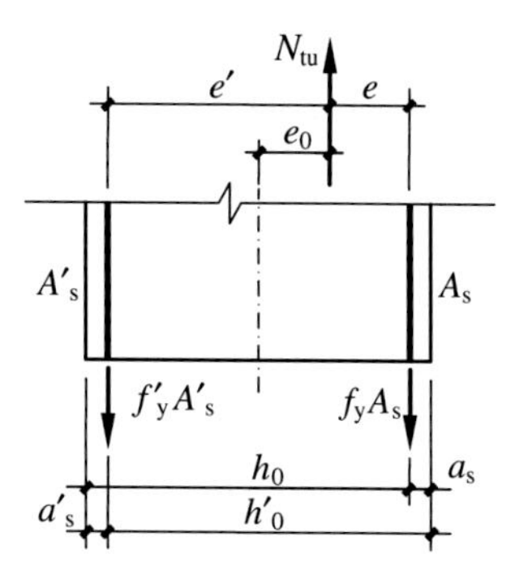

Fig. 6.28 Stress distribution across a small eccentrically tensioned section

the section cracks. For small e_0, the crack will rapidly cut through the member transversely. For big e_0, the compressive stress in the original compression zone will become tensile stress for equilibrium, because the concrete in the cracked tension zone quits working and the member is finally transversely cracked.

Therefore, for a small eccentricity section in the ultimate state, the concrete across the full section is cracked and the longitudinal bars are all yielding, as shown in Fig. 6.28.

From Fig. 6.28, we can write the equilibrium equations as

$$N_{tu} = f_yA_s + f'_yA'_s \tag{6.122}$$

$$N_{tu}e = f'_yA'_s(h_0 - a'_s) \tag{6.123}$$

$$N_{tu}e' = f_yA_s(h'_0 - a_s) \tag{6.124}$$

where e is the distance from the axial force to the resultant force of A_s

$$e = \frac{h}{2} - e_0 - a_s \tag{6.125}$$

e' is the distance from the axial force to the resultant force of A'_s

$$e' = \frac{h}{2} + e_0 - a'_s \tag{6.126}$$

6.11.2 *Ultimate Tension Capacities of Large Eccentrically Tensioned Sections*

If the axial force lies outside the segment connecting A_s and A'_s, the concrete near the axial force is under tension, while the concrete far away from the axial force is under compression. The tensile stress in the concrete near the axial force increases with the axial force and cracks when the tensile strength of concrete is reached. But there is always a compression zone in the section. Otherwise, the section cannot maintain equilibrium.

When the axial force is big enough to yield to the reinforcement near the axial force, the propagation of the crack will further reduce the compression zone, which leads to the increase of the compressive stress in the concrete, and finally, the concrete is crushed when the strain in the extreme compression fiber reaches ε_{cu}. It can be observed that this failure mode is similar to that of large eccentricity compression sections. However, if too many steel bars are placed at the face near the axial force and too few steel bars are placed at the face far away from the axial force, it is possible that the concrete far away from the axial force is crushed first, while the reinforcement near the axial force has not yielded, similar to that of an over-reinforced flexural section.

Based on Fig. 6.29, which shows the simplified stress distribution across a large eccentrically tensioned section, we can write the equilibrium equations as

$$N_{tu} = f_y A_s - \alpha_1 f_c bx - f'_y A'_s \tag{6.127}$$

$$N_{tu} e = \alpha_1 f_c bx\left(h_0 - \frac{x}{2}\right) + f'_y A'_s (h_0 - a'_s) \tag{6.128}$$

where e is the distance from the axial force to the resultant force of A_s

$$e = e_0 - \frac{h}{2} + a_s \tag{6.129}$$

When $x < 2\,a'_s$, assuming $x = 2\,a'_s$ and taking moments about A'_s yield

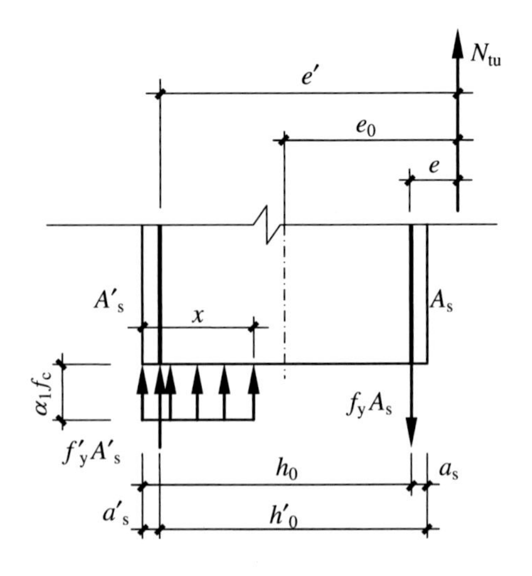

Fig. 6.29 Stress distribution across a large eccentrically tensioned section

$$N_{tu}e' = f_y A_s (h'_0 - a_s) \quad (6.130)$$

$$e' = \frac{h}{2} + e_0 - a'_s \quad (6.131)$$

When $x > \xi_b h_0$, the stress in the tension steels A_s should be calculated using Eq. (6.48).

6.12 Applications of the Ultimate Capacity Equations for Eccentrically Tensioned Members

6.12.1 *Design of Small Eccentrically Tensioned Sections*

The design of small eccentricity sections determines A_s and A'_s when the sectional dimensions, material properties and N_t and e_0 (or N_t and M) are given.

To guarantee the safety of the section, $N_t \leqslant N_{tu}$ must be satisfied. So the basic design equations are

$$N_t e = f'_y A'_s (h_0 - a'_s) \quad (6.132)$$

$$N_t e' = f_y A_s (h'_0 - a_s) \quad (6.133)$$

A'_s and A_s can be determined directly using Eqs. (6.132) and (6.133).

When the section is symmetrically reinforced, A_s can be calculated first by using Eq. (6.132), then take $A'_s = A_s$. It should be noted that A'_s usually does not yield in the ultimate state. And $A_s \geqslant \rho_{min} A$ and $A'_s \geqslant \rho'_{min} A$ should be checked, where A is the area of the section.

6.12.2 *Evaluation of Ultimate Capacities of Existing Small Eccentrically Tensioned Sections*

The evaluation of ultimate capacities of existing small eccentrically tensioned sections is to determine N_{tu} when the sectional dimensions, material properties, and reinforcement are given.

N_{tu} is the minimum of the results calculated by Eqs. (6.122), (6.123), and (6.124). If either A_s or A'_s is smaller than the minimum reinforcement ratio, the cracking load should be taken as N_{tu}. The calculation method of the cracking load of small eccentrically tensioned sections is similar to that of axially tensioned sections and will not be detailed herein for brevity.

6.12.3 Design of Large Eccentrically Tensioned Sections

The design of large eccentrically tensioned sections is to determine A_s and A'_s when the sectional dimensions, material properties and N_t and e_0 (or N_t and M) are given.

According to the design principle of $N_t \leqslant N_{tu}$, the design equations are

$$N_t = f_y A_s - \alpha_1 f_c bx - f'_y A'_s \tag{6.134}$$

$$N_t e = \alpha_1 f_c bx\left(h_0 - \frac{x}{2}\right) + f'_y A'_s (h_0 - a'_s) \tag{6.135}$$

To minimize the steel content, the same method in the design of large eccentrically tensioned sections can be used here, i.e., $\xi = \xi_b$. Then, A'_s can be determined by Eq. (6.135) and A_s by Eq. (6.134).If $A'_s < \rho'_{min}bh$, take $A'_s = \rho'_{min}bh$ and solve x by using Eq. (6.135), and then calculate A_s by using Eq. (6.134). If $x < 2\,a'_s$, take $x = 2\,a'_s$ and $A_s = \frac{N_t e'}{f_y (h'_0 - a_s)}$.

If $A_s < \rho_{min}\,bh$, take $A_s = \rho_{min}bh$.

If the section is symmetrically reinforced, that is, $f'_y A'_s = f_y A_s$, $x = -\frac{N_t}{\alpha_1 f_c b}$ from Eq. (6.134) is negative, which is obviously irrational. We can get $A_s = \frac{N_t e'}{f_y (h'_0 - a_s)}$ by assuming $x = 2\,a'_s$.

6.12.4 Evaluation of Ultimate Capacities of Existing Large Eccentrically Tensioned Sections

There are only two unknowns in Eqs. (6.127) and (6.128). So the ultimate capacity of existing large eccentrically tensioned sections can be determined by solving the two equations simultaneously. If $A_s < \rho_{min}bh$, the cracking load should be taken as the ultimate capacity of the section. If $x > \xi_b h_0$, Eq. (6.48) should be added to the solution. The calculation method of the cracking load of large eccentrically tensioned sections is similar to that of flexural sections and will not be detailed herein for brevity.

Example 6.15 An eccentrically tensioned member is subjected to N_t = 900 kN and M = 126 kN·m. The dimensions of the member are $b \times h$ = 300 mm × 400 mm, $a_s = a'_s$ = 35 mm. The concrete strength is f_c = 14.3 N/mm², f_t = 1.50 N/mm². The strength and modulus of elasticity of the steels are $f'_y = f_y$ = 300 N/mm², $E_s = 2 \times 10^5$N/mm². Determine A'_s and A_s.

Solution

1. Determine whether it is a large eccentrically tensioned section or not

$$e_0 = \frac{M}{N_t} = \frac{126}{900} = 0.14\,\text{m} = 140\,\text{mm} < \frac{h}{2} - a_s = 200 - 35 = 165\,\text{mm}$$

The axial force is applied between A_s' and A_s, and it is a small eccentrically tensioned section.

2. Calculate A_s' and A_s

$$e = \frac{h}{2} - e_0 - a_s = \frac{400}{2} - 140 - 35 = 25\,\text{mm}$$

$$e' = \frac{h}{2} + e_0 - a_s' = \frac{400}{2} + 140 - 35 = 305\,\text{mm}$$

$$A_s' = \frac{Ne}{f_y\left(h_0' - a_s'\right)} = \frac{900 \times 10^3 \times 25}{300 \times (365 - 35)}$$

$$= 227\,\text{mm}^2 < \rho'_{\min} bh = \max\begin{cases} 0.2bh = & 240\,\text{mm}^2 \\ 0.45\frac{f_t}{f_y} = & 270\,\text{mm}^2 \end{cases}$$

Take $A_s' = 270\,\text{mm}^2$

$$A_s = \frac{Ne'}{f_y\left(h_0' - a_s\right)} = \frac{900 \times 10^3 \times 305}{300 \times (365 - 35)} = 2773\,\text{mm}^2 > \rho_{\min} bh = 0.45\frac{f_t}{f_y}$$
$$= 270\,\text{mm}^2$$

Example 6.16 A rectangular water tank has a 400-mm-thick wall. $a_s = a_s' = 40$ mm. The concrete strength is $f_c = 14.3$ N/mm^2, $f_t = 1.50$ N/mm^2. The strength and modulus of elasticity of the steels are $f_y' = f_y = 300$ N/mm^2, $E_s = 2 \times 10^5$ N/mm^2. If the maximum moment per unit width (meter) in the horizontal direction at the mid-span of the wall is $M = 320$ kN·m and the corresponding tensile force is $N_t = 400$ kN, determine A_s' and A_s.

Solution

1. Determine whether it is a large eccentrically tensioned section or not

$$e_0 = \frac{M}{N_t} = \frac{320}{400} = 0.8\,\text{m} = 800\,\text{mm} > \frac{h}{2} - a_s = \frac{400}{2} - 40 = 160\,\text{mm}$$

It is a large eccentrically tensioned section.

2. Calculate A'_s and A_s

$$e = e_0 - \frac{h}{2} + a_s = 800 - \frac{400}{2} + 40 = 640\,\text{mm}$$

Taking $\xi = \xi_b = 0.55$ from Eq. (6.135), we have

$$\begin{aligned} A'_s &= \frac{N_t e - \alpha_1 f_c b h_0^2 (\xi - 0.5\xi^2)}{f'_y (h_0 - a'_s)} \\ &= \frac{400 \times 10^3 \times 640 - 14.3 \times 1000 \times 360^2 \times (0.55 - 0.5 \times 0.55^2)}{300 \times (360 - 40)} = -5031\,\text{mm}^2 < 0 \end{aligned}$$

Take $A'_s = \rho'_{\min} bh = 0.002 \times 1000 \times 400 = 800\,\text{mm}^2$, and use $\phi16@250$ (A'_s = 804 mm^2).

Now, the problem becomes determining A_s with A'_s known. From Eq. (6.135), we have

$$400 \times 10^3 \times 640 = 14.3 \times 1000 \times 360^2 \times (\xi - 0.5\xi^2) + 300 \times 804 \times (360 - 40)$$
$$\xi^2 - 2\xi + 0.193 = 0$$

Solving the equation yields

$$\xi = 0.102, x = \xi h_0 = 36.7\,\text{mm} < 80\,\text{mm}, \text{ take } x = 80\,\text{mm}.$$
$$e' = \frac{h}{2} + e_0 - a'_s = \frac{400}{2} + 800 - 40 = 960\,\text{mm}$$
$$A_s = \frac{N_t e'}{f_y (h'_0 - a_s)} = \frac{400 \times 10^3 \times 960}{300 \times (360 - 40)} = 4000\,\text{mm}^2$$
$$> 0.45 \frac{f_t}{f_y} bh = 0.45 \times \frac{1.5}{300} \times 1000 \times 400 = 900\,\text{mm}^2, \text{OK.}$$

Questions

6.1 Give engineering examples of eccentrically compressed members and eccentrically tensioned members except those mentioned in this chapter, five for each.

6.2 Compare the stress and strain distributions of eccentrically compressed members and flexural members, and state their common points and differences.

6.3 Why can't the brittle small eccentricity failure be avoided by limiting reinforcement ratio?

6.4 Symmetrical reinforcement pattern will increase the steel content in an eccentrically compressed member. Why is this pattern still widely used in engineering practice?

6.5 How do we distinguish large or small eccentrically compressed failure?

6.6 How do we consider the influence of the slenderness ratio on the ultimate capacity of eccentrically compressed members in the basic design equations?
6.7 Describe the relationship between the axial force and the bending moment for large and small eccentrically compressed sections based on the interaction diagram of N_{cu} and M_u. What is the bending capacity maximum?
6.8 What are the applications of the interaction diagram of N_{cu} and M_u?
6.9 Please explain why the moment magnifying coefficient η_s decreases with the increase of e_i in $\eta_s = 1 + \frac{1}{1300\frac{e_i}{h_0}}\left(\frac{l_0}{h}\right)^2\zeta_c$ if all other conditions are kept the same?
6.10 How do we calculate the ultimate capacities of large and small eccentrically compressed members of a rectangular concrete section?
6.11 Explain from equilibrium why there must be a compression zone in a large eccentrically tensioned section.
6.12 Why would a reinforcement far from the axial force yield first in eccentrically compressed members? What is the failure mode? How do we prevent such a failure mode?
6.13 Why should the additional eccentricity e_a be introduced?

Problems

6.1 An eccentrically compressed column is a rectangular section. The dimensions of the section are $b \times h = 400$ mm × 600 mm, $a_s = a_s' = 40$ mm, and $l_0 = 2.9$ m. The strength of the concrete is f_c = 16.7 N/mm². The properties of the reinforcement are $f_y' = f_y$ = 300 N/mm² and $E_s = 2 \times 10^5$ N/mm². A_s' = 603 mm² (3ϕ16), and A_s = 1521 mm² (4ϕ22). When e_0 = 50, 100, 150, 200, 250, 300, 350, 400, 450, and 500 mm, determine the ultimate capacities of the member, respectively, by using the simplified method; then, draw the interaction diagram of N_{cu} and M_u.
6.2 In Problem 6.l, when N_c = 0, 500, 1000, 1500, 2000, 2500, 3000, 3500, and 4000 kN, calculate M_u, respectively, and draw the interaction diagram of N_{cu} and M_u.
6.3 An eccentrically compressed column is a rectangular section. The dimensions of the section are $b \times h = 400$ mm × 600 mm, $a_s = a_s' = 40$ mm, and $l_0 = 4.6$ m. The strength of the concrete is f_c = 16.7 N/mm². The properties of the reinforcement are $f_y' = f_y$ = 300 N/mm², $E_s = 2 \times 10^5$ N/mm². If the axial compressive force N_c = 1200 kN and the moment M = 600 kN·m are applied on the column and the section is asymmetrically reinforced,

1. Determine A_s and A_s'.
2. If 4ϕ20 compression steel bars have already been placed (A_s' = 1257 mm²), determine A_s.
3. Compare the calculation results and analyze the difference.

6.4 An eccentrically compressed column is of rectangular section. The dimensions of the column are $b \times h = 500$ mm $\times$ 800 mm, $a_s = a'_s = 40$ mm, and $l_0 = 12.5$ m. The strength of the concrete is $f_c = 14.3$ N/mm^2. The properties of the reinforcement are $f'_y = f_y = 300$ N/mm^2 and $E_s = 2 \times 10^5$ N/mm^2. If the axial force $N_c = 1800$ kN and the moment $M = 1080$ kN·m are applied on the member and the section is asymmetrically reinforced, determine A_s and A'_s.

6.5 An eccentrically compressed column is of rectangular section. The dimensions of the column are $b \times h = 400$ mm $\times$ 600 mm, $a_s = a'_s = 45$ mm, and $l_0 = 5.6$ m. The strength of the concrete is $f_c = 14.3$ N/mm^2. The properties of the reinforcement are $f'_y = f_y = 300$ N/mm^2 and $E_s = 2 \times 10^5$ N/mm^2. If the axial force $N_c = 3200$ kN and the moment $M = 100$ kN·m are applied and the section is asymmetrically reinforced,

1. Determine A_s and A'_s.
2. If 3ϕ20 compression steel bars have already been placed ($A'_s = 942$ mm^2), calculate A_s.
3. Compare the calculation results and analyze the difference.

6.6 An eccentrically compressed column is of rectangular section. The dimensions of the column are $b \times h = 500$ mm $\times$ 800 mm, $a_s = a'_s = 45$ mm, and $l_0 = 4.6$ m. The strength of the concrete is $f_c = 14.3$ N/mm^2. The properties of the reinforcement are $f'_y = f_y = 300$ N/mm^2 and $E_s = 2 \times 10^5$ N/mm^2. If the axial force $N_c = 7000$ kN and the moment $M = 175$ kN·m are applied on the column and the section is asymmetrically reinforced, determine A_s and A'_s.

6.7 In Problem 6.4, if the section is symmetrically reinforced, determine $A_s = A'_s$.

6.8 In Problem 6.6, if the section is symmetrically reinforced, determine $A_s = A'_s$.

6.9 An I section column in a structural laboratory has the dimensions of $b = 120$ mm, $h = 800$ mm, $b'_f = b_f = 400$ mm, $h'_f = h_f = 130$ mm, $a_s = a_s' = 40$ mm, and $l_0 = 6.8$ m. The strength of the concrete is $f_c = 14.3$ N/mm^2. The properties of the reinforcement are $f'_y = f_y = 300$ N/mm^2. If the axial force $N_c = 1000$ kN and the moment $M = 400$ kN·m are applied on the column, and the section is symmetrically reinforced, determine $A_s = A'_s$.

6.10 An I section column in a single-story factory has the sectional dimensions shown in Fig. 6.30, i.e., $l_0 = 6.8$ m, $a_s = a'_s = 40$ m. The strength of the concrete is $f_c = 14.3$ N/mm^2. The properties of the reinforcement are $f'_y = f_y = 300$ N/mm^2. The column may be subjected to three groups of internal forces: (1) $N_c = 503.3$ kN and $M = 246.0$ kN·m; (2) $N_c = 740.0$ kN and $M = 294.0$ kN·m; (3) $N_c = 1040.0$ kN and $M = 312.0$ kN·m. Determine $A_s = A'_s$ (taking $h'_f = h_f = 162$ mm) if the section is symmetrically reinforced.

6.11 The wall thickness of a sewage pipe is 500 mm. $a_s = a'_s = 50$ mm. Under the pressure of the water inside the pipe and the soils outside the pipe, the axial tensile force and the moment applied on the section per unit width (meter) in the longitudinal direction are $N_t = 900$ kN and $M = 900$ kN·m. The strength

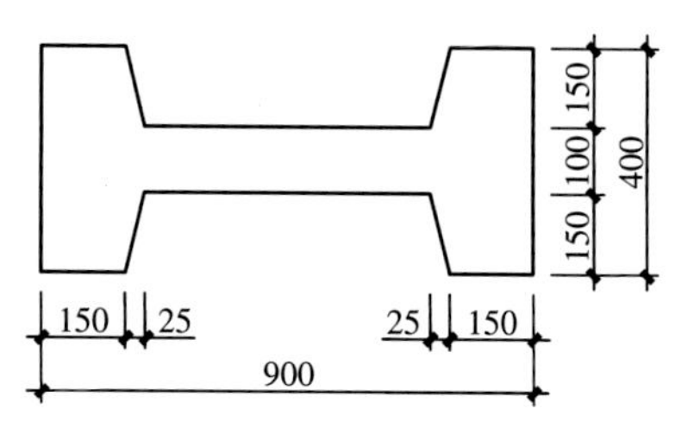

Fig. 6.30 Problem 6.10

of the concrete is $f_c = 16.7\ \text{N/mm}^2$. The strength of the reinforcement is $f'_y = f_y = 300\ \text{N/mm}^2$. Determine A_s and A'_s.

6.12 Detail the calculation steps for the design of eccentrically compressed sections with biaxial bending.

6.13 Detail the calculation steps for the evaluation of ultimate capacities of eccentrically compressed sections with biaxial bending.

6.14 Detail the calculation steps for the design of eccentrically compressed members with a circular section.

6.15 Detail the calculation steps for the evaluation of ultimate capacities of eccentrically compressed members with a circular section.

Chapter 7
Shear

7.1 Engineering Applications and Reinforcement

Reinforced concrete structural members such as beams, columns, tie bars, and walls are subject to two types of failure: (1) normal section failure by bending and (2) inclined section failure by shear. Generally, bending moment and shear force appear in a member simultaneously. However, which type of failure will happen depends on which internal force exceeds the corresponding bearing capacity of the member.

To prevent normal section failure, members are generally reinforced by longitudinal bars, which are parallel to the axes of the members. The mechanical property and analysis of the members in normal section failure have been detailed in Chaps. 4–6.

Inclined section failure includes inclined shear failure and inclined bending failure. To prevent inclined shear failure, stirrups should be placed perpendicular to the member axis. Sometimes longitudinal bars are also bent up at a certain angle to the member axis to resist shear force, as shown in Fig. 5.6d, e. To prevent inclined bending failure, calculations or detailing measures are necessary.

There are two types of stirrups, i.e., open stirrups (Fig. 7.1a) and closed stirrups (Fig. 7.1b). Closed stirrups can effectively restrain the transverse deformation of the concrete in the compression zone and can be used to resist torque. So closed stirrups are most frequently used in rectangular sections. For in situ cast T section beams, open stirrups can be adopted because transverse steel bars are already in place at the flange top. The hooked anchorage of closed stirrups should be bent into 135° rather than 90°. The straight segment of the hooked anchorage should not be less than 50 mm and $5d$, where d is the stirrup diameter, as shown in Fig. 7.1e.

The number of the vertical segments of a stirrup is called the leg number. Frequently used stirrups include single-legged (Fig. 7.1c), double-legged (Fig. 7.1a, b), and four-legged (Fig. 7.1d) stirrups.

X. Gu et al., *Basic Principles of Concrete Structures*,
DOI 10.1007/978-3-662-48565-1_7

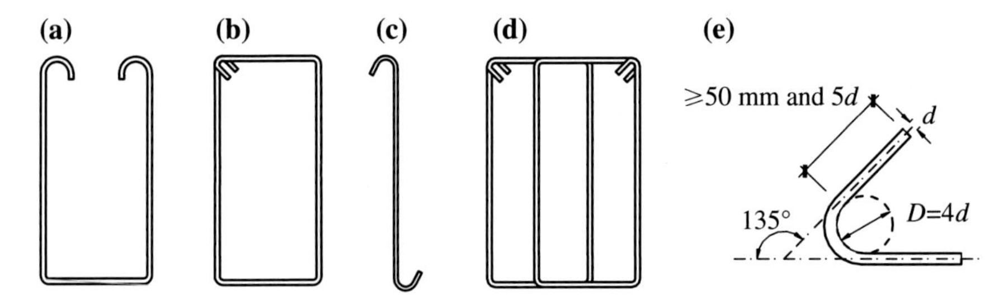

Fig. 7.1 Stirrup types and hooked anchorage. **a** Open stirrup (double leg), **b** closed stirrup (double leg), **c** single-legged stirrup, **d** four-legged stirrup, **e** hooked anchorage

In addition to resisting shear force, stirrups also serve as the supporting points of longitudinal bars. Therefore, closed stirrups should be used if calculated longitudinal bars are placed in the compression zone. The spacing between stirrups should be no more than 400 mm and 15*d*, where *d* is the minimum diameter of the longitudinal steel bars. And the stirrup diameter should be no less than 1/4 of the largest diameter of the longitudinal steel bars. If there are more than five compression steel bars with a diameter larger than 18 mm placed in one layer, the spacing between stirrups should be no more than 10*d*. If the sectional width of a beam is larger than 400 mm and there are more than three compression steel bars in one layer, or even if the sectional width is no more than 400 mm, but more than four compression steel bars are placed in one layer, complex stirrups should be used.

Double-legged stirrups are frequently used in beams with a sectional width less than 350 mm. If the sectional width is no less than 350 mm or there are more than five compression steel bars in one layer, four-legged stirrups should be used. At most, every other compression steel bar should be positioned at the stirrup corner. Single-legged stirrups can be used when the beam section is less than 150 mm wide or when they work as tie bars.

When a bent-up bar is used, the longitudinal anchorage length outside the bent-up end should be at least 20*d* in the tension zone and 10*d* in the compression zone, where *d* is the diameter of the bent-up bar. The bent-up angle can be 45° or 60°. The outermost two longitudinal steel bars at the bottom of a beam should not be bent upward and the outermost two longitudinal steel bars at the beam top should not be bent downward. Insufficiently anchored reinforcement prevents the use of a bent-up bar.

7.2 Behavior of Flexural Members Failing in Shear

Stirrups and bent-up bars in reinforced concrete flexural members are called web reinforcement. If a beam is reinforced by both longitudinal steel bars and web reinforcement, it is called the beam with web reinforcement. If a beam is reinforced by longitudinal steel bars only, it is called the beam without web reinforcement. Since flexure failure always happens to reinforced concrete slabs earlier than shear failure, web reinforcement is not usually necessary in the slabs. We will discuss the shear behavior of reinforced concrete beams without web reinforcement first.

7.2.1 Behavior of Beams Without Web Reinforcement

7.2.1.1 Stress State Prior to the Occurrence of Inclined Cracks

If the loads applied on a reinforced concrete beam are small and no crack appears, the beam can be visualized as a homogeneous elastic body, in which the longitudinal steel bars can be transformed into concrete of an equivalent cross-sectional area according to the ratio of the elastic moduli of the two materials. For a simply supported beam with two concentrated loads symmetrically applied (Fig. 7.2), according to the mechanics of materials, the flexural stress σ and the shear stress τ at an arbitrary point of any cross section in the shear–flexure regions AB or CD can be expressed, respectively, as follows:

$$\sigma = \frac{My}{I_0} \tag{7.1}$$

$$\tau = \frac{VS_0}{I_0 b} \tag{7.2}$$

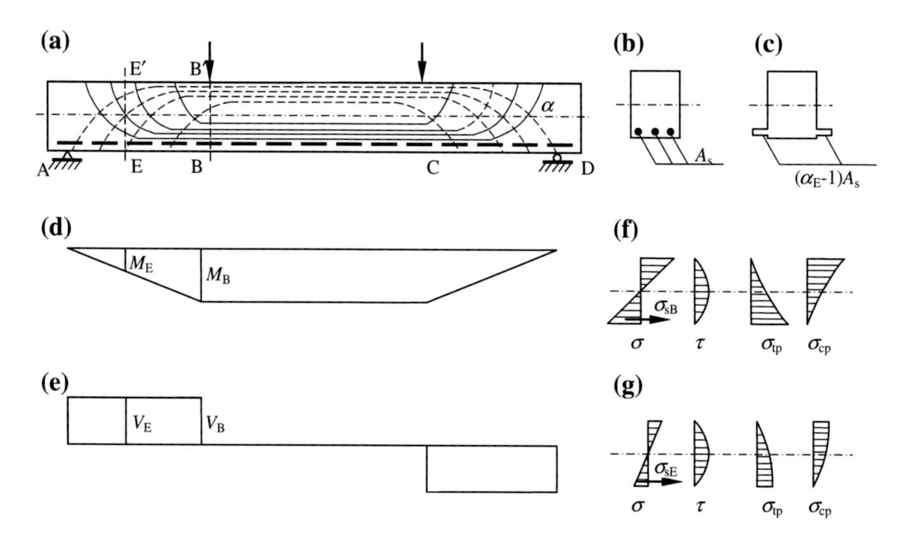

Fig. 7.2 Stress state prior to the occurrence of diagonal cracks in a beam without web reinforcement. **a** Principal stress trajectories. **b** Cross section. **c** Transformed section. **d** Diagram of moment. **e** Diagram of shear. **f** Stress of cross section B–B′. **g** Stress of cross section E–E′

where

M and V the bending moment and shear force acting upon the cross section, respectively;

I_0 the moment of inertia of the transformed cross section;

S_0 the first moment about the centroidal axis of the part of the cross-sectional area lying farther from the centroidal axis than the point when the shear stresses are being calculated;

b the width of the cross section; and

y the distance from the considered fiber to the centroid of the transformed cross section.

Due to the combined action of flexural stress and shear stress, the resulting principal tensile stress σ_{tp} and principal compressive stress σ_{cp} can be expressed as follows:

$$\sigma_{tp} = \frac{\sigma}{2} + \frac{1}{2}\sqrt{\sigma^2 + 4\tau^2} \tag{7.3}$$

$$\sigma_{cp} = \frac{\sigma}{2} - \frac{1}{2}\sqrt{\sigma^2 + 4\tau^2} \tag{7.4}$$

The inclination of the principal tensile stress to the longitudinal axis of the beam, α, can be determined by the following formula

$$\tan 2\alpha = -\frac{2\tau}{\sigma} \tag{7.5}$$

The trajectories of the principal stresses in the beam are shown in Fig. 7.2a, which intersect with the neutral axis at the angle of 45°. Figure 7.2f, g show the flexural stresses, shear stresses, principal tensile stresses, and principal compressive stresses of two cross sections, BB′ and EE′, respectively. By comparing Fig. 7.2f with Fig. 7.2g, it is found that the principal tensile stress σ_{tp} in the extreme fiber of cross section BB′ is larger than that of cross section EE′. And because the extreme fiber of cross section BB′ is nearly under uniaxial tension in the horizontal direction, it will be ruptured if the stress in it is larger than the tensile strength of concrete and vertical cracks will appear. With the increase of the loads, for an arbitrary cross section in the shear–flexure region, if the principal tensile stresses at any point below the centroidal axis exceed the tensile strength of concrete under biaxial loading, concrete will crack. Because the direction of the principal tensile stresses in the shear–flexure region is not perpendicular to the neutral axis, this kind of crack is called the inclined crack.

Most frequently encountered inclined cracks are formed by the development of short vertical cracks, which initially appear at the bottom of cross sections in the shear–flexure region. Furthermore, inclined cracks can also appear near the neutral axis in the web of a beam, and this type of inclined crack is called the web-shear crack, which will sometimes occur in beams with large flanges and thin webs.

7.2.1.2 Development of Cracks and Failure Modes of Beams Without Web Reinforcement

Once inclined cracks appear in a beam without web reinforcement, the length and width of the cracks will increase with the load. And new cracks will continuously appear, which finally result in inclined failure of the beam.

To describe the failure modes of a beam without web reinforcement, the term *shear-span ratio* must be first introduced. Shear-span ratio is a dimensionless factor, which reflects the relative magnitude of the applied bending moment M to the shear force V, and can be expressed as

$$\lambda = \frac{M}{Vh_0} \tag{7.6}$$

where h_0 is the effective depth of the cross section considered.

For a simply supported beam subjected to concentrated loads (Fig. 7.3), the shear-span ratio of the cross section B_{left} is

$$\lambda_{\text{B-left}} = \frac{M_{\text{B}}}{V_{\text{B-left}}h_0} = \frac{V_{\text{A}}a}{V_{\text{A}}h_0} = \frac{a}{h_0} \tag{7.7}$$

where a is the distance from the first concentrated load to the nearest support, i.e., the length of the shear span.

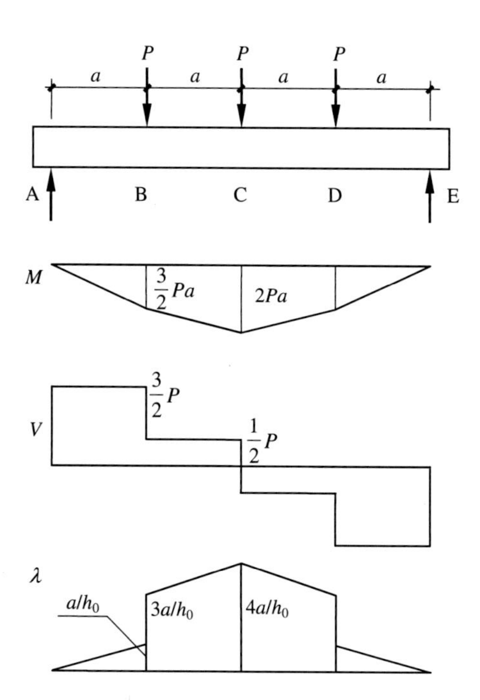

Fig. 7.3 Shear-span ratios for a beam under concentrated loads

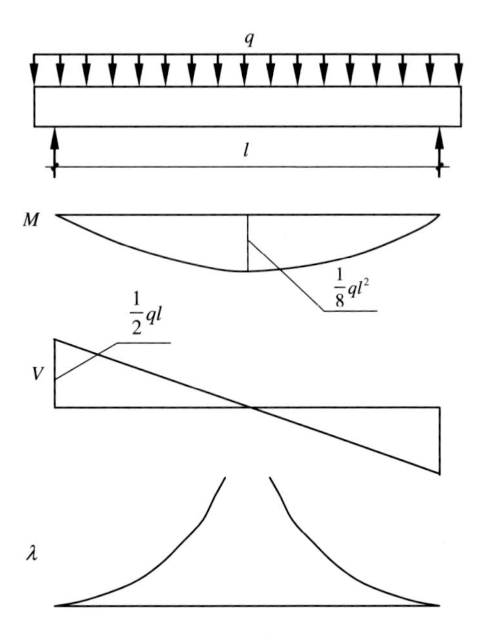

Fig. 7.4 Shear-span ratio of a beam under uniformly distributed load

Equation 7.7 indicates that for the cross section right under the first concentrated load, the shear-span ratio can be calculated either by Eqs. 7.6 or 7.7, which give the same result. But Eq. 7.7 is not applicable for the cross section under the second or the third concentrated load, and Eq. 7.6 should be employed. For a beam subjected to uniformly distributed loads (Fig. 7.4), Eq. 7.6 rather than Eq. 7.7 should be used, because the bending moments and shear forces vary along the beam span.

If a beam is asymmetrically loaded, because the ratios of the bending moment to the shear force of the left and the right cross sections at the loading point are different, the shear-span ratios should be calculated separately. The shear-span ratio defined in Eq. 7.6 is called a generalized shear-span ratio, which is defined in Eq. 7.7 as the computation shear-span ratio.

For different shear-span ratios λ, beams without web reinforcement will fail in different modes, which will be illustrated with a beam without web reinforcement under two symmetric concentrated loads as an example (Fig. 7.5). If $\lambda < 1$ (Fig. 7.5a), the concrete between the supports and the nearer acting points of the concentrated loads is similar to that of an inclined short column. Inclined cracks initiate at the web of the beam and propagate toward the loading points and the supports. With the increase of the loads, the cracks increase in number as well and are parallel to each other. Finally, the crush of the concrete between two major diagonal cracks indicates the inclined failure of the beam, which is called inclined compression failure.

If $1 \leqslant \lambda \leqslant 3$ (Fig. 7.5b), once the inclined cracks appear, they will continuously propagate toward the nearer load point with the widths becoming larger and larger. A critical crack, which exhibits the maximum width and length among all inclined

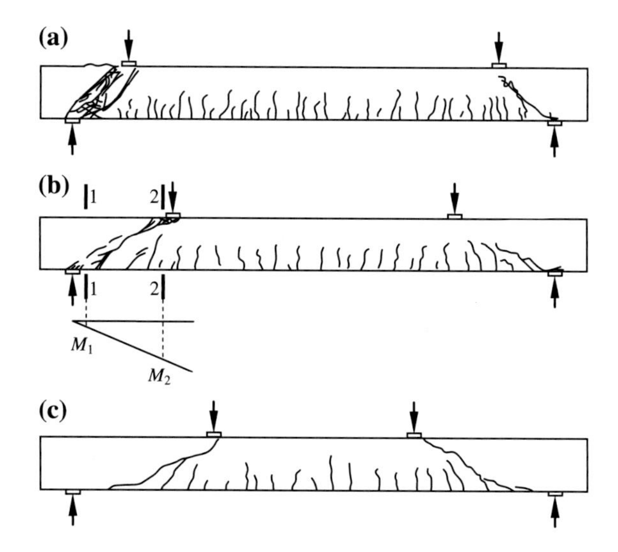

Fig. 7.5 Main failure modes of beams without web reinforcement

cracks, is formed. When the concrete in the shear-compression zone on top of the critical crack is crushed, the beam fails. This failure mode is called shear-compression failure.

If $\lambda > 3$ (Fig. 7.5c), once the inclined crack appears, it will rapidly propagate to the nearer load point. And the beam is split diagonally into two parts. This failure mode is called the inclined splitting failure or inclined tension failure.

Except for the three main failure modes (inclined compression failure, shear-compression failure, and inclined splitting failure or inclined tension failure) illustrated in Fig. 7.5a–c), beams without web reinforcement may diagonally fail in other patterns. As shown in Fig. 7.5b, the bending moment M_1 at cross section 1–1 is small before cracking, so the tensile stresses in the longitudinal bars are also small. But after cracking, the bending moment at cross section 1–1 becomes M_2, and the tensile stresses in the longitudinal bars significantly increase. If the anchorage length is not enough, the longitudinal bars will be pulled out, which is called anchorage failure.

7.2.1.3 Mechanism of Shear Resistance for Beams Without Web Reinforcement

After the occurrence of inclined cracks in beams without web reinforcement, the stress state of the beams will change essentially. Take a beam without web reinforcement failed by shear-compression as an example (Fig. 7.6). The vertical cracks in the pure-bending segment and the inclined cracks in the shear–flexure segments will split the beam into a comb-like structure with the longitudinal bars connecting the teeth at the bottom.

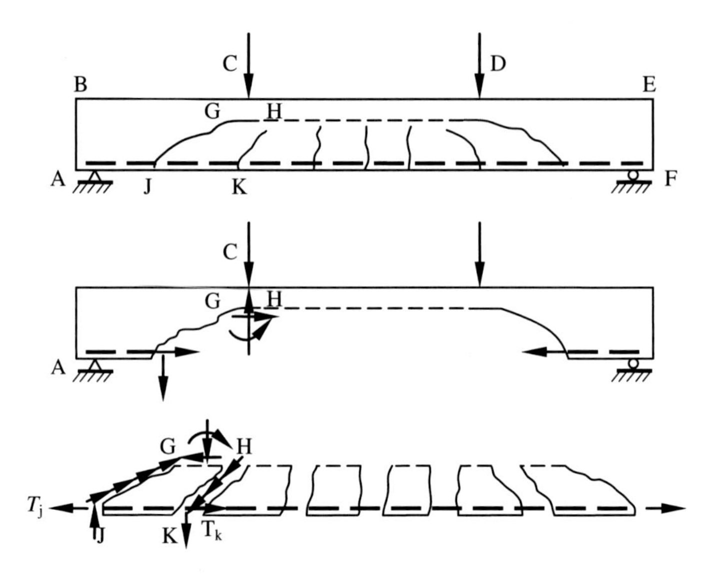

Fig. 7.6 Mechanism of shear resistance for beams without web reinforcement after the occurrence of inclined cracks

It has been indicated by experimental studies that a critical inclined crack will eventually form among all inclined cracks, and it separates the beam into two interconnected parts. The upper part is analogous to a two-hinge arch of a variable cross section; the longitudinal bars perform as a tension rod, and the arch and the beam have the same supports. The lower part is further divided by cracks into a series of comb-teeth, each of which can be considered as a cantilever beam with the end connected to the inner surface.

Take the tooth GHKJ as an example (Fig. 7.6). The end GH, which is connected to the arch, is equivalent to the fixed end of a cantilever beam. The end JK, which is equivalent to the free end, is subjected to unequal tension forces by the longitudinal reinforcement at points J and K. The tension force at point J is less than that at point K. The difference between the tension forces at points J and K caused by the distinct bending moments at sections J and K can be considered as the applied external force on the tooth and will place the cantilever beam under the combined action of bending and shear. Therefore, the flexure and shear on a tooth is due to the shear force in the beam, that is, part of the shear force in a beam is resisted by the cantilever action of the teeth.

At the initial stage of the development of inclined cracks, the tension difference is large, because the bonding between reinforcement and concrete is still good, and the shear force is mainly resisted by the cantilever action of the teeth. At the later stage, i.e., when the beam is approaching shear-compression failure, the tension force T_j is close to the tension force T_k due to the poor bonding. Consequently, the cantilever action of the teeth becomes weak and the shear force in the beam will be mainly resisted by the arch action.

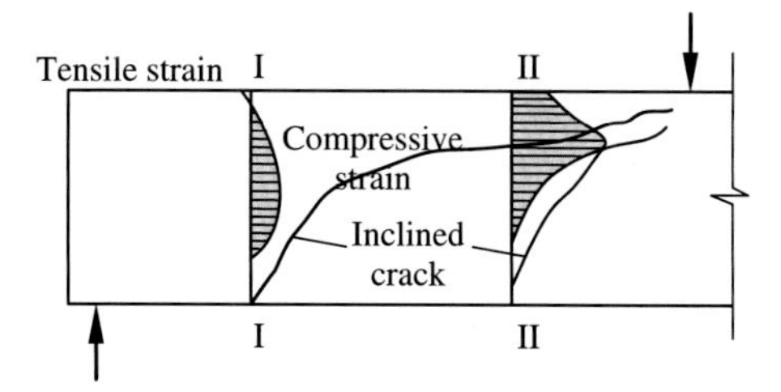

Fig. 7.7 Strain distributions along normal sections after the occurrence of inclined cracks

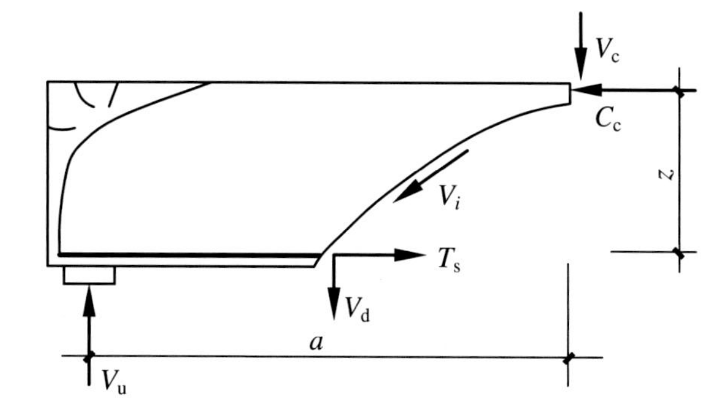

Fig. 7.8 Forces on the free body separated by a main diagonal crack

The strain distributions along normal sections of a beam after the occurrence of inclined cracks are shown in Fig. 7.7, which indicates that the plane section assumption is not valid anymore. From the force equilibrium in the vertical direction, it is found that the shear force in beams without web reinforcement is resisted by the following: (1) the concrete in the shear-compression zone, V_c; (2) the vertical component of the interlock of aggregates and friction along the inclined crack interface, V_i; and (3) the dowel action of longitudinal reinforcement, V_d. Among the three resistances, V_i and V_d are difficult to estimate. With the development of the cracks, V_i gradually decreases and V_d gradually increases, which keeps $V_i + V_d$ unchanged or only slightly changed. Therefore, the contribution of the concrete in the shear-compression zone is dominant (Fig. 7.8).

7.2.1.4 Factors Influencing the Shear Capacity of Beams Without Web Reinforcement

1. Shear-span ratio
   It has been experimentally indicated that the larger the shear-span ratio is, the lower the shear capacity will be for a beam without web reinforcement under a concentrated load (Fig. 7.9).

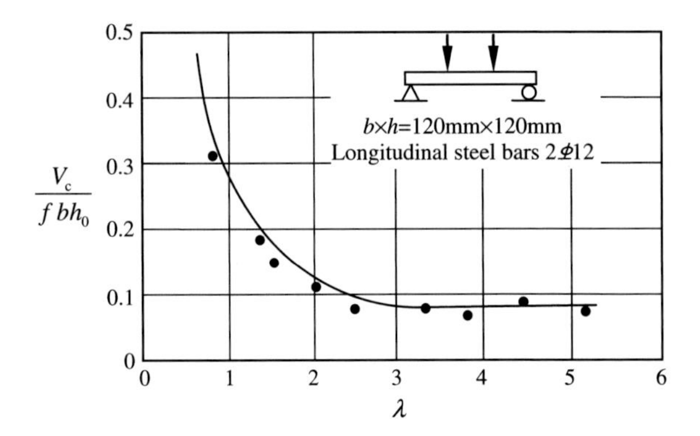

Fig. 7.9 Influence of shear-span ratio on the shear capacity of a beam without web reinforcement

According to the interaction between flexure and shear, when the shear-span ratio is large, even if the shear capacity of the beam is low, the bending moment in the beam increases more rapidly, and the beam will actually fail in flexure.

2. Concrete strength
   It is found from experimental results, as shown in Fig. 7.10, that if both the shear-span ratio and the steel content are constant, the shear capacity of a beam without web reinforcement will linearly increase with the cube strength of concrete, f_{cu}. However, the slopes of the linear relationships are different for different shear-span ratios.
3. Longitudinal reinforcement ratio
   Figure 7.11 shows the influence of longitudinal reinforcement ratio ρ on the shear capacity of a beam without web reinforcement, which is approximately in a linear relationship. The more the amount of longitudinal reinforcement is used, the higher the shear capacity of a beam will be. It can be explained by the fact

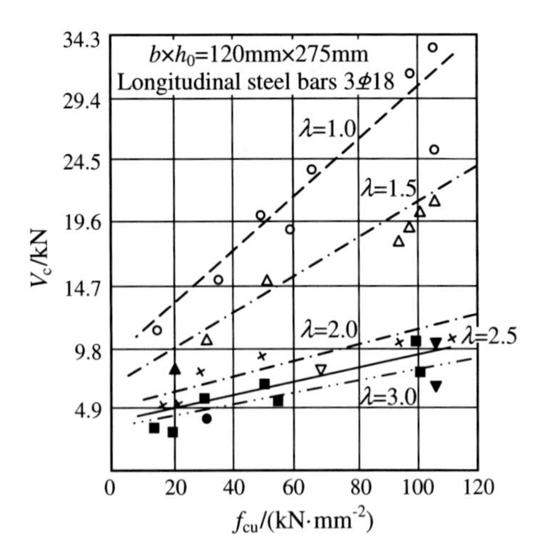

Fig. 7.10 Influence of concrete strength on the shear capacity of a beam without web reinforcement

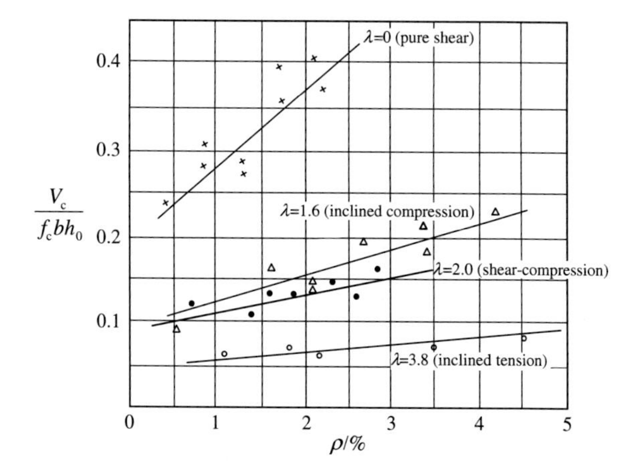

Fig. 7.11 Influence of longitudinal reinforcement ratio on the shear capacity of a beam without web reinforcement

that the longitudinal reinforcement can restrain the propagation of inclined cracks to some extent, so that the cross-sectional area of the shear-compression zone over inclined cracks is large, and consequently the shear capacity raises. Furthermore, the longitudinal reinforcement itself can provide a certain amount of shear resistance via the so-called *dowel action*. When the shear-span ratio λ is small, the dowel action will be significant, which means that ρ will greatly affect the shear capacity; however, when the shear-span ratio λ is large, the beam will fail in inclined tension, and the influence of ρ is limited.

In addition to the above factors, the shear capacity of a beam may also be influenced by the cross-sectional area, the loading pattern (such as loading on the beam top or the side surface) and the member type (simply supported beam, continuous beam), etc.

7.2.1.5 Analysis of Inclined Sections of Beams Without Web Reinforcement

1. Calculation diagram and equilibrium equations

   Take the most frequent shear-compression failure as an example. Take the part ABCGJ on the left side of the main inclined crack as the free body (Fig. 7.12), where surface CG is a section of the arch and the tie bar (longitudinal reinforcement) connects the arch springing at point J. Because the relative depth of the shear-compression zone is small, the shear stress and compressive stress along the section CG can be assumed uniformly distributed. If the dowel action and the interlock action are neglected, there are four forces acting on the free body, which are the reaction force at the support V_c, the tension in the longitudinal reinforcement $\sigma_s A_s$, the compressive force on the shear-compression

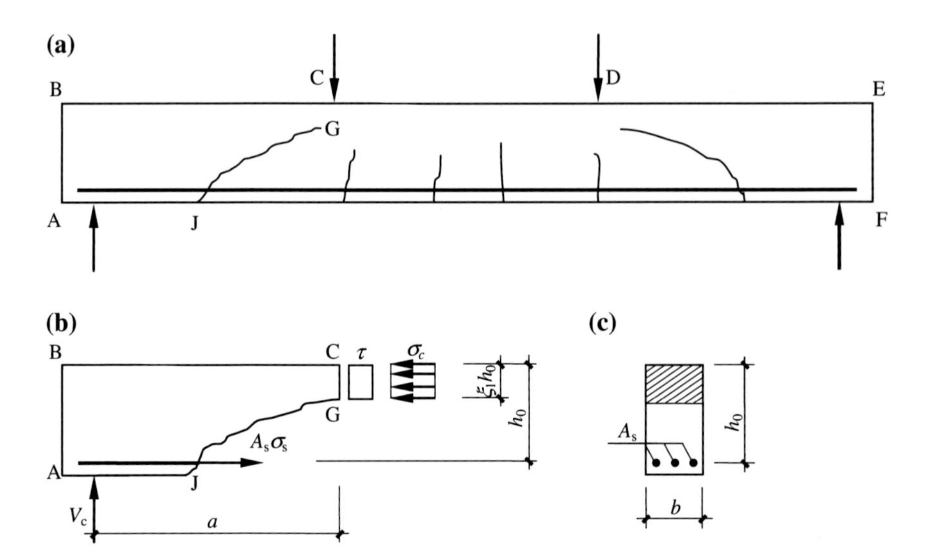

Fig. 7.12 Shear resistance calculation diagram of a beam without web reinforcement

zone $\sigma_c b\xi_1 h_0$ ($\xi_1 h_0$ is the height of the shear-compression zone) and the shear force $\tau b\xi_1 h_0$.

From the force equilibriums in the vertical direction and the horizontal direction and the moment equilibrium about the centroid of the longitudinal reinforcement at section *CG*, it can be determined that

$$\sigma_c b\xi_1 h_0 = \sigma_s A_s \tag{7.8}$$

$$V_c = \tau b\xi_1 h_0 \tag{7.9}$$

$$V_c a = \sigma_c b\xi_1 h_0 \left(h_0 - \frac{1}{2}\xi_1 h_0 \right) \tag{7.10}$$

where

V_c	the ultimate shear force carried by the inclined section of a beam without web reinforcement;
a	the shear span;
b and h_0	the sectional width and effective height, respectively;
σ_s and A_s	the tensile stress and sectional area of the longitudinal reinforcement, respectively;
ξ_1	the relative depth of the shear-compression zone; and
σ_c and τ	the compressive stress and shear stress along the shear-compression zone, respectively.

2. Strength criterion of concrete
   The concrete in the shear-compression zone is under the combined action of the compressive stress and the shear stress, so corresponding strength criterion should be adopted. Substituting Eq. (7.9) into Eq. (7.10), we have

$$\frac{\tau}{f_c} = \frac{1 - 0.5\xi_1}{\lambda} \cdot \frac{\sigma_c}{f_c} \tag{7.11}$$

From previous analysis, it can be found that when $\lambda < 1$, beams without web reinforcement will generally fail in inclined compression with the shear-compression zone depth being very small and ξ_1 being nearly zero; when λ is large, ξ_1 will increase up to ξ_b for under-reinforced beams. That is to say, when λ varies from 1 to 5, ξ_1 will change within $0 \sim \xi_b$. So by taking linear interpolation according to the above rules, ξ_1 can be expressed by λ. If $\xi_b = 0.55$, then Eq. (7.11) will become

$$\frac{\tau}{f_c} = \left(\frac{1.06875}{\lambda} - 0.06875\right)\frac{\sigma_c}{f_c} \tag{7.12}$$

The experimental curve for the strength criterion of concrete under shear and compression proposed by Hirai Yoshika can be employed here as shown in Eq. (7.13).

$$\frac{\tau}{f_c} = \sqrt{0.0089 + 0.095\frac{\sigma_c}{f_c} - 0.104\left(\frac{\sigma_c}{f_c}\right)^2} \tag{7.13}$$

Obviously, the intersection points of the loading curves represented by Eq. (7.12) and the strength criterion curve represented by Eq. (7.13) give the corresponding shear stresses and compressive stresses at inclined section failure for different shear-span ratios (Fig. 7.13). Linear regression of the intersection points corresponding to $\lambda = 2 \sim 5$ gives

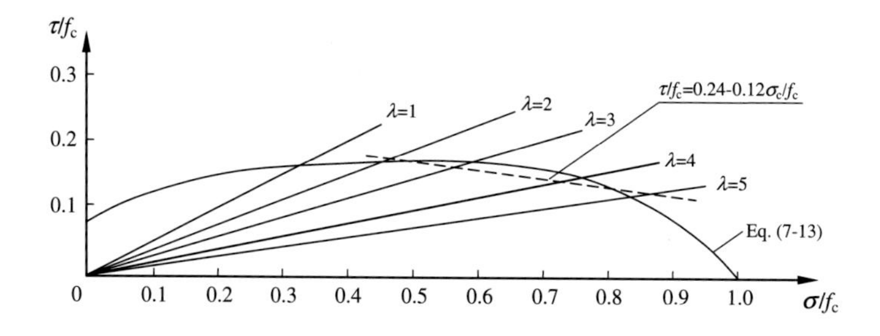

Fig. 7.13 Relationship between shear stress and compressive stress and shear-compression strength criterion

$$\frac{\tau}{f_c} = 0.24 - 0.12\frac{\sigma_c}{f_c} \tag{7.14}$$

Equation (7.14) can be taken as the shear-compression strength criterion of concrete for shear-span ratio from 2 to 5.

3. Shear capacity

From the above equilibrium conditions and strength criterion, the shear capacity of an inclined section can be derived. From Eq. (7.8) and considering $\rho = A_s/(bh_0)$, we have

$$\xi_1 = \frac{\sigma_s}{\sigma_c}\rho \tag{7.15}$$

Substituting Eqs. (7.14) and (7.15) into Eq. (7.9) gives

$$V_c = (0.24f_c - 0.12\sigma_c)b\frac{\sigma_s}{\sigma_c}\rho h_0$$

After rearrangement, we have

$$\frac{\sigma_c}{f_c} = \frac{0.24}{\frac{V_c}{\sigma_c b h_0}\cdot\frac{\sigma_c}{\sigma_s \rho} + 0.12} \tag{7.16}$$

Substituting Eq. (7.15) into Eq. (7.10) gives

$$V_c a = \sigma_c b\frac{\sigma_s}{\sigma_c}\rho h_0\left(h_0 - \frac{\sigma_s \rho}{\sigma_c}\cdot\frac{h_0}{2}\right) \tag{7.17}$$

Substituting Eq. (7.16) into Eq. (7.17) and canceling σ_c, we have

$$V_c = \left(\frac{0.24 - 0.06\frac{\sigma_s \rho}{f_c}}{0.5 + 0.24\lambda\frac{f_c}{\sigma_s \rho}}\right)f_c b h_0 \tag{7.18}$$

When the longitudinal reinforcement yields ($\sigma_s = f_y$), the shear capacity of the inclined section can be obtained by

$$V_c = \left(\frac{0.24 - 0.06\frac{f_y \rho}{f_c}}{0.5 + 0.24\lambda\frac{f_c}{f_y \rho}}\right)f_c b h_0 \tag{7.19}$$

7.2.2 *Experimental Study on Beams with Web Reinforcement*

The sectional dimensions, reinforcement, and loading scheme of a simply supported beam specimen with web reinforcement are shown in Fig. 7.14. The yield strength of the longitudinal reinforcement and web reinforcement are 340 N/mm^2 and 377 N/mm^2, respectively. The prism strength of the concrete is 17 N/mm^2.

The specimen behaved very differently before (Stage I) and after (Stage II) in the appearance of the main inclined crack. In Stage I, when the load was small, no cracks happened to the beam and the stresses in the longitudinal reinforcement and web reinforcement were both very small (Fig. 7.15). With the increase of the load, vertical cracks appeared in the pure-bending segment (segment CD) first; then, inclined cracks occurred in the shear–flexure segments (segment AC and BD).

After several inclined cracks merged into a main diagonal crack, the beam entered Stage II. The concrete quit working due to cracking, and the stress σ_{sv} in the web reinforcement across the main inclined crack increased rapidly, which was manifested by an apparent slope change in the load-stress curve (Fig. 7.15). With the increase of the load, the stress in the web reinforcement continued increasing; the main inclined crack propagated continuously toward the loading point (generally the side surface of the loading plate nearer to the support), and thus, the sectional area of the shear-compression zone over the crack tip was gradually reduced. When the stress in the web reinforcement reached 337 N/mm^2, the

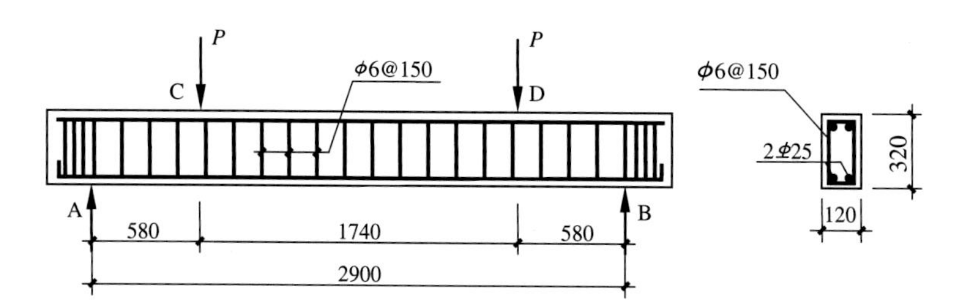

Fig. 7.14 A beam specimen with web reinforcement

Fig. 7.15 Load versus stress of web reinforcement

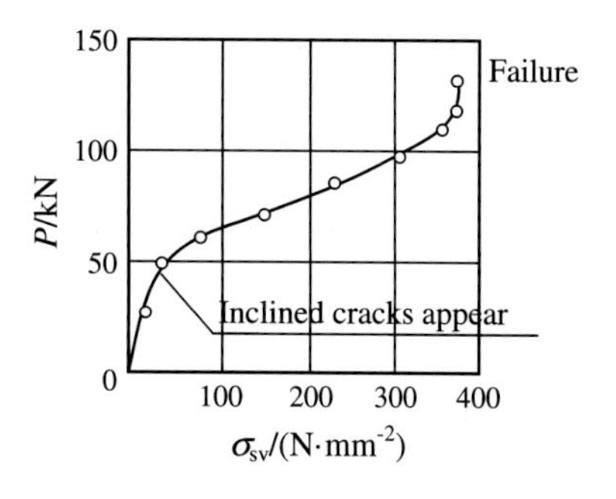

concrete in the shear-compression zone was crushed and Stage II was over. This kind of failure is called shear-compression failure. The failure pattern is similar to that of the beam without web reinforcement shown in Fig. 7.5b. The shear-compression failure generally happens to beams with a shear-span ratio of $1 \leqslant \lambda \leqslant 3$ and with a web reinforcement content that is appropriate.

When $\lambda < 1$ or even $1 \leqslant \lambda \leqslant 3$, but the web reinforcement is too much, the concrete between the support and the loading point acts like an inclined short column. Inclined cracks start from the beam web and propagate toward both the support and the loading point. With the increase of the load, more and more inclined cracks appear and the concrete in the beam web is crushed. This kind of failure is called inclined compression failure. The failure pattern is similar to that of the beam without web reinforcement shown in Fig. 7.5a. The web reinforcement is generally not yielding at the beam failure.

When $\lambda > 3$ and the web reinforcement ratio is small, once the inclined crack appears, it propagates rapidly to the loading point and splits the beam into two parts. This kind of failure is called inclined tension failure. The failure pattern is similar to that of the beam without web reinforcement shown in Fig. 7.5c. The concrete in the shear-compression zone is not crushed at the beam failure. The shear resistance of the beam depends on the tensile strength of the concrete, which results in an apparently lower shear capacity than that of the beam failed at shear-compression.

Except for these three failure modes (shear-compression failure, inclined compression failure, and inclined tension failure), local crushing failure or longitudinal reinforcement anchorage failure, etc., may also happen to the beam in different conditions.

7.2.3 Shear Resistance Mechanism of Beams with Web Reinforcement

If stirrups and/or bent-up reinforcement are provided in beams, the stress states and the failure modes will change dramatically compared to beams without web reinforcement. Stresses in stirrups are small prior to the occurrence of inclined cracks, which means that the web reinforcement has little influence on the cracking loads at that moment. After the occurrence of inclined cracks, stresses in stirrups that are intersected with the inclined cracks increase remarkably and restrain the development of inclined cracks; these stirrup stresses also increase the interlock forces of aggregates along interfaces of cracks as well, so that the cracks are scattered into several fine cracks. Because the longitudinal bars are secured by closed stirrups, the dowel action in the longitudinal bars increases, and correspondingly the shear–flexure capacities of the cross sections of the beams increase.

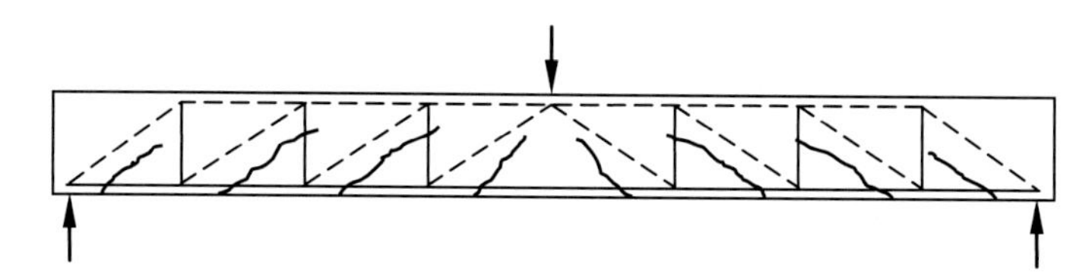

Fig. 7.16 Shear resistance mechanism of a beam with web reinforcement after cracking (*dashed lines* stand for compression chords; *solid lines* stand for tensile rods or chords)

The shear resistance mechanism of a beam with web reinforcement can be explained by several models, although the calculated results of shear capacities are not quite the same. Some models are too complicated for practical use.

After occurrence of the inclined cracks, shear resistance mechanism of a beam with web reinforcement can be explained in a planar truss model (Fig. 7.16). The top and bottom longitudinal bars act as the chords of the truss, and the stirrups act as vertical tie rods. Concrete between the inclined cracks is equivalent to diagonal struts. Meanwhile, the stirrups can also reduce the relative slip along the inclined cracks, and postpone the development of splitting cracks along the longitudinal reinforcement due to poor bond.

As stated above, the number of stirrups used has significant effect on the shear–flexure failure mode and shear–flexure capacity of a cross section. It can be seen from Fig. 7.17 that the shear–flexure capacity of a cross section increases linearly with the stirrup density [ρ_{sv} is the web reinforcement ratio defined in Eq. (7.31)]. The concrete strength, shear-span ratio, and ratio of longitudinal reinforcement all have a certain amount of influence on the capacity of the beam; however, compared to beams without web reinforcement, the influence of the last two factors is less significant.

Fig. 7.17 Influence of web reinforcement content on the shear capacity of a beam with web reinforcement

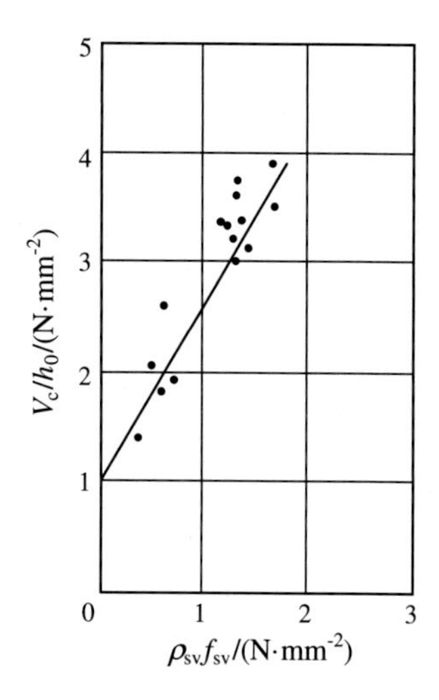

If the stirrups are not enough, after the stirrups that are intersected with the inclined cracks yield, the concrete in the shear-compression region will reach its ultimate strength, and the element will fail in the mode of shear-compression due to combined action of shear and compression.

If the stirrups are too many, the stirrups do not yield, the concrete between the inclined cracks fails in inclined compression due to the high level of principal compressive stress.

If the stirrups are too few, once the inclined cracks appear, the stirrups will immediately yield, because they cannot resist tensile forces that were originally resisted by concrete, so that the stirrups cannot restrain the development of inclined cracks anymore. The behavior of the beams is similar to that of the beams without web reinforcement, and inclined tension failure will occur if the shear-span ratio is large.

7.2.4 Analysis of Flexure–Shear Sections of Beams with Web Reinforcement

7.2.4.1 Calculation Diagram

Take a simply supported beam subjected to one concentrated load in the span as an example. After inclined cracks appear, the beam can be modeled as a planar truss with the angles between the chords and the inclined struts being indeterminate. Assume that all the longitudinal reinforcement concentrates in the top and the bottom of the beam in the spacing of h_{cor}.

Take an element A from the flexure–shear segment of the beam. The element consists of a whole diagonal crack. The length and height of the element are h_{cor} cot α and h_{cor}, respectively, where α is the inclined angle of the diagonal crack. On the left surface of the segment, there are shear force V and bending moment M. On the right surface, the shear force is still V, but the bending moment increases to $M + Vh_{cor} \cot \alpha$.

7.2.4.2 Basic Equations

As shown in Fig. 7.18, the shear force V on the left surface can be decomposed into a horizontal tensile force N and an inclined compressive force D, where

$$N = V \cot \alpha \tag{7.20}$$

$$D = \frac{V}{\sin \alpha} \tag{7.21}$$

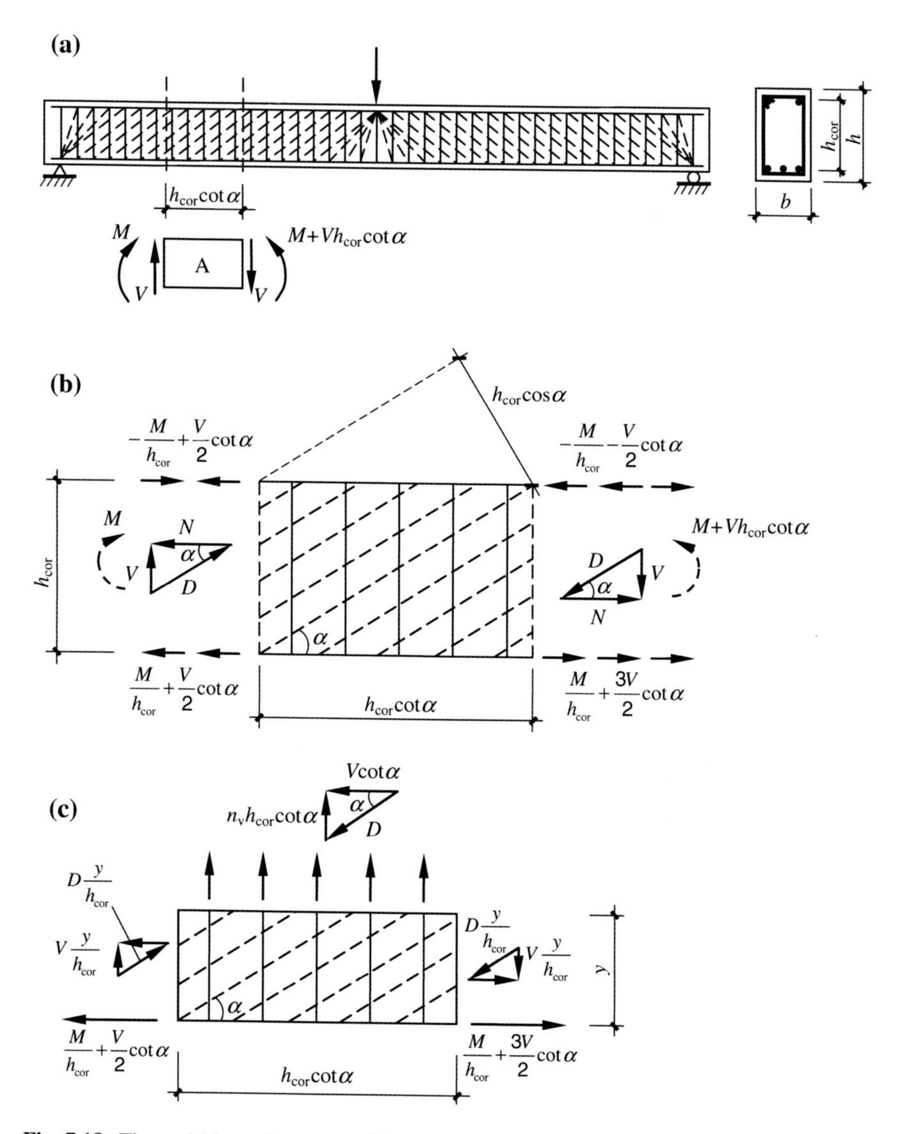

Fig. 7.18 The variable-angle truss model

The stresses in the concrete diagonal struts are

$$\sigma_d = \frac{D}{bh_{cor}\cos\alpha} = \frac{V}{bh_{cor}\cos\alpha\sin\alpha} \tag{7.22}$$

The horizontal tensile force N is equally assigned to the top and the bottom longitudinal reinforcement, i.e., $(V/2)\cot\alpha$. The bending moment is equivalent to a horizontal couple, whose value is M/h_{cor}. Therefore, the internal forces sustained by the top and bottom longitudinal reinforcement are

$$N_t = -\frac{M}{h_{cor}} + \frac{V}{2}\cot\alpha \tag{7.23}$$

$$N_b = \frac{M}{h_{cor}} + \frac{V}{2}\cot\alpha \tag{7.24}$$

Similarly, on the right surface, we have

$$N_t' = -\frac{M + Vh_{cor}\cot\alpha}{h_{cor}} + \frac{V}{2}\cot\alpha = -\frac{M}{h_{cor}} - \frac{V}{2}\cot\alpha \tag{7.25}$$

$$N_b' = \frac{M + Vh_{cor}\cot\alpha}{h_{cor}} + \frac{V}{2}\cot\alpha = \frac{M}{h_{cor}} + \frac{3V}{2}\cot\alpha \tag{7.26}$$

As shown in Fig. 7.18, the beam element A is in equilibrium under the action of the above four horizontal forces and the inclined compression D (Eq. 7.21).

If we further cut the beam element A at a distance y from the beam bottom and take the lower part as a free body, then there is a horizontal shear force V cot α acting along the top surface of the free body (Fig. 7.18c). This shear force can be decomposed into a vertical component n_v h_{cor} cot α and an inclined compression D in the concrete diagonal struts, where n_v is the vertical force in unit length. From the force triangle, we have

$$n_v h_{cor} = V\tan\alpha \tag{7.27}$$

The vertical force is sustained by uniformly distributed stirrups. When the stirrups yield, we have

$$n_v s = A_{sv} f_{yv} \tag{7.28}$$

where

A_{sv} the sectional area of the stirrups;
f_{yv} the yield strength of the stirrups; and
s the stirrup spacing.

Substituting Eq. (7.28) into Eq. (7.27) gives the ultimate shear capacity of the beam

$$V_u = f_{yv}\frac{A_{sv}}{s}h_{cor}\cot\alpha \tag{7.29}$$

If unit length of stirrups has the same force bearing capacity as unit length of longitudinal bars, i.e., equal strength reinforcement, α = 45°. However, unequal strength reinforcement is more general in shear–flexure sections and if unit length of stirrups can carry smaller force than unit length of longitudinal bars, $\alpha < 45°$.

The shear capacity of the beam with web reinforcement, i.e., Eq. (7.29), is derived from the truss model, which assumes that the stirrups sustain all of the shear force. The shear capacities estimated by Eq. (7.29) are smaller than corresponding experiment results. Therefore, the shear resistance contribution of the concrete should also be counted in engineering practice.

7.2.5 Practical Calculation Equations for Shear Capacities of Beams with Web Reinforcement

7.2.5.1 Equations Regressed from Experimental Data

From the foregoing analysis, it can be found that Eqs. (7.29) and (7.19) only consider the shear resistance contributions of the stirrups ($V_u = V_s$) and the concrete ($V_u = V_c$), respectively. So it is natural to assume that a combination of Eqs. (7.19) and (7.29) will give a more accurate estimation of the beam shear capacities.

$$V_u = \alpha_c f_c b h_0 + \alpha_{sv} \frac{f_{yv} A_{sv}}{s} h_0 \tag{7.30}$$

where α_c and α_{sv} are parameters to be determined from experimental results.

The next task would be to define the reinforcement ratio of the stirrups as

$$\rho_{sv} = \frac{n A_{sv1}}{bs} \tag{7.31}$$

where

n the number of stirrup legs in one cross section;
A_{sv1} sectional area of one stirrup leg;
s stirrup spacing; and
b beam sectional width.

To consider the influence of the concrete strength, two dimensionless variables $V_u/(f_c b h_0)$ and $\rho_{sv} f_{yv}/f_c$ are chosen in regression analysis.

For simply supported beams under concentrated loads, based on the lower bound of the experimental results on specimens without web reinforcement of different shear-span ratios, α_c can be taken as $0.2/(\lambda + 1.5)$, that is

$$\frac{V_u}{f_c b h_0} = \frac{0.2}{\lambda + 1.5} \tag{7.32}$$

In Eq. (7.32), when $\lambda < 1.4$, $\lambda = 1.4$; when $\lambda > 3$, $\lambda = 3$.

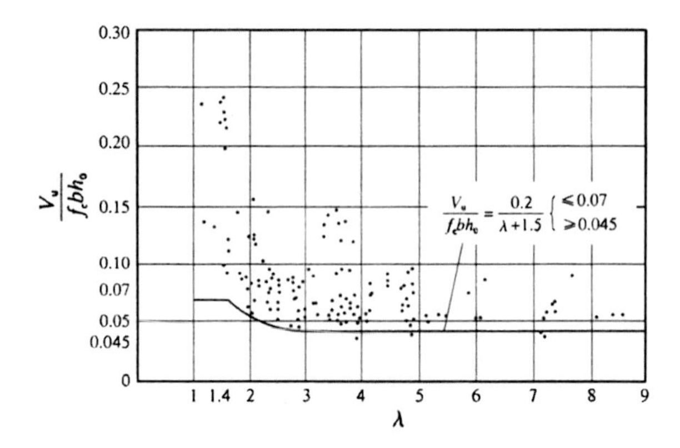

Fig. 7.19 Shear capacities versus shear-span ratios for beams without web reinforcement under concentrated loads

From experimental results of beams with stirrups under concentrated loads, α_{sv} can be determined as 1.25. Therefore, for beams under concentrated loads, Eq. (7.30) becomes (Fig. 7.19)

$$V_u = \frac{0.2}{\lambda + 1.5} f_c b h_0 + 1.25 f_{yv} \frac{A_{sv}}{s} h_0 \tag{7.33}$$

where A_{sv} = the total sectional area of all stirrup legs in one cross section, i.e.,

$$A_{sv} = n A_{sv1} \tag{7.34}$$

Figure 7.20 shows the comparison of Eq. (7.33) and experimental data.

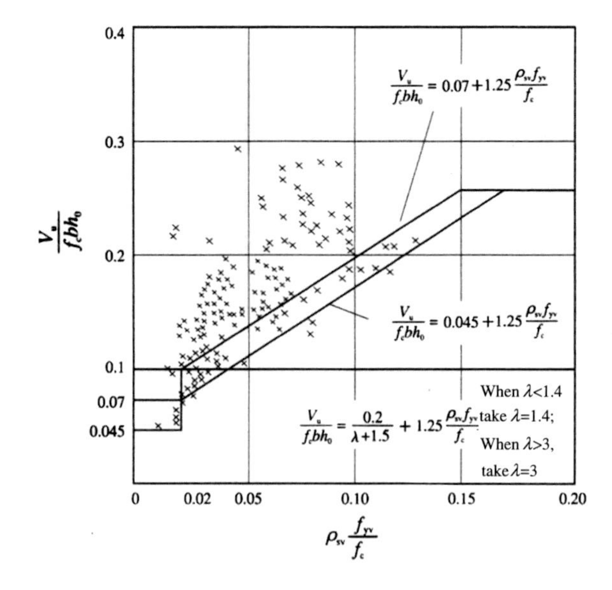

Fig. 7.20 Shear capacities versus reinforcement ratios of stirrups for beams under concentrated loads

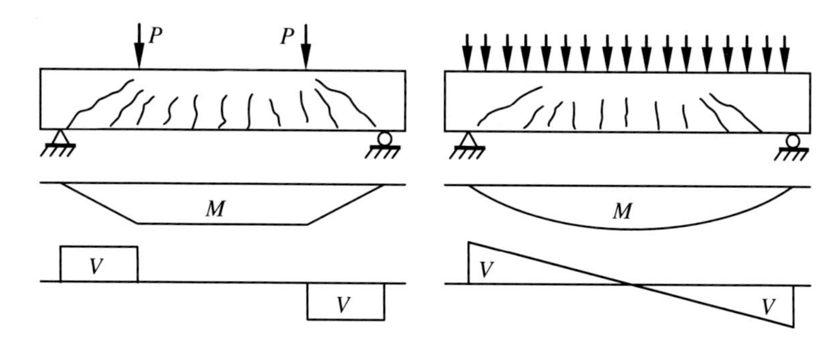

Fig. 7.21 Comparison of loading characteristics of beams under concentrated loads and under uniformly distributed loads

If the stirrups are too many, diagonal compression failure will happen. The concrete is crushed before the stirrups yield. So an upper bound of the shear capacity is stipulated to prevent this situation.

$$V_{\mathrm{u,max}} = 0.25 f_{\mathrm{c}} b h_0 \tag{7.35}$$

If the stirrups are too few, diagonal tension failure may happen. So a minimum reinforcement ratio of the stirrups should be enforced, i.e., $\rho_{\mathrm{sv}} \geqslant \rho_{\mathrm{sv,min}}$.

For beams under uniformly distributed loads, α_{c} = 0.07 and α_{sv} = 1.5 can be similarly determined from experimental results. So Eq. (7.30) becomes

$$V_{\mathrm{u}} = 0.07 f_{\mathrm{c}} b h_0 + 1.5 f_{\mathrm{yv}} \frac{A_{\mathrm{sv}}}{s} h_0 \tag{7.36}$$

For a beam under concentrated loads, both the bending moment and the shear force in the sections beneath the loading points are the largest (Fig. 7.21). Shear-compression failure generally occurs in these sections, because the normal stress and shear stress in the compression zone are maxima. However, for a beam under a uniformly distributed load, the largest shear force is at the cross section above the support, while the largest bending moment happens in the mid-span section. So shear-compression failure does not happen at the cross section subjected to the largest shear force, but generally occurs at the section 1/4 span from the support, where the bending moment and the shear force are both large. When the beam span–depth ratio l/h_0 is small, the failure position is close to the mid-span; when l/h_0 is large, the failure position is close to the support.

7.2.5.2 Shear Capacities of Continuous Beams

Continuous beams are subjected both positive and negative bending moments. The moment diagram of a continuous beam under concentrated loads is shown in Fig. 7.22. According to the definition of shear-span ratio, we have

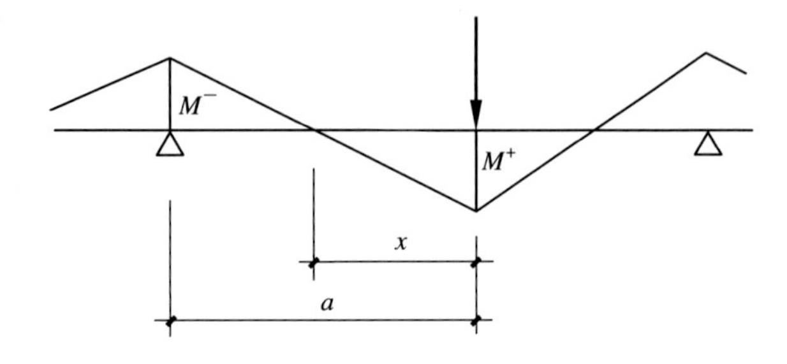

Fig. 7.22 Shear-span ratio of a continuous beam

$$\lambda = \frac{M}{Vh_0} = \frac{Vx}{Vh_0} = \frac{x}{h_0} \tag{7.37}$$

From the similarity of triangles in Fig. 7.22, we have

$$x = \frac{|M^+|}{|M^-| + |M^+|} a = \frac{a}{1 + n'} \tag{7.38}$$

where $n' = |M^-/M^+|$. Substituting Eq. (7.38) into Eq. (7.37) gives

$$\lambda = \frac{a}{h_0(1 + n')} \tag{7.39}$$

The general shear-span ratio of the continuous beam is less than its calculated shear-span ratio a/h_0. The larger the shear-span ratio is, the lower the shear capacity will be. So it is conservative to obtain the shear capacities of continuous beams by using the calculation shear-span ratio and Eq. (7.36).

7.2.5.3 Shear Capacity Formulae in GB 50010

Different codes adopt different formulae for shear capacity calculation. In this section, we focus our attention on *Code for Design of Concrete Structures* (GB 50010).

In order to raise the safety level in shear resistance design, the coefficients for stirrups in Eqs. (7.33) and (7.36) are changed from 1.25 and 1.5 to 1.0. Meanwhile, the application of high-strength concrete makes the concrete strength grade vary in a larger scope. Because f_t instead of f_c is in a good linear relation with the strength grade, and concrete fails in tension under shear force, f_t is adopted in current GB 50020 with corresponding parameter changes. Experiments have proved the validity of this change.

1. Shear capacities of beams if only stirrups are provided
   For common flexural elements of rectangular, T- and I-shaped cross sections subjected to uniformly distributed loads, if only stirrups are provided, the shear capacity formula can be expressed as follows:

$$V_u = V_{cs} = 0.7 f_t b h_0 + f_{yv} \frac{A_{sv}}{s} h_0 \tag{7.40}$$

   where

   - f_t the tensile strength of concrete;
   - b the width of the rectangular cross section, or width of the web of T- and I- sections;
   - h_0 the effective depth of the cross sections; and
   - f_{yv} the tensile strength of stirrups.

   The meanings of the other variables are the same as those in previous equations.
   For isolated beams of rectangular sections subjected to concentrated loads (including the case where several types of loads are applied, among which the shear force at the sections above the supports caused by concentrated loads counts for more than 75 % of the total shear force), the formula for shear capacity is as follows:

$$V_u = V_{cs} = \frac{1.75}{\lambda + 1.0} f_t b h_0 + f_{yv} \frac{A_{sv}}{s} h_0 \tag{7.41}$$

   where λ = the shear-span ratio for the cross section under consideration. Take $\lambda = a/h_0$, where a = the distance from the concentrated load to the nearer support of joint edge. If $\lambda < 1.5$, take $\lambda = 1.5$; and if $\lambda > 3$, take $\lambda = 3$. Stirrups in the beam segment between the concentrated loads and the supports should be uniformly spaced. The isolated beam mentioned above is such a beam that is not cast with its floor monolithically.
   If only stirrups are provided, the cross sections to be checked include the support edges and where the stirrup spacing or the web width is changed at the cross-sectional area (Fig. 7.23).
2. Shear capacities of beams with both stirrups and bent-up bars
   For flexural members of rectangular, T- and I-shaped cross sections, when both stirrups and bent-up bars are provided, the shear capacity should be evaluated as

$$V_u = V_{cs} + V_b = V_{cs} + 0.8 f_y A_{sb} \sin\alpha \tag{7.42}$$

   where

   - A_{sb} the cross-sectional area of the reinforcement bent up in the same plane;
   - f_y the yield strength of the bent-up reinforcement; and

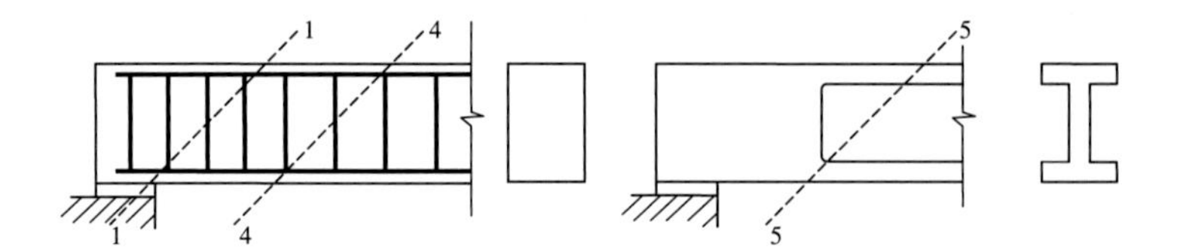

Fig. 7.23 Cross sections to be checked if only with stirrups (1–1 inclined sections at the support edge; 4–4 inclined sections where cross-sectional area or stirrup spacing is changed; 5–5 inclined section where web width is changed)

α the angle between the bent-up reinforcement and the longitudinal axis of the beam. It is usually measured as 45° for cross sections of a depth no more than 800 mm and 60° for cross sections of any depth greater than 800 mm.

The reduction factor 0.8 in Eq. (7.42) is based on the consideration that the intersection of the bent-up reinforcement with the inclined cracks is so close to the shear-compression zone that the reinforcement strength cannot be fully developed before the beam fails.

If bent-up reinforcement is provided, the cross sections at bending points should be checked (Fig. 7.24).

3. Shear capacities of slabs

Web reinforcement is not necessary in slabs and the shear capacity of a slab can be calculated as

$$V_u = 0.7\ \beta_h f_t b h_0 \tag{7.43}$$

$$\beta_h = \left(\frac{800}{h_0}\right)^{\frac{1}{4}} \tag{7.44}$$

where β_h = a factor considering the influence of the section depth, and h_0 can be taken as 800 mm if h_0 is < 800 mm; furthermore, if h_0 is > 2000 mm, h_0 = 2000 mm.

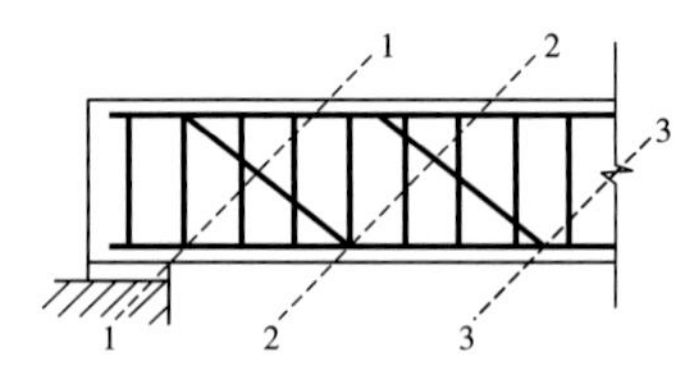

Fig. 7.24 Cross sections to be checked when bent-up reinforcement is provided (1–1 inclined section at the support edge; 2–2, 3–3 inclined sections at bending points)

4. Limits of shear capacities
   Above-mentioned formulae for shear capacities of beams are derived based on shear-compression failure along inclined cross sections. Therefore, they apply only under certain conditions. To ensure the suitability of these formulae, upper bound and lower bound should be specified.
   It is indicated by experimental results that the effect of web reinforcement to raise the shear capacity is limited if the cross-sectional area of a beam is too small, and when the beam will fail in inclined compression. So, for flexural members of rectangular, T- and I-shaped sections, the maximum shear capacities are specified in GB 50010 as follows:
   If $h_w/b \leqslant 4$,

$$V_{u,max} = 0.25\ \beta_c f_c b h_0 \tag{7.45}$$

   If $h_w/b \geqslant 6$,

$$V_{u,max} = 0.2\ \beta_c f_c b h_0 \tag{7.46}$$

   If $4 < h_w/b < 6$, $V_{u,max}$ is determined by linear interpolation,
   where

   β_c a factor that considers the influence of concrete strength. β_c can be taken as 1.0 when the concrete grade is lower than C50 and 0.8 when the concrete grade is C80. In other cases, it can be obtained via linear interpolation;
   f_c the compressive strength of concrete;
   b the width of a rectangular cross section, or width of the web of T- and I- sections;
   h_0 the effective depth of the cross section; and
   h_w web depth of the cross section. h_w can be taken as the effective depth for rectangular sections, the effective depth minus the flange depth for T section, and the net web depth for I section, respectively.

   For simply supported flexural members of T- or I-shaped cross sections, the coefficient 0.25 in Eq. (7.45) can be replaced by 0.3 if the application is supported by practical experience. For members of inclined tensile surface, the controlling conditions on cross sections can be relaxed if the application is supported by practical experience.
   The lower bound of the shear capacity is specified as

$$V_{u,min} = V_c \tag{7.47}$$

   where V_c = the shear capacity of a beam without web reinforcement.
5. The minimum reinforcement ratio of stirrups and the maximum stirrup spacing
   The shear capacity of a beam without web reinforcement can be taken as the lower bound of the shear capacity of the beam, as shown in Eq. (7.47).

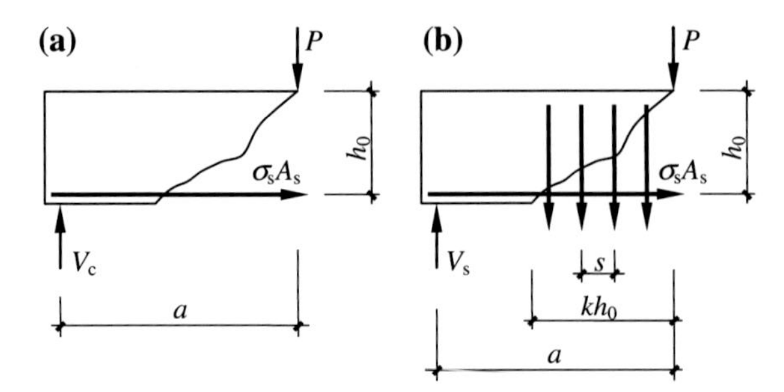

Fig. 7.25 The minimum reinforcement ratio of stirrups in a beam. **a** Without web reinforcement. **b** With web reinforcement

However, a beam without web reinforcement may fail in the mode of inclined tension failure, which is very brittle. In addition, the stirrups intersected with an inclined crack will restrict the development of the crack. So, in the design of flexural reinforced concrete members (except for slabs), the minimum reinforcement ratio of stirrups should be guaranteed.

For the beam without web reinforcement, as shown in Fig. 7.25a, when $\lambda = a/h_0 = 3$, the beam will fail in the mode of inclined tension failure. According to Eq. (7.41), we have

$$V_c = \frac{1.75}{\lambda + 1.0} f_t b h_0 = 0.44 f_t b h_0 \tag{7.48}$$

If the stirrups are lightly provided in the beam, the shear will be borne totally by the stirrups intersected with the inclined crack upon the initiation of the crack, as shown in Fig. 7.25b. The maximum shear capacity of stirrups intersected with the inclined crack is

$$V_s = f_{yv} \frac{A_{sv}}{s} k h_0 \tag{7.49}$$

where kh_0 = the length of the longitudinal projection of the inclined crack.

To avoid inclined tension failure, it should be guaranteed at least that $V_s = V_c$. So, the minimum reinforcement ratio of stirrups in a beam is

$$\rho_{sv,\,min} = \frac{A_{sv}}{bs} = \frac{0.44}{k} \cdot \frac{f_t}{f_{yv}} \tag{7.50}$$

When k = 1−3, we have

$$\rho_{sv,min} = (0.11 \sim 0.44) \frac{f_t}{f_{yv}} \tag{7.51}$$

Table 7.1 Maximum spacing between stirrups in beams s_{max}

Depth of beam, h/mm	$V > 0.7 f_t b h_0$	$V \leqslant 0.7 f_t b h_0$
$150 < h \leqslant 300$	150	200
$300 < h \leqslant 500$	200	300
$500 < h \leqslant 800$	250	350
$h > 800$	300	400

GB 50010 proposes that in the design of flexural reinforced concrete members (except for slabs), in order to guarantee that the shear capacity of a member is not smaller than its lower bound, the reinforcement ratios of stirrups should be checked as

$$\rho_{sv} \geqslant \rho_{sv,min} = 0.24 f_t / f_{yv} \tag{7.52}$$

Taking k in Eq. (7.50) as a value that is a little bit larger than 2, we can get Eq. (7.52).

From Fig. 7.25b, it can be seen that, to guarantee the intersection of stirrups with the inclined crack, the stirrup spacing generally should not be larger than the effective depth of section, that is, $s \leqslant h_0$. Table 7.1 lists the maximum stirrup spacing values s_{max} for reinforcement concrete beams under different conditions as proposed by GB 50010.

7.3 Applications of Shear Capacity Formulae for Flexural Members

7.3.1 Inclined Section Design Based on Shear Capacity

For this type of problem, usually the cross-sectional dimensions (b, h, h_0), computational span l_0, and clear span l_n of the beam, material strengths (f_c, f_t, f_y), and loading conditions are known, and A_{sv}/s, A_{sb} (or A_{sv1}, s, A_{sb}) are required to be calculated. To prevent inclined section failure under the given loading conditions, the shear capacities of the checked cross sections should not be lower than the applied shear force, that is, $V_u \geqslant V$. The following steps should be followed:

(1) Calculate the shear forces V on the cross sections to be designed (Figs. 7.23 and 7.24).
(2) Check $V \leqslant V_c$. If it is satisfied, web reinforcement is provided according to the detailing rather than calculation. Otherwise, go to the next steps.
(3) Check $V \leqslant V_{u,max} = (0.2\text{–}0.25)\, \beta_c f_c b h_0$ as per Eqs. (7.45) or (7.46). If it is not satisfied, go back to adjust the cross-sectional dimensions. Otherwise, continue the design.

(4) If both stirrups and bent-up bars are used, choose A_{sb} first. Then, calculate A_{sv}/s by using Eq. (7.42) with the assumption of $V = V_u$. If only stirrups are used, calculate A_{sv}/s by using Eqs. (7.40) or (7.41) with the assumption $V = V_u$.
(5) Check $\rho_{sv} = \frac{A_{sv}}{bs} \geqslant \rho_{sv,min} = 0.24 f_t / f_{yv}$. If it is satisfied, go to the next step; otherwise, take $\rho_{sv} = \rho_{sv,min}$ and continue the design.
(6) Choose the stirrup diameter and thus get A_{sv1}. Then, obtain s from A_{sv}/s. s must meet the requirement on the maximum spacing between stirrups in Table 7.1. Or one can determine s first from Table 7.1; then, get A_{sv1} from A_{sv}/s.

When stirrups are not required according to the calculation results, if the beam depth h is greater than 300 mm, stirrups should be provided along the full length of the beam; if h ranges between 150 mm and 300 mm, stirrups can be provided only at the ends, whose lengths are equal to one-fourth of the span. However, if concentrated loads are applied at mid-span, stirrups should be provided along the full length. If $h < 150$ mm, stirrups are not necessary.

For beams with a depth greater than 800 mm, the stirrup diameter should not be less than 8 mm. For beams with a depth not greater than 800 mm, the stirrup diameter should not be less than 6 mm.

When bent-up reinforcement is required according to calculation results, the distance between the starting point a and the ending point b of two adjacent rows of bent-up bars as shown in Fig. 7.25 should not be greater than the specified maximum spacing between stirrups for $V > 0.7 f_t b h_0$ (Fig. 7.26).

Example 7.1 For a simply supported reinforced concrete beam, the calculation span is 6 m, the net span is 5760 mm, the cross-sectional dimensions $b \times h = 200 \text{ mm} \times 500 \text{ mm}$. The applied uniformly distributed load is $p = 25.5$ kN/m (including self-weight). The concrete is of grade C20 ($f_c = 9.6 \text{ N/mm}^2$, $f_t = 1.10 \text{ N/mm}^2$), and the stirrup is HPB300 ($f_{yv} = 270 \text{ N/mm}^2$). Calculate the required stirrups for the beam.

Solution It is known that $f_c = 9.6 \text{ N/mm}^2$, $f_t = 1.10 \text{ N/mm}^2$, and $f_{yv} = 270 \text{ N/mm}^2$.

The effective depth of the cross section is $h_0 = h - 35 \text{ mm} = 500 - 35 = 465$ mm.

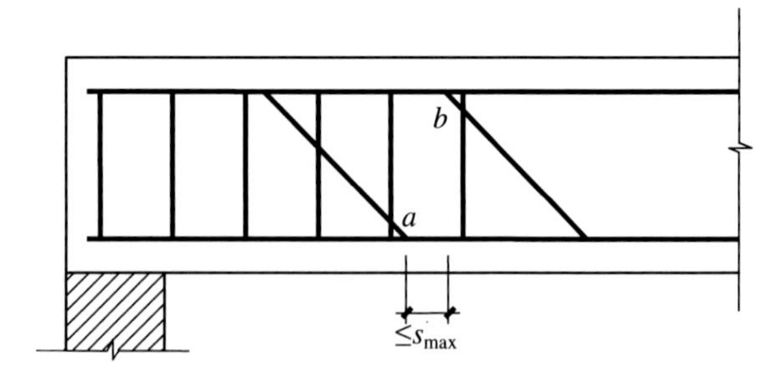

Fig. 7.26 Spacing between adjacent bent-up reinforcement bars

1. Calculate the shear force of the cross section
   Consider the cross section at the support and the maximum shear force for the net span is

$$V = \frac{1}{2} \times 5.76 \times 25.5 = 73.44\,\mathrm{kN}$$

2. Check if the calculation is required for stirrups

$$0.7 f_t b h_0 = 0.7 \times 1.1 \times 200 \times 465 = 71610\,\mathrm{N} < V = 73.44\,\mathrm{kN}$$

   Therefore, the required number of stirrups should be determined according to the calculation.
3. Check limits on the cross section
   Because $h_w/b < h/b = 500/200 < 4$, Eq. (7.45) is used:
   $0.25\ \beta_c f_c b h_0 = 0.25 \times 1.0 \times 9.6 \times 200 \times 465 = 223200$ N = 223.2 kN > V = 73.44 kN, OK.
4. Calculate the required stirrups
   According to Eq. (7.40), take $V = V_u$,

$$73.44 \times 10^3 = 71610 + 270 \times \frac{A_{sv}}{s} \times 465$$

$$\frac{A_{sv}}{s} = 0.0146$$

   The corresponding reinforcement ratio of stirrups is $\rho_{sv} = \frac{A_{sv}}{bs} = \frac{0.0146}{200} = 0.000073$.
5. Check the minimum steel ratio of stirrups

$$\rho_{sv,min} = 0.24 \times \frac{1.1}{270} = 0.00098 \geqslant \rho_{sv} = 0.000073$$

   So stirrups should be provided according to the minimum reinforcement ratio. Taking $\frac{A_{sv}}{bs} = \rho_{sv,min}$, we have

$$\frac{A_{sv}}{s} = b\rho_{sv,min} = 200 \times 0.00098 = 0.1956\,\mathrm{mm}$$

6. Select stirrups
   Stirrups with the diameter of 6 mm are chosen. $A_{sv1} = \pi 6^2/4 = 28.27\ \mathrm{mm}^2$. Use double-legged stirrup. $A_{sv} = 2A_{sv1} = 2 \times 28.27 = 56.54\ \mathrm{mm}^2$. Therefore, the stirrup spacing s should be

$$s = \frac{A_{sv}}{0.1956\,\mathrm{mm}} = \frac{56.54\,\mathrm{mm}^2}{0.1956\,\mathrm{mm}} = 289\,\mathrm{mm}$$

Further considering the detailing requirement on stirrup spacing, we have s = 200 mm. Finally, the provided stirrup is $\phi6@200$.

Conversely, if we determine s first and then determine A_{sv1}, the same results can be obtained.

Example 7.2 A simply supported beam is the same as that in Example 7.1, but is subjected to uniformly distributed load (including self-weight) p = 43 kN/m. The longitudinal bars based on the requirement of flexural capacity are chosen as $3\phi25$ (f_y = 300 N/mm^2). Concrete grade is C20, grade of stirrup is still HPB300 (f_{yv} = 270 N/mm^2). Calculate the required web reinforcement according to the shear capacity requirement of the inclined cross section.

Solution We already know that f_c = 9.6 N/mm^2, f_t = 1.10 N/mm^2, f_{yv} = 270 N/mm^2, h_0 = 465 mm, and f_y = 300 N/mm^2.

1. Calculate the shear force

$$V = \frac{1}{2} \times 43 \times 5.76 = 123.84\,\text{kN}$$

According to Example 7.1, the limits on cross section are obviously satisfied.

2. Both bent-up bars and stirrups provided (cross section I-I)

One $\phi25$ longitudinal steel bar (A_{sb} = 490.9 mm^2) is bent-up at a 45° angle. The upper bending point is 50 mm from the support edge, as shown in Fig. 7.27.

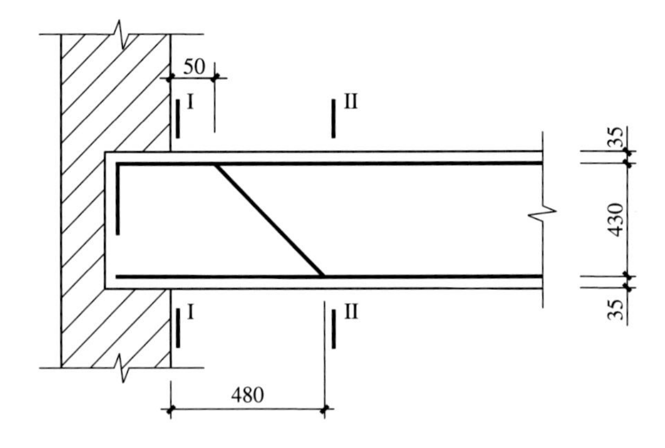

Fig. 7.27 Bent-up position and cross sections to be checked

The shear resistances of the bent-up reinforcement and the concrete are

$$\begin{aligned} V_{cb} &= 0.7 f_t b h_0 + 0.8 f_y A_{sb} \sin \alpha_s \\ &= 0.7 \times 1.1 \times 200 \times 465 + 0.8 \times 300 \times 490.9 \times \sin 45^\circ \\ &= 71610 + 83308 = 154918\,\text{N} \\ &= 154.9\,\text{kN} > V = 123.84\,\text{kN} \end{aligned}$$

So, in the bent-up segment stirrups, satisfying the detailing requirement is enough.

$$\frac{A_{sv}}{bs} = \rho_{sv,min} = 0.24 \frac{f_t}{f_{yv}} = 0.24 \times \frac{1.10}{270} = 0.00098$$

Choose $\phi 6$, that is, $A_{sv1} = 28.27\ \text{mm}^2$, $A_{sv} = 2 \times 28.27 = 56.54\ \text{mm}^2$

$$s = \frac{A_{sv}}{0.00098\,b} = \frac{56.54}{0.1956} = 289\,\text{mm}$$

Thus, stirrups of $\phi 6@200$ are provided.

3. Check the shear capacity at the section of the bottom bending point (cross section II-II)
   Shear force at the bottom bending point (cross section II-II) is

$$V_{II} = \frac{\frac{5760}{2} - 480}{\frac{5760}{2}} \times 123.84 = 103.20\,\text{kN}$$

Shear capacity of the cross section is

$$\begin{aligned} V_{cs} &= 71610 + 270 \times \frac{56.54}{200} \times 465 = 71610 + 35493 = 107103\,\text{N} \\ &= 107.10\,\text{kN} > V_{II}, \text{OK.} \end{aligned}$$

Therefore, stirrups of $\phi 6@200$ are used in the full beam length.

Example 7.3 A simply supported reinforced concrete beam is subjected to a uniformly distributed load of p = 13 kN/m (including self-weight) and two concentrated loads (each is P = 140 kN), as shown in Fig. 7.28. The dimensions of the cross section are $b \times h$ = 200 mm × 600 mm. Use C30 concrete (f_c = 14.3 N/mm^2, f_t = 1.43 N/mm^2). The grade of longitudinal reinforcement is HRB335 (f_y = 300 N/mm^2), and the grade of stirrups is HPB300 (f_{yv} = 270 N/mm^2). Calculate the required stirrups in the beam.

Fig. 7.28 Example 7.3

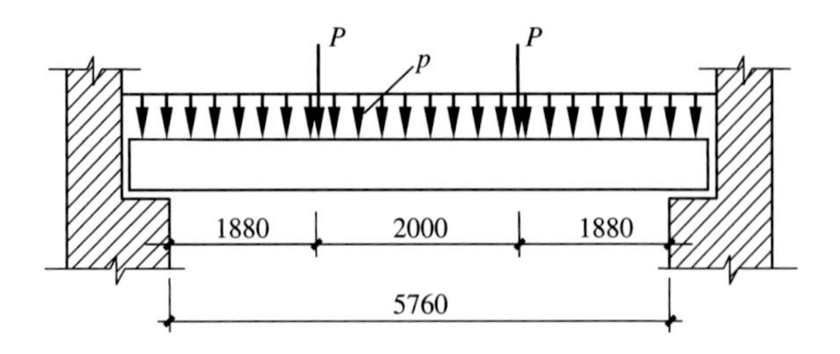

Solution

1. Calculate the shear force
   The shear force at the support edge caused by the distributed load is $V_p = 13 \times 5.76/2 = 37.44$ kN.
   The shear force at the support edge caused by the concentrated loads is $V_P = 140$ kN.
   Total shear force $V = 37.44 + 140 = 177.44$ kN
   The shear force caused by the concentrated loads counts for 140/177.44 = 0.789 = 78.9 % > 75 % of the total shear force; therefore, the shear capacity in the segment between the support edge and the concentrated load should be calculated according to the concentrated load case.
2. Check the limits on the cross section
   The load is large, so the longitudinal reinforcement is probably arranged in two layers. $h_0 = 600 - 60 = 540$ mm
   $0.25 f_c b h_0 = 0.25 \times 14.3 \times 200 \times 540 = 386100$ N = 386.1 kN > $V = 177.44$ kN, OK.
3. Calculate the required stirrups in the shear–flexure segment
   The shear-span ratio $\lambda = \frac{a}{h_0} = \frac{1880}{540} = 3.48 > 3.0$
   Take $\lambda = 3.0$. From Eq. (7.41) and $V = V_u$, we have

$$V = 177.44 \times 10^3 = \frac{1.75}{3+1} \times 1.43 \times 200 \times 540 + 270 \times \frac{A_{sv}}{s} \times 540$$

$$\frac{A_{sv}}{s} = 0.7536$$

The corresponding reinforcement ratio of stirrups is $A_{sv}/(bs) = 0.7536/200 = 0.003768$.
The minimum reinforcement ratio of stirrups is

$$\rho_{sv,min} = 0.24 \frac{f_t}{f_{yv}} = 0.24 \times \frac{1.43}{270} = 0.001271 < 0.003768$$

Stirrups should be provided by calculation. Double-legged stirrups with the diameter of 8 mm ($A_{sv} = 100.53$ mm^2) are chosen; then,

$$s = \frac{A_{sv}}{0.7536} = \frac{100.53}{0.7536} = 133.40\,\text{mm}$$

take s = 120 mm, that is, stirrups $\phi 8@120$ are provided in the shear–flexure segment.

4. Calculate the required stirrups in the mid-segment
In the mid-segment between the two concentrated loads, the maximum shear force, which was caused by the distributed load only, is

$$V_1 = \frac{1}{2} \times 13 \times 2 = 13.00\,\text{kN}$$

whereas $0.7\,f_t b h_0 = 0.7 \times 1.43 \times 200 \times 540 = 108108$ N = 108.1 kN > V_1 = 13.00 kN; therefore, the stirrups in this segment can be provided according to the detailing requirement. Finally $\phi 8@300$ is taken.

Example 7.4 For a reinforced concrete beam, the loads and corresponding shear force diagram are shown in Fig. 7.29. The cross-sectional dimensions are $b \times h$ = 200 mm × 500 mm. The concrete grade is C30 (f_c = 14.3 N/mm^2, f_t = 1.43 N/mm^2). Stirrups of HPB300 (f_{yv} = 270 N/mm^2) are used. Calculate the required stirrups.

Solution f_c = 14.3 N/mm^2, f_t = 1.43 N/mm^2, f_{yv} = 270 N/mm^2

1. Check the dimension of the cross section
$0.25\,f_c b h_0 = 0.25 \times 14.3 \times 200 \times 465 = 332475$ N = 332.48 kN > 119.17 kN, OK.
2. Calculate the required stirrups in each segment

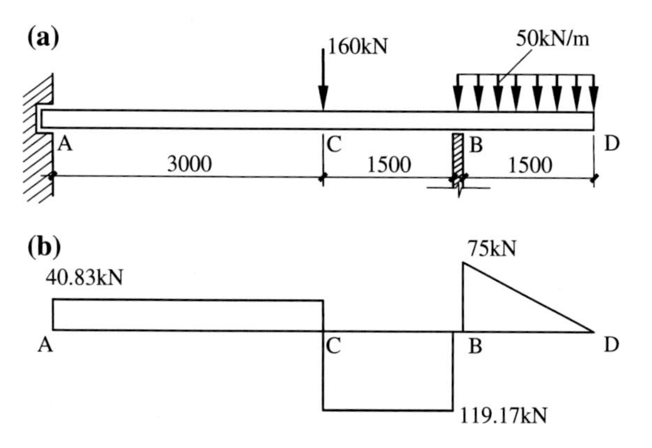

Fig. 7.29 Example 7.4

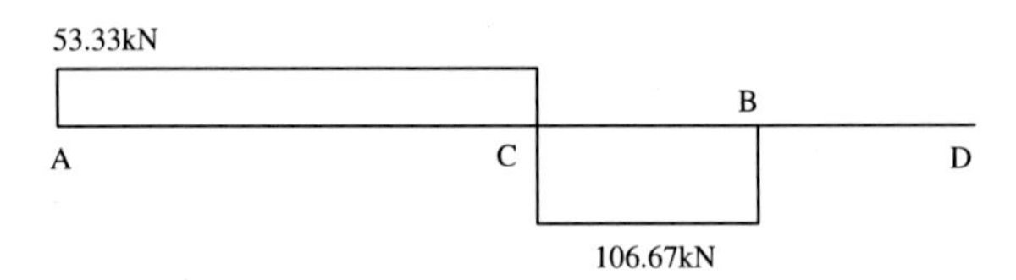

Fig. 7.30 Shear force caused by the concentrated load (Example 7.4)

1. Segment BD
   $0.7 f_t b h_0 = 0.7 \times 1.43 \times 200 \times 465 = 93093$ N = 93.093 kN > 75 kN (shear force value at the right side of support *B*); therefore, stirrups in segment *BD* can be provided according to the detailing requirement, that is, $\phi 6@300$.
   The shear force diagram due to the concentrated load is shown in Fig. 7.30. It can be seen that no matter the load is at support *A* or at the left side of support *B*, the shear forces caused by the concentrated load both count for more than 75 % of the total shear force at the corresponding cross section. Therefore, the shear capacity formula corresponding to the concentrated load should be used.
2. Segment AC
   The shear-span ratio is $\lambda = \frac{a}{h_0} = \frac{3000}{465} = 6.54 > 3.0$, take $\lambda = 3.0$.

$$\frac{1.75}{\lambda + 1} f_t b h_0 = \frac{1.75}{3 + 1} \times 1.43 \times 200 \times 465 = 58183\,\text{N}$$
$$= 58.183\,\text{kN} > 40.83\,\text{kN}$$

   Therefore, stirrups can be provided according to the detailing requirement in this segment, that is, $\phi 6@300$.
3. Segment CB
   The shear-span ratio is

$$\lambda = \frac{a}{h_0} = \frac{1500}{465} = 3.23 > 3.0,$$

   which leads to

$$\frac{1.75}{\lambda + 1} f_t b h_0 = 58183\,\text{N} = 58.183\,\text{kN} < 119.17\,\text{kN}.$$

   The calculation for the stirrup number is required. Hence,

$$\frac{A_{sv}}{s} = \frac{V - V_c}{f_{yv} h_0} = \frac{119170 - 58183}{270 \times 465} = 0.4858$$

   The corresponding reinforcement ratio of stirrups is

$$\rho_{sv} = \frac{A_{sv}}{bs} = \frac{0.4858}{200} = 0.002429 > \rho_{sv,min} = 0.24\frac{f_t}{f_{yv}} = 0.24 \times \frac{1.43}{270}$$
$$= 0.001271, \text{OK.}$$

Double-legged stirrups with the diameter of 6 mm (A_{sv} = 56.54 mm^2) are chosen, then $s = \frac{56.54}{0.4858} = 116.39\,\text{mm}$. Stirrups in the segment are $\phi 6@100$.

Example 7.5 For a simply supported reinforced concrete beam of rectangular section, the span and loads are shown in Fig. 7.31 (self-weight is included in the distributed load). The dimensions of the cross section are $b \times h$ = 200 mm × 600 mm (h_0 = 540 mm). C20 concrete (f_c = 9.6 N/mm^2, f_t = 1.1 N/mm^2) and HPB300 stirrups (f_{yv} = 270 N/mm^2) are used. Calculate the required stirrup amount.

Solution

(1) Diagrams for the total shear force and the shear force due to the concentrated load are shown in Fig. 7.31. It can be seen that the shear forces at both supports caused by the concentrated loads count for more than 75 % of the corresponding total shear forces. Therefore, the formula for concentrated loads should be used.
(2) Check the limits on the cross section
$0.25 f_c b h_0 = 0.25 \times 9.6 \times 200 \times 540 = 259200$ N = 259.20 kN > 148 kN, OK.
(3) Determine the required stirrups

Fig. 7.31 Example 7.5

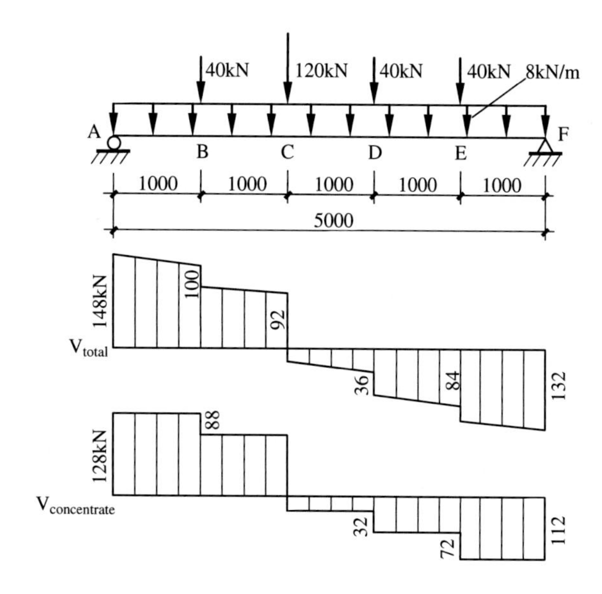

(1) At support A (segment AB),

$$1.5 < \lambda = \frac{a}{h_0} = \frac{1000}{540} = 1.852 < 3.0$$

$$\frac{1.75}{\lambda+1} f_t b h_0 = \frac{1.75}{1.852+1} \times 1.1 \times 200 \times 540 = 72896\,\text{N}$$
$$= 72.896\,\text{kN} < 148\,\text{kN}$$

Calculation is required to determine the stirrup number.

$$\frac{A_{sv}}{s} = \frac{V - V_c}{f_{yv} h_0} = \frac{148000 - 72896}{270 \times 540} = 0.5151$$

The corresponding reinforcement ratio of stirrups is

$$\rho_{sv} = \frac{A_{sv}}{bs} = \frac{0.5151}{200} = 0.002576 > \rho_{sv,\min} = 0.24 \frac{f_t}{f_{yv}} = 0.24 \times \frac{1.1}{270}$$
$$= 0.000978, \text{OK}$$

Choose double-legged stirrups with the diameter of 8 mm (A_{sv} = 100.53 mm^2), then $s = \frac{100.53}{0.5151} = 195.17\,\text{mm}$. So stirrups of $\phi 8@200$ are provided in segment AB. Even though the actual spacing is little bit larger than the calculated one, the result is still reasonable because the absolute relative error is smaller than 5 %.

(2) Segment BC
$\lambda = \frac{a}{h_0} = \frac{2000}{540} = 3.704 > 3.0$, take λ = 3.0.

$$\frac{1.75}{\lambda+1} f_t b h_0 = \frac{1.75}{3+1} \times 1.1 \times 200 \times 540 = 51975\,\text{N} = 51.975\,\text{kN} < 100\,\text{kN}$$

Calculation is required to determine the stirrup amount.

$$\frac{A_{sv}}{s} = \frac{V - V_c}{f_{yv} h_0} = \frac{100000 - 51975}{270 \times 540} = 0.3294$$

The corresponding reinforcement ratio of stirrups is

$$\rho_{sv} = \frac{A_{sv}}{bs} = \frac{0.3294}{200} = 0.001647 > \rho_{sv,\min} = 0.24 \frac{f_t}{f_{yv}} = 0.24 \times \frac{1.1}{270}$$
$$= 0.000978, \text{OK}$$

Choose double-legged stirrups with the diameter of 8 mm (A_{sv} = 100.53 mm^2), then $s = \frac{100.53}{0.3294} = 305.19\,\text{mm}$. So stirrups of $\phi 8@300$ are provided in segment BC.

(3) Segment CD
The shear-span ratio is obviously greater than 3, so take $\lambda = 3$. Because V_c = 51.975 kN > 36 kN, stirrups are provided to satisfy the detailing requirement. So stirrups of $\phi 8@300$ are provided in segment CD.

(4) Segment DE
$\lambda = 3.704 > 3.0$, take $\lambda = 3.0$. V_c = 51.975 kN < 84 kN, calculation is required.

$$\frac{A_{sv}}{s} = \frac{V - V_c}{f_{yv}h_0} = \frac{84000 - 51975}{270 \times 540} = 0.2197$$

The corresponding reinforcement ratio of stirrups is

$$\rho_{sv} = \frac{A_{sv}}{bs} = \frac{0.2197}{200} = 0.001098 > \rho_{sv,min} = 0.000978, \text{OK}$$

Choose double-legged stirrups with the diameter of 8 mm (A_{sv} = 100.53 mm^2), then $s = \frac{100.53}{0.2197} = 457.58\,\text{mm}$. So stirrups of $\phi 8@300$ are provided in segment DE.

(5) Segment EF

$$\lambda = 1.852$$

$$\frac{A_{sv}}{s} = \frac{V - V_c}{f_{yv}h_0} = \frac{132000 - 72896}{270 \times 540} = 0.4054$$

Choose double-legged stirrups with the diameter of 8 mm (A_{sv} = 100.53 mm^2), then $s = \frac{100.53}{0.4054} = 247.99\,\text{mm}$. So stirrups of $\phi 8@250$ are provided in segment EF (with an error of -0.8 %, OK).

7.3.2 *Shear Capacity Evaluation of Inclined Sections of Existing Members*

For this type of problem, usually the cross-sectional dimensions (b, h, h_0), strengths of materials (f_c, f_t, f_y), reinforcement (A_{sv}/s, A_{sb}), and loading conditions are known, and the shear capacities V_u of inclined cross sections are required to be calculated. The following steps are generally followed:

(1) Determine whether λ should be considered based on the loading conditions and the positions of the cross sections to be checked.
(2) When $\rho_{sv} \leqslant \rho_{sv,min}$ or when $\rho_{sv} > \rho_{sv,min}$, but $s > s_{max}$ (the maximum spacing between stirrups), $V_u = V_c$. Otherwise, go to the next step.
(3) Calculate the shear capacity by Eq. (7.53)

$$V_u = \min[(0.2\text{–}0.25)\beta_c f_c b h_0, V_{cs} + V_b] \tag{7.53}$$

Examples are omitted for brevity.

7.3.3 Discussion on Shear Forces for the Design of Beams

7.3.3.1 Shear Forces of Support Edges

Figures 7.23 and 7.24 give the different inclined critical sections in different conditions. These critical sections should be checked both in the design of a new beam and in the safety assessment of an existing beam.

In real practices, for section 1-1 shown in Fig. 7.23 or Fig. 7.24, the maximum shear force, that is, the calculated shear force at the edge of a support, is usually used for the sectional design or the sectional evaluation. However, in a beam loaded on the top and supported on the bottom as shown in Fig. 7.32a, the closest inclined crack that can occur to the supports will extend outward from the supports at roughly 45° to form the compression fans. Loads on the top of the beam within a distance h_0 or h_0' from a support will be transmitted to the support directly by the compression fan and will not affect the stresses in the stirrups crossing the cracks.

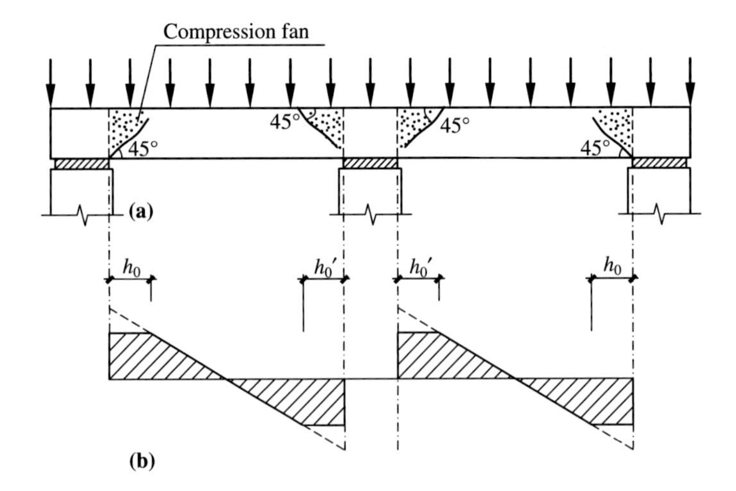

Fig. 7.32 Shear forces of the support edges. **a** Beam. **b** Shear force

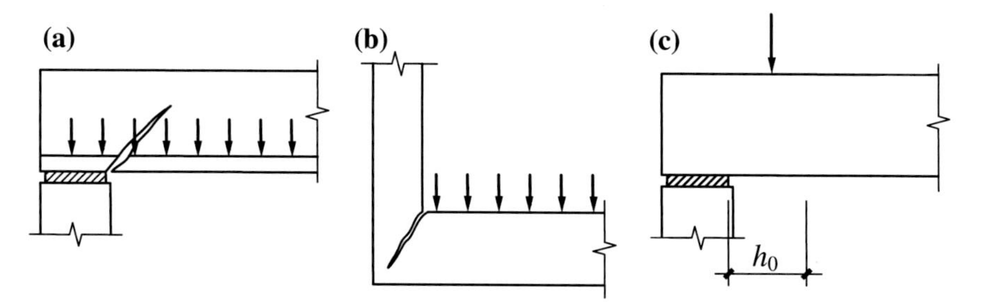

Fig. 7.33 Shear forces of the special supports. **a** Beam loaded on tension flange. **b** Beam supported by tension force. **c** Beam with concentrated load close to support

So, if the inclined angle of cracks is assumed to be 45°, the shear force at the section located at a distance of h_0 or h_0' from a support can be used to perform the design or the safety assessment for section 1-1, as shown in Fig. 7.23 or Fig. 7.24. Of course, in sectional design, the safer results will be obtained if the shear force at a support edge is used for section 1-1. But in the safety assessment of the section 1-1, totally different assessment results may be obtained by using different shear forces. It should be pointed out that for the cases shown in Fig. 7.33, only the shear force at the edge of the support can be used in the design or the safety assessment of the section

7.3.3.2 Shear Forces of Tapered Beams

Figure 7.34a shows the internal forces of a segment taken from a tapered beam. The moment at the end of the segment M_1 (or M_2) can be represented by two horizontal components of compressive force on concrete and tensile force on steel bars, C_1 and T_1 (or C_2 and T_2), separated by the lever arm γh_{01} (or γh_{02}): $M_1 = C_1\gamma h_{01} = T_1\gamma h_{01}$ ($M_2 = C_2\gamma h_{02} = T_2\gamma h_{02}$). When the height of the section increases with the increase of moment, two vertical components of compressive force on concrete and tensile force on steel bars will decrease the shear force on the section as shown in Fig. 7.34b. Otherwise, the shear force on the section will be increased as shown in Fig. 7.34c.

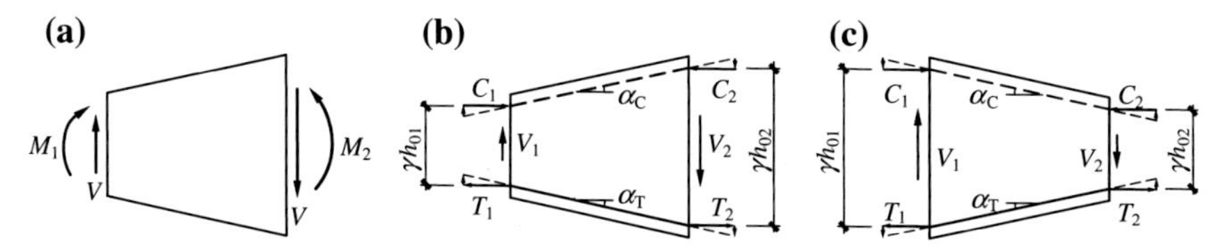

Fig. 7.34 Shear forces of segments in tapered beams. **a** Forces on segment of beam. **b** Shear forces resisted by stirrups and concrete (section increases with the moment increase). **c** Shear forces resisted by stirrups and concrete (section decreases with the moment increase)

Based on the forces at the left section of the segment shown in Fig. 7.34a and b, we have

$$V_1 = V - C_1 \tan \alpha_c - T_1 \tan \alpha_T \tag{7.54}$$

Substituting $C_1 = T_1 = \frac{M_1}{\gamma h_{01}}$ and letting $\alpha = \alpha_C + \alpha_T$ give the following general equation:

$$V_1 = V - \frac{|M|}{\gamma h_0} \tan \alpha \tag{7.55}$$

where

V_1 shear force resisted by stirrups and concrete in a critical section of a tapered beam;

$|M|$ absolute value of the moment on a critical section of a tapered beam;

γ coefficient of lever arm; and

α the sum of α_c, the inclined angle of the line joining the centroids of the compression stress blocks in concrete; α_T, the angle between the tensile force of steel bars and the horizontal line. α is positive if the lever arm γh_0 increases in the same direction as $|M|$ increases.

The examples for the use of Eq. (7.55) are shown in Fig. 7.35.

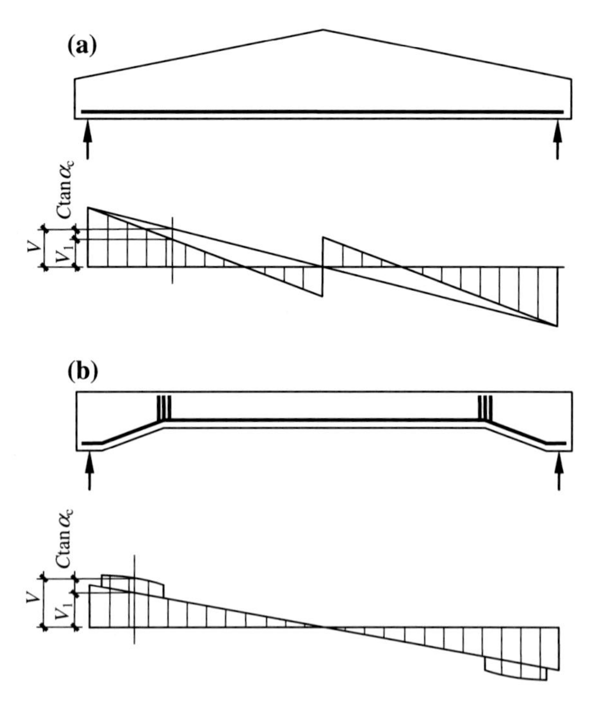

Fig. 7.35 Examples of V_1. **a** Roof beam under evenly distributed load. **b** Haunched simply supported beam under evenly distributed load

7.4 Measures to Ensure the Flexural Capacities of Inclined Cross Sections in Flexural Members

7.4.1 Flexural Capacities of Inclined Cross Sections

Take a free body along an inclined cross section as shown in Fig. 7.36. By taking the moment about the resultant force point in the compression zone, the flexural capacity of the inclined cross section can be expressed as

$$M_{\mathrm{u}}^{\mathrm{Inclined}} = f_{\mathrm{y}} A_{\mathrm{s1}} z + f_{\mathrm{y}} A_{\mathrm{sb}} z_{\mathrm{sb}} + \sum f_{\mathrm{yv}} A_{\mathrm{sv}} z_{\mathrm{sv}} \tag{7.56}$$

The corresponding flexural capacity of the normal cross section is

$$M_{\mathrm{u}}^{\mathrm{Normal}} = f_{\mathrm{y}} A_{\mathrm{s1}} z + f_{\mathrm{y}} A_{\mathrm{sb}} z \tag{7.57}$$

By comparing Eqs. (7.56) and (7.57), it can be found that $M_{\mathrm{u}}^{\mathrm{Inclined}} > M_{\mathrm{u}}^{\mathrm{Normal}}$ if $z_{\mathrm{sb}} \geqslant z$. That is, if the flexural capacity of the normal cross section is satisfactory according to the calculation results, the flexural capacity of the inclined cross section is generally conservative. However, the insufficient anchorage of the longitudinal reinforcement at supports and improper bent-up or cutoff of reinforcement will result in flexural failure of the inclined cross section.

Therefore, the flexural capacity of the inclined cross section is not necessary to be checked if the detailing requirements on bent-up, cutoff, and anchorage at supports are fulfilled.

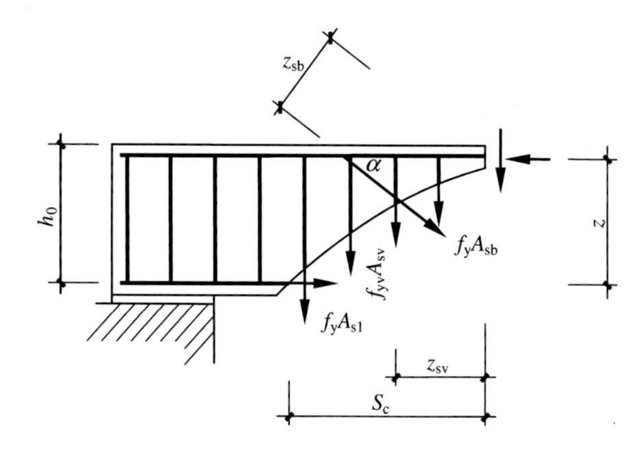

Fig. 7.36 Calculation diagram for the flexural capacity of an inclined cross section

7.4.2 Moment Capacity Diagram

The moment capacity diagram is also called the material diagram, which shows the maximum bending moment M_R resisted by the actually provided longitudinal reinforcement at each normal cross section along the beam length. The M_R diagram is used in determining where to bend up or cut off the reinforcement.

The total M_R at each cross section can be determined by the flexural capacity analysis of the normal section. M_R is actually the sum of contributions of each piece of reinforcement. The contribution of the *i*th bar to M_R, M_{Ri} can be approximately estimated by the ratio of the bar sectional area A_{si} to the total area of all the bars A_s, that is

$$M_{Ri} = \frac{A_{si}}{A_s} M_R \quad (7.58)$$

For instance, for a simply supported beam as shown in Fig. 7.37, because all the longitudinal bars are through the full length, the moment capacity diagram, M_R, is a trapezoid (abcd). In the beam mid-span, all longitudinal bars can be fully utilized, so the resistance moment is the maximum and the same at all cross sections. At the beam ends, the reinforcement needs a length l_a for anchorage. In the length l_a, the resistance moment gradually decreases toward the ends and becomes zero at point a or b. If the bond stress between the reinforcement and the concrete is uniformly distributed, the resistance moment at the beam ends will vary linearly. Figure 7.37 also presents the moment that each bar can resist. It can be seen that the strengths of the three bars are fully utilized at point 1 in the mid-span; the strengths of bars ① and ② are fully utilized at point 2, but theoretically, bar ③ is no longer necessary beyond point 2 (toward the support). Similarly, the strength of bar ① is fully utilized at point 2, but bar ② is no longer needed outside point 3. Therefore, points 1, 2, and 3 are referred to as *fully utilized points* of bars ①, ②, and ③, respectively; points 2, 3, and a are referred to as *unnecessary points* of bars ①, ②, and ③, respectively.

Figure 7.38 shows the moment capacity diagram of a simply supported beam with bent-up bars provided. Bar ③ is bent up at cross sections E and F. The influence of the bent-up bars on the normal section flexural capacity can be calculated according to the following assumptions. Take the bent-up bar at the left end

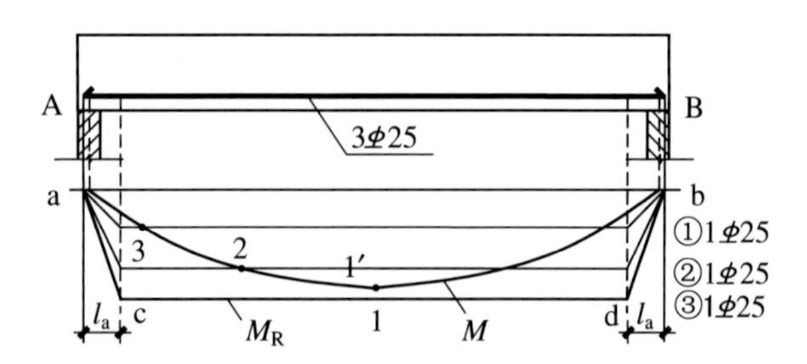

Fig. 7.37 Moment capacity diagram

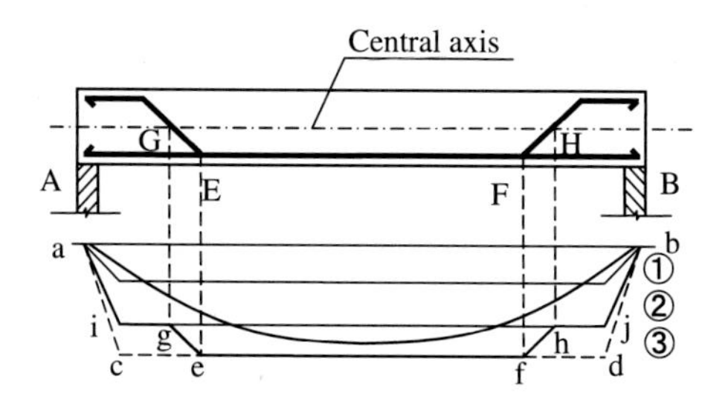

Fig. 7.38 Moment capacity diagram if bent-up bars are provided

of the beam for example. Suppose the bent-up bar intersects with the axis of the beam at point G, and at the left side of point G, the bent-up bar has no contribution to the positive bending moment. At point *E*, the bar is fully utilized. Between points G and E, the contribution of the bar to the positive bending moment can be determined by linear interpolation (line ge in Fig. 7.38).

In structural designs, the moment capacity diagram should envelope the moment diagram. And the closer the two diagrams are, the more economical the beam is.

7.4.3 Detailing Requirements to Ensure the Flexural Capacities of Inclined Sections with Bent-up Bars

The following three requirements should be satisfied when the longitudinal bars are to be bent up: (1) ensure the flexural capacities of normal sections, that is, $M_R \geqslant M$, the moment capacity diagram should envelope the moment diagram; (2) ensure the shear capacities of inclined sections; and (3) ensure the flexural capacities of inclined sections.

The last requirement is ensured by the detailing requirements. For a beam shown in Fig. 7.39, the longitudinal bars of area A_s are provided according to the flexural capacity of the normal section CC′. A bar (or a row of bars) of area A_{sb} is bent up at point K, and the remaining longitudinal bars, $A_{s1} = A_s - A_{sb}$, extend into the support. The potential inclined crack is represented by JH as shown in Fig. 7.39.

If we consider ABCC′ a free body, and take the moment about the resultant force point O at cross section CC′, the flexural capacity of the normal section is

$$M_u^{\text{Normal}} = f_y A_s z = f_y A_{s1} z + f_y A_{sb} z \tag{7.59}$$

where f_y is the yield strength of the longitudinal bars.

Next, take the free body ABCHJ on the left side of the inclined crack and also take the moment about point O, the flexural capacity of the inclined cross section, which can be expressed as

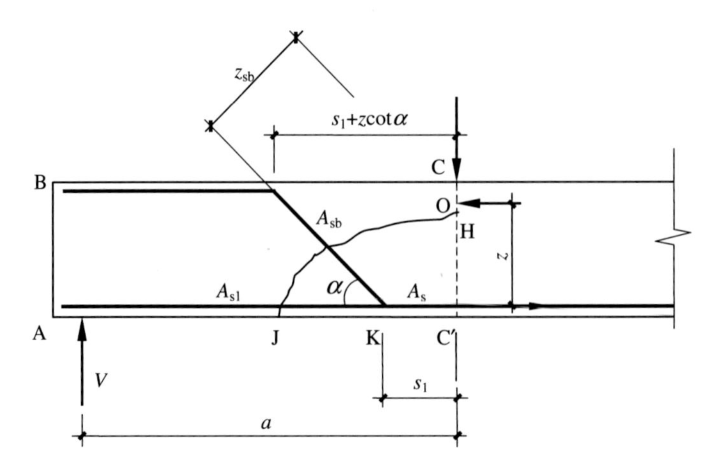

Fig. 7.39 Influence of bent-up bars on flexural capacity of an inclined section

$$M_{\mathrm{u}}^{\mathrm{Inclined}} = f_y A_{s1} z + f_y A_{sb} z_{sb} \tag{7.60}$$

If it is required that the flexural capacity of the inclined section should not be less than that of the normal section, we have

$$z_{sb} \geqslant z \tag{7.61}$$

The above condition can be satisfied by controlling the horizontal distance s_1 in Fig. 7.39. s_1 is the distance from the bent-up point to the fully utilized point. According to the geometric relationship in Fig. 7.39, it can be derived that

$$\begin{aligned} \frac{z_{sb}}{s_1 + z\cot\alpha} &= \sin\alpha \\ z_{sb} &= s_1 \sin\alpha + z\cos\alpha \end{aligned} \tag{7.62}$$

Substituting Eq. (7.62) into Eq. (7.61) yields

$$\begin{aligned} s_1 \sin\alpha + z\cos\alpha &\geqslant z \\ s_1 &\geqslant (\csc\alpha - \cot\alpha) z \end{aligned} \tag{7.63}$$

Take $z = (0.77 \sim 0.91)\,h_0$. When $\alpha = 45°$ and $60°$, Eq. (7.63) gives $s_1 \geqslant (0.319 \sim 0.372)\,h_0$ and $s_1 \geqslant (0.445 \sim 0.525)\,h_0$, respectively. So it can be taken that

$$s_1 \geqslant \frac{1}{2} h_0 \tag{7.64}$$

Equation (7.64) means that if the distance from the bent-up point to the fully utilized point of bars is not less than $h_0/2$, the flexural capacity of the inclined cross section is automatically satisfied.

7.4.4 Detailing Requirements to Ensure the Flexural Capacities of Inclined Sections When Longitudinal Bars Are Cut off

Longitudinal bars can be cut off where they are no longer needed to resist tensile forces or where the remaining bars are adequate to do so. The location of points where longitudinal bars are no longer needed is called the theoretical cutoff point, which is the unnecessary point in the moment capacity diagram. If longitudinal bars are cut off at the theoretical cutoff point, the flexural capacity of the normal section may not be sufficient as shown in Fig. 7.40. In addition, the sudden increase of the tensile stress in concrete due to stress concentration at the cutoff point will usually result in a flexure–shear-inclined crack. At this time, the remaining bars have been fully utilized; hence, they can only resist the bending moment at the theoretical cutoff point. However, after the occurrence of the inclined crack, the longitudinal bars at the theoretical cutoff point will resist the tensile stress caused by the bending moment at the tip of the inclined crack, which is larger than that at the theoretical cutoff point. (This phenomenon reflects the effects of shear on tensile stresses in longitudinal bars). So the stress in the longitudinal bars at the theoretical cutoff point exceeds the yield strength and inclined flexural failure will happen. Therefore, if bars are cut off at the theoretical cutoff point, flexural failure will occur along the inclined cross section.

To avoid the flexural failure at the normal section, longitudinal bars should be cut off after they extend out to the fully utilized point for a certain length l_a. To avoid the flexural failure along the inclined cross section, longitudinal bars should

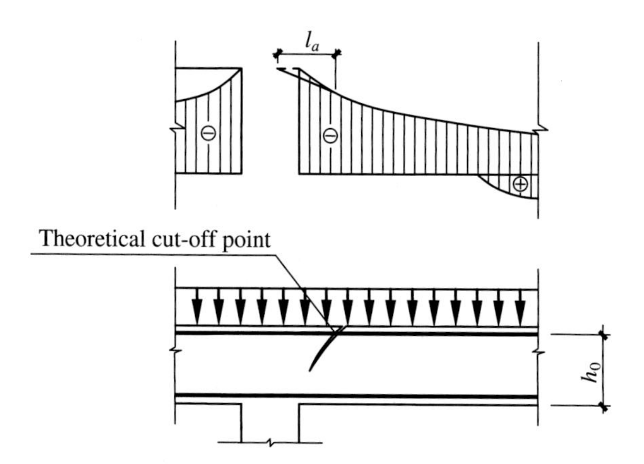

Fig. 7.40 Cutoff bars at the theoretical cutoff point

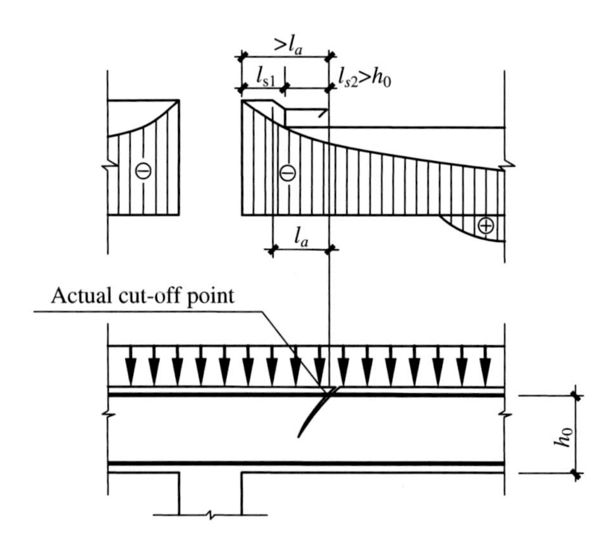

Fig. 7.41 Cutoff of reinforcement

be cut off after they extend past the theoretical cutoff point for a certain length l_{s2}, as shown in Fig. 7.41. If the inclined angle of an inclined crack appeared at the actual cutoff point to the axis of the beam, it would rest at 45°, the maximum distance from the actual cutoff point to the tip of the crack may reach h_0, so it is usually required that $l_{s2} > h_0$. Because the remaining longitudinal bars at the actual cutoff point have not exhausted their strength, they can still resist part of the increased bending moment due to the inclined crack. Besides, the stirrups that are intersected with the inclined crack can also resist part of the increased bending moment.

Detailed requirements on the cutoff of longitudinal bars are not the same in different codes. Take GB 50010 as an example. It is specified that the longitudinal tensile reinforcement to resist the negative bending moment at the cross sections above the supports should not be cut off in the tension zone. If the reinforcement must be cut off, the following requirements should be met:

(1) If $V \leqslant 0.7 f_t b h_0$, according to the calculation of the flexural capacity of normal sections, the bars should extend past the cross section at an unnecessary point for a length not less than $20d$, and it also should extend past the cross section at a fully utilized point for a length not less than $1.2l_a$.
(2) If $V > 0.7 f_t b h_0$, according to the calculation of the flexural capacity of normal sections, the bars should extend past the cross section at unnecessary points for a length not less than h_0 and $20d$, whichever is larger; and it also should extend out the cross section at a fully utilized point for a length not less than $1.2l_a + h_0$.
(3) If the cutoff point determined according to the above requirements is still located inside the tension zone corresponding to the negative bending moment, according to the calculation of the flexural capacity of normal sections, the bars should extend past the cross section at an unnecessary point for a length not less than $1.3h_0$ and $20d$, whichever is larger; and the bars should

also extend past the cross section at a fully utilized point for a length not less than $1.2l_a + 1.7h_0$.

When $V \leqslant 0.7 f_t bh_0$, the beam may not crack, the effects of shear on longitudinal bars can be neglected, and the requirements for the cutoff point can be relaxed appropriately.

In reinforced concrete cantilever beams, at least 2 bars should extend to the free end of the beams and be bent down to not less than $12d$; the remaining bars at the top of the beam should not be cut off, and they can only be bent down at a 45° or 60° angle to the axis of the beams; moreover, the requirements on the bent-up bars should also be satisfied.

Generally, the longitudinal bars at the bottom of the beam can be bent up but should not be cut off.

7.4.5 Illustration of Bent-up and Cutoff of Bars

An example for bent-up and cutoff of longitudinal bars in a reinforced concrete beam is illustrated in Fig. 7.42. Note the contribution of the bent-up bar to the moment capacity diagram when resisting the negative bending moment.

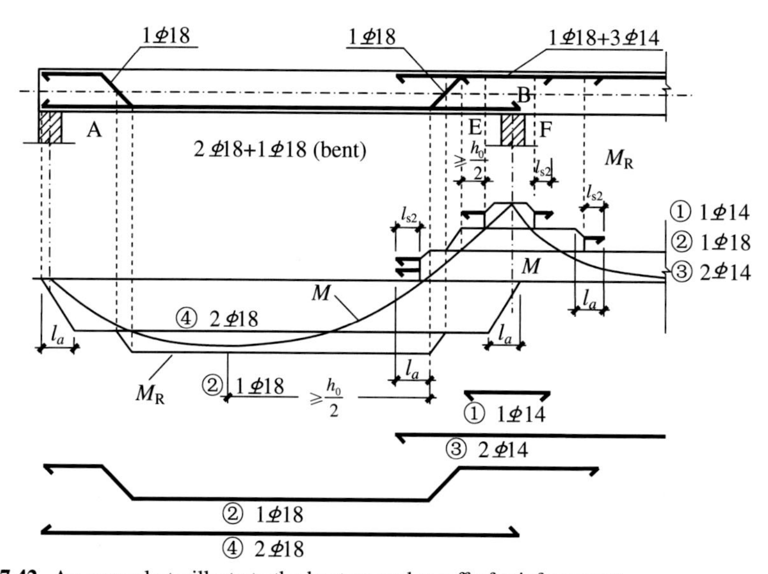

Fig. 7.42 An example to illustrate the bent-up and cutoff of reinforcement

7.4.6 Anchorage of Longitudinal Reinforcement at the Supports

As shown in Fig. 7.43, the bending moments at cross section A near the support are M_A and M_B before and after the occurrence of the inclined crack, respectively. Obviously, $M_B > M_A$; therefore, the tensile force in the reinforcement will significantly increase after the concrete cracks. If the anchorage length of the longitudinal reinforcement inside the support, l_{as}, is not sufficient, the reinforcement may be pulled out, and the anchorage failure happens. Therefore, the value of l_{as} should be specified.

In GB 50010, for simply supported and continuous slabs, the longitudinal bars at the bottom of the slabs should extend into the supports not less than $5d$ and are preferred to extend to the center line of the supports, where d is the diameter of the longitudinal reinforcement.

For simply supported and continuous beams, the longitudinal bars at the bottom of the beams should extend into the supports for a length l_{as}, which should satisfy the following requirements:

When $V \leqslant 0.7\ f_t bh_0$, $l_{as} \geqslant 5d$; when $V > 0.7\ f_t bh_0$, $l_{as} \geqslant 12d$ for deformed reinforcement and $l_{as} \geqslant 15d$ for plain reinforcement, where d is the maximum diameter of the longitudinal reinforcement.

If the anchorage of the longitudinal bars cannot satisfy these requirements, other effective anchorage measures should be taken, such as welding the reinforcement to steel anchorage plates or embedded parts at beam ends.

In GB 50010, it is not required that the longitudinal bars should extend to the outside of the centerline of the supports. From Fig. 7.43, it can be seen that if the reaction force of a support is located in the center of the support, and anchorage failure may happen because of insufficient flexural capacity if the longitudinal bars do not extend to the outside of the centerline of the support, so it is suggested by the authors of this book that the longitudinal bars should extend to the outside of the centerline of the supports.

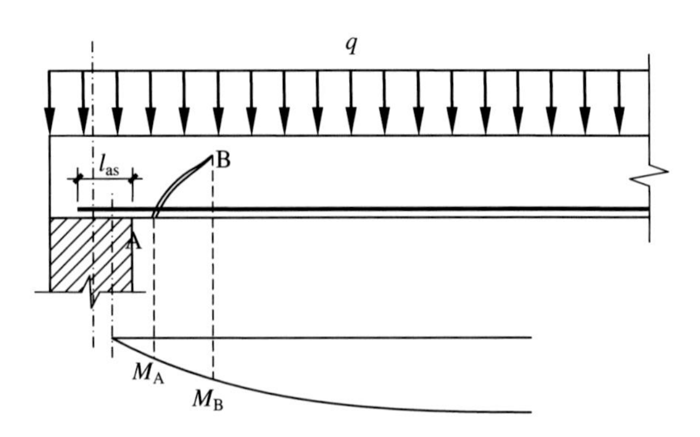

Fig. 7.43 Anchorage of reinforcement at the support

7.5 Shear Capacities of Eccentrically Loaded Members

7.5.1 Experimental Results

Figure 7.44 shows a continuous beam specimen subjected to antisymmetric loading. The stress state in the central segment is similar to that of a real frame column. As shown in Fig. 7.45, three typical failure modes may happen to the specimens.

7.5.1.1 Inclined Tension Failure

If the height–depth ratio of the column is large ($H/h_0 = 2 \sim 3$), the reinforcement ratio of stirrups is small and the axial compression is not high. A through inclined crack along the direction of the principal compressive stress splits the specimen into two parts in a brittle way. The shear capacities of members in this failure mode are low. Even if the height–depth ratio is not large, specimens without stirrups may still fail in inclined tension.

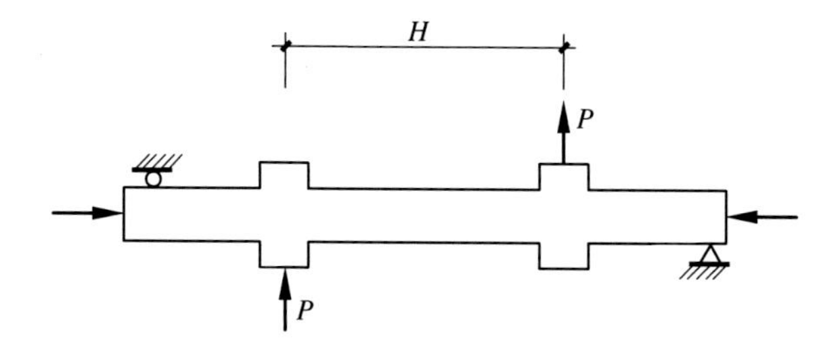

Fig. 7.44 Shear test of an eccentrically load member

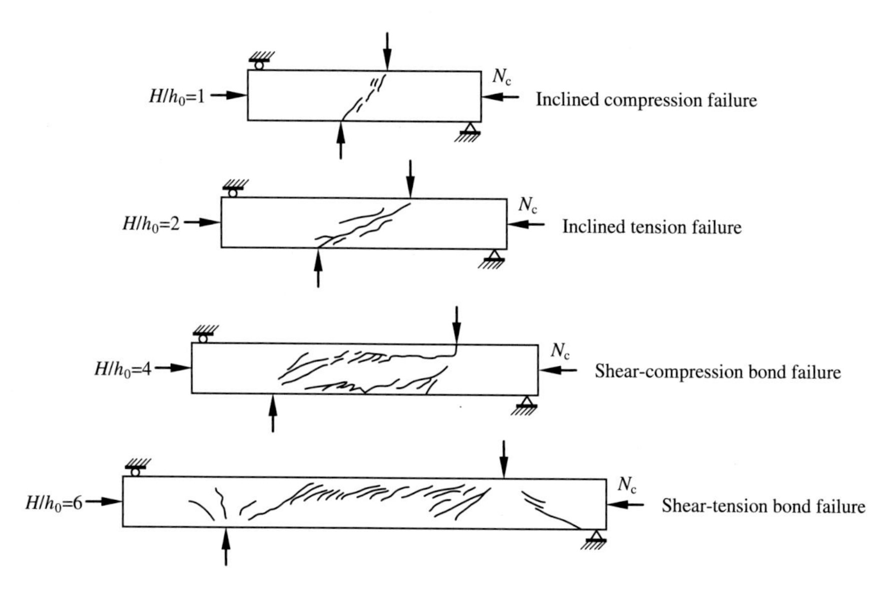

Fig. 7.45 Shear failures of eccentrically loaded members

7.5.1.2 Inclined Compression Failure

If the height–depth ratio of the column is small, but the axial compression is large or the reinforcement ratio of stirrups is high, several inclined cracks parallel to one another appear at the loading points and an inclined concrete strut is formed. Finally, a brittle failure will happen due to strut crushing.

7.5.1.3 Shear-Compression Bond Failure or Shear-Tension Bond Failure

If $H/h_0 = 3 \sim 5$, two nearly parallel web-shear-inclined cracks will occur one after another at the ends of the members. Then, tearing cracks along the longitudinal bars will appear due to bond failure. Finally, the concrete crushing at the diagonal crack tips indicates failure. If the reinforcement ratio of stirrups is low, it is possible that the stirrups yield before the crushing of concrete in the compression zone, and the members fail in shear tension. Shear-tension failure exhibits a certain ductility (different from inclined tension failure), and the concrete in the compression zone will not be crushed at failure (different from shear-compression failure).

Experimental results have indicated that the axial compression is beneficial to the shear capacities of members, because the occurrence and development of inclined cracks are postponed, and a larger depth is kept in the shear-compression zone, which increases the shear resistance of concrete.

Generally, axial compression will decrease the angles of inclined cracks and axial tension works in a reverse fashion. So axial tension is always adverse to shear resistance. Moderate axial compression is beneficial to shear resistance. However, too much axial compression will make the members fail in compression, which is not good for shear resistance.

Based on the experimental results of identical specimens failed by shear, the capacity envelop can be drawn in an N-V diagram. The flexural capacity envelop can be plotted in the same diagram if the flexural capacity of the normal section M is transferred to V (=M/a), where a is the shear span. The intersection of the two envelops is the failure load curve of the members (Fig. 7.46). It can be seen that when the axial load varies from tension to compression, the members undergo eccentric tension, shear tension, shear flexure, shear-compression, and small eccentricity compression.

The characteristics of eccentrically tensioned members are as follows: (1) Under the axial tension, an initial vertical crack through the specimen is formed first; (2) when the lateral force is applied, the inclined crack may develop independently and pass through the vertical crack or just develop from the extension of the vertical crack tip; (3) The inclined crack width and the inclined angle are large; (4) At the inclined crack tip, there a small shear-compression zone may appear; hence, the shear capacity is very small.

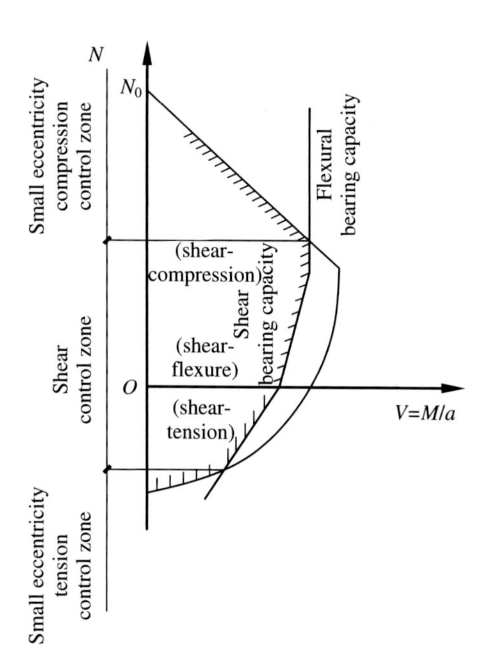

Fig. 7.46 Failure envelop of members under the combined action of axial force and shear

7.5.2 *Factors Influencing Shear Capacities of Eccentrically Loaded Members*

7.5.2.1 Height–Depth Ratio

Firstly, the axial compression ratio is defined as $N_c/(f_c bh)$. When the axial compression ratio and the web reinforcement ratio are constant, the shear capacities of eccentrically loaded members decrease significantly with the increase of the height–depth ratio, H/h_0 (Fig. 7.47). However, the deformation abilities of the

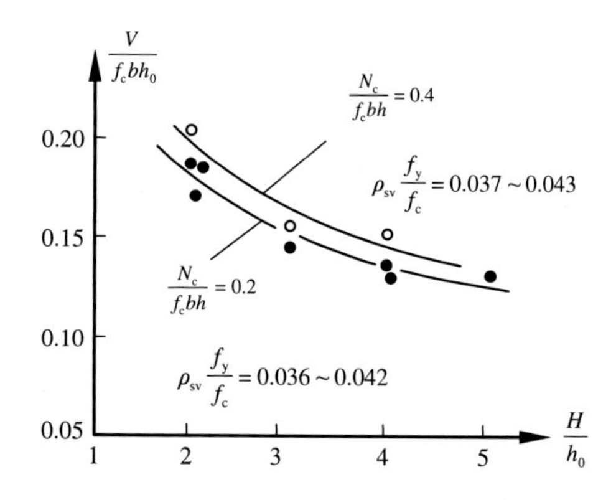

Fig. 7.47 Influence of height–depth ratio on shear capacities of members

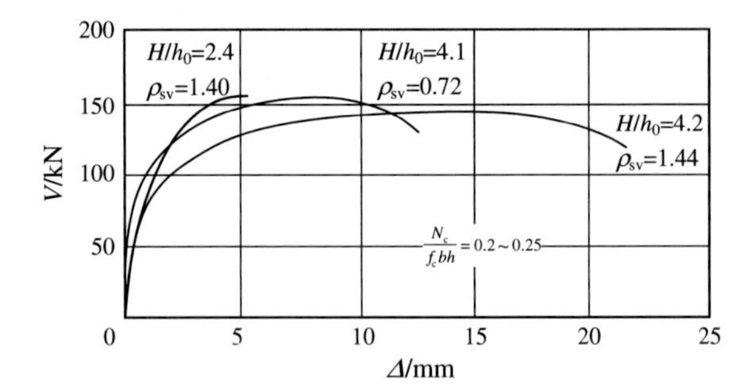

Fig. 7.48 Influence of height–depth ratio on deformation capacities of members

members are increased (Fig. 7.48). Besides, the magnitude of the axial load also plays a role in the influence of the height–depth ratio on the failure modes.

7.5.2.2 Axial Compression Ratio

Figure 7.49 shows the variation of the shear capacity of a specimen with the axial compression ratio. When the axial compression ratio is low ($N_c/(f_c bh) < 0.5$), the shear capacity increases with the increase of the axial compression ratio. This is because for members with a large height–depth ratio, the increase of the axial compression ratio will increase the compression area of the concrete; for members with a small height–depth ratio, the increase of the axial compression ratio will decrease the principal tensile stress in the concrete.

When the axial compression ratio is between 0.5 and 0.8, the shear capacity barely changes. On the one hand, when the axial force is increased to change the members from a large eccentrically loaded status to small eccentrically loaded status, both the tensile and compressive reinforcement yield and restrain the shear strengths of stirrups. Moreover, the micro-cracks inside the concrete weaken the shear capacity of the concrete. On the other hand, the increase of the axial compression reduces the principal tensile stress in the concrete. The above-mentioned beneficial and seemingly injurious factors cancel out, so the shear capacity is nearly constant in this stage.

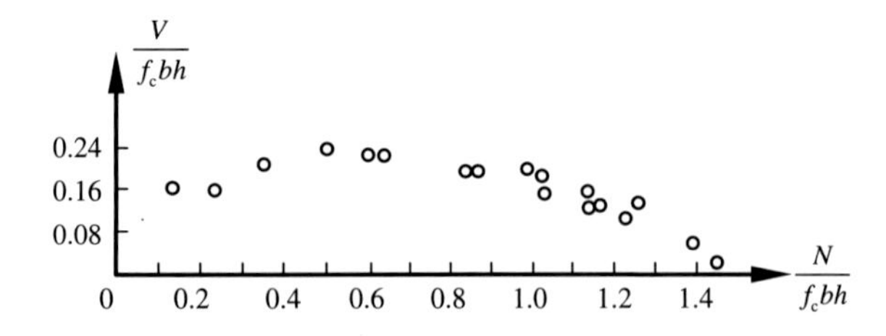

Fig. 7.49 Influence of axial compression ratio on shear capacities of members

When the axial compression ratio is larger than 0.8, the development of the cracks parallel to the member axis segregates the member into many small columns, and the sum of those shear capacities are lower than that of the member as a whole, as shown by the descending stage of the curve in Fig. 7.49.

Experimental results show that a certain amount of axial compression can restrain the occurrence and development of inclined cracks, enhance the aggregates interlock, and increase the depth of the shear-compression zone, so as to increase the shear capacities of the members. However, as shown in Fig. 7.49, the benefits of the axial compression to the shear capacities of the cross sections are limited. After the axial compression ratio $N_c/(f_c bh)$ exceeds 0.5, the shear capacity will gradually decrease and the failure pattern will transfer to small eccentricity compression.

7.5.2.3 Reinforcement Ratio of Stirrups

In column segments adjacent to the inflection point, at which the bending moment is zero, the stresses in the stirrups intersecting with inclined cracks are extremely nonuniform. The width of the inclined cracks and the stresses in the stirrups is small near the loading point due to the normal stress σ. The stirrup stresses near the inflection point are also low. However, between the loading point and the inflection point, the width of the inclined cracks and the stresses in the stirrups are large. Due to the inflection point, only part of the stirrups is stressed to yield. The shear resistance of the stirrups is lower than that of a member without inflection point.

If the stirrups are numerous and closely spaced, the width and spacing of the inclined cracks will decrease and the shear capacity and deformability will be high. Similar to flexural members, if the reinforcement ratio of stirrups, A_{sv}/bs, is raised to a certain number, the failure mode will transfer from the yielding of stirrups to the crushing of concrete, i.e., from shear-compression failure to inclined compression failure.

7.5.3 Calculation of Shear Capacities of Eccentrically Compressed Members

It is specified in GB 50010 that for eccentrically loaded members of rectangular shape, e.g., T and I sections, the shear capacities of inclined cross sections should be calculated by

$$V_u = \frac{1.75}{\lambda + 1} f_t b h_0 + f_{yv} \frac{A_{sv}}{s} h_0 + 0.07 N_c \tag{7.65}$$

where

λ shear-span ratio of calculated sections; and

N_c axial compression on members. If $N_c > 0.3 f_cA$, take $N_c = 0.3 f_cA$, where A is the cross-sectional areas of members.

The shear-span ratio of calculated sections should be determined as follows:

(1) For frame columns in all types of structures, it is suggested to take $\lambda = M/(Vh_0)$. For columns in frame structures, if the inflection point is within the story height, it can be taken as $\lambda = H_n/(2h_0)$; if $\lambda < 1$, take $\lambda = 1$; if $\lambda > 3$, take $\lambda = 3$, where M is the bending moment applied on the cross sections corresponding to the shear force V, and H_n is the clear height of the column.

(2) For other eccentrically loaded members, when uniformly distributed loads are applied, take $\lambda = 1.5$. When concentrated loads or several other types of loads are applied, the shear forces at the supports or joint edges caused by the concentrated loads count for more than 75 % of the total shear forces; hence, calculation is based on $\lambda = a/h_0$. Furthermore, if $\lambda < 1.5$, take $\lambda = 1.5$; if $\lambda > 3$, take $\lambda = 3$, where a is the distance from the concentrated load to the nearby supports or joint edges.

The upper bound of the shear capacity can still be calculated by Eqs. (7.45) or (7.46). And the lower bound of the shear capacity can be evaluated by

$$V_{u,\min} = \frac{1.75}{\lambda + 1} f_t b h_0 + 0.07 N_c \tag{7.66}$$

In design, if the shear force satisfies the following condition, the calculation of the shear capacity is not required, and the stirrups can be provided according to detailing requirements.

$$V \leqslant \frac{1.75}{\lambda + 1} f_t b h_0 + 0.07 N_c \tag{7.67}$$

7.5.4 Calculation of Shear Capacities of Eccentrically Tensioned Members

The axial tension increases the principal tensile stresses in a member. The angle between the inclined crack and the longitudinal axis of the member is enlarged, which will reduce the depth of the shear-compression zone at failure. When the axial tension is large, no shear-compression zone exists. In a word, the axial tension is adverse to the shear capacity of the member, and the decreased magnitude of the shear capacity will increase with the increase of the axial tension.

Therefore, the shear capacities of eccentrically tensioned members of rectangular, e.g., T and I sections, can be calculated by

$$V_u = \frac{1.75}{\lambda + 1} f_t b h_0 + f_{yv} \frac{A_{sv}}{s} h_0 - 0.2 N_t \tag{7.68}$$

where

λ shear-span ratio of calculated sections, determined in the same way as that for eccentrically compressed members; and

N_t axial tension on members.

In Eq. (7.68) V_u should not be less than $f_{yv} A_{sv} h_0 / s$. And the minimum value of $f_{yv} A_{sv} h_0 / s$ is 0.36 $f_t b h_0$.

The upper and lower bounds of the shear capacities of eccentrically tensioned members and the restrictions in design are similar to those for eccentrically compressed members.

7.5.5 *Shear Capacities of Columns of Rectangular Sections Under Bidirectional Shear*

Figure 7.50 shows a cross section of a reinforced concrete column under bidirectional shear. Assuming V_{ux0} and V_{uy0} are the shear capacities of the cross section when subjected to unidirectional shear in x and y directions, respectively, from previous argument we have

$$V_{ux0} = \frac{1.75}{\lambda_x + 1} f_t b h_0 + f_{yv} \frac{A_{svx}}{s} h_0 + 0.07 N_c \tag{7.69}$$

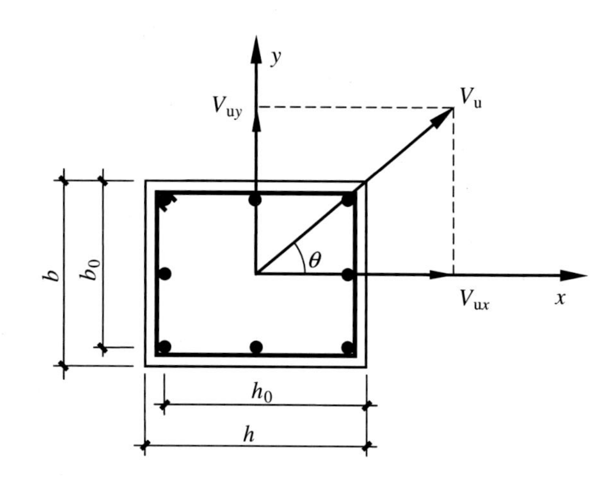

Fig. 7.50 Cross section of a reinforced concrete column under bidirectional shear

$$V_{uy0} = \frac{1.75}{\lambda_y + 1} f_t b h_0 + f_{yv} \frac{A_{svy}}{s} h_0 + 0.07 N_c \tag{7.70}$$

where

λ_x, λ_y computed shear-span ratios in x and y directions, which can be determined in the same way as that for eccentrically compressed members; and

A_{svx}, A_{svy} sectional areas of all stirrup legs in x and y directions in the calculated cross section.

When the cross section is subjected to unidirectional shear, the maximum shear resistances in x and y directions are

$$V_{ux0,\max} = 0.25\, \beta_c f_c b h_0 \tag{7.71}$$

$$V_{uy0,\max} = 0.25\, \beta_c f_c b h_0 \tag{7.72}$$

When the shear force is applied unidirectionally, the stirrups perpendicular to the force direction are not stressed. This is greatly different from bidirectional shear, in which case all stirrup legs are stressed. Experimental results indicate that the shear capacity envelop of columns under bidirectional shear is approximately an ellipse. Assume V_u is the shear capacity of a column under bidirectional shear, the angle between V_u and x axis is θ and the components of V_u in x and y directions are V_{ux} and V_{uy}, respectively, then

$$\left(\frac{V_{ux}}{V_{ux0}}\right)^2 + \left(\frac{V_{uy}}{V_{uy0}}\right)^2 = 1 \tag{7.73}$$

$$\begin{cases} V_{ux} = V_u \cos\theta \\ V_{uy} = V_u \sin\theta \end{cases} \tag{7.74}$$

Substituting Eq. (7.74) into Eq. (7.73) gives

$$V_u = \frac{1}{\sqrt{\frac{\cos^2\theta}{V_{ux0}^2} + \frac{\sin^2\theta}{V_{uy0}^2}}} \tag{7.75}$$

Similarly, there are also upper and lower bounds on the shear capacities of columns under bidirectional shear. Readers are invited to derive them independently (without help).

7.5.6 *Shear Capacities of Columns of Circular Sections*

When reinforced concrete columns are reinforced as in Fig. 7.51, the shear capacities can still be expressed as

$$V_{\mathrm{u}} = V_{\mathrm{c}} + V_{\mathrm{s}} \tag{7.76}$$

where
V_c shear resistance of the concrete; and
V_s shear resistance of the stirrups.

Analysis of the experimental results at home and abroad yields

$$V_{\mathrm{c}} = \frac{1.0}{\lambda + 1.5} f_{\mathrm{t}} D^2 + 0.083 N_{\mathrm{c}} \tag{7.77}$$

When $N_c > 0.3 f_c A$, take $N_c = 0.3 f_c A$. λ is determined in the same manner as shown in Eq. (7.65).

Assume the inclined crack is 45° to the longitudinal axis of the column, the stirrups intersecting with the inclined crack all yield at failure, and the spacing between stirrups s is small compared with the diameter D' of the circle formed by the center line of the stirrup, as shown in Fig. 7.52. Project the tensile forces in the stirrups intersecting with the inclined crack onto a horizontal plane; then, the sum of the components of the projected tensile forces in the shear force direction is the shear resistance of the stirrups, that is

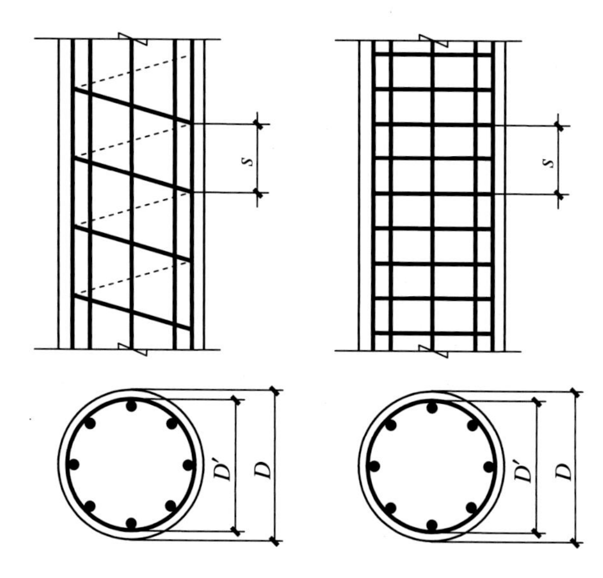

Fig. 7.51 Reinforcement in a column with a circular cross section

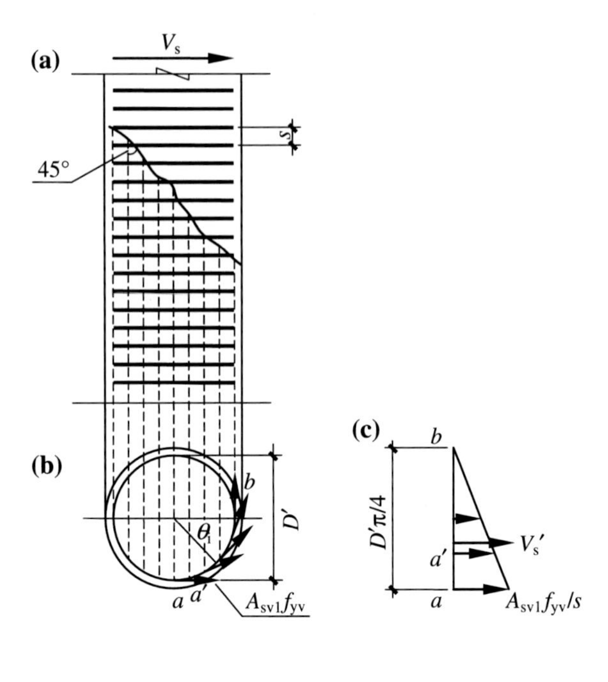

Fig. 7.52 Contribution of stirrups to column shear resistance

$$V_s = \sum A_{sv1} f_{yv} \sin\theta_i \tag{7.78}$$

The computation of θ_i is complex, so appropriate simplification is necessary. When s is small, $aa' \approx s$, the force intensities in the shear force direction at points a and b are $q_a = A_{sv1}f_{yv}/s$ and $q_b = 0$, respectively. If we assume a quarter of circumference is straightened out and the force intensities between points a and b are linearly distributed (Fig. 7.52), then

$$V_s' = \frac{1}{2}\frac{A_{sv1}f_{yv}}{s}\frac{D'\pi}{4} = \frac{\pi A_{sv1}f_{yv}}{8s}D' \tag{7.79}$$

For the whole circumference,

$$V_s = \frac{\pi A_{sv1}f_{yv}}{2s}D' \tag{7.80}$$

Substituting Eqs. (7.77) and (7.80) into Eq. (7.76) gives the shear capacity formula for columns of circular sections as

$$V_u = \frac{1.0}{\lambda + 1.5}f_t D^2 + \frac{\pi}{2}\frac{A_{sv1}f_{yv}}{s}D' + 0.083\,N_c \tag{7.81}$$

where D' is the diameter of the circle formed by the centerline of the stirrup and A_{sv1} is the section area of a single leg of the stirrups. The upper and lower bounds of the shear capacities are

$$V_{u,max} = 0.25\,\beta_c f_c D^2 \tag{7.82}$$

$$V_{u,max} = \frac{1.0}{\lambda + 1.5} f_c D^2 + 0.083\,N_c \tag{7.83}$$

7.6 Applications of Shear Capacity Formulae for Eccentrically Loaded Members

The section design and shear capacity evaluation of eccentrically loaded members are similar to those of flexural members except the axial force should be considered herein. So detailed calculation steps will be explained in two examples in this section.

Example 7.6 A frame column is shown in Fig. 7.53. The cross-sectional dimensions are $b \times h$ = 300 mm × 400 mm, and the clear height of the column is H_n = 3 m. The concrete is C30 (f_c = 14.3 N/mm^2, f_t = 1.43 N/mm^2). The stirrups are HPB300 (f_{yv} = 270 N/mm^2), and the longitudinal bars are HRB335 (f_y = 300 N/mm^2). After the combination of internal forces, the bending moments at the column ends are M = 120 kN m; the axial compressive force is N_c = 600 kN, and the shear forces are V = 180 kN. The method for determining the required stirrups for the column is as follows.

Solution

1. Check the cross-sectional dimensions

$$0.25 f_c b h_0 = 0.25 \times 14.3 \times 300 \times 365 = 391{,}463\,\text{N} > 180{,}000\,\text{N, OK}$$

2. Calculate the stirrups
   Shear-span ratio $\lambda = \frac{H_n}{2h_0} = \frac{3000}{2 \times 365} = 4.110 > 3.0$, take λ = 3.0
   Check the upper bound of the axial compressive force:

$$0.3 f_c b h = 0.3 \times 14.3 \times 300 \times 400 = 514{,}800\,\text{N} < N_c = 600{,}000\,\text{N}$$

Therefore, take N_c = 514800 N. Then,

$$V_{cN} = \frac{1.75}{\lambda + 1} f_t b h_0 + 0.07\,N_c = \frac{1.75}{3+1} \times 1.43 \times 300 \times 365 + 0.07 \times 514{,}800$$
$$= 104{,}542\,\text{N} < V = 180{,}000\,\text{N}$$

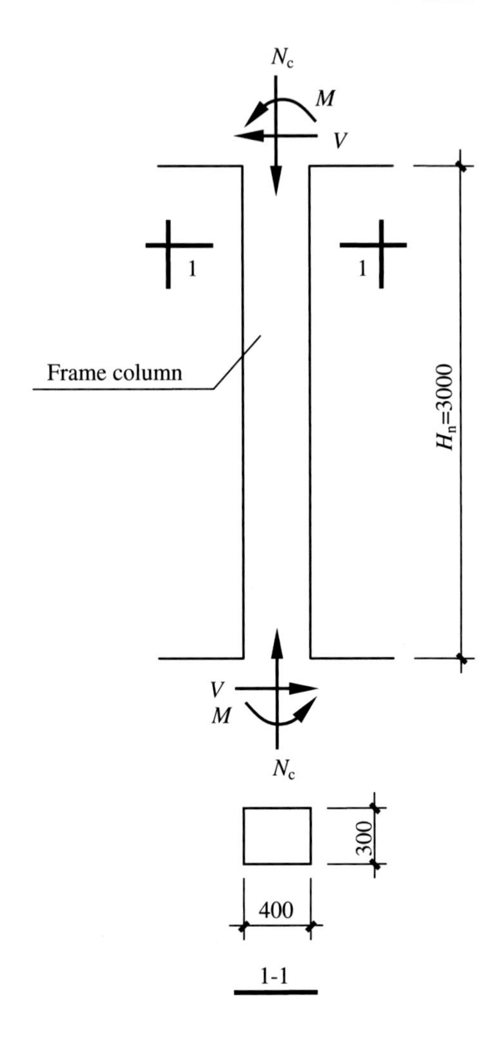

Fig. 7.53 Example 7.6

Calculation of the number of stirrups needed can be done by

$$\frac{A_{sv}}{s} = \frac{V - V_{cN}}{f_{yv}h_0} = \frac{180{,}000 - 104{,}542}{270 \times 365} = 0.7657$$

If choosing double-legged stirrups with a diameter of 8 mm (A_{sv} = 100.53 mm^2); then, $s = \frac{100.53}{0.7657} = 131.29\,\text{mm}$. So $\phi 8@125$ is OK.

Example 7.7 For a bottom chord in a reinforced concrete truss, the cross-sectional dimensions are $b \times h$ = 240 mm × 300 mm, and a 100 kN load is suspended at the center of a 6 m-long segment. The applied tensile force is N_t = 70 kN. The concrete is C30 (f_c = 14.3 N/mm^2, f_t = 1.43 N/mm^2), the stirrups are HPB300 (f_y = 270 N/mm^2), and the longitudinal reinforcement is HRB335 (f_y = 300 N/mm^2). Calculate the required stirrups in the bottom chord of the truss.

Solution

1. To calculate the internal forces
   Maximum bending moment is $M = \frac{1}{4} \times 100 \times 6 = 150\,\text{kN} \cdot \text{m}$
   Shear force is V = 50 kN
2. To calculate the number of stirrups
   Shear-span ratio $\lambda = \frac{a}{h_0} = \frac{3000}{265} = 11.32 > 3.0$, take λ = 3.0

$$\begin{aligned} V = 50{,}000 &= \frac{1.75}{\lambda+1} f_t b h_0 + f_{yv} \frac{A_{sv}}{s} h_0 - 0.2N_t \\ &= \frac{1.75}{3+1} \times 1.43 \times 240 \times 265 + 270 \times 265 \times \frac{A_{sv}}{s} - 0.2 \times 70{,}000 \\ &= 25{,}790 + 71{,}550 \frac{A_{sv}}{s} \end{aligned}$$

The above equation makes the following solution possible.

$$\frac{A_{sv}}{s} = \frac{50{,}000 - 25{,}790}{71{,}550} = 0.3384\,\text{mm}$$

Check the low bound of stirrups:

$$\begin{aligned} 0.36 f_t b h_0 &= 0.36 \times 1.43 \times 240 \times 265 = 32{,}741\,\text{N} \\ > f_{yv} \frac{A_{sv}}{s} h_0 &= 71{,}550 \times 0.3384 = 24{,}213\,\text{N}. \end{aligned}$$

Therefore, stirrups should be provided according to $f_{yv} \frac{A_{sv}}{s} h_0 \geqslant 0.36 f_t b h_0$. Then

$$f_{yv} \frac{A_{sv}}{s} h_0 = 715{,}500 \frac{A_{sv}}{s} \geqslant 0.36 f_t b h_0 = 32{,}741\,\text{N}$$

that is, $\frac{A_{sv}}{s} \geqslant \frac{32{,}741}{71{,}550} = 0.4576\,\text{mm}^2/\text{mm}$
Choose double-legged stirrups with the diameter of 6 mm (A_{sv} = 56.55 mm^2), then $s \leqslant \frac{56.55}{0.4576} = 123.58\,\text{mm}$. So ϕ6@125 is OK, the error is within 5 %.

7.7 Shear Performance of Deep Flexural Members and Structural Walls

7.7.1 Shear Performance of Deep Flexural Members

7.7.1.1 Detailing Requirements

The reinforcement in deep flexural members is shown in Fig. 5.42. The minimum reinforcement ratios for deep beams are listed in Table 5.2, in which the reinforcement ratios of longitudinal tensile bars, ρ, of horizontal distribution bars, ρ_{sh}, and of vertical distribution bars, ρ_{sv}, are defined, respectively, as follows:

$$\rho = \frac{A_s}{bh}, \quad \rho_{sh} = \frac{A_{sh}}{bs_v}, \quad \rho_{sv} = \frac{A_{sv}}{bs_h} \tag{7.84}$$

where

A_{sv} total cross-sectional area of the vertical distribution reinforcement provided in one vertical cross section;

s_h horizontal spacing of vertical distribution reinforcement;

A_{sh} total cross-sectional area of horizontal distribution reinforcement provided in one horizontal cross section; and

s_v vertical spacing of horizontal distribution reinforcement.

As l_0/h decreases, the failure mode of deep flexural members transfers from shear-compression to inclined compression, and the contribution of concrete to the shear capacity gradually increases.

As to the contribution of shear reinforcement, if l_0/h is larger (close to 5.0), only vertical distribution reinforcement contributes to the shear resistance, which is similar to common shallow beams; but if l_0/h is small, only horizontal distribution reinforcement makes limited contribution to the shear resistance. When l_0/h changes from large to small, the contributory change of each material will transition smoothly.

It should be noted that because the contributions of the horizontal and vertical distribution reinforcement in deep beams to shear capacity are limited, the required shear capacity should be satisfied mainly by adjusting the cross-sectional dimensions and concrete grade.

7.7.1.2 Calculation of Shear Capacities of Deep Flexural Members

For deep flexural members of rectangular, T- and I-shaped cross sections under uniformly distributed loads, if the vertical and horizontal distribution reinforcements are provided, the calculation formula for the shear capacities of inclined cross sections is

$$V_{\mathrm{u}} = 0.7\frac{(8 - l_0/h)}{3}f_{\mathrm{t}}bh_0 + 1.25\frac{(l_0/h - 2)}{3}f_{\mathrm{yv}}\frac{A_{\mathrm{sv}}}{s_{\mathrm{h}}}h_0 + \frac{(5 - l_0/h)}{6}f_{\mathrm{yh}}\frac{A_{\mathrm{sh}}}{s_{\mathrm{v}}}h_0 \tag{7.85}$$

For deep flexural members under concentrated loads (or if several kinds of loads are applied, but the shear forces at the supports caused by the concentrated loads count for more than 75 % of the total shear force), the shear capacities of inclined cross sections are evaluated by

$$V_{\mathrm{u}} = \frac{1.75}{\lambda + 1}f_{\mathrm{t}}bh_0 + \frac{(l_0/h - 2)}{3}f_{\mathrm{yv}}\frac{A_{\mathrm{sv}}}{s_{\mathrm{h}}}h_0 + \frac{(5 - l_0/h)}{6}f_{\mathrm{yh}}\frac{A_{\mathrm{sh}}}{s_{\mathrm{v}}}h_0 \tag{7.86}$$

where λ = computation shear-span ratio. If $l_0/h \leqslant 2.0$, take $\lambda = 0.25$; if $2.0 < l_0/h < 5.0$, take $\lambda = a/h_0$, where a is the distance from the concentrated load to the nearby support; the upper bound of λ is $(0.92\ l_0/h - 1.58)$ and the low bound is $(0.42l_0/h - 0.58)$, where l_0/h is the span–depth ratio, and if $l_0/h < 2.0$, take $l_0/h = 2.0$.

The maximum shear capacities of the inclined cross sections of deep flexural members are

$$\text{If } h_{\mathrm{w}}/b \leqslant 4, \quad V_{\mathrm{u,max}} = \frac{1}{60}(10 + l_0/h)\beta_{\mathrm{c}}f_{\mathrm{c}}bh_0 \tag{7.87}$$

$$\text{If } h_{\mathrm{w}}/b \geqslant 6, \quad V_{\mathrm{u,max}} = \frac{1}{60}(7 + l_0/h)\beta_{\mathrm{c}}f_{\mathrm{c}}bh_0 \tag{7.88}$$

If $4 < h_{\mathrm{w}}/b < 6$, the value can be determined by linear interpolation.

7.7.2 *Shear Performance of Structural Walls*

If the longer dimension of a concrete beam's cross section (length) is four times greater than the shorter dimension (width), the vertical eccentrically loaded member can be called a structural wall or a shear wall.

Shear walls (or wall panels) in structures may be eccentrically tensioned or eccentrically compressed, and are usually subjected to (horizontal) shear. Therefore, the calculation of shear capacities of inclined cross sections is required.

For flanged shear walls, the effective width of the flange can be taken as the minimum among the following four values: spacing of shear walls, width of the flanged wall between door (or window) openings, depth of shear wall plus 6 times the depth of each flange wall on both sides; and 1/10 of the total height of the wall panel.

The maximum shear capacities of inclined cross sections in a shear walls is

$$V_{u,max} = 0.25\,\beta_c f_c bh \tag{7.89}$$

where

β_c influence coefficient of concrete strength;

b width of rectangular cross section or web width of T and I sections (depth of wall); and

h depth of the cross section (length of wall).

The calculating method used to determine the shear capacities of shear walls is as follows. If a seismic design is necessary, specific requirements for that parameter must also be satisfied.

7.7.2.1 Shear Capacities of Eccentrically Compressed Walls

Shear capacities of the inclined cross sections in eccentrically compressed walls should be calculated according to the following formula:

$$V_u = \frac{1}{\lambda - 0.5}\left(0.5 f_t b h_0 + 0.13\,N_c \frac{A_W}{A}\right) + f_{yv}\frac{A_{sh}}{s_v} h_0 \tag{7.90}$$

where

N_c axial compressive force applied on the shear wall. If $N_c \geqslant 0.2\, f_c bh$, take $N_c = 0.2\, f_c bh$;

A cross-sectional area of the shear wall. The effective area of flange can be calculated after the effective width of flange has been determined by the foregoing introduction;

A_w cross-sectional area of the web in the shear wall of T or I section. For shear wall of rectangular section, take $A_w = A$;

A_{sh} total cross-sectional area of horizontal distribution reinforcement provided in a horizontal cross section;

s_v vertical spacing of horizontal distribution reinforcement; and

λ shear-span ratio of calculated section, $\lambda = \frac{M}{Vh_0}$.

If $\lambda < 1.5$, take $\lambda = 1.5$; if $\lambda > 2.2$, take $\lambda = 2.2$. If the distance between the calculated section and the bottom of the wall is less than $h_0/2$, λ should be calculated based on the bending moment and the shear force at the cross section $h_0/2$ from the wall bottom.

In section design, if the shear force is not greater than the first term in the right-hand side of Eq. (7.90), the horizontal distribution reinforcement should be provided according to the detailing requirements.

7.7.2.2 Shear Capacities of Eccentrically Tensioned Walls

Shear capacities of the inclined cross sections in eccentrically tensioned walls should be calculated according to the following formula:

$$V_u = \frac{1}{\lambda - 0.5}\left(0.5 f_t b h_0 - 0.13 N_t \frac{A_W}{A}\right) + f_{yv} \frac{A_{sh}}{s_v} h_0 \tag{7.91}$$

where N_t is the axial tensile force applied on the shear wall. Definitions for other notations are the same as previously mentioned. If the calculated value is less than $f_{yv} A_{sh} h_0 / s_v$, take it as $f_{yv} A_{sh} h_0 / s_v$.

7.7.2.3 Detailing Requirements

The grade of the concrete used in reinforced concrete shear walls is preferred to be not less than C20. The thickness of the wall should not be less than 140 mm. For shear wall structures, the preferred thickness of walls is not less than 1/25 of the floor height; for frame-shear wall structures, the preferred thickness of walls is not less than 1/20 of the floor height. When precast floor slabs are used, requirements for the supporting length of precast slabs on walls and for continuous vertical reinforcement in full height of the walls should be under consideration when determining the thickness of the wall.

The reinforcement ratios of the horizontal distribution reinforcement ρ_{sh} and the vertical distribution reinforcement ρ_{sv} can be calculated by Eq. (7.84).

Both ρ_{sh} and ρ_{sv} should not be less than 0.2 %. For shear walls of importance, the reinforcement ratios of both the horizontal and vertical distribution reinforcement should be properly increased. At locations of high thermal stress or large shrinkage, the reinforcement ratio of horizontal distribution reinforcement should also be increased.

The diameter of the horizontal and vertical distribution reinforcement in shear walls should not be less than 8 mm, and the spacing should not be greater than 300 mm.

For shear walls with a thickness greater than 160 mm, double-layer fabrics of distribution reinforcement should be provided. For shear walls at important positions of a structure, even if the thickness is not greater than 160 mm, it is also preferred to provide double-layer fabrics of distribution reinforcement. The two fabrics should be placed near the two faces of the shear walls and linked by ties. The diameter of the ties is preferred to be not less than 6 mm, and the spacing is preferred to be not greater than 600 mm.

7.8 Shear Transfer Across Interfaces Between Concretes Cast at Different Times

There are many situations in reinforced concrete structures where it is necessary to transfer shear across interfaces between concretes cast at different times. Figure 7.54 shows a simply supported reinforced concrete beam. When the beam is precast and the slab is cast in situ, there is an interface between the precast beam and the cast in situ slab. When a load is applied on that beam, which is usually called the composite reinforced concrete beam, shear stress will be generated across the interface between different concretes as shown in Fig. 7.54a. If the shear can be transferred efficiently across the interface, the new cast slab and the precast beam will work well together.

The shear-friction concept provides a simple and powerful model to analyze situations such as the composite reinforced concrete beam shown in Fig. 7.54. As shown in Fig. 7.55, when shear V_h is big enough, concrete will crack along the interface. Shear V_h will cause not only a shear displacement but also a widening of the crack. This crack opening will cause tension in the reinforcement crossing the crack balanced by compressive stresses in concrete across the crack. The shear

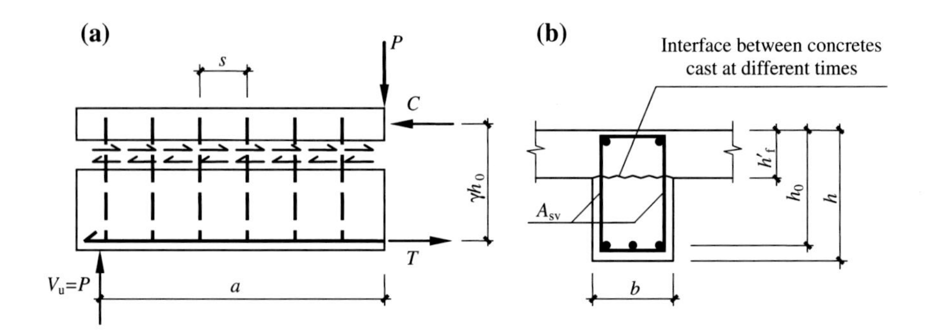

Fig. 7.54 Shear transfer across the interface between cast in situ slab and precast beam

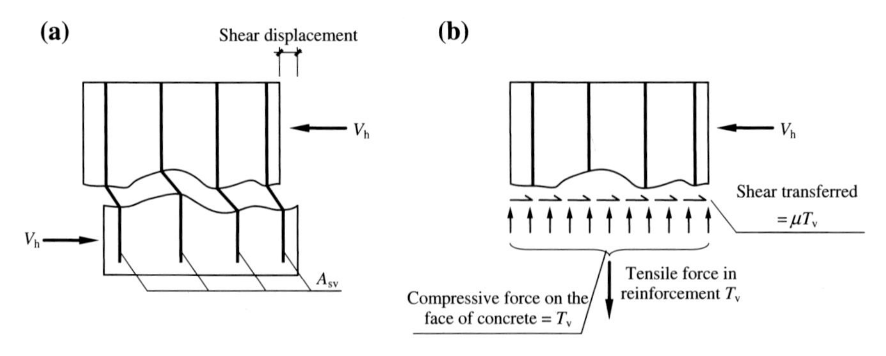

Fig. 7.55 Shear friction concept

stresses on the concrete face are assumed to be related to the compressive force on the face by a coefficient of friction μ. The maximum shear transfer capacity is assumed to be reached when the reinforcement crossing the crack yields leading to Eq. (7.92).

$$V_{\mathrm{hu}} = \mu T_{\mathrm{v}} = \mu A_{\mathrm{sv}} f_{\mathrm{yv}} \tag{7.92}$$

where

μ coefficient of friction between concrete, which varies from 0.6 to 1.0;
A_{sv} cross-sectional area of reinforcement crossing the crack or the interface; and
f_{yv} yield strength of reinforcement.

For the composite reinforced concrete beam shown in Fig. 7.54, the maximum shear transfer capacity of the interface is $V_{\mathrm{hu}} = C$. Based on the equilibrium condition, we get

$$C = \frac{M}{\gamma h_0} = \frac{V_{\mathrm{u}} a}{\gamma h_0} \tag{7.93}$$

$$C = V_{\mathrm{hu}} = \mu \frac{A_{\mathrm{sv}}}{s} \cdot a \cdot f_{\mathrm{yv}} \tag{7.94}$$

Combining Eqs. (7.93) and (7.94) yields

$$V_{\mathrm{u}} = \mu \gamma h_0 \frac{A_{\mathrm{sv}} f_{\mathrm{yv}}}{s} \tag{7.95}$$

Based on Eq. (7.95), taking into account the contribution of concrete, Chinese code (GB 50010) recommends the calculation equation for the shear capacities of interfaces in composite reinforced concrete structures, which is

$$V_{\mathrm{u}} = 1.2 f_{\mathrm{t}} b h_0 + 0.85 f_{\mathrm{yv}} \frac{A_{\mathrm{sv}}}{s} h_0 \tag{7.96}$$

where f_{t} = small tensile strength of concretes cast at different times.

Questions

7.1 What types of cracks will appear on the simply supported beam shown in Fig. 7.56? Sketch the crack patterns.

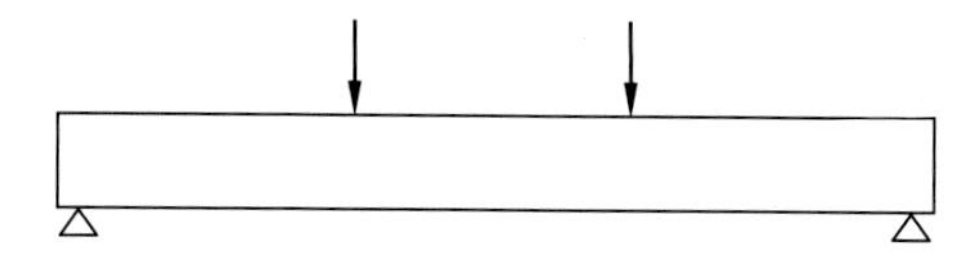

Fig. 7.56 Question 7.1

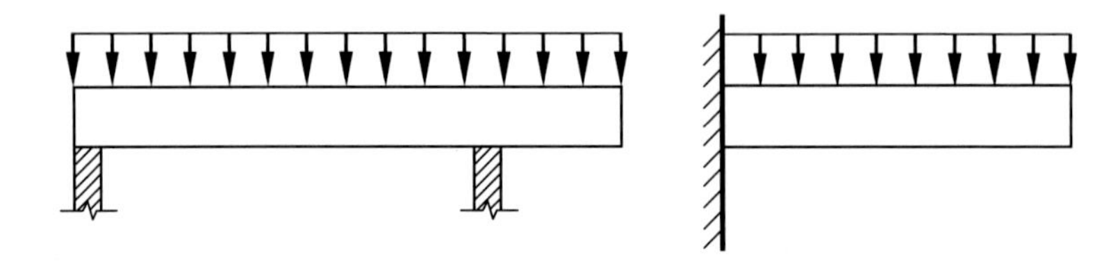

Fig. 7.57 Question 7.4

7.2 The formula of shear capacity along an inclined section takes the form that the shear capacity equals the sum of concrete resistance and reinforcement resistance. Does this mean that these two terms are not correlated?
7.3 Why do inclined cracks appear in reinforced concrete beams under loading?
7.4 Sketch the potential inclined cracks and the directions of propagation of the cantilever beams in Fig. 7.57.
7.5 Why does the flexural failure happen along inclined section? How can we prevent this failure mode?
7.6 What are the main failure patterns along inclined cross sections of simply supported beams with or without web reinforcement? When do they occur? And what are their failure characteristics?
7.7 When should the shear-span ratio be used in the shear capacity design of beams?
7.8 Why should the maximum spacing of stirrups be specified?
7.9 What is the difference between generalized shear-span ratio and computed shear-span ratio? Why can the computed shear-span ratio be used in calculation of shear capacity?
7.10 What are the main factors that will significantly influence the capacities along inclined cross sections in beams with web reinforcement?
7.11 What is the moment capacity diagram? What is the relationship between the moment capacity diagram and the design moment diagram?
7.12 What is the shear resistance mechanism in a simply supported beam with web reinforcement after the occurrence of inclined cracks?
7.13 Can the shear capacity definitely be raised if more stirrups are provided? Why?
7.14 How would you prevent inclined compression failure in design?
7.15 What is the difference in the stress state of a simply supported beam without web reinforcement before and after the occurrence of inclined cracks?
7.16 How would you define the influence of the axial forces on the shear resistance of a reinforced concrete column?
7.17 Why is it not correct to simply combine the shear resistance in two perpendicular directions in the calculation of shear capacity of a column under bidirectional shear?
7.18 List the differences between shear reinforcement in deep beams and shear reinforcement in normal beams.

7.19 Besides composite reinforced concrete beams, can you tell us other engineering examples where shear transfers across interfaces between concretes cast at different times?

Problems

7.1 For a reinforced concrete beam, the cross-sectional dimensions are $b \times h$ = 200 mm × 500 mm. a_s = 35 mm. The concrete used is C20 (f_c = 9.6 N/mm^2, f_t = 1.10 N/mm^2). The applied shear force $V = 1.2 \times 10^5$ N. Calculate the required stirrups (f_{yv} = 270 N/mm^2).

7.2 All other conditions are the same as those in Problem 7.1 except the shear forces are $V = 6.2 \times 10^4$ N and $V = 2.8 \times 10^5$ N, respectively. Calculate the required stirrups.

7.3 A reinforced concrete beam subjected to uniformly distributed load q = 40 kN/m (including self-weight) is shown in Fig. 7.58. C20 concrete (f_c = 9.6 N/mm^2, f_t = 1.10 N/mm^2) is used. The cross-sectional dimensions are $b \times h$ = 200 mm × 400 mm. Calculate the required stirrups at cross sections A_{right}, B_{left}, and B_{right} (f_{yv} = 270 N/mm^2).

7.4 For a simply supported beam as shown in Fig. 7.59, a uniformly distributed load q = 70 kN/m (including self-weight) is applied. C20 concrete (f_c = 9.6 N/mm^2, f_t = 1.10 N/mm^2) is used. The strengths of longitudinal

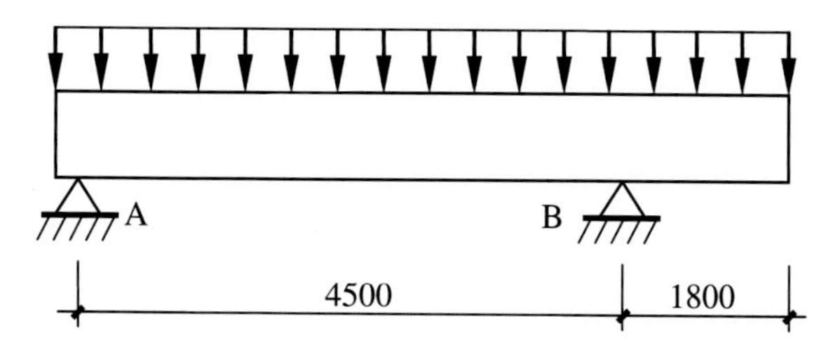

Fig. 7.58 Problem 7.3

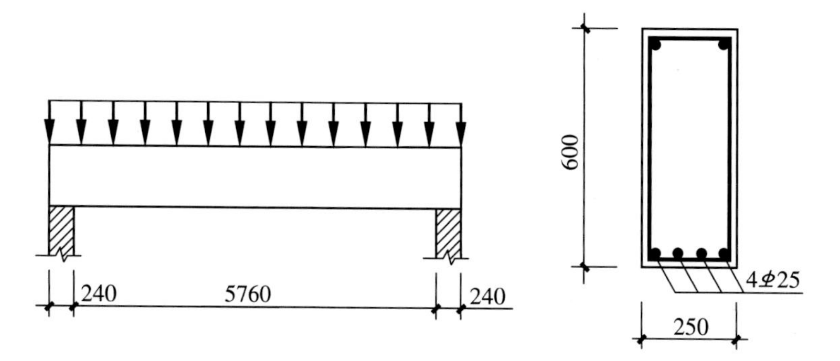

Fig. 7.59 Problem 7.4

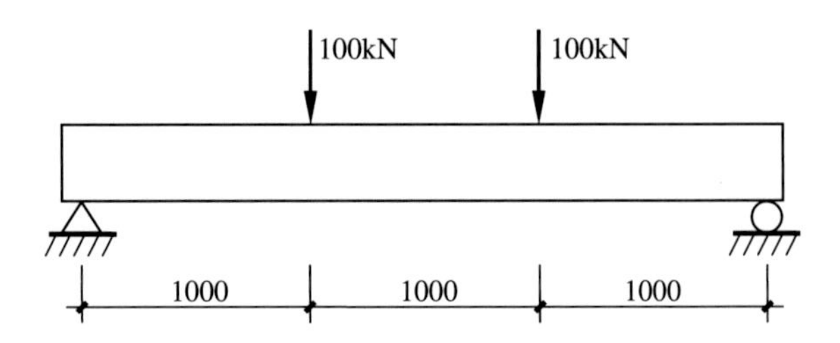

Fig. 7.60 Problem 7.5

reinforcement and stirrups are f_y = 300 N/mm^2 and f_{yv} = 270 N/mm^2, respectively. Calculate the following: (1) the required stirrups if bent-up bars are not provided, (2) the required stirrups if existing longitudinal bars are utilized as bent-up bars, and (3) the bent-up bars if stirrups of $\phi 8@200$ are provided.

7.5 A simply supported beam of rectangular cross section is shown in Fig. 7.60. The cross-sectional dimensions are $b \times h$ = 200 mm × 400 mm. Concrete of C20 (f_c = 9.6 N/mm^2, f_t = 1.10 N/mm^2) is used. The strengths of longitudinal reinforcement and stirrups are f_y = 300 N/mm^2 and f_{yv} = 270 N/mm^2, respectively. If ignoring the self-weight of the beam, try to calculate the following: (1) the required longitudinal tensile bars, (2) the required stirrups (without bent-up bars), and (3) the required stirrups if longitudinal bars are utilized as bent-up bars.

7.6 For a simply supported reinforced concrete beam as shown in Fig. 7.61, a uniformly distributed load q = 88.8 kN/m (including self-weight) is applied. Concrete of C20 (f_c = 9.6 N/mm^2, f_t = 1.10 N/mm^2) is used. The strengths of longitudinal reinforcement and stirrups are f_y = 300 N/mm^2 and f_{yv} = 270 N/mm^2, respectively. Is the beam safe?

7.7 A simply supported beam is subjected to loads as shown in Fig. 7.62, of which the uniformly distributed load has already included self-weight. The concrete used is C20 (f_c = 9.6 N/mm^2, f_t = 1.10 N/mm^2). The strengths of longitudinal reinforcement and stirrups are f_y = 300 N/mm^2 and f_{yv} = 270 N/mm^2, respectively. Calculate the required stirrups.

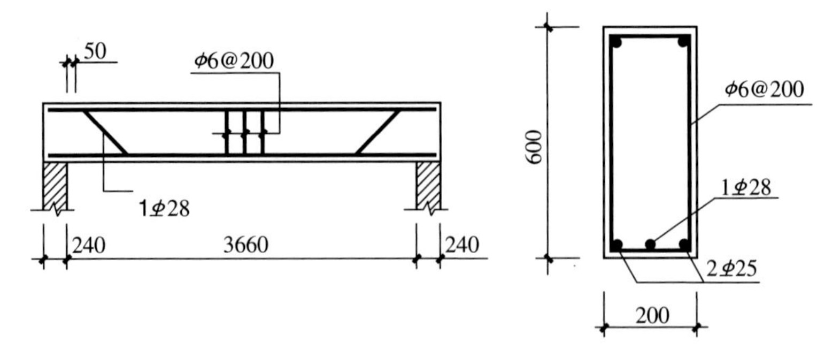

Fig. 7.61 Problem 7.6

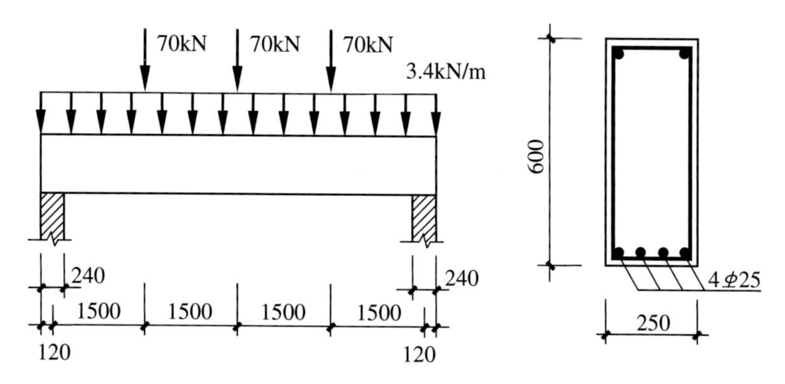

Fig. 7.62 Problem 7.7

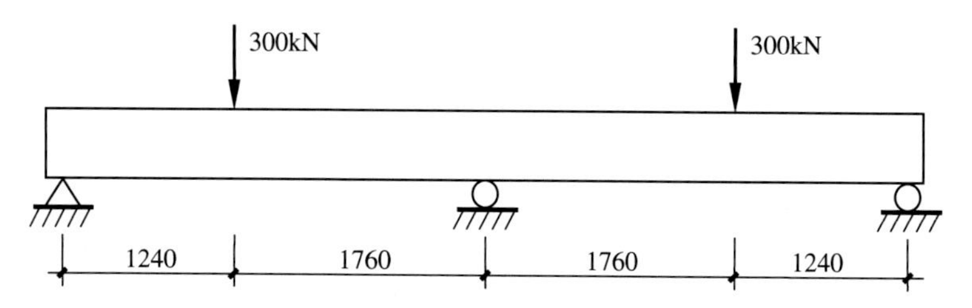

Fig. 7.63 Problem 7.8

7.8 A two-span continuous beam with the cross-sectional dimensions of $b \times h = 250$ mm × 500 mm is shown in Fig. 7.63. The concrete used is C20 (f_c = 9.6 N/mm^2, f_t = 1.10 N/mm^2). The strengths of longitudinal reinforcement and stirrups are f_y = 300 N/mm^2 and f_{yv} = 270 N/mm^2. Calculate the flexural and shear reinforcement (ignore the self-weight).

7.9 A simply supported reinforced concrete beam is shown in Fig. 7.64. C20 concrete (f_c = 9.6 N/mm^2, f_t = 1.10 N/mm^2) is used. The strengths of longitudinal reinforcement and stirrups are f_y = 300 N/mm^2 and

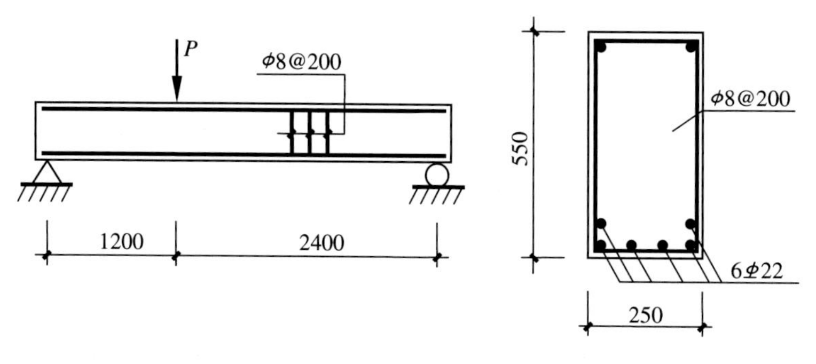

Fig. 7.64 Problem 7.9

$f_{yv} = 270$ N/mm^2, respectively. If ignoring the self-weight of the beam and the contribution of erection bars, try to calculate the maximum load P_{max} that the beam can resist and determine the failure mode.

7.10 An eccentrically loaded reinforced concrete frame column has a rectangular section of $b \times h = 400$ mm $\times$ 600 mm. $H_n = 3.1$ m, $a_s = a_s' = 40$ mm. C30 concrete ($f_c = 14.3$ N/mm^2, $f_t = 1.43$ N/mm^2) is used. Stirrups are HPB300 ($f_{yv} = 270$ N/mm^2), and longitudinal reinforcement is HRB400 ($f_y = 360$ N/mm^2). An axial force $N_c = 1500$ kN and a shear force $V = 282$ kN are applied on the column. Try to calculate the required stirrups.

7.11 All other conditions are the same as those in Problem 7.10 except the circular cross section of the same area is used. Calculate the required stirrups and compare with the result of Problem 7.10.

7.12 Detail the design steps of reinforced concrete columns of rectangular cross sections under bidirectional shear.

7.13 Detail the steps for the calculation of shear capacities of reinforced concrete columns of rectangular cross sections under bidirectional shear.

Chapter 8
Torsion

8.1 Engineering Applications and Reinforcement Detailing

Torsion is one of the basic loading types of structural members. The spandrel beam in frame structures (Fig. 8.1a) and the canopy beam (Fig. 8.1b) are two typical examples of torsional members. Torsion seldom occurs by itself. Generally, there are also bending moments and shearing forces. But it is helpful to study the mechanical behavior of reinforced concrete members under pure torsion for the best understanding of mechanical behavior of reinforced members under combined torsion, shear, and bending. So, in this chapter, pure torsion-reinforced concrete members are studied first.

If the magnitude of torques resisted by members is independent of the torsional rigidities or the torsional moment is required for the equilibrium of the structure, it is called equilibrium torsion. For instance, the canopy beam shown in Fig. 8.1b is a typical example of equilibrium torsion. Obviously, no matter how the torsional rigidity of the beam varies, the torque on it is constant and it can be determined by equilibrium conditions (herein only members of uniform cross section are considered).

If the magnitude of torques resisted by members is caused by the continuity between members in statically indeterminate structures and is dependent on the torsional rigidities of members, then it is called compatibility torsion. For instance, the spandrel beam in the frame shown in Fig. 8.1a is a typical example of compatibility torsion. In this case, if the torsional rigidity of the spandrel beam is reduced by cracking, then the torque on it will also decrease. Therefore, even if the spandrel beam is not designed for torsion, the load carrying capacity of the beam may still be sufficient at the expense of cracking and large deformation.

For common members of rectangular cross sections and cross sections composed of rectangles, longitudinal reinforcement and stirrups are installed to resist torques. The stirrups must be in closed form and put along the periphery of the cross

X. Gu et al., *Basic Principles of Concrete Structures*,
DOI 10.1007/978-3-662-48565-1_8

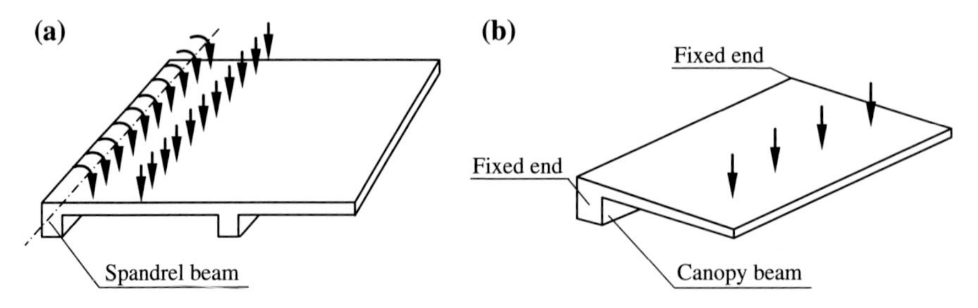

Fig. 8.1 Typical examples of torsional members. **a** Spandrel beam in a frame structures. **b** Canopy beam

sections. The hook ends of the stirrups should be set at the angle of 135°, and the extension length should be 10 times larger than the stirrup diameter. The longitudinal reinforcement should be uniformly distributed along the stirrup except for their placement at the four corners of the stirrup, as shown in Fig. 8.2. Note that the inside two legs of the four-legged stirrup can only resist shear rather than torsion.

In statically indeterminate structures, the spacing between stirrups provided due to compatibility torsion is preferred to be not larger than $0.75b$, where b is the width of the beam web.

The spacing between longitudinal reinforcement along the periphery of the cross section should not be larger than 200 mm and the length of the short dimension of the cross section. The longitudinal reinforcement designed to resist the torques should be anchored in the supports as other tensile steel bars.

Detailing of reinforcement for thin-walled tube sections, such as the box sections, can be found in specific textbooks.

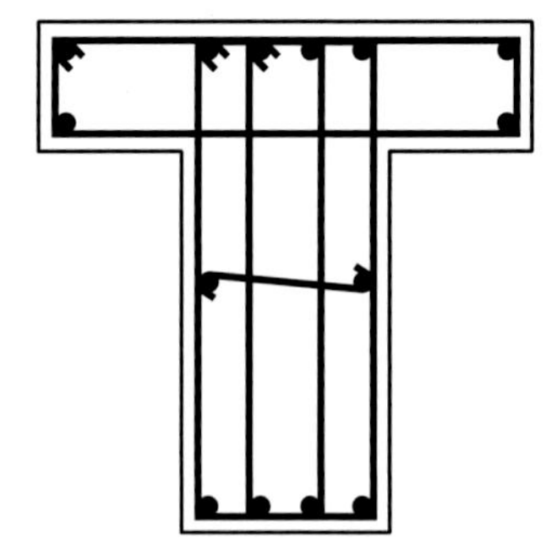

Fig. 8.2 Reinforcement of torsional members

8.2 Experimental Results of Members Subjected to Pure Torsion

Although real members are generally under the combined action of flexure, shear, and torsion, research on pure torsional members is still meaningful. First, this kind of research can reveal the loading characteristics of torsional members. Secondly, the design of torsional members in the early days was based on the experimental results of pure torsional members.

Torques are exerted at the two ends of plain concrete members or reinforced concrete members of rectangular cross sections provided with longitudinal bars and stirrups, which results in a pure torsion state in the members (Fig. 8.3).

For plain concrete members, the linear relationship between torque and twist angle is found at the beginning of loading. The principal tensile stress at the midpoint of the longer side of a member eventually causes cracking that spirals around the member, as shown by the line of ABCD in Fig. 8.4. This crack causes the failure of member immediately.

For reinforced concrete members, it is found from the recorded torque-twist angle curve (Fig. 8.5) that the behavior of the members conforms to the elastic torsion theory prior to cracking; hence, the reinforcement stresses are low, and the curve is almost a straight line. The initial crack happens at approximately midpoint of the longer dimension of the cross section. The inclination of the crack is 45° to the longitudinal axis of the member. Under loading, this crack propagates in a spiral way and maintains its 45° angle to the longitudinal axis. Meantime, more and more spiral cracks appear (Fig. 8.6).

After cracking, the members can still bear the torques; however, the torsional rigidities of the members drop dramatically, which is indicated in Fig. 8.5 by an apparent slope change. A new resistance mechanism is formed in the members after cracking. The concrete is in compression and the longitudinal bars and the stirrups are in tension. The establishment of this mechanism must be accompanied by a certain deformation, so there is a yield plateau in the curve shown in Fig. 8.5. As the torque continues to increase, the mechanism remains unchanged and the stresses in the concrete and reinforcement increase continuously until the final failure.

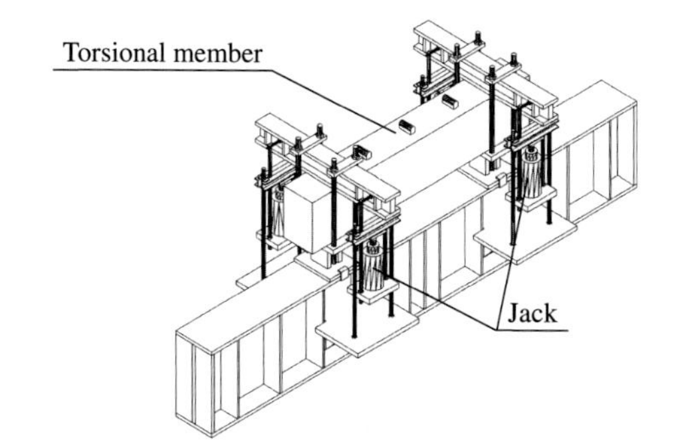

Fig. 8.3 Experimental setup for pure torsional members at Tongji University

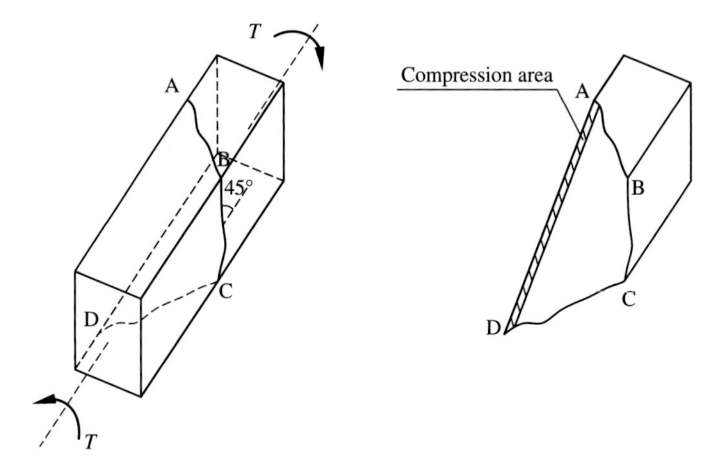

Fig. 8.4 Failure mode of pure torsional plain concrete members

Fig. 8.5 Typical torque-twist angle curve

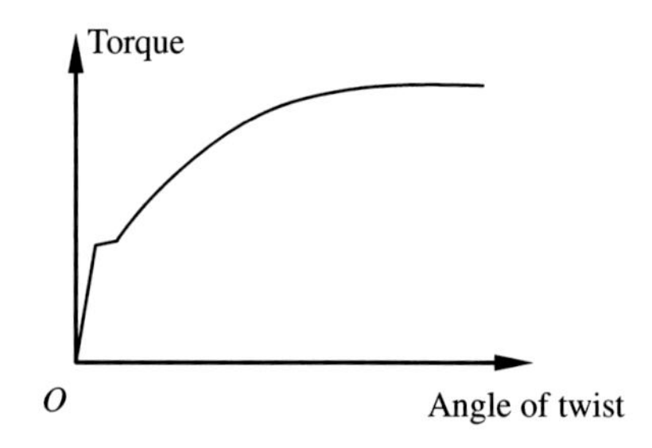

Fig. 8.6 Typical surface crack pattern at failure

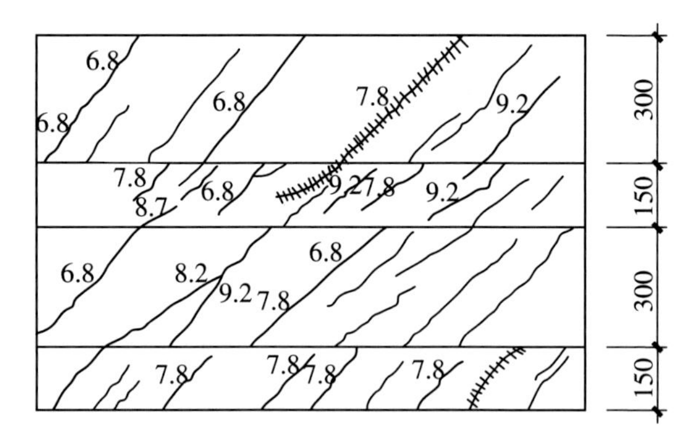

For members with different contents of longitudinal bars and stirrups, different failure modes may happen as follows:

1. Tension failure
   When stirrups and/or longitudinal bars are not sufficient or the spacing between stirrups is too large, once the member cracks, the reinforcement cannot resist the tensile stress transferred from the concrete, so the member fails immediately. The failure is brittle and appears with little warning. Torsional capacity of this type of member is almost equal to that of a plain concrete member. The requirements for the minimum content of stirrups and longitudinal bars together with the maximum spacing between stirrups are specified in design code to avoid this failure mode.
2. Skew bending failure
   When the contents of stirrups and longitudinal bars are appropriate, the member will not immediately fail after the occurrence of initial cracks. Several spiral cracks will appear on the surface at an angle of about 45° to the axis of the member. As the stirrups and longitudinal bars that intersect the spiral cracks yield, the width of the cracks increases. Finally, the member fails due to the crushing of concrete between inclined cracks. The total process exhibits a certain ductility with obvious warning before failure, which is a dominant failure mode that is permitted in practical design.
3. Compression failure
   When stirrups and longitudinal bars are overly provided, before either of them yield, the member suddenly fails due to the crushing of concrete. To avoid this brittle failure mode in practical design, the minimum cross-sectional area is usually specified in codes.
4. Partially over-reinforced failure
   If either longitudinal bars or stirrups are overly provided, the overly provided reinforcement will not yield at failure. Although partially over-reinforced failure exhibits ductility to some extent, the ductility is not as much as that of the under-reinforced failure. To avoid this failure mode, the ratio of longitudinal bars to stirrups is specified.

The behavior of members of high strength concrete (f_{cu} = 77.2–91.9 N/mm^2) without web reinforcement under pure torsion is nearly the same as that of common concrete but more brittle. The inclination of cracks is steeper. The cracked surface is flat and most of the aggregates are snapped. The cracking load is close to the failure load for members of high strength concrete.

The crack development and failure process of high strength concrete members with web reinforcement are similar to those of common concrete but the inclination of cracks is steeper.

Excluding the above-mentioned failure modes, the corners of members subjected to torsion may spall off. From the space truss model, which will be discussed in

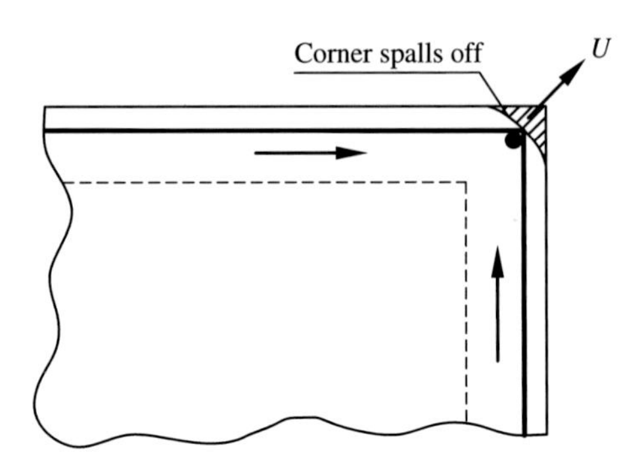

Fig. 8.7 Corners of torsion members spall off

Sect. 8.4, intersected struts at the corner of a cross section will produce a radial force U that pushes the corner out (Fig. 8.7). If no closely spaced stirrups or rigid corner longitudinal reinforcement exists to resist this force, the corner will drop. Experiments show that only stirrups with spacing less than or equal to 100 mm or corner longitudinal reinforcement of large diameter can reliably prevent this kind of failure when the torsion-induced shear stress is large. In addition, because the corner spalls off, the ends of stirrups should be anchored inside the concrete with hooks placed at a 135° angle (Fig. 8.2).

8.3 Cracking Torque for Members Under Pure Torsion

To avoid tension failure, the torsional capacities of reinforced members should be at least greater than the capacities of the members of plain concrete, whereas the torsional capacities of plain concrete members are equal to the cracking torques. Therefore, calculation of the cracking torques of members is required to provide a basis for determining the minimum amount of torsional reinforcement. Before cracking, the stress of the reinforcement is very low, so it is usually neglected in the calculation. Structural members to resist torsion in building engineering are usually solid. However, hollow members such as box beams are usually used in bridge engineering. In this section, both solid and hollow members will be discussed.

8.3.1 Solid Members

For elastic materials, the cracking torque should be calculated according to elastic theory. For plastic materials, plastic theory should be used to determine the cracking torque. The softening characteristic of the stress–strain relationship of concrete under tension indicates that the concrete is actually an elasto-plastic material.

Thus, there are two ways to determine cracking torques. One is based on elastic theory, where calculated results are enlarged to consider the plasticity of concrete. Another is based on plastic theory, where calculated results are discounted. ACI code uses the first method while Chinese codes prefer the second method.

8.3.1.1 Method Based on Elastic Theory

The distribution of shear stresses in a torsional member of rectangular cross section is shown in Fig. 8.8a. The shear stresses at the section's corners are zero, and the maximum shear stress $\tau_{\max}$ appears at the center of the longer dimension of the cross section, which can be visualized by using the soap-film analogy. The value of $\tau_{\max}$ is

$$\tau_{\max} = \frac{T}{\alpha' b^2 h} \tag{8.1}$$

where

b, h the short and long dimensions of the cross section; and

α' the shape factor, whose value is approximately 1/4.

It is shown in Fig. 8.8b that the concrete at the center of the longer dimension is under pure shear, i.e., at this point, the concrete is subjected to not only principal tensile stress σ_1, but also principal compressive stress of equivalent value σ_2 in the

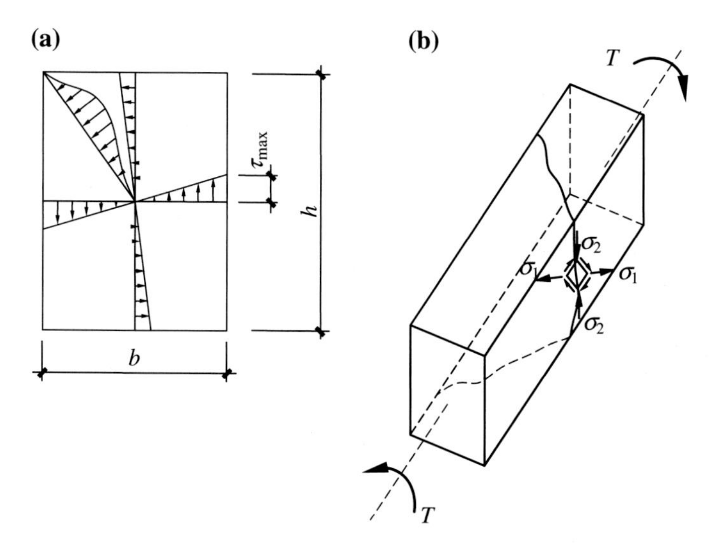

Fig. 8.8 Shear stress of a torsional member with rectangular section. **a** Shear stress distribution of the section from elastic analysis. **b** Principal stresses at the midpoint of the longer side of the section

perpendicular direction. From the strength curve of concrete under biaxial stress shown in Fig. 2.27, it can be seen that when $\sigma_1 = \sigma_2$, the tensile strength of concrete will be reduced by about 10 %. In other words, the tensile strength of concrete under pure shear is about 0.9 f_t, where f_t is the tensile strength of concrete under uniaxial tension, so

$$T_{cr,e} = 0.9\frac{f_t b^2 h}{4} \tag{8.2}$$

To consider the plasticity of concrete, the shape factor 1/4 in Eq. (8.2) is changed to 1/3 as shown in Eq. (8.3), which gives a larger estimation of the cracking torque compared with the elastic results.

$$T_{cr} = 0.9\frac{f_t b^2 h}{3} \tag{8.3}$$

For T- and I sections, the cracking torque can be conservatively taken as the sum of the cracking torques of all component rectangles. The way to divide the compound cross section into rectangles should make $\Sigma b^2 h$ maximum. The corresponding cracking torque is as follows:

$$T_{cr} = 0.9 f_t \frac{\sum b^2 h}{3} \tag{8.4}$$

8.3.1.2 Method Based on Plastic Theory

If the full cross section is assumed in plasticity, the torsion-induced shear stresses reach τ_{max} everywhere in the section. Dividing the rectangular section into four portions, Fig. 8.9 shows how the resultant forces of each portion form two pairs of couples that are cracking torsion of the section as shown in Eq. (8.5).

$$T_{cr,p} = 2(F_1 d_1 + F_2 d_2) = \tau_{max}\frac{b^2}{6}(3h - b) \tag{8.5}$$

where

h, b the dimensions of the cross section, $h > b$.

For ideal plastic material, τ_{max} can be assumed as the tensile strength herein. Then, the plastic cracking torque is as follows:

$$T_{cr,p} = f_t W_t \tag{8.6}$$

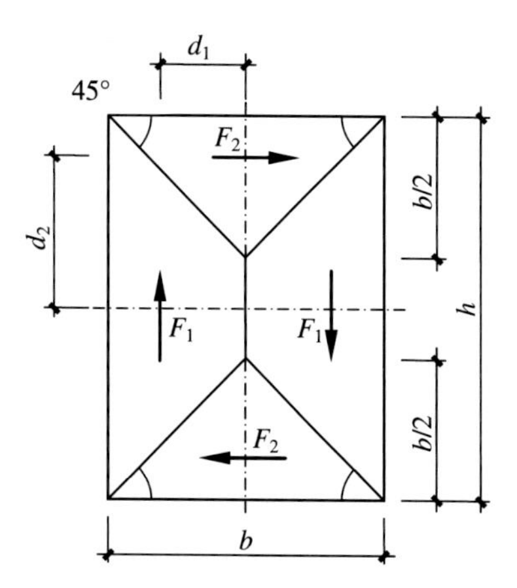

Fig. 8.9 Division and plastic internal forces of a rectangular section under pure torsion

where

f_t the tensile strength of concrete; and

W_t the plastic torsional modulus of the cross section, which can be calculated by Eq. (8.7).

$$W_t = \frac{b^2}{6}(3h - b) \tag{8.7}$$

Equation (8.7) can also be derived by using the sand-heap analogy. Heaping the sand freely on the rectangular section of a torsional member, the sole shape of the sand heap can be finally formed as shown in Fig. 8.10. According to the geometric shape of the sand heap, its volume can be calculated as follows:

$$V_R = \frac{1}{2}\theta\frac{b}{2}b(h - b) + \frac{1}{3}b\theta\frac{b}{2} = \frac{b^2}{12}(3h - b)\theta \tag{8.8}$$

where

θ the inclined angle of the sand-heap surface.

Defining $W_t = 2V_R/\theta$ yields Eq. (8.7).

Because concrete is not an ideal plastic material, the cracking torque should be appropriately discounted. Experimental results show that the reduction factor is 0.7 for high strength concrete and 0.8 for low strength concrete. So for safety, the cracking torque is taken as

$$T_{cr} = 0.7f_tW_t \tag{8.9}$$

For rectangular sections, W_t can be calculated by Eq. (8.7). For compound sections such as T-, and I sections, the plastic torsional modulus can be obtained by sand-heap analogy. The plastic torsional modulus of a compound cross section is

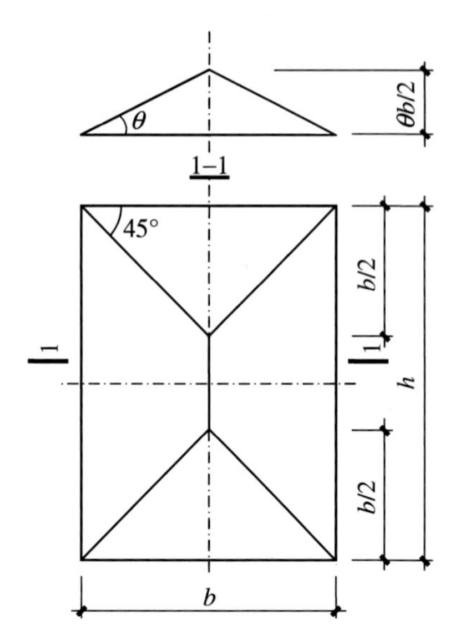

Fig. 8.10 Sand heap on a rectangular section

obviously larger than the sum of the plastic torsional moduli of the component rectangles, because the heaped sand at the intersections of the rectangles is apparently more than that when the rectangles separate.

The sand-heap analogy of cross sections composed of several rectangles is illustrated in Fig. 8.11. It is really complex to calculate the sand-heap volume. To simplify, we assume that the heap 1′2′3′ at the intersection is moved to fill the end 123. Then, the volume of the sand heap on the flange is as follows:

$$V_{\mathrm{f}} = \theta' \frac{h_{\mathrm{f}}^2}{4}(b_{\mathrm{f}} - b) \tag{8.10}$$

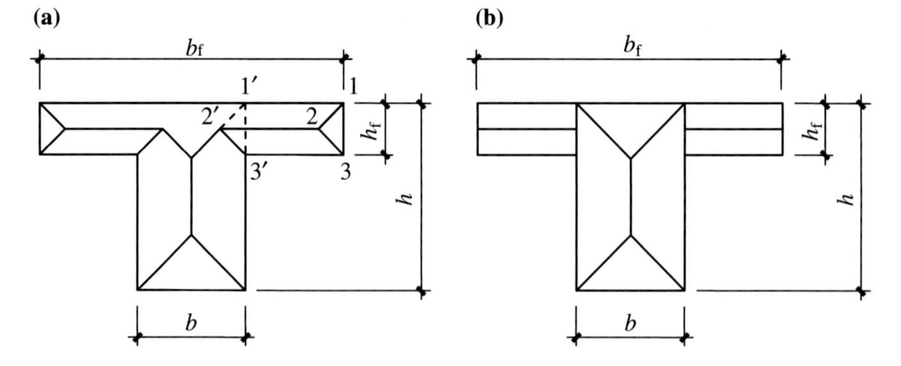

Fig. 8.11 Sand-heap analogy of a compound section

where θ' = the inclined angle of the sand heap. Thus, the torsional modulus of the flange shown in Fig. 8.11 is given as follows:

$$W_{\mathrm{tf}} = 2V_{\mathrm{f}}/\theta' = \frac{h_{\mathrm{f}}^2}{2}(b_{\mathrm{f}} - b) \tag{8.11}$$

Experimental results show that the effective overhung flange length should not exceed three times that of the flange thickness.

The torsional modulus of a compound cross section composed of several rectangles is given as:

$$W_{\mathrm{t}} = \sum W_{\mathrm{t}i} \tag{8.12}$$

The minimum content of torsional reinforcement should be provided to ensure that the ultimate torque of a cross section is larger than the cracking torque.

Example 8.1 A T section beam (Fig. 8.12) has the dimensions of $b_{\mathrm{f}}' = 650\,\mathrm{mm}$, $h_{\mathrm{f}}' = 120\,\mathrm{mm}$, b = 250 mm, and h = 500 mm. The tensile strength of concrete is $f_{\mathrm{t}} = 1.5\ \mathrm{N/mm^2}$. Calculate the cracking torque by using elastic and plastic methods, respectively.

Solution

1. Based on elastic method
   From Eq. (8.4), we have

$$\begin{aligned} T_{\mathrm{cr}} &= 0.9 f_{\mathrm{t}} \frac{\sum x^2 y}{3} \\ &= 0.9 \times 1.5 \times \frac{2 \times 120^2 \times 200 + 250^2 \times 500}{3} = 1.665 \times 10^7\ \mathrm{N{\cdot}mm} \end{aligned}$$

2. Based on plastic method
   From Eq. (8.12), we have

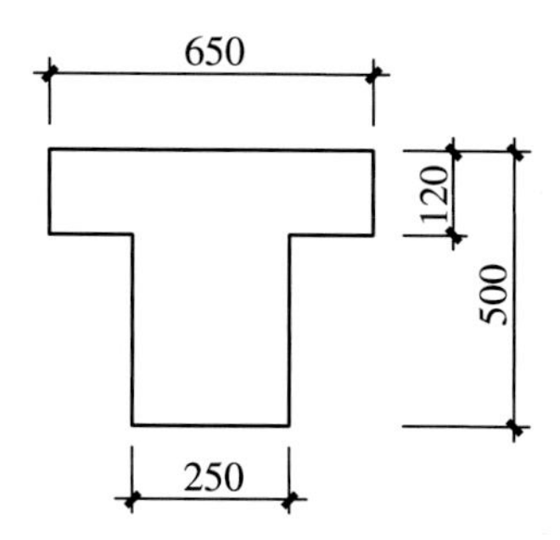

Fig. 8.12 Example 8.1

$$W_t = W_{tw} + W_{tf} = \frac{250^2}{6} \times (3 \times 500 - 250) + \frac{120^2}{2} \times (650 - 250)$$
$$= 1.59 \times 10^7 \text{mm}^3$$

And from Eq. (8.9), we have

$$T_{cr} = 0.7 \times 1.5 \times 1.59 \times 10^7 = 1.670 \times 10^7 \text{ N·mm}$$

Comparing the results based on different methods, it can be concluded that the modified elastic method and the modified plastic method can almost give the same result for the beam in Example 8.1.

8.3.2 *Hollow Members*

8.3.2.1 Method Based on Thin-Walled Tube Theory

Figure 8.13a shows a thin-walled tube with a continuous wall subjected to pure torsion about its longitudinal axis. An element cut from the tube wall ABCD is shown in Fig. 8.13b. The thicknesses of the wall along sides AB and CD are t_1 and t_2, respectively. Shear forces caused by the applied torque T on the four sides of the element are V_{AB}, V_{BC}, V_{CD}, and V_{DA}, respectively. From $\Sigma F_x = 0$, it can be found that $V_{AB} = V_{CD}$. Because $V_{AB} = \tau_1 t_1 \mathrm{d}x$ and $V_{CD} = \tau_2 t_2 \mathrm{d}x$, we get $\tau_1 t_1 = \tau_2 t_2$, where τ_1 and τ_2 are the shear stresses acting on sides AB and CD, respectively. The product τt is defined as the shear flow q.

Two smaller elements cut at corner B and corner C in Fig. 8.13b, respectively, are shown in Fig. 8.13c and d. For the equilibrium of the elements, $\tau_1 = \tau_3$ and $\tau_2 = \tau_4$. Thus, at points B and C on the perimeter of the tube, $\tau_3 t_1 = \tau_4 t_2$. This shows that shear flow keeps constant around the perimeter of the tube for a given applied torque T.

Figure 8.13a shows the shear force acting on the length ds of wall, qds. The moment of this force about the centroid longitudinal axis of the tube is rqds, where r is the distance measured from the mid-place of the tube wall to the centroid axis. Integrating the moment around the perimeter gives the torque in the tube:

$$T = \oint_{u_{cor}} q \times r \times \mathrm{d}s = q \oint_{u_{cor}} r\mathrm{d}s = 2qA_{cor} \tag{8.13}$$

where

A_{cor} the area enclosed by the shear flow; and

u_{cor} the perimeter of the tube section.

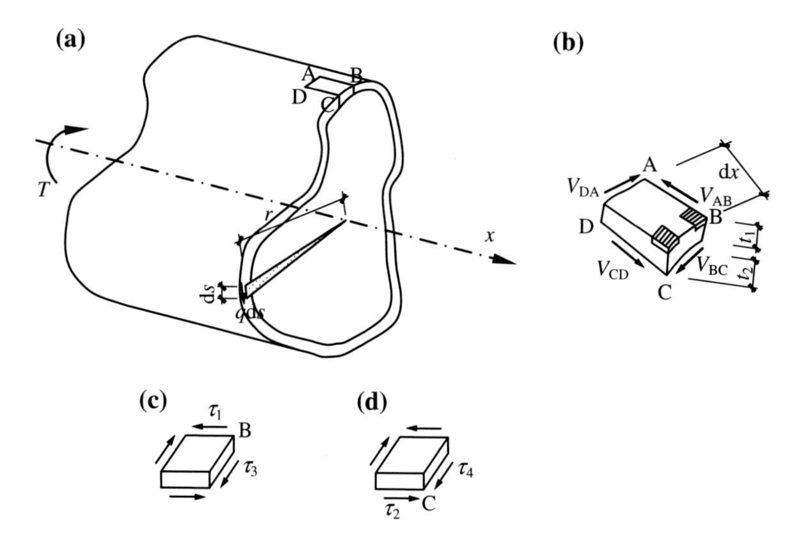

Fig. 8.13 Shear stresses in a thin-walled tube

Substituting q by τt gives

$$\tau = \frac{T}{2A_{\text{cor}}t} \tag{8.14}$$

Equation (8.14) shows that the maximum shear stress inside the tube under a given torque T initiates at the point where the thickness of the tube is the thinnest. Obviously, the stress state of concrete everywhere in the tube is biaxial tension and compression caused by shear stress. From the strength curve for the concrete subjected to biaxial stress shown in Fig. 2.27, it can be concluded that the concrete will crack when $\tau_{\max} = 0.9f_t$. So, we get

$$T_{\text{cr}} = 0.9f_{\text{t}} \times 2A_{\text{cor}}t_{\min} = 1.8f_{\text{t}}A_{\text{cor}}t_{\min} \tag{8.15}$$

where

$t_{\min}$ the minimum thickness of the thin-walled tube.

8.3.2.2 Method Based on Plastic Theory

The cracking torque of a box-section beam can also be calculated by using Eq. (8.9) considering the plasticity of concrete. But the plastic modulus of the section should be calculated with a method different from that for the rectangular section or I- or T-sections. Similar to that for the rectangular or I- or T sections, the equation to calculate W_t for a box section can also be derived by using the sand-heap analogy. But the sand cannot be heaped up directly on the box section when deriving the equation.

For the box section shown in Fig. 8.14a, the sand is heaped up firstly on the area enclosed by the outer perimeter of the section as shown in Fig. 8.14b. The volume of sand heap is given as:

$$V_h = \frac{b_h^2}{12}(3h_h - b_h)\theta \tag{8.16}$$

where

b_h, h_h the dimensions for the short and the long sides of the section.

Then, the sand is also heaped up on the area enclosed by the inner perimeter of the section as shown in Fig. 8.14c. The volume of the sand heap in this case is given as follows:

$$V_w = \frac{(b_h - 2t_w)^2}{12}[3h_w - (b_h - 2t_w)]\theta \tag{8.17}$$

The plastic modulus of the box section is given as:

$$W_t = 2(V_h - V_w)/\theta = \frac{b_h^2}{6}(3h_h - b_h) - \frac{(b_h - 2t_w)^2}{6}[3h_w - (b_h - 2t_w)] \tag{8.18}$$

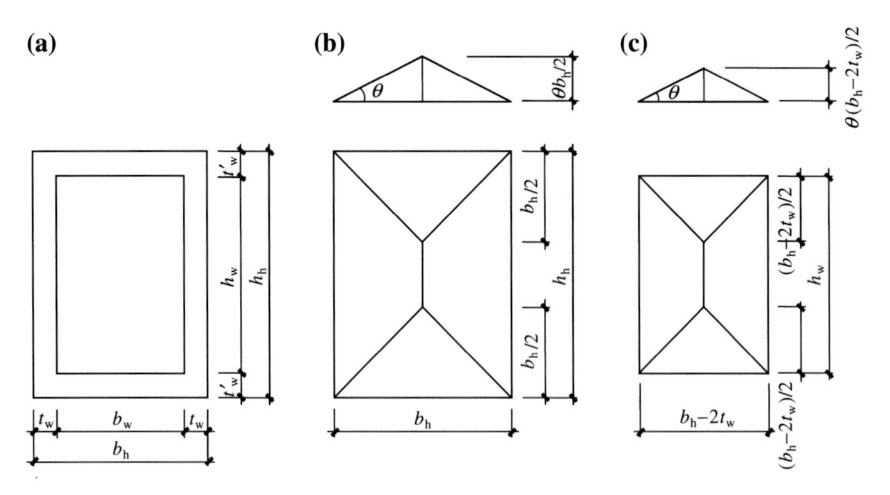

Fig. 8.14 Sand-heap analogy of a box section. **a** Dimension of the enclosed section ($t_w \leqslant t'_w$). **b** Sand heap on the area enclosed by the outer perimeter of the section. **c** Sand heap on the area enclosed by the inner perimeter of the section

Fig. 8.15 Example 8.2

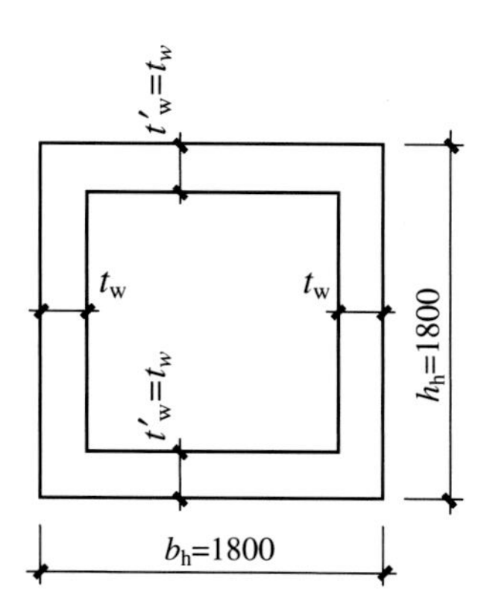

where

h_w the clear height of the web; and

t_w the thickness of the web.

Example 8.2 The dimension of the section for a box-section beam is shown in Fig. 8.15, f_t = 2.1 MPa. When t_w = 300, 400, 500, 600, 700, 800, and 900 mm, calculate the corresponding torques by using the method based on the thin-walled tube theory and the plastic theory, respectively.

Solution

1. Based on the thin-walled tube theory
   According to Eq. (8.15), the results are listed in the following table (Table 8.1).
2. Based on the plastic theory
   According to Eqs. (8.9) and (8.18), the results are listed in the following table (Table 8.2).

Comparing the corresponding results from different methods, it can be seen that the calculated results by the method based on the thin-walled tube theory are larger. This is because the cracking torques may be overestimated by the method based on the thin-walled tube theory when the thickness of the wall is large. Equations (8.9)

Table 8.1 Results based on the thin-walled tube theory

f_t/(N·mm^{-2})	t_w/mm	A_{cor}/mm^3	T_{cr}/(N·mm)
2.1	300	2.250×10^6	2.552×10^9
	400	1.960×10^6	2.964×10^9
	500	1.690×10^6	3.194×10^9
	600	1.440×10^6	3.266×10^9
	700	1.210×10^6	3.202×10^9
	800	1.000×10^6	3.204×10^9
	900	0.810×10^6	2.756×10^9

Table 8.2 Results based on the plastic theory

f_t/(N·mm^{-2})	t_w/mm	W_t/mm^3	T_{cr}/(N·mm)
2.1	300	1.368×10^9	2.011×10^9
	400	1.611×10^9	2.368×10^9
	500	1.773×10^9	2.606×10^9
	600	1.872×10^9	2.752×10^9
	700	1.923×10^9	2.827×10^9
	800	1.941×10^9	2.853×10^9
	900	1.944×10^9	2.858×10^9

and (8.18) are recommended by Code GB 50010 in China for the calculation of cracking torque of a box-section beam.

8.4 Calculation of Torsional Capacities for Members of Rectangular Sections Subjected to Pure Torsion

So far, the torsional capacities of warped sections in reinforced concrete members have been calculated by two quite different theories—the space truss analogy and the skew bending theory, both of which will be introduced as follows.

8.4.1 Space Truss Analogy

E. Rausch proposed a space truss model in 1929, in which the angles between the inclined web members and the longitudinal axis were constant (45°). However, the inclination of inclined cracks varies with the volumes and strengths of stirrups and longitudinal bars. Setting a constant angle of 45° is equivalent to apply additional constraints on members, which makes the calculation unreliable. However, because the coefficients in the final calculation formulae are determined by experiments and discounted by safety consideration, we can assume a constant angle of 45° and not run the risk of unreliability or error. Nevertheless, the constant angle space truss model does not reflect the influence of the volumes and strengths of longitudinal bars and stirrups on the inclined angle and the ultimate torsion capacity, so the reliability is not uniform.

In 1968, P. Lampert and B. Thuerlimann came up with the variable angle space truss model. This model overcame the above-mentioned shortcomings and will be detailed herein.

The variable angle space truss model makes the following assumptions: (1) The original solid cross section is simplified as a tubular section, as shown in Fig. 8.16a. The concrete in the tubular section was split by spiral cracks, which led to many compression struts with the inclination of α; (2) the longitudinal bars and stirrups

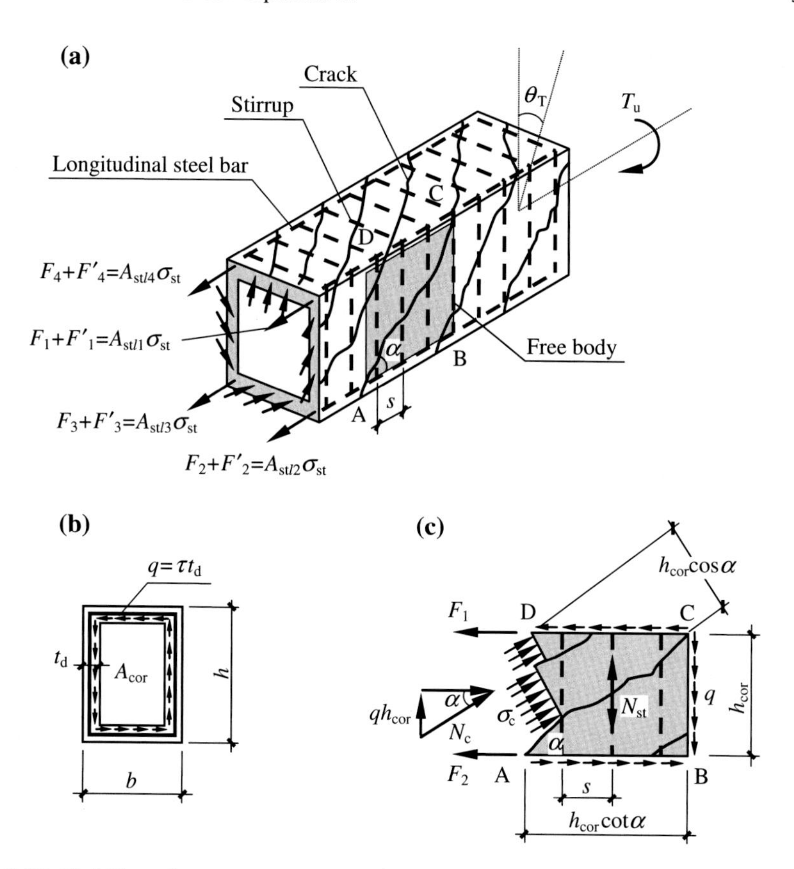

Fig. 8.16 Variable angle space truss analogy. **a** Variable angle space truss model. **b** Shear flow in thin wall of the section. **c** Free body of side wall in the truss

serve as tensile members and form the space truss with the compression struts; and (3) the dowel action of the bars is ignored.

Under these assumptions, if the concept of shear flow is introduced and only the average compressive stress is considered for the concrete diagonal struts, the problem becomes statically determinate.

Based on Eq. (8.13), the shear flow q induced by a torque T_u in the tubular section can be expressed as follows:

$$q = \tau t_d = \frac{T_u}{2A_{cor}} \tag{8.19}$$

where

A_{cor} the area enclosed by the center line of the tubular section wall. It is herein taken as the area enclosed by the connection line of centroids of longitudinal bars at corners of the section, i.e., $A_{cor} = b_{cor}h_{cor}$ (Fig. 8.16b);

τ the shear stress due to torsion; and
t_d the thickness of the wall of tubular section.

A side wall of the box-section ABCD which includes a complete diagonal crack is taken from the space truss as a free body (Fig. 8.16c). Figure 8.16c also shows the shear flow q induced forces in the truss members. For the diagonal struts, the average compressive stress is σ_c and the resultant of the compressive forces is N_c. The horizontal component of N_c is balanced by the tensile forces in the longitudinal bars and the vertical component of N_c is balanced by the shear stress on the right side of the free body. Then, we have the following equation:

$$F_1 + F_2 = qh_{cor} \cot \alpha \quad (8.20)$$

Similar analyses of the other three walls of the tubular section give the following equations:

$$F_1' + F_4' = qb_{cor} \cot \alpha \quad (8.21)$$

$$F_4 + F_3 = qh_{cor} \cot \alpha \quad (8.22)$$

$$F_3' + F_2' = qb_{cor} \cot \alpha \quad (8.23)$$

The resultant tensile force of longitudinal reinforcement obtained from all side walls is given as:

$$F_1 + F_1' + F_2 + F_2' + F_3 + F_3' + F_4 + F_4' = q \cot \alpha \cdot 2(b_{cor} + h_{cor}) = qu_{cor} \cot \alpha \quad (8.24)$$

where
$u_{cor} = 2(b_{cor} + h_{cor})$ is the perimeter corresponding to the area A_{cor} enclosed by the shear flow.

If the content of the longitudinal bars is appropriate, the bars yield prior to failure. Then,

$$A_{stl} f_y = qu_{cor} \cot \alpha \quad (8.25)$$

where
A_{stl} the total cross-sectional area of all longitudinal bars.

Cutting the free body ABCD from the diagonal crack and considering the equilibrium of the upper part, if N_{st} is the tensile force in one stirrup leg and s is the spacing between stirrups, we have

$$\frac{N_{st}h_{cor}\cot\alpha}{s} = qh_{cor} \tag{8.26}$$

If the content of the stirrups is also appropriate, the stirrups yield prior to the failure. Then

$$A_{st1}f_{yv}\frac{h_{cor}\cot\alpha}{s} = qh_{cor} \tag{8.27}$$

where

A_{st1} the cross-sectional area of one stirrup leg.

It can be seen from Eq. (8.27) that the angle α of the diagonal struts in each side wall will be the same if an equal amount of stirrups, A_{st1}, is contained in each side wall.

If q is canceled out from Eqs. (8.25) and (8.27), then

$$\cot\alpha = \sqrt{\frac{f_y A_{stl}s}{f_{yv}A_{st1}u_{cor}}} = \sqrt{\zeta} \tag{8.28}$$

$$\text{where} \quad \zeta = \frac{f_y A_{stl}s}{f_{yv}A_{st1}u_{cor}} \tag{8.29}$$

where ζ is called the reinforcement strength ratio of the longitudinal bars to the stirrups in a torsional member. If the longitudinal bars are asymmetrically provided in a cross section, the less reinforced side can be used for calculation purposes.

Because the reinforcement almost fails to contribute to the section prior to cracking, the initial inclined crack is at an angle of about 45°. However, when the torque is close to the torsion capacity, the inclination of the critical diagonal crack is controlled by the reinforcement strength ratio ζ. This means that when $\zeta \neq 1$, as the torque increases after cracking, inclined cracks continue to propagate and the angle of the diagonal struts varies correspondingly until the critical inclination α is reached. Canceling out α from Eqs. (8.25) and (8.27) gives the following:

$$q = \sqrt{\frac{A_{stl}f_y A_{st1}f_{yv}}{su_{cor}}} \tag{8.30}$$

Substituting Eqs. (8.30) into (8.19) yields

$$T_u = 2A_{cor}\sqrt{\zeta}\frac{A_{st1}f_{yv}}{s} \tag{8.31}$$

Equation (8.31) reflects the influence of the reinforcement on the torsional capacity. For thin-walled members of arbitrary cross sections, similar formulae can be derived.

The above conclusions were obtained based on the assumption that both longitudinal bars and stirrups yield at the ultimate state. When ζ is too large or too small, the longitudinal bars or stirrups may not yield. Experimental results show that if α is between 30° and 60° and the corresponding ζ is between 3 and 0.333 according to Eq. (8.29), the two kinds of steel bars yield at the member failure provided that the contents of longitudinal reinforcement and stirrups are appropriate. To limit the crack width of the members under service loads, α should satisfy the following condition:

$$3/5 \leqslant \tan\alpha \leqslant 5/3 \tag{8.32}$$

$$\text{or} \quad 0.36 \leqslant \zeta \leqslant 2.778 \tag{8.33}$$

To avoid compression failure, the maximum content of reinforcement must be limited.

8.4.2 Skew Bending Theory

Skew bending theory for members under torsion was suggested by H.H. Лессцг in 1959, based on the experimental results of reinforced concrete members under combined torsion, shear, and flexure. The theory is also called ultimate equilibrium theory on a skew failure surface.

The skew failure surface confined by a spiral crack ABCD is shown in Fig. 8.17, where the shade area near side AD is in compression. The compression zone is usually very small, and its depth can be approximately taken as twice that of the concrete cover of the longitudinal bars. The angle between the inclined crack and the axis of the member is still α. If longitudinal bars and stirrups are appropriately provided, both of them that intersect with the inclined cracks can yield.

According to the static equilibrium, the same equation for the ultimate torque T_u as previously mentioned can be derived.

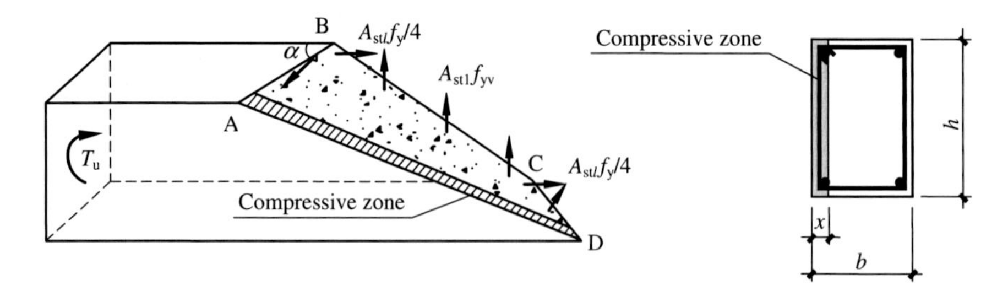

Fig. 8.17 Calculation diagram of skew bending theory

The shortcoming of the skew bending theory lies in the difficulty of choosing a skew failure surface for nonrectangular sections, whereas the space truss analogy can be more easily used in the torsional analysis of members of irregular sections.

8.4.3 Calculation Method in GB 50010

The formula for torsion resistance of reinforced concrete members of rectangular sections subjected to pure torsion specified in GB 50010 is:

$$T_u = 0.35 f_t W_t + 1.2\sqrt{\zeta}\frac{f_{yv} A_{st1} A_{cor}}{s} \quad (8.34)$$

where the first term on the right-hand side of the equation is the torsion resistance of concrete, and the second term is the torsion resistance of reinforcement. A_{cor} is the area enclosed by the inside profile of the stirrup, which can be calculated by subtracting the thickness of concrete cover from the cross-sectional dimensions.

The maximum torsional capacity is given as:

$$\text{when} \quad h_0/b \leqslant 4, T_{u,max} = 0.2\beta_c f_c W_t \quad (8.35)$$

$$\text{when} \quad h_0/b = 6, T_{u,max} = 0.16\beta_c f_c W_t \quad (8.36)$$

when $4 < h_0/b < 6$, the value is determined by linear interpolation. β_c is the influence factor of concrete strength. If the concrete grade is not higher than C50, then $\beta_c = 1.0$; If the concrete grade is C80, then $\beta_c = 0.8$; if the concrete grade is between C50 and C80, then β_c is determined by linear interpolation.

The minimum torsional capacity is given as:

$$T_{u,min} = 0.7 f_t W_t \quad (8.37)$$

It is assumed in Eq. (8.34) that the ultimate torque equals the sum of concrete resistant torque and reinforcement resistant torque. The basic form of the formula is based on the plastic ultimate torque formula for concrete and the ultimate torque formula from variable angle space truss analogy. Then, the factors 0.35 and 1.2 are determined experimentally for the concrete term and the reinforcement term, respectively. Figure 8.18 compares experimental data with the predictions by Eq. (8.34).

The factor in front of the reinforcement term should be 2.0 according to the space truss analogy, but is taken as 1.2 in the code instead. This is because: (1) The concrete resistant torque has already been considered by the first term in Eq. (8.34); (2) the calculation of A_{cor} in the code formula is based on the area enclosed by the inside profile of the stirrup, whereas A_{cor} in the space truss model is based on the area enclosed by the connection lines of the centroids of the longitudinal bars at

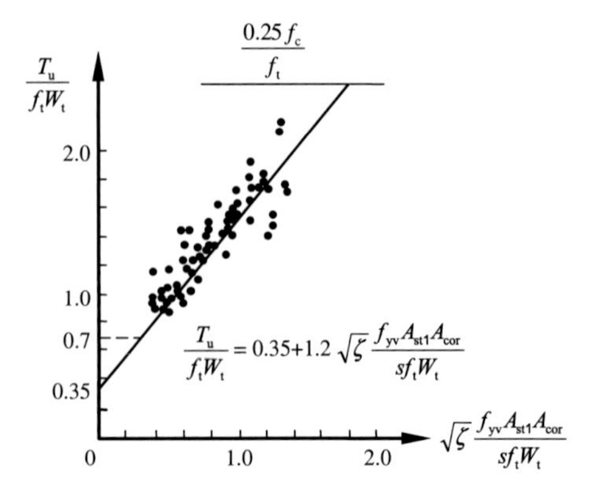

Fig. 8.18 Comparison of Eq. (8.34) and experimental data

the corners of the section; and (3) the experimental data, based on which the code formula is derived, includes a portion of over-reinforced members.

Equation (8.9) is derived for uncracked members. However, members are severely cracked in the ultimate state, so the concrete contribution to the torsion resistance in Eq. (8.34) is taken as half of the cracking torque, i.e., the factor in front of the concrete term is 0.35.

Equation (8.33) shows that when ζ is between 0.36 and 2.778, both the longitudinal bars and the stirrups can yield; meanwhile, the maximum width of cracks can be limited to an allowable value under the service load. GB 50010 conservatively specifies $0.6 \leqslant \zeta \leqslant 1.7$. And if $\zeta > 1.7$, take $\zeta = 1.7$.

The steel ratio of torsional longitudinal bars ρ_{tl} is defined as follows:

$$\rho_{tl} = \frac{A_{stl}}{bh} \tag{8.38}$$

where

b the width of rectangular section, or web width of T- or I sections;
h the depth of cross section; and
A_{stl} the total cross-sectional area of longitudinal torsion bars.

The minimum steel ratio of longitudinal bars for members under pure torsion is given as:

$$\rho_{tl,\min} = 0.85\frac{f_t}{f_y} \tag{8.39}$$

The minimum steel ratio of stirrups is given as:

$$\rho_{st,\min} = \frac{A_{st}}{bs} = 0.28\frac{f_t}{f_{yv}} \tag{8.40}$$

where $A_{st} = 2A_{st1}$.

8.5 Calculation of Torsional Capacities for Members of I-, T-, and Box Sections Subjected to Pure Torsion

8.5.1 *Method Based on the Space Truss Analogy*

Similar to the space truss analogy for rectangular sections introduced in Sect. 8.4.1, the same equation as Eq. (8.31) for member of I-, T-, and box sections subjected to pure torsion can be derived to consider the contribution of reinforcement to the torsional bearing capacities of the members. The main difference between the rectangular sections and the I-, T-, or box sections is the calculation of A_{cor}. Figure 8.19 gives the calculation examples in different cases.

However, in the design of I- or T section members subjected to pure torsion, it is difficult to arrange the reinforcement bars and stirrups between the web and flange (or flanges) after they have been determined by using Eq. (8.31). So, simplification of the method is needed.

8.5.2 *Method in GB 50010*

Similar to the method of cracking torque calculation, the ultimate torque of a compound cross section composed of several rectangles can be conservatively calculated by summing the ultimate torques of the component rectangles, that is

$$T_{\mathrm{u}} = \sum T_{\mathrm{u}i} \tag{8.41}$$

where

$T_{\mathrm{u}i}$ the ultimate torque of the *i*th rectangle.

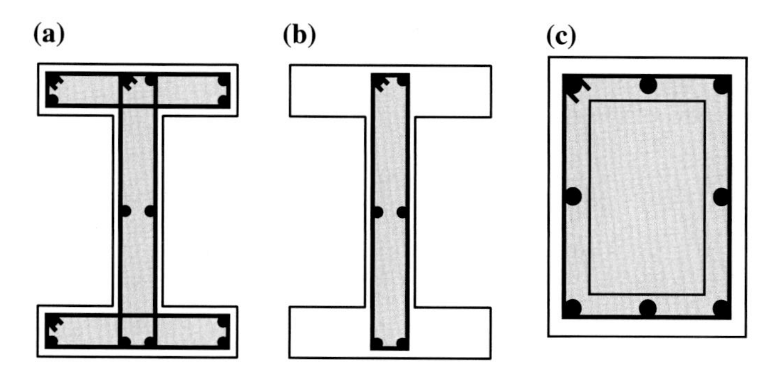

Fig. 8.19 Calculation examples of A_{cor} for I- (T-) and box sections (A_{cor} = shaded area)

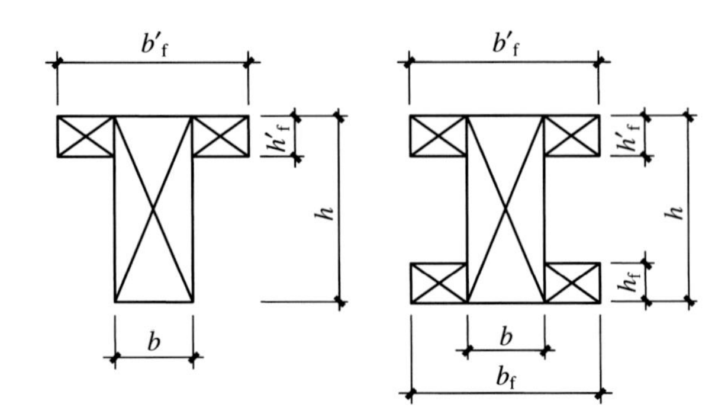

Fig. 8.20 Optimum subdivision of compound cross sections

Obviously, there are many ways to decompose a compound section into several rectangles. The optimum way is to maximize T_u, which is shown in Fig. 8.20.

The effective overhung length of the flange in a T- or I section should not be larger than three times that of the flange depth.

After the cross section has been subdivided, the torque resisted by each component rectangle is proportional to its torsional rigidity (or elastic/plastic torsional modulus).

The plastic torsional moduli of web W_{tw}, compressive flange W'_{tf}, and tensile flange W_{tf} can be calculated by Eqs. (8.7) and (8.11), respectively, that is

$$W_{tw} = \frac{b^2}{6}(3h - b) \tag{8.42}$$

$$W'_{tf} = \frac{h'^2_f}{2}(b^2_f - b) \tag{8.43}$$

$$W_{tf} = \frac{h^2_f}{2}(b_f - b) \tag{8.44}$$

where

h the section height;
b the web width;
h_f, h'_f the thicknesses of tensile and compressive flanges; and
b_f, b'_f the widths of tensile and compressive flanges.

The overall plastic torsional modulus of a compound cross section is

$$W_t = W_{tw} + W'_{tf} + W_{tf} \tag{8.45}$$

After subdivision of an I- or T section, A_{cor} and u_{cor} can be calculated according to the dimensions of the rectangles shown in Fig. 8.20.

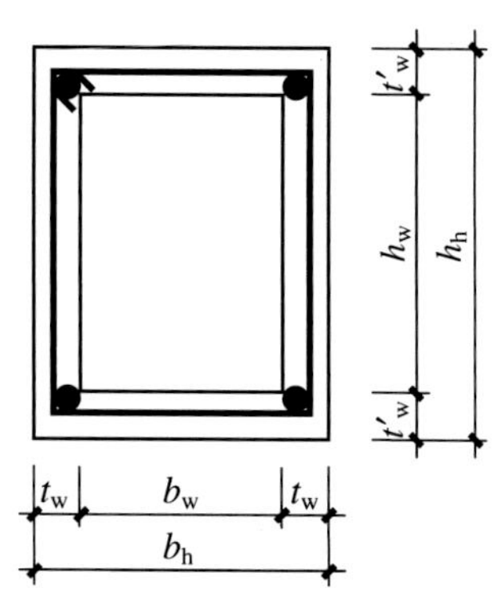

Fig. 8.21 Box section ($t_w \leqslant t'_w$)

For the box section shown in Fig. 8.21, the torsional capacity can be calculated by

$$T_u = 0.35W_t\left(\frac{2.5t_w}{b_h}\right)f_t + 1.2\sqrt{\zeta}\frac{A_{st1}f_{yv}}{s}A_{cor} \tag{8.46}$$

where 2.5 t_w/b_h = the modified coefficient to the contribution of concrete. If 2.5 $t_w/b_h > 1.0$, take 2.5 $t_w/b_h = 1.0$. W_t is calculated by using Eq. 8.18.

8.6 Applications of Calculation Formulae for Torsional Capacities of Members Subjected to Pure Torsion

8.6.1 *Cross-Sectional Design*

For this type of problems, usually the cross-sectional dimensions (b, h, h_0; or b, h, h_0, h'_f, b'_f; or b, h, h_0, h_f, b_f, h'_f, b'_f; or b_h, h_h, t_w, t'_w), strengths of the materials (f_c, f_t, f_y, f_{yv}) and the torque applied on a member are known, and the reinforcement (A_{stl}, A_{st1}) and s are to be calculated. To ensure member safety under the given torque T, the torsional capacity of the warped section should not be less than the torque applied, that is, $T_u \geqslant T$. The design steps are detailed as follows:

1. Check $T \leqslant (0.16\text{–}0.2)\ \beta_c W_t f_c$ to avoid over-reinforced failure. If this condition cannot be satisfied, the cross section should be enlarged.
2. When $T \leqslant T_{u,min}$, the longitudinal reinforcement and stirrups can be provided by the detailing requirements without any calculation.
3. If it is a compound cross section, subdivide the section into several rectangles and calculate the torque that each rectangle can resist.
4. Choose $\zeta = 1.0\text{–}1.3$.
5. Calculate A_{st1}/s by corresponding torsional capacity formulae assuming $T = T_u$.
6. Check $\rho_{st} = \frac{A_{st}}{bs} \geqslant 0.28\frac{f_t}{f_{yv}}$. If not satisfied, take $\frac{A_{st}}{bs} = 0.28\frac{f_t}{f_{yv}}$.

7. Calculate A_{stl} with ζ and A_{st1} from Eq. (8.29).
8. Check $\rho_{stl} = \frac{A_{stl}}{bh} \geqslant 0.85\frac{f_t}{f_y}$. If not satisfied, take $\frac{A_{stl}}{bh} = 0.85\frac{f_t}{f_y}$.

Note: the longitudinal reinforcement should be placed uniformly along the cross-sectional perimeter except at the four corners. Otherwise, local over-reinforcement will make the design unsafe.

Example 8.3 For a reinforced concrete beam of rectangular cross section, the cross-sectional dimensions are $b \times h = 250$ mm $\times$ 500 mm, and the design torque is T = 12 kN·m. The concrete uses C20 (f_c = 9.6 N/mm^2, f_t = 1.10 N/mm^2). The longitudinal bars are HRB335 (f_y = 300 N/mm^2) and the stirrups are HPB300 (f_{yv} = 270 N/mm^2). Calculate the amount of longitudinal bars and stirrups.

Solution The plastic torsional modulus of the cross section is given as:

$$W_t = \frac{250^2}{6} \times (3 \times 500 - 250) = 1.302 \times 10^7 \text{mm}^3$$

1. Check the cross-sectional dimensions
   From Eq. (8.35), we have

$$\begin{aligned} 0.2f_c W_t &= 0.2 \times 9.6 \times 1.302 \times 10^7 = 2.500 \times 10^7 \text{N·mm} \\ &= 25.00 \, \text{kN·m} > T = 12 \, \text{kN·m}, \text{ OK.} \end{aligned}$$

2. Check if the reinforcement should be calculated
   From Eq. (8.37), we have

$$\begin{aligned} 0.7f_t W_t &= 0.7 \times 1.10 \times 1.302 \times 10^7 = 1.003 \times 10^7 \text{ N·mm} \\ &= 10.03 \, \text{kN·m} < T \end{aligned}$$

   So, calculation is required to determine the reinforcement.
3. Calculate the stirrups
   Take ζ = 1.0. A_{cor} = 450 × 200 = 90000 mm^2. If $T = T_u$ is assumed, then from Eq. (8.34), we have

$$\frac{A_{st1}}{s} = \frac{T - 0.35f_t W_t}{1.2\sqrt{\zeta} f_{yv} A_{cor}} = \frac{12 \times 10^6 - 0.35 \times 1.10 \times 1.302 \times 10^7}{1.2 \times 1 \times 270 \times 90000} = 0.240 \, \text{mm}$$

   If choose $\phi 6$ (A_{st1} = 28.27 mm^2), $s = \frac{28.27}{0.240} = 117.79$ mm. So, the stirrups of $\phi 6@100$ are chosen.
   Check the minimum content of stirrups

$$\rho_{st} = \frac{56.54}{250 \times 100} = 0.00226 > 0.28\frac{f_t}{f_{yv}} = 0.28 \times \frac{1.1}{270} = 0.00114, \text{ OK.}$$

4. Calculate longitudinal bars
   From Eq. (8.29) and $u_{cor} = 2 \times (450 + 200) = 1300$ mm, we have

$$A_{stl} = \frac{\zeta f_{yv} u_{cor}}{f_y} \cdot \frac{A_{st1}}{s} = \frac{1.0 \times 270 \times 1300}{300} \times 0.240 = 280.80\,\text{mm}^2$$

   From the detailing requirements, choose 6ϕ10 (A_{stl} = 471 mm^2). The reinforcement of the cross section is shown in Fig. 8.22.

Example 8.4 For a reinforced concrete beam of T section shown in Fig. 8.23, b = 250 mm, h = 500 mm, b'_f = 500 mm, h'_f = 150 mm, and torque T = 14.59 kN·m. The concrete uses C20 (f_c = 9.6 N/mm^2, f_t = 1.10 N/mm^2). The longitudinal bars use HRB335 (f_y = 300 N/mm^2) and the stirrups use HPB300 (f_{yv} = 270 N/mm^2). Calculate the required amount of longitudinal bars and stirrups.

Solution

1. Calculate the plastic torsional modulus

$$W_{tw} = \frac{b^2}{6}(3h - b) = \frac{250^2}{6} \times (3 \times 500 - 250) = 1.302 \times 10^7\,\text{mm}^3$$
$$W'_{tf} = \frac{h'^2_f}{2}(b'_f - b) = \frac{150^2}{2} \times (500 - 250) = 2.813 \times 10^6\,\text{mm}^3$$
$$W_t = W_{tw} + W'_{tf} = 1.302 \times 10^7 + 2.813 \times 10^6 = 1.583 \times 10^7\,\text{mm}^3$$

2. Check the cross-sectional dimensions

$$0.2 f_c W_t = 0.2 \times 9.6 \times 1.583 \times 10^7 = 3.040 \times 10^7\,\text{N·mm} = 30.40\,\text{kN·m} > T = 14.59\,\text{kN·m, OK.}$$

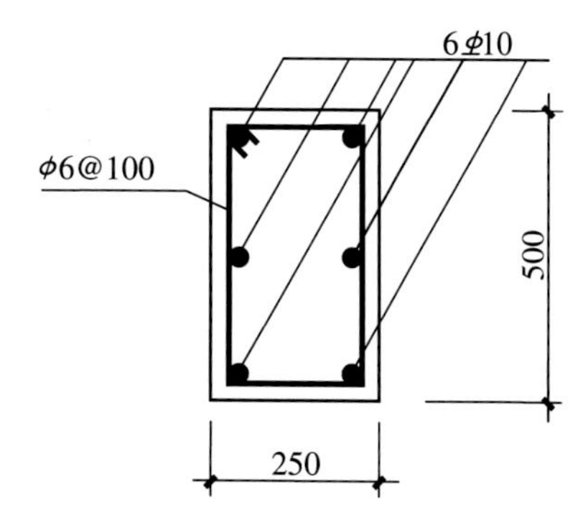

Fig. 8.22 Reinforcement of Example 8.3

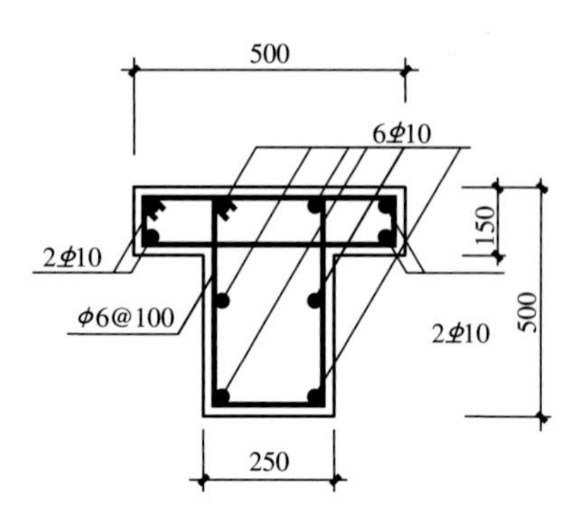

Fig. 8.23 Reinforcement of Example 8.4

3. Check if calculation is required for reinforcement

$$0.7f_tW_t = 0.7 \times 1.10 \times 1.583 \times 10^7 = 1.219 \times 10^7\,\text{N·mm} = 12.19\,\text{kN·m} < T \\ = 14.59\,\text{kN·m}$$

Therefore, calculation is required to determine the reinforcement.

4. Distribution of the torque

$$T_w = \frac{W_{tw}}{W_t}T = \frac{1.302 \times 10^7}{1.583 \times 10^7} \times 14.59 = 12.00\,\text{kN·m}$$

$$T_f' = \frac{W_{tf}'}{W_t}T = \frac{2.813 \times 10^6}{1.583 \times 10^7} \times 14.59 = 2.59\,\text{kN·m}$$

5. Calculate the stirrups and longitudinal bars in the web
   The calculation is the same as that in Example 8.3. Reinforcement is also the same.
6. Calculate the stirrups and longitudinal bars in the flange
   Choose $\zeta = 1.0$. $A_{cor}' = 100 \times 200 = 20,000\,\text{mm}^2$. If $T = T_u$ is assumed, then from Eq. (8.34), we have

$$\frac{A_{st1}'}{s} = \frac{T_f' - 0.35f_tW_{tf}'}{1.2\sqrt{\xi}f_{yv}A_{cor}'} = \frac{2.59 \times 10^6 - 0.35 \times 1.10 \times 2.813 \times 10^6}{1.2 \times 1 \times 270 \times 20000} = 0.233\,\text{mm}$$

To match the stirrups provided in the web, the stirrups in the flange also choose $\phi 6@100$. Because the short dimension of the flange is 150 mm and the total overhung flange length is 250 mm, the minimum content of stirrups is obviously satisfied.

Calculate the longitudinal bars in the flange: From Eq. (8.29) and $u_{cor}' = 2 \times (100 \times 200) = 600\,\text{mm}$, we have

$$A'_{stl} = \frac{\zeta f_{yv} u'_{cor}}{f_y} \cdot \frac{A'_{st1}}{s} = \frac{1.0 \times 270 \times 600}{300} \times 0.233 = 125.82\,\text{mm}^2$$

Choose 4ϕ10 ($A'_{stl} = 314\,\text{mm}^2$). The reinforcement of the cross section is shown in Fig. 8.23.

8.6.2 *Evaluation of Torsional Capacities of Existing Members*

In this type of problems, the cross-sectional dimensions (b, h, h_0; or b, h, h_0, h'_f, b'_f; or b, h, h_0, h_f, b_f, h'_f, b'_f; or b_h, h_h, t_w, t'_w), strengths of materials (f_c, f_t, f_y, f_{yv}), and reinforcement (A_{stl}, A_{st1}, s) are known. Calculate T_u. Two cases will be considered herein.

8.6.2.1 Rectangular or Box Section

1. Determine the area of longitudinal reinforcement to resist torsion based on the principle that the longitudinal bars should be uniformly distributed or biaxially symmetrical.
2. Check $\frac{A_{st}}{bs} \geqslant 0.28\frac{f_t}{f_{yv}}$ and $\frac{A_{stl}}{bh} \geqslant 0.85\frac{f_t}{f_y}$. If either condition is not satisfied, $T_u = 0.7 f_t W_t$.
3. Calculate ζ. If $\zeta > 1.7$, take $\zeta = 1.7$.
4. Calculate T_u by corresponding capacity formulae, and $T_u \leqslant (0.16 - 0.2)\beta_c W_t f_c$ should be satisfied.

8.6.2.2 I- or T Section

Subdivide the compound section into several rectangles, calculate T_{ui}, then $T_u = \Sigma T_{ui}$. The detailed steps are similar to those for rectangular sections and will be omitted for brevity.

Example 8.5 For a member of T section as shown in Fig. 8.24, the cross-sectional dimensions are $b'_f = 400\,\text{mm}$, $h'_f = 120\,\text{mm}$, $b = 250$ mm, $h = 500$ mm. For concrete, $f_c = 13.5\ \text{N/mm}^2$, $f_t = 1.5\ \text{N/mm}^2$. Some longitudinal bars are HRB335 ($f_y = 335\ \text{N/mm}^2$). The stirrups and other longitudinal bars are HPB300 (f_y or $f_{yv} = 300\ \text{N/mm}^2$). Calculate the torsional capacity of the section.

Fig. 8.24 Example 8.5

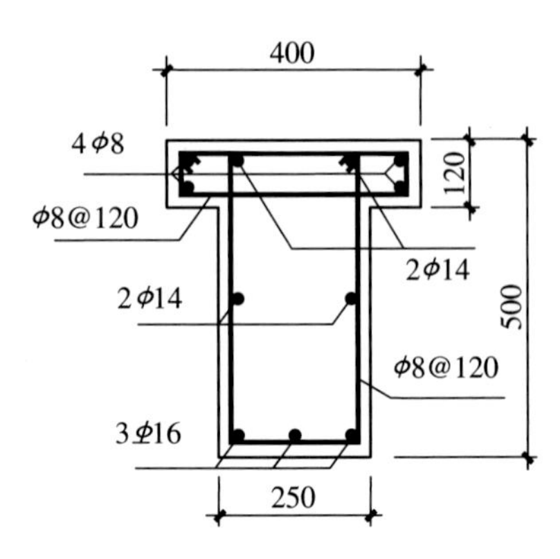

Solution Plastic torsional modulus of the cross section

$$W_{\mathrm{tw}} = \frac{b^2}{6}(3h - b) = \frac{250^2}{6}(3 \times 500 - 250) = 1302.1 \times 10^4\,\mathrm{mm}^3$$
$$W'_{\mathrm{tf}} = \frac{h_{\mathrm{f}}^{\prime 2}}{2}(b'_{\mathrm{f}} - b) = \frac{120^2}{2}(400 - 250) = 108 \times 10^4\,\mathrm{mm}^3$$
$$W_{\mathrm{t}} = W_{\mathrm{tw}} + W'_{\mathrm{tf}} = (1302.1 + 108) \times 10^4 = 1410.1 \times 10^4\,\mathrm{mm}^3$$

1. Torsional capacity of the web, T_{wu}
   The longitudinal reinforcement is not symmetrically provided in the web. Only the symmetric portion contributes to the torsion resistance. Therefore, the torsional longitudinal bars are 6ϕ14, and the total area of the longitudinal bars is $A_{\mathrm{st}l}$ = 923 mm^2. The area of a stirrup leg is A_{st1} = 50.3 mm^2, and the spacing between the stirrups is s = 120 mm. b_{cor} = 200 mm, h_{cor} = 450 mm, A_{cor} = 200 × 450 = 90,000 mm^2, u_{cor} = 2 × (200 + 450) = 1300 mm.

$$\frac{A_{\mathrm{st}}}{bs} = \frac{2 \times 50.3}{250 \times 120} = 0.0034 > 0.28\frac{f_{\mathrm{t}}}{f_{\mathrm{yv}}} = 0.28 \times \frac{1.5}{300} = 0.0014$$
$$\frac{A_{\mathrm{st}l}}{bh} = \frac{923}{250 \times 500} = 0.0074 > 0.85\frac{f_{\mathrm{t}}}{f_{\mathrm{y}}} = 0.85 \times \frac{1.5}{300} = 0.0043$$
$$\zeta = \frac{f_{\mathrm{y}} A_{\mathrm{st}l} s}{f_{\mathrm{yv}} A_{\mathrm{st1}} u_{\mathrm{cor}}} = \frac{300 \times 923 \times 120}{300 \times 50.3 \times 1300} = 1.694 < 1.7$$

$$\begin{aligned} T_{\mathrm{wu}} &= 0.35 f_{\mathrm{t}} W_{\mathrm{tw}} + 1.2\sqrt{\zeta_{\mathrm{w}}}\frac{f_{\mathrm{yv}} A_{\mathrm{st1}} A_{\mathrm{cor}}}{s} \\ &= 0.35 \times 1.5 \times 1302.1 \times 10^4 + 1.2 \times \sqrt{1.694} \times \frac{300 \times 50.3 \times 90000}{120} \\ &= 2.4511 \times 10^7\,\mathrm{N{\cdot}mm} \end{aligned}$$

2. Torsional capacity of the flange, T_{fu}
   Longitudinal bars are 4ϕ8, of which the area is $A'_{stl} = 201\,\text{mm}^2$. The area of a stirrup leg is $A'_{st1} = 50.3\,\text{mm}^2$. The spacing between the stirrups is $s' = 120$ mm, $b'_{cor} = 100\,\text{mm}$, $h'_{cor} = 70\,\text{mm}$, $A'_{cor} = 70 \times 100 = 7000\,\text{mm}^2$, and $u'_{cor} = 2 \times (100 + 70) = 340\,\text{mm}$.

$$\frac{A_{st}}{b's'} = \frac{2 \times 50.3}{150 \times 120} = 0.0056 > 0.28\frac{f_t}{f_{yv}} = 0.28 \times \frac{1.5}{300} = 0.0014$$

$$\frac{A'_{stl}}{b'h'} = \frac{201}{150 \times 120} = 0.0112 > 0.85\frac{f_t}{f_y} = 0.85 \times \frac{1.5}{300} = 0.0043$$

$$\zeta' = \frac{f_y A'_{stl} s'}{f_{yv} A'_{st1} u'_{cor}} = \frac{300 \times 201 \times 120}{300 \times 50.3 \times 340} = 1.410 < 1.7$$

$$T_{fu} = 0.7 f_t W'_{tf} + 1.2\sqrt{\zeta'}\frac{f_{yv} A_{st1}}{s'} A'_{cor} = 0.7 \times 1.5 \times 108 \times 10^4$$
$$+ 1.2 \times \sqrt{1.410} \times \frac{300 \times 50.3}{120} \times 7000 = 2.388 \times 10^6\,\text{N}\cdot\text{mm}$$

Torsional capacity of the whole cross section

$$T_u = T_{wu} + T_{fu} = 2.2875 \times 10^7 + 2.388 \times 10^6 = 2.526 \times 10^7\,\text{N}\cdot\text{mm}$$
$$= 25.26\,\text{kN}\cdot\text{m}$$

8.7 Experimental Results on Members Under Combined Torsion, Shear, and Flexure

The stress states in members under the combined action of bending moment M, shear force V, and torque T are very complicated. Each combination ratio of M, V, and T gives a distinct failure result. If plotting all the failure results in M, V, T space, we can obtain a closed failure surface. The points right on the surface represent failure states, whereas the points inside the space enclosed by the failure surface indicate safe states.

Research has indicated that stress states are to a large extent controlled by the torsion–moment ratio, $\psi = T/M$, and torsion–shear ratio, $\chi = T/(Vb)$, where b is the section width. For different values of torsion–moment ratio and torsion–shear ratio, the members will fail in the following three modes:

1. *Bending-type failure* (Fig. 8.25a). This occurs when the torque–moment ratio ψ is small. The inclined crack due to torsion first appears on the bottom surface, which is under flexural tension, then propagates to the side faces. No crack can be observed on the top surface, which is under flexural compression. At failure,

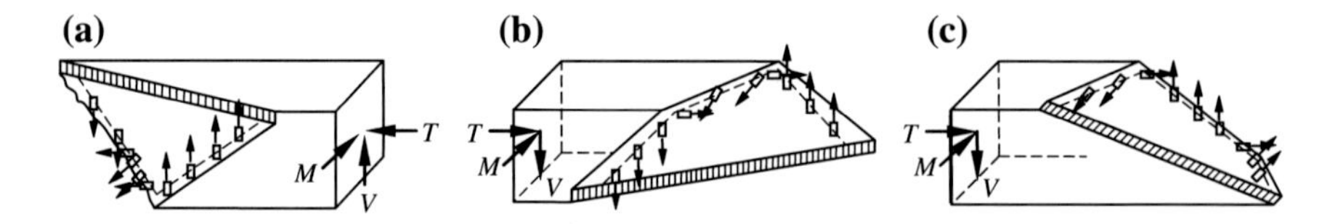

Fig. 8.25 Failure modes of members under combined torsion, shear, and moment. **a** Bending failure mode. **b** Torsion failure mode. **c** Shear–torsion failure mode

both the longitudinal bars and the stirrups that intersect with the spiral cracks yield in tension. The member's top section is under compression.

2. *Torsion failure mode* (Fig. 8.25b). This occurs when both the torsion–moment ratio ψ and the torsion–shear ratio χ are large, and the longitudinal reinforcement at the top is less substantial than at the bottom. Because the bending moment is small, the flexure-induced compressive stress at the member's upper section is not sufficient to counteract the torsion-induced tensile stress. Longitudinal bars at both the top and the bottom of the member are in tension. Furthermore, because the longitudinal reinforcement is not symmetrically provided, the tensile stresses in the top bars are even larger than those in the bottom bars. Therefore, torsional inclined cracks first appear at the top of the member and spread to the side surfaces, whereas the member bottom remains in compression.
3. *Shear–torsion failure mode* (Fig. 8.25c). This occurs when the bending moment is small and the shear and torque are dominant. As this mode of failure occurs, both the shear force and the torque cause shear stress in the cross section. The superposition of the shear stress from different sources makes the shear stress increase on one section side and decrease on the opposite side. Therefore, torsional inclined cracks first appear on the side surface with larger shear stress and then propagate to the top and bottom surfaces. The other side surface is in compression. At failure, both the longitudinal bars and the stirrups that intersect with the spiral cracks yield.

In addition to the above-mentioned three failure modes, shear-type failure may also occur if the shear force is large and the torque is small, which is similar to the shear-compression failure.

8.8 Bearing Capacities of Members Under Combined Torsion, Shear, and Flexure

8.8.1 Bearing Capacities of Members Under Combined Torsion and Flexure

The bearing capacities of members under combined torsion and flexure can be analyzed by the space truss analogy or skew bending theory. For calculation of the flexural capacities, it is assumed that the internal lever arms along the longitudinal bars are constant and equal to the distance between the chords in a space truss. It is also assumed that this lever arm is independent of the amount of reinforcement used. A parabolic torsion–moment interaction curve can be derived by both the space truss analogy and the skew bending theory. In the space truss analogy, torque can be accurately calculated based on the correct internal lever arms; in the skew bending theory, correct results can be obtained only for pure bending if the proper internal lever arm is in place. Based on these results, the interaction curve will be in good agreement with experimental results.

If the longitudinal bars yield at the flexural tension side, the interaction curve is given as:

$$\left(\frac{T_{\mathrm{u}}}{T_{\mathrm{u0}}}\right)^2 = r\left(1 - \frac{M_{\mathrm{u}}}{M_{\mathrm{u0}}}\right) \tag{8.47}$$

If the longitudinal bars yield at the flexural compression side, the interaction curve is given as:

$$\left(\frac{T_{\mathrm{u}}}{T_{\mathrm{u0}}}\right)^2 = 1 + r\frac{M_{\mathrm{u}}}{M_{\mathrm{u0}}} \tag{8.48}$$

where

T_{u}, M_{u} the ultimate torque and bending moment, respectively;

T_{u0}, M_{u0} the ultimate torque corresponding to the pure torsion and the ultimate bending moment corresponding to the pure flexure, respectively; and

r the ratio of the yield force of the tensile reinforcement to that of the compressive reinforcement.

$$r = \frac{A_{\mathrm{s}} f_{\mathrm{y}}}{A'_{\mathrm{s}} f'_{\mathrm{y}}} \tag{8.49}$$

In the interaction curves shown in Fig. 8.26, in the region where the compression reinforcement yields in tension, the tensile stresses in the bars decrease with the increase in the bending moment, so that the yielding of the compression reinforcement is delayed and the torsional capacity is correspondingly raised. In the region where the tension reinforcement yields, the yielding of the tension reinforcement will accelerate with the increase in the bending moment, so that the torsional capacity is reduced. Obviously, these relationships are derived under the condition that the failure initiates from the yielding of the reinforcement. Therefore, the cross section of the member should not be too small nor should the reinforcement be too heavily provided, to ensure that the concrete will not be crushed prior to the yielding of the reinforcement.

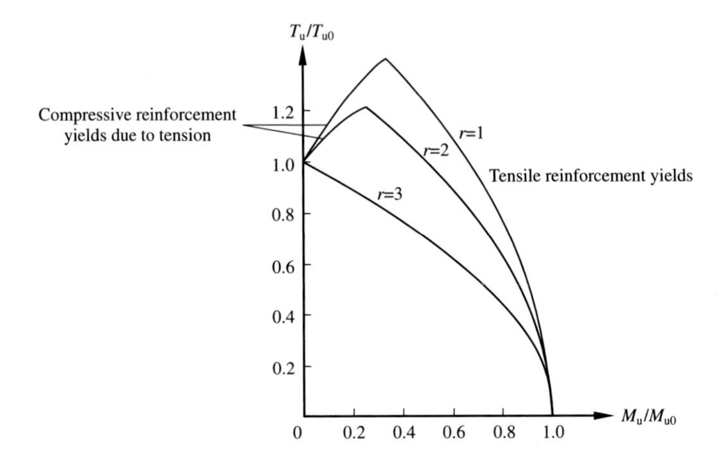

Fig. 8.26 Torsion–flexure interaction curves

When the shear force is low, GB 50010 allows the torsion–flexure coupling to be ignored, and the flexural and torsional capacities can be calculated independently according to the pure flexure and the pure torsion, respectively.

8.8.2 Bearing Capacities of Members Under Combined Torsion and Shear

From previous analysis, we know that the shear and torsional capacities can be expressed as follows:

$$V_u = V_c + V_s \tag{8.50}$$

$$T_u = T_c + T_s \tag{8.51}$$

where

V_s, T_s the reinforcement contributions to the shear resistance and the torsion resistance, respectively; and

V_c, T_c the concrete contributions to the shear resistance and the torsion resistance, respectively.

If the shear resistance and the torsion resistance were calculated independently, the concrete would be considered twice. To avoid this irrationality, the interaction of the shear and the torsion should be taken into account for the concrete, whereas the superposition form of the concrete term and the reinforcement term can still be accepted in the formulae of shear and torsional capacities. Detailed discussion is as follows:

8.8.2.1 Distributed Loads

In cases where the beams of rectangular cross sections are under uniformly distributed loads together with the beams of I- and T sections under arbitrary loads, the reinforcement contribution to the shear resistance, V_s, and to the torsion resistance, T_s, in Eqs. (8.50) and (8.51) can directly copy the corresponding terms in the previously mentioned formulae:

$$V_s = f_{yv}\frac{A_{sv}}{s}h_0 \tag{8.52}$$

$$T_s = 1.2\sqrt{\zeta}\frac{f_{yv}A_{st1}A_{cor}}{s} \tag{8.53}$$

Shear–torsion interaction should be considered in the expressions of V_c and T_c in Eqs. (8.50) and (8.51). It has been indicated by experimental results that the shear–torsion interaction curves for members with or without web reinforcement are in the shape of a quarter of a circle (Fig. 8.27a, b). For the sake of simplicity in practical use, the shear–torsion interaction of concrete is represented by a trilinear relationship as shown in Fig. 8.27c. If the independently calculated concrete resistances

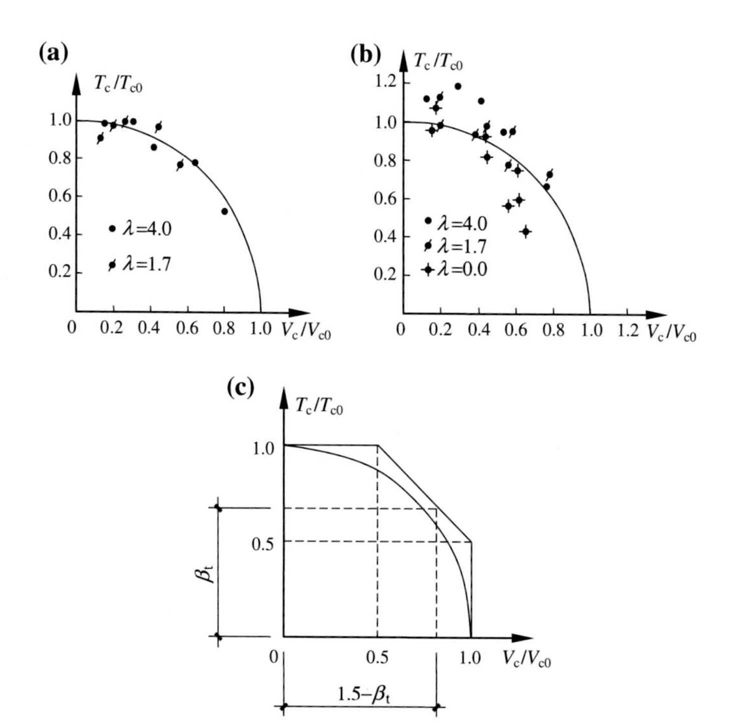

Fig. 8.27 Shear–torsion interaction curve for concrete. **a** Members without web reinforcement. **b** Members with web reinforcement. **c** Simplified model

under pure shear and under pure torsion are $0.7 f_t b h_0$ and $0.35 f_t W_t$, respectively, V_c in Eq. (8.50) and T_c in Eq. (8.51) can be expressed as follows:

$$V_c = 0.7(1.5 - \beta_t) f_t b h_0 \tag{8.54}$$

$$T_c = 0.35 \beta_t f_t W_t \tag{8.55}$$

where
β_t the torsional capacity reduction factor for concrete.

Using T_c and T_{c0} to represent the concrete torsional capacities under combined torsion and shear and under pure torsion, respectively, and using V_c and V_{c0} to represent the concrete shear capacities under combined torsion and shear and under pure shear, respectively, from the trilinear curve in Fig. 8.27c, we have

$$\text{when} \quad \frac{V_c}{V_{c0}} \leqslant 0.5 \quad \frac{T_c}{T_{c0}} = 1.0 \tag{8.56}$$

$$\text{when} \quad \frac{T_c}{T_{c0}} \leqslant 0.5 \quad \frac{V_c}{V_{c0}} = 1.0 \tag{8.57}$$

$$\text{when} \quad \frac{V_c}{V_{c0}} > 0.5 \quad \text{and} \quad \frac{T_c}{T_{c0}} > 0.5, \quad \frac{T_c}{T_{c0}} + \frac{V_c}{V_{c0}} = 1.5 \tag{8.58}$$

In Eq. (8.58), define

$$\beta_t = \frac{T_c}{T_{c0}} \tag{8.59}$$

$$\text{then} \quad \frac{V_c}{V_{c0}} = 1.5 - \beta_t \tag{8.60}$$

Substituting Eq. (8.59) into Eq. (8.60) gives the following:

$$\beta_t = \frac{1.5}{1 + \frac{V_c/V_{c0}}{T_c/T_{c0}}} \tag{8.61}$$

Substituting $T_{c0} = 0.35 f_t W_t$ and $V_{c0} = 0.7 f_t b h_0$ into Eq. (8.51) yields the following:

$$\beta_t = \frac{1.5}{1 + 0.5 \frac{V_c W_t}{T_c b h_0}} \tag{8.62}$$

If the reduction factor, β_t, calculated from the above formula is less than 0.5, takes $\beta_t = 0.5$, that is, then the influence of torques on the shear resistance of concrete is neglected. If β_t is greater than 1.0, take β_t = 1.0, that is, then the influence of shear forces on the torsion resistance of concrete is neglected.

8.8.2.2 Concentrated Loads

For independent members of rectangular cross sections under concentrated loads (including the case where several types of loads are applied, and the shear forces at the supports or connection edges caused by the concentrated loads account for more than 75 % of the total shear forces), the contribution of the reinforcement to the shear resistance, V_s, in Eq. (8.50) and the contribution of the reinforcement to the torsion resistance, T_s, in Eq. (8.51) are still taken as the aforementioned values.

V_c and T_c in Eqs. (8.50) and (8.51) are taken as follows:

$$V_c = \frac{1.75}{\lambda + 1}(1.5 - \beta_t) f_t b h_0 \tag{8.63}$$

$$T_c = 0.35 \beta_t f_t W_t \tag{8.64}$$

β_t also satisfies Eq. (8.61). Substituting $T_{c0} = 0.35\, f_t W_t$ and $V_{c0} = [1.75/(\lambda + 1)]$ $f_t b h_0$ gives the following:

$$\beta_t = \frac{1.5}{1 + 0.2(\lambda + 1)\frac{V_c}{T_c}\frac{W_t}{b h_0}} \tag{8.65}$$

8.8.3 *Capacity Calculation of Members Under Combined Torsion, Shear, and Flexure*

8.8.3.1 Calculation Principles

For members under combined torsion, shear, and flexural, the flexural and torsional capacities are calculated separately, but the shear–torsion interaction should be considered for concrete terms. For I- and T sections, the flanges are generally treated as under pure torsion and the webs are under combined torsion and shear.

8.8.3.2 Minimum Content of Stirrups

Theoretically, the minimum content of stirrups for members under combined torsion, shear, and flexure should take the interaction of shear and torsion into account. For example, it can be expressed as follows:

$$\rho_{\mathrm{svt,min}} = 0.02[1 + 1.75(2\beta_{\mathrm{t}} - 1)]\frac{f_{\mathrm{c}}}{f_{\mathrm{yv}}} \geqslant 0.28\frac{f_{\mathrm{t}}}{f_{\mathrm{yv}}} \tag{8.66}$$

When $\beta_t < 0.5$, take $\beta_t = 0.5$, and Eq. (8.66) gives the minimum content of stirrups for members under pure shear. When $\beta_t > 1$, take $\beta_t = 1$, and Eq. (8.66) gives the minimum content of stirrups for members under pure torsion.

However, for the sake of simplicity and considering practical experience, the minimum content of stirrups can be taken as follows:

$$\rho_{\mathrm{svt,min}} = 0.28\frac{f_{\mathrm{t}}}{f_{\mathrm{yv}}} \tag{8.67}$$

8.8.3.3 Minimum Content of Longitudinal Bars

The minimum content of longitudinal bars for members under combined torsion, shear, and flexure is taken as the superposition of the minimum contents of longitudinal bars for flexural members and torsional members. The minimum content of longitudinal bars for torsional members is taken as follows:

$$\rho_{\mathrm{st}l,\mathrm{min}} = \frac{A_{\mathrm{st}l,\mathrm{min}}}{bh} = 0.6\sqrt{\frac{T_{\mathrm{u}}}{V_{\mathrm{u}}b}} \cdot \frac{f_{\mathrm{t}}}{f_{\mathrm{y}}} \tag{8.68}$$

If $T_u/(V_u b) > 2.0$, take $T_u/(V_u b) = 2.0$, b is the web width of a beam cross section.

8.8.3.4 Maximum Capacity of Cross Sections Under Combined Torsion and Shear

For members under combined torsion, shear, and moment, to ensure that the concrete is not crushed prior to the yielding of the reinforcement, the bearing capacities considering shear–torsion interaction should satisfy the following:

$$\text{when} \quad h_{\mathrm{w}}/b \leqslant 4, \frac{V_{\mathrm{u}}}{bh_0} + \frac{T_{\mathrm{u}}}{0.8W_{\mathrm{t}}} \leqslant 0.25\beta_{\mathrm{c}}f_{\mathrm{c}} \tag{8.69}$$

$$\text{when} \quad h_{\mathrm{w}}/b = 6, \frac{V_{\mathrm{u}}}{bh_0} + \frac{T_{\mathrm{u}}}{0.8W_{\mathrm{t}}} \leqslant 0.2\beta_{\mathrm{c}}f_{\mathrm{c}} \tag{8.70}$$

when $4 < h_w/b < 6$, the value is obtained by linear interpolation,
where

b the web width; and

h_w the web depth. For a rectangular cross section, h_w is taken as the effective depth; For a T section, h_w is taken as the effective depth minus the thickness of the flange; and for an I section, h_w is taken as the clear depth of the web.

8.8.3.5 Minimum Capacity of Cross Sections Under Combined Torsion and Shear

When a member is lightly reinforced, the shear–torsional bearing capacity of the section is given as:

$$\frac{V_u}{bh_0} + \frac{T_u}{W_t} = 0.7f_t \tag{8.71}$$

8.9 Applications of Capacity Formulae for Members Under Combined Torsion, Shear, and Moment

8.9.1 Cross-Sectional Design

For this type of problems, usually the cross-sectional dimensions (b, h, h_0; or b, h, h_0, h'_f, b'_f; or b, h, h_0, h_f, b_f, h'_f, b'_f; or b_h, h_h, t_w, t'_w), the strength of materials (f_c, f_t, f_y, f_{yv}), and the bending moment M, shear force V, and torque T applied on the members are known. The required longitudinal bars and stirrups are to be calculated. To ensure that members do not fail under given loads, it is required that the flexural, shear, and torsional capacities are greater than the corresponding internal forces, that is $M_u \geqslant M$, $V_u \geqslant V$, and $T_u \geqslant T$.

According to the calculation principles for the capacities of the members under combined torsion, shear, and flexure, the flexural capacities are calculated individually, and for the shear and torsional capacities, the interaction between the two should be considered for the concrete term in the formulae. Next, taking members of rectangular cross sections subjected to combined torsion, shear, and flexure as an example, we will detail the design steps.

1. Check $\frac{V}{bh_0} + \frac{T}{0.8W_t} \leqslant (0.2 - 0.25)\beta_c f_c$. If this condition cannot be satisfied, the cross section should be enlarged.
2. Check $\frac{V}{bh_0} + \frac{T}{W_t} \leqslant 0.7f_t$. If yes, the reinforcement can be provided according to the detailing requirement. Otherwise, the reinforcement should be calculated.
3. Calculate β_t by Eqs. (8.62) or (8.65), in which V_c and T_c can be replaced by V and T, respectively.
4. Calculate the shear and torsional reinforcement by corresponding capacity formulae assuming $T = T_u$ and $V = V_u$.
5. Calculate the flexural reinforcement by assuming it is a singly reinforced section.
6. Check the minimum contents of longitudinal bars and stirrups. If neither condition is satisfied, the corresponding reinforcement should be provided according to the minimum content.
7. Detailing of reinforcement. Note that the longitudinal bars to resist the bending moment should be placed in the tension zone and the longitudinal bars to resist the torque should be uniformly placed along the perimeter of the cross section.

Example 8.6 The cross section is the same that as in Example 8.3. The loads applied on the section include a torque T = 12 kN·m, a bending moment M = 90 kN·m, and a shear force due to distributed load V = 90 kN. Calculate the required longitudinal bars and stirrups.

Solution

1. Check the cross-sectional dimensions

$$\frac{V}{bh_0}+\frac{T}{0.8W_t}=\frac{90000}{250\times 465}+\frac{12\times 10^6}{0.8\times 1.302\times 10^7}=1.926\,\text{N/mm}^2$$
$$<0.25b_c f_c=0.25\times 1.0\times 9.6=2.4\,\text{N/mm}^2,\ \text{OK}$$

2. Check whether the shear and torsional reinforcement should be calculated

$$\frac{V}{bh_0}+\frac{T}{W_t}=1.696\,\text{N/mm}^2>0.7f_t=0.7\times 1.1=0.77\,\text{N/mm}^2$$

   The shear and torsional reinforcement should be provided according to calculation.
3. Calculate the torsional reinforcement

$$\beta_t=\frac{1.5}{1+0.5\frac{VW_t}{Tbh_0}}=\frac{1.5}{1+0.5\times\frac{90000\times 1.302\times 10^7}{12\times 10^6\times 250\times 465}}=1.056>1.0$$

   Take β_t = 1.0, i.e., the influence of shear on the torsional capacity of the concrete is not considered. Take ζ = 1.0, then according to the results of Example 8.3, $\frac{A_{st1}}{s}=0.240\text{mm}$; A_{stl} = 280.80 mm^2.
4. Calculate the shear reinforcement
   From Eqs. (8.50), (8.52), and (8.54), we have

$$V=V_c+V_s=0.7(1.5-\beta_t)f_t bh_0+f_{yv}\frac{A_{sv}}{s}h_0$$

   then

$$\frac{A_{sv}}{s}=\frac{nA_{sv1}}{s}=\frac{V-0.7(1.5-\beta_t)f_t bh_0}{f_{yv}h_0}$$
$$=\frac{90000-0.7\times(1.5-1.0)\times 1.1\times 250\times 465}{270\times 465}=0.360$$

   take n = 2, then $\frac{A_{sv1}}{s}=\frac{0.360}{2}=0.180$.
5. Calculate the flexural reinforcement
   Take it as a singly reinforced rectangular cross section

$$f_c bx = f_y A_s, M = f_c bx\left(h_0 - \frac{1}{2}x\right)$$

$$9.6 \times 250x = 300A_s$$

$$90 \times 10^6 = 9.6 \times 250x \times (465 - 0.5x)$$

Solving the equation set yields x = 89.20 mm, A_s = 713.61 mm^2.

6. Total contents of longitudinal bars and stirrups

(1) Calculate the total contents of longitudinal bars and stirrups
Cross-sectional area of the longitudinal bars at the top $= \frac{1}{3}A_{stl} = \frac{280.80}{3} = 93.60\,\text{mm}^2$
Cross-sectional area of the longitudinal bars at the mid-height = 93.60 mm^2
Cross-sectional area of the longitudinal bars at the bottom

$$= \frac{1}{3}A_{stl} + A_s = 93.60 + 713.61 = 807.21\,\text{mm}^2$$

Content of stirrups $= \frac{A_{st1}}{s} + \frac{A_{sv1}}{s} = 0.240 + 0.180 = 0.420\,\text{mm}.$

(2) Check minimum contents of longitudinal bars and stirrups
$A_{s,min}$ = 0.002bh = 0.002 × 250 × 500 = 250 mm^2

$$\frac{45f_t}{100f_y}bh = \frac{45 \times 1.1}{100 \times 300} \times 250 \times 500 = 206.3\,\text{mm}^2$$

A_s = 713.61 mm^2 > $A_{s,min}$ = 250 mm^2, OK.
For torsional longitudinal bars, T_u and V_u in Eq. (8.68) are replaced by T and V, respectively, then

$$A_{stl,min} = 0.6\sqrt{\frac{T}{Vb}} \cdot \frac{f_t}{f_y} \cdot bh = 0.6 \times \sqrt{\frac{12}{90 \times 0.25}} \times \frac{1.1}{300} \times 250 \times 500$$
$$= 200.83\,\text{mm}^2$$

The torsional longitudinal bars are placed in three layers, the area for each layer is 200.83/3 = 66.94 mm^2 < 93.60 mm^2, OK.
For shear and torsional stirrups

$$\frac{A_{sv}}{bs} = \frac{2}{b}\left(\frac{A_{st1}}{s} + \frac{A_{sv1}}{s}\right) = \frac{2}{250} \times 0.420 = 0.00336$$

$$> 0.28\frac{f_t}{f_{yv}} = 0.28 \times \frac{1.1}{270} = 0.00114, \text{OK.}$$

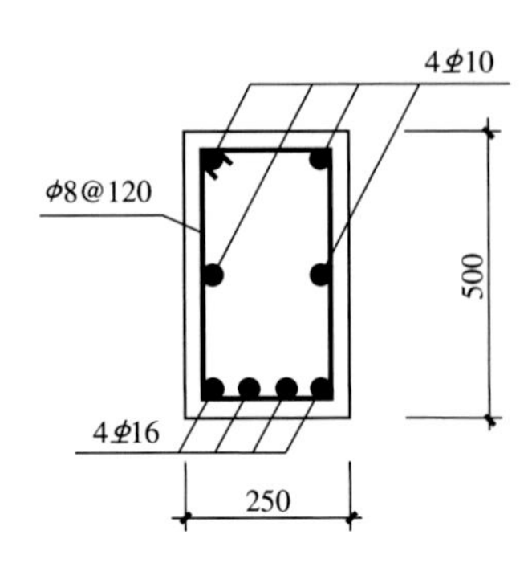

Fig. 8.28 Reinforcement for Example 8.6

(3) Reinforcement
For the longitudinal bars at the top, 2⌀10 (157 mm^2) is chosen; the longitudinal bars at the mid-height are 2⌀10, and the longitudinal bars at the bottom are 4⌀16 (804 mm^2, error is less than 5 %). For the stirrups, choose ϕ8 (50.3 mm^2/leg), and the spacing is calculated as $s = \frac{50.3}{0.420} = 119.76\,\text{mm}$, take s = 120 mm; therefore, the stirrups are ϕ8@120. The reinforcement of the cross section is shown in Fig. 8.28.

8.9.2 Capacity Evaluation of Members Under Combined Torsion, Shear, and Flexure

For this type of problem, usually the cross-sectional dimensions (b, h, h_0; or b, h, h_0, h_f', b_f'; or b, h, h_0, h_f, b_f, h_f', b_f'; or b_h, h_h, t_w, t_w'), the reinforcement, the strengths of materials (f_c, f_t, f_y, f_{yv}), and the loading types are known. M_u, V_u, and T_u are to be evaluated. The detailed steps for members of rectangular cross sections are as follows:

1. Determine the flexure and torsional longitudinal reinforcement A_s and A_{stl}.
2. Calculate M_u from A_s.
3. Choose ζ and calculate A_{st1}.
4. Determine A_{sv1} from the total stirrups and A_{st1}.
5. Assume $0.5 \leqslant \beta_t \leqslant 1.0$.
6. Calculate V_c by Eq. (8.54) and T_c by Eq. (8.55).
7. Calculate V_u by Eq. (8.50) and T_u by Eq. (8.51).
8. Check $\frac{V_u}{bh_0} + \frac{T_u}{0.8W_t} \leqslant (0.2 - 0.25)\beta_c f_c$. If it is not satisfied, adjust the results and recalculate.
9. Check $A_{stl} \geqslant 0.6\sqrt{\frac{T_u}{V_u b}} \cdot \frac{f_t}{f_y} bh$. If it is not satisfied, ignore the contribution of longitudinal bars to the torsional capacity.
10. Check $A_{svt} \geqslant 0.28 \frac{f_t}{f_{yv}} bs$. If it is not satisfied, ignore the contribution of stirrups to the shear and torsional capacities.

Example 8.7 The cross-sectional dimensions, strength of materials, and reinforcement are the same as those in Example 8.6. The shear force is mainly caused by uniformly distributed loads. Calculate the flexural capacity M_u, torsional capacity T_u, and shear capacity V_u.

Solution

1. Calculate M_u
   From Fig. 8.28, we know that

$$A_s = 804 - 157 = 647\,\text{mm}^2 > 0.2\,\%\,bh = 250\,\text{mm}^2$$

$$A_s = 804 - 157 = 647\,\text{mm}^2 > 0.45\frac{f_t}{f_y}bh = 206.3\,\text{mm}^2,\text{OK.}$$

$$9.6 \times 250x = 300 \times 647$$

$$x = 80.875\,\text{mm}$$

$$\xi_b = \frac{0.8}{1 + \frac{f_y}{0.0033E_s}} = \frac{0.8}{1 + \frac{300}{0.0033 \times 2.0 \times 10^5}} = 0.55$$

x = 80.875 mm ⩽ $\xi_b h_0$ = 0.55 × 465 = 255.75 mm, and it is under-reinforced.

$$M_u = f_c bx\left(h_0 - \frac{x}{2}\right) = 9.6 \times 250 \times 80.875 \times \left(465 - \frac{80.875}{2}\right)$$
$$= 82{,}407{,}581\,\text{N} \cdot \text{mm} = 82.408\,\text{kN}\cdot\text{m}$$

2. Calculate V_c and T_c
   Take ζ = 1.3. From Fig. 8.28, A_{stl} = 3 × 157 = 471 mm^2
   From $\zeta = \frac{f_y A_{stl} s}{f_{yv} A_{st1} u_{cor}}$, we get

$$A_{st1} = \frac{f_y A_{stl} s}{f_{yv} u_{cor} \zeta} = \frac{300 \times 471 \times 120}{70 \times 1300 \times 1.3} = 37.160\,\text{mm}^2$$
$$A_{sv1} = 50.3 - 37.160 = 13.140\,\text{mm}^2, A_{sv} = 2 \times 13.140 = 26.280\,\text{mm}^2$$

Assume β_t = 1.0

$$V_c = 0.7(1.5 - \beta_t)f_t bh_0 = 0.7 \times 0.5 \times 1.1 \times 250 \times 465 = 4.476 \times 10^4\,\text{N}$$
$$T_c = 0.35\beta_t f_t W_t = 0.35 \times 1.0 \times 1.1 \times 1.302 \times 10^7 = 5.013 \times 10^6\,\text{N}\cdot\text{mm}$$

3. Calculate V_u and T_u

$$V_u = V_c + V_s = 4.476 \times 10^4 + 270 \times \frac{26.280}{120} \times 465 = 72255\text{N} = 72.255\,\text{kN}$$

$$T_u = T_c + T_s = 5.013 \times 10^6 + 1.2 \times \sqrt{1.3} \times \frac{270 \times 37.160 \times 90000}{120}$$
$$= 15,308,647\,\text{N·mm} = 15.309\,\text{kN·m}$$

4. Check for the maximum capacity and minimum content of stirrups

$$\frac{V_u}{bh_0} + \frac{T_u}{0.8W_t} = \frac{72255}{250 \times 465} + \frac{15.309 \times 10^6}{0.8 \times 1.302 \times 10^7} = 2.091\,\text{N/mm}^2$$
$$< 0.25\beta_c f_c = 0.25 \times 1.0 \times 9.6 = 2.4\,\text{N/mm}^2, \text{OK.}$$

$$\frac{T_u}{V_u b} = \frac{15.309 \times 10^6}{72255 \times 250} = 0.847$$

$$A_{stl} = 471\,\text{mm}^2 > 0.6 \times \sqrt{0.847} \times \frac{1.1}{300} \times 250 \times 500 = 253\,\text{mm}^2, \text{ OK.}$$

$$A_{svt} = 100.6\,\text{mm}^2 > 0.28 \times \frac{1.1}{270} \times 250 \times 120 = 34.2\,\text{mm}^2, \text{ OK.}$$

Comparing the results with those of Example 8.6, it can be found that the calculated torsional capacity based on the actual reinforcement (15.309 kN·m) is greater than the torque in Example 8.6 (12 kN·m). This is because the actual amount of torsional longitudinal bars is greater than the calculated amount. Correspondingly the shear capacity (73.868 kN) is lower than that of Example 8.6 (90 kN). In addition, since both ζ and β_t are arbitrarily chosen, V_u and T_u obtained in this example is not the unique solution.

8.10 Capacities of Members Under Combined Torsion, Shear, Flexure, and Axial Force

8.10.1 Capacities of Members with Rectangular Sections Under Combined Torsion, Shear, Flexure, and Axial Compression

From the development of cracks (Fig. 8.6), it can be found that the failure of members under torsion initiates from the inclined crack caused by torsion-induced shear stress. Similar to members under shear, axial compression on members under torsion can restrain the occurrence and propagation of inclined cracks to some extent, but too large an axial compression will change the failure mode. Experimental results show that axial compression can significantly influence the stress in longitudinal reinforcement and improve the aggregate interlock and dowel action at inclined cracks, so appropriate axial compression can increase the shear capacity of members. GB 50010 has considered this favorable influence and proposed the following equations:

$$V_{\mathrm{u}} = (1.5 - \beta_{\mathrm{t}})\left(\frac{1.75}{\lambda + 1}f_{\mathrm{t}}bh_0 + 0.07N_{\mathrm{c}}\right) + f_{\mathrm{yv}}\frac{A_{\mathrm{sv}}}{s}h_0 \tag{8.72}$$

$$T_{\mathrm{u}} = \beta_{\mathrm{t}}\left(0.35f_{\mathrm{t}} + 0.07\frac{N_{\mathrm{c}}}{A}\right)W_{\mathrm{t}} + 1.2\sqrt{\zeta}f_{\mathrm{yv}}\frac{A_{\mathrm{st1}}A_{\mathrm{cor}}}{s} \tag{8.73}$$

where

- λ the shear-span ratio of the calculated cross section, whose value can be taken as that in Eq. (7.41);
- β_{t} can be calculated by Eq. (8.62) without considering the axial compression; and
- N_{c} the axial compression on the members. If $N_{\mathrm{c}} > 0.3\, f_{\mathrm{c}}A$, take $N_{\mathrm{c}} = 0.3\, f_{\mathrm{c}}A$.

In cross-sectional design, when $T \leqslant (0.175f_{\mathrm{t}} + 0.035N_{\mathrm{c}}/A)\, W_{\mathrm{t}}$, the influence of the torque can be neglected.

8.10.2 *Capacities of Members with Rectangular Sections Under Combined Torsion, Shear, Flexure, and Axial Tension*

Axial tension will always weaken the members, so GB 50010 has proposed the following equations:

$$V_{\mathrm{u}} = (1.5 - \beta_{\mathrm{t}})\left(\frac{1.75}{\lambda + 1}f_{\mathrm{t}}bh_0 - 0.2N_{\mathrm{t}}\right) + f_{\mathrm{yv}}\frac{A_{\mathrm{sv}}}{s}h_0 \tag{8.74}$$

$$T_{\mathrm{u}} = \beta_{\mathrm{t}}\left(0.35f_{\mathrm{t}} - 0.2\frac{N_{\mathrm{t}}}{A}\right)W_{\mathrm{t}} + 1.2\sqrt{\zeta}f_{\mathrm{yv}}\frac{A_{\mathrm{st1}}A_{\mathrm{cor}}}{s} \tag{8.75}$$

The calculation of β_{t} does not consider the influence of axial tension. When the value of the right-hand side of Eq. (8.74) is smaller than $f_{\mathrm{yv}}\frac{A_{\mathrm{sv}}}{s}h_0$; take it as $f_{\mathrm{yv}}\frac{A_{\mathrm{sv}}}{s}h_0$. When the value of the right-hand side of Eq. (8.75) is smaller than $1.2\sqrt{\zeta}f_{\mathrm{yv}}\frac{A_{\mathrm{st1}}A_{\mathrm{cor}}}{s}$, take it as $1.2\sqrt{\zeta}f_{\mathrm{yv}}\frac{A_{\mathrm{st1}}A_{\mathrm{cor}}}{s}$.

Questions

8.1 What kind of members will be subjected to torsion in practical engineering?

8.2 How many stages are there for members of a rectangular cross section under pure torsion from the beginning of loading to failure? What is each stage featured by?

8.3 What is the similarity and difference of the cracks between a member under pure torsion and a member under shear?
8.4 What is the relationship between the inclination of the cracks and the direction of the applied torque for a member of rectangular cross section subjected to pure torsion?
8.5 What are the failure modes and characteristics for members under pure torsion?
8.6 What is the equilibrium torsion? What is the compatibility torsion? Provide some practical examples.
8.7 How is the plastic torsional modulus, W_t, of a rectangular cross section derived? How would you calculate W_t for T- and I-shaped cross sections?
8.8 What is ζ? Why is the value of ζ limited? What is the influence of ζ on failure modes?
8.9 Where will the first crack take place in a rectangular cross section member under pure torsion?
8.10 Consider two members of the same dimensions—one is of high strength concrete and the other is of ordinary concrete. What is the difference in the failure modes between the two members under pure torsion?
8.11 Why does cross section corner spall off under torsion? How would you prevent this kind of failure?
8.12 What is a partly over-reinforced member?
8.13 What is the basis in determining the minimum torsional reinforcement?
8.14 What are the basic assumptions of space truss analogy?
8.15 What are the features of the flexure–torsion interaction curve? How will a curve vary with the reinforcement of longitudinal bars?
8.16 What is the physical meaning of β_t in the formula for calculating torsional capacity? What is represented in the expression of β_t? What factors are considered in this formula?
8.17 What should be taken into consideration in the placement of longitudinal bars and stirrups in torsional members?
8.18 Are the stress states and detailing requirement of torsional stirrups and shear stirrups the same? Why?
8.19 What are the influence of axial compression and axial tension on the shear–torsion capacities of members? Why?

Problems

8.1 For a rectangular cross-sectional member under pure torsion, the cross-sectional dimensions are $b \times h = 300 \times 500$ mm. The longitudinal bars are 4ϕ14 ($f_y = 300$ N/mm^2), and stirrups are ϕ8@150 ($f_{yv} = 270$ N/mm^2). The concrete is C25 ($f_c = 11.9$ N/mm^2, $f_t = 1.27$ N/mm^2). Calculate the maximum torque that the cross section can resist.
8.2 For a reinforced concrete member subjected to pure torsion $T = 4940$ N m, and the cross-sectional dimensions are $b \times h = 200 \times 400$ mm. The concrete is C30 ($f_c = 14.3$ N/mm^2, $f_t = 1.43$ N/mm^2). Grade I reinforcement

(f_y = 210 N/mm^2, f_{yv} = 210 N/mm^2) is used. Calculate the required reinforcement.

8.3 For a reinforced concrete member under combined torsion and flexure, the cross-sectional dimensions are $b \times h = 200 \times 400$ mm. The bending moment is M = 55 kN·m and the torque is T = 9 kN·m. C25 concrete (f_c = 11.9 N/mm^2, f_t = 1.27 N/mm^2) is used. The stirrups are grade I reinforcement (f_{yv} = 270 N/mm^2) and the longitudinal bars use grade II reinforcement (f_y = 300 N/mm^2). Calculate the required reinforcement.

8.4 For a member of cross section $b \times h = 250 \times 600$ mm, according to calculation, the internal forces on the cross section include a bending moment M = 142 kN·m, a shear force V = 97 kN, and a torque T = 12 kN·m. The concrete is C30 (f_c = 14.3 N/mm^2, f_t = 1.43 N/mm^2), Grade I stirrups (f_{yv} = 270 N/mm^2), and grade II longitudinal bars (f_y = 300 N/mm^2) are used. Calculate the required reinforcement (Shear force is mainly caused by uniformly distributed loads).

8.5 For a member of T section subjected to combined torsion, shear, and flexure (shear force is mainly caused by uniformly distributed loads), the cross-sectional dimensions are $b_f' = 400\,\text{mm}$, $h_f' = 80\,\text{mm}$, b = 200 mm, and h = 450 mm. The reinforcement of this section is shown in Fig. 8.29. The internal forces include a bending moment M = 54 kN·m, a shear force V = 42 kN, and a torque T = 8 k N m. The concrete is C20 (f_c = 9.6 N/mm^2, f_t = 1.10 N/mm^2), and the reinforcement is of grade I (f_y = 270 N/mm^2, f_{yv} = 270 N/mm^2). Check whether the cross section can resist the given internal forces (a_s = 35 mm).

8.6 For a reinforced concrete member under combined torsion and shear, the cross-sectional dimensions are $b \times h = 250 \times 500$ mm. The shear force is V = 80 kN and the torque is T = 8 k N·m. C30 concrete (f_c = 14.3 N/mm^2, f_t = 1.43 N/mm^2) and grade I reinforcement (f_y = 270 N/mm^2, f_{yv} = 270 N/mm^2) are used. Calculate the required reinforcement (Shear force is mainly caused by uniformly distributed loads).

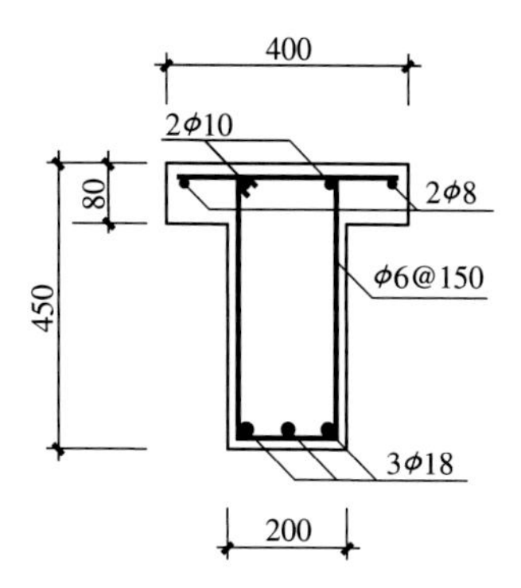

Fig. 8.29 Problem 8.5

8.7 For a reinforced concrete member under combined torsion and moment, the cross-sectional dimensions are $b \times h$ = 200 mm × 400 mm. The applied forces are a bending moment M = 54 kN·m and a torque T = 9.7 kN·m. C20 concrete (f_c = 9.6 N/mm^2, f_t = 1.10 N/mm^2) and grade I reinforcement (f_y = 270 N/mm^2, f_{yv} = 270 N/mm^2) are used. Reinforcement of the section is shown in Fig. 8.30. Check whether the member can resist the given internal forces (a_s = 35 mm).

8.8 For a reinforcement concrete member of I-shaped cross section under pure torsion T = 8.5 kN·m, the cross-sectional dimensions are shown in Fig. 8.31. C20 concrete (f_c = 9.6 N/mm^2, f_t = 1.10 N/mm^2) and grade I reinforcement (f_y = 270 N/mm^2, f_{yv} = 270 N/mm^2) are used. Calculate the torques resisted by the web, compression flange, and tension flange, respectively. And calculate the required stirrups and longitudinal bars in the web.

8.9 A spandrel beam in a reinforced concrete frame is subjected to uniformly distributed loads. The cross-sectional dimensions are $b \times h$ = 250 mm × 400 mm. The internal forces at the support include a torque T = 8 kN·m, a bending moment M = 45 kN·m (the top side of the section is in tension), and a shear force V = 46 kN. C20 concrete (f_c = 9.6 N/mm^2, f_t = 1.10 N/mm^2) and grade I reinforcement (f_y = 270 N/mm^2, f_{yv} = 270 N/mm^2)

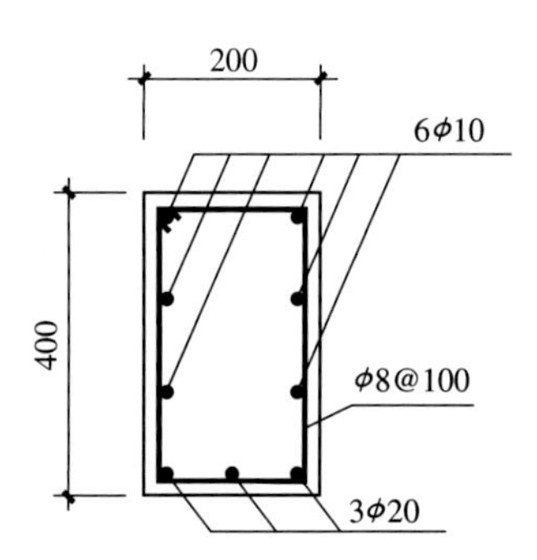

Fig. 8.30 Problem 8.7

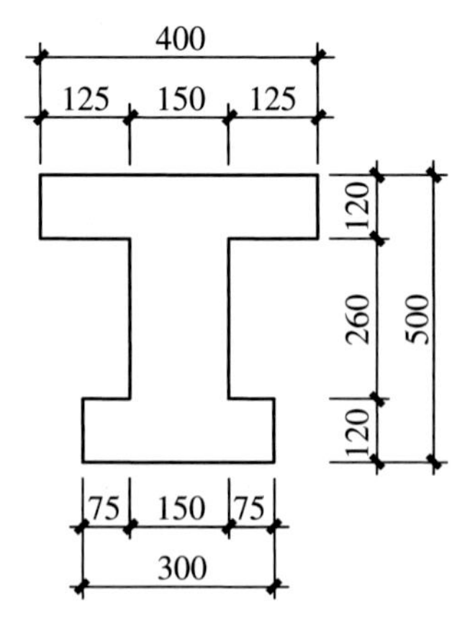

Fig. 8.31 Problem 8.8

Fig. 8.32 Problem 8.10

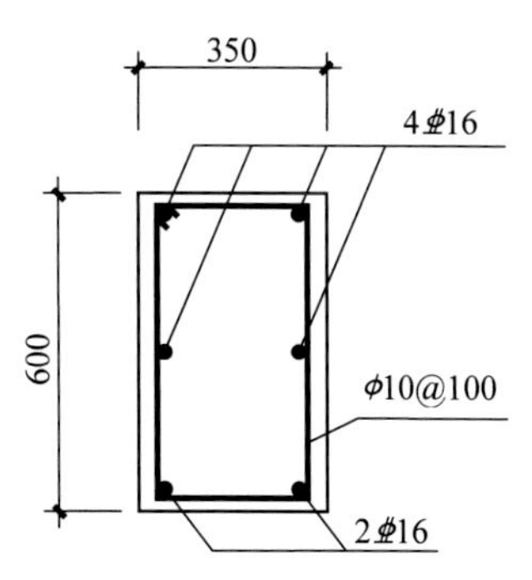

are used. Calculate the required reinforcement in the cross section under combined torsion, shear, and flexure and draw the reinforcement diagram.

8.10 For a member of rectangular cross section under pure torsion, the cross-sectional dimensions and the reinforcement are shown in Fig. 8.32. The concrete is C30 (f_c = 14.3 N/mm^2, f_t = 1.43 N/mm^2). Six longitudinal bars with a diameter of 16 mm are of grade III (f_y = 400 N/mm^2). The stirrups are grade I (f_{yv} = 210 N/mm^2). Calculate the maximum torque that the member can resist.

8.11 Detail the design steps of reinforced concrete members under combined torsion, shear, flexure, and axial compression/tension.

8.12 Detail the capacity evaluation steps of reinforced concrete members under combined torsion, shear, flexure, and axial compression/tension.

Chapter 9
Punching Shear and Bearing

9.1 Punching Shear

9.1.1 Punching Shear Failure in Slabs

9.1.1.1 Failure Features of Punching Shear

Various slabs in reinforced concrete structures, such as flat floor slabs, spread footings, and pile caps (Fig. 1.4), may fail in the patterns shown in Fig. 9.1, under local compressions, which are perpendicular to the slabs and of high intensity. When the failure happens, circumferential cracks will appear on the top and bottom surfaces, and a truncated pyramid surrounded by the circumferential cracks will (or tends to) be separated from the surrounding part in the force direction. This type of failure is called punching shear failure, and the local load is called punching shear load; the dropped-out part is called punching shear failure pyramid.

It is observed that the punching shear failure patterns are different in different types of slabs. According to the distance from the circumferential crack to the edge of the local load, the inclinations of the pyramidal surface may reach 90°, i.e., failure happens along columns, and 40°–60° for footings. The punching shear failure patterns for pile caps are very complicated, and they are significantly influenced by the arrangement of piles and columns.

9.1.1.2 Punching Shear Process

Punching shear failure is inherently related to flexure. The members that failed in punching shear can also be classified as flexural members according to their mechanical behavior. Generally, both the bending moment and the shear force are critical in the vicinity of the truncated pyramid. As concrete slabs fail in punching

X. Gu et al., *Basic Principles of Concrete Structures*,
DOI 10.1007/978-3-662-48565-1_9

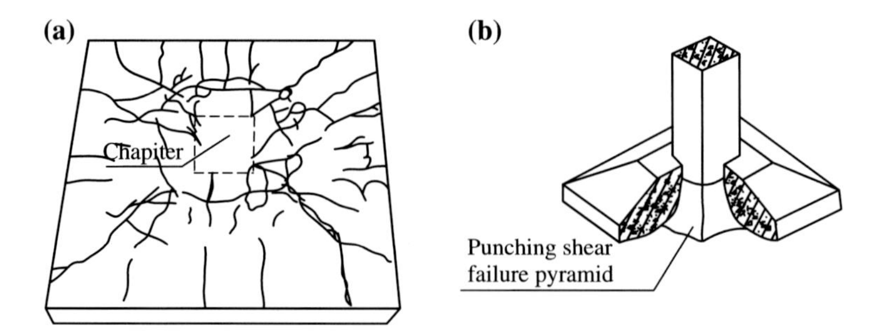

Fig. 9.1 Failure patterns of punching shear. **a** Slab–column connection. **b** Tapered footing

shear, bending cracks, which are perpendicular to the slabs, will usually accompany the appearance of the punching shear cracks.

The development of cracks on the flexural tensile surface of the slab in a slab–column connection is illustrated in Fig. 9.2. Flexural cracks first appear around the perimeter of the column at about 20 % of the ultimate load, and then, radial cracks appear at the corners of the column and extend to the edges of the slab. The pattern of flexural cracks is almost stable at about 70 % of the ultimate load. According to the strain measured inside the slab, the inclined cracks corresponding to punching failure first take place near the mid-height of the slab and gradually spread to the top and bottom surfaces of the slab. The sign of punching failure appears quite late and is shown on the surfaces of the slab. When circumferential cracks are observed on the tensile surface of the slab, this means that the slab has actually failed in punching shear. Therefore, punching shear failure seems to happen suddenly without any warning. Because flexural cracks are always noticed prior to punching shear cracks, the flexural cracks are called primary cracks and the punching shear cracks are referred to as secondary cracks.

Compared to flexural failure, the ultimate deflection of punching shear failure is smaller. Once punching shear failure occurs, the load-bearing capacity drops sharply and the failure is brittle. Experimental results show that not all the longitudinal bars across the punching failure pyramidal surface can yield, that is, the flexural bearing capacity of the slab has not yet been fully reached.

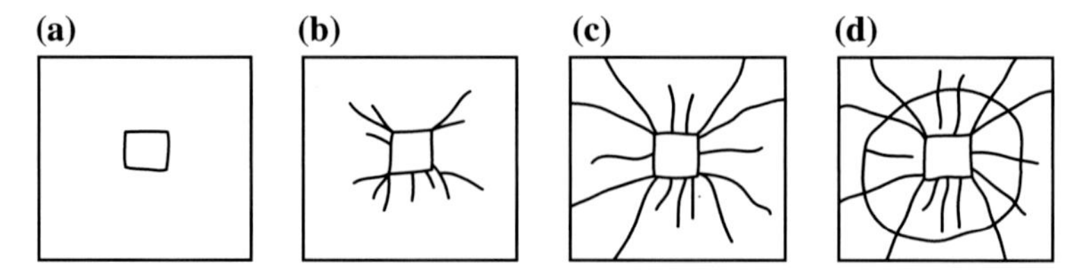

Fig. 9.2 Punching shear failure in a slab–column connection (P_u is the ultimate load of the slab–columns connection). **a** 22 % P_u. **b** 30 % P_u. **c** 70 % P_u. **d** 100 % P_u

Punching shear failure essentially belongs to a shear-type failure, but it is somewhat different from the shear failure in a plane member. The latter involves an inclined crack extending across the entire width of the member, whereas the former involves a truncated pyramid around the local load. Punching shear sometimes is also called two-way shear in foreign references due to its 3D failure characteristics. The failure modes are classified as inclined compression, shear compression, and inclined tension, which are similar to shear failure in beams.

9.1.1.3 Factors Affecting Punching Shear Capacity

A great amount of domestic and foreign experiments have shown that the punching shear capacities of concrete slabs are mainly affected by the following factors:

1. Concrete strength
   Punching shear capacity increases with concrete strength. Foreign experimental results have shown that a linear relationship exists between the punching shear capacity and the square root of the concrete cylinder's compressive strength $\sqrt{f_c'}$. Domestic research indicates that the punching shear capacity is linearly proportional to concrete tensile strength f_t.
2. Effective depth of slab
   If the magnitude and area of a local load are given, the most direct and effective way to raise the punching shear capacity of a slab is to increase the depth of the slab.
3. Loading area
   When the area of a local load is square-shaped, the punching shear capacity depends on the ratio of the loading area width to the effective depth of the slab. For a given depth of the slab, the punching shear capacity is linear to the ratio. The shape of the loading area will also affect the punching shear capacity. When the loading areas have equal perimeters, the punching shear capacity of a circular area will be greater than that of a square area. Research has shown that the punching shear capacity of a square loading area is equal to that of a circular loading area with the diameter being 1.2 times the width of the square. Also, the square loading area has higher punching shear capacity than the rectangular loading area of the same perimeter. This is mainly because shear stresses in slabs along square or rectangular loading areas distribute unevenly as shown in Fig. 9.3.
4. Size effect
   Experimental results have shown that punching shear capacities decrease as the size of specimens increases if the dimensions of the specimens are scaled and the specimens are made from concrete of the same grade. In other words, size affects punching shear.
5. Flexural reinforcement
   It is reported that flexural reinforcement also contributes to punching shear capacity. According to experimental results of circular plates with circumferential reinforcement, the punching shear capacity raises with the increase of the

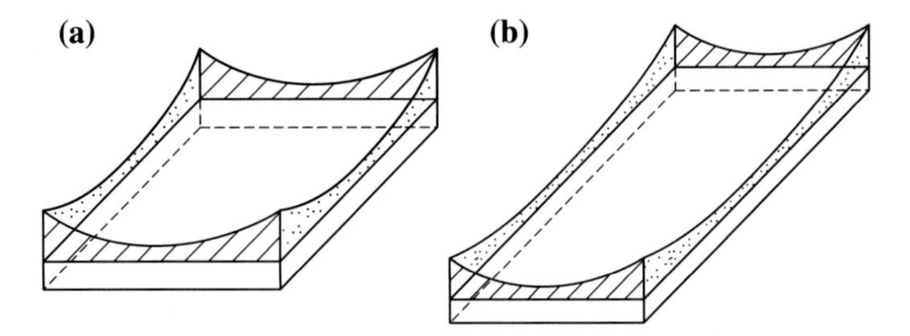

Fig. 9.3 Shear stress distribution in slabs along local loads. **a** Square area. **b** Rectangular area

content of the circumferential reinforcement. The arrangement of the flexural reinforcement will also affect the punching shear capacity. Comparing with the case that equal amount of flexural reinforcing steel bars is uniformly spaced in the slabs, the punching shear capacity will decrease if the reinforcing steel bars are closely spaced in the strips through columns.

6. Boundary conditions
   If the rotation at the boundary of a slab is restrained, the punching shear capacity of the slab will increase. Comparing the experimental results of slabs with rigid restraints to those free from restraints, it is found that the boundary restraints can increase the punching shear capacities of slabs. And the lower the content of reinforcement is, the more significant the increase is. However, the restraints will reduce the ductility of the members.
   Experimental results on punching shear capacities are usually obtained from isolated slab–column connection specimens. However, the boundary conditions of the slabs in this type of isolated connections are quite different from the slabs in real slab–column structural systems. Tests carried out on the whole structure to understand the punching shear capacities of slabs are too expensive. Therefore, an economic and realistic testing scheme for punching shear capacities still needs to be developed.
7. Punching shear-span ratio
   Punching shear-span ratio λ can be defined as the ratio of the net distance between the edge of the punching load and the edge of the support at the perimeter of the slab, a, to the effective depth, h_0. Test results show that the punching shear-span ratio significantly affects punching shear capacity. Figure 9.4 shows that the punching shear capacity will greatly increase with the decrease of the punching shear-span ratio.
   According to the failure patterns observed from specimens of different punching shear-span ratios, it can be assumed that the shape of the punching pyramids varies with the punching shear-span ratios as shown in Fig. 9.5. If the ratio is small ($\lambda < 1$), the shape of the punching pyramids is controlled by the ratio, so that the angles of the pyramidal surfaces vary with the ratios and are greater than 45°. If the ratios are large ($\lambda \geqslant 1$), the shape of the punching pyramids is independent of the ratios, so that the angles of the pyramidal surfaces are equal to 45°.

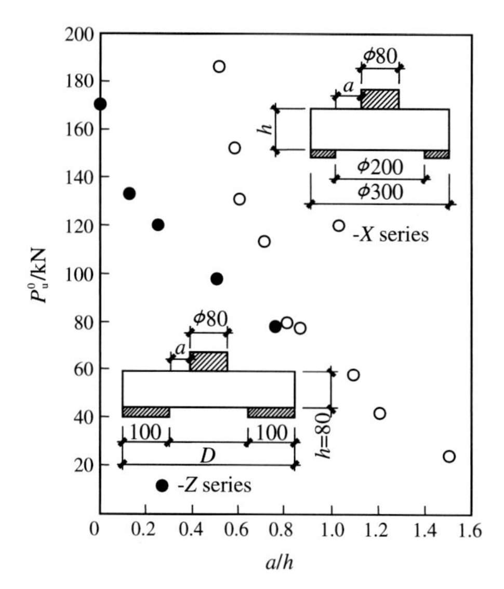

Fig. 9.4 Influence of punching shear-span ratio on punching shear capacity

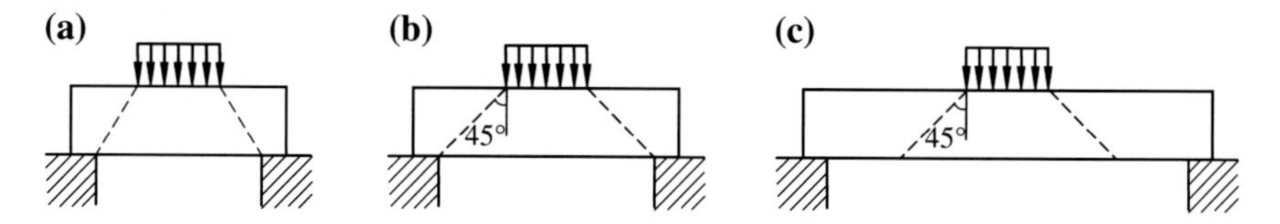

Fig. 9.5 Punching pyramids for different punching shear-span ratios. **a** $\lambda < 1$. **b** $\lambda = 1$. **c** $\lambda > 1$

Test results have shown that punching shear reinforcement makes almost no contribution to the punching shear capacity of slabs when the punching shear-span ratio is small.

9.1.2 Measures to Increase Punching Shear Capacities of Members

9.1.2.1 Column Capital and Drop Panel

For slab–column connections in flat-plate structures (the upper floor in Fig. 1.4), punching shear capacities of slabs can be effectively raised by enlarging the loading area or increasing the depth of the slabs, which can be realized by setting up column capitals (Fig. 9.6) and/or drop panels (Fig. 9.7).

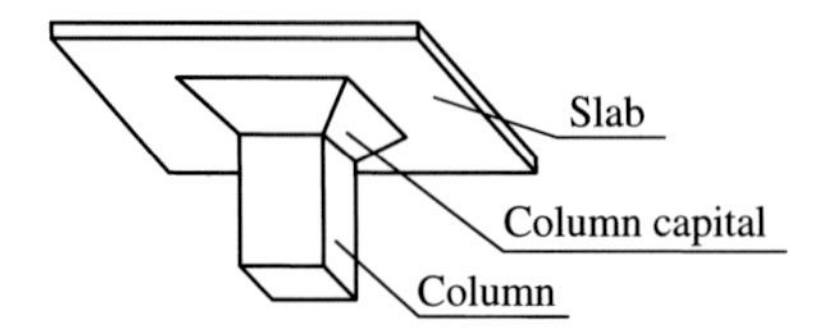

Fig. 9.6 Column capital

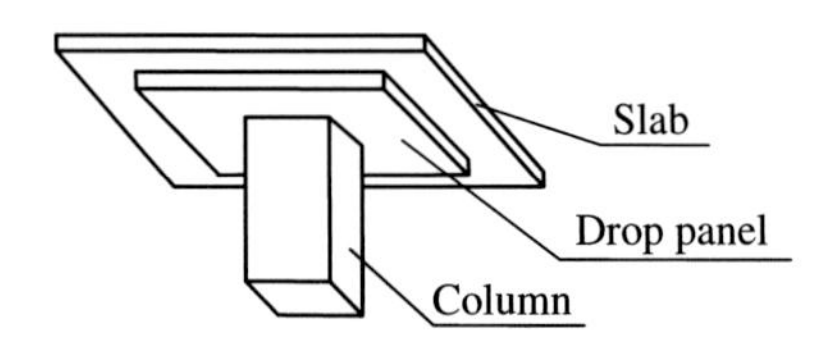

Fig. 9.7 Drop panel

Column capitals can be in arbitrary shapes according to the architects' requirement. It is generally accepted that the local loads spread from the columns in 45° truncated pyramids to the slabs (Fig. 9.5c). Therefore, if the inclination of the side surface of a column capital is larger than 45° or the depth of a drop panel is larger than the extended length, the stresses in the column capital and drop panel are usually very small. The reinforcement in the column capitals and drop panels can be placed according to the detailing requirements without any calculation (Fig. 9.8).

9.1.2.2 Punching Shear Reinforcement

In engineering practice, if the slab depth is limited and column capitals are not preferred due to architectural effects and space saving, the best choice is to provide

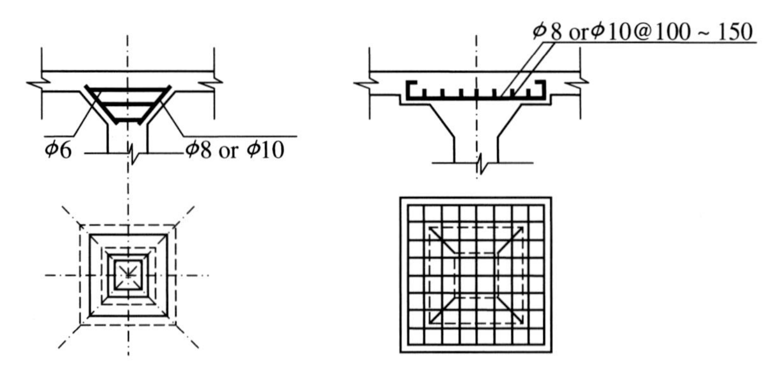

Fig. 9.8 Reinforcement in column capitals and drop panels

punching shear reinforcement in the potential punching shear failure region. The frequently used types of punching shear reinforcement include stirrups and bent-up bars, as shown in Fig. 9.9.

According to domestic and foreign experiments on slabs provided with punching shear reinforcement, the following conclusions can be drawn:

1. Punching shear capacity will raise with the increase of the amount of punching shear reinforcement in slabs.
2. The type and arrangement of punching shear reinforcement have a significant effect on the punching shear capacities. If properly detailed, both bent bars and stirrups can effectively increase the punching shear capacities. It is preferred to arrange the punching shear reinforcement in the vicinity of punching loads and across the area with the highest potential for inclined cracks. The anchorage of punching shear reinforcement is very important because poor anchorage will not benefit to the strength development of the reinforcement.
3. Appropriately reinforced slabs exhibit better ductility at failure. The ultimate deflections are larger than those of slabs without punching shear reinforcement. The magnitude of increases in deflections is proportional to the amount of punching shear reinforcement.

To facilitate the placement of longitudinal reinforcement and reduce the work of passing longitudinal reinforcement through stirrups, domestic and foreign researchers investigate the possible type improvement of punching shear reinforcement. Figure 9.10 illustrates two new types of punching shear reinforcement, which have not been widely accepted.

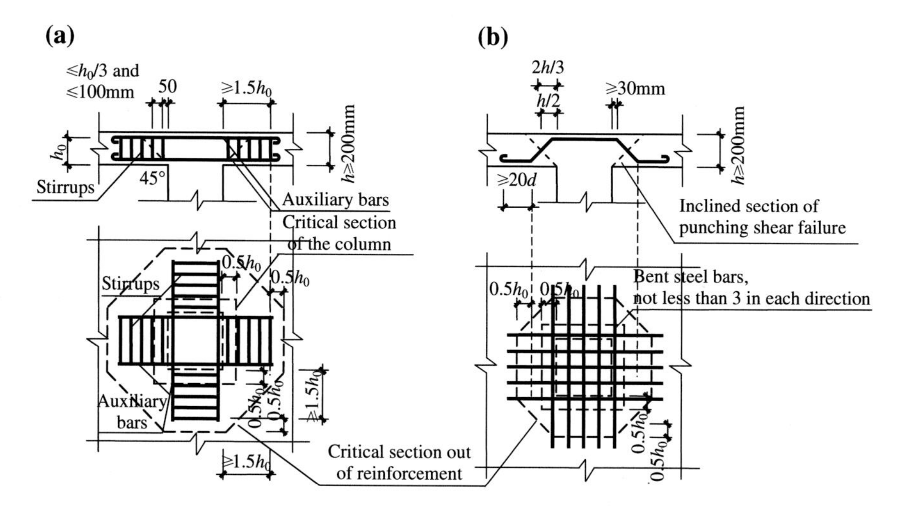

Fig. 9.9 Punching shear reinforcement specified in GB 50010. **a** Stirrups. **b** Bent bars

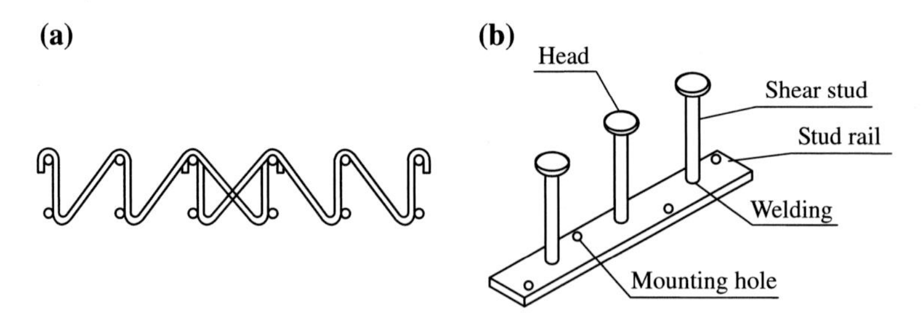

Fig. 9.10 Two types of punching shear reinforcement. **a** Zigzag-form steel bar. **b** Header shear studs

9.1.3 Calculation of Punching Shear Capacities

Different calculation methods of punching shear capacities are given in different codes. For the sake of simplicity, only the evaluation approach in China's code for design of concrete structures (GB 50010) is introduced in this section.

9.1.3.1 Punching Shear Capacities of Slabs Without Punching Shear Reinforcement

The punching shear problem is essentially two-way shear of slabs. Referring to the calculation method of shear capacities of inclined sections for one-way shear slabs (Eq. 7.43) and, assuming that, for concrete slabs without stirrups or bent bars to resist the punching shear (Fig. 9.11), the inclination of the pyramidal surface is 45°, and the punching shear capacities can be estimated by

$$F_{lu} = 0.7\beta_{\mathrm{h}} f_{\mathrm{t}} u_{\mathrm{m}} h_0 \tag{9.1}$$

where

f_{t} tensile strength of concrete;

u_{m} perimeter of the critical section or calculation section, i.e., the calculation perimeter at a distance of $h_0/2$ to the edge of loading area (Fig. 9.11);

h_0 effective depth of cross section; and

β_{h} coefficient considering the influence of sectional depth. If $h \leqslant 800$ mm, $\beta_{\mathrm{h}} = 1.0$; if $h \geqslant 2000$ mm, $\beta_{\mathrm{h}} = 0.9$. If $200 < \beta_{\mathrm{h}} < 800$ mm, β_{h} is determined by linear interpolation.

Because of the requirements for the placement of conduits and ducts, openings are sometimes needed in the vicinity of columns. These openings will decrease the punching shear capacities of the slabs. Based on experimental results, the punching shear capacities of slabs with openings are based on the formulae for slabs without openings, but the critical perimeter is discounted to consider the influence of openings.

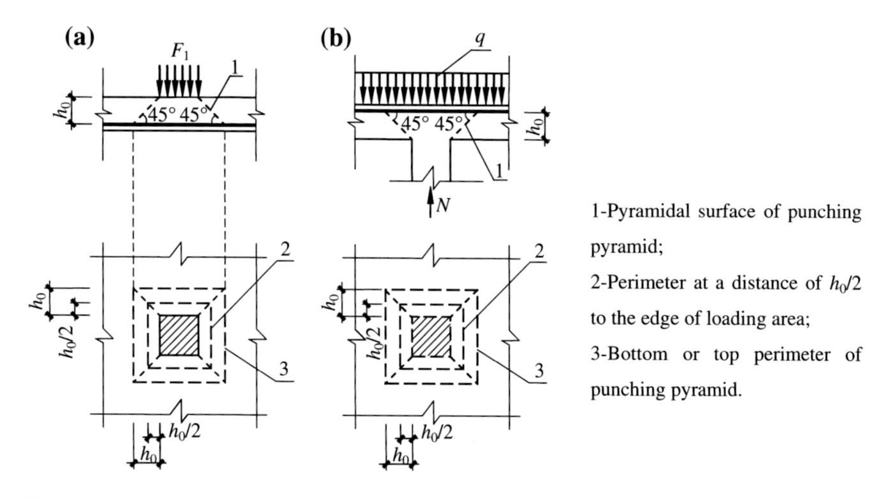

Fig. 9.11 Assumed punching pyramid in GB 50010. **a** Local load. **b** Concentrated reaction force

According to GB 50010, when Eq. (9.1) is used to calculate the punching shear capacity of a slab with an opening located at a distance not greater than 6 times the effective slab depth from the area edge of the concentrated load or the reaction, the length bounded by the straight lines connecting the load area center and the corner points of the opening should be deducted from the critical perimeter (Fig. 9.12). If $l_1 > l_2$, l_2 is replaced by $\sqrt{l_1 l_2}$, and if the center of the opening is near the edge of the column and the maximum width of the opening is less than 1/4 width of the column or 1/2 depth of slab, whichever is smaller, the influence of the opening on the perimeter, u_{m}, can be ignored.

For irregular shapes of the punching load area, such as L-shape, cross, or triangle, the critical perimeters can be referred to in Fig. 9.13.

Figure 9.3 shows that stresses distribute unevenly along square or rectangular load areas. The stress at corners is larger than that in the middle of either side, which is more evident for long sides of rectangular cross sections. In order to take the effect of nonuniformity on the punching shear capacities into consideration, an

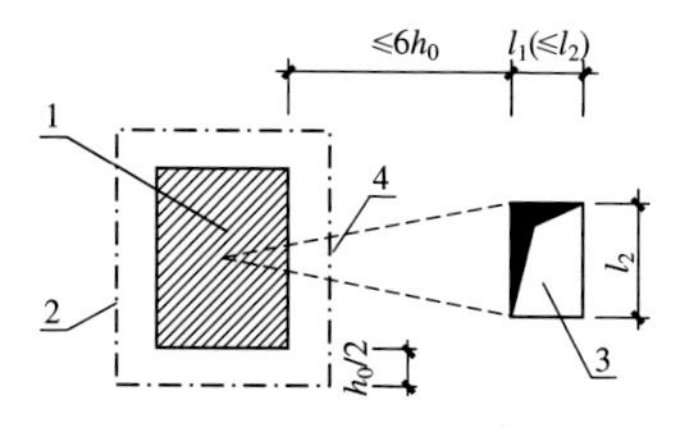

Fig. 9.12 Critical perimeter in the vicinity of an opening

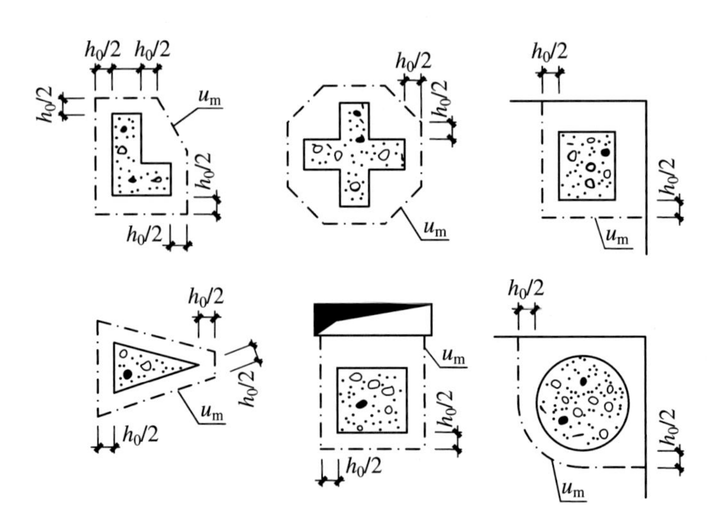

Fig. 9.13 Critical perimeters under irregular punching loads

area shape coefficient of local loads or concentrated reaction forces η_1 is introduced for Eq. (9.1), which is expressed by Eq. (9.2).

$$\eta_1 = 0.4 + 1.2/\beta_s \tag{9.2}$$

where β_s = the ratio of long dimension to short dimension of a rectangular local or concentrated loading area. β_s is preferred to be no larger than 4; if $\beta_s < 2$, take $\beta_s = 2$; for circular area, take $\beta_s = 2$; for other irregular areas, β_s is calculated based on Fig. 9.14.

If the area of local loads or reaction forces is very large, the nonuniformity of shear stress along rectangular or square areas is more evident. In order to consider the negative effect of the nonuniformity, a coefficient, η_2, reflecting the influence of

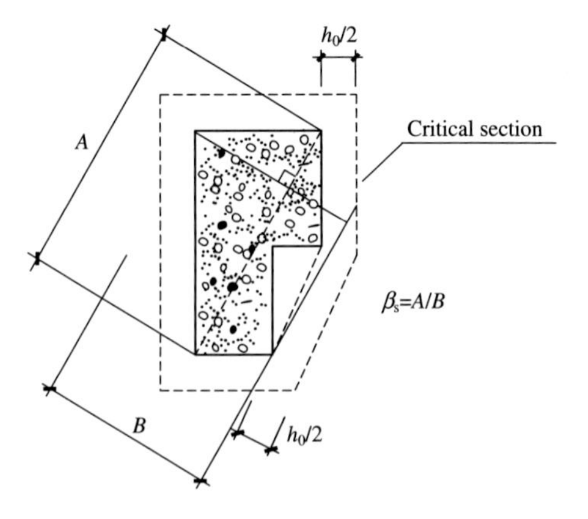

Fig. 9.14 Calculation of β_s for irregular areas

the ratio of the perimeter of the critical section to the effective height of a slab is adopted when $u_m/h_0 \geqslant 20$, as expressed by Eq. (9.3).

$$\eta_2 = 0.5 + \frac{\alpha_s h_0}{4u_m} \tag{9.3}$$

where α_s = the coefficient considering the column type in flat-plate structures. For central columns, take α_s = 40; for edge columns, take α_s = 30; and for corner columns, take α_s = 20.

Since both coefficients η_1 and η_2 are used to reflect the negative effect of unevenly distributed shear stresses, an integrated correction coefficient $\eta = \min(\eta_1, \eta_2)$ is introduced. Consequently, Eq. (9.1) is expressed by

$$F_{lu} = 0.7\beta_h f_t \eta u_m h_0 \tag{9.4}$$

9.1.3.2 Punching Shear Capacities of Slabs with Punching Shear Reinforcement

For concrete slabs with punching shear reinforcement, the punching shear capacities can be estimated by Eq. (9.5) or Eq. (9.6).

If only stirrups are provided

$$F_{lu} = 0.5 f_t \eta u_m h_0 + 0.8 f_{yv} A_{svu} \tag{9.5}$$

and if only bent bars are provided

$$F_{lu} = 0.5 f_t \eta u_m h_0 + 0.8 f_y A_{sbu} \sin\alpha \tag{9.6}$$

where

A_{svu} total area of stirrups intersecting with a pyramidal surface that is 45° to the slab;

A_{sbu} total area of bent-up bars intersecting with pyramidal surface that is 45° to the slab;

f_{yv} tensile strength of stirrups, which should not be greater than 360 N/mm^2;

f_y tensile strength of bent-up bars; and

α angle between bent-up bars and the slab.

In Eqs. (9.5) and (9.6), the coefficients of the concrete terms are taken as 0.5, which are 0.7 times the corresponding coefficient 0.70 in Eq. (9.4). This is because the concrete becomes severely cracked when punching shear reinforcement is fully utilized, and the concrete contribution to the punching shear capacity is mainly from the aggregate interlock and the dowel action of the reinforcement. In addition, because the depth of the slabs with punching shear reinforcement is generally not large, the coefficient β_h is not considered in Eqs. (9.5) and (9.6).

For slabs with punching shear reinforcement, the maximum punching shear capacity is

$$F_{lu,\max} = 1.2 f_t \eta u_m h_0 \tag{9.7}$$

For slabs with punching shear reinforcement, the punching shear failure will possibly take place just outside the region with punching shear reinforcement. In that case, the area bounded by punching shear reinforcement with effective anchorage at the bottom of the slab can be visualized as the loading area, and the critical perimeter is taken at a distance of $0.5h_0$ from the new loading area, i.e., the perimeter of the critical section out of the reinforcement (as shown in Fig. 9.9). Then, the punching shear capacity can be estimated by Eq. (9.4), which is applied in slabs without punching shear reinforcement.

Detailing requirements on punching shear reinforcement are listed as follows:

(1) The slab depth should not be less than 200 mm;
(2) The required stirrups and corresponding auxiliary bars should be placed in the punching pyramid and the region extending out from the column perimeter to a distance of $1.5h_0$, as shown in Fig. 9.9a. The stirrups should be in closed form with a diameter of not less than 6 mm and a spacing not greater than $h_0/3$ and 100 mm;
(3) The required bent-up bars should be placed in the punching pyramid, and the inclination is between 30° and 45° depending on the slab depth as shown in Fig. 9.9b. The bent-up bars should intersect with the pyramidal surface of the punching pyramid, and the intersecting points should be located in (1/2–2/3) h from the column perimeter. The diameter of the bent bars should not be less than 12 mm, and the number of bars should not be less than three in each direction.

Example 9.1 For the center column–slab connection shown in Fig. 9.15, the column section is a square of width $b = 400$ mm. The slab depth is $h = 200$ mm, so the effective depth is $h_0 = 175$ mm. C30 concrete ($f_t = 1.43$ N/mm^2) is used. The total

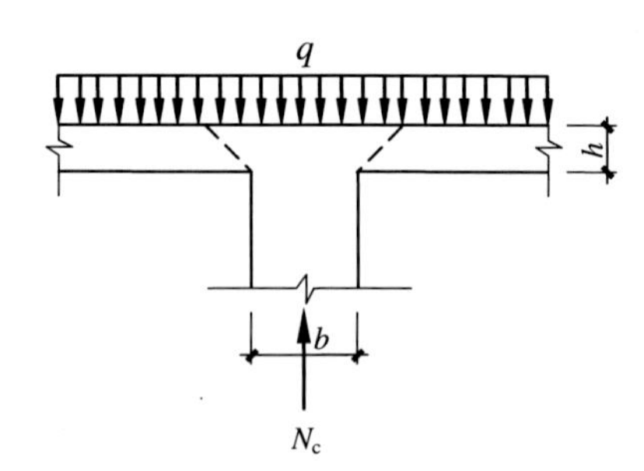

Fig. 9.15 Column–slab connection (Example 9.1)

uniformly distributed load (including self-weight) on the floor is $q = 12$ kN/m^2, and the axial compression in the column is $N_c = 700$ kN. Check the punching shear capacity of the connection.

Solution

1. Without punching shear reinforcement
The punching pyramid is presented in Fig. 9.15 in dashed lines. The local load is

$$F_l = N_c - q(b + 2h_0)^2 = 700 - 12 \times (0.4 + 2 \times 0.175)^2 = 693.3\,\text{kN}$$

The critical perimeter is $u_m = 4(b + h_0) = 4 \times (400 + 175) = 2300\,\text{mm}$

$$\beta_s = \frac{400}{400} = 1 < 2,\ \text{take}\ \beta_s = 2$$
$$\alpha_s = 40$$
$$\eta_1 = 0.4 + \frac{1.2}{\beta_s} = 1.0$$
$$\eta_2 = 0.5 + \frac{\alpha_s h_0}{4u_m} = 0.5 + \frac{40 \times 175}{4 \times 2300} = 1.261$$
$$\eta = \min(\eta_1, \eta_2) = 1.0$$

The punching shear capacity of the slab is

$$F_{lu} = 0.7 f_t \eta u_m h_0 = 0.7 \times 1.43 \times 1.0 \times 2300 \times 175 = 4.029 \times 10^5\,\text{N} = 402.9\,\text{kN} < F_l$$

So, the punching shear capacity of the slab is not satisfactory.
2. With punching shear reinforcement

$$F_{lu,\max} = 1.2 f_t \eta u_m h_0 = 1.2 \times 1.43 \times 1.0 \times 2300 \times 175 = 6.907 \times 10^5\,\text{N} = 690.7\,\text{kN} < F_l = 693.3\,\text{kN}$$

Therefore, the punching shear capacity cannot be satisfied by providing punching shear reinforcement.
3. Using column capital
If the width $B = 750$ mm is assumed for the column capital, the punching shear capacity can be calculated according to the punching pyramid shown in Fig. 9.16 in dashed lines as

$$F_l = N_c - q(B + 2h_0)^2 = 700 \times 10^3 - 12 \times 10^{-3} \times (750 + 2 \times 175)^2 = 685{,}500\,\text{N}$$
$$u_m = 4 \times (750 + 175) = 3700\,\text{mm}$$

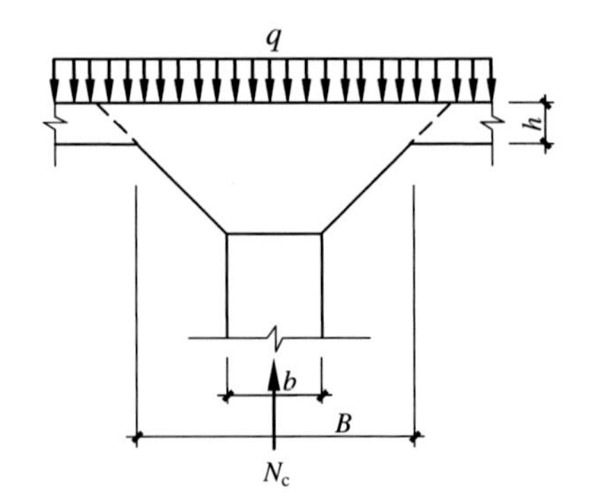

Fig. 9.16 Column–slab connection with column capital (Example 9.1)

$$\beta_s = \frac{750}{750} = 1 < 2, \text{ take } \beta_s = 2$$
$$\alpha_s = 40$$
$$\eta_1 = 0.4 + \frac{1.2}{\beta_s} = 1.0$$
$$\eta_2 = 0.5 + \frac{\alpha_s h_0}{4u_m} = 0.5 + \frac{40 \times 175}{4 \times 3700} = 0.973$$
$$\eta = \min(\eta_1, \eta_2) = 0.973$$
$$F_{lu} = 0.7 f_t \eta u_m h_0 = 0.7 \times 1.43 \times 0.973 \times 3700 \times 175 = 630{,}648\,\text{N} < F_l$$

The punching shear capacity is still not satisfactory. Increase the width of the column capital to B = 900 mm, and then

$$F_l = N_c - q(B + 2h_0)^2 = 700 \times 10^3 - 12 \times 10^{-3} \times (900 + 2 \times 175)^2 = 681{,}300\,\text{N}$$
$$u_m = 4 \times (900 + 175) = 4300\,\text{mm}$$

$$\beta_s = \frac{900}{900} = 1 < 2, \text{ take } \quad \beta_s = 2$$
$$\alpha_s = 40$$
$$\eta_1 = 0.4 + \frac{1.2}{\beta_s} = 1.0$$
$$\eta_2 = 0.5 + \frac{\alpha_s h_0}{4u_m} = 0.5 + \frac{40 \times 175}{4 \times 4300} = 0.907$$
$$\eta = \min(\eta_1, \eta_2) = 0.907$$

$$F_{lu} = 0.7 f_t \eta u_m h_0 = 0.7 \times 1.43 \times 0.907 \times 4300 \times 175$$
$$= 683{,}200\,\text{N} > F_l, \text{ OK.}$$

9.1.4 Eccentric Punching Shear Problems

In the aforementioned punching shear problems, the punching load is axial compression, which results in symmetric stress distribution on the pyramidal surface. If the local loads that cause punching shear failure in slabs are eccentric, the stress on the pyramidal surface must be asymmetrically distributed to balance the eccentricity-induced bending moment. The moment resisted by the asymmetric stress distribution is called unbalanced moment. In practice, the punching shear with eccentricity is more frequently encountered.

In 1960, J. Di Stasio and M.P. Van Buren proposed a model to calculate the shear stress on the critical section of a slab under the eccentric local load. This model assumes a linear distribution of shear stress on the critical perimeter under the combined axial compression and flexure (Fig. 9.17). For slab–column structures, suppose a portion of the moment transferred from the column to the slab, γ_v, is resisted by the shear stress on the critical perimeter, and the remaining portion of the moment, $1-\gamma_v$, is resisted by the flexure of the column and the reinforced slab strip 1.5 times the slab depth on both sides of the column. The fraction of moment that is transferred by shear stresses on the critical section, γ_v, is determined by

$$\gamma_v = 1 - \frac{1}{1+2/3\sqrt{\beta_{cr}}} \tag{9.8}$$

where β_{cr} the ratio of the dimension of the critical perimeter in the direction parallel to the bending moment to that in the direction perpendicular to the moment (Fig. 9.17), that is

$$\beta_{cr} = \frac{c_1+h_0}{c_2+h_0} \tag{9.9}$$

where
h_0 the effective depth of the slab.

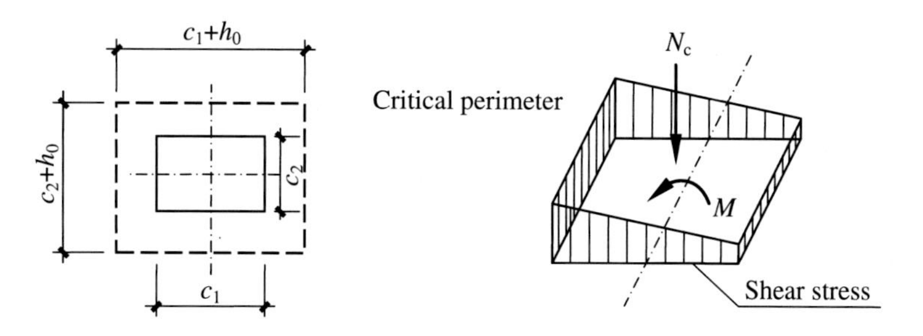

Fig. 9.17 Combined shear and moment transfer model

Under the combined axial compression and bending moment, the shear stress at an arbitrary point on the critical perimeter can be calculated by

$$v_c = \frac{N_c}{A_c} \pm \frac{\gamma_v M c_v}{J_c} \tag{9.10}$$

where

N_c punching load;

M moment transferred at the centroid of the critical perimeter;

A_c area enclosed by critical perimeter;

$$A_c = 2h_0(c_1 + c_2 + 2h_0) \tag{9.11}$$

c_v distance from the centroid of the critical perimeter to the point where the shear stress is to be calculated; and

J_c polar moment of inertia of the critical perimeter about the centroid.

$$J_c = \frac{2h_0(c_1 + h_0)^3}{12} + \frac{2(c_1 + h_0)h_0^3}{12} + 2h_0(c_2 + h_0)\left(\frac{c_1 + h_0}{2}\right)^2 \tag{9.12}$$

If the maximum shear stress in the perimeter reaches the value corresponding to the axial punching shear failure, or the reinforced slab strip of an effective width $c_2 + 3h$ reaches its flexural bearing capacity, the ultimate capacity of the eccentric punching shear is deemed as being attained.

The combined shear and moment transfer model has been accepted by ACI 318 code.

For slab–column connections in eccentric punching shear, the method in the ACI 318 code was adopted by GB 50010. For the unbalanced moment transferred by shear stress, when Eqs. (9.3), (9.5), and (9.6) are used to calculate the punching shear capacities, the local load is replaced by an equivalent concentrated reaction force F_{leq}. The method to calculate F_{leq} can be referred to in the Appendix of the GB 50010 code, which is the design value of equivalent concentrated reaction used for the design of slab–column connection, which will not be detailed herein.

For footings under eccentric punching shear, the calculation method suggested in the GB 50010 code is simple and applicable. For rectangular footings under columns of rectangular cross sections, as shown in Fig. 9.18, the punching shear capacities at the sections where (1) the column intersects with the foundation and (2) one step intersects with another step should be calculated by following equations in GB 50010:

$$F_{lu} = 0.7\beta_h f_t b_m h_0 \tag{9.13a}$$

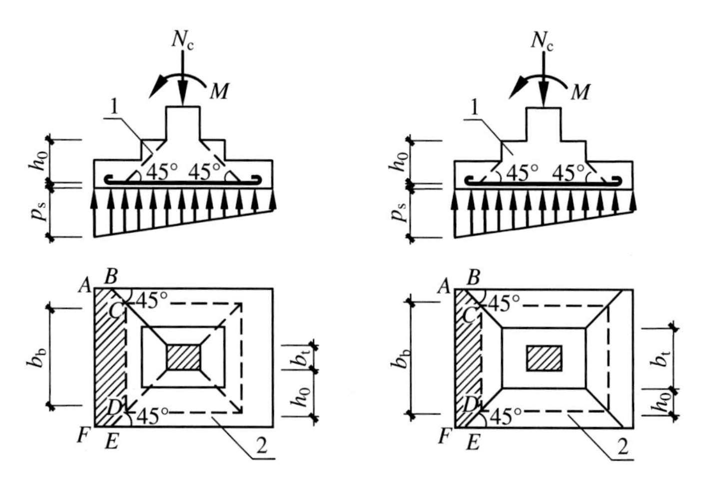

Fig. 9.18 Calculation sections for punching shear capacities of stepped footings. *1* Critical pyramidal surface. *2* Bottom perimeter of punching pyramid

$$F_l = p_s A \tag{9.13b}$$

$$b_m = \frac{b_t + b_b}{2} \tag{9.13c}$$

where

h_0 the effective depth of punching pyramid in a footing;

β_h the same as that in Eq. (9.1);

p_s a subgrade reaction on the unit bottom area of a footing under design load. If the load is eccentric, it takes the maximum unit reaction. The reaction due to self-weight of the foundation and weight of backfill can be deducted from the total reaction;

A the area of the polygon tributary to punching load as shown by the shaded area *ABCDEF* in Fig. 9.18;

b_t the top width of the critical punching pyramid. For the calculation of punching shear capacity at the column–footing intersection, it is taken as the column width, and for the calculation of punching shear capacity at the section where two steps intersect, it is taken as the width of the upper step; and

b_b the bottom width of the critical punching pyramid. For the calculation of punching shear capacity at the column–footing intersection, it is taken as the column width plus twice the effective depth of foundation, and for the calculation of punching shear capacity at the section where two steps intersect, it is taken as the width of the upper step plus twice the effective depth of the corresponding footing.

Example 9.2 For a tapered footing under column as shown in Fig. 9.19, the applied axial load on the top surface of the foundation is N_{c1} = 1647 kN. The bending moment is M_1 = 616.6 kN·m, and the shear force is V_1 = 79.36 kN. C30 concrete and HPB300 reinforcement are used. The effective depth of the footing at the column edge is h_0 = 1165 mm, and the effective depth at where the depth of the steps changes is h_0 = 765 mm. The depth of the soil over the bottom surface of the footing is 1.8 m, and the average density of the soil and the foundation is γ_m = 20 kN/m^2. Calculate the punching shear capacity of the foundation.

Solution The bending moment on the bottom surface of foundation is expressed as

$$M = 616.6 + 79.36 \times 1.2 = 711.8\,\mathrm{kN \cdot m}$$

After the self-weight of foundation and backfill is deducted, the axial load on the bottom surface of the foundation is

$$N_c = 1647 - 4.8 \times 3.6 \times 1.8 \times 20 = 1025\,\mathrm{kN}$$

The subgrade reaction on the unit area of the bottom surface of the foundation can be calculated as

$$\begin{matrix} p_{s,\max} \\ p_{s,\min} \end{matrix} = \frac{N_c}{A} \pm \frac{M}{W} = \frac{1025}{4.8 \times 3.6} \pm \frac{711.8}{\frac{1}{6} \times 3.6 \times 4.8^2} = \begin{cases} 110.8\,\mathrm{kN/m^2} \\ 7.812\,\mathrm{kN/m^2} \end{cases}$$

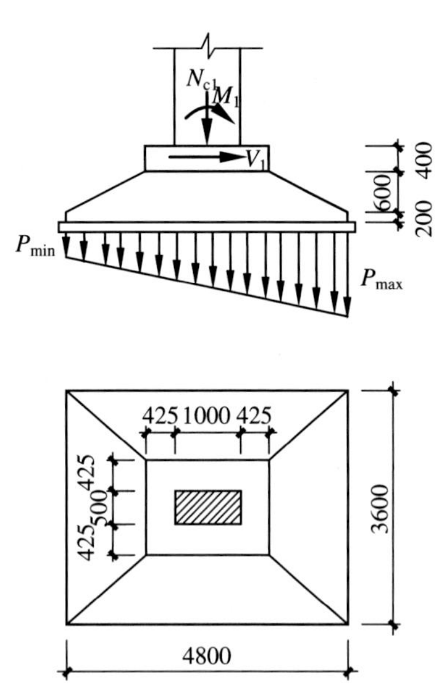

Fig. 9.19 Example 9.2

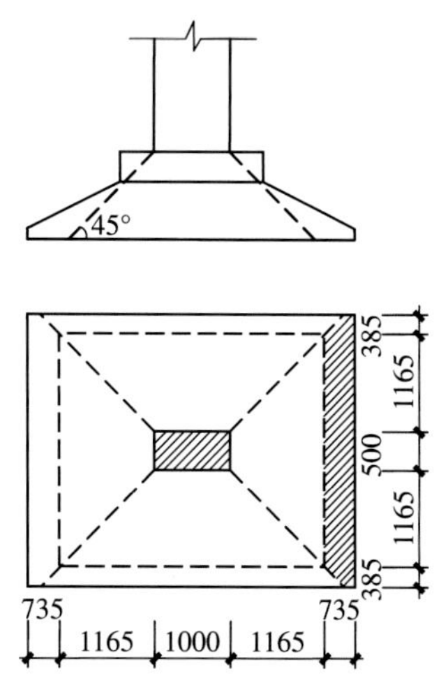

Fig. 9.20 Punching pyramid at the column edge (Example 9.2)

For the calculation of the punching capacity at the column edge, the area of the punching load, as shown by the shaded area in Fig. 9.20, is

$$A = 3600 \times 735 - 385 \times 385 = 2.498 \times 10^6\,\text{mm}^2 = 2.498\,\text{m}^2$$

then $F_l = p_{\text{s,max}}A = 110.8 \times 2.498 = 276.8\,\text{kN}$

$$\beta_\text{h} = 1.0 - \frac{1200 - 800}{2000 - 800} \times (1.0 - 0.9) = 0.9667$$

$$\begin{aligned} 0.7\beta_\text{h} f_\text{t} b_\text{m} h_0 &= 0.7 \times 0.9667 \times 1.43 \times \frac{500 + (3600 - 2 \times 385)}{2} \times 1165 \\ &= 1.877 \times 10^6\,\text{N} = 1877\,\text{kN} > F_l \end{aligned}$$

Therefore, the punching capacity is satisfactory. Similarly, it can be proved that the punching shear capacity at the section where the depths of the steps change is also satisfactory, but the detailed calculation procedure is omitted here for brevity.

9.2 Bearing

Bearing is a common loading state in reinforced concrete structures. The loads transferred from trusses, beams, or slabs to columns, walls, or piers directly or indirectly through the base plate, and the local compression exerted by anchorage devices on the ends of posttensioned prestressed concrete elements (which will be discussed in Chap. 10) are all considered bearing problems.

9.2.1 Mechanism of Bearing Failure

The stress state in the vicinity of a local load is very complex. The distributions of stresses in the longitudinal direction, σ_z, and in the radial direction, σ_r, in a concrete strut under a local load are shown in Fig. 9.21. It is found that the high local compressive stress, σ_z, at the bearing surface gradually becomes uniformly distributed compressive stress across the whole transverse section after a certain transition length, which is approximately equal to the depth of the cross section, $2b$. The transverse stress σ_r changes from compressive stress to tensile stress to zero in the transition length and gets the maximum tensile value at about $(0.5\text{–}1.0)b$ from the bearing surface.

The stress state in the vicinity of the bearing end can be divided into three regions: In Region I, the concrete right beneath the local load expands laterally under the vertical compressive stress. But the lateral expansion is restrained by surrounding concrete, so the concrete is in 3D compression. In Region II, circumferential tensile stress is produced in the surrounding concrete due to the lateral pressure of the core concrete, so the surrounding concrete is in 2D or 3D tension–compression. In Region III, the concrete bounded by horizontal tensile stress and the trajectory of principal compressive stress is in 3D tension–compression. The

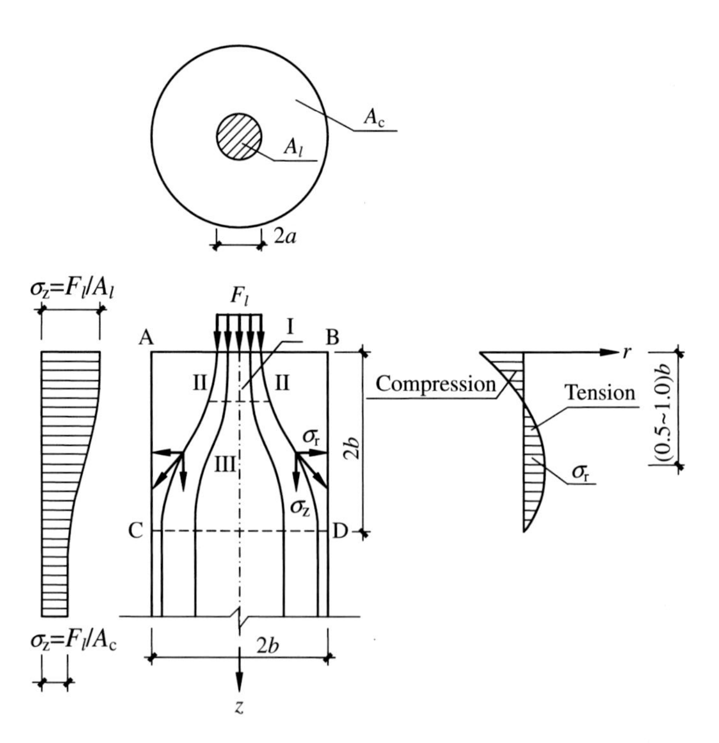

Fig. 9.21 Stress states in the vicinity of bearing area

specific partition of the regions and the magnitude of the stress values as well as the failure patterns mainly depend on the ratio of the sectional area A_c to the bearing area A_l (A_c/A_l). When A_c/A_l is small (generally less than 9), splitting failure is dominant; when A_c/A_l is large (generally larger than 36), indentation failure is more apparent.

Experimental study conducted by Hawkins in 1968 demonstrated the bearing failure mechanism more directly. As the local compressive load F_l increased, crack 1 appeared first under the bearing area due to the effect of σ_r; then, it propagated toward the surface, and crack 2 was generated. Afterward, the resulting conical wedge among the cracks squeezed the peripheral concrete out, causing circumferential tension in the surrounding concrete. If A_c was comparatively small, crack 3 would appear and final failure occurred. If A_c was comparatively large, the wedge might be crushed till failure.

The compressive stress of concrete corresponding to bearing failure is much higher than the axial compressive strength of concrete. Neglecting different bearing failure modes, define the maximum compressive stress in the bearing area (within the region of A_l) as the local compressive strength of concrete and also introduce a strength enhancement factor, β_l, which is defined as the ratio of the compressive strength of concrete under bearing to the axial compressive strength of concrete. It has been shown by experimental results that β_l raises with the increase of A_c/A_l and is approximately proportional to $\sqrt{A_c/A_l}$ (Fig. 9.23).

The strength enhancement factor, β_l, also depends on the relative position of the bearing area within the cross section. When the bearing area is not concentric with the cross section, but instead located close to the edges or corners, the compressive strength of concrete under bearing will increase slightly compared to the axial compressive strength of concrete.

The effective measures to raise the bearing capacities include the following: (1) setting base plates of sufficient stiffness under the local loads to enlarge the bearing area; (2) increasing the concrete grade; and (3) providing indirect reinforcement, including wire fabrics or spirals, as shown in Fig. 9.25, to resist transverse tensile stress, so that the propagation of longitudinal splitting cracks (crack 3 in Fig. 9.22) is restrained and a 3D compressive strength of concrete in Region I shown in Fig. 9.21 is fully developed.

9.2.2 *Calculation of Bearing Capacities*

Because of the complexity of the stress state in the region near the bearing areas, no widely accepted theoretical model has been established for the calculation of bearing capacities. Generally, a half-theoretical and half-empirical method is adopted for the calculation of bearing capacities.

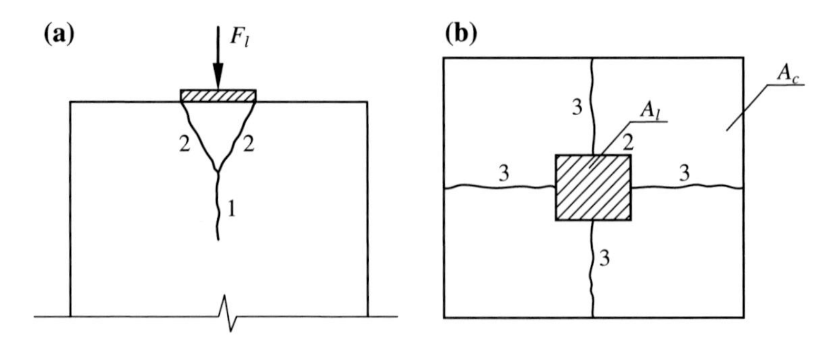

Fig. 9.22 Failure mode of concrete specimens under local compression. **a** Front view. **b** Birds-eye view

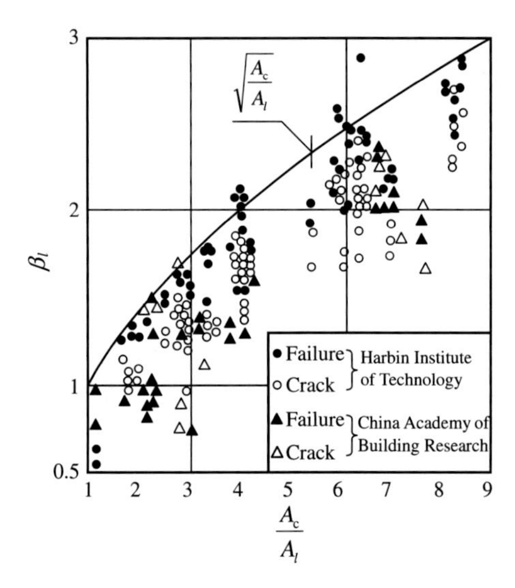

Fig. 9.23 Relationship between strength enhancement factor and A_c/A_l

9.2.2.1 Calculation of Bearing Capacities of Elements Without Indirect Reinforcement

For reinforced concrete elements without indirect reinforcement, the bearing capacities can be calculated by

$$F_{lu} = 0.9\beta_c\beta_l f_c A_{ln} \tag{9.14}$$

where

β_c the influential factor of concrete strength. If the concrete grade is no more than C50, take β_c as 1.0; if the concrete grade is C80, take β_c as 0.8, while the other values are taken by linear interpolation; and

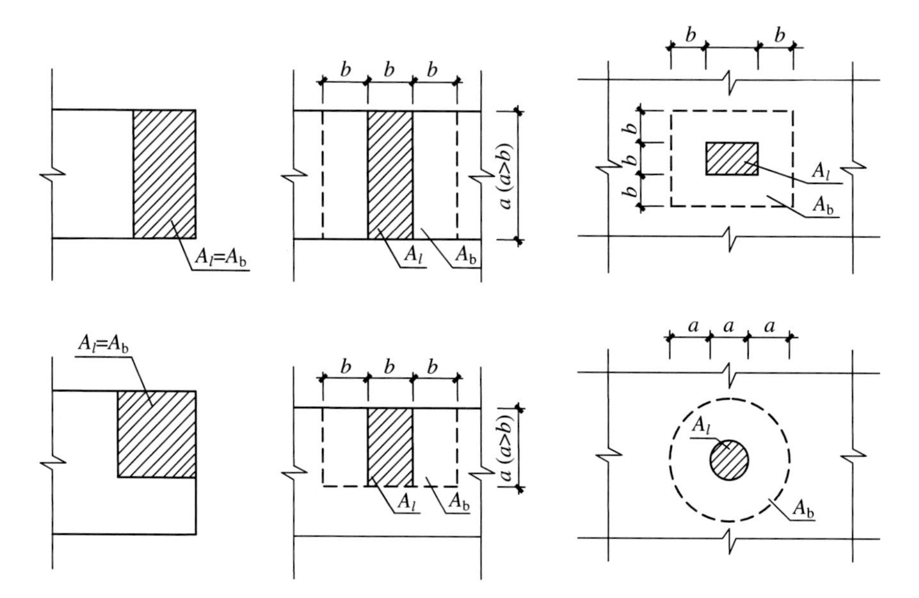

Fig. 9.24 Tributary bottom areas A_b for calculation of bearing capacities

A_{ln} the net bearing area of concrete. For posttensioned prestressed concrete elements, which will be introduced in Chap. 10, the areas of conduits or grooves should be deducted.

The notation β_l in Eq. (9.14) is a strength enhancement factor and defined as Eq. (9.15) based on experimental results.

$$\beta_l = \sqrt{\frac{A_b}{A_l}} \tag{9.15}$$

where

A_l the bearing area of concrete; and
A_b the tributary bottom area for calculation of the bearing capacity.

A_b can be determined according to Fig. 9.24 for common cases. It should be noted that the tributary bottom areas in Fig. 9.24 are symmetric and concentric with the bearing areas.

9.2.2.2 Calculation for Bearing Capacities of Members with Indirect Reinforcement

If the bearing capacity does not satisfy the requirement of Eq. (9.14), indirect reinforcement can be provided to raise the bearing capacity.

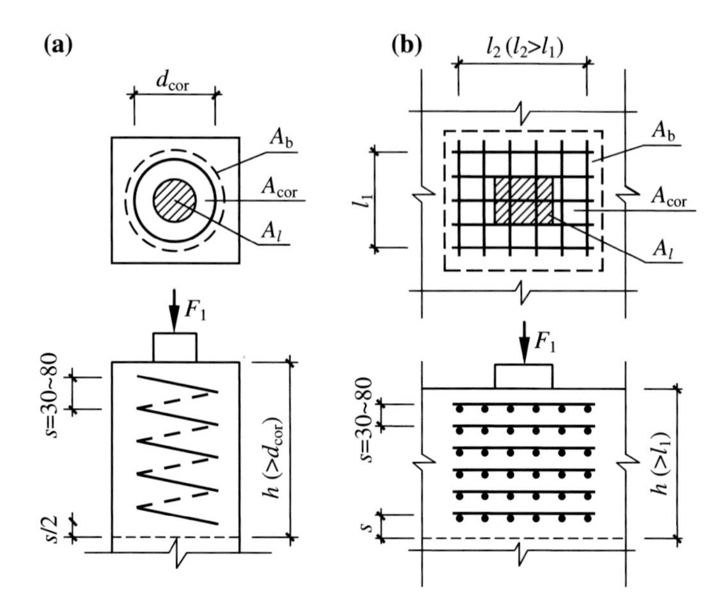

Fig. 9.25 Indirect reinforcement in vicinity of bearing area. **a** Spiral bars. **b** Wire fabrics

1. Detailing requirements for indirect reinforcement
   Generally, there are two types of indirect reinforcement, i.e., spirals and wire fabrics, as shown in Fig. 9.25. Indirect reinforcement should be provided in the region of length h as shown in Fig. 9.25. For a connection with columns, h should not be less than 15 times the diameter of the longitudinal steel bars. The wire fabrics provided should not be less than four pieces, and the spiral bars provided should not be less than four circles.
2. Maximum bearing capacities
   For reinforced concrete structural members with indirect reinforcement, the maximum bearing capacities for the region under local compression is

$$F_{l\mathrm{u}} = 1.35\beta_{\mathrm{c}}\beta_l f_{\mathrm{c}} A_{l\mathrm{n}} \tag{9.16}$$

   where the notations are the same as those specified in Eqs. (9.14) and (9.15).
3. Calculation of bearing capacities
   When concrete is confined by indirect reinforcement, its compressive behavior would be improved. The compressive strength improvement is attributed to the lateral compressive stress σ_{r} provided by the reinforcement in tension. The compressive strength would be improved by $4\sigma_{\mathrm{r}}$ based on Eq. (4.32).
   For spiral reinforcement, the volume content is defined as

$$\rho_{\mathrm{v}} = \frac{4A_{\mathrm{ss1}}}{d_{\mathrm{cor}}s} \tag{9.17}$$

where

A_{ss1} the section area of a single spiral leg;
d_{cor} the diameter of core concrete confined by spirals; and
s the pitch of spirals.

Hence, it is obtained from Eq. (4.35) that the strength of concrete within spirals is increased by $2\alpha f_{yv}\rho_v$, i.e., the concrete strength is improved from f_c to $f_c + 2\alpha f_{yv}\rho_v$. By using Eq. (9.14), the bearing capacities can be expressed by Eq. (9.18) when spirals are adopted and the core area $A_{cor} \geqslant A_l$ (Fig. 9.25).

$$F_{lu} = 0.9(\beta_c \beta_l f_c + 2\alpha \rho_v \beta_{cor} f_{yv}) A_{ln} \quad (9.18)$$

where

β_{cor} the enhancement factor for bearing capacity due to indirect reinforcement. It can still be calculated by Eq. (9.15), but A_b should be replaced by A_{cor}. If $A_{cor} > A_b$, take $A_{cor} = A_b$;
A_{cor} the area of concrete core bounded by the internal surface of spirals, which should be concentric with A_l;
α the reduction factor for restraint of indirect reinforcement to concrete. If the concrete grade is not larger than C50, take the value as 1.0. If the concrete grade is C80, take the value as 0.85. The other values are determined by linear interpolation; and
ρ_v the volume content of indirect reinforcement, which is the volume of indirect reinforcement contained in the unit volume of core concrete.

The other notations are the same as those in Eqs. (9.14) and (9.15).

If wire fabrics are used (Fig. 9.25b), the volume content can be defined similarly as

$$\rho_v = \frac{n_1 A_{s1} l_1 + n_2 A_{s2} l_2}{A_{cor} s} \quad (9.19)$$

where

n_1, A_{s1} the number of bars and the cross-sectional area of a single bar, respectively, along the l_1 direction of a wire fabric;
n_2, A_{s2} the number of bars and the cross-sectional area of a single bar, respectively, along the l_2 direction of a wire fabric; and
s the spacing of indirect reinforcement.

When wire fabrics yield, the lateral compressive stresses applied to concrete in the directions l_1 and l_2 are expressed by

$$\sigma_1 = \frac{n_1 A_{s1} f_{yv}}{l_2 s} \quad (9.20)$$

$$\sigma_2 = \frac{n_2 A_{s2} f_{yv}}{l_1 s} \quad (9.21)$$

If cross-sectional areas of wire fabrics in a unit length of two directions are approximate (less than 1.5 times), the lateral compressive stresses provided by wire fabrics in the two directions are considered to be equal and can be calculated as

$$\sigma_r = \frac{\sigma_1 + \sigma_2}{2} = \frac{n_1 A_{s1} l_1 + n_2 A_{s2} l_2}{2 A_{cor} s} f_{yv} = \frac{1}{2} \rho_v f_{yv} \quad (9.22)$$

When introducing the reduction coefficient α due to the effect of concrete strength, it can be deduced from Eq. (4.32) that the strength of concrete within fabrics is improved from f_c to $f_c + 2\,\alpha f_{yv} \rho_v$. Similarly, Eq. (9.18) can be derived from Eq. (9.14), where A_{cor} is the core area of concrete within wire fabrics. This method is also adopted by the *Code for Design of Concrete Structures* (GB 50010).

Example 9.3 For a posttensioned prestressed tie rod in a prestressed concrete truss, as shown in Fig. 9.26, two prestressed tendons are threaded through a conduit with a diameter of 52 mm and anchored at the ends of the member by an anchorage with a diameter of 100 mm and a base plate 11 mm thick. The concrete grade for the truss is C40 (f_c = 19.1 N/mm^2). The design value of the load applied by each anchorage at the ends is 360 kN. Check the bearing capacities at the ends of the member (details of prestressed concrete structures are given in Chap. 10).

Solution The anchorages are of circular cross sections. Because they are so close to each other and the area of the surrounding concrete is limited, it is difficult to determine the tributary bottom area for the calculation of the bearing capacity according to the aforementioned "concentric, symmetric" principle. To simplify the calculation, the circular anchorages are transformed to equivalent square ones of the same area.

Because the anchorages do not act on the concrete directly, the bearing area should be amplified according to the 45° rigid angle in the steel base plate. Therefore, the width of the equivalent square for an anchorage is

$$b = \sqrt{\frac{\pi}{4}(100 + 2 \times 11)^2} = 108.1\,\text{mm}$$

If the tributary bottom area for each anchorage is determined separately, two areas will partially overlap. Therefore, the area enclosed by the common perimeter of the two squares, as shown in Fig. 9.27, is taken as the bearing area, that is

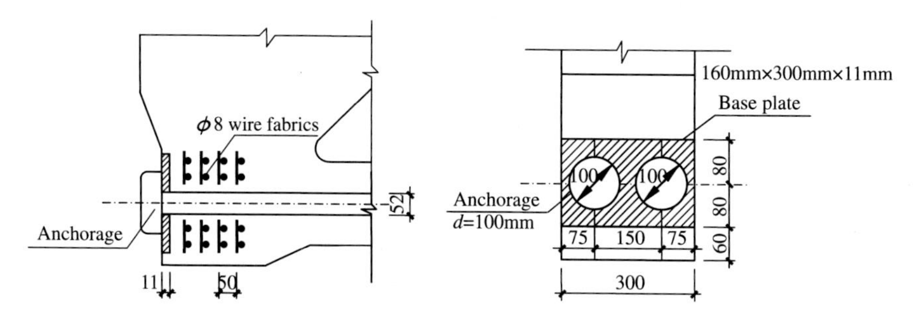

Fig. 9.26 Prestressed concrete tie rod in a truss (Example 9.3)

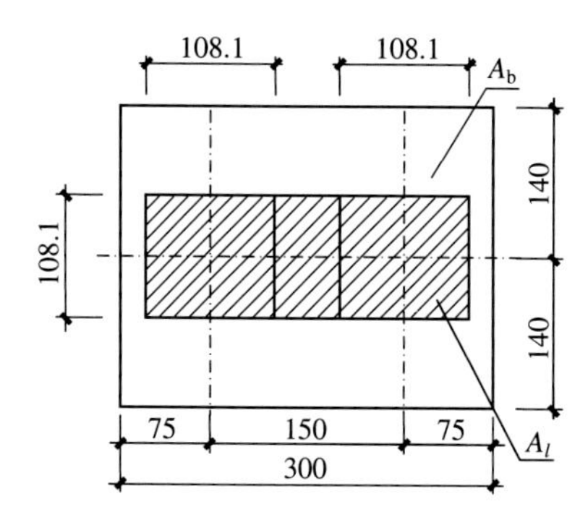

Fig. 9.27 Calculation of A_l and A_b (Example 9.3)

$$A_l = (108.1 + 150) \times 108.1 = 2.790 \times 10^4\,\text{mm}^2$$

Limited by the actual dimension of the member, the tributary bottom area for the calculation of the bearing capacity is

$$A_b = 300 \times (160 + 2 \times 60) = 8.4 \times 10^4\,\text{mm}^2$$

The enhancement factor for the bearing capacity is

$$\beta_l = \sqrt{\frac{A_b}{A_l}} = \sqrt{\frac{8.4 \times 10^4}{2.790 \times 10^4}} = 1.735$$

Deducting the conduit area, the net bearing area will be

$$A_{ln} = 2.790 \times 10^4 - 2 \times \frac{\pi}{4} \times 52^2 = 2.365 \times 10^4\,\text{mm}^2$$

$$\begin{aligned} F_{lu,\max} &= 1.35\beta_l f_c A_{ln} = 1.35 \times 1.735 \times 19.1 \times 2.365 \times 10^4 = 1.058 \times 10^6\,\text{N} \\ &= 1058\,\text{kN} > F_l = 2 \times 360 = 720\,\text{kN} \end{aligned}$$

Therefore, the cross-sectional area is satisfactory.

$$F_{lu} = 0.9\beta_l f_c A_{ln} = 0.9 \times 1.735 \times 19.1 \times 2.365 \times 10^4 = 7.054 \times 10^5\,\text{N}$$
$$= 705.4\,\text{kN} < F_l$$

Indirect reinforcement is required.

$\phi 8$ bars (f_{yv} = 270 N/mm^2) are welded to wire fabrics with a spacing of 50 mm. Then, the volume content of the indirect reinforcement is expressed as

$$\rho_v = \frac{n_1 A_{s1} l_1 + n_2 A_{s2} l_2}{A_{cor} s} = \frac{6 \times 50.3 \times 140 + 4 \times 50.3 \times 280}{140 \times 280 \times 50} = 0.0503$$

The enhancement factor for the bearing capacity with indirect reinforcement is expressed as

$$\beta_{cor} = \sqrt{\frac{A_{cor}}{A_l}} = \sqrt{\frac{140 \times 280}{2.790 \times 10^4}} = 1.185$$

$$\begin{aligned} F_{lu} &= 0.9(\beta_l f_c + 2\rho_v \alpha \beta_{cor} f_{yv}) A_{ln} \\ &= 0.9 \times (1.735 \times 19.1 + 2 \times 0.0503 \times 1.0 \times 1.185 \times 270) \times 2.365 \times 10^4 \\ &= 13.905 \times 10^5\,\text{N} = 1390.5\,\text{kN} > F_l \end{aligned}$$

After indirect reinforcement is provided, the bearing capacity is satisfactory.

Questions

9.1 What is the characteristic of punching shear failure?
9.2 What factors will affect the punching shear capacity?
9.3 How can punching shear capacities be increased by using column capitals and drop panels?
9.4 What are the main types of punching shear reinforcement?
9.5 What is the failure mechanism of bearing failure?
9.6 What are the main types of indirect reinforcement? How does the reinforcement affect the bearing capacity?

Problems

9.1 For a slab–column connection of the same geometry as that in Example 9.1, the axial compressive load is N = 600 kN. If the punching shear reinforcement uses stirrups and bent-up bars, respectively, what is the required punching shear reinforcement? Draw the reinforcement diagram.
9.2 For the same slab–column connection as that in Example 9.1, the axial compression is N = 350 kN. An opening of 200 mm × 200 mm is in the vicinity of the column, as shown in Fig. 9.28. Check whether the punching shear capacity is satisfactory.

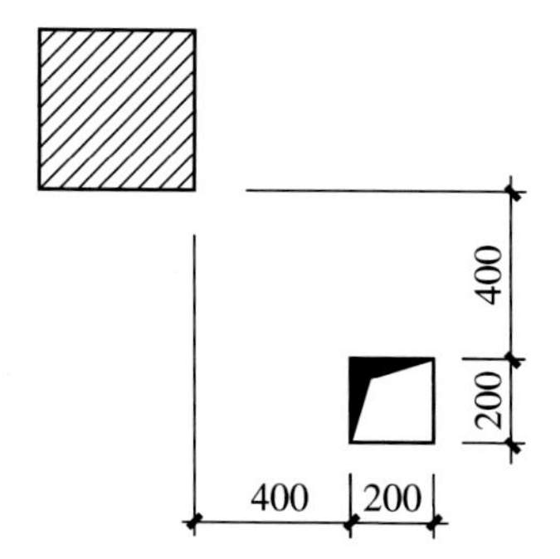

Fig. 9.28 Problem 9.2

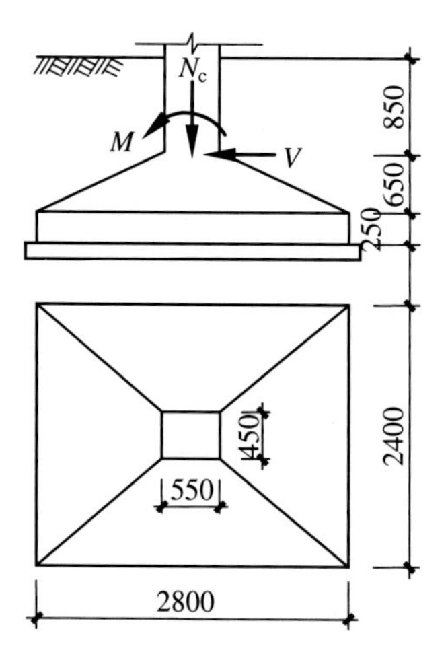

Fig. 9.29 Problem 9.3

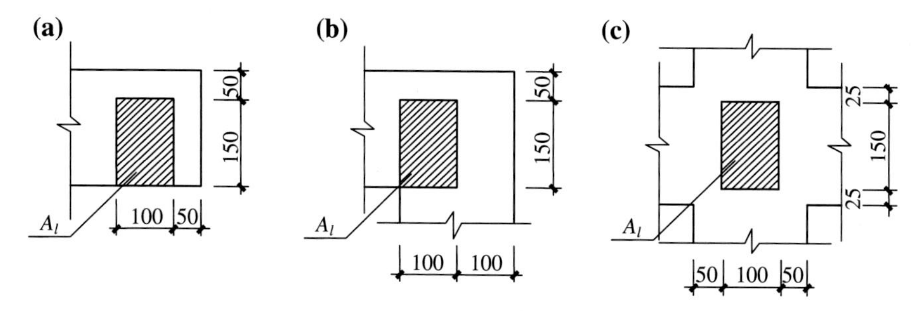

Fig. 9.30 Problem 9.4

9.3 A tapered footing under a column is subjected to an axial load $N_c = 750$ kN, a bending moment $M = 100$ kN·m, and a shear force $V = 25$ kN (Fig. 9.29). The top surface of the footing is above groundwater level. The average density of the backfill over the bottom surface of the footing and the self-weight of the footing is 20 kN/m^2. The concrete grade of the footing is C25 ($f_c = 11.9$ N/mm^2, $f_t = 1.27$ N/mm^2). The effective depth of the footing at the perimeter of the column is $h_0 = 865$ mm. Check whether the depth of the footing is satisfactory.

9.4 Local loads are applied at various positions of concrete cross sections, as shown in Fig. 9.30. For each case, draw the tributary bottom area A_b and calculate the enhancement factor β_l.

Chapter 10
Prestressed Concrete Structures

10.1 Basic Concepts and Materials

10.1.1 Characteristics of Prestressed Concrete Structures

Prestressed concrete is made by introducing compressive stresses (prestress) into areas where an external load will produce tensile stresses. Before the established compressive stresses (prestress) offset, the concrete is not subjected to any tensile stresses.

This process can be visualized by examining the simply supported beam shown in Fig. 10.1. The beam is prestressed by exerting an eccentric compressive force N at the beam ends (Fig. 10.1a) before the beam is subject to any external load. This will cause the beam section to be subjected to compressive stresses at the bottom fiber and tensile stresses at the top fiber. Under the action of the external load P, the stress distribution in the beam is the well-known pattern shown in Fig. 10.1b. Since the beam at this stage generally works within the linear elastic range, the principle of superposition can be applied. The superposition of the two aforementioned stress states will result in the stress state in Fig. 10.1c. It can be seen that, depending on the relative magnitudes of the compressive and tensile stresses produced previously at the bottom fiber, the final stress at the bottom fiber can be either compressive or tensile. Even if this stress is tensile, it is still much smaller than the tensile stress produced by external load without any prestress.

Therefore, if the final tensile stress is less than the tensile strength of the concrete, the beam will not crack.

The principle of prestress can also be illustrated by common examples. Figure 10.2 shows a pile of books. The books in loose state will not even be able to carry their own weight. But after being tightly tied by a string, they are prestressed in compression so that they can carry not only their own weight but also some extra books.

X. Gu et al., *Basic Principles of Concrete Structures*,
DOI 10.1007/978-3-662-48565-1_10

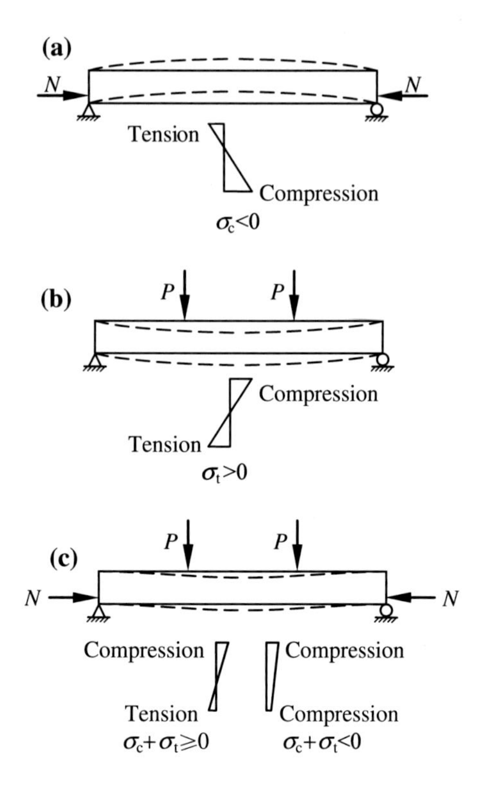

Fig. 10.1 Simply supported prestressed beam

Another example is the wooden bucket shown in Fig. 10.3. The loop will produce compressive stresses in the circumferential direction to compensate for the tensile stresses produced by the water pressure.

Compared with ordinary concrete structures, prestressed concrete structures have the following characteristics:

1. Improving the serviceability behavior of a structure. Prestress can keep the structure free from cracking under external load or make the crack width much smaller. Prestress can also increase the stiffness of the structure, which together with the camber caused by prestress will significantly reduce the deflection of the structure. When the camber is excessive, the structure may even show upward deflection.
2. Reducing the section height and self-weight of the member. This is especially effective for heavily loaded large-span structures, because prestress will effectively increase the limit on span-to-depth ratio.
3. Using high-strength steel more effectively. In ordinary concrete structures, the use of high-strength rebars is usually limited by the requirements on crack width and deflections. In prestressed concrete structures, the high-strength rebars can be prestressed so that they can generally reach their yield strengths at the ultimate limit state of the structure.

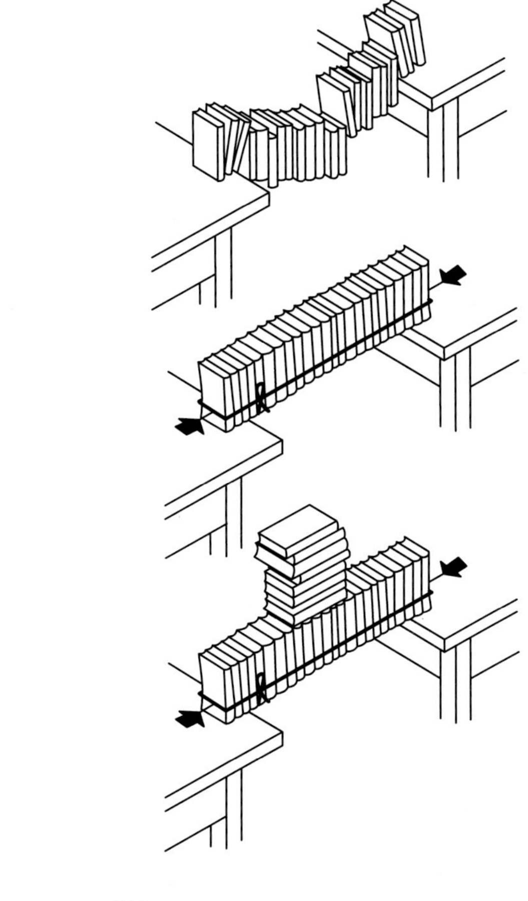

Fig. 10.2 Tied books

Fig. 10.3 Looped bucket

4. Showing better crack closing and deformation restoration properties. After a prestressed structure is unloaded, the cracks can usually be closed and the deformation restored. This will increase the section stiffness and make the structure more durable.
5. Increasing shear strength. Prestress can suppress the appearance of inclined cracks and hence make the shear-compression zone bigger, resulting in higher shear strength of the member. In addition, the web width of a prestressed beam can be designed thinner to further reduce the self-weight.

6. Increasing the fatigue strength. Prestress can effectively reduce the magnitude of stress cycles in rebars. The stress state in the concrete can also be improved with respect to fatigue resistance. These improvements help increase the fatigue life of a structure.
7. Showing good economy. For those structures that can be prestressed, prestressing can save 20–40 % of concrete and 30–60 % of longitudinal reinforcement. And the cost of a prestressed concrete structure is usually half that of the corresponding steel structure.

However, the unit price of materials used in prestressed concrete structures is usually higher. The design and construction of prestressed concrete structures are generally more complicated. And there are still unsolved problems, which need further research work.

10.1.2 Definition of Degree of Prestress

The prestress ratio PPR, which was first put forward by A.E. Naaman, is defined as

$$\mathrm{PPR} = \frac{M_{\mathrm{u,p}}}{M_{\mathrm{u,p+s}}} \tag{10.1}$$

where

$M_{\mathrm{u,p}}$ resistance moment provided by the tendons; and
$M_{\mathrm{u,p+s}}$ resistance moment provided by both the tendons and the ordinary steel rebars.

When the tendons and steel rebars reach their yield strengths, the above formula can be expressed as

$$\mathrm{PPR} = \frac{A_{\mathrm{p}} f_{\mathrm{py}} \left(h_{\mathrm{p0}} - \frac{x}{2}\right)}{A_{\mathrm{p}} f_{\mathrm{py}} \left(h_{\mathrm{p0}} - \frac{x}{2}\right) + A_{\mathrm{s}} f_{\mathrm{y}} \left(h_0 - \frac{x}{2}\right)} \tag{10.2}$$

where

A_{p} and A_{s} areas of tendons and steel rebars, respectively;
f_{py} and f_{y} tensile strengths of tendon and steel rebars, respectively;
h_{p0} and h_0 distance from the tendon center and the steel rebar center to the extreme fiber in the compression zone, respectively; and
x height of the concrete compression zone.

If $h_{p0} = h_0$, then Eq. (10.2) becomes

$$\mathrm{PPR} = \frac{A_p f_{py}}{A_p f_{py} + A_s f_y} \tag{10.3}$$

The degree of prestress of a flexural section is defined as

$$\lambda_p = \frac{M_0}{M} \tag{10.4}$$

where

M_0 decompression moment, which is the moment that makes the prestress of the extreme tension fiber zero; and

M moment at the critical section under characteristic loading.

Similarly, the degree of prestress of a tensioned member is defined as

$$\lambda_p = \frac{N_{t0}}{N_t} \tag{10.5}$$

where

N_{t0} decompression axial tensile force, which makes the prestress at the critical section zero; and

N_t axial tensile force caused by characteristic loading combination.

10.1.3 Grades and Classification of Prestressed Concrete Structures

In the *Recommendations for the Design of Partially Prestressed Concrete Structures* issued by China Civil Engineering Society in 1986, reinforced concrete is classified into fully prestressed, partially prestressed, and reinforced categories according to different degrees of prestressing.

Partially prestressed concrete is further classified into classes A and B. Class A includes members in which the tensile stresses in the precompressed zone shall not exceed the allowable values. Class B includes members in which the tensile stresses in the precompressed zone are allowed to exceed the allowable values, but the crack width cannot exceed the corresponding allowable value.

The degree of prestress can vary between fully prestressed concrete and reinforced concrete. Therefore, the degree of prestress can be used as a unifying condition for the classification of prestressed concretes. If $\lambda_p \geqslant 1.0$, the concrete is fully prestressed; if $\lambda_p = 1.0$, the concrete is nonprestressed; and if $0 < \lambda_p < 1.0$, the concrete is partially prestressed.

The degree of prestress can also be expressed by the stress ratio K_{f0}, which is defined as

$$K_{f0} = \frac{\sigma_{pc}}{\sigma_t} \quad (10.6)$$

where

σ_{pc} effective precompressive stress in concrete; and

σ_t tensile stress in concrete caused by service loads.

10.1.4 Types of Prestressed Concrete Structures

10.1.4.1 Classification by Prestressing Methods

The two basic prestressing methods are pretensioning and posttensioning. The process of pretensioning is as follows (Fig. 10.4a). First, the tendons are tensioned to the required stresses on a pretensioning bed or a steel form and anchored. Then, the concrete is cast. When the concrete gets sufficient strength (usually at least 70 % of its design strength), the tendons are cut. And the concrete is prestressed by the contraction of the tendons. In this case, the transfer of prestress is mainly by the bond between tendon and concrete. Sometimes, additional anchors are provided to help with the transfer of prestress.

Pretensioning can be done either between two abutments or on a steel form. In the former case, the tendons are usually straight, but sometimes, the tendons can be bent in either a vertical or horizontal direction.

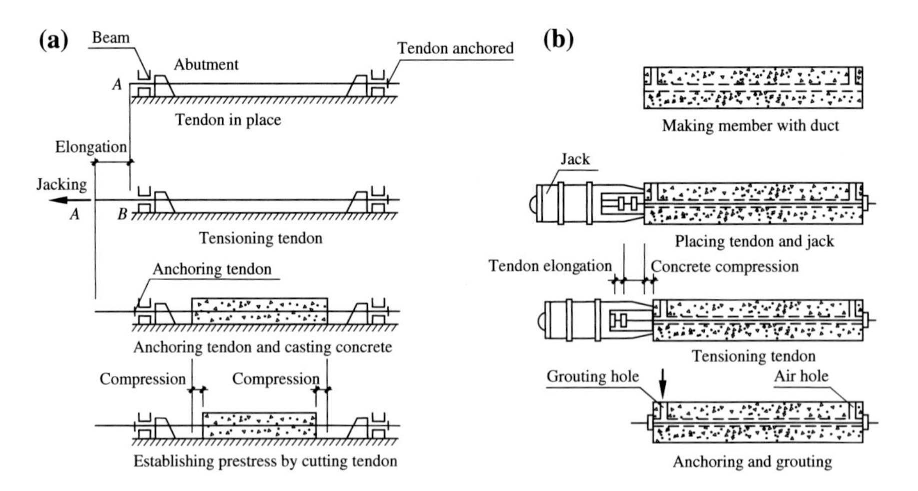

Fig. 10.4 **a** Pretensioning and **b** posttensioning

The distance between abutments in a tensioning bed is usually 80–100 m or even longer. Such a tensioning bed is widely used in various prefabrication factories due to its simplicity and natural curing; however, the use of steel forms is more efficient and takes less room.

Compared with posttensioning, pretensioning is simpler and more efficient. The quality of products made by pretensioning is usually better and more stable. The cost is also less because anchorage of members is not needed.

The process of posttensioning is as follows (Fig. 10.4b). First, the member is cast with ducts. Then, it is cured. When the concrete of the member gets sufficient strength (usually not less than 70 % of its design strength), the tendons are put through the ducts. The tendon in a duct is usually anchored at one end and tensioned by jacking at the other end (the tendon can also be tensioned at both ends). After tensioning, the tensioned end is also anchored. The ducts are then grouted with pressure. Obviously, in this case, the prestress is transferred to the concrete by the anchors at the two ends of the tendon.

The method of posttensioning is particularly suitable for tensioning very large members on a construction site. Because permanent anchorages are used, the cost is usually higher.

10.1.4.2 Classification by Degree of Prestress

According to the degree of prestress, prestressed concrete structures can be classified as fully prestressed and partially prestressed.

Fully prestressed concrete structures are structures in which no tensile stress will occur in critical normal sections under the most unfavorable loads.

Partially prestressed concrete structures are structures in which tensile stresses in concrete in critical normal sections will not exceed the given limit or when the crack width is within that limit.

10.1.4.3 Classification by Prestressing Systems

According to different prestressing systems used, prestressed concrete structures can be classified as structures with internal or external prestressing, structures with bonded or unbonded tendons, structures with a pretensioned compression zone, and structures with preflex of shaped steel.

Concrete structures with internal prestressing are structures in which the tendons are put inside the members. This is the normal way of prestressing. The usual pretensioning and posttensioning belong to this category.

Concrete structures with external prestressing are structures in which the tendons are put outside the members as shown in Fig. 10.5.

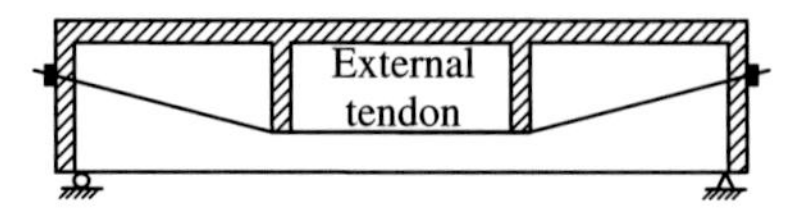

Fig. 10.5 Structures with external prestressing

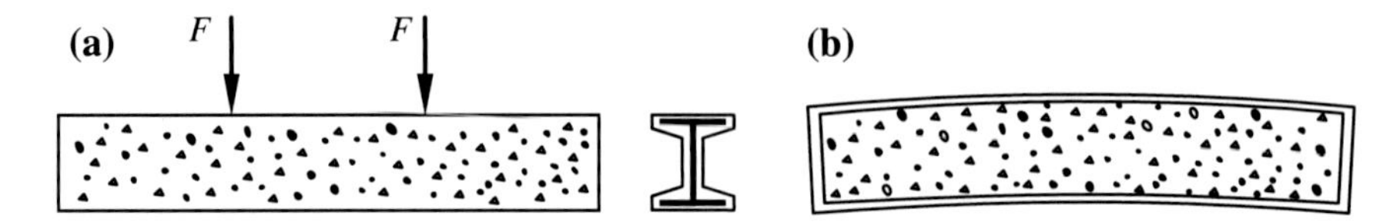

Fig. 10.6 Prestressing by preflex. **a** Shaped steel preflex and concrete casting. **b** Establishing prestress by releasing corresponding loads

Prestressing tendons are usually bonded to the surrounding concrete. For example, pretensioned members and the usual posttensioned members with grouting are members with bonded tendons.

Structures could also be prestressed with unbonded tendons in the case of posttensioning. In this case, the tendons are specially made by first being greased and then wrapped in a plastic sheet. The tendons so made are cast in concrete together with steel rebars. After the concrete gathers sufficient strength, the tendons are tensioned and anchored as usual but without grouting.

Usually, the prestressing is accomplished by stretching the tendons, but the concrete in the compression zone can also be pretensioned by other external forces. In this case, the concrete may be all precompressed or partially precompressed in the tension zone.

Prestressing can also be accomplished by the preflexing of shaped steel. In this case, a shaped steel member is bent before the concrete is cast as shown in Fig. 10.6a. Then, the concrete is cast around the bent shaped steel. After the concrete gains sufficient strength, the forces that have been causing preflex of the shaped steel are released. The rebounding of the shaped steel will then produce the required prestress in the concrete (Fig. 10.6b).

10.1.5 Materials

10.1.5.1 Concrete

The concrete used in prestressed concrete structures should satisfy the following requirements.

1. High strength
   High-strength concrete together with high-strength tendons can effectively reduce the member section dimension and, hence, the weight of a structure. Furthermore, higher prestress can be established in high-strength concrete and results in higher resistance to cracking. For pretensioned members, the high-strength concrete can increase the bond strength between the tendon and the concrete. For posttensioned members, the high-strength concrete will enable the anchorage region to withstand higher local compression.
2. Less shrinkage and creep
   This can reduce the prestress loss due to shrinkage and creep.
3. Higher early-age strength
   This can enable the jacking to be performed at an early age after the concrete is cast, resulting in more efficient construction.
   In the selection of strength grade of concrete, the construction method (pretensioning or posttensioning), span length of members, service conditions (with or without vibrating load), and types of tendons should be carefully considered. GB 50010 specifies that the grade of concrete used in prestressed concrete structures should not be less than C30 with a preference for a concrete grade not less than C40.

10.1.5.2 Tendons

Tendons should be of high strength in order to achieve a satisfactory prestressing effect. High-strength tendons can enable higher tensile stress to be established in the tendons so as to increase crack resistance. The tendons should also possess the required degrees of plasticity and weldability as well as good cold workability. Tendons used in pretensioned members must have a good bond with the concrete.

Commonly used types of tendons mainly include steel strands, steel wires, and heat-treated steel rebars. The nonprestressed steel rebars are expected to be made of HRB400 and HRB335 grades. HPB300 and RRB400 steel rebars can also be used. However, RRB400 rebars are not generally used in important members or in members meant to stand under fatigue load.

In recent years, tendons made of fiber-reinforced plastics (FRP) have been investigated internationally as a substitute for steel tendons. The main advantage of FRP tendons is that they do not corrode. FRP tendons will not be covered in this book.

The main tendons are as follows:

1. Heat-treated bars
   They are made by subjecting hot-rolled ribbed bars to such quality modification procedures as quenching and tempering. According to the shapes, they can be classified into those with and without longitudinal rib. The types of heat-treated rebar include $40Si_2Mn$, $48SiCr_2Mn$, and $45Si_2Cr$ with the tensile strengths as

high as 1230 N/mm^2. The tendons are delivered in coiled form, relieving the procedures of welding and cold drawing. Therefore, heat-treated rebars are widely used.

2. Stress-relief steel wires
   Stress-relief steel wires are made by cold-drawing, high-carbon steel coil to relieve the stresses. According to the shapes, they can be classified into indented wires, helically ribbed wires, and plain wires. Such steel wires are high in carbon content with low ultimate elongation of about 2–6 %. The strength of such wires can be as high as 1860 N/mm^2, which means they are mainly used in large members.
3. Steel strands
   They are made by twisting several steel wires together, such as a 3- or 7-wire strand. The characteristic tensile strength of a steel strand can be as high as 1960 N/mm^2. The merit of steel strands is that they are convenient in construction. Hence, they are mainly used in large posttensioned members.

The different types of tendons shown above have their own characteristics, which should be comprehensively weighed in reasonable tendon selection.

10.2 Methods of Prestressing and Anchorage

10.2.1 Methods of Prestressing

There are many methods of prestressing. The most commonly used method is by tensioning the tendon. The tendon cannot only exert compressive stresses on concrete but withstand tension as well. The main prestressing methods are outlined below.

10.2.1.1 Direct Tensioning

The methods of direct tensioning steel bars can be further divided into the pretensioning and posttensioning methods mentioned previously. In cases where the tendons are unbonded, the tendons can be easily retensioned and even changed in certain cases.

The pretensioning and posttensioning methods can be combined in some very large-span members. In this case, pretensioning is mainly for counteracting the self-weight of the member and the stresses induced by the process of hoisting, transportation, and erection, whereas posttensioning is mainly for counteracting the dead and live loads imposed on the structure after erection.

There is a method of posttensioning with self-anchorage. The self-anchorage is achieved by using a cone-shaped duct end, onto which the tendon is anchored by cast in situ concrete, thus no steel anchorages are needed. In this case, special rigs are used, by which the tendons can be jacked and temporarily anchored. Then, the concrete in the anchorage cone is cast. When the concrete gains sufficient strength, the tendons are cut and the concrete in the cone will anchor the tendon through bond action. The rigs are then removed and can be reused. The process is shown in Fig. 10.7.

The method of posttensioning with self-anchorages has been used in roof trusses spanning 15–30 m, in crane beams with rated cargo weight below 2000 kN and span lengths between 6–12 m, and in roof beams. Test results and engineering applications have shown that the self-anchorages have good properties and can function well. The disadvantage is that the process is very complicated and takes a long time. Also, the detailing at the member ends becomes complicated as the number of tendons to be anchored increases, which can result in cracking at the ends if proper measures are overlooked. These can be improved by using epoxy

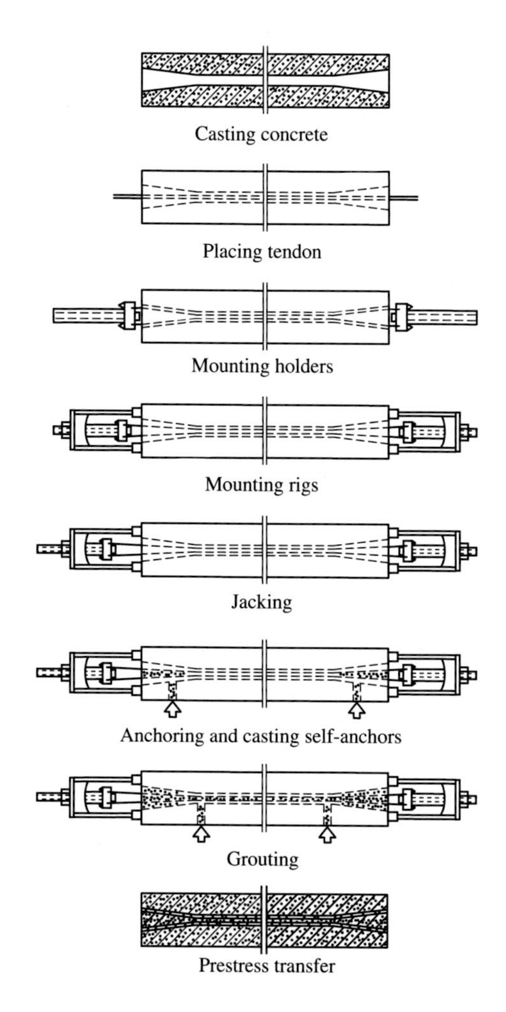

Fig. 10.7 Posttensioning with self-anchorages

resin mortar instead of concrete in the cone to achieve better bond and higher strength at an early age.

10.2.1.2 Electric Heating

This method is based on the principle that the tendon will elongate under heating and contract under cooling. The elongation of the tendon is achieved by passing strong electric current through the tendon. The strong current will do work through the resistance of the tendon and thus heat the tendon. When the elongation has reached the required value, the tendon is anchored to the concrete and the electric current is switched off. The cooling of the tendon will cause it to contract, and this will in turn produce the required compressive stress in the concrete.

Compared with the normal mechanical methods of tensioning, tensioning by electric heating has the merits of being simple, convenient, fast, and efficient. It can be easily used in curved members such as circular cistern and tanks and in frame members, which are high above ground. However, the accuracy of the prestress produced in this way is difficult to control. Therefore, this method may not be suitable for members with a high level of crack control. What is more, before batch production, jacks should be used to calibrate the level of prestress, so that the concrete quality with electric heating method can be guaranteed.

10.2.1.3 Continuous Gyration

This method is usually used for prestressing circular cisterns or similar structures. In this case, a platform with tendon coils is made to rotate about the axis of the circular section; meanwhile, the structure is being compressed by the tendons on the circumference of the structure.

10.2.1.4 Self-prestressing

Concrete made with self-stressing cement will expand in the hardening process. This expansion is constrained by the use of rebars in concrete, and thus, the concrete is subjected to precompression and the rebars to pretension. Such self-prestressing method has been used, for example, in the production of self-stressing concrete pipes.

10.2.1.5 Direct Compression

A member can be compressed by direct action of the jacking force at the two ends of the member. Obviously, external supports are needed in this case to provide the necessary reactions. The jacking can be accomplished by inserting circular thin

Fig. 10.8 Circular thin jack

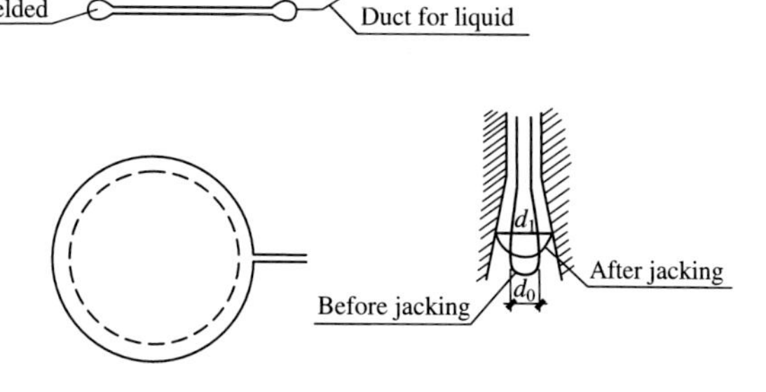

jacks (Fig. 10.8) between the members. The thin jacks can be taken out afterward or left in the structure. This method has been used in prestressing airplane runways and circular cisterns. In the latter case, the walls of the cistern are first assembled with outside circular rebars and then prestressed.

10.2.2 Anchorages and Clamps

Anchorages and clamps are tools used to anchor the tendons in the process of manufacturing prestressed concrete members. Among them, those that can be removed after the member has been made are called clampers; those that cannot be removed and hence must be left as part of the member are called anchorages. Sometimes, they can all be called anchorages. The working principles of anchorages are based on friction, clamping, and bearing. As has been stated, anchorages must be used with posttensioned members and may be used in pretensioned members to help anchoring the tendons.

Anchorages must be sufficiently safe and rigid to guarantee their proper functions. Anchorages should also make the slippage of tendons that they anchor as small as possible. Furthermore, anchorages should be sufficiently simple, convenient to use, and economical with less material consumption.

The most frequently used anchorages are described below.

10.2.2.1 Screw Anchorages and Welded Head Anchorages

The screw anchorage is composed of an end bolt and a nut as shown in Fig. 10.9a. The bolt is usually welded to the tendon, and it can be connected to the jack as well. After tensioning, the nut is tightened to enable the end of the member under the nut to be subjected to compression through a steel base plate.

The welded head is made by welding three short steel bars of 50–60 mm in length around the tendon at 120° pitch angles. The short steel bars can then be anchored to the concrete through a 15–20-mm-thick steel plate, as shown in

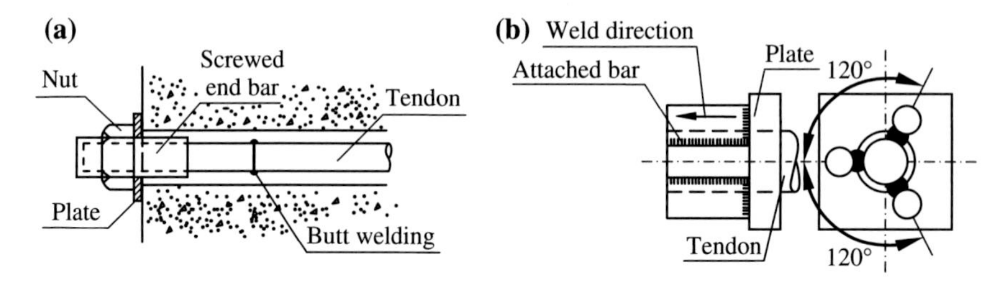

Fig. 10.9 **a** Screw anchorage and **b** welded head anchorage

Fig. 10.9b. The short bars are usually made of the same material as the tendon. Such anchorages are suitable for 12–40-mm-diameter tendons.

These are the commonly used anchorages to anchor a single tendon of large diameter. The welded head is suitable for the anchored end, while the screw anchorage is suitable for the jacked end.

10.2.2.2 Clamping-Type Anchorages

The clamping-type anchorage is composed of a clamping ring, an anchoring plate, and a gasket (Fig. 10.10). The clamping ring can be divided into two or three parts by slots in the ring. The anchoring plate is with holes in the shape of a cone. The tendon passing the clamping ring is then anchored in the cone-shaped hole by wedging during the contraction process of the tendon after being jacked. The slots in the ring can be parallel or at an angle to the tendon axis.

The names of the clamping-type anchorages that are often used in China include OVM, HVM, QM, STM, XM, XYM, and YM. These are mainly for anchoring the seven-wire strand with a wire diameter of $\phi 4$ or $\phi 5$. The number of strands that can be anchored ranges from one to several dozens. Auxiliaries of the clamping-type anchorages include anchored end anchorages and connectors.

10.2.2.3 Wedged Last Anchorage

The wedged last anchorage is also called a Freyssinet anchorage. It is suitable for anchoring a steel strand composed of 12–24 wires that are 5 mm in diameter. It

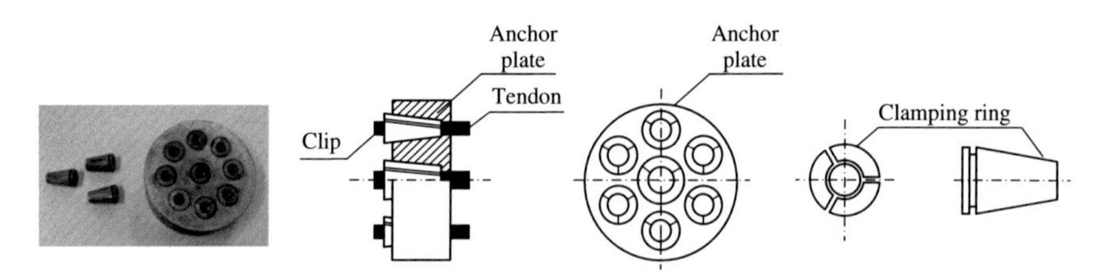

Fig. 10.10 Clamping-type anchorages

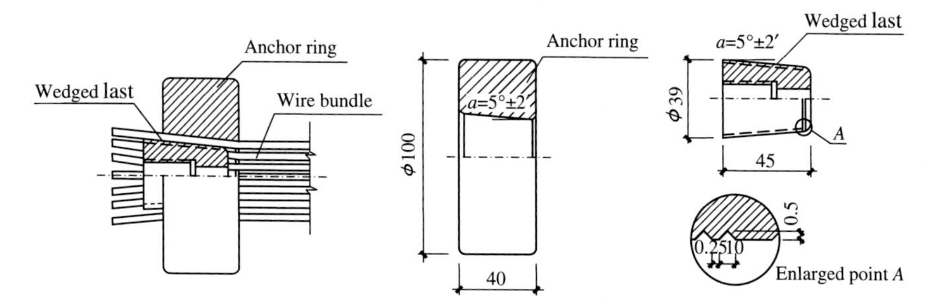

Fig. 10.11 Wedged last anchorage

consists of a ring and a wedged last as shown in Fig. 10.11. The tensioning is performed by a special double action jack that simultaneously tensions the tendons and presses the wedged last into the ring, making the wires tightly clamped between the wedged last and the ring. The tendons are thus anchored by friction.

10.2.2.4 Anchorages for Wire Bundles with Cold-Headed Ends

This type of anchorages is composed of an anchoring cup and a nut as shown in Fig. 10.12. The cup has holes for the wires to pass through, which allows them to be anchored by their cold-headed ends. The nut is for anchoring the whole set against the concrete through a steel plate. Obviously, this type of anchorage is more reliable with heavy tonnage and easy operation. It is suitable for anchoring wire bundles; however, it requires more accurate cutting of the wires. Furthermore, the end of the duct in the member must be enlarged, and a special device for cold heading the tendon ends is needed.

10.2.3 Profiles of Posttensioned Tendons

In posttensioned construction, tendon profiles are usually of parabolic shape considering the shape of a moment diagram of a beam under distributed load. The tendon profile for a simply supported beam is usually a single parabolic curve with the maximum eccentricity located at mid-span (Fig. 10.13).

Taking the right half of the curve as the objective, the equation of the curve in coordinate system *xoy* is

$$y = e_0\left(\frac{x}{l_0}\right)^2 \tag{10.7}$$

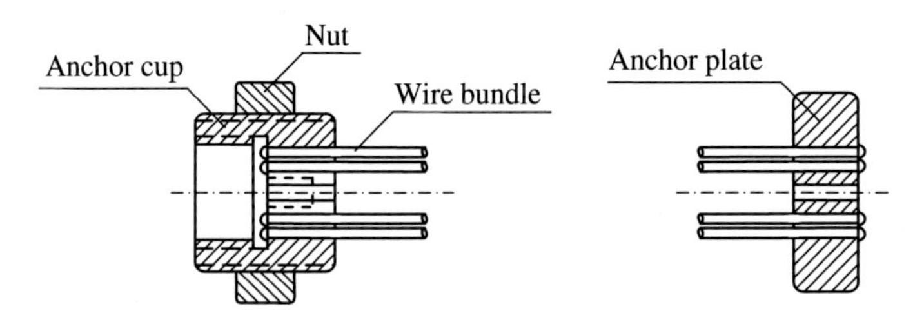

Fig. 10.12 Anchorages for wire bundle with cold-headed ends

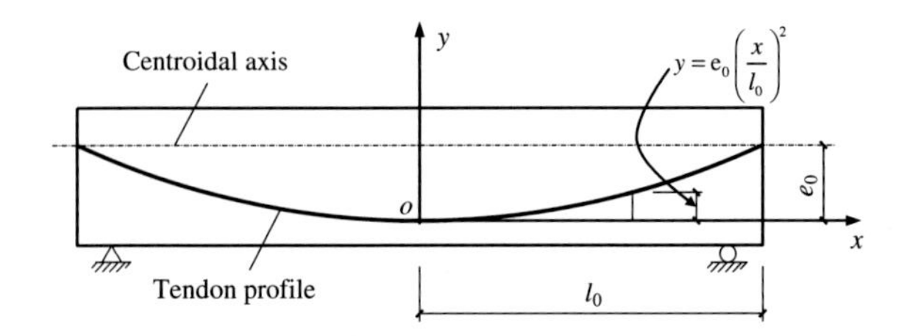

Fig. 10.13 Parabolic tendon profile for a simply supported beam

where

e_0 maximum eccentricity of the tendon; and

l_0 length of the parabolic segment

At $x = 0$, the radius of curvature is

$$r = \frac{l_0^2}{2e_0} \tag{10.8}$$

At the right end of the tendon, the slope of curve is

$$\text{Slope} = \frac{2e_0}{l_0} \tag{10.9}$$

Tendon profiles in continuous beams can be described as a series of parabolic segments with concave segments in the spans and convex segments over the supports, as shown in Fig. 10.14a. Figure 10.14b illustrates how a series of parabolic segments fit together. At maximum eccentricity e_1, both parabola 1 and parabola 2 have zero slopes and hence are compatible. To keep parabola 2 and parabola 3 compatible either, their slopes at inflection point must be equal, which means

$$\frac{2(e_1 + e_2 - h_2)}{(\lambda - \beta)l_1} = \frac{2h_2}{\beta l_1} \tag{10.10}$$

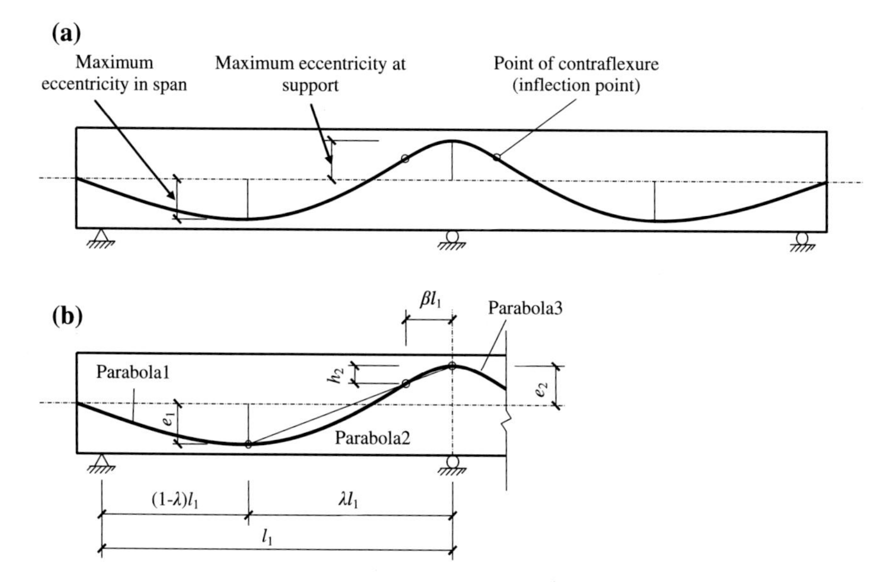

Fig. 10.14 Parabolic tendon profile for a continuous beam

Thus, the inflection point must be located at a distance, h_2, below the highest point, where

$$h_2 = \frac{\beta}{\lambda}(e_1 + e_2) \tag{10.11}$$

The inflection point must therefore lie on a straight line connecting the points of maximum eccentricities as shown in Fig. 10.14b.

10.2.4 Control Stress σ_{con} at Jacking

The control stress σ_{con} is the permissible maximum stress in a tendon at the time of jacking. It is measured as the stress obtained by dividing the total tension indicated by the dial gauge of the jack by the area of the tendon.

The choice of the control stress value will directly affect properties of the prestressed member. If the control stress is too low, the prestress will also be very low after the various losses of prestress have taken place, which will not effectively increase the resistance to cracking of the member and, therefore, result in lower stiffness of the member. However, a control stress that is too high may cause the following problems:

Table 10.1 Allowable control stress values $[\sigma_{con}]$

Steel type	$[\sigma_{con}]$
Steel wires and steel strands	$0.75f_{ptk}$
Heat-treated bars	$0.85f_{pyk}$

Notes (1) The term f_{ptk} in the table stands for the characteristic value of the ultimate strength of tendons, and f_{pyk} is the characteristic value of the yield strength of heat-treated bars
(2) The listed values of $[\sigma_{con}]$ in the table are allowed to be increased by $0.05f_{ptk}$ or $0.05f_{pyk}$ in the following cases
Tendons are put in the potential compression zone under service load condition in order to increase the resistance to cracking during construction
The increase in $[\sigma_{con}]$ is to partially compensate for the prestress losses due to stress relaxation, friction, and batch jacking, as well as the temperature difference between tendons and jacking abutments

1. Tensile stress in the concrete due to prestress may be induced during the construction stage. This tensile stress may cause cracking of concrete and, in the case of posttensioning, may cause the end of the member to be subjected to a bearing failure.
2. The cracking load may be too close to the ultimate load, causing little warning before failure. The ductility of the member may be poor.
3. In order to reduce the prestress loss, overtensioning is sometimes used. Such overtensioning may cause yielding or even rupturing of the tendon if the control stress is too high.

It follows that the choice of the control stress depends on the properties of the tendon steel. The degree of plasticity of strand and destressed wires is better for they will show obvious yield plateau; therefore, their control stress may be set higher. In contrast, the heat-treated bars are generally poor in plasticity; therefore, their control stresses are usually set lower.

The allowable control stress, $[\sigma_{con}]$, as specified by code GB 50010 is given in Table 10.1. These values are generally not to be exceeded.

In order to guarantee the minimum effect of prestressing, $[\sigma_{con}]$ is specified in the GB 50010 code not to be less than $0.4f_{ptk}$.

10.3 Prestress Losses

The prestress established by jacking will gradually become smaller in the process of construction and service due to the nature of construction and material properties. This reduction in tendon stress after jacking is called prestress loss. Many factors will cause prestress losses. Some factors such as shrinkage and creep of concrete as well as relaxation of tendon will vary with time and environment. These factors can even influence one another. Therefore, accurate calculation of the various prestress losses is a complicated task. In engineering applications, it is usually assumed that the prestress losses due to different reasons can be calculated separately, and the total prestress loss is equal to the summation of the individual prestress losses. Hence, the calculation of the different types of prestress losses is given below.

10.3.1 Prestress Loss σ_{l1} Due to Anchorage Deformation

When the tension in a tendon is transferred to the anchorage, the tendon will shorten due to the deformation of the anchorage set in the form of gaps being pressed and slippage. The amount of shortening, a, can be determined for the various types of anchorages as listed in Table 10.2. Hence, the prestress loss due to such shortening, σ_{l1}, can be expressed by the following formula for straight tendons:

$$\sigma_{l1} = \frac{a}{l} E_{\mathrm{P}} \tag{10.12}$$

where

l distance between the anchored end and the jacked end; and

E_{p} modulus of elasticity of the tendon.

Obviously, the prestress loss due to anchorage deformation is considered only for the jacked end. There is no such loss at the anchored end, because the anchorage at this end has already been pressed tightly during the jacking process.

For structures assembled with several segments, the gaps between the segments will cause prestress losses. Such losses can be calculated by assuming 1 mm shortening for each gap if the gap is filled by concrete or mortar.

The prestress loss due to anchorage deformation in the case of a curved tendon will be discussed later.

The measures to reduce the prestress loss due to anchorage deformation are as follows:

Table 10.2 Shortening a of tendon due to anchorage deformation (mm)

Types of anchorages		a
Bearing-type anchorages (e.g., anchorages for wire bundle with extended ends)	Gap due to a nut	1
	Gap of each extra gasket	1
Wedge-type anchorages (e.g., steel wedge anchorages for steel wires)		5
Clamping-type anchorages	With pressing pressure	5
	Without pressing pressure	6–8

Notes (1) The values of shortening in the table can also be determined by measurement
(2) The values of shortening of other types of anchorages should be determined by measurement

1. Select anchorages that will deform less or make the tendon slip less. Use as few gaskets as possible, for with each extra gasket, the shortening is increased by 1 mm.
2. Increase the length of the platform, that is, increase the distance between the two abutments. When the length of the platform is 100 m or longer in pretensioning production, the loss due to anchorage deformation becomes so small that it can be omitted.

10.3.2 Prestress Loss σ_{l2} Due to Friction Between Tendon and Duct

In the case of posttensioning with a straight tendon, friction will be generated due to crookedness of the duct, error in duct position, curvature of the tendon (may be caused by eccentricity in connections and accidental bent, etc.), roughness of tendon surface, etc. The effect of such friction will be greater with greater distance from the jacked end. The situation becomes worse with geometrically curved tendons/ducts, for the intended curvature will tighten the tendon against the duct. The prestress loss due to such friction is denoted as σ_{l2}.

Now let us analyze the prestress loss due to friction. An infinitesimal segment dx of the tendon is taken out as a free body (Fig. 10.15). The friction can be thought to be caused by two components:

1. The first component of friction results from the intended change of angle of the tendon profile. Let κ be the frictional influencing factor due to crookedness per unit length of the tendon, the values of which can be looked up from Table 10.3. Then, we have

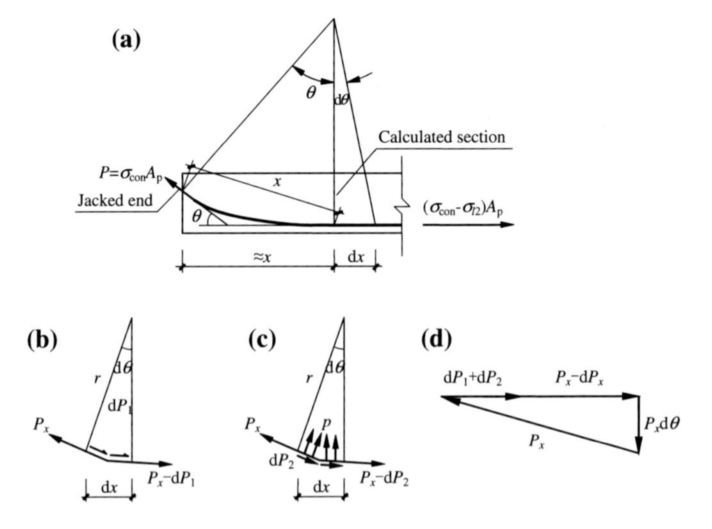

Fig. 10.15 Prestress loss due to friction

Table 10.3 Coefficients of friction κ and μ

Type of duct	κ	μ	
		Steel wires and steel strands	Heat-treated bar
Embedded corrugated metal tube	0.0015	0.25	0.5
Embedded corrugated plastic tube	0.0015	0.15	–
Embedded steel tube	0.0010	0.30	–
Formed by core withdrawal	0.0014	0.55	0.60
Unbonded tendon	0.0040	0.09	–

Notes Friction coefficients can also be determined by measurement

$$\mathrm{d}P_1 = -\kappa P_x r \mathrm{d}\theta \approx -\kappa P_x \mathrm{d}x \tag{10.13}$$

where P_x = the tensile force in the tendon.

2. The second component of friction is due to the intended curvature of the tendon. Let the angle corresponding to the infinitesimal segment $\mathrm{d}x$ be $\mathrm{d}\theta$. By radial equilibrium of the forces acting on the segment, the line load p exerted by the tendon on the concrete can be derived as (Fig. 10.15c, d)

$$p = -P_x \mathrm{d}\theta \tag{10.14}$$

The frictional force $\mathrm{d}P_2$ acting on the segment is thus

$$\mathrm{d}P_2 = -\mu P_x \mathrm{d}\theta \tag{10.15}$$

where μ = the coefficient of friction.

Now, the total infinitesimal friction force $\mathrm{d}P_x$ can be derived as

$$\mathrm{d}P_x = \mathrm{d}P_1 + \mathrm{d}P_2 = -\kappa P_x \mathrm{d}x - \mu P_x \mathrm{d}\theta \tag{10.16}$$

Upon integration, we have

$$\int_P^{P_x} \frac{\mathrm{d}P_x}{P_x} = -\int_0^x \kappa \mathrm{d}x - \int_0^\theta \mu \mathrm{d}x \tag{10.17}$$

Hence,

$$\Delta P = P - P_x = P\left(1 - \frac{1}{e^{\kappa x + \mu\theta}}\right) \tag{10.18}$$

Dividing both sides of the above equation by the tendon area, we have the expression for the prestress loss due to friction:

$$\sigma_{l2} = \sigma_{\mathrm{con}}\left(1 - \frac{1}{e^{\kappa x + \mu\theta}}\right) \tag{10.19}$$

when $\kappa x + \mu\theta \leqslant 0.3$, σ_{l2} can be approximately expressed as

$$\sigma_{l2} = \sigma_{\mathrm{con}}(\kappa x + \mu\theta) \tag{10.20}$$

where

x distance in meters between the calculated section and the jacked end; and

θ angle in radians between the tangent of the tendon at the jacked end and the tangent of the tendon at the calculated section.

The coefficient of friction μ can be looked up from Table 10.3.

If the shape of the tendon is an arc of a circle with its central angle not greater than 30°, the variation of tendon stress due to friction between tendon and duct at the time of jacking can be approximated by the straight line ABC in Fig. 10.16. At the end of jacking, the tendon is anchored, and at the same time, the tendon will shorten. This shortening of the tendon will cause friction between the tendon and the duct in a direction opposite to the friction at jacking. The shortening of tendon will cause reduction in tendon stress, while the reversed friction will cause the stress loss due to tendon shortening to become smaller with greater distance from the jacked end. Therefore, there generally exists a distance l_{f} from the jacked end at which the stress loss due to shortening of tendon is zero. This length l_{f} is called the influencing length due to reversed friction. The stress variation within the influencing length is shown as straight line A′B in Fig. 10.16.

Now, we shall derive the expression for l_{f} and calculate the prestress loss σ_{l1} due to anchorage deformation.

Let the prestress loss due to friction be $\sigma_{l2} = \sigma_{\mathrm{con}}(\kappa x + \mu\theta)$. Because θ increases with x, σ_{l2} can be approximately expressed as a linear function of x. Therefore, it is assumed that

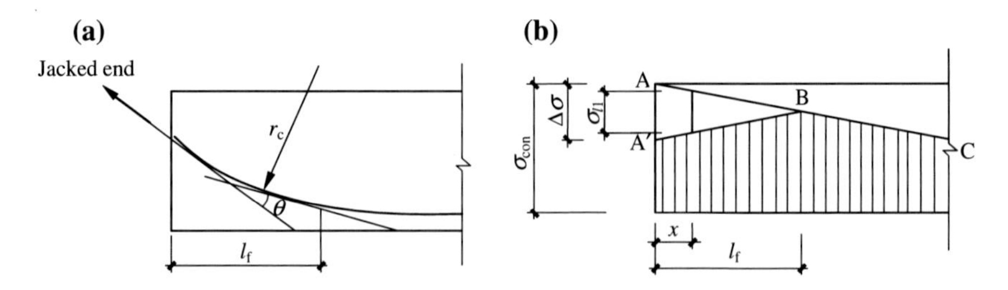

Fig. 10.16 Prestress loss in circular tendon due to anchorage deformation. **a** Cirular tendon. **b** Distribution of σ_{l1}

$$\lambda x = \kappa x + \mu\theta \tag{10.21}$$

where

$$\lambda = \frac{\kappa x + \mu\theta}{x} = \kappa + \mu\frac{\theta}{x} \tag{10.22}$$

Because the tendon shortens due to anchorage deformation, the tendon stress at the jacked end is reduced from point A to point A′ (Fig. 10.16) by a value of $\Delta\sigma$. The tendon stress σ_1 at an arbitrary point x along line AB can be calculated by the control stress σ_{con} minus the stress loss due to friction as:

$$\sigma_1 = \sigma_{con}[1 - (\kappa x + \mu\theta)] = \sigma_{con}(1 - \lambda x) = \sigma_{con} - \lambda x\sigma_{con} \tag{10.23}$$

Let the coefficient of friction in the reversed direction be the same as that in the positive direction. With the symmetry so produced, the tendon stress σ_2 at point x along line A′B can be calculated by the tendon stress ($\sigma_{con} - \Delta\sigma$) at point A′ plus the prestress loss due to friction, i.e.,

$$\sigma_2 = \sigma_{con} - \Delta\sigma + \lambda x\sigma_{con} \tag{10.24}$$

At point B where $x = l_f$, $\sigma_1 = \sigma_2$ by definition. Hence, we have

$$\sigma_{con} - \lambda l_f\sigma_{con} = \sigma_{con} - \Delta\sigma + \lambda l_f\sigma_{con} \tag{10.25}$$

Upon solving l_f, we have

$$l_f = \frac{\Delta\sigma}{2\lambda\sigma_{con}} \tag{10.26}$$

The stress loss $\Delta\sigma$ can be derived as follows. The anchorage deformation a is distributed within distance l_f resulting in average shortening strain a/l_f. Therefore, the average prestress loss is $(a/l_f)E_p$. Because the prestress loss varies linearly within the segment indicated by l_f, it follows that the prestress loss at the middle point of the segment must be equal to the average prestress loss. Hence, $\Delta\sigma$ must be twice this average prestress loss. Thus, $\Delta\sigma = 2(a/l_f)E_p$. Substituting $\Delta\sigma$ in Eq. (10.26), we have

$$l_f = \frac{\Delta\sigma}{2\lambda\sigma_{con}} = \frac{2\frac{a}{l_f}E_p}{2\lambda\sigma_{con}}$$

Hence,

$$l_f = \sqrt{\frac{aE_p}{\lambda\sigma_{con}}} = \sqrt{\frac{aE_p}{\sigma_{con}\left(\kappa + \frac{\mu\theta}{x}\right)}} \tag{10.27}$$

Suppose the tendon is of a circular arc with its central angle θ less than 30°, then we have

$$\frac{\theta}{x} = \frac{1}{r_c} \tag{10.28}$$

where

r_c the corresponding radius of curvature of the arc given in meters.

Substituting $\frac{\theta}{x}$ in Eq. (10.27), and let the units of all lengths be meter except a (the unit of a is mm), we get

$$l_f = \sqrt{\frac{aE_p}{1000\sigma_{con}\left(\frac{\mu}{r_c} + \kappa\right)}} \tag{10.29}$$

Therefore, from Eq. (10.26), we have

$$\Delta\sigma = 2\sigma_{con}\lambda l_f = 2\sigma_{con} l_f \left(\frac{\mu}{r_c} + \kappa\right) \tag{10.30}$$

The prestress loss due to shortening of tendon caused by anchorage deformation at any point x ($x \leqslant l_f$) can be obtained by the linear relationship in x

$$\sigma_{l1} = 2\sigma_{con} l_f \left(\frac{\mu}{r_c} + \kappa\right)\left(1 - \frac{x}{l_f}\right) \tag{10.31}$$

Measures to reduce the prestress loss due to friction include the following:

1. Jacking at both ends of the tendon can be performed for long members. In this case, the duct length used in calculation can be reduced by half. The effect of this measure in reducing the friction loss is obvious as shown in Fig. 10.17a and Fig. 10.17b. However, this measure will increase σ_{l1}.

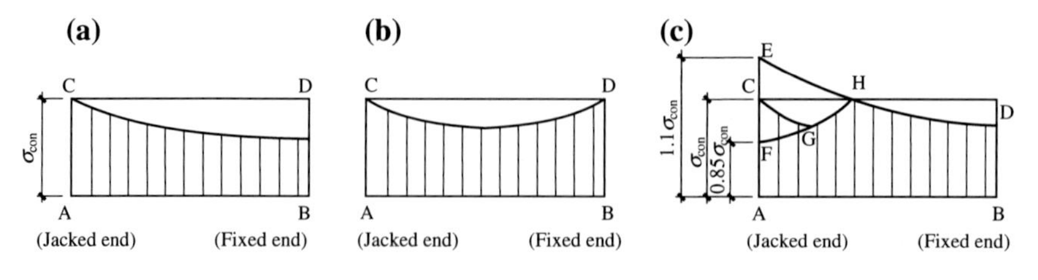

Fig. 10.17 Effect in reducing friction loss by jacking at two ends and by overtensioning. **a** Jacking at A. **b** Jacking at A and B. **c** Overtensioning

2. Overtensioning can be used as shown in Fig. 10.17c. The process of overtensioning can be:

$$0 \longrightarrow 1.1\sigma_{\rm con} \xrightarrow{\text{Stop for 2 min}} 0.85\sigma_{\rm con} \xrightarrow{\text{Stop for 2 min}} \sigma_{\rm con} \qquad (10.32)$$

When the end A is overtensioned by 10 %, the distribution of tendon stress is represented by curve EHD in the figure. When the tensioning stress is lowered to $0.85\sigma_{\rm con}$, the distribution of tendon stress is given by curve FGHD as the tendon is now affected by reversed friction. When the tensioning end A finally reaches $\sigma_{\rm con}$ again, the distribution is given by curve CGHD which is obviously more uniform with less prestress loss.

10.3.3 Prestress Loss σ_{l3} Due to Temperature Difference

Steam curing is often used to reduce the time needed in the production of pretensioned members. In Fig. 10.18, when the temperature is raised by steam curing form t_0 to t_1, the concrete has not fully hardened, and the tendon can expand freely resulting in prestress loss due to the tendon being fixed at the abutments. The stress in the tendon is $\sigma'_{\rm con}$. When the temperature drops to normal, the concrete has hardened and bonded to the tendon resulting in the stress in the tendon to be $\sigma''_{\rm con} < \sigma_{\rm con}$ produced, previously being permanently nonrecoverable deformation. So, the prestress loss due to temperature difference is $\sigma_{l3} = \sigma_{\rm con} - \sigma''_{\rm con}$.

Let the temperature difference between the tendon and the abutment platform be Δt (in °C) as the member is being steam cured, and if the coefficient of expansion of the tendon be 0.00001/°C, then σ_{l3} can be derived as

$$\begin{aligned}\sigma_{l3} &= \varepsilon_{\rm s}E_{\rm p} = \frac{\Delta l}{l}E_{\rm p} = \frac{0.00001l\Delta t}{l}E_{\rm p} = 0.00001E_{\rm p}\Delta t \\ &= 0.00001 \times 2.0 \times 10^5\Delta t = 2\Delta t\left(\text{N/mm}^2\right)\end{aligned} \qquad (10.33)$$

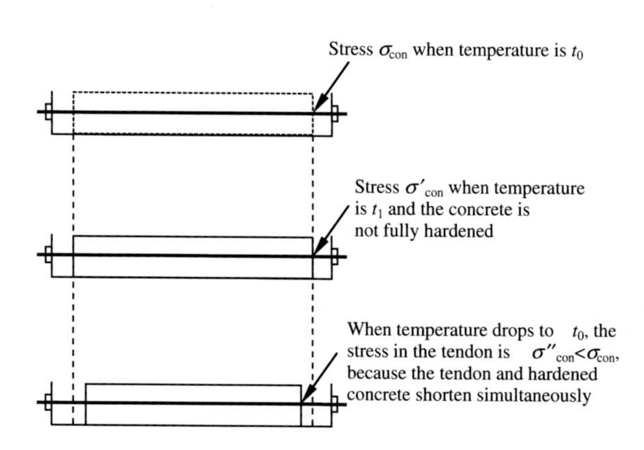

Fig. 10.18 Prestress loss due to temperature difference

Measures to reduce this temperature loss include the following:

1. A process of cure in which the temperature rises two times can be adopted. That is, the curing temperature is first risen to a lower level, which is usually less than 20 °C before the concrete is hardened and bonded to the tendon. After the concrete reaches certain strength, e.g., C7.5–C10, the temperature is further raised to a higher curing temperature at which time the concrete and the tendon have bonded together allowing no prestress loss due to temperature difference.
2. The tendon can be tensioned in a steel form which is also subjected to the curing steam. Thus, there is no temperature difference between the tendon and the form, and hence, no prestress loss will be caused due to this process.

10.3.4 Prestress Loss σ_{l4} Due to Tendon Stress Relaxation

Steel bars generally show creep properties under high stress. The related phenomenon that the stress in the steel bar decreases with increasing time as the total strain of the bar is held at a constant high level (e.g., by fixing the two ends on the abutments) is called stress relaxation of the steel bar. Prestress loss due to such relaxation of tendon is called relaxation loss and denoted as σ_{l4}.

The formulae given in the code GB 50010 for calculating σ_{l4} are as follows, which are based on test results.

For heat-treated steel bars, the formula

$$\sigma_{l4} = 0.04\sigma_{\text{con}} \tag{10.34}$$

is used in case of single tensioning, and the formula

$$\sigma_{l4} = 0.03\sigma_{\text{con}} \tag{10.35}$$

is used in case of overtensioning.

For steel wires and strands with ordinary relaxation properties, the formula

$$\sigma_{l4} = 0.4\psi\left(\frac{\sigma_{\text{con}}}{f_{\text{ptk}}} - 0.5\right)\sigma_{\text{con}} \tag{10.36}$$

is used, in which $\psi = 1$ for single tensioning and $\psi = 0.9$ for overtensioning.

For steel wires and strands with low relaxation properties, the formula

$$\sigma_{l4} = 0.125\left(\frac{\sigma_{\text{con}}}{f_{\text{ptk}}} - 0.5\right)\sigma_{\text{con}} \tag{10.37}$$

Table 10.4 Coefficients for time variation of relaxation loss

Time (day)	2	10	20	30	⩾40
Coefficient of relaxation loss	0.50	0.77	0.88	0.95	1.0

is used when $\sigma \leqslant 0.7 f_{\text{ptk}}$, and the formula

$$\sigma_{l4} = 0.2\left(\frac{\sigma_{\text{con}}}{f_{\text{ptk}}} - 0.575\right)\sigma_{\text{con}} \tag{10.38}$$

is used when $0.7 f_{\text{ptk}} < \sigma_{\text{con}} < 0.8 f_{\text{ptk}}$.

When the above formulae for calculating prestress loss due to relaxation are used, the process of tensioning should conform to the requirements given by the national standard *Code for Quality Checking of Construction of Concrete Structures* (GB 50204).

The coefficients given in Table 10.4 can be used to multiply the σ_{l4} value obtained previously to get the relaxation loss corresponding to a given time after tensioning.

The following factors influence the degree of relaxation loss:

10.3.4.1 Time

The development of relaxation loss is faster at the initial stage such that the relaxation loss within the first two days can be as high as 50 % of the total relaxation loss. The development of relaxation loss will be slower afterward.

10.3.4.2 Type of Steel

The relaxation loss of heat-treated steel bars is less than that of steel wires and strands.

10.3.4.3 Control Stress

The higher the control stress is, the greater will be the amount of relaxation loss.

The main measure to reduce relaxation loss is overtensioning. The process for this overtensioning is as follows. First, the tensioning stress is raised to (1.05–1.1) σ_{con} and kept at this level for 3–5 min. Then, the stress is released. After that the stress is finally raised to σ_{con}. The reason for this process to reduce the relaxation loss is that the relaxation caused by short-term high stress can be as large as the relaxation caused by long-term low stress. Therefore, most of the relaxation loss has taken place in the overtensioning process. The relaxation value of a steel bar depends on the initial stress. When the initial stress is less than $0.7f_{py}$, the relationship between relaxation and initial stress is linear; when the initial stress is greater than $0.7f_{py}$, the relationship becomes nonlinear since the relaxation increases more rapidly.

10.3.5 Prestress Loss σ_{l5} Due to Creep and Shrinkage of Concrete

Concrete will shrink when hardened in air under normal temperatures. The action of prestress will make the concrete creep in the direction of compression. That will cause the member and the tendon to shorten, resulting in prestress loss. Although creep and shrinkage are two different phenomena, their effects in producing prestress loss are usually taken into account as a whole due to their similar influencing factors and similar laws of variation.

The prestress loss due to creep and shrinkage of the tendon in the tension zone is denoted as σ_{l5} and that of the tendon in the compression zone is denoted as σ'_{l5}. They can be calculated as follows.

For pretensioned members,

$$\sigma_{l5} = \frac{60 + 340\frac{\sigma_{pc}}{f'_{cu}}}{1 + 15\rho} \tag{10.39}$$

$$\sigma'_{l5} = \frac{60 + 340\frac{\sigma'_{pc}}{f'_{cu}}}{1 + 15\rho'} \tag{10.40}$$

For posttensioned members,

$$\sigma_{l5} = \frac{55 + 300\frac{\sigma_{pc}}{f'_{cu}}}{1 + 15\rho} \tag{10.41}$$

$$\sigma'_{l5} = \frac{55 + 300\frac{\sigma'_{pc}}{f'_{cu}}}{1 + 15\rho'} \tag{10.42}$$

In the above formulae, σ_{pc} and σ'_{pc} are the normal compressive stresses in concrete at the resultant points of the tendons in the tension and compression zone, respectively. In this case, only the prestress loss before prestress transfer (first batch loss) is considered, and the corresponding σ_{l5} and σ'_{l5} in the nonprestressed rebar should be taken as zero. The values of σ_{pc} and σ'_{pc} must not exceed $0.5f'_{cu}$, where f'_{cu} is the cube strength of concrete at the time of prestress transfer. Furthermore, if σ'_{pc} is tensile, the values of σ'_{pc} in Eqs. (10.40) and (10.42) should be taken as zero. In the calculation of σ_{pc} and σ'_{pc}, the self-weight of the member should be considered according to the actual condition during the process of manufacture.

The terms ρ and ρ' in the above formulae are the reinforcement ratios of the total steel bars including both prestressed and nonprestressed steel bars in the tension and compression zones, respectively. For pretensioned members,

$$\rho = \frac{A_p + A_s}{A_0}, \ \rho' = \frac{A'_p + A'_s}{A_0} \tag{10.43}$$

and for posttensioned members,

$$\rho = \frac{A_p + A_s}{A_n}, \ \rho' = \frac{A'_p + A'_s}{A_n} \tag{10.44}$$

where

A_0 the area of the transformed section, and A_n is the net area of the concrete section.

For members with symmetric reinforcement of both tendons and steel rebars, $\rho = \rho'$ is taken. In this case, the reinforcement ratio is taken as half of the total steel area.

Equations (10.39)–(10.42) indicate the following:

1. The relationship between σ_{l5} and σ_{pc}/f'_{cu} is linear with the implication that linear creep is assumed. That is why the condition $\sigma_{pc} < f'_{cu}$ must be satisfied. Otherwise, the prestress loss will increase out of proportion.
2. The σ_{l5} of a posttensioned member is lower than that of a pretensioned member, for some of the shrinkage loss of the former has taken place when the member is prestressed by tensioning.

The above equations are suitable for environments with usual humidity. In case that the environment is so dry that the annual average humidity is less than 40 %, the σ_{l5} and σ'_{l5} values should be increased by 30 %.

If the structural member is important, its final σ_{l5} and σ'_{l5} values can be calculated by using the following method. For tendons in the tension zone,

$$\sigma_{l5} = \frac{0.9\alpha_{Ep}\sigma_{pc}\varphi_{\infty} + E_p\varepsilon_{\infty}}{1 + 15\rho} \tag{10.45}$$

where σ_{pc} = the concrete compressive stress produced by prestressing at points in the tension zone at which points the resultant of tendons acts. The value of σ_{pc} must not be greater than 0.5 f'_{cu}. In the case of a simply supported beam, the average values of σ_{pc} at mid-span and 1/4 span point can be taken. In the case of a

Table 10.5 Final values of creep coefficient of concrete

Annual average relative humidity RH		$40\ \% \leqslant RH \leqslant 70\ \%$				$70\ \% \leqslant RH \leqslant 90\ \%$			
Theoretical thickness $2A/\mu$ (mm)		100	200	300	⩾600	100	200	300	⩾600
Age of concrete at prestressing t_0 (day)	3	4.83	4.09	3.57	3.09	3.47	2.95	2.60	2.26
	7	4.35	3.89	3.44	3.01	3.12	2.80	2.49	2.18
	10	4.06	3.77	3.37	2.96	2.91	2.70	2.42	2.14
	14	3.73	3.62	3.27	2.91	2.67	2.59	2.35	2.10
	28	2.90	3.20	3.01	2.77	2.07	2.28	2.15	1.98
	60	1.92	2.54	2.58	2.54	1.37	1.80	1.82	1.80
	90	1.45	2.12	2.27	2.38	1.03	1.50	1.60	1.68

Notes (1) The age of concrete at prestressing can be taken as 3–7 days for pretensioned members and 7–28 days for posttensioned members
(2) The term A is the section area of the member, and μ is that part of the perimeter of the section which is in contact with air
(3) This table is suitable for concrete made by ordinary silicate cement or rapid hardening cement. The values in the table are based on C40 concrete, and for C50 or higher concrete, $\sqrt{\frac{32.4}{f_c}}$ should be multiplied to the values
(4) This table is suitable for average temperature varying seasonally between −20 °C and +40 °C
(5) Linear interpolation can be used if the values of the theoretical thickness and the age of concrete at prestressing fall between the values listed in the table

continuous beam or a frame, the average values of σ_{pc} at several representative sections can be taken.

In the above equation, the term φ_∞ is the final creep coefficient of concrete; ε_∞ is the final shrinkage strain of concrete; and α_{Ep} is the ratio of tendon modulus to concrete modulus.

If no reliable data can be used, the values of φ_∞ and ε_∞ given in Tables 10.5 and 10.6 can be adopted.

For tendons in the compression zone,

$$\sigma'_{l5} = \frac{0.9\alpha_{Ep}\sigma'_{pc}\varphi_\infty + E_p\varepsilon_\infty}{1 + 15\rho'} \tag{10.46}$$

where σ'_{pc} = the concrete compressive stress produced by prestressing at points in the compression zone at which points the resultant of tendons acts. The value of σ'_{pc} must not be greater than 0.5 f'_{cu}. If σ'_{pc} is tensile, then $\sigma'_{pc} = 0$ is taken.

The term ρ' in the above equation is the reinforcement ratio including both tendons and rebars. For pretensioned members, $\rho' = (A'_p + A'_s)/A_0$; and for post-tensioned members, $\rho' = (A'_p + A'_s)/A_n$.

It should be noted that for members with tendon area A'_p and rebar area A'_s in the compression zone, the σ_{pc} and σ'_{pc} in Eqs. (10.45) and (10.46) should be calculated according to the total prestress of the section.

Table 10.6 Final values of shrinkage strain of concrete

Annual average relative humidity RH		$40\ \% \leqslant RH \leqslant 70\ \%$				$70\ \% \leqslant RH \leqslant 90\ \%$			
Theoretical thickness $2\,A/\mu$ (mm)		100	200	300	$\geqslant$600	100	200	300	$\geqslant$600
Age of concrete at prestressing t_0 (day)	3	3.51	3.14	2.94	2.63	2.78	2.55	2.43	2.23
	7	3.00	2.68	2.51	2.25	2.37	2.18	2.08	1.91
	10	2.80	2.51	2.35	2.10	2.22	2.04	1.94	1.78
	14	2.63	2.35	2.21	1.97	2.08	1.91	1.82	1.67
	28	2.31	2.06	1.93	1.73	1.82	1.68	1.60	1.47
	60	1.99	1.78	1.67	1.49	1.58	1.45	1.38	1.27
	$\geqslant$90	1.85	1.65	1.55	1.38	1.46	1.34	1.28	1.17

Notes The same as those in Table 10.5

Table 10.7 Coefficients for time variation of prestress loss

Time (day)	Coefficient of loss due to creep and shrinkage	Time (day)	Coefficient of loss due to creep and shrinkage
2	–	60	0.50
10	0.33	90	0.60
20	0.37	180	0.75
30	0.40	365	0.85
40	0.43	1095	1.00

The variation of prestress loss due to creep and shrinkage of concrete can be obtained as the product of the final values of σ_{l5} and σ'_{l5} and the coefficients listed in Table 10.7.

Measures for reducing prestress loss due to creep and shrinkage of concrete include the following:

1. Use high-grade cement, reduce cement content, reduce water cement ratio, and adopt dry hardening concrete.
2. Adopt aggregates that are well graded and make good vibration of the fresh concrete to increase the compactness of the concrete.
3. Cure the concrete with care to reduce shrinkage of concrete.

10.3.6 *Prestress Loss σ_{l6} Due to Local Deformation Caused by Pressure*

Circular members are often with helical tendons. In this case, the tendon will press the concrete resulting in reduction in the diameter of the member. This will make

Table 10.8 Combination of prestress losses into batches

Combination of prestress losses	Pretensioned members	Posttensioned members
Losses before prestress transfer (first batch) σ_{lI}	$\sigma_{l1}+\sigma_{l2}+\sigma_{l3}+\sigma_{l4}$	$\sigma_{l1}+\sigma_{l2}$
Losses after prestress transfer (second batch) σ_{lII}	σ_{l5}	$\sigma_{l4}+\sigma_{l5}+\sigma_{l6}$

Notes The distribution of σ_{l4} between the first batch and the second batch can be determined according to real conditions if this distinguishing is needed

the tendon shorten resulting in prestress loss, which is denoted as σ_{l6}. The prestress loss so produced is in reverse proportion to the diameter d of the member. The simplified calculation of σ_{l6} is as follows.

$$\text{If } d \leqslant 3\,\text{m}, \quad \sigma_{l6} = 30\,\text{N/mm}^2 \tag{10.47}$$

$$\text{If } d > 3\,\text{m}, \quad \sigma_{l6} = 0. \tag{10.48}$$

10.3.7 Combination of Prestress Losses

The above prestress losses are combined into batches to facilitate the calculation. The method of combination given in the code GB 50010 is shown in Table 10.8, wherein the prestress losses are combined into two batches. The first batch of the prestress loss includes those losses that take place before the prestress transfer. The second batch includes those take place after prestress transfer.

The actual prestress loss could be higher than the calculated value. Therefore, if the calculated prestress loss is less than $\sigma_{l\min}$ = 100 N/mm^2 in the case of pretensioning or less than $\sigma_{l\min}$ = 80 N/mm^2 in the case of posttensioning, these values should be adopted, respectively.

10.4 Properties of the Zone for Prestress Transfer

10.4.1 Transfer Length and Anchorage Length of Pretensioned Tendons

It is through the bond between the tendon and the concrete that the force in the tendon is transferred to the concrete in the case of pretensioning. This transfer is within a short distance called transfer length at the end of the member. Let the segment of the tendon within a distance x from the member end be taken out as a

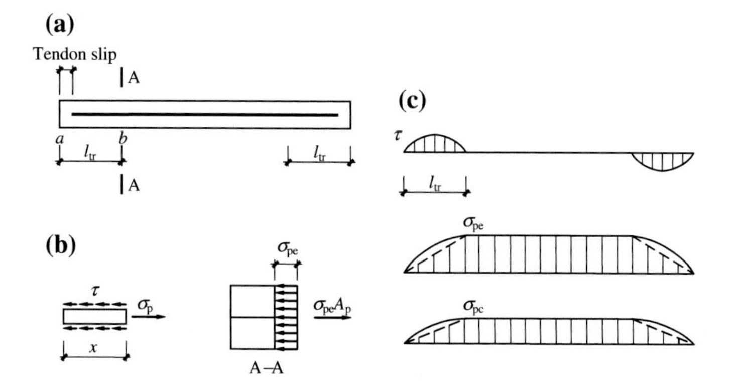

Fig. 10.19 Analysis of tendon releasing. **a** Tendon slip. **b** Bond stress and stress of section A–A. **c** Distribution of stresses

free body (Fig. 10.19). At the time of prestress transfer, the tendon will shorten and tend to slip with respect to the concrete. At the end point a, the prestress of the tendon is zero. Within the member, the shortening of the tendon is resisted by the concrete, resulting in compressive stresses in the concrete and bond stresses between the tendon and the concrete. With increasing value of x, the accumulation of the bond stresses is greater, and hence, the prestress σ_{pe} in the tendon and the prestress σ_{pc} in the concrete are also greater. When x reaches a certain length l_{tr} (the distance between sections a and b in Fig. 10.19a), the bond stress within the distance l_{tr} will be in equilibrium with the tension $\sigma_{pe}A_p$ in the tendon. Therefore, the required effective prestress σ_{pe} in the tendon can only be established for x that are beyond the length l_{tr}, that is, beyond section b. This length l_{tr} is therefore called the transfer length of the tendon; and the portion of the member between sections a and b is called the self-anchoring zone of the pretensioned member.

A linear variation of prestress in the self-anchoring zone can be assumed (Fig. 10.19c). For any calculation within the zone in which prestress level is used, such as checking of resistance to cracking, the reduced prestress in the zone must be considered. The transfer length is given by the formula

$$l_{tr} = \alpha_v \frac{\sigma_{pe}}{f'_t} d \tag{10.49}$$

where

σ_{pe} effective prestress of tendon;
d nominal diameter of the tendon;
α_v tendon's shape coefficient (Table 3.1); and
f'_t tensile strength of concrete at the time of prestress transfer.

If the release of tendon is sudden, the end region of the tendon could be debonded. Therefore, the starting point of the self-anchoring zone should be moved inwardly by a distance of 0.25 l_{tr}.

10.4.2 Anchorage Zone of Posttensioned Members

The tendon force in the case of posttensioned member is transferred to the concrete through the anchorage zone. The concrete under the anchorage plate is subjected to local compression. Such local compression can cause cracking in the anchorage zone. Hence, the strength of the anchorage zone must be checked by using the method given in Chap. 9. One thing to be noted is that the design value of the local compressive force should be taken as $F_l = 1.2\ \sigma_{con} A_p$.

10.5 Analysis of Members Subjected to Axial Tension

10.5.1 Characteristics of Pretressed Members Subjected to Axial Tension

In order to understand the characteristics of prestressed concrete members subjected to axial tension, experiment study was conducted on both prestressed concrete members and normal concrete members. The geometric and physical parameters of normal concrete members were all the same as the prestressed concrete members except the steel bars. Figure 10.20 presents the load–deflection curves (N_t-δ) of the

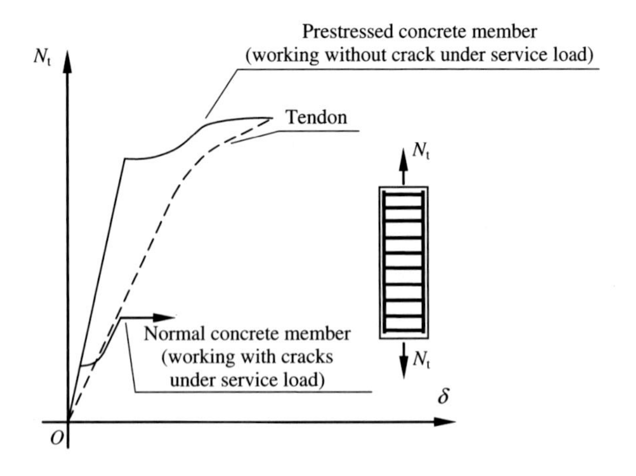

Fig. 10.20 N_t-δ curves of both prestressed concrete member and normal concrete member subjected to axial tension

test specimens. It is indicated that the N_t-δ relationship is approximately linear before cracking; the stress in tendons increases rapidly and the member goes into the nonlinear branch after cracking. The ultimate bearing capacity depends on the amount and strength of tendons. The prestressed concrete member has an initial stiffness almost the same as normal concrete members, but a much higher cracking load. The experiment reflects the characteristics of prestressed concrete members under axial tensile force. The whole process from tendon jacking to member failure of a member subjected to axial tension can be divided into two stages: the construction stage and the loading stage. These two stages will be analyzed in this section separately.

10.5.2 Pretensioned Members Subjected to Axial Tension

10.5.2.1 Construction Stage

The tendon is first jacked against the abutments. When the concrete reaches sufficient strength, the tendon is released. The force in the tendon so released will act on the whole transformed section of the member. This will not only cause precompression in the concrete but also cause reduction in the tendon force due to elastic shortening of the tendon. The method of self-balance of the section can be used for the analysis.

1. Jacking the tendon
   The tendon is jacked against the abutments to the control stress σ_{con}. Let the tendon area be A_p, then the total tension is $\sigma_{con}A_p$. Now, the concrete has not yet been cast.
2. After first batch of prestress loss
   At this moment, the concrete has not been subjected to compression yet. After the jacking was completed, the concrete was cast and cured, during which the first batch of prestress loss σ_{lI} has taken place. Now, the tendon stress is reduced from σ_{con} to $\sigma_{pe} = \sigma_{con} - \sigma_{lI}$. The concrete stress $\sigma_{pc} = 0$ because the tendon is not released yet.
3. Releasing the tendon
   When the tendon reaches sufficient strength (usually above 75 % of design strength), the tendon is released. Now that the tendon and the concrete have bonded together, the concrete is subjected to compression and the tendon stress is reduced due to its shortening.

The process of releasing the tendon is equivalent to applying the force $N_{pI} = (\sigma_{con} - \sigma_{lI})A_p$ on the transformed section A_0 (Fig. 10.21). The concrete stress so caused is

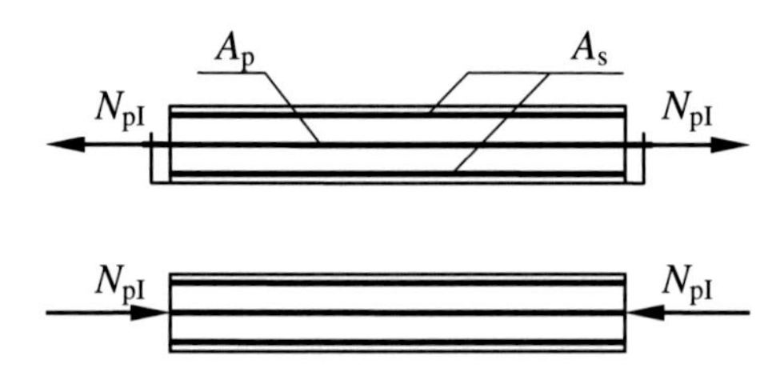

Fig. 10.21 The process of tendon releasing

$$\sigma_{pcI} = \frac{N_{pI}}{A_0} \tag{10.50}$$

Let $\alpha_E = E_s/E_c$ and $\alpha_{Ep} = E_p/E_c$, where E_s, $E_{p,}$ and E_c are the elastic moduli of steel rebars, tendon, and concrete, respectively, and the compressive stress in the steel rebars is

$$\sigma_{sI} = \alpha_{Es}\sigma_{pcI} \tag{10.51}$$

and the tensile stress in the tendon is reduced by an amount of

$$\Delta\sigma_p = \alpha_{Ep}\sigma_{pcI} \tag{10.52}$$

4. After second batch of prestress loss
   After the second batch of prestress loss has taken place, the stresses in the tendon and in the concrete are further reduced. Let the concrete stress be σ_{pcII}, then σ_{pcII} can be calculated in the same way as σ_{pcI} except that σ_{lI} is now replaced by the total loss σ_l. Hence, we have

$$N_{pII} = (\sigma_{con} - \sigma_l)A_p \tag{10.53}$$

and

$$\sigma_{pcII} = \frac{N_{pII}}{A_0} \tag{10.54}$$

Now, the tendon stress becomes

$$\sigma_{pII} = \sigma_{con} - \sigma_l - \alpha_{Ep}\sigma_{pcII} \tag{10.55}$$

and the compressive stress in the steel rebars is

$$\sigma_{sII} = \alpha_E\sigma_{pcII} \tag{10.56}$$

The term σ_{pcII} is also called the effective prestress established in the concrete.

10.5.2.2 Loading Stage

1. The destressing state
   The tensile load that causes exact zero stress in the concrete is called the destressing load N_{t0}. Now, the load is acting on the whole transformed section; hence,

$$N_{t0} = \sigma_{pcII} A_0 \tag{10.57}$$

   The tendon stress now becomes

$$\sigma_{p0} = \sigma_{con} - \sigma_l - \alpha_{Ep}\sigma_{pcII} + \alpha_{Ep}\sigma_{pcII} = \sigma_{con} - \sigma_l \tag{10.58}$$

2. The cracking load
   If the load is further increased beyond the destressing load N_{t0}, tensile stress will be induced in the concrete. The load corresponding to the state at which the tensile stress of concrete is equal to the characteristic tensile strength f_t is the cracking load N_{tcr}. Hence,

$$N_{tcr} = \left(\sigma_{pcII} + f_t\right) A_0 \tag{10.59}$$

   The cracking load of a prestressed member is obviously much greater than the nonprestressed member, because σ_{pcII} is usually much greater than f_t.
3. The ultimate load
   After cracking, the load is totally resisted by the tendon and the rebars. When the tendon and the rebars reach their tensile strengths, the member reaches its ultimate strength

$$N_{tu} = f_{py} A_p + f_y A_s \tag{10.60}$$

10.5.3 Posttensioned Members Subjected to Axial Tension

10.5.3.1 Construction Stage

1. Before jacking
   The process before jacking includes casting the concrete and curing. These are considered not to produce any stress in the member.
2. Jacking
   As the tendon is jacked against the concrete, the concrete is compressed, producing friction loss σ_{l2} in tendon stress. Hence, the tendon stress becomes $\sigma_p = \sigma_{con} - \sigma_{l2}$.

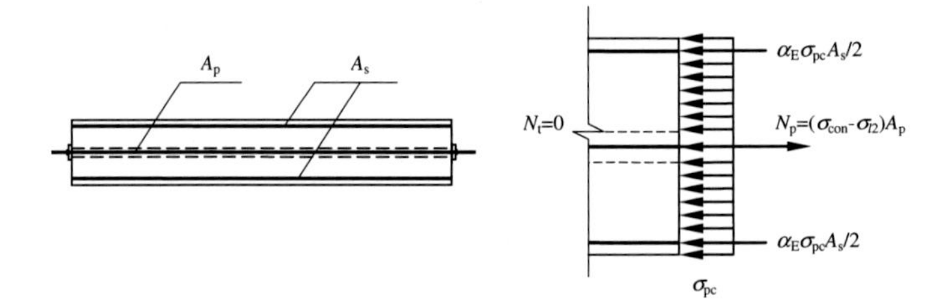

Fig. 10.22 Equilibrium of forces on section at jacking

Let the concrete stress be σ_{pc}, then the rebar stress is $\sigma_s = \alpha_E\sigma_{pc}$. Equilibrium of the section requires (Fig. 10.22)

$$(\sigma_{con} - \sigma_{l2})A_p = \sigma_{pc}A_c + \alpha_E\sigma_{pc}A_s \tag{10.61}$$

where A_c = the area of concrete in the section. Hence, we have

$$\sigma_{pc} = \frac{(\sigma_{con} - \sigma_{l2})A_p}{A_c + \alpha_E A_s} = \frac{(\sigma_{con} - \sigma_{l2})A_p}{A_n} = \frac{N_P}{A_n} \tag{10.62}$$

where $A_n = A_c + \alpha_E A_s$.

3. After first batch of prestress loss
   After jacking, the tendon is anchored against the member. The first batch of prestress loss has now taken place, which also includes the loss due to anchorage deformation. Let the concrete stress be σ_{pcI}, then the stress in the rebar becomes $\sigma_{sI} = \alpha_E\sigma_{pcI}$. From the equilibrium of the stresses in the section, we have

$$\sigma_{pcI} = \frac{(\sigma_{con} - \sigma_{lI})A_p}{A_c + \alpha_E A_s} = \frac{N_{pI}}{A_n} \tag{10.63}$$

   where $N_{pI} = (\sigma_{con} - \sigma_{lI})A_p$.
4. After second batch of prestress loss
   Now, the losses due to relaxation, creep, and shrinkage have taken place. That is, the total prestress loss has taken place. The tendon stress becomes $\sigma_{con} - \sigma_l$. Because the tendon force is in equilibrium with the compression of the concrete, we have

$$\sigma_{pcII} = \frac{(\sigma_{con} - \sigma_{lI} - \sigma_{lII})A_p}{A_c + \alpha_E A_s} = \frac{N_{pII}}{A_n} \tag{10.64}$$

where $N_{pII} = (\sigma_{con} - \sigma_l)A_p$.

10.5.3.2 Loading Stage

1. The destressing state
   Again, the load that causes zero stress in the concrete is

$$N_{t0} = \sigma_{pcII} A_0 \tag{10.65}$$

   The rebar stress is zero now, but the tendon stress becomes

$$\sigma_{p0} = \sigma_{con} - \sigma_l + \alpha_{Ep} \sigma_{pcII} \tag{10.66}$$

2. The cracking load
   The load N_{tcr} that causes cracking of the concrete corresponds to the concrete stress f_t. Hence,

$$N_{tcr} = N_{t0} + f_t A_0 = (\sigma_{pcII} + f_t) A_0 \tag{10.67}$$

   The tendon stress now is

$$\sigma_p = \sigma_{con} - \sigma_l + \alpha_{Ep} (\sigma_{pcII} + f_t) \tag{10.68}$$

3. Ultimate load
   The ultimate load of the member again is

$$N_{tu} = f_{py} A_p + f_y A_s \tag{10.69}$$

because now only the tendon and the rebar resist the load.

10.5.4 Comparison Between Pretensioned and Posttensioned Members and Discussion

From the above analyses, we have the following conclusions:

1. The loading stage formulae, i.e., the formulae for N_{t0}, $N_{tcr,}$ and N_{tu}, are identical in the form for both pretensioned and posttensioned members. However, the values for effective prestress σ_{pcII} are different. The construction stage formulae for both cases are almost the same except that A_0 is used for pretensioning, while A_n is used for posttensioning.
2. The tendon stress is always at high levels. In addition, the variation in tendon stress before cracking of concrete is very small. In contrast, the concrete is subjected to only compression before the load reaches N_{t0}. Therefore, the advantages of both materials are fully utilized.

3. The cracking load N_{tcr} of a prestressed member is usually much greater than the cracking load of a nonprestressed member, with the accompanying phenomenon of the cracking load setting up closer to the ultimate load. The latter phenomenon can be problematic in certain cases.
4. The ultimate bearing capacity of a prestressed member is equal to the ultimate bearing capacity of the corresponding nonprestressed member with the same materials and section dimensions.

10.6 Design of Members Subjected to Axial Tension

In the design of a prestressed member, the whole process from construction to service to ultimate state must be considered. Therefore, the design can also be divided into construction stage and loading stage. These include the calculation of ultimate bearing capacity as well as a number of separate status checks on crack resistance, crack width if cracked, bearing capacity of member, and local bearing strength of the anchorage zone (if applicable) at jacking or when releasing tendons.

10.6.1 Design for the Loading Stage

10.6.1.1 Ultimate Bearing Capacity

At ultimate limit state, the total load is resisted by tendon and rebars only (Fig. 10.23), and they all have yielded. Let the design axial tensile load be N_t. Then, the requirement for the design is

$$N_t \leqslant N_{tu} = f_{py}A_p + f_yA_s \tag{10.70}$$

where

f_{py} yield strength of the tendon; and

f_y yield strength of the rebar.

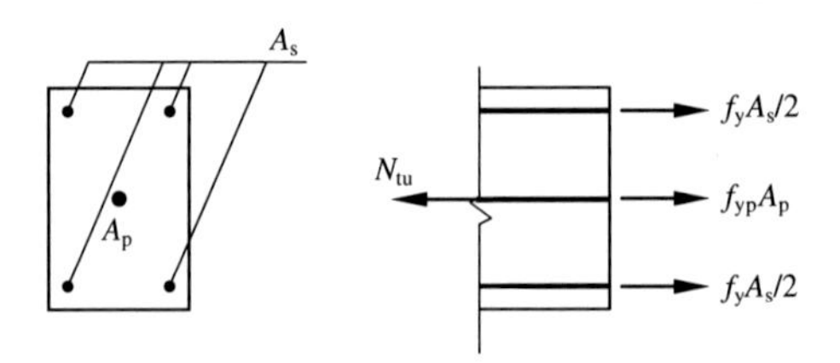

Fig. 10.23 Forces on section at ultimate limit state

10.6.1.2 Crack Resistance Checking

The main purpose of prestressing is to increase the cracking load. Therefore, checking for crack resistance is one of the main tasks in the design. However, since this is a status check task that falls in the field of serviceability, details on the check requirements will be presented in Chap. 11.

10.6.2 Design for Construction Stage

The construction stage design should include all critical events during construction. However, in this section, only the state of prestress transfer is considered. This corresponds to jacking in the case of posttensioned members and releasing the tendons in the case of pretensioned members. At prestress transfer, the prestressing force is the highest, while the strength of the concrete is usually lower than its design strength. Therefore, the relevant strengths must be checked:

10.6.2.1 Check on Member Strength at Prestress Transfer

The compressive stress σ_{cc} in the concrete caused by prestress transfer must satisfy the following condition

$$\sigma_{cc} \leqslant 0.8 f_c' \tag{10.71}$$

where

f_c' compressive strength of concrete prism at the time of prestress transfer.

In the case of pretensioned members, the first batch of prestress loss σ_{lI} is considered to have taken place in the calculation of σ_{cc}. Therefore,

$$\sigma_{cc} = \frac{(\sigma_{con} - \sigma_{l1})A_p}{A_0} \tag{10.72}$$

In the case of posttensioned members, no prestress loss is considered in the calculation of σ_{cc}. Therefore,

$$\sigma_{cc} = \frac{\sigma_{con} A_p}{A_n} \tag{10.73}$$

Note that in the above two equations, A_0 is used for pretensioning and A_n for posttensioning cases.

10.6.2.2 Check on Local Bearing Strength of the Anchorage Zone

The method given in Chap. 9 can be used for this check. In this case, the design compressive force acting on the local area is taken as $F_l = 1.2\,\sigma_{con}A_p$.

10.6.3 Steps for the Design

The steps for the design of a prestressed member subjected to axial tension are as follows:

1. Determine the section dimension, the strengths and the elastic moduli of concrete, tendons, and rebars, the percentage of concrete strength at prestress transfer, the tendon control stress, the construction method (pretensioning or posttensioning), the internal forces caused by external loading, the coefficient of structural importance, etc.
2. Determine A_p and A_s according to the ultimate bearing capacity requirement.
3. Calculate the prestress loss σ_l.
4. Calculate the effective compressive prestress σ_{pcII}.
5. Check the crack resistance or the crack width under service load conditions (by using the relevant contents in Chap. 11). If this is not satisfied, go back to step 1 and adjust initial parameters.
6. Check the construction stage. For pretensioned members, check the concrete strength requirement at tendon releasing; for posttensioned members, check the concrete strength requirement and the local bearing strength requirement at jacking. If any of these are not satisfied, go back to step 1 and adjust the initial parameters until all requirements are satisfied.

Example 10.1 Design the prestressed concrete lower chord of a 24-m roof truss. The design conditions are as given in Table 10.9.

Solution

1. Loading stage calculation
   From Eq. (10.65), we have

$$A_p = \frac{\gamma_0 N_t - f_y A_s}{f_{py}} = \frac{1.1(408{,}000 + 196{,}000) - 300 \times 452}{1110} = 476\ \text{mm}^2$$

   Take $24\phi^P5$ (A_p = 471 mm^2) and the result is shown in Fig. 10.24c.

Table 10.9 Design conditions of Example 10.1

Materials	Concrete	Tendon	Rebar
Type and strength grade	C40	$\phi^{P}5$	HRB335
Section	250 × 160 (mm) with ducts 2ϕ50	–	4 ϕ12 (A_s = 452 mm^2) as required by detailing
Strength (N·mm^{-2})	$f_c = 19.1, f_t = 1.71, f_{tk} = 2.39$	$f_{py} = 1110$ $f_{ptk} = 1570$	$f_y = 300$ $f_{yk} = 335$
Elastic modulus (N·mm^{-2})	$E_c = 3.25 \times 10^4$	$E_{ps} = 2.05 \times 10^5$	$E_s = 2 \times 10^5$
Prestressing method	Posttensioning at only one end with overtensioning; button head anchors for wires adopted. Ducts are made by withdrawal of pressured rubber tube		
Control stress	$\sigma_{con} = 0.75 f_{ptk} = 0.75 \times 1570 = 1177.5\ \mathrm{N/mm^2}$		
Concrete strength at jacking	$f'_{cu} = 40\ \mathrm{N/mm^2}, f'_c = 26.8\ \mathrm{N/mm^2}$		
Internal forces in the lower chord	Characteristic tension N_{tk} = 340 kN and design tension N_t = 408 kN due to dead load		
	Characteristic tension N_{tk} = 140 kN and design tension N_t = 196 kN due to live load		
	Quasi-permanent coefficient for live load is 0		
*Structural importance coefficient	γ_0 = 1.1; Crack control grade: generally no cracking		

Notes In the application of Eq. (10.70) for a real project, the equation is usually adjusted by using the structural importance coefficient to consider the different safety requirements for different kinds of structures. For example, in Chinese code GB 50010, Eq. (10.70) is modified as $\gamma_0 N_t = N_{tu} = f_y A_s + f_p A_p$

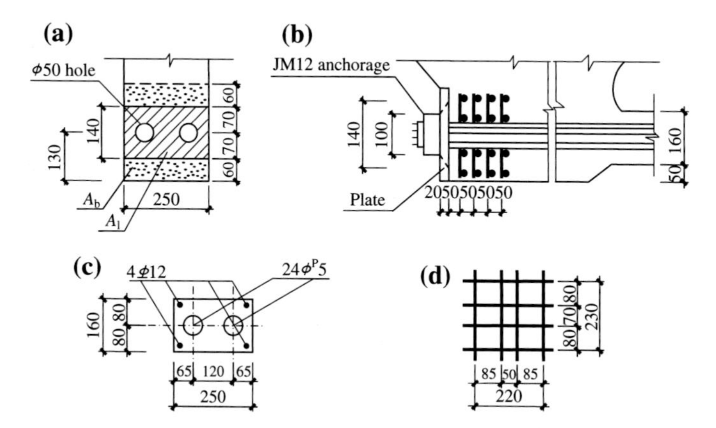

Fig. 10.24 The lower chord of a roof truss. **a** Compression area. **b** End joint of the lower chord. **c** Reinforcement of the section. **d** Rebar mesh

2. Prestress loss and effective prestress

(1) Geometric properties of the section (Fig. 10.24)

$$\alpha_E = \frac{E_s}{E_c} = \frac{2.0 \times 10^5}{3.25 \times 10^4} = 6.15, \ \alpha_{Ep} = \frac{E_p}{E_c} = \frac{2.05 \times 10^5}{3.25 \times 10^4} = 6.31$$

$$\begin{aligned} A_n &= A_c + \alpha_E A_s = 250 \times 160 - 2 \times \frac{3.14}{4} \times 50^2 - 452 + 6.15 \times 452 \\ &= 38,401 \text{ mm}^2 \end{aligned}$$

$$A_0 = A_n + \alpha_{Ep} A_p = 38,403 + 6.31 \times 471 = 41,373 \text{ mm}^2$$

(2) Prestress losses

$$\sigma_{con} = 1000 \text{ N/mm}^2$$

Loss due to anchorage deformation:
With 1 mm gap of nut and 1 mm gap of plate, thus

$$\sigma_{l1} = \frac{a}{l} E_p = \frac{2}{24,000} \times 2.05 \times 10^5 = 17.1 \text{ N/mm}^2$$

Loss due to friction:
At the anchored end, l = 24 m and θ = 0°. Thus,

$$\sigma_{l2} = \sigma_{con}(\kappa x + \mu\theta) = 1000 \times (0.0014 \times 24 + 0) = 33.6 \text{ N/mm}^2$$

The first batch of loss is

$$\sigma_{lI} = \sigma_{l1} + \sigma_{l2} = 17.1 + 33.6 = 50.7 \text{ N/mm}^2$$

Loss due to relaxation:

$$\begin{aligned} \sigma_{l4} &= 0.4\psi \left(\frac{\sigma_{con}}{f_{ptk}} - 0.5 \right) \sigma_{con} = 0.4 \times 0.9 \times \left(\frac{1000}{1570} - 0.5 \right) \times 1000 \\ &= 49.3 \text{ N/mm}^2 \end{aligned}$$

Loss due to creep and shrinkage:
The concrete stress after first batch of loss is

$$\sigma_{pcI} = \frac{(\sigma_{con} - \sigma_{l1})A_p}{A_n} = \frac{(1000 - 50.7) \times 471}{38,401} = 11.64 \text{ N/mm}^2$$

Other relevant quantities are:

$$\frac{\sigma_{pc}}{f'_{cu}} = \frac{11.64}{40} = 0.29 < 0.5$$

$$\rho = \frac{A_s + A_p}{A_n} = \frac{452 + 471}{38,403} = 0.024$$

Hence,

$$\sigma_{l5} = \frac{55 + 300\left(\frac{\sigma_{pc}}{f'_{cu}}\right)}{1 + 15\rho} = \frac{55 + 300 \times 0.29}{1 + 15 \times 0.024} = 104.4\ \text{N/mm}^2$$

The second batch of loss is

$$\sigma_{lII} = \sigma_{l4} + \sigma_{l5} = 49.3 + 104.4 = 153.7\ \text{N/mm}^2$$

The total loss is

$$\sigma_l = \sigma_{lI} + \sigma_{lII} = 50.7 + 153.7 = 204.4\ \text{N/mm}^2 > 80\ \text{N/mm}^2$$

Therefore, the effective prestress in concrete is

$$\sigma_{pcII} = \frac{(\sigma_{con} - \sigma_l)A_p}{A_n} = \frac{(1000 - 204.4) \times 471}{38,401} = 9.76\ \text{N/mm}^2$$

3. Check on crack resistance
   See Example 11.2 in Chap. 11.
4. Check on construction stage

(1) Concrete stress check: because of 5 % overtensioning, the maximum jacking force is

$$N_p = 1.05\sigma_{con}A_p = 1.05 \times 1000 \times 471 = 494{,}550\ \text{N}$$

The corresponding concrete stress is

$$\sigma_{cc} = \frac{N_p}{A_n} = \frac{494{,}550}{38{,}401} = 12.88\ \text{N/mm}^2 < 0.8f'_c = 0.8 \times 26.8 = 21.44\ \text{N/mm}^2$$

(2) Local bearing strength check: the diameter of the anchorage is 100 mm. The thickness of the plate is 20 mm. The local bearing area can in principle be calculated as the area of the anchorage plus the enlarged part obtained by assuming a 45° proliferation angle. In the calculation, the area of the two circles can be approximated by a rectangle as shown in Fig. 10.24a.

Local bearing area:

$$A_l = 250 \times (100 + 2 \times 20) = 35{,}000\ \text{mm}^2$$

Base area:

$$A_\text{b} = 250 \times (140 + 2 \times 60) = 65{,}000\ \text{mm}^2$$

$$\beta_l = \sqrt{\frac{A_\text{b}}{A_l}} = \sqrt{\frac{65{,}000}{35{,}000}} = 1.36$$

Design value of local compressive force:

$$F_l = 1.2\sigma_\text{con}A_\text{p} = 1.2 \times 1000 \times 471 = 565{,}200\ \text{N} = 565.2\ \text{kN}$$

Net local bearing area:

$$A_{l\text{n}} = 35{,}000 - 2 \times \pi/4 \times 50^2 = 31{,}075\ \text{mm}^2$$

Check local bearing strength:

$$\begin{aligned} F_{l\text{u}} &= 0.9\beta_\text{c}\beta_l f_\text{c} A_{l\text{n}} = 0.9 \times 1.0 \times 1.36 \times 19.1 \times 31{,}075 \\ &= 726{,}483.8\,\text{N} \approx 726.5\,\text{kN} > F_l = 565.2\,\text{kN, OK} \end{aligned}$$

Four lateral indirect rebar meshes made of $\phi 6$ welded bars with spacing s = 50 mm are taken as shown in Fig. 10.22d.

Area of mesh core:

$$A_\text{cor} = 220 \times 230 = 50{,}600\,\text{mm}^2 < A_\text{b} = 650{,}00\,\text{mm}^2$$

Volumetric reinforcement ratio of the meshes:

$$\rho_\text{v} = \frac{n_1 A_{s1} l_1 + n_2 A_{s2} l_2}{A_\text{cor} s} = \frac{4 \times 28.3 \times 220 + 4 \times 28.3 \times 230}{50{,}600 \times 50} = 0.02 > 0.5\,\%, \text{OK.}$$

10.7 Analysis of Prestressed Flexural Members

10.7.1 Characteristics of Pretressed Flexural Members

Similar to prestressed concrete members subjected to tensile loading, the whole process from tendon jacking to a flexural member failure can be divided into two stages: the construction stage and the loading stage. Each stage also includes several branches. Figure 10.25 shows the load–deflection curve of a posttensioned concrete beam under two concentrated loads P. Point ① represents the camber with stress losses and self-weight neglected (camber means the deflection of the beam caused by tendon jacking). Point ② indicates the camber with self-weight neglected but stress losses considered. Indeed, Points ① and ② are theoretically estimated, and Point ③ is the real camber. The whole cross section of the beam is in compression at Point ④. When the stress at the bottom edge of the beam equals to zero, it comes to Point ⑤. The beam cracks at Point ⑥ and reaches the ultimate bearing capacity at Point ⑧. It is shown from Fig. 10.25 that the cracking moment of a prestressed concrete beam approaches to the ultimate bearing capacity and the deflection at failure is reduced significantly due to the camber.

10.7.2 Pretensioned Flexural Members

10.7.2.1 Construction Stage

First, the tendons are jacked against their abutments. After the concrete has been cast and has reached a certain strength (usually equal to or greater than 75 % of the design strength), the tendons are released. As shown in Fig. 10.26, the prestressing force $N_{\mathrm{pI}} = (\sigma_{\mathrm{con}} - \sigma_{II})A_{\mathrm{p}} + (\sigma'_{\mathrm{con}} - \sigma'_{II})A'_{\mathrm{p}}$, so released force is acting in the opposite direction (as compression) on the whole transformed section composed of concrete, tendons, and rebars. For simplicity, the rebars are not shown in Fig. 10.26.

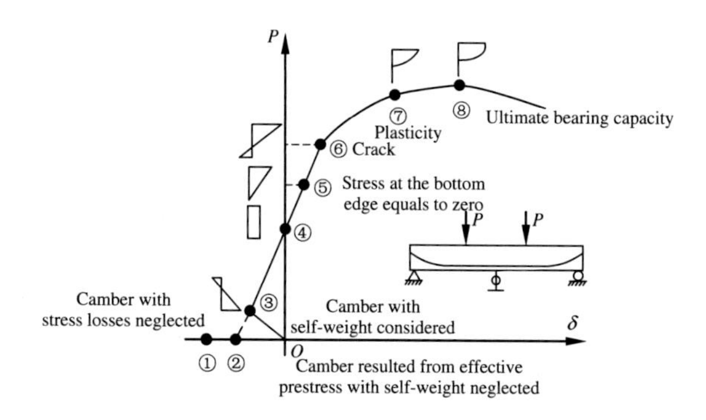

Fig. 10.25 P-δ curve of a prestressed flexural member

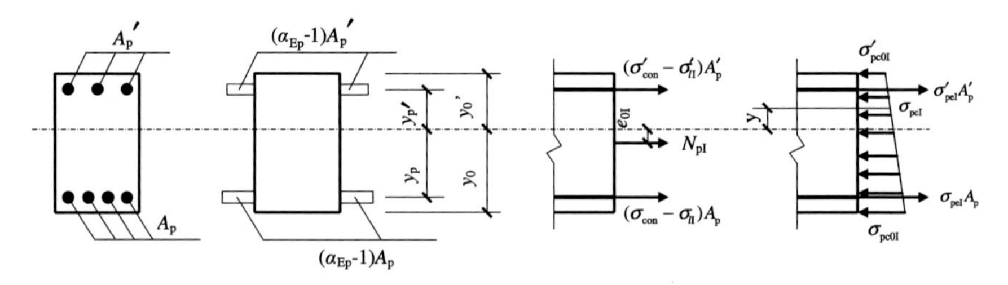

Fig. 10.26 Transformed section and distribution of stress in the section after releasing tendons

Hence, the compressive stress in the concrete fiber at a distance y_0 from the neutral axis (positive for downward distance) of the beam becomes (Fig. 10.26)

$$\sigma_{pcI} = \frac{N_{pI}}{A_0} \pm \frac{N_{pI} e_{0I}}{I_0} y \tag{10.74a}$$

$$e_{0I} = \frac{(\sigma_{con} - \sigma_{lI}) A_p y_p - (\sigma'_{con} - \sigma'_{lI}) A'_p y'_p}{N_{pI}} \tag{10.74b}$$

where

A_0 and I_0 the sectional area and the moment of inertia of the transformed section; and

y_p and y'_p the distances from the resultants of tendon forces in tension and compression zones to the centroid of the transformed cross section, respectively.

The stresses in the tendons are

$$\sigma_{peI} = \sigma_{con} - \sigma_{lI} - \alpha_{Ep} \sigma_{pcI} \tag{10.75}$$

$$\sigma'_{peI} = \sigma'_{con} - \sigma'_{lI} - \alpha_{Ep} \sigma'_{pcI} \tag{10.76}$$

After the second batch of prestress loss has taken place, the stress in the concrete and tendons can be calculated in the same way as above except that σ_{lI} is replaced by σ_l. That is, now N_{pI} is replaced by N_{pII}, where

$$N_{pII} = (\sigma_{con} - \sigma_l) A_p + (\sigma'_{con} - \sigma'_l) A'_p \tag{10.77a}$$

$$e_{0II} = \frac{(\sigma_{con} - \sigma_l) A_p y_p - (\sigma'_{con} - \sigma'_l) A'_p y'_p}{N_{pII}} \tag{10.77b}$$

Thus, the compressive stress of concrete at the fiber represented by y becomes

$$\sigma_{\mathrm{pcII}} = \frac{N_{\mathrm{PII}}}{A_0} \pm \frac{N_{\mathrm{PII}} e_0}{I_0} y_0 \tag{10.78}$$

The stresses in the tendons are

$$\sigma_{\mathrm{peII}} = \sigma_{\mathrm{con}} - \sigma_l - \alpha_{\mathrm{Ep}} \sigma_{\mathrm{pcII}} \tag{10.79}$$

$$\sigma'_{\mathrm{peII}} = \sigma'_{\mathrm{con}} - \sigma'_l - \alpha_{\mathrm{Ep}} \sigma'_{\mathrm{pcII}} \tag{10.80}$$

σ_{pcII} and σ'_{pcII} are the effective prestress terms established in the concrete in the tension and compression zone, respectively.

10.7.2.2 The Loading Stage

1. The destressing moment M_0
   The external moment is resisted by the whole transformed section. The moment that causes zero stress in the extreme precompressed fiber of concrete is called the destressing moment M_0. Let the y coordinate of the bottom fiber be y_0. If we denote the compressive prestress of this fiber as σ_{pc0II}, we have

$$\sigma_{\mathrm{pc0II}} = \frac{M_0}{I_0} y_0 \tag{10.81}$$

   Then, by definition, we have

$$M_0 = \frac{\sigma_{\mathrm{pc0II}} I_0}{y_0} \tag{10.82}$$

2. Cracking moment M_{cr}
   The cracking moment is the moment that causes cracking of the extreme precompressed concrete fiber. Therefore, we have

$$M_{\mathrm{cr}} = \left(\sigma_{\mathrm{pc0II}} + \gamma f_{\mathrm{t}}\right) \frac{I_0}{y_0} \tag{10.83}$$

   where γ = a coefficient to consider the plasticity of concrete in the tension zone. Theoretically, γ can be calculated by letting the right side of Eq. 5.16 equal to the right side of Eq. 5.14. But it is not easy to get such a general equation in a real application.

Table 10.10 Basic value of the coefficient to consider concrete in the tension zone γ_m

Item	①	②	③		④		⑤
Shape of the section	Rectangular section	T section with compressive flange	Symmetrical I section or box section		T section with tensile flange		Circular or annular section
			$b_f/b \leqslant 2$	$b_f/b > 2$ $h_f/h < 0.2$	$b_f/b \leqslant 2$	$b_f/b > 2$ $h_f/h < 0.2$	
γ_m	1.55	1.5	1.45	1.35	1.50	1.40	1.6–0.24 r_1/r

Notes (1) For I sections, if , the value between Item ② and Item ③ can be used; otherwise, the value between Item ③ and Item ④ can be used
(2) For box sections, b stands for the summation of all of the widths of the webs
(3) r_1 stands for the inner radius of an annular section. For circular sections, take $r_1 = 0$

An experienced equation was recommended in the *Code for Design of Concrete Structures* (GB 50010), which is expressed as Eq. (10.84).

$$\gamma = \left(0.7 + \frac{120}{h}\right)\gamma_m \tag{10.84}$$

where

γ_m the basic value of the coefficient to consider the plasticity of concrete in the tension zone, which can be found in Table 10.10; and

h the height of the section (mm). Take $h = 400$ when $h < 400$, and take $h = 1600$ when $h > 1600$; for circular or annular sections, take $h = 2r$, where r is the radius of a circular section or the outer radius of an annular section.

Equation (10.85) was recommended in the *Code for Design of Highway Reinforced Concrete and Prestressed Concrete Bridges and Culverts* (JTG62-2012).

$$\gamma = \frac{2S_0}{W_0} \tag{10.85}$$

where, S_0 = the area moment for the sectional area above or below the centroid of the transformed section about the centroid.

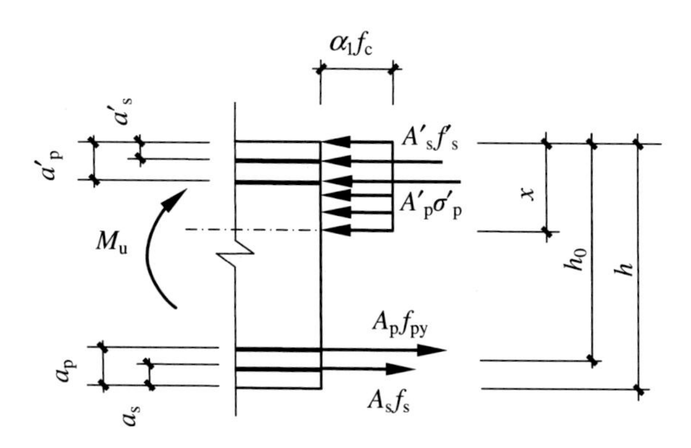

Fig. 10.27 Ultimate limit state

3. Ultimate moment M_u
   If the section is under-reinforced, all longitudinal tendons and rebars in the tension zone will have yielded at ultimate limit state. The calculation model for this case is shown in Fig. 10.27 (the rebars are added in the figure), from which we have the following equilibrium equations:

$$\alpha_1 f_c bx = f_y A_s + f_{py} A_p - f'_y A'_s - \sigma'_p A'_p \tag{10.86}$$

$$M_u = \alpha_1 f_c bx\left(h_0 - \frac{x}{2}\right) + f'_y A'_s\left(h_0 - a'_s\right) + \sigma'_p A'_p(\mathrm{h}_0 - a'_p) \tag{10.87}$$

In the above equations, it is assumed that the compression rebars also yield at the ultimate state. This, of course, should be checked.

10.7.3 Posttensioned Flexural Members

10.7.3.1 Construction Stage

In the case of posttensioned members, the prestressing force N_p is directly applied on the concrete section A_n, which only includes the concrete and the rebar. This is illustrated in Fig. 10.28. For simplicity, the rebars are not shown in Fig. 10.28.

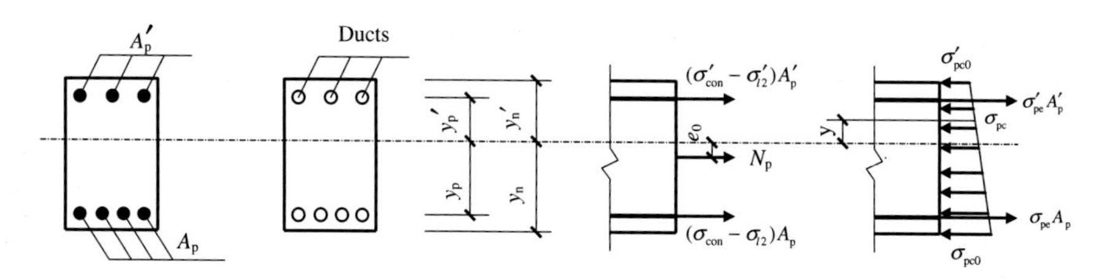

Fig. 10.28 Section and stress distribution of the section when jacking the tendons

From equilibrium, the stresses in the concrete can be expressed as

$$\sigma_{\mathrm{pc}} = \frac{N_{\mathrm{p}}}{A_{\mathrm{n}}} + \frac{N_{\mathrm{p}} e_0}{I_{\mathrm{n}}} y \tag{10.88}$$

where A_{n}, I_{n}, e_0, and y are the corresponding geometric quantities associated with the concrete section only including the concrete and the rebars. In addition, A_{n} and I_{n} are the area and moment of inertia of the concrete section, respectively; e_0 is the eccentricity of the resultant tendon force as measured from the center of the concrete section; and y is the distance between the concrete section center and the layer of concrete considered (downward as positive).

After the first batch of prestress loss occurs, the force in the tendon becomes

$$N_{\mathrm{pI}} = (\sigma_{\mathrm{con}} - \sigma_{lI})A_{\mathrm{p}} + (\sigma'_{\mathrm{con}} - \sigma'_{lI})A'_{\mathrm{p}} \tag{10.89}$$

The compressive stress in the concrete becomes

$$\sigma_{\mathrm{pcI}} = \frac{N_{\mathrm{pI}}}{A_{\mathrm{n}}} \pm \frac{N_{\mathrm{pI}} e_{0\mathrm{I}}}{I_{\mathrm{n}}} y \tag{10.90}$$

After the second batch of prestress loss occurs, the force in the tendon becomes

$$N_{\mathrm{pII}} = (\sigma_{\mathrm{con}} - \sigma_l)A_{\mathrm{p}} + (\sigma'_{\mathrm{con}} - \sigma'_l)A'_{\mathrm{p}} \tag{10.91}$$

and the compressive stress in the concrete becomes

$$\sigma_{\mathrm{pcII}} = \frac{N_{\mathrm{pII}}}{A_{\mathrm{n}}} \pm \frac{N_{\mathrm{pII}} e_{0\mathrm{II}}}{I_{\mathrm{n}}} y \tag{10.92}$$

Let y_{n} be the y coordinate corresponding to the extreme compression fiber of concrete (usually the bottom fiber of the section), then the effective prestress established in the concrete fiber is

$$\sigma_{\mathrm{pcII}} = \frac{N_{\mathrm{pII}}}{A_{\mathrm{n}}} + \frac{N_{\mathrm{pII}} e_{0\mathrm{II}}}{I_{\mathrm{n}}} y_{\mathrm{n}}. \tag{10.93}$$

10.7.3.2 The Loading Stage

The formulae for the analysis of the loading stage of a posttensioned beam are essentially the same in form as those for a pretensioned beam. In both cases, the beams reach a similar state, i.e., the effective prestress σ_{pcII} is established. From this state onward, the loading processes are the same; therefore, identical formulae are used for both cases.

1. The destressing moment M_0
   Similar to the case of a pretensioned beam, the destressing moment M_0, i.e., the moment that makes the stress of the extreme precompression fiber zero, is:

$$M_0 = \frac{\sigma_{\mathrm{pc0II}} I_0}{y_0} \tag{10.94}$$

2. The cracking moment M_{cr}
   The formula for cracking moment M_{cr} is also identical in form to that of the pretensioning case. Thus,

$$M_{\mathrm{cr}} = \left(\sigma_{\mathrm{pc0II}} + \gamma f_{\mathrm{t}}\right) \frac{I_0}{y_0} \tag{10.95}$$

3. The ultimate moment M_{u}
   The ultimate moment M_{u} is expressed by the same set of equations as that in the case of a pretensioned beam.

10.8 Design of Prestressed Flexural Members

Design of a prestressed concrete flexural member normally includes the design of normal sections, the check on crack resistance of normal sections, the design of inclined sections based on shearing capacities, the crack resistance of inclined sections, and the check on the construction stage.

10.8.1 Design of Normal Sections

The design for a prestressed normal section generally follows the same principles as found in the design of an ordinary nonprestressed section. But there are several special issues to be noted. In the following part, these special issues will be analyzed and the design method will be presented.

10.8.1.1 The Relative Height of Compression Zone at Balanced Failure

The tendon stress corresponding to a state at which the stress of the concrete at the very location of the tendon equals zero is defined as σ_{p0}, and the corresponding tendon strain is defined as $\varepsilon_{\mathrm{p0}}$. According to the tradition, if the tendon is located in the tension zone, the term σ_{p0} is used; otherwise, the term σ'_{p0} is used. Usually, the

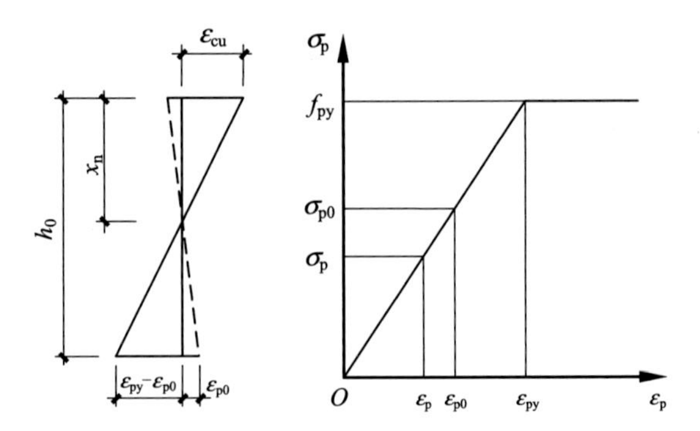

Fig. 10.29 Strain profile for tendon with yield point

group of tendons in the tension zone is treated as one tendon represented by their center line (or the resultant of the group of tendons); in this case, the term σ_{p0} (or σ'_{p0}) represents the stress in the group of tendons when the stress of the concrete at the center of the group is zero.

The derivation of the relative height of the compression zone at balanced failure, ξ_b, for tendons with (evident) yield point is different from tendons without (evident) yield point; however, the common principle is the same. That is, at the state σ_{p0}, the tendon has already been subjected to a stress σ_{p0}. Therefore, the remaining strain that can be developed before yielding is only the difference between the yielding strain and the strain at the state σ_{p0}.

The state σ_{p0} corresponds to a zero stress in the concrete. From this state, the concrete at the extreme compression fiber will develop ultimate strain ε_{cu} at balanced failure. Also from this state, the tendon which has already been subjected to the stress σ_{p0} will only be able to develop the strain $\varepsilon_{py} - \varepsilon_{p0}$ at balanced failure, where ε_{py} is the strain at yielding, and $\varepsilon_{p0} = \sigma_{p0}/E_p$.

If the external load is zero, the concrete at the level of the tendon is subjected to compression. Under the load that causes the tendon stress σ_{p0}, the strain profile is shown in Fig. 10.29 by the vertical line showing zero strain in the concrete. At balanced failure, the strain profile is shown in Fig. 10.29 by the line corresponding to tensile strain $\varepsilon_{py} - \varepsilon_{p0}$ in the tendon. Thus, from Fig. 10.29, we have

$$\xi_b = \frac{\beta_1}{1 + \frac{f_{py} - \sigma_{p0}}{E_p \varepsilon_{cu}}} \tag{10.96}$$

If the tendon does not have a yield point, then $\varepsilon_{py} = \varepsilon_{0.2} = f_{py}/E_p + 0.002$. The states corresponding to zero load, σ_{p0}, and balanced failure, as shown in Fig. 10.30, are similar to those in Fig. 10.29. Thus, from Fig. 10.30, we have

$$\xi_b = \frac{\beta_1}{1 + \frac{0.002}{\varepsilon_{cu}} + \frac{f_{py} - \sigma_{p0}}{E_p \varepsilon_{cu}}} \tag{10.97}$$

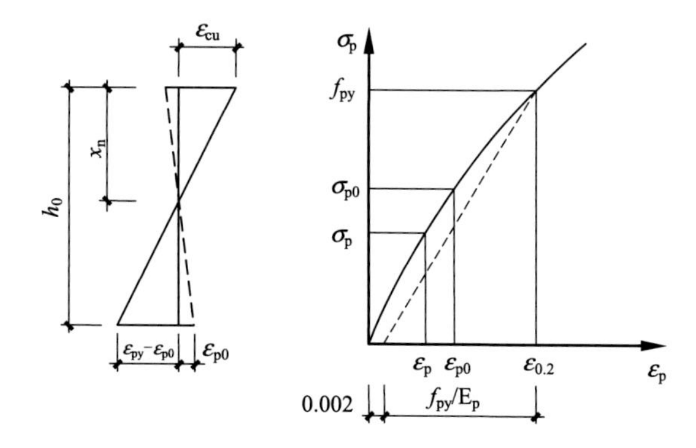

Fig. 10.30 Strain profile for tendon without yield point

If there are several layers of tendons or rebars, each of them will have its own ξ_b value. These values should be calculated separately, and usually, the smallest value ξ_b is used.

10.8.1.2 Stress in Tendons or Rebars at Arbitrary Layer

Obviously, the calculation of the stress of a tendon or a rebar at an arbitrary layer in the section at ultimate state is needed to calculate the ultimate strength. This can be done as follows.

Take the *i*th tendon as an example (Fig. 10.31). Let the position of the tendon measured from the extreme compression fiber of concrete be h_{0i}, the stress of the

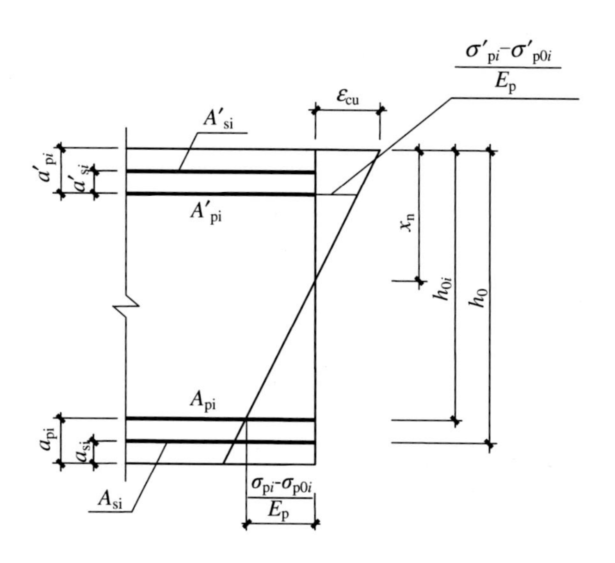

Fig. 10.31 Strain profile at ultimate state

tendon be $\sigma_{\rm pi}$, and its stress corresponding to zero concrete stress be $\sigma_{\rm p0i}$. From Fig. 10.31, we have

$$\frac{\varepsilon_{\rm cu}+\frac{\sigma_{\rm p\it i}-\sigma_{\rm p0\it i}}{E_{\rm p}}}{h_{0i}}=\frac{\varepsilon_{\rm cu}}{x_{\rm n}} \quad (10.98)$$

Substituting $x=\beta_1 x_{\rm n}$ into Eq. (10.98) yields

$$\sigma_{\rm p\it i}=\varepsilon_{\rm cu}E_{\rm p}\left(\frac{\beta_1 h_{0i}}{x}-1\right)+\sigma_{\rm p0\it i} \quad (10.99)$$

If the steel bar is not prestressed, then we obtain the rebar stress at ultimate state as:

$$\sigma_{\rm s\it i}=\varepsilon_{\rm cu}E_{\rm s}\left(\frac{\beta_1 h_{0i}}{x}-1\right) \quad (10.100)$$

If the approximate method (linear interpolation) is used, the above two equations become

$$\sigma_{\rm p\it i}=\frac{f_{\rm py}-\sigma_{\rm p0\it i}}{\xi_{\rm b}-\beta_1}\left(\frac{x}{h_{0i}}-\beta_1\right)+\sigma_{\rm p0\it i} \quad (10.101)$$

$$\sigma_{\rm s\it i}=\frac{f_{\rm y}}{\xi_{\rm b}-\beta_1}\left(\frac{x}{h_{0i}}-\beta_1\right) \quad (10.102)$$

The above two equations are those used in the usual design calculations. In these equations, $\sigma_{\rm pi}$ and $\sigma_{\rm si}$ are the stress in the tendon and the rebar of the *i*th layer, respectively, with positive value for tension; h_{0i} is the distance between the center of the *i*th layer steel and the extreme compression fiber of concrete; x is the height of the equivalent rectangular stress block acting on the compression zone; and $\sigma_{\rm p0i}$ is the tendon stress of the *i*th layer of the tendon corresponding to a state at which the stress in the concrete located at the center of the tendon is zero.

The above two equations are applicable only when the steel has not yielded. That is, in the case of a tendon, the following must be satisfied:

$$\sigma_{\rm p0\it i}-f'_{\rm py}\leqslant\sigma_{\rm p\it i}\leqslant f_{\rm py} \quad (10.103)$$

where $f'_{\rm py}>0$ is the design value of yield compressive stress (i.e., the compressive strength of the tendon usually not greater than 360 N/mm^2). If $\sigma_{\rm p\it i}>f_{\rm py}$, take $\sigma_{\rm p\it i}=f_{\rm py}$; if $\sigma_{\rm pi}<\sigma_{\rm p0\it i}-f'_{\rm py}$, take $\sigma_{\rm pi}=\sigma_{\rm p0\it i}-f'_{\rm py}$. If the *i*th layer of tendon is in the compression zone of the section, the term $\sigma_{\rm p0\it i}$ is replaced by $\sigma'_{\rm p0\it i}$ according to the tradition.

In the case of a rebar, the following must be satisfied:

$$-f'_{y} \leqslant \sigma_{si} \leqslant f_{y} \tag{10.104}$$

where $f'_{y} > 0$ is the compressive strength of the rebar. If $\sigma_{si} > f_{y}$, take $\sigma_{si} = f_{y}$; if $\sigma_{si} < (-f'_{y})$, take $\sigma_{si} = (-f'_{y})$.

10.8.1.3 Design Formulae for Flexural Bearing Capacities of Normal Sections

At ultimate limit state, the tension steels in an under-reinforced section will yield. That is, the tendons and the rebars in the tension zone will yield. The rebars in the compression zone normally will also yield. However, the tendon in the compression zone normally will not yield in compression, because its σ_{p0} value usually is very great. As is indicated by Eq. (10.103), the tendon strain can only be reduced from σ_{p0}/E_{p} by a value not greater than f'_{py}/E_{p} because of the limited ultimate compressive strain of the concrete (usually taken as 0.002 in this case). Therefore, the stress of a tendon in the compression zone at ultimate state cannot be less than $\sigma'_{p0} - f'_{py}$, which in many cases is still tensile.

With the above analyses, we are now ready to establish the equilibrium equations of the section at ultimate limit state. The calculation model is shown in Fig. 10.32, from which the following equilibrium equations can be derived:

$$\sum X = 0, \quad \alpha_{1} f_{c} b x = f_{y} A_{s} - f'_{y} A'_{s} + f_{py} A_{p} + \sigma'_{p} A'_{p} \tag{10.105}$$

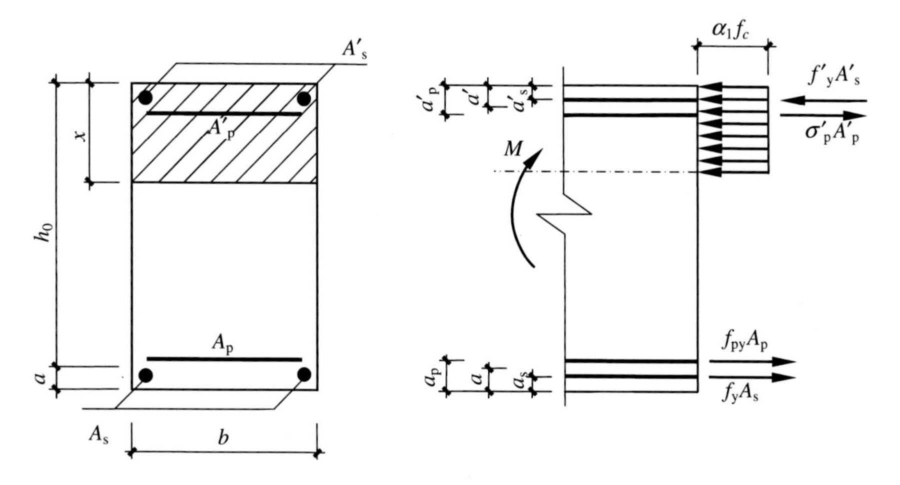

Fig. 10.32 Section and equilibrium at ultimate limit state

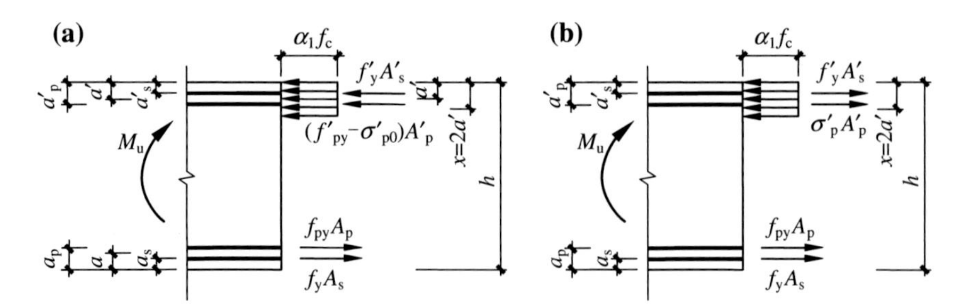

Fig. 10.33 Methods for case $x < 2a'$

$$\sum M = 0, \quad M \leqslant M_u = \alpha_1 f_c bx\left(h_0 - \frac{x}{2}\right) + f'_y A'_s(h_0 - a'_s) - \sigma'_p A'_p\left(h_0 - a'_p\right) \tag{10.106}$$

where M is the moment acting on the section at ultimate limit state; a' is the distance between the resultant of the tendons and the rebars in the compression zone and the edge of the compression zone. If there is no tendon in the compression zone or if the tendon stress σ'_p is tensile, a' should be replaced by a'_s. Meanwhile, a'_s and a'_p are the positions of the resultants of rebars and tendons in the compression zone, respectively, measured as distances from the edge of the compression zone; and σ'_p is the stress of the tendon in the compression zone at ultimate limit state given by the expression

$$\sigma'_p = \sigma'_{p0} - f'_{py} \tag{10.107}$$

Equations (10.105) and (10.106) require the following conditions on height x of the rectangular stress block:

$$x \leqslant \xi_b h_0 \tag{10.108}$$

$$x \geqslant 2a' \tag{10.109}$$

In case that $x < 2'_a$, the flexural strength of the section should be calculated as follows. If σ'_p is compressive, take $x = 2'_a$ as shown in Fig. 10.33a. Then, we have

$$M \leqslant M_u = f_{py}A_p\left(h - a_p - a'\right) + f_y A_s(h - a_s - a') \tag{10.110}$$

If σ'_p is tensile, take $x = 2'_a$ as shown in Fig. 10.33b. Then, we have

$$M \leqslant M_u = f_{py}A_p\left(h - a_p - a'_s\right) + f_y A_s\left(h - a_s - a'_s\right) + \sigma'_p A'_p\left(a'_p - a'_s\right) \tag{10.111}$$

The above method is suitable for rectangular sections and T sections with only tension flanges.

For I sections and T sections with compressive flanges, the flexural strength calculation is similar to nonprestressed section's calculation. That is, if either the condition

$$f_y A_s + f_{py} A_p \leqslant \alpha_1 f_c b'_f h'_f + f'_y A'_s - \sigma'_p A'_p \tag{10.112}$$

or the condition

$$M \leqslant M_u = \alpha_1 f_c b'_f h'_f \left(h_0 - \frac{h'_f}{2} \right) + f'_y A'_s (h_0 - a'_s) - \sigma'_p A'_p \left(h_0 - a'_p \right) \tag{10.113}$$

is satisfied, the section belongs to a Type I T section. In this case, the section can be calculated as the rectangular section with width b'_f (Fig. 10.33a) by using the following formulae

$$\alpha_1 f_c b'_f x = f_y A_s - f'_y A'_s + f_{py} A_p + \sigma'_p A'_p \tag{10.114}$$

$$M \leqslant M_u = \alpha_1 f_c b'_f x \left(h_0 - \frac{x}{2} \right) + f'_y A'_s (h_0 - a'_s) - \sigma'_p A'_p \left(h_0 - a'_p \right) \tag{10.115}$$

If the above condition is not satisfied, the section is classified as a Type II T section with $x > h'_f$. The corresponding formulae are (Fig. 10.34b) as

$$\sum X = 0, \quad \alpha_1 f_c \left[bx + \left(b'_f - b \right) h'_f \right] = f_y A_s - f'_y A'_s + f_{py} A_p + \sigma'_p A'_p \tag{10.116}$$

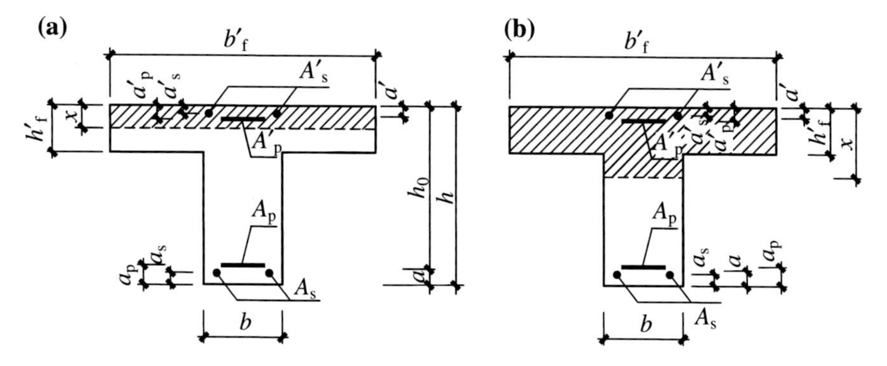

Fig. 10.34 The case of T sections

$$\sum M = 0,$$
$$M \leqslant M_u = \alpha_1 f_c bx\left(h_0 - \frac{x}{2}\right) + \alpha_1 f_c (b'_f - b)\left(h_0 - \frac{h'_f}{2}\right)h'_f + f'_y A'_s (h_0 - a'_s) - \sigma'_p A'_p \left(h_0 - a'_p\right) \tag{10.117}$$

where
h'_f and b'_f the height and effective width of the compressive flange, respectively.

The following conditions on the range of x are used in conjunction with the above equations:

$$x \leqslant \xi_b h_0 \tag{10.118}$$

$$x \geqslant 2a' \tag{10.119}$$

It can be seen that the tendons in the compression zone will generally not contribute much to the ultimate bearing capacity. They are there mainly for the purpose of controlling cracks during the construction stage.

Although there are some differences in the formulae for the ultimate bearing capacity between the prestressed and nonprestressed sections, the design formulae for the flexural strength of prestressed sections are generally similar.

10.8.2 Design of Inclined Sections

Prestressing is generally beneficial to the shear resistance of a beam since prestressing can help resist the appearance and development of cracks. The shear compression zone is also made larger by prestressing, which results in a higher contribution to the shear resistance by concrete.

Figure 10.35 shows the experimental results about the effect of prestress level on the shear resistance, where σ_{pc0} is the compressive prestress in concrete at the center of the transformed section. Test results indicate that the increase in shear resistance

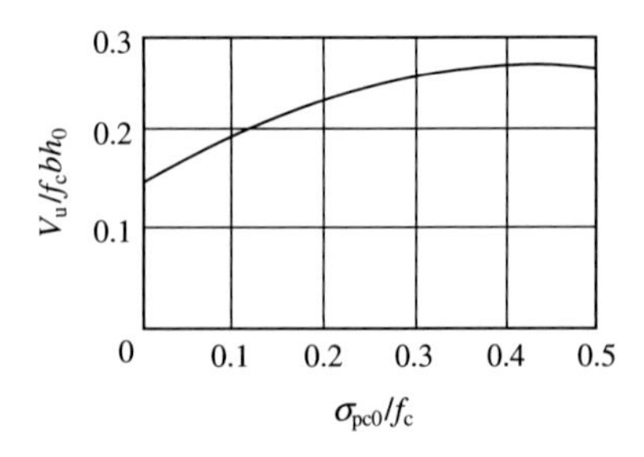

Fig. 10.35 Effect of prestress on shear resistance

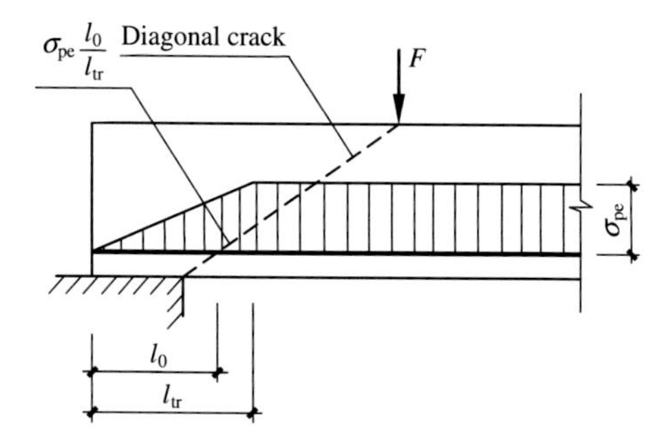

Fig. 10.36 Variation of prestress within zone of transfer

depends on the level of prestress as well as the acting point of the prestressing force. It has been found that if the ratio σ_{pc0}/f_c exceeds $0.3 \sim 0.4$, the beneficial effect of the prestress will be reduced.

Therefore, the design formula for the shear resistance of prestressed members is

$$V \leqslant V_u = V_{cs} + V_p = 0.7 f_t b h_0 + f_{yv} \frac{A_{sv}}{s} h_0 + 0.05 N_{p0} \tag{10.120}$$

where $V_p = 0.05 N_{p0}$ and N_{p0} is the resultant of all tendons and rebars when the stress in the concrete is zero, and the meaning of the other symbols is the same as previously defined. There is a limit on N_{p0}. That is, if $N_{p0} > 0.3 f_c A_0$, then take $N_{p0} = 0.3 f_c A_0$.

Within the zone of prestress transfer of a pretensioned member, N_{p0} should be calculated according to the assumed linear distribution of prestress, which has been indicated in Sect. 10.4.1 of this chapter (Fig. 10.36). This leads to the formula

$$N_{p0} = \sigma_{pe} \frac{l_0}{l_{tr}} A_p \tag{10.121}$$

where l_{tr} is calculated according Eq. (10.49) and l_0 is the distance between the point of intersection of the tendon and the inclined cracks and the end of the member.

In the following cases, the term V_p is taken as zero: (1) The moment caused by N_{p0} is in the same direction as the external moment. (2) The beam is continuous. (3) The beam is simply supported, but cracking is allowed.

If the beam is with bent-up bars and tendons, the design formula is

$$V \leqslant V_u = V_{cs} + V_p + 0.8 f_y A_{sb} \sin \alpha_s + 0.8 f_{py} A_{pb} \sin \alpha_p \tag{10.122}$$

where

A_{sb} and A_{pb} the areas of the bent-up bars and tendons, respectively, which are effective for shear resistance (i.e., within the same section at which the shear capacity is calculated); and

α_s and α_p the angles of inclination of the bent-up bars and the bent-up tendons, respectively.

For a simply supported beam subjected to a concentrated load (including the case in which the shear caused by concentrated load at the support section is greater than or equal to 75 % of the total shear at that section), the term V_{cs} should adopt the following form

$$V_{cs} = \frac{1.75}{\lambda + 1.0} f_t b h_0 + f_{yv} \frac{A_{sv}}{s} h_0 \tag{10.123}$$

where λ = the shear-span ratio.

In order to avoid inclined compression failure, the section dimension should meet the following conditions:

$$V \leqslant 0.25 \beta_c f_c b h_0 \text{ as } \frac{h_w}{b} \leqslant 4.0 \tag{10.124}$$

$$V \leqslant 0.2 \beta_c f_c b h_0 \text{ as } \frac{h_w}{b} \geqslant 6.0 \tag{10.125}$$

$$\text{Linear interpolation is adopted as } 4.0 \leqslant \frac{h_w}{b} \leqslant 6.0 \tag{10.126}$$

$$\text{If the condition } V \leqslant 0.7 f_t b h_0 + 0.05 N_{p0} \tag{10.127}$$

is satisfied for the usual prestressed flexural members of rectangular, T section or I section, or if the condition

$$V \leqslant \frac{1.75}{\lambda + 1.0} f_t b h_0 + 0.05 N_{p0} \tag{10.128}$$

is satisfied for a simply supported beam subjected to concentrated loading, the calculation of the shear capacity is not required, and the stirrups can be provided only based on detailing requirements.

The range of applications of the above formulae, the critical section positions, and the steps for the calculations are the same as those for nonprestressed members.

10.8.3 Serviceability Checks

The serviceability checks of prestressed flexural members include the crack resistance check of normal sections, the crack resistance check of inclined sections, the crack width check if cracking is allowed, and the deflection check. These contents can be found in Chap. 11 of this book.

10.8.4 Check on the Construction Stage

The eccentricity of each tendon's center in a flexural member together with the negative moment caused by self-weight during the process of transportation and erection may cause cracking, especially at the top fiber of the member. Such cracking will reduce the stiffness of the member and affect the crack resistance during service of the member. For this and other reasons, the construction stage must be checked for all critical events.

For members in which cracking is not allowed during the process of construction including manufacture, transportation, and erection, or for members that are subjected to compression over the whole section during prestressing, the stresses at the edges of the section due to the actions of prestressing, self-weight, and construction load (including dynamic coefficient if necessary) must satisfy the following conditions:

$$\sigma_{\mathrm{ct}} \leqslant 1.0 f_{\mathrm{t}}' \tag{10.129}$$

$$\sigma_{\mathrm{cc}} \leqslant 0.8 f_{\mathrm{c}}' \tag{10.130}$$

where

f_{t}' and f_{c}' the tensile and compressive strengths of the concrete at the time of construction corresponding to the concrete cube strength f_{cu}', respectively;

σ_{cc} and σ_{ct} the compressive and tensile stresses at the edges of control sections corresponding to the construction stage.

The terms σ_{cc} and σ_{ct} are calculated by using the following equations.

For pretensioned members $$\sigma_{\mathrm{cc}} = \sigma_{\mathrm{pc0I}} + \frac{N}{A_0} + \frac{M}{W_0} \tag{10.131a}$$

$$\sigma_{\mathrm{ct}} = \left| \sigma_{\mathrm{pc0I}}' + \frac{N}{A_0} - \frac{M}{W_0} \right| \tag{10.131b}$$

For post-tensioned members $$\sigma_{\mathrm{cc}} = \sigma_{\mathrm{pc0I}} + \frac{N}{A_{\mathrm{n}}} + \frac{M}{W_{\mathrm{n}}} \tag{10.132a}$$

$$\sigma_{\mathrm{ct}} = \left| \sigma_{\mathrm{pc0I}}' + \frac{N}{A_{\mathrm{n}}} - \frac{M}{W_{\mathrm{n}}} \right| \tag{10.132b}$$

where

σ_{pc0I} and σ'_{pc0I} the normal compressive and tensile stress in concrete, respectively, due to prestressing (positive for compression);

N and M the axial force and moment, respectively, on the control section due to the self-weight and construction load; and

W_0 and W_{n} the section moduli for the corresponding section edges.

For members in which cracking is allowed during the construction process if there is no tendon in the pretensioned zone, the normal stresses at the edges of the section should satisfy the following conditions:

$$\sigma_{\mathrm{ct}} \leqslant 2.0 f'_{\mathrm{t}} \tag{10.133}$$

$$\sigma_{\mathrm{cc}} \leqslant 0.8 f'_{\mathrm{c}} \tag{10.134}$$

The method for local bearing strength checking of the anchorage zone is the same as that for members subjected to axial tension.

For prestressed members in hoisting and transportation, the negative bending moment induced by self-weight will crack the concrete in the tension zone, decrease the stiffness of the members, and influence the crack resistance in service. Checking in this stage is the same as that in the construction stage, except in the calculation of σ_{cc} and σ_{ct} where one more term is added to consider the negative bending moment caused by self-weight, M_{q}, which is 1.5 $M_{\mathrm{q}}y_0/I_0$ for pretensioned members and 1.5 $M_{\mathrm{q}}y_{\mathrm{n}}/I_{\mathrm{n}}$ for posttensioned members. The number 1.5 in the term is the dynamic coefficient.

10.8.5 Steps for Design of Prestresed Flexural Members

The following steps are generally followed in the design:

(1) Determine the section dimension, as well as the strengths and elastic moduli of the concrete, the tendons, and the rebars. Determine the strength of concrete at the time of prestress transfer, the control stress of the tendons, and the construction method (pretensioning or posttensioning). Choose the initial section area of the tendons and the rebars. Look up the structural importance coefficient and calculate the internal forces of the structure.
(2) Calculate the prestress loss σ_l.
(3) Calculate the effective prestress σ_{pcII}.
(4) Calculate the bearing capacities of normal sections at ultimate limit state. If the bearing capacity requirement is not satisfied, go back to step (1), adjust the parameters, and do the above steps again.
(5) Check the crack resistance and crack width (if applicable) of the normal sections under serviceability conditions. If any of the requirements are not satisfied, go back to step (1).

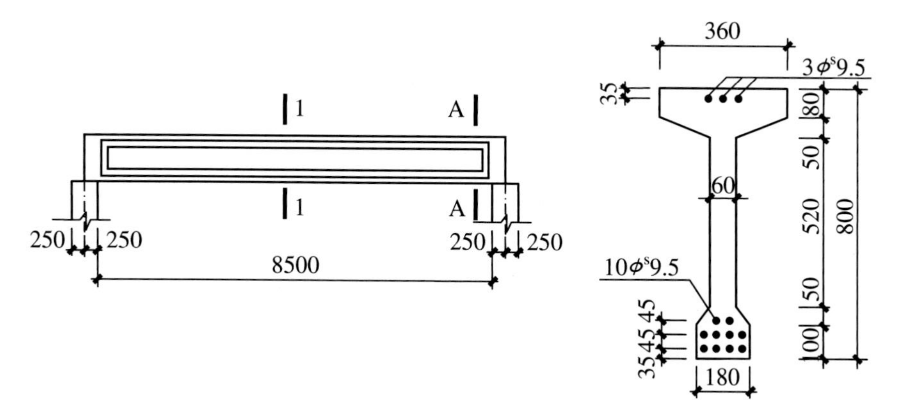

Fig. 10.37 Figure of Example 10.2

(6) Check the shear capacity at ultimate limit state. If this is not satisfied, adjust the parameters and redo the calculation.
(7) Check the resistance to inclined cracking under serviceability conditions. If this is not satisfied, adjust the parameters and redo the calculation.
(8) Check the deflections. If this is not satisfied, adjust the parameters and redo the calculation.
(9) Check the construction stage including manufacture, transportation, lifting, and erection. If any of these are not satisfied, adjust the parameters and redo the calculation.

Example 10.2 A prestressed concrete beam has a clear span of 8.5 m. Its section dimensions and reinforcing details are as shown in Fig. 10.37. The tendons are pretensioned on a 100-m-long casting bed with overtensioning. The tendons are released when the concrete reaches 90 % of its design strength. The beam is subjected to a uniformly distributed load q = 38.6 kN/m. Check the strengths and the bearing capacities of the beam for all stages.

Solution

1. Data for calculation

(1) Concrete: Choose C40 grade. Concrete cover is 25 mm.

$$f_{cu} = 40\,\text{N/mm}^2, f'_{cu} = 0.9 \times 40 = 36\,\text{N/mm}^2$$
$$f_c = 26.8\ \text{N/mm}^2, f'_c = 0.9 \times 26.8 = 24.12\ \text{N/mm}^2$$
$$f_t = 2.39\ \text{N/mm}^2, f'_t = 0.9 \times 2.39 = 2.15\ \text{N/mm}^2$$
$$E_c = 3.25 \times 10^4\ \text{N/mm}^2.$$

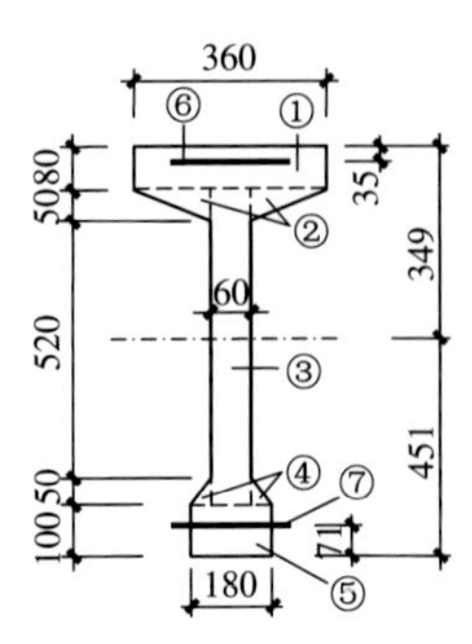

Fig. 10.38 Section geometry of Example 10.2

(2) Tendon and reinforcement: Choose ϕ^{s} 9.5 tendon

$f_{py} = 1220\,\text{N/mm}^2, f_{ptk} = 1720\,\text{N/mm}^2, f'_{py} = 390\,\text{N/mm}^2$,
$E_p = 1.95 \times 10^5\,\text{N/mm}^2, A'_p = 212.5\,\text{mm}^2 (3\phi^{s}9.5)$,
$A_p = 708.5\ \text{mm}^2\ (10\phi^{s}9.5)$.
HPB300 steel is used for stirrups with $f_y = 270\ \text{N/mm}^2$.

(3) Control stress: $0.7f_{ptk} = 0.7 \times 1720 = 1204\ \text{N/mm}^2$, take $\sigma_{con} = \sigma'_{con} = 880\ \text{N/mm}^2$.

(4) Others: a 100-m-long casting bed with no anchorage loss and with temperature difference $\Delta t = 20\ °\text{C}$ due to curing.

2. Internal force calculation
Effective span $l = l_0 + a = 8.5 + 0.25 = 8.75\,\text{m}$.
Maximum moment and shear:

$$M_{max} = \frac{1}{8} \times 38.6 \times 8.75^2 = 369.41\,\text{kN m}$$

$$V_{max} = \frac{1}{2} \times 38.6 \times 8.75 = 164.05\,\text{kN}$$

3. Geometric properties of the section
With the division and numbering of the section as shown in Fig. 10.38, the following can be obtained as follows:
The modulus ratio is expressed as $a_{Ep} = \frac{E_p}{E_c} = \frac{1.95 \times 10^5}{3.25 \times 10^4} = 6.0$
The distance a_p between the center of the lower tendons and the bottom edge of the section is expressed as:

$$a_p = \frac{4 \times 35 + 4 \times 80 + 2 \times 125}{10} = 71\,\text{mm}$$

$$h_0 = 800 - 71 = 729\,\text{mm}$$

Table 10.11 Geometric parameters of Example 10.2

No.	A_i/mm^2	a_i/mm	$S_i = A_i a_i$/mm^3	y_i/mm	$A_i y_i^2$/mm^4	I_i/mm^4
(1)	28,800	760	21,888,000	310	276,768 × 10^4	1536 × 10^4
(2)	7500	703	5,273,000	253	48,006.8 × 10^4	104 × 10^4
(3)	37,200	410	15,252,000	40	5952 × 10^4	119,164 × 10^4
(4)	3000	117	351,000	333	33,266.7 × 10^4	42 × 10^4
(5)	18,000	50	900,000	400	288,000 × 10^4	1500 × 10^4
(6)	1063	765	813,195	315	10,548 × 10^4	
(7)	3543	71	251,553	379	50,892 × 10^4	
	99,106		44,728,748		713,434 × 10^4	122,346 × 10^4

The distance from the center of the section to the bottom of the section:

$$y_0 = \frac{\sum S_i}{\sum A_i} = \frac{44{,}728{,}748}{99{,}106} = 451\,\text{mm}$$

The distance from the center of the section to the top of the section:

$$y_0' = 800 - 451 = 349\,\text{mm}$$

Moment of inertia of the section:

$$I_0 = \sum A_i y_i^2 + \sum I_i = 835{,}780 \times 10^4\,\text{mm}^4$$

The details of the calculation are listed in Table 10.11.
In Table 10.11, a_i is the distance from the center of area A_i to the bottom of the section; y_i is the distance from the center of area A_i to the center of the section; and I_i is the moment of inertia of area A_i with respect to its own center.

4. Prestress loss

(1) Loss due to anchorage deformation

$$\sigma_{l1} = \sigma_{l1}' = 0 \quad \text{according to the given condition.}$$

(2) Loss due to temperature difference

$$\sigma_{l3} = \sigma_{l3}' = 2\Delta t = 2 \times 20 = 40\,\text{N/mm}^2$$

(3) Loss due to relaxation

Overtensioning is used. Therefore,

$$\sigma_{l4} = 0.035\sigma_{\text{con}} = 0.035 \times 880 = 30.8\ \text{N/mm}^2 = \sigma_{l4}'$$

(4) First batch of prestress loss and related quantities
With 50 % of relaxation loss grouped into first batch, we have

$$\sigma_{lI} = \sigma_{l3} + 0.5\sigma_{l4} = 40 + 0.5 \times 30.8 = 55.4\,\text{N/mm}^2 = \sigma'_{lI}$$

The resultant of tendons after first batch of loss:

$$\begin{aligned} N_{\text{pI}} &= (\sigma_{\text{con}} - \sigma_{lI})A_{\text{p}} + \left(\sigma'_{\text{con}} - \sigma'_{lI}\right)A'_{\text{p}} \\ &= (880 - 55.4) \times 708.5 + (880 - 55.4) \times 212.5 = 759{,}456.6\,\text{N} \end{aligned}$$

Distance from the position of resultant of tendons to center of section:

$$\begin{aligned} e_{0\text{I}} &= \frac{(\sigma_{\text{con}} - \sigma_{lI})A_{\text{p}}y_{\text{p}} - \left(\sigma'_{\text{con}} - \sigma'_{lI}\right)A'_{\text{p}}y'_{\text{p}}}{N_{\text{pI}}} \\ &= \frac{(880 - 55.4) \times 708.5 \times 379 - (880 - 55.4) \times 212.5 \times 315}{759{,}456.6} \\ &= 218.9\,\text{mm} \end{aligned}$$

Compressive prestress in concrete at centers of A_{p} and A'_{p}:

$$\sigma_{\text{pcI}} = \frac{N_{\text{pI}}}{A_0} + \frac{N_{\text{pI}}e_{\text{p0I}}}{I_0}y_{\text{p0}} = \frac{759{,}456.6}{99{,}106} + \frac{759{,}456.6 \times 218.9}{835{,}780 \times 10^4} \times 379 = 15.20\,\text{N/mm}^2$$

$$\sigma'_{\text{pcI}} = \frac{N_{\text{pI}}}{A_0} - \frac{N_{\text{pI}}e_{\text{p0I}}}{I_0}y'_{\text{p0}} = \frac{759{,}456.6}{99{,}106} - \frac{759{,}456.6 \times 218.9}{835{,}780 \times 10^4} \times 315 = 1.39\,\text{N/mm}^2$$

(5) Loss due to creep and shrinkage

$$\rho = \frac{A_{\text{p}}}{A_0} = \frac{708.5}{99{,}106} = 0.00715, \quad \rho' = \frac{A'_{\text{p}}}{A_0} = \frac{212.5}{99{,}106} = 0.00214$$

$$\frac{\sigma_{\text{pcI}}}{f'_{\text{cu}}} = \frac{15.2}{0.9 \times 40} = 0.422, \quad \frac{\sigma'_{\text{pcI}}}{f'_{\text{cu}}} = \frac{1.39}{0.9 \times 40} = 0.0386$$

$$\sigma_{l5} = \frac{60 + 340\frac{\sigma_{\text{pcI}}}{f'_{\text{cu}}}}{1 + 15\rho} = \frac{60 + 340 \times 0.422}{1 + 15 \times 0.00715} = 183.77\,\text{N/mm}^2$$

$$\sigma'_{l5} = \frac{60 + 340\frac{\sigma'_{\text{pcI}}}{f'_{\text{cu}}}}{1 + 15\rho'} = \frac{60 + 340 \times 0.0386}{1 + 15 \times 0.00214} = 70.85\,\text{N/mm}^2$$

(6) Second batch and total losses
Second batch losses:

$$\sigma_{lII} = 0.5\sigma_{l4} + \sigma_{l5} = 0.5 \times 30.8 + 183.77 = 199.17\,\text{N/mm}^2$$

$$\sigma'_{lII} = 0.5\,\sigma'_{l4} + \sigma'_{l5} = 0.5 \times 30.8 + 70.85 = 86.25\,\text{N/mm}^2$$

Total prestress losses:

$$\sigma_l = \sigma_{lI} + \sigma_{lII} = 55.4 + 199.17 = 254.57\,\text{N/mm}^2 > 100\,\text{N/mm}^2$$
$$\sigma'_l = \sigma'_{lI} + \sigma'_{lII} = 55.4 + 86.25 = 141.65\,\text{N/mm}^2 > 100\,\text{N/mm}^2$$

5. Calculation of bearing capacity of normal section

$$\sigma_{p0} = \sigma_{con} - \sigma_l = 880 - 254.57 = 625.43\,\text{N/mm}^2$$
$$\sigma'_{p0} = \sigma'_{con} - \sigma'_l = 880 - 141.65 = 738.35\,\text{N/mm}^2$$
$$\sigma'_p = \sigma'_{p0} - f'_{py} = 738.35 - 390 = 348.35\,\text{N/mm}^2 \quad \text{(tension)}$$

$$x = \frac{f_{py}A_p + \sigma'_p A'_p}{f_c b'_f} = \frac{1220 \times 708.5 + 348.35 \times 212.5}{26.8 \times 360}$$
$$= 97.26\,\text{mm} < h'_f = 80 + \frac{50}{2} = 105\,\text{mm}$$

Hence, the neutral axis is in the flange, and it is a Type I T section.

$$\xi = \frac{x}{h_0} = \frac{97.26}{729} = 0.133 \leqslant \xi_b = \frac{0.8}{1 + \frac{0.002}{\varepsilon_{cu}} + \frac{f_{py} - \sigma_{p0}}{0.0033E_s}}$$
$$= \frac{0.8}{1 + \frac{0.002}{0.0033} + \frac{1220 - 625.43}{0.0033 \times 1.95 \times 10^5}} = 0.316$$

$$M_u = 588.09\,\text{kN·m} \geqslant M_{max} = 369.41\,\text{kN·m}, \quad \text{OK.}$$

6. Bearing capacity of inclined section
Check on section size:

$$h_w = 520\,\text{mm} \qquad \frac{h_w}{b} = \frac{520}{60} = 8.67 > 6.0$$

$$0.2\beta_c f_c b h_0 = 0.2 \times 26.8 \times 60 \times (800 - 71)$$
$$= 234{,}446\,\text{N} = 234.45\,\text{kN} > V_{max} = 164.05\,\text{kN}$$

Calculate stirrups:

$$0.7f_tbh_0 + \min\{0.05N_{p0},\ 0.05(0.3f_cA_0)\} = 0.7 \times 2.39 \times 60 \times 729 + 30001$$
$$= 103178\text{N} < 164050\text{N}$$

Hence, stirrups are needed. Take $\phi 8@150$ with two legs, we have

$$\begin{aligned} V &= 0.7f_tbh_0 + f_{yv}\frac{A_{sv}}{s}h_0 + V_P \\ &= 103,178 + 270 \times \frac{2 \times 50.24}{150} \times 729 \\ &= 235,028\,\text{N} = 235.03\,\text{kN} > V_{max} = 164.05\,\text{kN},\ \text{OK} \end{aligned}$$

7. Strength during construction and crack resistance

   (1) At prestress transfer

$$N_{pI} = 759.457\,\text{kN} \quad e_{0I} = 218.9\,\text{mm}$$

   Stress of concrete at top edge of section:

$$\begin{aligned} \sigma_{ct} &= \frac{N_{pI}}{A_0} - \frac{N_{pI}e_{0I}}{I_0}y_0' = \frac{759{,}457}{99{,}106} - \frac{759{,}457 \times 218.9}{835{,}780 \times 10^4} \times 349 \\ &= 0.80\,\text{N/mm}^2(\text{compression}) < f_t' = 2.15\,\text{N/mm}^2,\ \text{OK} \end{aligned}$$

   Stress of concrete at bottom edge of section:

$$\begin{aligned} \sigma_{cc} &= \frac{N_{pI}}{A_0} + \frac{N_{pI}e_{p0I}}{I_0}y_0 = \frac{759{,}457}{99{,}106} + \frac{759{,}457 \times 218.9}{835{,}780 \times 10^4} \times 451 \\ &= 16.63\,\text{N/mm}^2 < 0.8f_c' = 19.30\,\text{N/mm}^2,\ \text{OK} \end{aligned}$$

   (2) At hoisting

   Self-weight of the beam according to Table 10.11:

$$g = (0.0288 + 0.0075 + 0.0372 + 0.0003 + 0.018) \times 25 = 2.36\,\text{kN/m}$$

   Let the lifting point be 0.7 m from the beam end, then

$$M_q = \frac{1}{2}gl^2 = \frac{1}{2} \times 2.36 \times 0.72 = 0.58\,\text{kN}\cdot\text{m}$$

Obtain stress of concrete at top edge by considering the dynamic coefficient 1.5:

$$\sigma_{ct} = \frac{N_{pI}}{A_0} - \frac{N_{pI}e_{0I}}{I_0}y_0' - \frac{1.5M_q}{I_0}y_0' = \frac{759{,}457}{99106}$$
$$- \frac{759{,}457}{835{,}780 \times 10^4} \times 349 - \frac{1.5 \times 0.58 \times 10^6}{835780 \times 10^4} \times 349$$
$$= 0.68\,\text{N/mm}^2 \text{ (compressive stress), OK.}$$

Obtain stress of concrete at bottom edge with

$$\sigma_{cc} = \frac{N_{pI}}{A_0} + \frac{N_{pI}e_{0I}}{I_0}y_0 + \frac{1.5M_q}{I_0}y_0 = \frac{759{,}457}{99{,}106}$$
$$+ \frac{759{,}457}{835{,}780 \times 10^4} \times 451 + \frac{1.5 \times 0.58 \times 10^6}{835{,}780 \times 10^4} \times 451$$
$$= 16.68\,\text{N/mm}^2 < 0.8f_c' = 19.30\,\text{N/mm}^2, \text{ OK.}$$

10.9 Statically Indeterminate Prestressed Structures

For statically determinate prestressed structures, the stress resultants inside the structures caused by prestress can be determined by the tendon force and its eccentricity. However, for posttensioned statically indeterminate structures, the stress resultants inside the structures caused by prestress are influenced by not only the tendon force and its eccentricity, but also the imposed deformation (restraint action). Figure 10.39 shows an example of a posttensioned two-span beam with a straight, eccentric tendon. Moment inside the beam caused by the tendon force only is shown in Fig. 10.39b. Tensioning of the tendon would cause the beam to try to lift off the center support. The restraint forces (Fig. 10.39c) required to hold the beams on the support would cause moments in the beam (Fig. 10.39d). These restraint moments will vary linearly between supports and are often referred to as "Secondary moments." Finally, the moments caused by prestress in the two-span beam are shown in Fig. 10.39e, which are the basic resultants for the analysis or the design of a statically indeterminate prestressed structure (two-span beam).

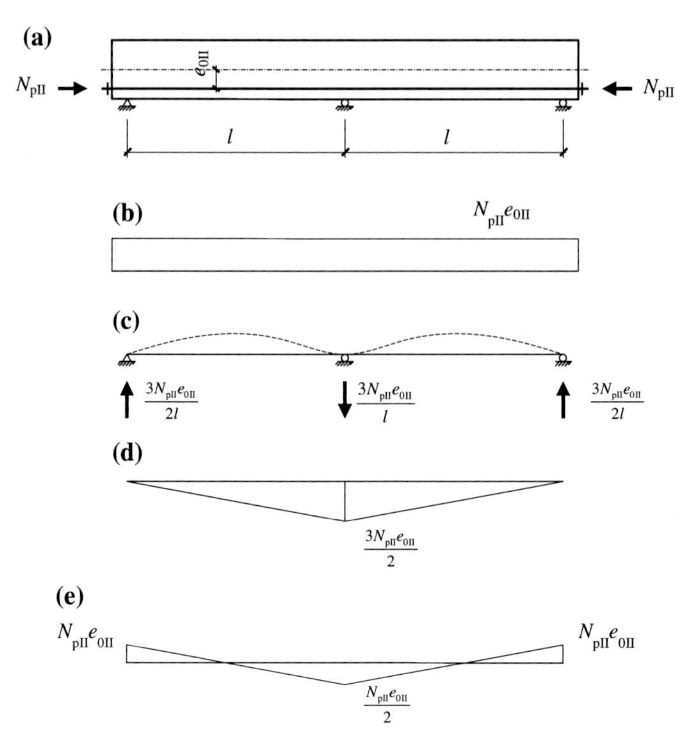

Fig. 10.39 Restraint action caused by a posttensioning two-span beam. **a** Beam. **b** Moment caused by prestress. **c** Resraint forces. **d** Secondary moment. **e** Moment in the beam

10.10 Detailing for Prestressed Concrete Members

10.10.1 Detailing for Pretensioned Members

10.10.1.1 Spacing of Tendons and Concrete Cover

In a pretensioned member, the tendons rely on the bond with concrete to develop prestress. Therefore, the concrete surrounding the tendon must be carefully specified in terms of the spacing of tendons and concrete cover to tendons.

The minimum thickness of concrete cover for tendons is the same as that for rebars.

The clear spacing between adjacent tendons, i.e., the minimum clear distance between two adjacent tendons, must not be less than the values specified in Table 10.12, in which d_{p} is the diameter of a steel bar used as tendon; and d_{pe} is the equivalent diameter of bundled bars as the tendon, $d_{\mathrm{pe}} = \sqrt{n}d_{\mathrm{p}}$, where n is the number of bars in the bundle.

Table 10.12 Minimum clear spacing between pretensioned tendons (mm)

Type of tendon	Steel wire	3-wire strand	7-wire strand
Absolute value	15	20	25
Relative value	1.25 d_p or 2.5 d_{pe}		

10.10.1.2 Detailing of the End Regions of Pretension Members

The ends of pretensioned members should be strengthened, so that on the one hand, the end part of the members will not crack under the high prestressing force, and on the other hand, the end part can be effectively restrained to enable a better transfer of the prestressing force. There are several details depending on the types of member, which are shown in the following summary of reinforcements.

1. Helical reinforcement
   Helical reinforcement is suitable for a single steel bar or a bundle of bars. It is made by providing helically formed steel rings around the tendon at the end (Fig. 10.40). The length of the helical reinforcement is not less than 150 mm and has at least 4 rings. The helical reinforcement can provide constraint on the concrete within the ring so that the strength of the constrained concrete is increased to avoid local compression failure in the form of cracks.
2. Support plate with inserted bars
   In case it is difficult to provide helical reinforcement, but a support plate exists at the end of the beam, inserted bars can be added to the plate and anchored into the end region to play a role similar to the helical reinforcement, i.e., to constrain the end region of the concrete surrounding the tendon. To this end, the inserted bars must be provided so that the tendon passes through the gaps between every two rows of the inserted bars. The total number of the inserted bars must not be less than 4 with the length of each inserted bar not less than 120 mm (Fig. 10.41). This detailing measure has proved effective especially in the end region of roof slabs.
3. Steel meshes
   If the end region is large and has many tendons, it can be difficult to provide helical rings for each tendon separately. In this case, steel meshes can be used to enhance the local bearing strength of the end region. Such steel meshes are

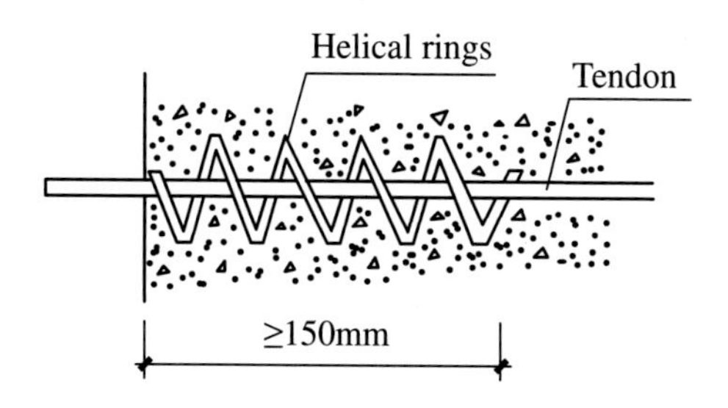

Fig. 10.40 Helical reinforcement

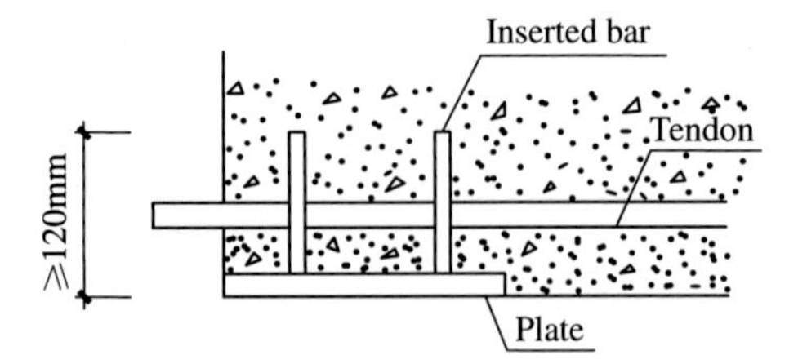

Fig. 10.41 Support plate with inserted bars

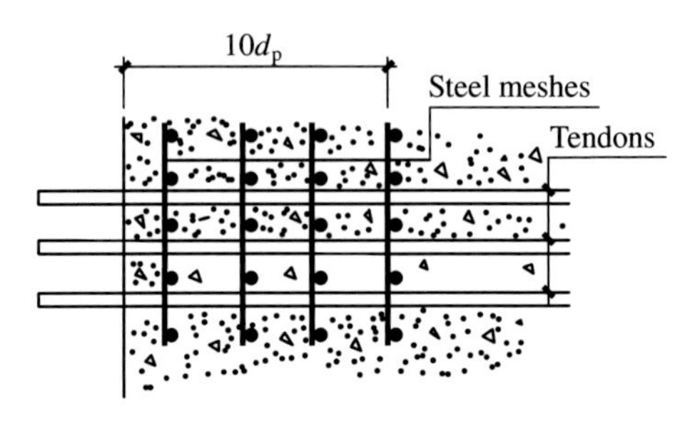

Fig. 10.42 Steel meshes

usually formed by welding or tying steel bars of small diameter. A number of 3–5 steel meshes are usually provided with enough width to cover the region of local compression at a depth not less than 10 times the diameter d_p of the tendon (Fig. 10.42).

4. Lateral detailing bars
This measure is shown in Fig. 10.43 and suitable for slablike members. In this case, the previous measures are difficult to use. Hence, providing additional lateral bars is more effective because these bars can also increase the local bearing strength and thereby limit the development of cracks in the end region. The number of added lateral bars should not be less than 2, and the lateral bars should be arranged within a range of 100 mm in width at the end of the slab (Fig. 10.43).

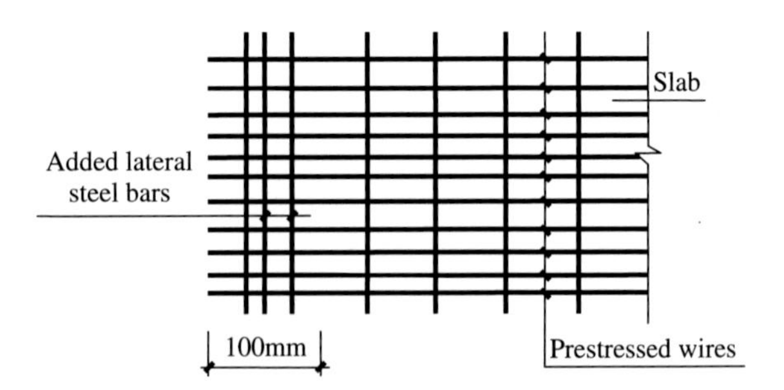

Fig. 10.43 Added lateral bars

10.10.1.3 The Technology of Releasing Tendons

The measures given previously could solve the problem associated with the end region of a pretensioned member; however, the method for releasing the tendons also plays an important role in alleviating the end-region problem and controlling the strength of the member at the time of tendon releasing in general.

Usually, cutting or sawing is used to release tendons. Such a method will cause a sudden release of the tendon, which will in turn produce impact on the end of the member. The impact so produced could be enough to produce cracks at the end of the member so that the anchoring of the tendon and the prestress transfer could be affected. As a result, the forces in the tendon are released suddenly, and the starting point of the prestress transfer should be moved inwardly into the member by a length of $0.25l_{tr}$, where l_{tr} is the length of prestress transfer.

The order by which the tendons are released is also very important. If the tendons are not released in a symmetric way, the greater eccentricity so produced could cause cracking of the concrete. Therefore, the order of releasing tendons should be checked and strictly specified.

If the tendons are released slowly and gradually, the stress state in the end region will be much improved so that the problems of local compression and cracking are greatly alleviated. There are many ways in which a gradual release of the tendon can be realized. One of them is to slowly displace the abutment at which the tendons are anchored so that the stresses in the tendons are gradually reduced. The tendons are then cut at a low stress level.

10.10.2 Detailing for Posttensioned Members

10.10.2.1 Detailing of Duct

Steel wires and strands are usually employed as tendons in ducts in posttensioned members. Therefore, the ducts must be sufficiently large in diameter to accommodate the tendons. In addition, the spacing between the ducts and the concrete cover to the duct must be enough to enable proper action of the tendons. These requirements are often critical in controlling the design due mainly to the limited dimension of the member section. In general, the following requirements should be met.

1. Diameter of ducts

   The diameter of a duct should be greater than the outer diameter of the prestressing wire bundle, or prestressing strand, or any part of the anchorage that must be put in the duct, by at least 10–15 mm. And the cross-sectional area of the duct is preferred to be 3–4 times that of the tendon. This is the minimum requirement for the proper operation of putting these things in the duct.

2. Arrangement of ducts in the end face of members
   When arranging ducts in the end surface of a member, such factors as the dimension of the anchorage sets, the dimension of the jacking devices, and the local bearing strength of the concrete must be carefully considered. Often it is needed to enlarge the end part of the member to meet all these requirements and avoid unexpected difficulties in the jacking process.
3. Duct spacing and concrete cover
   The clear horizontal spacing between the ducts is suggested to be no less than 50 mm and 1.25 times the diameter of coarse aggregates. The concrete cover to the duct (the clear distance between the duct and the edge of the member) is preferred to be at least 30 mm and should not be less than half of the duct diameter.
4. Duct spacing and concrete cover of frame beams
   In a frame beam, the moments are usually negative at supports and positive at mid-spans. As a result, the tendons are usually curved accordingly. The ducts for such tendons must have a clear spacing, that is, at least 1.5 times the duct diameter in the horizontal direction and one time the duct diameter in the vertical direction with a preferred diameter 1.25 times larger than the diameter of coarse aggregates. The concrete cover to the ducts is suggested to have a clearance of at least 40 mm from the sides of the beam and at least 50 mm from the bottom of the beam.
5. Treatment of camber
   Camber is to make the beam deflect upward before the load is applied. Camber is often needed to compensate for excessive deflection especially in the case of beams with large spans. When camber is used, the ducts must also be cambered to avoid secondary stresses that are not considered in the design.
6. Grouting and the position of air outlets
   Grouting is to put cement mortar into the ducts usually by pressure so that the tendons are protected from corrosion and capable of transferring the forces to the concrete through the bond action. Therefore, inlets for the grout and outlets for air and grout must be provided in the form of holes made of ducts. The spacing of these holes should not be greater than 20 m.

10.10.2.2 Reinforcement in Member End

To make sure that the tendons will not be overcrowded in the end region of the member to such a degree as to cause local compression failure or cracking, the following detailing measures are often asserted.

1. Part of tendons bent up
   Part of the tendons are usually bent up near the support region of such members as prestressed roof girders and crane beams. Doing so not only reduces the harmful stress concentration and construction difficulties caused by congestion of tendons at the bottom of the beam, but also reduces the likelihood of cracking by making the principal tensile stresses smaller near the supports. Furthermore,

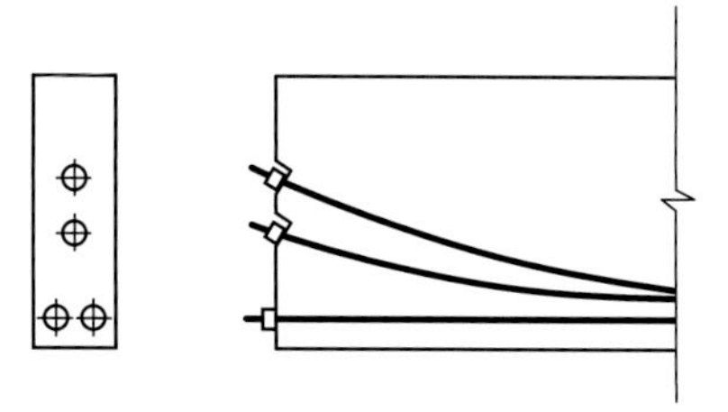

Fig. 10.44 Tendons bent up at member end

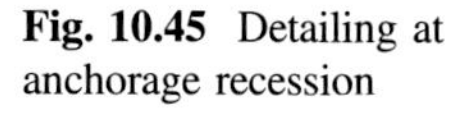

Fig. 10.45 Detailing at anchorage recession

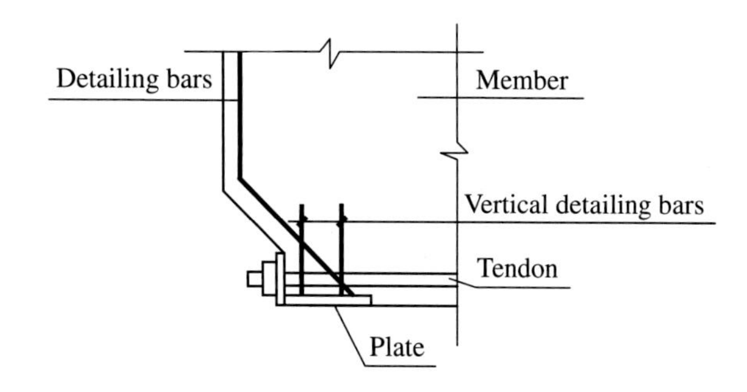

the load-carrying capacities of the support region are basically unaffected because of the smaller values of moment near the support (Fig. 10.44).

2. Detailing at anchorage recessions
   Recessions are often made at the position of the anchorages. In this case, detailing rebars in the bent form should be provided. These together with the inserted bars on the support plate will constrain the anchorage zone (Fig. 10.45)
3. Detailing at welded support
   A prestressed member is often welded to the support. In this case, stresses due to constrain in the longitudinal direction could be induced by such actions as creep and shrinkage of concrete as well as temperature change. The situation is more serious if the member is longer. Therefore, sufficient longitudinal rebar should be provided in the member to resist such actions.

10.10.2.3 Measures for Reinforcing the End of Posttensioned Member

Measures must be taken to carefully detail the end of a posttensioned member owing to the great prestressing forces there. The main aspects are as follows.

1. Inserted plate
   An inserted steel plate must cover that part of the end which is under the pressure of the anchorages or the jacking devices so that these huge forces will not act directly on the concrete.

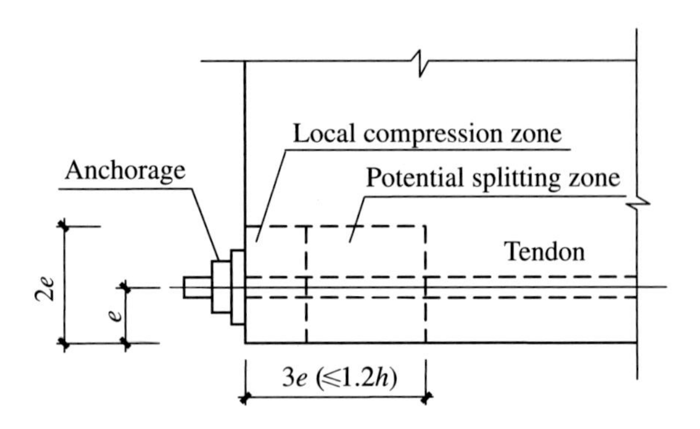

Fig. 10.46 Potential splitting zone

2. Check on local bearing strength
   The local bearing strength of the end region must be checked by using the corresponding formulae and the indirect reinforcement provided as required. The corresponding volumetric reinforcement ratio should not be less than 0.5 %.
3. Reinforcement to prevent splitting of ducts
   The stress concentration at the end of the member together with the limited end dimension could be to such a degree as to cause splitting of the ducts outside of the anchorage zone which are not strengthened by the local bearing requirement. Therefore, additional indirect reinforcement should be provided in this zone, which is outside the zone determined by local bearing requirements. This zone has a height of $2e$ and a length of $3e$ (the length is not greater than 1.2 h), where e is the distance from the resultant tendon forces (due to local compression) to the edge of the member, and h is the section height at the end (Fig. 10.46). The corresponding volumetric reinforcement ratio should also be no less than 0.5 %. The area of reinforcement can be calculated by

$$A_s \geqslant 0.18\left(1-\frac{l_l}{l_b}\right)\frac{F_l}{f_{yv}} \quad (10.135)$$

   where

   F_l resultant force of the local compression at the member end;
   l_l and l_b length of side or diameter of A_l and A_b, in the member height direction, respectively. A_l and A_b can be calculated based on Fig. 9.24; and
   f_{yv} tensile strength of the reinforcement.

4. Additional vertical rebar
   If the tendons cannot be uniformly arranged over the end surface of the member, they are usually arranged in the lower part of the surface or arranged in both the lower and upper part of the surface. In this case, the eccentricity of the tendons

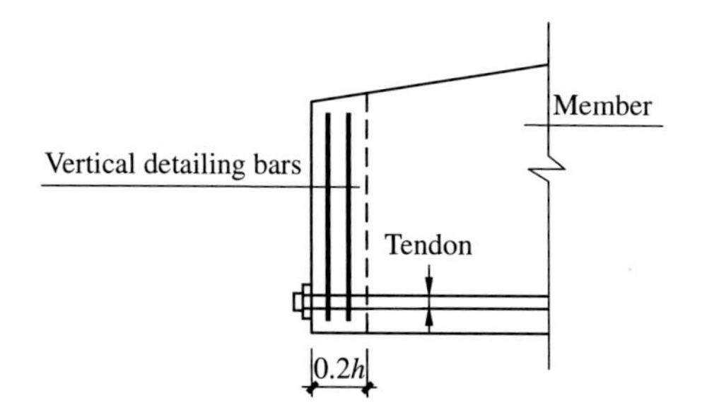

Fig. 10.47 Additional vertical rebar

could cause tensile stresses in the middle part of the section and even cause cracking. Therefore, additional vertical rebars should be provided within a 0.2-h-thick layer near the surface. These rebars can be in the form of closed stirrups, welded steel meshes, or other types (Fig. 10.47). The area A_{sv} of the section of the vertical rebar is determined by the following formulae:

$$A_{sv} \geqslant \left(0.25 - \frac{e}{h}\right) F_l / f_{yv} \tag{10.137}$$

If $e > 0.2\ h$, appropriate vertical rebars should also be provided accordingly.

10.10.2.4 Other Detailing Measures

The following measures should also be noted.

1. Reinforcement to resist inversed moment
   The eccentricity of tendons can cause inversed moment at prestress transfer. Such moment could cause cracking when the concrete strength is low and/or the service load is not acting. Therefore, appropriate rebars should be provided in the potential cracking zone to control such cracks.
2. Finite element analysis of the end region
   If the member is very important in stress control, the finite element method should be used, and its results should be considered in design. Laboratory testing can be performed as a check if necessary.
3. Choice of anchorages
   The anchorages must be reliable and conform to the specifications given in the standard *Anchorages, Clippers and Connectors for Tendons* (GB/T 14370). In addition, the proper treatment of the anchorage devices after the tendons are anchored should not be overlooked. The exposed part of these devices should be covered by concrete or protected by using proper measures against rusting and corrosion.

Questions

10.1 What are prestressed concrete structures?

10.2 What are the merits of prestressed concrete structures? Why must high-strength concrete and high-strength reinforcement be used in prestressed concrete structures?

10.3 What is a fully prestressed concrete structure? What is a partially prestressed concrete structure? What is the degree of prestressing?

10.4. What are pretensioned and posttensioned concrete structures?

10.5 What are the stress transfer mechanisms of different types of anchorages?

10.6 What are the requirements on materials in prestressed concrete structures?

10.7 What are the factors that will influence the allowable value of control stress? Why are these factors important?

10.8 How many kinds of prestress losses are there in prestressed concrete structures? How do we calculate them? And how do we reduce them?

10.9 What is the transfer length of prestress? How do we check the bearing capacity at the ends of posttensioned members?

10.10 Draw the internal force diagrams, stress diagrams at destressing state, cracking state, and ultimate limit state on cross sections of pretensioned and post-tensioned members under axial tension.

10.11 Where are tendons usually placed in flexural members? Illustrate it in a simply supported beam, a simply supported beam with overhung part and a two-span continuous beam.

10.12 Draw the internal force diagrams, stress diagrams at destressing state, cracking state, and ultimate limit state on cross sections of pretensioned and posttensioned members under flexure.

10.13 What is the difference in deriving equations for the relative depth of a balanced cross section in prestressed and ordinary flexural members?

10.14 Why is the shear capacity of a prestressed concrete beam higher than that of the corresponding ordinary reinforced concrete beam?

10.15 Why should the cracking resistance of inclined sections be checked in prestressed flexural members? Provide detailed checking steps.

10.16 What are the similarities and differences of the detailing requirements in prestressed concrete members and ordinary reinforced concrete members?

Problems

10.1 A pretensioned concrete member is under axial tension. The cross-sectional dimensions of the member are 200 mm × 200 mm. C40 concrete is used, and $9\phi^{H}9$ tendons are provided. The control stress is σ_{con} = 1000 N/mm^2 and f_{py} = 1110 N/mm^2. There is no ordinary reinforcement. The first and second batch of prestress losses are σ_{lI} = 68 N/mm^2 and σ_{lII} = 52 N/mm^2, respectively. Calculate (1) the prestress σ_c in construction; (2) the load needed to make the normal compressive stress in concrete zero; (3) the cracking load; and (4) the ultimate capacity of the member.

10.2 A 21-m-long posttensioned concrete member is under axial tension of $N = 700$ kN. The cross-sectional dimensions of the member are $b \times h = 250$ mm $\times$ 200 mm. C40 concrete and screw anchorage are used. And $4\phi12$ rebars are provided. The grade for tendons is ϕ^{T}. When concrete gets sufficient strength, the tendons are pretensioned at one end with overtensioning. The duct of 50 mm in diameter is formed by withdrawing a pressurized rubber tube. Calculate (1) the amount of tendons; (2) check the compression resistance of concrete in construction stage; and (3) check the bearing capacity of the anchorage zone in construction stage.

10.3 A posttensioned concrete I section beam is reinforced by straight-line tendons. There is no any ordinary reinforcement in the beam. If the tendons at the top and bottom of the beam are tensioned simultaneously, derive the expression of the area of tendons at the beam top when no tensile stress is allowed in the cross section.

10.4 A pretensioned concrete beam is subjected to a bending moment of $M = 10$ kN m. The sectional dimensions are $b \times h = 150$ mm $\times$ 300 mm. Tendons of $3\phi^{\mathrm{HM}}9$ are provided in the tension zone. The control stress is $\sigma_{\mathrm{con}} = 0.7f_{\mathrm{ptk}}, f_{\mathrm{ptk}} = 1270\,\mathrm{N/mm^2}$. Prestress losses are $\sigma_{l1} = 10.5\,\mathrm{N/mm^2}$, $\sigma_{l3} = 0, \sigma_{l4} = 19\,\mathrm{N/mm^2}$, and $\sigma_{l5} = 53\,\mathrm{N/mm^2}$. The concrete grade is C40. Check the flexural capacity of the beam; also, check the beam strengths in the construction stage when tendons are released, at which the concrete has attained 85 % of the designed strength.

Chapter 11
Serviceability of Concrete Structures

In the design of new concrete structures or in the appraisal of existing concrete structures, calculation for load-carrying capacity should be performed for all structural members in case any concrete members have reached their ultimate limit states due to strength failure or buckling. Concrete members can also reach their serviceability limit state, which would indicate a large deformation or undesirable vibrations or excessive crack width. Thus, to meet the functional requirements of the structures, it is necessary to control the deformation and the crack width and restrict their deviation within reasonable limits.

In the checking of crack width and deformation of members, appropriate combinations of load effects corresponding to different structural types such as buildings, harbors, and bridges should be considered.

11.1 Crack Width Control

11.1.1 Classification and Causes of Cracks in Concrete Structures

Concrete is a material composed of cement paste and aggregates (sand and stone). During the hardening of concrete, micropores and micro-cracks are evident. Micro-cracks can happen in cement paste, at the interfaces of cement paste and aggregates and in aggregates. Generally, the former two micro-cracks are dominant before the concrete is loaded. When loads are being applied on the concrete, the micro-cracks and micropores link together and propagate to form macro-cracks. Further development of the macro-cracks may fail the concrete. This chapter will focus on the macro-cracks that may influence the serviceability and durability of concrete structures.

There are many types of cracks with different characteristics and causes. They may have different influence on the functions of structures. So, it is necessary to distinguish the crack types and take corresponding measures.

X. Gu et al., *Basic Principles of Concrete Structures*,
DOI 10.1007/978-3-662-48565-1_11

11.1.1.1 Classification of Cracks

Cracks in concrete structures should be classified according to the following three factors:

(1) Occurrence time: in construction stage and in service.
(2) Causes of cracking: inappropriate materials, improper construction, plasticity of concrete, static loads, temperature change, shrinkage of concrete, corrosion of reinforcement, freeze–thaw cycle, differential settlement of foundation, seismic effect, fire, etc.
(3) Shapes, distributions, and regularities of cracks: map cracks, transverse cracks, longitudinal cracks, X-shaped cracks, etc.

11.1.1.2 Causes of Cracks

1. Cracks in construction stage
   Plastic cracks. They happen in the first several hours before the concrete hardens and can generally be observed about 24 h after the casting of concrete. There are two types of plastic cracks in concrete. One is plastic slumping crack, which is caused by the forms and reinforcement blocking the settlement of concrete under gravity and the concrete bleeding at its surface (Fig. 11.1). The deep and wide plastic slumping cracks follow the longitudinal direction of reinforcement and are one of the main causes of reinforcement corrosion, which can bring great harm to structures. Another one is map cracking, which is caused by rapid drying of the surfaces of structural members during big winds or high temperatures, as shown in Fig. 11.2. Map cracking generally occurs when the concrete cover is too thin.
   Heat-of-hydration cracks. They usually happen in large volume concrete structures, such as dams and water gates. The hardening of concrete produces a large amount of hydration heat so that the temperature in concrete is raised. If the temperature difference between the inside and outside of concrete is too big, the thermal strain may exceed the ultimate tensile strain of concrete causing cracks to appear. For members of ordinary dimensions, the cracks are usually

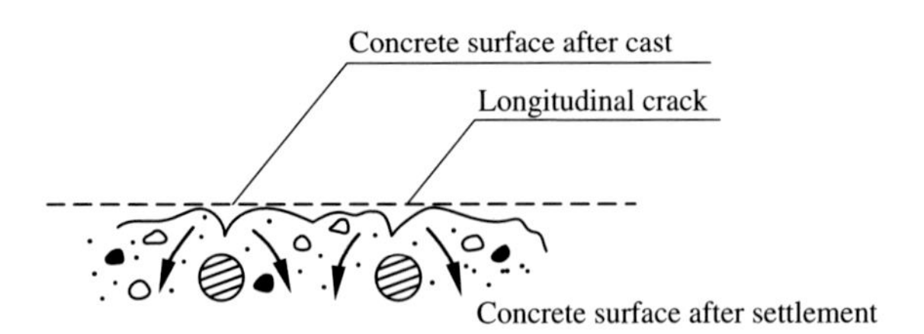

Fig. 11.1 Plastic slumping cracks

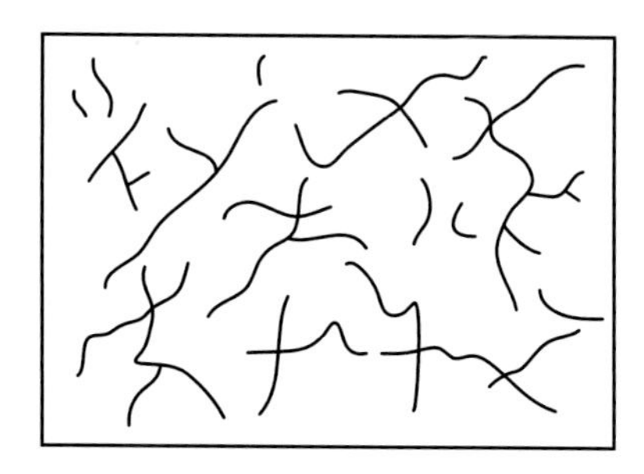

Fig. 11.2 Map cracking

perpendicular to the longitudinal axis. Sometimes, the cracks are only at the surface; sometimes, they are through the whole cross section.

Shrinkage cracks. Shrinkage cracks can occur when the volume change of concrete due to shrinkage is restrained, which is frequently met in a fixed beam or a beam with high steel content. This can also happen when new concrete is cast on old concrete or on top of a solid foundation or if concrete is improperly cured. The cracks are generally perpendicular to the longitudinal axis of the member and sometimes are wide and extend through the cross section.

Cracks due to poor construction quality. These cracks are mainly due to insufficient reinforcement, wrong position of negative reinforcement caused in early stage of construction. Cracks can be caused by trampling, removing supports too early, wrong prestressing, etc. In addition, if the concrete is not properly cured in construction, map cracks may happen at member surfaces in the initial set, but they are very shallow.

Freeze–thaw cracks. Freeze–thaw on early age concrete may produce cracks along the longitudinal bars and stirrups. These cracks are of different widths and may reach the longitudinal bars in depth.

2. Time-dependent cracks in service

   Longitudinal cracking due to reinforcement corrosion. For concrete structures in unfavorable environments, e.g., coastal buildings, marine structures, and structures subjected to high humidity and temperature, if the concrete cover is too thin, and especially if the compactness of concrete is poor, it is very easy for the reinforcement to corrode. Figure 11.3 shows the expansion of corrosion products in the cracking of concrete. The longitudinal cracks along the reinforcement speeds up the corrosion process, which may cause the whole concrete cover layer to drop out. These cracks are very harmful to the safety and durability of concrete structures.

   Cracks due to temperature change and shrinkage. For in situ cast frame beams, slabs, and bridge decks, the thermal deformations and shrinkage of concrete will be restrained by other members of high stiffness, so cracks happen. Thermal cracks also occur in structures subjected to high temperature, such as concrete chimneys and nuclear reactor vessels. Practical experience shows that there exists large transverse thermal stress in top slabs of box-section beams in

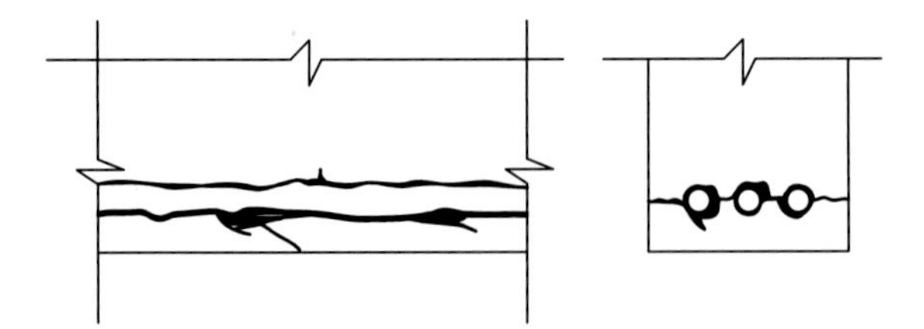

Fig. 11.3 Longitudinal cracks due to reinforcement corrosion

highway bridges. If prestress or sufficient reinforcement is not provided in the transverse direction, the top slabs will surely crack, and more cracks will develop with time (Fig. 11.4). When in situ cast concrete floor contracts due to low temperature or dryness, cracks may appear in the middle or in the corners of the floor.

Cracks due to differential settlement of foundations. When differential settlement of foundations under statically indeterminate structures happens, members in the structures may crack because their deformations are restrained. These cracks are frequently observed in buildings, and their cracks will evolve with the development of differential settlement.

Cracks due to Freeze–thaw, alkali–aggregate reaction, and corrosion by salt or acid. Attacked by freeze–thaw cycles, salt or acid, the concrete will crack and finally cause peeling of concrete. Alkali–aggregate reaction will cause map cracking on surfaces of concrete structural members.

3. Load-induced cracks in service

   Members will crack under external loads. Moreover, different loading conditions, e.g., flexure, shear, combined torsion, shear and flexure, and local loads, will produce different shapes and distributions of cracks, as discussed in previous chapters. In this chapter, we are only concerned with cracks caused by static loads.

   In summary, there are many reasons for concrete cracking. Practical experience indicates that the direct action of loads is generally not the main reason for

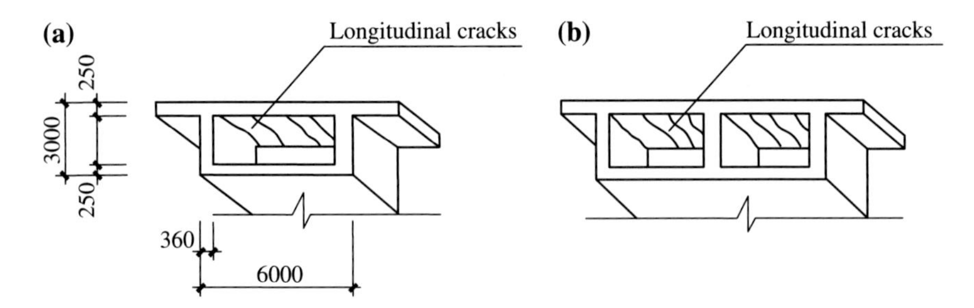

Fig. 11.4 Longitudinal temperature cracks in top slabs of box girder in highway bridges. **a** Single-chamber box beam. **b** Double-chamber box beam

producing cracks of excessive width if the structures are properly designed, constructed, and used. Many cracks are induced by a combination of several reasons, among which temperature changes and concrete shrinkage play important roles. Cracks due to restrained deformations usually occur in certain positions of a structure rather than just in tensile zones of individual members. These cracks should be controlled by reasonable structural arrangement and corresponding detailing requirements.

11.1.2 Purpose and Requirements of Crack Control

11.1.2.1 Purpose of Crack Control

The tensile strength of concrete is much lower than its compressive strength. Concrete will crack easily under small tensile stress. For example, cracks in the tension zone of a flexure member in service are very normal and hard to prevent.

One purpose of crack control is to ensure the durability of structures. When unacceptable wide cracks appear, gas, moisture, and chemical agents will penetrate concrete and lead to corrosion of the reinforcement. The corrosion of reinforcement (usually steel bars) will not only reduce the sectional area of the reinforcement, but also cause the concrete cover layer to drop out with the expansion of corrosion products. In recent years, high-strength reinforcement is more and more widely used. The stress and strain in the reinforcement are correspondingly increased, which results in larger crack widths and a more severe consequence if the reinforcement corrodes. So all structural design codes require crack width checking in normal sections of reinforced concrete members. And for cisterns that have specific requirements, calculation is indispensable in design to ensure that the concrete is uncracked. Actually, high-quality, good compactness, and sufficient concrete cover depth are more important to the durability of structures than the crack width control. Using high performance concrete and prestressing techniques can effectively improve the crack resistance of members.

On the other hand, experiments also show that the degree, scope, and development of reinforcement corrosion at cracks in normal sections are not as severe as expected. The speed of corrosion and even whether the reinforcement corrodes or not are not always directly related to the crack width. Another reason for crack control has to do with the aesthetic aspect as it relates to public concern. A special project about public response to cracking concrete reported that most people worry about cracks 0.3 mm in width.

11.1.2.2 Load Effect Combination

Before discussing crack control requirements and the calculation method, it is necessary to introduce the concept of load effect combinations. Different design codes have different stipulations on load effect combinations.

For example, two kinds of load effect combinations are given in *Load Code for the Design of Building Structures* (GB 50009) for checking the serviceability of structures with the consideration of time-related uncertainties of live loads and concrete creep.

1. Characteristic combination

   The characteristic combination of load effect is expressed as follows

$$S = S_{\mathrm{Gk}} + S_{Q1\mathrm{k}} + \sum_{i=2}^{n} \psi_{\mathrm{c}i} S_{Qi\mathrm{k}} \tag{11.1}$$

where

S_{Gk} load effect calculated according to the standard values of permanent or dead load G_{k};

$S_{Qi\mathrm{k}}$ load effect calculated according to the standard values of live load $Q_{i\mathrm{k}}$, among which $S_{Q1\mathrm{k}}$ is the predominant one; and

$\psi_{\mathrm{c}i}$ coefficients for live load combinations, which are given in *Load Code for the Design of Building Structures* (GB 50009).

2. Quasi-permanent combination

$$S = S_{\mathrm{Gk}} + \sum_{i=1}^{n} \psi_{\mathrm{q}i} S_{Qi\mathrm{k}} \tag{11.2}$$

where

$\psi_{\mathrm{q}i}$ quasi-permanent coefficients for live loads, which are given in *Load Code for the Design of Building Structures* (GB 50009).

Actually, the first combination considers the short-term effect of loads, whereas the second combination considers the long-term effect of loads. The short-term and long-term load effect combinations are also applied in the serviceability analysis of highway bridges.

For convenience, unless otherwise specified, all the load effect combinations in this chapter refer to those stipulated in GB 50009.

11.1.2.3 Grade and Requirement of Crack Control

The division of crack control grade for concrete members is mainly according to the requirements on function of structures, the effects of environmental conditions on reinforcement corrosion, corrosion sensitivity of different types of reinforcement, the duration of load action, and so on.

The crack control of concrete members is generally divided into three grades, which indicate different levels of strictness on crack control. The requirements on crack control will be introduced as follows with GB 50010 as an example. Note that the formulae are only suitable for cracks in normal sections. For crack control in inclined sections in prestressed concrete structures, one can refer to Sect. 11.2.

1. Grade I—Members in which cracks are strictly not allowed
   No tensile stress is allowed in the extreme tension fiber under the characteristic combination of load effects, i.e.,

$$\sigma_{\mathrm{ck}} - \sigma_{\mathrm{pcII}} \leqslant 0 \tag{11.3}$$

   where

   σ_{ck} normal stress in the extreme tension fiber at the checked cross section under the characteristic combination of load effects; and

   σ_{pcII} precompression stress in the extreme tension fiber at the checked cross section after deducting all prestress losses.

2. Grade II—Members in which cracks are generally not allowed
   The tensile stress in the extreme tension fiber under the characteristic combination of load effects should not be larger than the characteristic tensile strength of concrete, i.e.,

$$\sigma_{\mathrm{ck}} - \sigma_{\mathrm{pcII}} \leqslant f_{\mathrm{tk}} \tag{11.4}$$

3. Grade III—Members in which cracks are allowed
   The maximum crack width $w_{\max}$ should not exceed the limit value w_{lim}. $w_{\max}$ is calculated for reinforced concrete members using the quasi-permanent combination of load effects and considering the long-term effects of loads. $w_{\max}$ is calculated for prestressed concrete members using the characteristic combination of load effects and considering the long-term effects of loads.
   For prestressed concrete members in Class II environment (see Table 11.1), the stress in the extreme tension fiber under the quasi-permanent combination of load effects should also satisfy the following formula:

$$\sigma_{\mathrm{cq}} - \sigma_{\mathrm{pcII}} \leqslant f_{\mathrm{tk}} \tag{11.5}$$

   where

   σ_{cq} normal stress in the extreme tension fiber at the checked cross section under the quasi-permanent combination of load effects.

Table 11.1 Environment class for concrete structures

Environment class		Conditions
I		Dry, indoors; submerged in still water containing no erosive agents
II	a	Humid, indoors; outdoors in regions that are not very cold; in direct contact with water or soil containing no erosive agents in regions that are not very cold; in direct contact with water or soil containing no erosive agents below the freezing line in (severe) cold regions
	b	Alternate drying and wetting; water-level frequently changes; outdoors in (severe) cold regions; in direct contact with water or soil containing no erosive agents above freezing line in (severe) cold regions
III	a	Water-level frequently changes in winters in (severe) cold regions; exposed to deicing salt; exposed to sea wind
	b	Saline soil; in direct contact with deicing salt; seashore
IV		Seawater
V		Exposed to man-made or natural erosive agents

Table 11.2 Limit value of the maximum crack width

Environment class	Reinforced concrete structures		Prestressed concrete structures	
	Grade of crack control	w_{lim}/mm	Grade of crack control	w_{lim}/mm
I	III	0.3 (0.4)	III	0.2
II_a	III	0.2	III	0.1
II_b	III	0.2	II	–
III_a, III_b	III	0.2	I	–

The limit values of the maximum crack width stipulated in GB 50010 are listed in Table 11.2. The grade of crack control and the maximum crack width in the table can only be used in normal section checking.

For crack control of other structures, such as chimneys, silos and structures subjected to liquid pressure, one should refer to corresponding specific actions. Compared with those of buildings, the limit values for crack width in concrete members in highway bridges are smaller (see *Code for Design of Reinforced Concrete and Prestressed Concrete Highway Bridges and Culverts* [JTG D62-2004]).

11.2 Calculation of Cracking Resistance in Prestressed Concrete Members

Cracking resistance should be calculated at normal and inclined sections in prestressed concrete members. The concepts and methods in the calculation are nearly the same for all civil engineering structures. The method in GB 50010 will be introduced herein.

11.2.1 Cracking Resistance of Normal Sections

Equations (11.3)–(11.5) can be used in cracking resistance checking at normal sections in prestressed concrete structures. But σ_{ck} (or σ_{cq}) and σ_{pcII} should be determined first.

11.2.1.1 Normal Stress in Concrete

Prestressed concrete members are in elastic stage before cracking, so the normal stress in concrete can be calculated by formulae in *Mechanics of Materials*. Area of transformed section A_0 and corresponding section modulus W_0 should be used.

1. Axially tensioned members

$$\sigma_{ck} = \frac{N_k}{A_0} \tag{11.6}$$

$$\sigma_{cq} = \frac{N_q}{A_0} \tag{11.7}$$

2. Flexural members

$$\sigma_{ck} = \frac{M_k}{W_0} \tag{11.8}$$

$$\sigma_{cq} = \frac{M_q}{W_0} \tag{11.9}$$

where

N_k and M_k represent the axial tension and bending moment under the characteristic combination of load effects, respectively;

N_q and M_q represent the axial tension and bending moment under the quasi-permanent combination of load effects, respectively; and

A_0 and W_0 represent area and section modulus of the transformed section, respectively.

11.2.1.2 Precompression Stress

It is the stress in the extreme tension fiber at the checked cross section after deducting all prestress losses, which can be evaluated by corresponding equations in Chap. 10.

Example 11.1 Check the cracking resistance (Grade II crack control) of the post-tensioned beam of an I-shaped cross section as shown in Fig. 11.5. C55 concrete and $\phi^{P}5$ plain stress-relief steel wires are used. The bending moment at the mid-span of the beam is $M_k =$ kN·m. The following data has been obtained.

$$A_n = 3184.7 \times 10^2\,\text{mm}^2, I_n = 7041714 \times 10^4\,\text{mm}^4, y_n = 821.1\,\text{mm}$$

$$A_0 = 3394.7 \times 10^2\,\text{mm}^2, I_0 = 7847869 \times 10^4\,\text{mm}^4, y_0 = 790.5\,\text{mm}$$

$$\sigma_{con} = \sigma'_{con} = 1177.5\,\text{N/mm}^2$$

No. 1 tendon: $\sigma_l = 405.2$ N/mm^2; No. 2 and No. 3 tendons: $\sigma_l = 378.4$ N/mm^2; No. 4 and No. 5 tendons: $\sigma_l = 364.0$ N/mm^2; No. 8 to No. 11 tendons: $\sigma_l = 385.8$ N/mm^2; and No. 6 and No. 7 tendons: $\sigma_l = 248.0$ N/mm^2.

Solution

1. Calculate N_{pII} and e_{0II}

Tendon number	A_{pi}/mm^2	$(\sigma_{con} - \sigma_l)$/(N·mm^{-2})	$N_p = (\sigma_{con} - \sigma_l)A_{pi}$/N
1	353	772.3	272,622
2, 3	706	799.1	544,165
4, 5	706	813.5	574,331
8–11	1413	791.7	1,118,672
6, 7	471	929.5	437,795

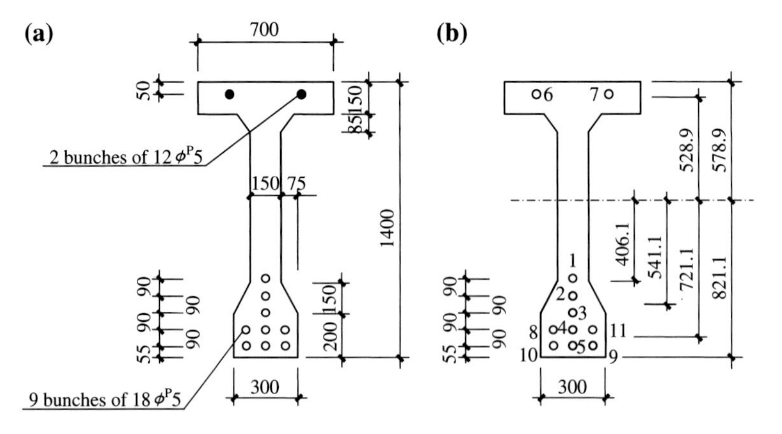

Fig. 11.5 Example 11.1

$$N_{\text{pII}} = \sum N_{\text{pII}i} = 2,947,585\,N = 2948\,\text{kN}$$

$$\begin{aligned} e_{0\text{II}} &= \sum N_{\text{pII}i} y_{\text{pn}i} \Big/ N_{\text{pII}} \\ &= (272.6 \times 406.1 + 564.2 \times 541.1 + 574.3 \times 721.1 \\ &\quad + 1118.7 \times 721.1 - 437.8 \times 528.9)/2948 = 477\,\text{mm} \end{aligned}$$

2. Calculate σ_{pcII}

$$\begin{aligned} \sigma_{\text{pcII}} &= \frac{N_{\text{pII}}}{A_{\text{n}}} + \frac{N_{\text{pII}} e_{0\text{II}}}{I_{\text{n}}} y_{\text{n}} = \frac{2948 \times 10^3}{3184.7 \times 10^2} + \frac{2948 \times 10^3 \times 477}{7041714 \times 10^4} \times 821.1 \\ &= 25.7\,\text{N/mm}^2 \text{ (compressive stress)} \end{aligned}$$

3. Calculate σ_{ck}

$$\sigma_{\text{ck}} = \frac{M_{\text{k}}}{W_0} = \frac{M_{\text{k}}}{I_0} y_0 = \frac{2600 \times 10^6}{7847869 \times 10^4} \times 790.5 = 26.2\,\text{N/mm}^2$$

4. Check cracking resistance

$$\sigma_{\text{ck}} - \sigma_{\text{pcII}} = 26.2 - 25.7 = 0.5\,\text{N/mm}^2 \leqslant f_{\text{tk}} = 2.74\,\text{N/mm}^2, \text{OK.}$$

Example 11.2 Check the cracking resistance of the bottom chord in Example 10.1.

Solution

$$\begin{aligned} N_{\text{k}} &= 340 + 140 = 480\,\text{kN} \\ \sigma_{\text{ck}} &= \frac{N_{\text{k}}}{A_0} = \frac{480 \times 10^3}{41373} = 11.60\,\text{N/mm}^2 \\ \sigma_{\text{ck}} - \sigma_{\text{pcII}} &= 11.60 - 9.76 = 1.84\,\text{N/mm}^2 < f_{\text{tk}} = 2.39\,\text{N/mm}^2, \text{OK.} \end{aligned}$$

Example 11.3 Check the cracking resistance at the normal section of the prestressed concrete beam in Example 10.2. The characteristic value of the uniformly distributed load on the beam is q = 30 kN/m. Grade II crack control is used. For calculation convenience, take f_{tk} = 2.39 N/mm^2.

Solution

The resultant force in the tendon after deducting all prestress losses is

$$\begin{aligned} N_{\text{pII}} &= (\sigma_{\text{con}} - \sigma_l)A_{\text{p}} + (\sigma'_{\text{con}} - \sigma'_l)A'_{\text{p}} \\ &= (880 - 254.57) \times 708.5 + (880 - 141.65) \times 212.5 = 600,017\,\text{N} \end{aligned}$$

The distance from the resultant force to the centroid of the transformed section is

$$e_{0\mathrm{II}} = \frac{(\sigma_{\mathrm{con}} - \sigma_l)A_{\mathrm{p}}y_{\mathrm{p}} - (\sigma'_{\mathrm{con}} - \sigma'_l)A'_{\mathrm{p}}y'_{\mathrm{p}}}{N_{\mathrm{pII}}} = 197.5\,\mathrm{mm}$$

The precompression stress at the bottom edge is

$$\sigma_{\mathrm{pcII}} = \frac{N_{\mathrm{pII}}}{A_0} + \frac{N_{\mathrm{pII}}e_{0\mathrm{II}}}{I_0}y = \frac{600,017}{99,106} + \frac{600,017 \times 197.5}{835,780 \times 10^4} \times 450 = 12.43\,\mathrm{N/mm^2}$$

The tensile stress in the extreme tension fiber under the characteristic combination of load effects is

$$M_{\mathrm{k}} = \frac{1}{8} \times (17 + 13) \times 8.75^2 = 287.11\,\mathrm{kN{\cdot}m}$$

$$\sigma_{\mathrm{ck}} = \frac{M_{\mathrm{k}}}{I_0}y = \frac{287.11 \times 10^6}{835780 \times 10^4} \times 450 = 15.46\,\mathrm{N/mm^2}$$

$$\sigma_{\mathrm{ck}} - \sigma_{\mathrm{pcII}} = 15.46 - 12.43 = 3.03 > f_{\mathrm{tk}} = 2.39\,\mathrm{N/mm^2}$$

The design can be modified by increasing the control stress, using higher grade concrete, etc.

11.2.2 Cracking Resistance of Inclined Sections

11.2.2.1 Requirements on Checking

For prestressed concrete members under combined flexure and shear, inclined cracks may occur since the principal tensile stress exceeds the tensile strength of concrete. The cracking resistance of inclined cracks is evaluated by checking the principal tensile stress and principal compressive stress under the characteristic combination of load effects. The reason to check principal compressive stress is that for concrete in biaxial stress states, the compressive stress in one direction will influence the tensile strength in another direction.

Cross sections at unfavorable positions in the beam span, e.g., cross sections that have been subjected to a larger bending moment and shear or a cross section that has undergone a sudden shape change, should be selected for cracking resistance checking. In addition, for the selected cross sections, the centroid of the transformed section and the positions where the section width changes, such as the intersections of flanges and web in an I-shaped section, should also be checked.

1. Principal tensile stress in concrete σ_{tp}

$$\text{For Grade I crack control, } \sigma_{tp} \leqslant 0.85 f_{tk} \tag{11.10}$$

$$\text{For Grade II crack control, } \sigma_{tp} \leqslant 0.95 f_{tk} \tag{11.11}$$

2. Principal compressive stress in concrete σ_{cp}

$$\text{For Grade I and II crack control, } \sigma_{cp} \leqslant 0.60 f_{ck} \tag{11.12}$$

11.2.2.2 Calculation of Principal Stresses in Flexural Members

Concrete before cracking can be treated as the homogeneous elastic material; the principal tensile and compressive stresses can be calculated by equations, which are introduced in courses of *Mechanics of Materials* as follows:

$$\begin{matrix}\sigma_{tp} \\ \sigma_{cp}\end{matrix} = \frac{\sigma_x + \sigma_y}{2} \pm \sqrt{\left(\frac{\sigma_x - \sigma_y}{2}\right)^2 + \tau^2} \tag{11.13}$$

$$\sigma_x = \sigma_{pcII} + \frac{M_k y_0}{I_0} \tag{11.14}$$

$$\tau = \frac{\left(V_k - \sum \sigma_{pe} A_{pb} \sin \alpha_p\right) S_0}{I_0 b} \tag{11.15}$$

where

- σ_x normal stress in the calculated fiber caused by precompression and bending moment;
- σ_y vertical compressive stress caused by concentrated load in characteristic value F_k;
- τ shear stress in the calculated fiber caused by shear force V_k and precompression in prestressing bent-up reinforcement;
- σ_{pcII} normal stress in the calculated fiber caused by precompression after deducting all prestress losses;
- y_0 distance from the centroid of the transformed section to the calculated fiber;
- I_0 moment of inertia of transformed section;
- V_k shear force calculated by characteristic combination of load effects;
- S_0 moment of the area of transformed section above the calculated fiber around the centroid of the transformed section;
- σ_{pe} effective prestress in prestressing bent-up reinforcement;

A_{pb} section area of prestressing bent-up reinforcement in the same row at the calculated cross section; and

α_p angle between the tangent of prestressing bent-up reinforcement and the longitudinal axis at the calculated cross section.

σ_x, σ_y, σ_{pcII}, and $\frac{M_k y_0}{I_0}$ in Eqs. (11.13) and (11.14) take positive values if they are tensile stresses and negative values if they are compressive stresses.

For pretensioned members, if the checked cross section is close to the member end and within the transfer length l_{tr}, the stresses variation within l_{tr} should be considered when using N_{pII} and e_{0II} to calculate σ_{pe} and σ_{pcII}.

Example 11.4 Check the cracking resistance at inclined sections in Example 10.2. For calculation convenience, take $q_k = 30$ kN/m, $f_{ck} = 26.8$ MPa.

Solution

The simply supported beam is subjected to uniformly distributed loads; thus, along the longitudinal direction, the maximum shear stress happens at the support edge (section A-A in Fig. 10.37). Moreover, along the section depth, sections 1-1, 2-2 and 3-3 have large principal stresses (Fig. 11.6), which means cracking resistance should be checked for all of these sections.

1. Normal stresses

   Section A-A in Fig. 10.37: The bending moment induced by the external load is $M \approx 0$, so the corresponding normal stress is $\sigma = 0$. But the normal stress caused by precompression is

$$\sigma_{pcII} = \frac{N_{pII}}{A_0} \pm \frac{N_{pII} e_{0II}}{I_0} y = 6.05 \pm 0.0142y$$

   Section 1-1 in Fig. 11.6: $\sigma_{pcII} = 6.05 - 0.0142 \times (350 - 130) = 2.93\,\text{N/mm}^2$
   Section 2-2 in Fig. 11.6: $\sigma_{pcII} = 6.05\,\text{N/mm}^2$

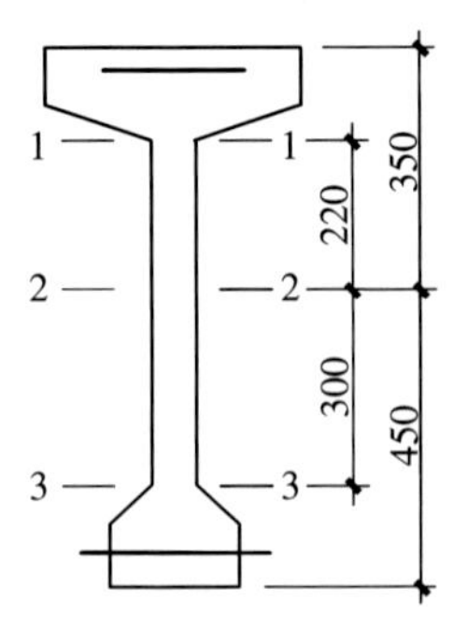

Fig. 11.6 Example 11.4

Section 3-3 in Fig. 11.6: $\sigma_{\mathrm{pcII}} = 6.05 + 0.0142 \times (450 - 150) = 10.31\,\mathrm{N/mm^2}$

2. Shear stresses

$$V_{\mathrm{k}} = \frac{1}{2} \times 30 \times 8.75 = 127.5\,\mathrm{kN}$$

$$\tau_{xy} = \frac{V_{\mathrm{k}} S_0}{b I_0} = \frac{127.5 \times 10^3 S_0}{60 \times 835{,}780 \times 10^4} = 2.54 \times 10^{-7} S_0$$

Section 1-1 in Fig. 11.6:

$$S_{1-1} = 28{,}800 \times (350 - 40) + 7500 \times \left(350 - 80 - \frac{50}{3}\right) + 60 \times 50 \times \left(350 - 80 - \frac{50}{2}\right) + 1063(350 - 35) = 11{,}897{,}595\,\mathrm{mm^3}$$

$$\tau_{1-1} = 3.02\,\mathrm{N/mm^2}$$

Section 2-2 in Fig. 11.6:

$$S_{2-2} = 11{,}897{,}595 + 60 \times (350 - 130) \times \frac{(350 - 130)}{2} = 13{,}349{,}595\,\mathrm{mm^3}$$

$$\tau_{2-2} = 3.39\,\mathrm{N/mm^2}$$

Section 3-3 in Fig. 11.6:

$$S_{3-3} = 11{,}897{,}595 - 60 \times 300 \times 300/2 = 9{,}197{,}595\,\mathrm{mm^3}$$

$$\tau_{3-3} = 2.34\,\mathrm{N/mm^2}$$

3. Principal stresses

$$\sigma_{\mathrm{tp}} = \frac{\sigma_{\mathrm{pcII}}}{2} + \sqrt{\left(\frac{\sigma_{\mathrm{pcII}}}{2}\right)^2 + \tau^2} \quad \sigma_{\mathrm{cp}} = \frac{\sigma_{\mathrm{pcII}}}{2} - \sqrt{\left(\frac{\sigma_{\mathrm{pcII}}}{2}\right)^2 + \tau^2}$$

Section 1-1 in Fig. 11.6:

$$\sigma_{tp} = 1.90\,\text{N/mm}^2\ \text{(tension)} \quad \sigma_{cp} = -4.83\,\text{N/mm}^2\ \text{(compression)}$$

Section 2-2 in Fig. 11.6:

$$\sigma_{tp} = 1.52\,\text{N/mm}^2\ \text{(tension)} \quad \sigma_{cp} = -7.57\,\text{N/mm}^2\ \text{(compression)}$$

Section 3-3 in Fig. 11.6:

$$\sigma_{tp} = 0.51\text{N/mm}^2\ \text{(tension)} \quad \sigma_{cp} = -10.82\text{N/mm}^2\ \text{(compression)}$$

Maximum principal tensile stress

$$\sigma_{tp,max} = 1.90\,\text{N/mm}^2 < 0.95 f_{tk} = 0.95 \times 2.39 = 2.27\,\text{N/mm}^2, \text{OK.}$$

Maximum principal compressive stress

$$\sigma_{cp,max} = 10.82\,\text{N/mm}^2 < 0.60 f_{ck} = 0.60 \times 26.8 = 16.08\,\text{N/mm}^2, \text{OK.}$$

11.3 Calculation of Crack Width in Normal Sections

11.3.1 Theories on Crack Width Calculation

The factors that affect the crack width include reinforcing bar type, bar stress, bar distribution, reinforcement ratio, bond strength, and loading conditions. So, it is difficult to establish a crack width equation for common use. In this section, the theories and methods for calculating the crack width of members under axial loads and flexural members will be discussed.

11.3.1.1 Bond-Slip Theory

This theory, which is based on experimental results of axially loaded members, assumes that the development of the cracks depends mainly on the quality of the bond between the concrete and the steel bars. When a crack occurs, the local bond near the crack fails, and the relative slip between the steel bars and the concrete is the crack width. Actually, the stress in concrete is supposed to be uniformly distributed along the member, and the strain obeys the plane section assumption (Fig. 11.7). Thus, we can first determine the spacing between cracks from the transfer law of bond stress and then get the formula for crack width, which is proportional to crack spacing.

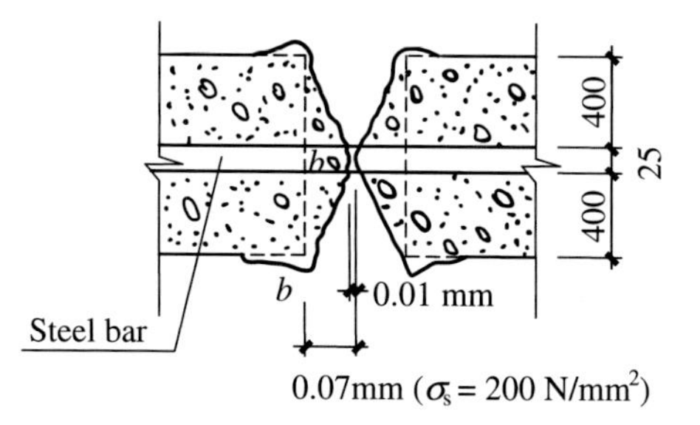

Fig. 11.7 Contracted deformation of tensioned concrete member after cracking

1. Crack spacing
   Before cracking, the tensile stresses and strains in the steel and the concrete along the longitudinal direction of a member in service are generally uniformly distributed. When the tensile strain in the concrete approaches the ultimate tensile strain, all cross sections are going to crack. However, because the mechanical properties of the concrete are not the same everywhere and local defects and micro-cracks are often induced by shrinkage and temperature change, the first (batch of) cracked samples based on any test site will occur in the weakest cross section (see section 1 in Fig. 11.8), so the cracking position is random. After cracking, the concrete at the cracked section quits working and the steel sustains all of the tensile force, which is signaled by the sudden increase of σ_s with an increment of $\Delta\sigma_s$. Meanwhile, the originally tensioned concrete contracts at both sides of the crack. A relative slip happens between the steel and the concrete. So, the crack has a certain width once it appears. The stress in the steel at the cracked cross section will be gradually transferred back to the concrete through bonding stress. With the increase of the distance from the cracked cross section, the bonding stress accumulates, which makes the tensile stress σ_s and strain ε_s in the steel correspondingly decrease, and the tensile stress and strain in the concrete gradually increase. At the distance l_{tr} from the cracked cross section, the strains in the steel and in the concrete are equal. The bonding stress and relative slip become zero. The stresses in the steel and in the concrete are the same as those before cracking.
   Obviously, the cross sections within l_{tr} from the cracked sections and within two adjacent cracks whose spacing is less than $2l_{tr}$ cannot crack anymore, because the tensile stress in the concrete at those cross sections cannot be accumulated through the bond to the ultimate value. Therefore, theoretically, the minimum crack spacing is l_{tr}, the maximum crack spacing is $2l_{tr}$, and the average value is $l_m = 1.5l_{tr}$.

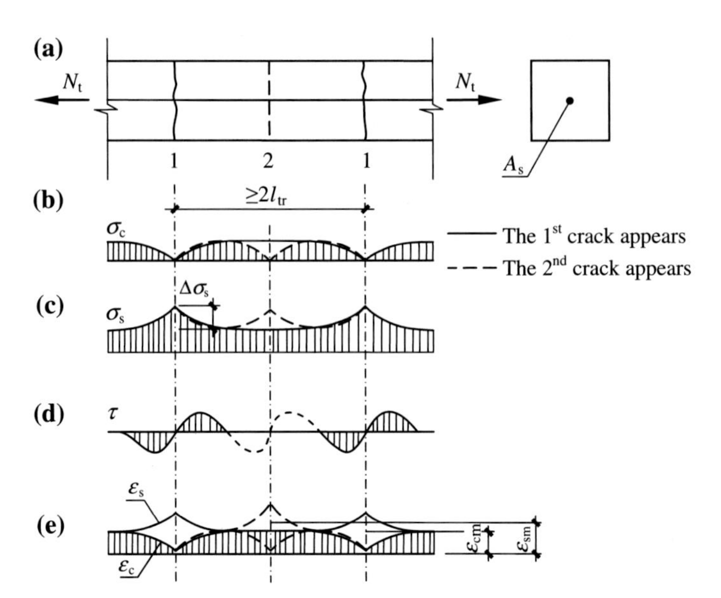

Fig. 11.8 Crack propagation process. **a** The first batch of crack appear. **b** Concrete stress distribution. **c** Steel stress distribution. **d** Bond stress distribution. **e** Strain distribution of the steel bar and concrete

With the increase of load, more and more cracks appear (see section 2 in Fig. 11.8). The variations of stresses and strains discussed in this section are common and will repeat until no more cracks appear. Consequently, the strains in the steel and the concrete are not uniform along the member.

The average crack spacing l_m can be calculated by equilibrium. If the cross-sectional area of the member and the reinforcement are A and A_s, respectively, the reinforcement diameter is d and the average bonding stress within l_m is τ_m; then, from the equilibrium on the free body as shown in Fig. 11.9, we can get

$$\Delta\sigma_s A_s = f_t A$$

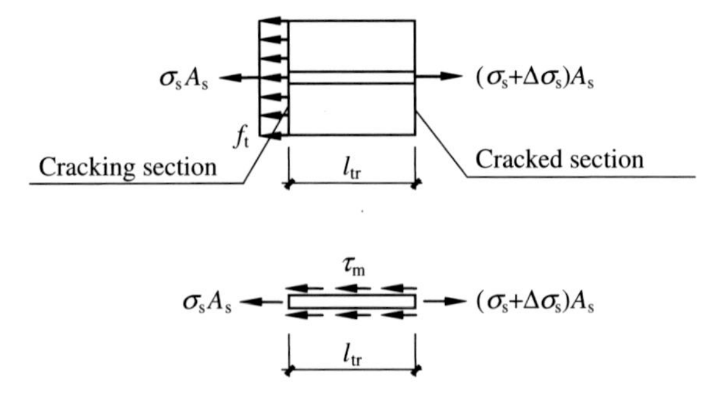

Fig. 11.9 Free body with development length of l_{tr}

and

$$\Delta\sigma_s A_s = \tau_m \pi\, d l_{tr}$$

So

$$l_{tr} = f_t A / \tau_m \pi\, d$$

$A_s = \pi\, d^2/4$, reinforcement ratio $\rho = A_s/A$, and the average crack spacing is

$$l_m = 1.5 l_{tr} = \frac{1.5}{4} \cdot \frac{f_t}{\tau_m} \cdot \frac{d}{\rho} = k_2' \frac{d}{\rho} \tag{11.16}$$

The value k_2' depends on τ_m and f_t. Experiments show that τ_m is proportional to f_t; therefore, their quotient is a constant, which means k_2' can be considered a constant. According to the bond-slip theory, the average crack spacing l_m is independent of the concrete strength. When the reinforcement type and the steel stress are constant, the variable to determine l_m is d/ρ, which is proportional to l_m.

Experiments and analyses indicate that Eq. (11.16) also applies to flexural and eccentrically loaded members. Considering that the cross sections of flexural and eccentrically loaded members cannot be in full tension, the reinforcement ratio ρ can be uniformly substituted by the effective reinforcement ratio ρ_{te}, which is calculated by using the sectional area of effectively tensioned concrete A_{te}. Then, Eq. (11.16) becomes

$$l_m = k_2 \frac{d}{\rho_{te}} \tag{11.17}$$

For reinforced concrete members

$$\rho_{te} = \frac{A_s}{A_{te}} \tag{11.18}$$

For prestressed concrete members

$$\rho_{te} = \frac{A_s + A_p}{A_{te}} \tag{11.19}$$

where

A_s cross-sectional area of nonprestressed steel bars in tension;
A_p cross-sectional area of tendons in tension; and
A_{te} effectively tensioned concrete area (Fig. 11.10).

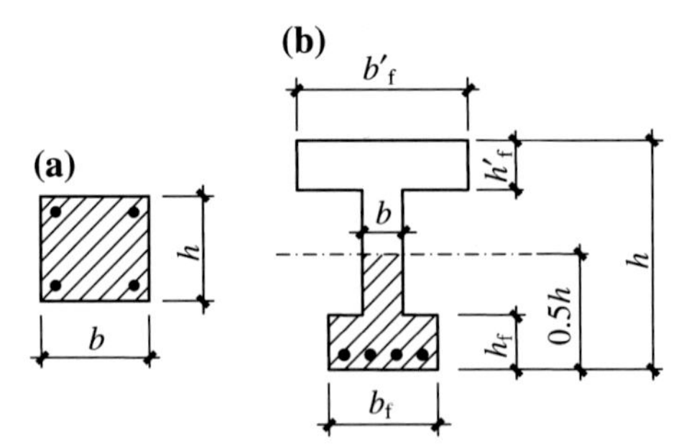

Fig. 11.10 Effectively tensioned concrete area. **a** Axially tensioned member. **b** Flexural and eccentrically loaded member

For axially tensioned members

$$A_{\rm te} = A = bh \tag{11.20}$$

For flexural and eccentrically loaded members

$$A_{\rm te} = 0.5bh + (b_{\rm f} - b)h_{\rm f} \tag{11.21}$$

2. Crack width
   Crack width is measured on the lateral surface of a structural member near the gravity center of the tension reinforcement.
   According to the bond-slip theory, the crack width is the relative slip between the steel and its surrounding concrete within two adjacent cracks. If the average steel strain within the average crack spacing $l_{\rm m}$ is $\varepsilon_{\rm sm}$ and the average concrete strain is $\varepsilon_{\rm cm}$, the average crack width can then be calculated as follows:

$$w_{\rm m} = (\varepsilon_{\rm sm} - \varepsilon_{\rm cm})l_{\rm m} = k'_{\rm w2}\varepsilon_{\rm sm}l_{\rm m} \tag{11.22}$$

where $k'_{\rm w2} = 1 - \varepsilon_{\rm cm}/\varepsilon_{\rm sm}$.
Let $\psi = \varepsilon_{\rm sm}/\varepsilon_{\rm s}$, where $\varepsilon_{\rm s}$ is the steel strain at the cracked section and ψ is a coefficient considering the nonuniform distribution of steel strain between cracks. And because $\varepsilon_{\rm s} = \sigma_{\rm s}/E_{\rm s}$, where $\sigma_{\rm s}$ is the steel stress at the cracked section and $E_{\rm s}$ is the modulus of elasticity of the steel, Eq. (11.22) can be further written as follows:

$$w_{\rm m} = k'_{\rm w2}\varepsilon_{\rm sm}l_{\rm m} = k'_{\rm w2}\psi\varepsilon_{\rm s}l_{\rm m} = k'_{\rm w2}\psi\frac{\sigma_{\rm s}}{E_{\rm s}}l_{\rm m} = k_{\rm w2}\psi\frac{\sigma_{\rm s}}{E_s}\frac{d}{\rho_{\rm te}} \tag{11.23}$$

After the probability distribution type of the crack width has been determined and the long-term effects of loads have been considered, the maximum crack width w_{max} can finally be calculated from the average crack width with the required degree of confidence.

11.3.1.2 Nonslip Theory

Experimental results show that the assumptions of strain distribution at cracked sections and crack shapes in the bond-slip theory are inconsistence with practical conditions. After cracking, the distribution of contraction deformation in concrete is illustrated in line b-b as shown in Fig. 11.7 (axially tensioned members) and Fig. 11.11 (flexural members). The crack width increases with distance from the steel surface; that is, the crack width at the steel surface is much less than that at the member surface. This means that the steel can restrain the concrete contraction due to the bond between the steel and the concrete and the contracted surface is not a plane. Under the steel stress in service (generally $\sigma_s = 160 - 210$ N/mm^2), the relative slip between the steel and the concrete is so small that it can be assumed zero at the steel surface. The crack width is controlled by the uneven contraction of the concrete, which depends mainly on the distance from the measured point to the nearest steel. Therefore, the crack width is mainly influenced by the concrete cover depth rather than d/ρ.

The average crack width w_m can be expressed as follows:

$$w_m = k_{w1} c \frac{\sigma_s}{E_s} \tag{11.24}$$

where

c distance from the point, at which the crack width is to be investigated, to the surface of the nearest reinforcement; and

k_{w1} a coefficient.

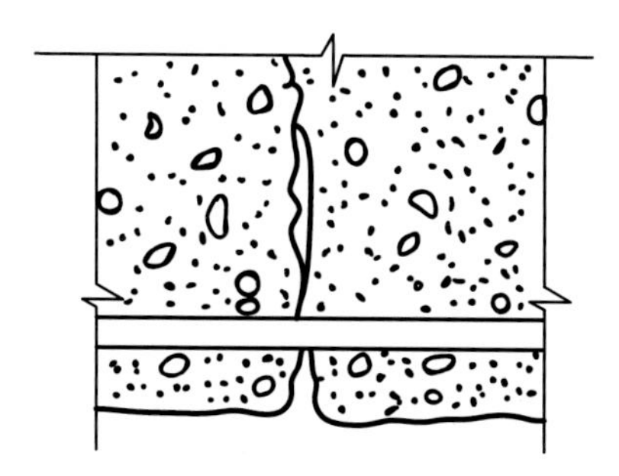

Fig. 11.11 Contracted deformation of a flexural concrete member after cracking

The maximum crack width w_{max} can also be calculated after the probability distribution type of the cracks and the degree of confidence have been determined and the long-term effects of loads have been considered.

The essence of this theory is that within allowable crack width, there is not any relative slip at the interface between the concrete and the steel, so it is called the nonslip theory, which conflicts with the bond-slip theory.

11.3.1.3 Combination of Bond-Slip Theory and Nonslip Theory

Neither the bond-slip theory nor the nonslip theory agrees well with experimental data. The bond-slip theory indicates that the average crack width w_m is proportional to d/ρ, and the proportional constant depends on the bond strength, which conflicts with experimental results. For example, the bond strength of ribbed reinforcement is 2–3 times that of plain reinforcement, but the average crack width in members with ribbed reinforcement is only 1.2–1.3 times that of plain reinforcement. In addition, Eq. (11.16) represents a straight line passing the origin. However, experiments show that when the reinforcement ratio is very large, d/ρ tends to be zero, and the average crack spacing l_m approaches a certain value rather than 0. This certain value is related with the concrete cover depth. The nonslip theory points out that the cracked width is mainly affected by the concrete cover depth. But actually the crack width is not zero at the steel surface. So it is natural for us to combine the two theories together.

Let us first analyze the influences of the concrete cover depth and the effectively restrained area by reinforcement.

1. Concrete cover

   The contraction deformations of concrete at cracked sections are not uniform. The crack width at the steel surface is much smaller than that at the member surface. This means that the concrete cover depth (more accurately the distance from the member surface to the nearest reinforcement) is the main factor that affects the crack width at the member surface.

   The four axially tensioned specimens are identical in reinforcement ratio and diameter of reinforcement but different in concrete cover depth (Fig. 11.12). Experiments on these specimens show that the average crack width (the value

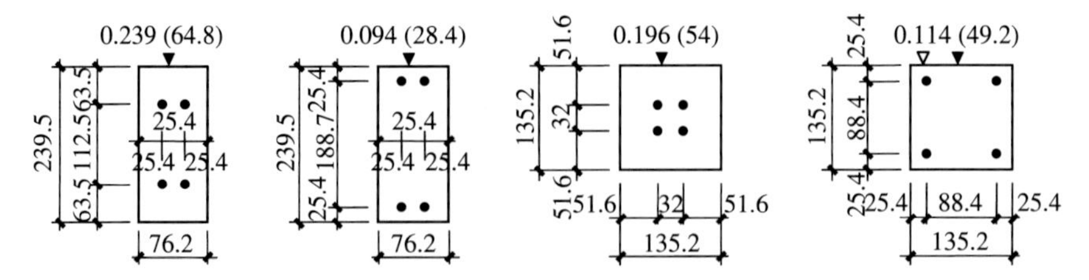

Fig. 11.12 Comparison of crack widths of members with different concrete covers

near "▼" in Fig. 11.12, unit: mm) is generally proportional to the distance (the value in parentheses near "▼" in Fig. 11.12, unit: mm) from the measured point to the nearest reinforcement.

2. Effectively restrained area by reinforcement
The development of cracks is induced by the contraction of the concrete surrounding the reinforcement. The reinforcement transfers the tensile stress to the concrete through the bonding stress. Thus, the concrete contraction must be restrained by the reinforcement. The area pertaining to this restraint is called the effectively restrained area by reinforcement. If this restraint is weakened or disappears, both the crack spacing and the crack width will increase. For the one-way slab of large reinforcement spacing shown in Fig. 11.13, the maximum crack width under external loads does not happen near reinforcement but between the reinforcement. The cracks near reinforcement are dense and thin. In a T section beam in Fig. 11.14, the maximum crack width happens in the web rather than near reinforcement or at the beam bottom. The cracks near reinforcement are also dense and thin. All the cracks mentioned above are branch-like, which indicates clearly that the restraint by reinforcement to concrete is limited in a certain range. Reducing reinforcement spacing in the slab and providing longitudinal reinforcement in the web of a T section beam can prevent these branch-like cracks and decrease the crack spacing and crack width (Fig. 11.15). Experiments show that restraint by reinforcement will apparently decrease if the spacing between longitudinal reinforcement is larger than 15 times the reinforcement diameter.

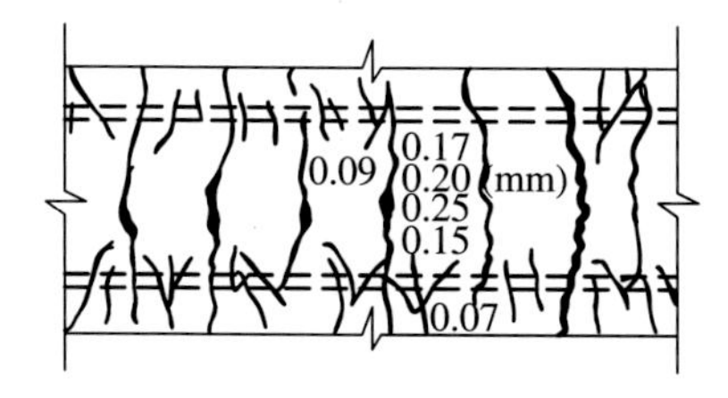

Fig. 11.13 Cracks in the bottom of a one-way slab

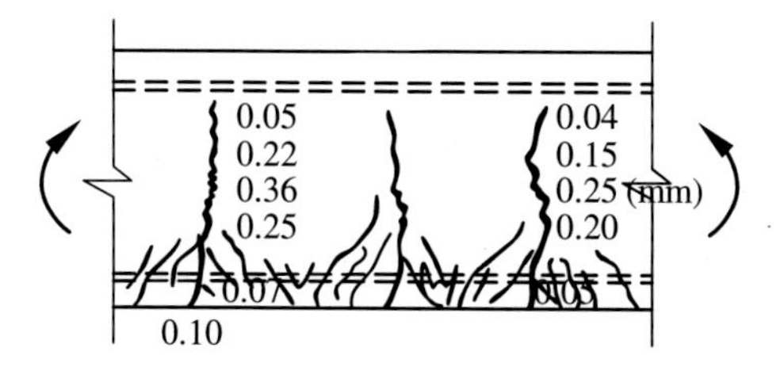

Fig. 11.14 Cracks in a T section beam with large height

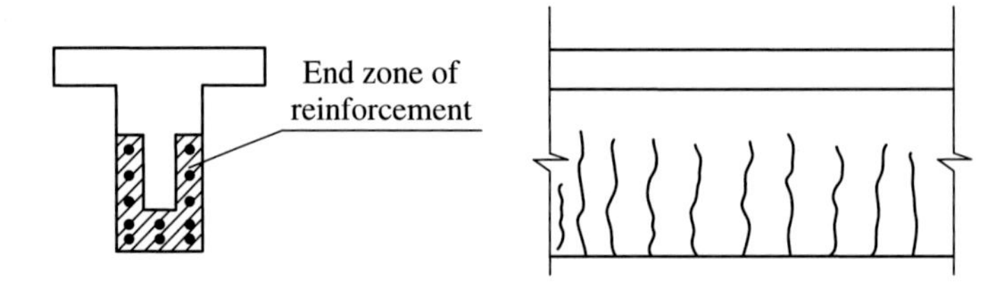

Fig. 11.15 Cracks of a T section beam with longitudinal reinforcement in the web

The effectively restrained area by reinforcement is similar in concept to the effectively tensioned area in the bond-slip theory. In real calculation, the former can be substituted by the latter (Eqs. 11.20 and 11.21).

3. Calculation formula

   After cracks appear, the uneven contraction deformation of concrete and the bond-slip between the steel and the concrete are inevitable. So the crack width is related not only to d/ρ but also to the concrete cover depth c. Further consideration of the influence of the effectively restrained area by reinforcement, the average crack width of a member can be calculated by

$$w_{\mathrm{m}} = k_{\mathrm{w}}\psi\frac{\sigma_{\mathrm{s}}}{E_{\mathrm{s}}}\left(k_1 c + k_2\frac{d}{\rho_{\mathrm{te}}}\right) \tag{11.25}$$

   where k_{w}, k_1, and k_2 are all different from those in previous formulae and should be determined by experiments.

11.3.2 Maximum Crack Width

There are two methods in calculating the maximum crack width in members under external loads. The first one is the semitheoretical and semiempirical method. The second one is mathematical statistics method.

11.3.2.1 Semitheoretical and Semiempirical Method

This method is actually a combination of the bond-slip theory and the nonslip theory. It can be found in GB 50010 and *Design code for hydraulic concrete structures* (SL/T191-96).

1. Calculation formula

   The maximum crack width can be obtained by multiplying the average crack width with amplification coefficient, which reflects the nonuniformity of crack distribution and the influence of long-term effects of loads.

It has been found that the experimental results of crack width are very scattering but basically obey the normal distribution for flexural and eccentrically compressed members. If the degree of confidence is taken as 95 %, the maximum crack width can be calculated by

$$w_{\max} = w_{\mathrm{m}}(1 + 1.645\delta) \tag{11.26}$$

where

δ coefficient of variation of crack width.

For flexural and eccentrically compressed members, $\delta = 0.4$, so $w_{\max} = 1.66\ w_{\mathrm{m}}$; for axially tensioned and eccentrically tensioned members, whose crack width shows a skewed distribution, $w_{\max}/w_{\mathrm{m}}$ is large, so we use $w_{\max} = 1.9\ w_{\mathrm{m}}$. Under long-term loads, the crack width increases with time due to further contraction and creep of concrete and slip creep between the steel and the concrete. After analysis, the amplification coefficient considering the influence of load-term loads is 1.5.

Moreover, from experimental results and practical experiences, it can be determined that $k_{\mathrm{w}} = 0.77$ for flexural and eccentrically compressed members or 0.85 for axially tensioned and eccentrically tensioned members, $k_1 = 1.9$ and $k_2 = 0.08$. And equivalent diameter d_{eq} is used to reflect the reinforcement type and diameter and the bonding properties between the steel and the concrete.

Considering all the above-mentioned influencing factors, the maximum crack width can be finally expressed as follows:

$$w_{\max} = \alpha_{\mathrm{cr}}\psi\frac{\sigma_{\mathrm{s}}}{E_{\mathrm{s}}}\left(1.9c + 0.08\frac{d_{\mathrm{eq}}}{\rho_{\mathrm{te}}}\right) \tag{11.27}$$

where

α_{cr} coefficient considering loading characteristics. For axially tensioned reinforced concrete members, $\alpha_{\mathrm{cr}} = 1.5 \times 1.90 \times 0.85 \times 1.1 = 2.7$; for eccentrically tensioned reinforced concrete members, $\alpha_{\mathrm{cr}} = 1.5 \times 1.90 \times 0.85 \times 1.0 = 2.4$; and for flexural and eccentrically compressed reinforced concrete members, $\alpha_{\mathrm{cr}} = 1.5 \times 1.66 \times 0.77 \times 1.0 = 1.9$. Using α_{cr} for prestressed concrete members has a little different from that for reinforced concrete members (see Table 11.3)

ψ coefficient considering uneven distribution of tensile strain in longitudinal reinforcement between cracks;

σ_{s} stress in longitudinal tension reinforcement in reinforced concrete members calculated by a quasi-permanent combination of load effects or equivalent stress in longitudinal tensile tendons in prestressed concrete members calculated by a characteristic combination of load effects;

E_{s} Modulus of elasticity of longitudinal tension reinforcement, but for prestressed concrete members take E_{p};

Table 11.3 Value of α_{cr}

Loading type	α_{cr}	
	Reinforced concrete members	Prestressed concrete members
Flexure, eccentric compression	1.9	1.5
Eccentric tension	2.4	–
Axial tension	2.7	2.2

c distance from the outside surface of the outmost layer of tension reinforcement to the extreme tension fiber (mm). When $c < 20$, take $c = 20$; when $c > 65$, take $c = 65$;

d_{eq} equivalent reinforcement diameter (mm), calculated by

$$d_{eq} = \frac{\sum n_i d_i^2}{\sum n_i \upsilon_i d_i} \tag{11.28}$$

d_i diameter of the ith-type longitudinal tension reinforcement;

n_i number of the ith-type longitudinal tension reinforcement;

v_i coefficient considering the relative bonding property of the ith-type longitudinal tension reinforcement (Table 11.4); and

ρ_{te} reinforcement ratio of the longitudinal tension reinforcement calculated by the effectively tensioned area of concrete A_{te}. When $\rho_{te} < 0.01$, take $\rho_{te} = 0.01$.

2. σ_s at cracked sections

   σ_s can be obtained by the equilibrium at the cracked section under axial tension or bending moment, which is calculated by the characteristic combination or quasi-permanent combination of load effects.

   (1) Axially tensioned members

$$\sigma_s = \frac{N - N_{p0}}{A_p + A_s} \tag{11.29}$$

where

N_{p0} resultant force of ordinary reinforcement and tendons when the normal stress in concrete is zero, $N_{p0} = N_{t0}$. For reinforced concrete members, $A_p = 0$, $N_{po} = 0$; and

N axial force applied on members. For reinforced concrete members, $N = N_q$, and N_q is calculated by using a quasi-permanent combination of load effects. For prestressed concrete members, $N = N_k$, and N_k is calculated by using a characteristic combination of load effects.

Table 11.4 Value of v_i

Reinforcement type	Ordinary reinforcement		Pretensioned tendons			Posttensioned tendons		
	Plain reinforcement	Ribbed reinforcement	Ribbed reinforcement	Helically ribbed reinforcement	Indented steel wires, steel strands	Ribbed reinforcement	Steel strands	Plain reinforcement
v_i	0.7	1.0	1.0	0.8	0.6	0.8	0.5	0.4

Notes v_i for epoxy-coated ribbed reinforcement should be the corresponding values in the table times 0.8

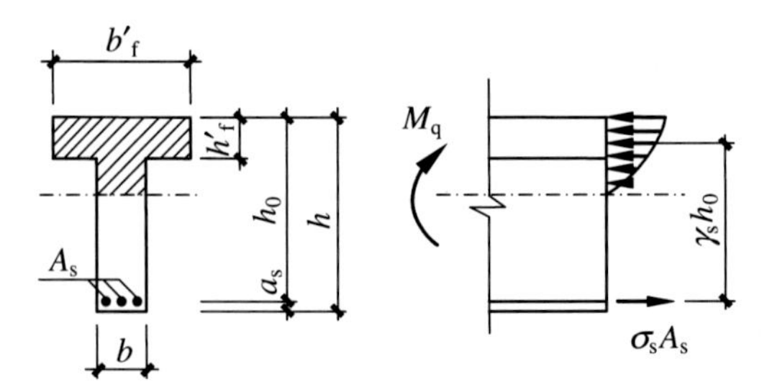

Fig. 11.16 Stress distribution of the cracked section of a flexural member

(2) Flexural members (Fig. 11.16)

$$\sigma_s = \frac{M_q}{A_s \gamma_s h_0} \tag{11.30}$$

where

M_q bending moment calculated by using a quasi-permanent combination of load effects; and

γ_s lever arm length factor at cracked sections.

$$\gamma_s = 1 - 0.4\frac{\sqrt{\alpha_E \rho}}{1 + 2\gamma'_f} \tag{11.31}$$

where

ρ reinforcement ratio of longitudinal tension reinforcement;

α_E ratio of modulus of elasticity of steel E_s to modulus of elasticity of concrete E_c; and

γ'_f strengthening factor of compressive flange

$$\gamma'_f = \frac{(b'_f - b)h'_f}{bh_0} \tag{11.32}$$

Under service loads, $M = (0.6\text{–}0.8)\ M_u$, and the beam is in Stage II. Experimental and theoretical analyses show that for frequently used concrete strengths and reinforcement ratios, the relative compression zone depth $\xi = x/h_0$ varies little; hence, γ_s is within 0.83–0.93 and can be approximately taken as $\gamma_s = 0.87$.

For prestressed flexural members,

$$\sigma_s = \frac{M_k - N_{p0}(z - e_p)}{(\alpha_p A_p + A_s)z} \tag{11.33}$$

where

M_k bending moment calculated by using a characteristic combination of load effects;

N_{p0} precompression or resultant force of all reinforcement when the normal stress in concrete is zero;

α_p equivalent deduction factor, for unbonded tendons, $\alpha_p = 0.3$ and for grouted posttensioned tendons, $\alpha_p = 1.0$;

e_p distance from resultant force of all reinforcement in the tension zone to N_{p0}; and

z distance from resultant force of all reinforcement in the tension zone to resultant force in the compression zone.

$$z = \left[0.87 - 0.12\left(1 - \gamma_f'\right)\left(\frac{h_0}{e}\right)^2\right] h_0 \quad (11.34)$$

$$e = e_p + \frac{M_k}{N_{p0}} \quad (11.35)$$

(3) Eccentrically tensioned reinforced concrete members

$$\sigma_s = \frac{N_q e'}{A_s\left(h_0 - a_s'\right)} \quad (11.36)$$

where e' = distance from the axial tensile force to the resultant force of longitudinal reinforcement in the compression zone or in the less tension side.

(4) Eccentrically compressed reinforced concrete members

$$\sigma_s = \frac{N_q(e - z)}{A_s z} \quad (11.37)$$

$$e = \eta_s e_0 + y_s \quad (11.38)$$

$$\eta_s = 1 + \frac{1}{4000 e_0/h_0}\left(\frac{l_0}{h}\right)^2 \quad (11.39)$$

where

z distance from resultant force of longitudinal tension reinforcement to the resultant force in the compression zone calculated by Eq. (11.34) and should be less than or equal to $0.87h_0$;

e_0 initial eccentricity under quasi-permanent combination of load effects, $e_0 = M_q/N_q$;

y_s distance from sectional gravity center to resultant force of longitudinal tension reinforcement; and

η_s amplification factor for eccentric compression in service. Equation (11.39) is derived by assuming that the curvature of the section in service is 1/3 of that in ultimate limit state. When $l_0/h \leqslant 14$, $\eta_s = 1.0$.

3. Coefficient for uneven distribution of tensile strain in reinforcement ψ
   It is the ratio of the average steel strain ε_{sm} between cracks to the steel strain at cracked sections. Actually, it also reflects the contribution of concrete between cracks to resist tension. Experiments indicate that ψ is related to concrete strength, reinforcement ratio, and steel strain values at cracked sections,

$$\psi = 1.1 - 0.65\frac{f_{tk}}{\rho_{te}\sigma_s} \tag{11.40}$$

   It is of no physical meaning if $\psi > 1$. So if $\psi > 1$, take $\psi = 1$. If ψ is too small, the concrete contribution will be overestimated. So if $\psi < 0.2$, take $\psi = 0.2$. Equation (11.40) applies to axially tensioned, flexural, and eccentrically loaded members.

Example 11.6 An axially tensioned bottom chord in a truss has the cross-sectional dimensions of $b \times h = 200$ mm $\times$ 160 mm. The concrete cover depth is $c = 20$ mm. 4ϕ16 ($A_s = 804$ mm^2) are provided. The concrete grade is C25 ($f_{tk} = 1.78$ N/mm^2). The axial force under the quasi-permanent combination of load effects is $N_q = 142$ kN. The crack width limit is $w_{lim} = 0.2$ mm. Check the crack width using the semitheoretical and semiempirical method.

Solution

$$\rho_{te} = \frac{A_s}{bh} = \frac{804}{200 \times 160} = 0.0251$$
$$\sigma_s = \frac{N_q}{A_s} = \frac{142,000}{804} = 177\text{N/mm}^2$$

$$\psi = 1.1 - 0.65\frac{f_{tk}}{\rho_{te}\sigma_s} = 1.1 - 0.65 \times \frac{1.78}{0.0251 \times 177} = 0.84$$

$$\begin{aligned} w_{max} &= \alpha_{cr}\psi\frac{\sigma_s}{E_s}\left(1.9c + 0.08\frac{d_{eq}}{\rho_{te}}\right) \\ &= 2.7 \times 0.84 \times \frac{177}{2.0 \times 10^5} \times \left(1.9 \times 20 + 0.08 \times \frac{16}{0.0251}\right) \\ &= 0.18\text{mm} < w_{lim} = 0.2\,\text{mm, OK.} \end{aligned}$$

Example 11.7 The cross-sectional dimensions of a T section beam are shown in Fig. 11.17. The bending moment on the section to be checked is $M_q = 440$ kN·m. The characteristic tensile strength of concrete is $f_{tk} = 1.54$ N/mm^2. The concrete

Fig. 11.17 Example 11.7

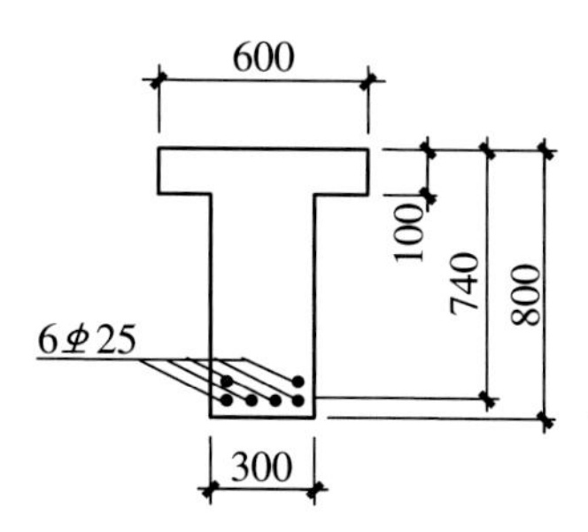

cover depth is $c = 25$ mm. 6⌀25 ($A_s = 2945$ mm^2) are provided in the tension zone. $E_s = 2.0 \times 105$ N/mm^2. Calculate the maximum crack width using the semitheoretical and semiempirical method.

Solution

$$\rho_{te} = \frac{A_s}{0.5bh} = \frac{2945}{0.5 \times 300 \times 800} = 0.0245$$

$$\sigma_s = \frac{M_q}{0.87A_s h_0} = \frac{440 \times 10^6}{0.87 \times 2945 \times 740} = 232\,\text{N/mm}^2$$

$$\psi = 1.1 - 0.65\frac{f_{tk}}{\rho_{te}\sigma_s} = 1.1 - 0.65 \times \frac{1.54}{0.0245 \times 232} = 0.917$$

$$w_{max} = \alpha_{cr}\psi\frac{\sigma_s}{E_s}\left(1.9c + 0.08\frac{d_{eq}}{\rho_{te}}\right)$$

$$= 1.9 \times 0.917 \times \frac{232}{2.0 \times 10^5} \times \left(1.9 \times 25 + 0.08 \times \frac{25}{0.0245}\right) = 0.26\,\text{mm}$$

11.3.2.2 Mathematical Statistics Method

This method represents one of the most popular methods created to calculate crack width. This is because that the semitheoretical and semiempirical method is very complicated, but the calculated values do not agree well with experimental data; on the other hand, the current stipulation on the crack width limit is still coarse.

The prerequisite of this method is to accumulate huge measured crack width data. After the main factors that influence crack width have been determined, the calculation formula for maximum crack width can be obtained from data analysis and regress. This method has been adopted in the *Code for Design of Highway of Reinforced Concrete and Prestressed Concrete Bridges and Culverts*

(JTG D62-2004), and *Design Code for Concrete Structures of Port and Waterway Engineering* (JTS151-2011). However, the calculation equations in the two codes are a little bit different. As an example, the method proposed in JTG D62-2004 is introduced in this section.

1. Main influencing factors
   Analysis of experiment results indicated that the following factors will influence the crack width of concrete members.

   (1) Crack width increases with the stress in tension reinforcement σ_s;
   (2) Crack width increases with the reinforcement diameter;
   (3) Crack width decreases with the reinforcement ratio;
   (4) Crack width increases with the concrete cover depth, so it is unfavorable for crack width control to increase the concrete cover depth. But some researches show that the thicker the concrete cover depth is, the lesser the corrosion of reinforcement is. Actually, the concrete cover depth cannot vary too much for ordinary structural members, so it can be neglected in calculation formulae;
   (5) Surface profile of reinforcement. The crack width of members provided with ribbed reinforcement is smaller than that with plain reinforcement;
   (6) Loading characteristics. Large crack width can be observed in members subjected to long-term loads or repeated loads;
   (7) Loading types include flexure, axial tension, etc.
   Tensile strength of concrete does not affect the crack width much.

2. Calculation formula

$$w_{\max} = C_1 C_2 C_3 \frac{\sigma_{ss}}{E_s}\left(\frac{30+d}{0.28+10\rho}\right)\text{(mm)} \tag{11.41}$$

where

C_1 coefficient related to surface profile of reinforcement. C_1 = 1.0 for ribbed reinforcement and C_1 = 1.4 for plain reinforcement;

C_2 coefficient related to long-term effects of loads or actions. $C_2 = 1 + 0.5\frac{N_l}{N_s}$, where N_l and N_s are internal forces (axial tension or bending moment) calculated by combinations of long-term load effects and short-term load effects, respectively. $C_2 = 1.0$ if the maximum crack width under short-term loads is to be calculated;

C_3 coefficient related to member type. C_3 = 1.15 for slab-like flexural members, and C_3 = 1.0 for flexural members with webs;

σ_{ss} stress in longitudinal tension reinforcement at crack sections calculated by combination of short-term load effects;

d diameter of longitudinal tension reinforcement (mm). If reinforcement of different diameters is provided, d should be replaced by $d_e = \frac{\sum n_i d_i^2}{\sum n_i d_i}$, where n_i is the number of the ith-type reinforcement in the tension zone; d_i is the nominal diameter of the ith-type ordinary reinforcement or the equivalent diameter of the *i*th-type steel wire bundle or steel strands $d_{pe} = \sqrt{n}d$. *n* is the number of steel wires or steel strands, and *d* is the nominal diameter of a single steel wire or steel strand; and

ρ reinforcement ratio of the longitudinal tension reinforcement in flexural members.

$$\rho = \frac{A_s + A_p}{bh_0 + (b_f - b)h_f} \tag{11.42}$$

When $\rho > 0.02$, take $\rho = 0.02$; when $\rho < 0.006$, take $\rho = 0.006$.

Example 11.8 The cross section of a prefabricated reinforced concrete bridge beam with the span of 19.5 m is shown in Fig. 11.18. C25 concrete and Grade II welded steel cages with longitudinal tension reinforcement of 8Φ32 + 2Φ16 are provided. $E_s = 2.0 \times 10^5$ N/mm^2. The bending moments at the mid-span are as follows: $M_G = 751$ kN·m by dead load, $M_{Q1} = 596.04$ kN·m by vehicle load (impact factor $\mu = 1.191$), and $M_{Q2} = 55.30$ kN·m by pedestrian load. $w_{lim} = 0.2$ mm. Check the maximum crack width under long-term loads using the mathematical statistics method.

Solution

$$C_1 = 1.0$$

$$M = M_G + \frac{M_{Q1}}{1.191} + M_{Q2} = 751 + \frac{596.04}{1.191} + 55.30 = 1306.75\,\text{kN}\cdot\text{m}$$

$$C_2 = 1 + 0.5\frac{M_G}{M} = 1 + 0.5 \times \frac{751}{1306.75} = 1.29$$

$$A_s = 804 \times 8 + 201 \times 2 = 6834\,\text{mm}^2$$

$$a_s = \frac{804 \times 8 \times 99 + 201 \times 2 \times 179}{6834} = 103.7\,\text{mm}$$

Fig. 11.18 Example 11.8

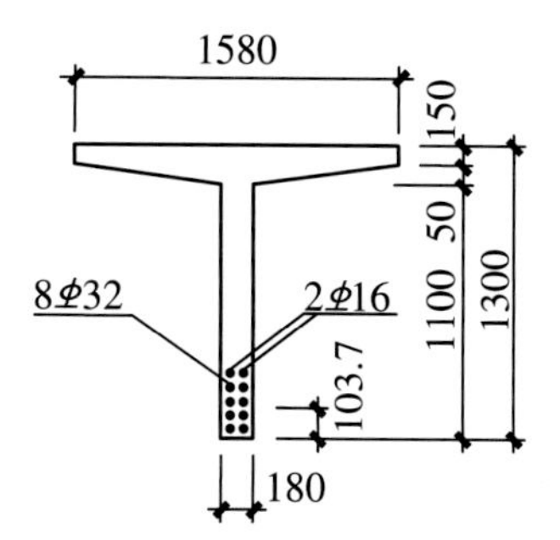

$$h_0 = h - a_s = 1300 - 103.7 = 1196.3\,\text{mm}$$

$$\sigma_{ss} = \frac{M}{\gamma_s h_0 A_s} = \frac{1306.75 \times 10^6}{0.87 \times 1196.3 \times 6834} = 183.7\,\text{N/mm}^2$$

$$d_e = \frac{2 \times 16^2 + 8 \times 32^2}{2 \times 16 + 8 \times 32} = 30.2\,\text{mm}$$

Take b = 180 mm, h_0 = 1196.3 mm, b_f = 0, h_f = 0

$$\rho = \frac{A_s}{bh_0 + (b_f - b)h_f} = \frac{6834}{180 \times 1196.3} = 0.0317 > 0.02$$

Take ρ = 0.02

$$\begin{aligned} w_{max} &= C_1 C_2 C_3 \frac{\sigma_{ss}}{E_s}\left(\frac{30+d}{0.28+10\rho}\right) \\ &= 1.0 \times 1.287 \times 1.0 \times \frac{183.7}{2.0 \times 10^5} \times \left(\frac{30+30.2}{0.28+10 \times 0.02}\right) \\ &= 0.15\,\text{mm} < w_{lim} = 0.2\,\text{mm, OK.} \end{aligned}$$

Example 11.9 If M_k = 600 kN·m, calculate the maximum crack width of the beam in Example 11.7 by using the mathematical statistics method.

Solution

$$C_1 = 1.0,\ C_3 = 1.0$$

$$C_2 = 1 + 0.5\frac{M_q}{M_k} = 1 + 0.5 \times \frac{440}{600} = 1.37$$

$$\sigma_s = 232\,\text{N/mm}^2$$

$$\rho = \frac{A_s}{bh_0} = \frac{2945}{300 \times 740} = 0.0133$$

$$\begin{aligned} w_{max} &= C_1 C_2 C_3 \frac{\sigma_{ss}}{E_s}\left(\frac{30+d}{0.28+10\rho}\right) \\ &= 1.0 \times 1.37 \times 1.0 \times \frac{232}{2.0 \times 10^5} \times \left(\frac{30+25}{0.28+10 \times 0.0133}\right) \\ &= 0.21\,\text{mm} \end{aligned}$$

Comparing the results of Examples 11.7 and 11.9, it can be found that the maximum crack widths calculated by different methods are not the same. But it should be noted that the crack width limits are also different for different methods. So corresponding values must be used in $w_{max} \leqslant w_{lim}$ to evaluate the crack width control of members.

11.4 Deflection Control

11.4.1 Purpose and Requirement of Deflection Control

11.4.1.1 Purpose

1. Ensure service functions of structures
   Excessive deformation will disrupt or even deprive functions of structures. For example, excessive deflection of the upper portion of a bridge will make the deck uneven, thereby impairing traffic safety could even cause the bridge deck to fail; outdoor floors (e.g., parking lots) or roofs may become ponds due to too large a deflection, which increases the risk of leakage. Excessive deflection of the floors in workshops requiring precise instruments will directly affect the product quality. Too much deflection in crane girders will not only disturb the normal operation of cranes, but will also increase the wear-down of rail components.
2. Aesthetic requirement and avoid panicking public
3. Avoid damaging nonstructural elements
   The so-called nonstructural members mean members only supporting self-weight or built by the detailing requirements. Excessive deformation of adjacent load-carrying members will fail the nonstructural members. For example, partitions made of brittle materials, such as hollow bricks or gypsum board, may be cracked and damaged by too big a deflection in supporting beams (Fig. 11.19), and excessive deflection of lintels will damage windows or doors beneath them. To avoid damages to nonstructural members is one of the main factors in determining deformation limits.
4. Avoid unfavorable influence to other members
   If a member deforms too much, its loading conditions may be different from those assumed in calculation and other members connected with it may also experience large deformation. Sometimes, even the transfer path, magnitude, and property of the load will be changed. For example, operation of cranes on girders that have excessive deformation may induce vibration of a factory building.

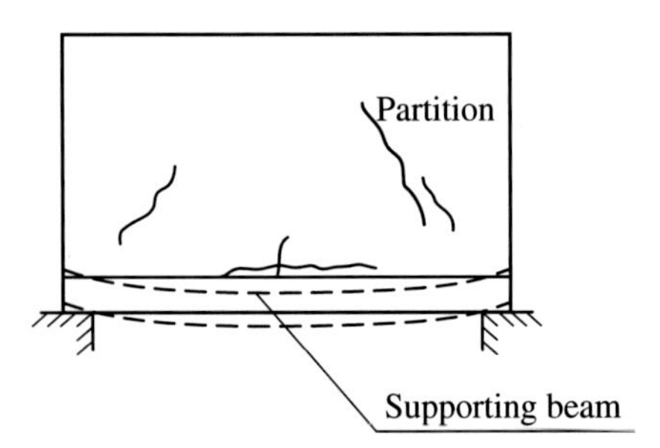

Fig. 11.19 Cracks in a partition due to excessive deformation of the supporting beam

11.4.1.2 Requirement

Currently, research on deformation control is limited to deflection control of flexural members, as well as trusses and arches in highway bridges. It is generally required that

$$f \leqslant f_{\rm lim} \tag{11.43}$$

Deflection limits are mainly determined by the above-mentioned purposes and practical experience. Tables 11.5 and 11.6 list the deflection limits for reinforced concrete flexural members in common civil and industrial structures and for reinforced concrete highway bridges and culverts.

Table 11.5 Deflection limits of flexural members in building structures

Member types	Deflection limit (using computation span l_0)
Crane girders	
Hand operated	$l_0/500$
Motor operated	$l_0/600$
Roof, floor, and stairs	
$l_0 < 7$ m	$l_0/200$ ($l_0/250$)
7 m $\leqslant l_0 < 9$ m	$l_0/250$ ($l_0/300$)
$l_0 > 9$ m	$l_0/300$ ($l_0/400$)

Notes (1) If members are cambered in fabrication, which is also allowed in service, the deflections are the calculated deflection values minus the camber values. For prestressed concrete members, cambers by precompression can also be deducted

(2) Values in parentheses are applied to members of strict deflection control

(3) For cantilever members, the deflection limits are twice the corresponding values in the table

Table 11.6 Limits on vertical deflections for members of highway bridges and culverts

Member types	Deflection limit
Mid-span of girders in girder bridge	$l/600$
Cantilever end of girders in girder bridge	$l_1/300$
Trusses and arches	$l/800$

Notes (1) l is the computation span, and l_1 is the cantilever length
(2) Values in the table can be increased by 20 % for vehicle or crawler loading tests
(3) If both positive and negative deflections happen in a single span, the calculated deflection should be the sum of the absolute values of both deflections

11.4.2 Deformation Checking for Reinforced Concrete Flexural Members

Deflection of reinforced concrete flexural members can be obtained by relevant formulae in the *Mechanics of Materials* class if the flexural stiffness has been determined. The calculation of flexural stiffness should be reasonable and can reflect the plasticity of members after cracking.

From the plane section assumption, the differential equation for the deflection curve of a homogeneous elastic beam is

$$\frac{d^2y(x)}{dx^2} = \frac{1}{r} = -\frac{M(x)}{EI} \quad (11.44)$$

where
$y(x)$ deflections;
r radius of curvature; and
EI flexural stiffness.

Solving this equation gives

$$f = S\frac{Ml_0^2}{EI} \quad (11.45)$$

where
M maximum bending moment;
S coefficient related to loading types and supporting conditions. For example, S = 5/48 for simply supported beams under a uniformly distributed load; and
l_0 computation span.

The flexural stiffness of a homogeneous elastic beam is a constant if the materials and dimensions have been selected. Deflection f is linearly proportional to bending moment M, as shown by the dashed line OA in Fig. 11.20.

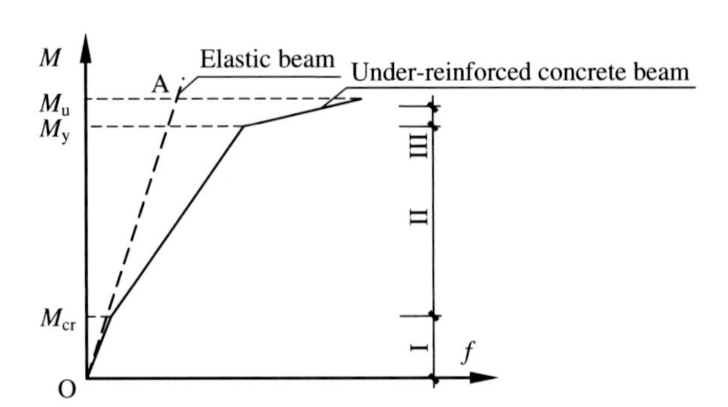

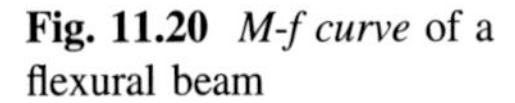
Fig. 11.20 *M-f curve* of a flexural beam

As noted in Chap. 5, under-reinforced concrete beams experience three stages from being loaded to failure (the solid line in Fig. 11.20). Before cracking ($M \leqslant M_{cr}$), the beam works elastically. The relationship of f and M is a straight line. If the flexural stiffness is EI, this straight line overlaps the dashed line OA. Once the concrete in the tension zone cracks ($M \geqslant M_{cr}$), the flexural stiffness decreases noticeably. And when the tension reinforcement yields ($M \geqslant M_y$), the flexural stiffness drops sharply. The above phenomena indicate that the flexural stiffness of a reinforced concrete beam is not constant and will be significantly affected by cracking. Because flexural members in service are working with cracks, Stage II should be based on for deformation or deflection calculation.

There are two kinds of methods in deformation calculation for reinforced concrete flexural members. One is the semitheoretical and semiempirical method, which assumes that the concrete bond between cracks still works. The other method neglects the contribution of concrete between cracks and takes the moment of inertia of a transformed cracked section as the basis for calculation. This method is actually based on elastic flexural stiffness, which gives us the formulae for flexural stiffness.

11.4.2.1 Semitheoretical and Semiempirical Method

1. Characteristics of strains in flexural members in service
   Experimental results show that when the cracks are stable, the tensile strains in reinforcement and compressive strains in concrete are unevenly distributed in longitudinal direction (Fig. 11.21). The strains are large at cracked sections, whereas they are small between cracks. Therefore, the neutral axis is actually a wavy line. Even in the pure bending segment, the neutral axis height x_n also varies.
   Coefficient $\psi = \varepsilon_{sm}/\varepsilon_s$ is still being used to reflect the uneven distribution of tensile strains in reinforcement between cracks. It also reflects the contribution of concrete between cracks to resist tension. The average steel stress σ_{sm} corresponds to ε_{sm}.

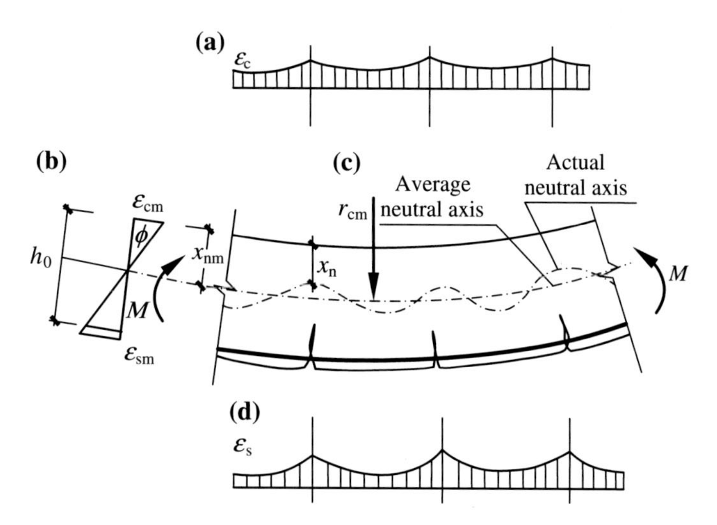

Fig. 11.21 Strain distribution of a flexural beam and location of the neutral axis. **a** Compressive strain distribution in concrete. **b** Average strain distribution on a section. **c** Location of the neutral axis. **d** Strain distribution in the steel bar

The magnitude of variation of compressive strain in concrete is much less than that in steel. The maximum value is not far from the average value ε_{cm}.
The average of neutral axis height x_{nm} is called the average height of neutral axis. The corresponding neutral axis and curvature are called the average neutral axis and average curvature, respectively.
Experiments show that the average steel strain ε_{sm} and the average concrete strain ε_{cm} comply with the plane section assumption (Fig. 11.21b).

2. Formulae for flexural stiffness

(1) Short-term flexural stiffness B_s
It can be derived by synthesizing geometrical relationship, constitutive law, and equilibrium conditions on a specific cross section.
Geometrical relationship: From the plane section assumption, the curvature of the cross section in a pure bending segment is

$$\phi = \frac{1}{r_{cm}} = \frac{\varepsilon_{sm} + \varepsilon_{cm}}{h_0} \tag{11.46}$$

where
r_{cm} the average radius of curvature.

Constitutive law: For member in service, the average steel stress σ_{sm} and the average steel strain ε_{sm} obey the Hooke's law, i.e., $\varepsilon_{sm} = \frac{\sigma_{sm}}{E_s}$. So,

$$\varepsilon_{\mathrm{sm}} = \psi \varepsilon_{\mathrm{s}} = \psi \frac{\sigma_{\mathrm{s}}}{E_{\mathrm{s}}} \tag{11.47}$$

In addition, because $\varepsilon_{\mathrm{cm}}$ does not represent a big difference from ε_{c} and $E'_{\mathrm{c}} = \nu E_{\mathrm{c}}$, $\varepsilon_{\mathrm{cm}}$ should be used to consider the plasticity of concrete; hence, we have

$$\varepsilon_{\mathrm{cm}} \approx \varepsilon_{\mathrm{c}} = \frac{\sigma_{\mathrm{s}}}{E'_{\mathrm{c}}} = \frac{\sigma_{\mathrm{s}}}{\nu E_{\mathrm{c}}} \tag{11.48}$$

Equilibrium condition: Figure 11.22a shows the real stress distribution along the cracked section, which can be replaced by an equivalent stress block (Fig. 11.22b) with the magnitude of $\omega\sigma_{\mathrm{c}}$, where ω is a coefficient. If the compression zone depth of the cracked section is assumed to be ξh_0, the lever arm of internal forces is $\gamma_{\mathrm{s}} h_0$, and the concrete stress in the compression zone can be obtained from the equilibrium conditions in Fig. 11.22b as follows:

$$\sigma_{\mathrm{c}} = \frac{M}{\xi \omega \gamma_{\mathrm{s}} b h_0^2} \tag{11.49}$$

And the stress of steel bars is

$$\sigma_{\mathrm{s}} = \frac{M}{A_{\mathrm{s}} \gamma_{\mathrm{s}} h_0} \tag{11.50}$$

Summarizing Eqs. (11.46)–(11.49) gives

$$\begin{aligned} \phi &= \frac{\varepsilon_{\mathrm{sm}} + \varepsilon_{\mathrm{cm}}}{h_0} = \frac{\psi \frac{\sigma_{\mathrm{s}}}{E_{\mathrm{s}}} + \frac{\sigma_{\mathrm{c}}}{\nu E_{\mathrm{c}}}}{h_0} = \frac{\psi \frac{M}{E_{\mathrm{s}} A_{\mathrm{s}} \gamma_{\mathrm{s}} h_0} + \frac{M}{\nu E_{\mathrm{c}} \xi \omega \gamma_{\mathrm{s}} b h_0^2}}{h_0} \\ &= M \left(\frac{\psi}{E_{\mathrm{s}} A_{\mathrm{s}} \gamma_{\mathrm{s}} h_0^2} + \frac{1}{\nu E_{\mathrm{c}} \xi \omega \gamma_{\mathrm{s}} b h_0^3} \right) \end{aligned} \tag{11.51}$$

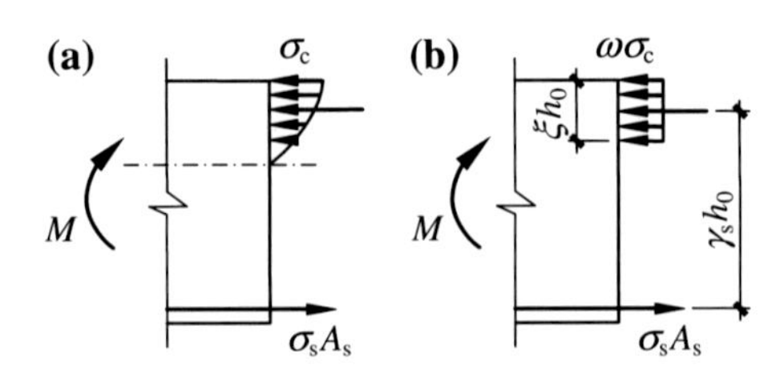

Fig. 11.22 Calculation schematic of the cracked section

Assuming $\zeta = \nu\xi\omega\gamma_s$, we can get

$$B_s = \frac{M}{\phi} = \frac{1}{\frac{\psi}{E_s A_s \gamma_s h_0^2} + \frac{1}{\zeta E_c b h_0^3}} = \frac{E_s A_s h_0^2}{\frac{\psi}{\gamma_s} + \frac{\alpha_E \rho}{\zeta}} \tag{11.52}$$

where ψ can be calculated by Eq. (11.40), α_E is the ratio of modulus of elasticity of steel to that of concrete, and ρ is the reinforcement ratio of longitudinal reinforcement. Obviously, once the dimensions and reinforcement ratio of a member have been determined, the first term in the denominator of Eq. (11.51) reflects the influence of uneven distribution of steel strain on the flexural stiffness. When M is small and σ_s is also small, which means the steel and the concrete are bonded well, so the uneven distribution of steel strain is not that significant. Concrete makes a big contribution to resisting tension. The value of ψ is small, and the value of B_s is large. When M is large, B_s is reduced. The second term in the denominator reflects the influence of concrete in the compression zone on flexural stiffness.

ζ synthetically reflects the effects of ν, ξ, ω, and γ_s. When M is small, ν and ξ are large, but γ_s and ω are small. When M is large, γ_s and ω are big. The variation of bending moment within service load range will not affect ζ much, so ζ can be assumed to be independent of bending moment. Instead, ζ is related to concrete strength, reinforcement ratio, and cross section type of the member.

From Eq. (11.49), we have

$$M = \sigma_c \xi \omega \gamma_s b h_0^2 = \nu E_c \varepsilon_c \xi \omega \gamma_s b h_0^2 = \zeta \varepsilon_{cm} E_c b h_0^2 \tag{11.53}$$

This means that ζ can be obtained from experiments because E_c, b, and h_0 of a specimen are known and ε_{cm} can be measured.

Analysis of experimental data yields

$$\frac{\alpha_E \rho}{\zeta} = 0.2 + \frac{6\alpha_E \rho}{1 + 3.5\gamma_f'} \tag{11.54}$$

where γ_f' = strengthening factor of compressive flange (Eq. 11.32). When $h_f' > 0.2h_0$, take $h_f' = 0.2h_0$. For rectangular cross section, $\gamma_f' = 0$.

Equation (11.54) will give a constant value if concrete strength and reinforcement ratio have been determined. Substituting Eq. (11.54) and $\gamma_s = 0.87$ into Eq. (11.52) yields

$$B_s = \frac{E_s A_s h_0^2}{1.15\psi + 0.2 + \frac{6\alpha_E \rho}{1 + 3.5\gamma_f'}} \tag{11.55}$$

(2) Flexural stiffness B

Under long-term loads, the flexural stiffness of reinforced concrete members will decrease with time so that the deflection will increase. Experiments showed that deflections of beams increased quickly in the first half year. Then, the increasing speed gradually decreased. After about a year, the deflection became stable with minute increase. The increase of ε_{cm} due to concrete creep is the main reason for deflection augment. Besides the slip creep between the concrete and the steel (especially for beams with small reinforcement ratio), which will make a portion of concrete quit working, the development of concrete plasticity in compression zone concrete and the contraction difference between tensioned and compressed concrete will also reduce the flexural stiffness. Therefore, all factors that influence the creep and shrinkage of concrete, such as components and their proportion in concrete, reinforcement ratio of compression reinforcement, loading time, and environmental conditions (temperature and humidity), will influence flexural stiffness.

The influence of long-term loads on the increase of deflection can be represented by θ, which is defined as follows:

$$\theta = \frac{f}{f_s} \tag{11.56}$$

where f and f_s = long-term deflection and short-term deflection, respectively.

θ can be determined by experiments, and the restraint of compression reinforcement on creep and shrinkage of concrete should be considered. When the reinforcement ratio of compression reinforcement $\rho' = \frac{A'_s}{bh_0} = 0$, $\theta = 2.0$; when $\rho' = \rho$, $\theta = 1.6$; and when $0 < \rho' < \rho$, θ can be determined by linear interpolation. For a T section with the flange in the tension zone, θ should be increased by 20 %.

M_q calculated by the quasi-permanent combination of load effects is a long-term load effect. Thus, if the deformation of a member is calculated by the quasi-permanent combination of load effects, we have

$$B = B_s/\theta \tag{11.57}$$

M_k calculated by a characteristic combination of load effects includes two effects. One is the long-term bending moment M_q produced by the dead load and dead part in the live load. Another is the short-term bending moment $M_k - M_q$ produced by the live part of the live load. The deflection of a beam induced by these two effects is illustrated in Fig. 11.23. If the deformation is calculated by the characteristic combination of load effects, it should be the sum of f'_l and f'_s as shown in Fig. 11.23. That is,

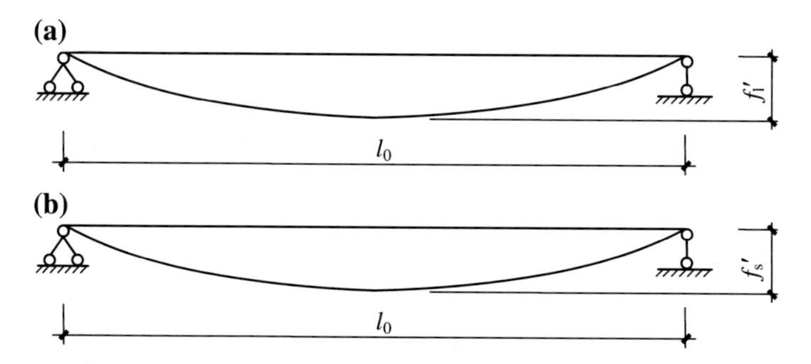

Fig. 11.23 Beam deflections under different loads. **a** Deformation caused by the dead load and dead part in the live load. **b** Deformation caused by the live part in the live load

$$f = f_l' + f_s' = \theta S \frac{M_q l_0^2}{B_s} + S \frac{M_k - M_q}{B_s} l_0^2 = S \frac{M_k}{B} l_0^2 \tag{11.58}$$

Thus, the flexural stiffness calculated by the characteristic combination of load effects and considering the influence of long-term load is

$$B = \frac{M_k}{M_q(\theta - 1) + M_k} B_s \tag{11.59}$$

GB 50010 adopts this calculation method and stipulates that the maximum deflection of reinforced concrete flexural members uses the quasi-permanent combination of load effects, whereas the maximum deflection of prestressed concrete flexural members uses the characteristic combination of load effects, and the influence of long-term load should be considered in both cases.

11.4.2.2 Simplified Calculation Method Based on Elastic Stiffness

In the semitheoretical and semiempirical method, the relationship between B_s and $\alpha_E \rho$ is a hyperbola, which makes the calculation complex. To simplify the calculation and be conservative, another method, which multiplies the elastic stiffness with a reduction factor, has been suggested and adopted by *Code for Design of Reinforced Concrete and Prestressed Concrete Highway Bridges and Culverts* (JTG D62-2004) and the *Design Code for Concrete Structures of Port and Waterway Engineering* (JTS151-2001). However, the calculation equations in the two codes are different. As an example, the method in JTG D62-2004 is introduced in this section.

The short-term flexural stiffness is calculated by

$$B_s = \frac{0.95E_cI_0}{\left(\frac{M_{cr}}{M_s}\right)^2 + \left[\left(1 - \frac{M_{cr}}{M_s}\right)^2\right]\frac{0.95I_0}{I_{0cr}}} \tag{11.60}$$

where
I_0 moment of inertia of transformed section;
I_{0cr} moment of inertia of transformed cracked section;
M_s bending moment calculated by combination of short-term load effect; and
M_{cr} cracking moment of section calculated by Eq. (5.16).

When the influence of long-term loads is considered, an amplification factor η_θ is multiplied to the result shown in Eq. (11.60). η_θ has the same physical meaning as θ in Eq. (11.57), but they are of different values. JTGD62 specifies that when the strength grade of concrete is less than C40, $\eta_\theta = 1.60$; when C40–C80 concrete is used, $\eta_\theta = 1.45 - 1.35$.

The transformed area and corresponding moment of inertia of a singly reinforced section can be referred to in Chap. 5. The transformed area of a cracked section is (Fig. 11.24)

$$A_{0cr} = bx + \alpha_E A_s \tag{11.61}$$

where the depth of compression zone can be obtained by

$$x = \frac{\alpha_E A_s}{b}\left(\sqrt{1 + \frac{2bh_0}{\alpha_E A_s}} - 1\right) \tag{11.62}$$

The corresponding moment of inertia is

$$I_{0cr} = \frac{1}{3}bx^3 + \alpha_E A_s(h_0 - x)^2 \tag{11.63}$$

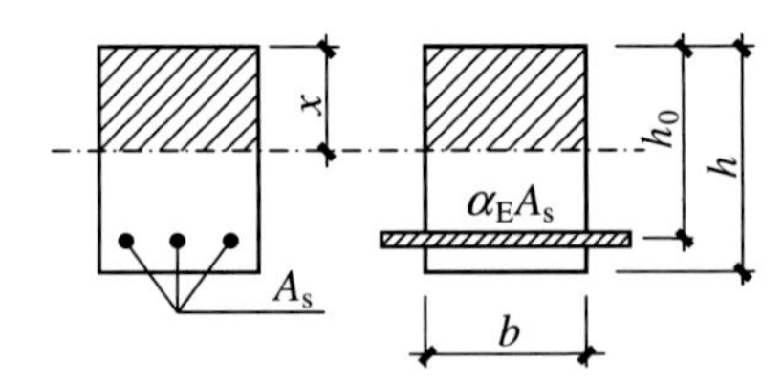

Fig. 11.24 Transformed area of a cracked section

11.4.2.3 Deformation Checking

After the flexural stiffness has been obtained, the deflection of flexural members can be calculated by using the flexural stiffness to replace EI in Eq. (11.45). Then, Eq. (11.43) is used to check whether the deflection is satisfactory.

1. Semitheoretical and semiempirical method
   The flexural stiffness represented by Eqs. (11.55), (11.57), or (11.59) is an average stiffness along the pure bending segment. But the bending moments in the shear–flexure segment of a reinforced concrete beam are not equal at any two cross sections. And the tension zones of different cross sections crack differently. As shown in Fig. 11.25, no normal cracks appear at cross sections close to the supports when $M < M_{cr}$. Therefore, the flexural stiffness of these cross sections is much larger than those of cracked cross sections at mid-span. Along the member, the flexural stiffness varies with steel content and steel stress. The cross section subjected to the maximum bending moment has the smallest flexural stiffness $B_{s,min}$. Theoretically, the deflection should be calculated by varied flexural stiffness rates, but it is too complicated. For simplification, the flexural stiffness rates of all cross sections in beam segments subjected to bending moments of one sign are equal in engineering design to flexural members of identical cross sections. Moreover, the flexural stiffness $B_{s,min}$ at the position of the maximum bending moment is taken for deflection calculation, as shown by the dashed line in Fig. 11.25a. This is the so-called minimum stiffness principle.
   According to the minimum stiffness principle, the calculated curvature $M/B_{s,min}$ near supports is larger than the real one (Fig. 11.25b). However, the above-mentioned stiffness calculation only considers the flexural deformation of normal sections. Since shear deformation also increases deflection, it should not be neglected in reinforced concrete beams when inclined cracks occur in shear–flexure segments. Moreover, the occurrence of inclined cracks increases the steel stress in shear–flexure segment. The influence of these factors on deflection

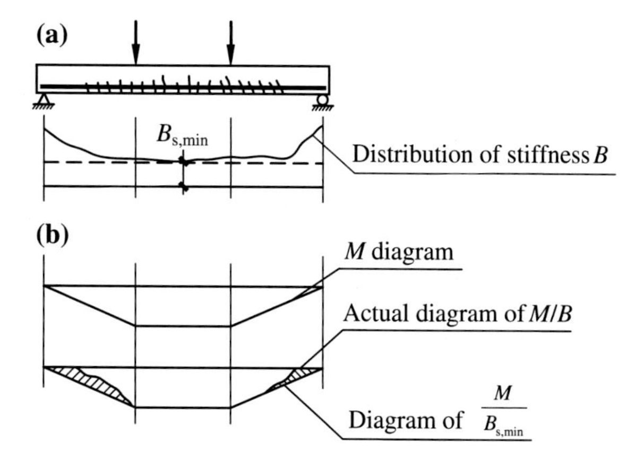

Fig. 11.25 Application of the principle of minimum stiffness. **a** Minimum stiffness. **b** Distribution of beam curvature

increase approximately offsets the calculation deviance due to using the minimum stiffness principle, which has been proven by experiments on 350 beams. Therefore, it is sound to calculate the deflection by the minimum stiffness principle.

2. Simplified method based on elastic stiffness

   When Eq. (11.60) is employed to calculate flexural stiffness, equal stiffness along the whole member is also assumed. The deflection of a reinforced concrete bridge includes two parts. One is induced by structure weight (dead load). Another is from static live load without considering impact. When the deflection produced by structure weight and vehicle load (neglect impact) exceeds $l_0/1600$, precamber should be set, the value of which is equal to the long-term deflection calculated by structural weight and a half vehicle load (neglect impact). When the deformation of highway bridges is being checked, if the move of vehicle load produces both positive and negative deflections on a single span, the maximum deflection should be the sum of the absolute values of both deflections.

11.4.2.4 Measures to Increase Flexural Stiffness

From the formulae, it can be found that the most efficient way to enhance flexural stiffness is to increase the section height h. In engineering practice, the deformation of members is controlled in advance by reasonably selecting section height-to-span ratio (h/l). When the dimension of cross sections cannot be modified, other measures, such as increasing the content of longitudinal reinforcement, using doubly reinforced sections, can also be adopted. In addition, using high performance concrete or applying prestress is efficient too.

Example 11.10 The hollow plate shown in Fig. 11.26a has a computation span of l_0 = 3.04 m. C20 concrete and 9ϕ6 are provided. The concrete cover depth is c = 10 mm. The bending moment calculated by quasi-permanent combination of load effects is M_q = 3.53 kN·m. Check the deformation of the plate according to the semitheoretical and semiempirical method.

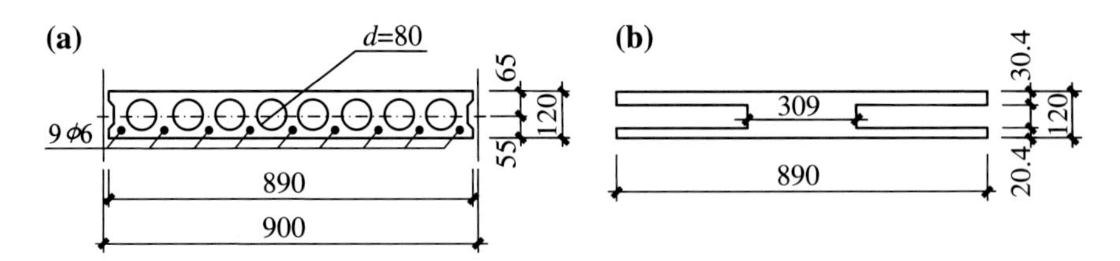

Fig. 11.26 Example 11.10

Solution

1. Calculation cross section
   Transform the cross section of the hollow plate to an I-shaped section, i.e., replace the circular holes by equivalent rectangular holes with the area, centroid and moment of inertia of the cross section unchanged.

$$\frac{\pi d^2}{4} = b_a h_a, \quad \frac{\pi d^4}{64} = \frac{b_a h_a^3}{12}$$

   Solving the equation set, we have b_a = 72.6 mm and h_a = 69.2 mm.
   The dimensions of the I-shaped section are (Fig. 11.26b)

$$b = 890 - 72.6 \times 8 = 309\,\text{mm}, \quad h_f' = 65 - \frac{69.2}{2} = 30.4\,\text{mm},$$

$$h_f = 55 - \frac{69.2}{2} = 20.4\,\text{mm}$$

2. Deflection checking is expressed as follows:

$$\alpha_E \rho = \frac{E_s}{E_c} \frac{A_s}{bh_0} = \frac{2.1 \times 10^5}{2.55 \times 10^4} \times \frac{28.3 \times 9}{309 \times 107} = 0.0634$$

$$\gamma_f' = \frac{(b_f' - b)h_f'}{bh_0} = \frac{(890 - 309) \times 30.4}{309 \times 107} = 0.534$$

$$\rho_{te} = \frac{A_s}{0.5bh + (b_f - b)h_f} = \frac{28.3 \times 9}{0.5 \times 309 \times 120 + (890 - 309) \times 20.4} = 0.00838$$

$$\sigma_{sq} = \frac{M_q}{0.87 h_0 A_s} = \frac{3.53 \times 10^6}{0.87 \times 107 \times 28.3 \times 9} = 148.9\,\text{N/mm}^2$$

$$f_{tk} = 1.5\,\text{N/mm}^2$$

$$\psi = 1.1 - 0.65 \frac{f_{tk}}{\rho_{te}\sigma_{sq}} = 1.1 - 0.65 \times \frac{1.5}{0.00838 \times 148.9} = 0.319$$

$$B_s = \frac{E_s A_s h_0^2}{1.15\psi + 0.2 + \frac{6\alpha_E \rho}{1 + 3.5\gamma_f'}}$$

$$= \frac{2.1 \times 10^5 \times 28.3 \times 8 \times 107^2}{1.15 \times 0.319 + 0.2 + \frac{6 \times 0.0634}{1 + 3.5 \times 0.534}} = 7.78 \times 10^{11}\,\text{N}\cdot\text{mm}^2$$

$$B = \frac{B_s}{\theta} = \frac{7.78 \times 10^{11}}{2} = 3.89 \times 10^{11}\,\text{N·mm}^2$$

$$\begin{aligned} f &= \frac{5}{48}\frac{M_q l_0^2}{B} = \frac{5}{48} \times \frac{3.53 \times 10^6 \times 3040^2}{3.89 \times 10^{11}} = 8.74\,\text{mm} < \frac{l_0}{200} \\ &= \frac{3040}{200} = 15.2\,\text{mm, OK.} \end{aligned}$$

Example 11.11 A flexural member of an I-shaped section is shown in Fig. 11.27. C30 concrete and Grade II reinforcement are used. The computation span is $l_0 = 11.7$ m, $M_q = 550$ kN·m. The deflection limit is $l_0/300$. Check the deformation of the member according to the semitheoretical and semiempirical method.

Solution

$$A_s = 2945\,\text{mm}^2, \quad h_0 = 1290 - 65 = 1225\,\text{mm}$$

$$\begin{aligned} A_{te} &= 0.5bh + (b_f - b)h_f \\ &= 0.5 \times 80 \times 1290 + (200 - 80) \times 130 \\ &= 67,200\,\text{mm}^2 \end{aligned}$$

$$\rho_{te} = \frac{A_s}{A_{te}} = \frac{2945}{67,200} = 0.0438$$

$$\rho = \frac{A_s}{bh_0} = \frac{2945}{80 \times 1225} = 0.0301$$

Fig. 11.27 Example 11.11

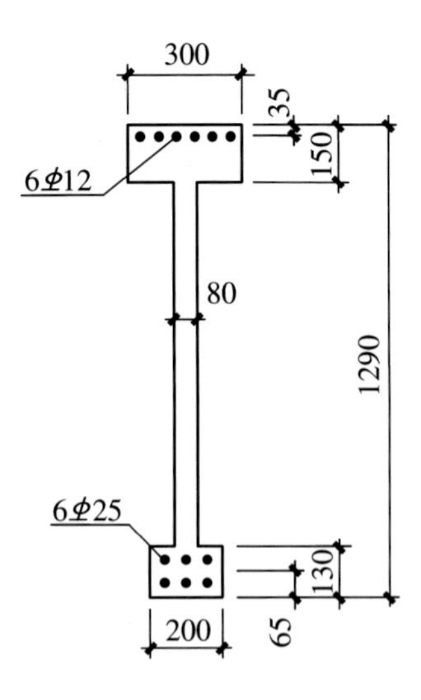

$$\alpha_E \rho = \frac{E_s}{E_c}\rho = \frac{2.0 \times 10^5}{3.0 \times 10^4} \times 0.0301 = 0.201$$

$$\sigma_{sq} = \frac{M_q}{0.87 h_0 A_s} = \frac{550 \times 10^6}{0.87 \times 1225 \times 2945} = 175\,\text{N/mm}^2$$

$$f_{tk} = 2\,\text{N/mm}^2$$

$$\psi = 1.1 - 0.65\frac{f_{tk}}{\rho_{te}\sigma_{sq}} = 1.1 - 0.65 \times \frac{2}{0.0438 \times 175} = 0.93$$

$$\gamma_f' = \frac{(b_f' - b)h_f'}{bh_0} = \frac{(300 - 80) \times 150}{80 \times 1225} = 0.337$$

$$\begin{aligned} B_s &= \frac{E_s A_s h_0^2}{1.15\psi + 0.2 + \frac{6\alpha_E \rho}{1+3.5\gamma_f'}} \\ &= \frac{2.0 \times 10^5 \times 2945 \times 1225^2}{1.15 \times 0.93 + 0.2 + \frac{6 \times 0.201}{1+3.5 \times 0.337}} = 485 \times 10^{12}\,\text{N}\cdot\text{mm}^2 \end{aligned}$$

$$A_s' = 678\,\text{mm}^2$$

$$\rho' = \frac{A_s'}{bh_0} = \frac{678}{80 \times 1225} = 0.0069$$

$$\frac{\rho'}{\rho} = \frac{0.0069}{0.301} = 0.229$$

$$\theta = 2.0 - 0.4\frac{\rho'}{\rho} = 2.0 - 0.4 \times 0.229 = 1.91$$

$$B = \frac{B_s}{\theta} = \frac{485 \times 10^{12}}{1.91} = 254 \times 10^{12}\,\text{N}\cdot\text{mm}^2$$

$$\begin{aligned} f &= \frac{5}{48}\frac{M_q l_0^2}{B} = \frac{5}{48} \times \frac{550 \times 10^6 \times 11{,}700^2}{254 \times 10^{12}} = 29.6\,\text{mm} < \frac{l_0}{300} = \frac{11{,}700}{300} \\ &= 39\,\text{mm, OK.} \end{aligned}$$

Example 11.12 The known conditions are the same as those in Example 11.8, where $I_0 = 64.35 \times 10^9\ \text{mm}^4$, $I_{cr} = 50.71 \times 10^9\ \text{mm}^4$, $W_0 = 7.55 \times 10^7\ \text{mm}^3$, and $E_c = 2.85 \times 10^4\ \text{N/mm}^2$. Check the deflection of the member.

Solution
From Example 11.8, we know that

$$M_s = 1306.75\,\text{kN m}, \quad \alpha_E = \frac{2.0 \times 10^5}{2.85 \times 10^4} = 7.02$$

The distance from the centroid to the bottom surface in the transformed section is shown as follows:

$$y_0 = \frac{I_0}{W_0} = \frac{64.35 \times 10^9}{7.55 \times 10^7} = 852.3\,\text{mm}$$

$$S_0 = 180 \times 852.3 \times \frac{852.3}{2} + 7.02 \times 6834 \times (852.3 - 103.7) = 1.013 \times 10^8\,\text{mm}^3$$

$$\gamma = 2S_0/W_0 = 2 \times 1.013 \times 10^8/7.55 \times 10^7 = 2.683$$

$$M_{cr} = \gamma f_{tk} W_0 = 2.683 \times 1.78 \times 7.55 \times 10^7 = 360.568 \times 10^6\,\text{N·mm}$$

$$\begin{aligned} B_s &= \frac{0.95E_c I_0}{\left(\frac{M_{cr}}{M_s}\right)^2 + \left[\left(1 - \frac{M_{cr}}{M_s}\right)^2\right] \frac{0.95E_c I_0}{I_{0cr}}} \\ &= \frac{0.95 \times 2.85 \times 10^4 \times 64.35 \times 10^9}{\left(\frac{360.568 \times 10^6}{1306.75 \times 10^6}\right)^2 + \left[\left(1 - \frac{360.568 \times 10^6}{1306.75 \times 10^6}\right)^2\right] \times \frac{0.95 \times 64.35 \times 10^9}{50.71 \times 10^9}} \\ &= 2.452 \times 10^{15}\,\text{N·mm} \end{aligned}$$

$$\begin{aligned} f &= \eta_\theta \frac{5}{48}\frac{M_s l^2}{B_s} = 1.6 \times \frac{5}{48} \times \frac{1306.75 \times 10^6 \times 19.5^2 \times 10^6}{2.452 \times 10^{15}} \\ &= 33.78\,\text{mm} > \frac{l_0}{600} = \frac{19.5 \times 10^3}{600} = 32.5\,\text{mm} \end{aligned}$$

This deflection is not satisfactory; the design needs to be modified, and the calculation of the precamber is omitted herein.

11.4.3 Deformation Checking for Prestressed Concrete Flexural Members

We take GB 50010 as an example to introduce the method of deformation checking for prestressed concrete flexural members. The calculation methods for deflection of other structures are very similar.

11.4.3.1 Short-Term Stiffness

1. Members with cracks not allowed

$$B_s = 0.85E_cI_0 \tag{11.64}$$

2. Members with cracks allowed
   For prestressed concrete flexural members, in which cracks appear in service, there is a bilinear relationship between bending moment and curvature. The intersection of the two straight lines corresponds to the cracking moment M_{cr}, as shown in Fig. 11.28.

Then the short-term stiffness can be obtained as follows:

$$B_s = \frac{E_cI_0}{\frac{1}{\beta_{0.4}} + \frac{\frac{M_{cr}}{M_k} - 0.4}{1 - 0.4}\left(\frac{1}{\beta_{cr}} - \frac{1}{\beta_{0.4}}\right)} \tag{11.65}$$

where $\beta_{0.4}$ and β_{cr} = the stiffness reduction factors when $\frac{M_{cr}}{M_k} = 0.4$ and 1.0, respectively. β_{cr} = 0.85. $\frac{1}{\beta_{0.4}}$ can be regressed from experimental data as follows:

$$\frac{1}{\beta_{0.4}} = \left(0.8 + \frac{0.15}{\alpha_E \rho}\right)(1 + 0.45\gamma_f) \tag{11.66}$$

Substituting β_{cr} and $\frac{1}{\beta_{0.4}}$ into Eq. (11.65) yields

$$B_s = \frac{0.85E_cI_0}{\kappa_{cr} + (1 - \kappa_{cr})\omega} \tag{11.67}$$

$$\omega = \left(1.0 + \frac{0.21}{\alpha_E \rho}\right)(1 + 0.45\gamma_f) - 0.7 \tag{11.68}$$

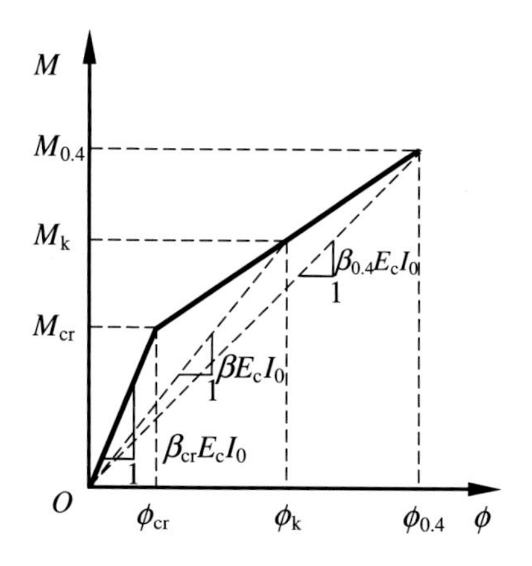

Fig. 11.28 Bilinear moment–curvature relationship of a bonded prestressed concrete beam

where

ρ reinforcement ratio of longitudinal tension reinforcement, $\rho = (A_s + A_p)/(bh_0)$;

γ_f ratio of tension flange area to effective area of web, $\gamma_f = \frac{(b_f - b)h}{bh_0}$; and

κ_{cr} ratio of M_{cr} to M_k, when $\kappa_{cr} > 1.0$, $\kappa_{cr} = 1.0$. M_{cr} is calculated by Eq. (10.83) or Eq. (10.95).

For members with cracks occurring in a pretension zone during precompression, B_s from Eq. (11.67) should be reduced by 10 %.

11.4.3.2 Flexural Stiffness *B*

B can be calculated by Eq. (11.59) by taking $\theta = 2.0$ to consider the influence of long-term load.

11.4.3.3 Deformation Checking

The deflection of a prestressed concrete flexural member is superimposed by the one induced by loads and another one by camber.

1. Deflection induced by loads f_1

$$f_1 = S\frac{M_k l_0^2}{B} \tag{11.69}$$

2. Deflection induced by camber f_2
   f_2 is produced by the total precompression N_{PII} of eccentricity $e_{0\mathrm{II}}$ and can be calculated by the relevant formula used by students in the class of *Mechanics of Materials*. The flexural stiffness is $B = E_c I_0$. Due to the long duration of the precompression, concrete creep in the precompression zone will increase the camber. The calculated camber should be multiplied by an amplification factor 2.0. At this time, N_{PII} and $e_{0\mathrm{II}}$ should be determined after all prestress losses have been deducted.
   f_2 can be obtained by a simply supported beam model, which has the span of l_0 and is subjected to bending moments ($=N_{\mathrm{PII}} e_{0\mathrm{II}}$) at both ends. Then

$$f_2 = 2.0 \times \frac{N_{\mathrm{pII}} e_{0\mathrm{II}} l_0^2}{8B} \tag{11.70}$$

3. Deflection checking can be accomplished by using the following equation:

$$f = f_1 - f_2 \leqslant f_{\mathrm{lim}} \tag{11.71}$$

Example 11.13 Check the deflection of the beam in Example 10.2. The beam is subjected to a uniformly distributed load of q = 30 kN/m, among which the permanent load is 17 kN/m, the live load is 13 kN/m, and the quasi-permanent coefficient is 0.65. The beam is required not to crack generally.

Solution

$$B_s = 0.85E_cI_0 = 23.33 \times 10^{13}\,\text{N}\cdot\text{mm}^2$$

$$B = \frac{M_k}{M_q(\theta - 1) + M_k} B_s = 12.62 \times 10^{13}\,\text{N}\cdot\text{mm}^2$$

$$f_1 = \frac{5}{384} \times \frac{ql^4}{B} = 27.1\,\text{mm}$$

$$f_2 = 2 \times \frac{N_{\text{pII}}e_{0\text{II}}l^2}{8E_cI_0} = 9.64\,\text{mm}$$

$$f_l = f_1 - f_2 = 27.1 - 9.64 = 17.46\,\text{mm}$$

$$\frac{f_l}{l} = \frac{17.46}{8.75 \times 10^3} = \frac{1}{501} < [f] = \frac{1}{250}, \text{OK.}$$

Questions

11.1 Why do we control crack width and deformation? Which loading stage should be the calculations of crack width and deformation of flexural members based on?

11.2 What is the idea of establishing crack width formula in semitheoretical and semiempirical method? What is the physical meaning of ψ?

11.3 Why is concrete cover depth a main factor influencing the widths of cracks on member surfaces? What are the other factors?

11.4 Explain the concept and practical meaning of effectively restrained area by reinforcement. How is it calculated?

11.5 Can we say that limiting $w_{\max}$ equals limiting the tensile stress in reinforcement of flexural members?

11.6 What are the requirements of cracking resistance checking at normal sections in prestressed concrete members? Why check principal tensile and principal compressive stresses during cracking resistance checking at inclined sections?

11.7 What is the idea of establishing short-term stiffness for flexural members by using the semitheoretical and semiempirical method? How can the characteristics of concrete be reflected?

11.8 Simply analyze the influence of bending moment, reinforcement ratio of longitudinal tension reinforcement, section shape, concrete strength, and section height on short-term stiffness of flexural members.

11.9 What are the characteristics of the calculation method for the deflection of prestressed concrete flexural members? What should be noted in calculating the camber of members in service?

Problems

11.1 A precast trough plate to be used in multistory industrial buildings is shown in Fig. 11.29. The computation span is $l_0 = 5.8$ m. Concrete grade is C25. $2\phi16$ tension steel bars are provided. The bending moments by characteristic combination of load effects and quasi-permanent combination of load effects are $M_k = 18$ kN·m and $M_q = 14$ kN·m, respectively. Check the crack width by using the semitheoretical and semiempirical method.

11.2 A T section simply supported beam of $l_0 = 6$ m (Fig. 11.30) is subjected to a uniformly distributed load. Concrete grade is C30. $6\phi25$ tension reinforcement (ribbed) and $2\phi20$ compression reinforcement are provided. The bending moments by characteristic combination of load effects and quasi-permanent combination of load effects are $M_k = 315.5$ kN·m and $M_q = 301.5$ kN·m, respectively. Check the crack width using the semitheoretical and semiempirical method.

11.3 A T section prefabricated reinforced concrete bridge girder has the standard span $l_0 = 20$ m. $4\phi16 + 8\phi32$ ($A_s = 7238$ mm^2), and the ribbed steel bars are provided. The effective height of cross sections is $h_0 = 1200$ mm. The web

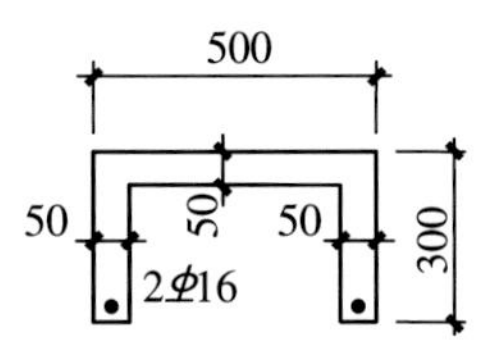

Fig. 11.29 Exercise 11.1

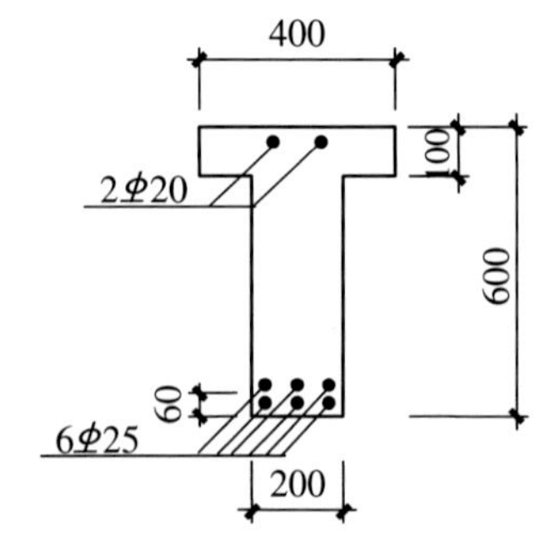

Fig. 11.30 Exercise 11.2

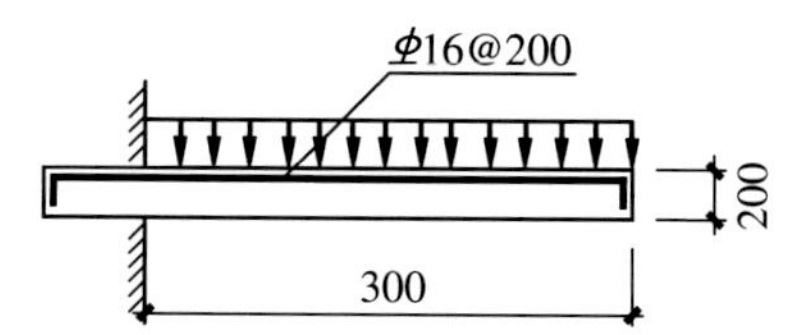

Fig. 11.31 Exercise 11.6

width is b = 128 mm. The steel stress under short-term static load (neglect impact) is σ_s = 197 N/mm^2. The dead load-induced bending moment accounts for 54.5 % of the total moment. The limit of the maximum crack width is w_{lim} = 0.25 mm. Calculate the maximum crack width under short-term load by using the mathematical statistics method and check the crack width control under long-term load.

11.4 Check whether the maximum deflection of the beam in Problem 11.2 satisfies the deflection limit of $l_0/200$.

11.5 Calculate the deflection of the trough plate in Problem 11.1 under long-term load.

11.6 The cantilever slab which is 200 mm thick shown in Fig. 11.31 is subjected to bending moments of M_k = M_q = 38.25 kN·m. The computation span is l_0 = 3 m. Concrete grade is C30. Φ16@200 ribbed steel bars are provided. Check the maximum deflection by using the semitheoretical and semiempirical method.

11.7 All conditions are the same as those in Problem 10.1. If it is a member that is generally not allowed to crack, calculate the maximum characteristic load effect that the member can sustain.

11.8 All conditions are the same as those in Problem 10.2. If it is a member that is generally not allowed to crack, and the axial tension by characteristic combination of load effects is N_k = 600 kN, check the crack resistance of normal sections in service.

11.9 All conditions are the same as those in Problem 10.4. If it is a member that is strictly not allowed to crack, and the bending moment by characteristic combination of load effects is M_k = 8 kN·m, check the crack resistance of normal sections in service.

Chapter 12
Durability of Concrete Structures

12.1 Influencing Factors

Irreversible changes can occur to concrete structures due to combined environmental and service loads. This is caused by complex physical and chemical actions in addition to direct mechanical loads. These changes include deterioration of materials, degradation of structural performance, and accumulation of damage. This process can affect the serviceability and bearing capacity of a structure and finally raise a threat to the safety of the structure. For instance, in marine atmosphere, chloride penetrates into concrete and leads to corrosion of embedded steel. This can not only cause expansive cracking of concrete cover and damage to serviceability, but also reduce the sectional area of the steel bars and therefore impair the safety of the structure.

Durability of a structure is defined as the ability of the structure to maintain safety and serviceability when exposed to its intended service environment without spending a lot of money to repair or rehabilitate.

Deterioration of durability for a structure is the degradation of structural performance arising from either intrinsic ingredients or external agents; thus, durability of a concrete structure depends on the property of the structure and the service environmental conditions as well, as shown in Fig. 12.1.

The internal deterioration of a concrete structure is mainly caused by its internal defects due to unreasonable design, unqualified materials, poor construction quality, and inappropriate use and maintenance of the structure. For instance, water, oxygen, and other aggressive agents can rapidly enter into concrete through micro-cracks as well as big and connective pores. Thin concrete cover and small bar spacing make it much easier for these agents to reach the bar surface. If seawater, sea sand, and admixtures containing chloride are used to mix the concrete during casting, the excessive chloride can lead to bar corrosion. If alkali-active aggregates are used, they can react with the alkali in concrete and lead to expansion cracks.

X. Gu et al., *Basic Principles of Concrete Structures*,
DOI 10.1007/978-3-662-48565-1_12

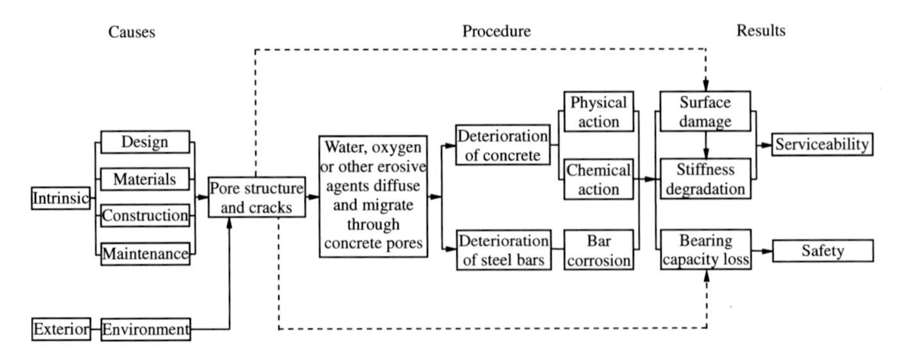

Fig. 12.1 Deterioration of concrete structures

Durability deterioration of concrete structures can also be caused by environmental attacks. For instance, carbon dioxide or acid rain can lead to neutralization of concrete and subsequent corrosion of embedded steel; acid, alkali, or salt in industrial area may lead to deterioration of concrete and bar corrosion; and chloride in marine atmosphere can penetrate into concrete and result in steel corrosion as well. During these corrosion processes, temperature, humidity, and oxygen concentration are crucial factors. In GB 50010, the working environment can be divided into five classes, as listed in Table 11.1.

12.2 Deterioration of Concrete

Deterioration of concrete structures always initiates from materials, i.e., either concrete or embedded steel. It could be intrinsic deterioration like alkali–aggregate reaction (AAR), but mostly, it is caused by environmental attacks, such as carbonation, frost attack, chemical attack, surface wear, and corrosion of embedded steel. These attacks are usually classified into two groups, including the physical actions and chemical actions although both actions are usually involved in one deterioration process. For example, corrosion of embedded steel in the concrete itself is a chemical reaction, but diffusion of oxygen and chloride in concrete, as an indispensable part of corrosion, is a physical process. Different forms of deterioration are summarized in Fig. 12.2.

The classification in Fig. 12.2 is based on the primary course; however, every kind of material deterioration cannot occur with simple environmental effect or internal course. For instance, external humidity is a necessary condition for intrinsic AAR. Additionally, the deterioration normally arises from combined actions. The deterioration of a concrete structure in marine atmosphere may be caused by frost attack, salt crystallization, steel corrosion, etc. In addition, the deterioration of a concrete structure under a deicing salt attack may result from crystallization and steel corrosion.

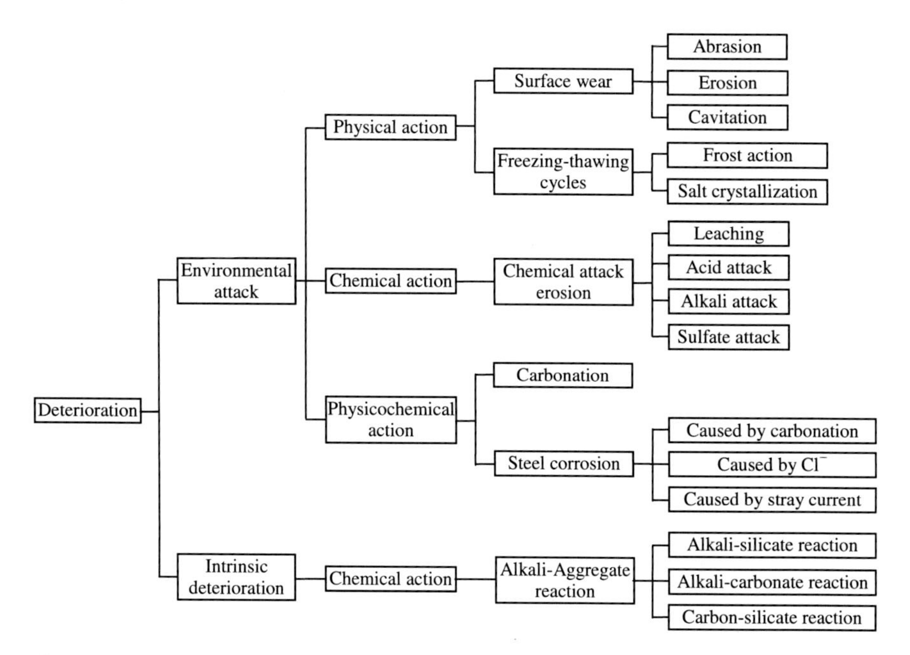

Fig. 12.2 Classification of causes of concrete deterioration

12.2.1 Carbonization

12.2.1.1 Mechanisms

Concrete neutralization refers to the decreasing process of PH value, which is caused by the reaction between alkali in concrete and acid gas or liquid from air, soil, or underground water, because normal air contains carbon dioxide gas (CO_2), which can react with hydroxides to form carbonate. This CO_2-induced reaction, also referred to as carbonization, is the most common kind of concrete neutralization.

Carbonization is a complex physicochemical process. After sufficient hydration of cement, $Ca(OH)_2$ and $3CaO{\cdot}2SiO_2{\cdot}3H_2O$ are produced. Pore solution tests show strong alkalinity with pH value of 12–13 in the presence of saturated $Ca(OH)_2$. Stable water films come into being between pore water and surrounding moisture and achieve hydrothermal balance. With CO_2 continuously diffusing into concrete and dissolving in pore solution, solid $Ca(OH)_2$ dissolves and diffuses toward carbonated areas simultaneously by concentration gradient; thus, in the carbonation front, CO_2 can react with $Ca(OH)_2$ and $3CaO{\cdot}2SiO_2{\cdot}3H_2O$ in solution and solid–liquid interface, respectively.

$$Ca(OH)_2 + CO_2 \rightarrow CaCO_3 + H_2O \quad (12.1)$$

$$(3CaO{\cdot}2SiO_2{\cdot}3H_2O) + 3CO_2 \rightarrow 3CaCO_3 + 2SiO_2 + 3H_2O \quad (12.2)$$

Carbonization, on the one hand, blocks the pores with $CaCO_3$ and other solid products, finding big pores and reducing porosity, thereby slowing down CO_2 diffusion and increasing concrete strength, compactness, and brittleness. On the other hand, it decreases $Ca(OH)_2$ concentration and pH value to 8.5–9.0 in completely carbonized regions and leads to corrosion of embedded steel. Figure 12.3 shows a physical model of the concrete carbonation process.

1. Water cement (W/C) ratio
   Pore structure and porosity of concrete mainly depend on W/C ratio, while the saturation degree of concrete pores (ratio of water volume to pore volume) relies on how much free water is left after hydration. Therefore, W/C ratio is one of the governing factors for the effective diffusion coefficient of CO_2 and the consequent carbonation rate, which increases with increasing W/C ratio.
2. Cement type and content
   Cement type and content in concrete are also crucial to carbonization rate. This is because the content of minerals in cement varies with cement type, which decides how many hydrates are available in unit volume of concrete.
   Concrete with less cement will suffer a relatively quick carbonization. This is because less CO_2 can be consumed per unit volume of concrete with less cement. If admixtures are used, the carbonization will be even faster because fewer hydrates are produced. Therefore, Portland cement is most favorable to carbonization, while fly-ash cement, Portland Pozzolana cement, and Portland

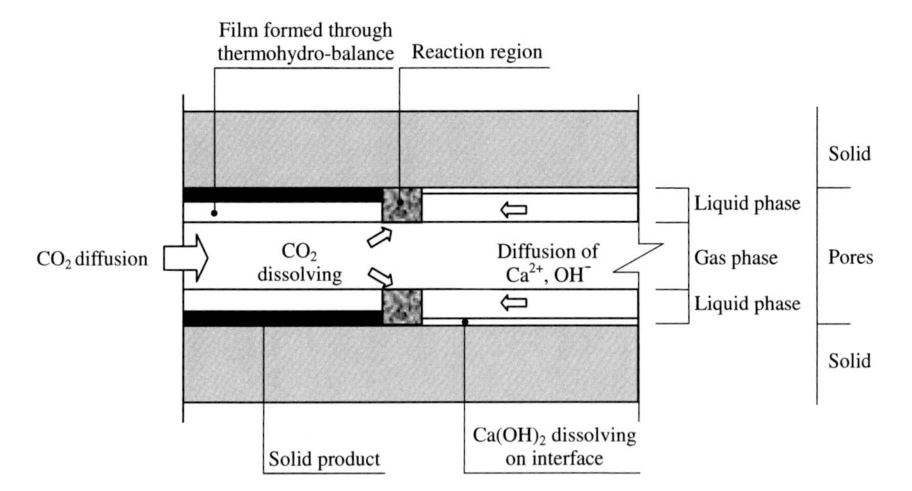

Fig. 12.3 Model of concrete carbonation process

blast furnace slag cement vary greatly from the premixed Portland cement that predominates the market.

3. Aggregate type and size
   Large aggregates make it easier for CO_2 to diffuse through relatively weak aggregate–matrix interface. Many lightweight aggregates contain pozzolanic component, and this can react with $Ca(OH)_2$ during curing. Some silica aggregates also consume $Ca(OH)_2$ through AAR and accelerate carbonization.
4. Admixtures
   Both water-reducing agents and air-entraining agents can remarkably reduce the effective coefficient of CO_2 diffusion and the subsequent carbonization rate. This is mainly because the former can reduce the water content and thus lead to decrease of porosity, and the latter can reduce the capillary suction by introducing numerous enclosed pores into the concrete.
5. Curing condition and time
   The amount of hydrates that can be carbonated largely depends on the curing condition and time. Inappropriate curing at early age of concrete can lead to insufficient hydration of cement and thereby accelerate the concrete carbonization process.
6. Concrete strength
   Concrete strength can be used to evaluate its carbonization resistance, since the concrete strength reflects porosity and compactness of concrete to some extent. Normally, high-strength concrete has a relatively small carbonization rate.
7. CO_2 concentration
   CO_2 penetrates into concrete more effectively under greater concentration gradient. Normally, CO_2 concentration is 0.03 % in suburban areas, 0.04 % in urban areas, and 0.1 % in indoor conditions.
8. Relative humidity
   Relative humidity is another major environmental factor. Saturation degree of pore water is affected by the environmental humidity according to the thermohydrobalance, which on the one hand affects CO_2 diffusion rate and on the other hand provides necessary solution and solid–liquid interface for the required chemical reaction.
   In excessively high-moisture surroundings, carbonization nearly stops. This is because CO_2 diffuses very slowly in concrete that is nearly water saturated. Under dry conditions, although CO_2 enters into concrete rapidly, the carbonization still develops very slowly without the necessary liquid phase. The most favorable relative humidity for concrete carbonization is between 70 and 80 %.
9. Temperature
   Temperature rise can accelerate chemical reaction and diffusion of CO_2 and thus accelerate the carbonization process. Alternation of temperature has a similar effect on the concrete carbonization process.

10. Surface coating
    Surface coating acts as a barrier to carbonization. It covers the concrete surface and blocks surface pores and micro-cracks. Some coating materials such as cement mortar or gypsum mortar contain active components that can react with CO_2 and, therefore, retard the carbonization process.
11. Stress status
    Research on concrete carbonization has mainly focused on the materials, and concrete carbonization in existing structures always develops under different stress status. Small compressive stress facilitates the compactness of concrete by retarding the diffusion of CO_2 and carbonation accordingly, but high stress can lead to micro-crack expansion and the subsequent carbonization acceleration. A relatively small tensile stress of less than 0.3 f_t is considered to have no obvious effect on carbonization, whereas larger tensile stress can lead to a remarkably accelerated carbonation and crack formations.

12.2.1.2 Determination of Carbonization Depth

CO_2 reacts with $Ca(OH)_2$ in concrete to form $CaCO_3$. However, it cannot go through the barrier of uncarbonated concrete until all the $Ca(OH)_2$ is converted, at which time it begins to move into the next layer of uncarbonated concrete. Therefore, the highest $CaCO_3$ content and the lowest $Ca(OH)_2$ content are found in carbonated regions near the surface. However, the situation is reversed in uncarbonated concrete. In between, there is a partially carbonated region where the pH value normally ranges between 9 and 12.5, which gives a good explanation for why steel corrosion starts before the concrete cover is fully carbonated.

Carbonization depth is usually determined by either X-rays or chemical agents. The partially carbonated region can be determined in a similar way with X-rays, but special equipment is needed. For this reason, carbonization depth tests are normally conducted in a laboratory for precise measurements. Chemical agents are normally used on-site because they can be easily applied. A phenolphthalein indicator is the most common agent used to determine carbonization depth as it distinguishes the dividing line of pH value equal to 9 and presents the uncarbonated area by changing it to pink. Another agent called Rainbow Indicator can differentiate pH value ranging from 5 to 13 by showing different colors, which also allow the partially carbonated region to be determined.

Figure 12.4 shows the content of $Ca(OH)_2$ and $CaCO_3$ in concrete measured by X-rays. Obviously, there is a partially carbonated region in which $Ca(OH)_2$ content increases and $CaCO_3$ content decreases by depth.

12.2.1.3 Carbonization Depth Prediction

Various models have been developed to predict the carbonization depth. These models can be divided into two classes: (1) a theoretical model derived

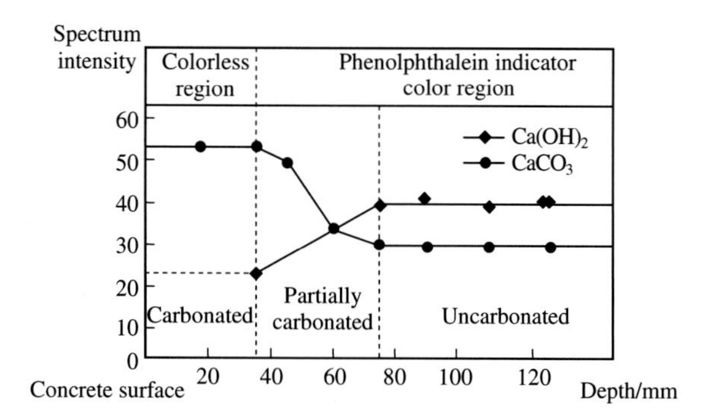

Fig. 12.4 Concentration variation of $Ca(OH)_2$ and $CaCO_3$

mathematically based on the carbonization mechanism, and (2) a more empirical numerical model derived from experimental results. The second mechanism-based numerical model proposed by Tongji University can be used to predict both the carbonization depth and partially carbonated region, respectively, as

$$x_{\mathrm{c}} = 839(1 - \mathrm{RH})^{1.1}\sqrt{\frac{W/\gamma_{\mathrm{c}}C - 0.34}{\gamma_{\mathrm{HD}}\gamma_{\mathrm{c}}C}n_0} \cdot \sqrt{t} \tag{12.3}$$

$$x_{\mathrm{hc}} = 1.017 \times 10^4(0.7 - \mathrm{RH})^{1.82}\sqrt{\frac{W/C - 0.31}{C}} \tag{12.4}$$

where

x_{c} and x_{hc}	carbonization depth and partially carbonated depth in mm, respectively;
n_0	CO_2 volume concentration;
W and C	water and cement content in kg/m^3, respectively;
t	carbonization time measured in days;
RH	environmental relative humidity;
γ_{HD}	coefficient of cement hydration degree and equals 1.0 when the curing time is more than 90 days or 0.85 for the curing time of 28 days. Linear interpolation is used for curing time in between; and
γ_{c}	coefficient of cement type and equals 1.0 for Portland cement and 1.0 minus admixture content for other types of cement.

In normal air, it can take several years and sometimes decades for concrete that is several centimeters thick to be fully carbonated, depending on the compactness of the concrete. Thus, accelerated carbonization tests are commonly employed to avoid a long experiment cycle in which high CO_2 concentration is applied. Similar to Eq. (12.3), and in various other models, it is understood that the carbonization depth is proportional to the square root of CO_2 concentration n_0 and carbonization time t. This idea is also used in the design of accelerated carbonization tests. For instance, given the carbonization depth x_c^0 with time t_0 and CO_2 concentration n_0^1 in an acceleration test, the carbonization depth of the same specimen in natural environment with time t and CO_2 concentration n_0 can be calculated by Eq. (12.5). Based on this principle, the carbonization depth x_c at time t can also be calculated through the on-site measured value x_c^0 at t_0.

$$x_c = \sqrt{\frac{n_0 t}{n_0^1 t_0}} \cdot x_c^0 \tag{12.5}$$

12.2.2 Frost Action

As water turns to ice at its freezing point, there is an increase in volume of about 9 %. When porous material like concrete is saturated with water, this expansion upon freezing can force pore solution into the surrounding gel pores and finally lead to disruption of the material if its strength is insufficient to resist the hydrostatically induced pressure. Unless the material is more than 91.7 % saturated, disruption will not occur since there is empty space to accommodate the expansion. The water in concrete pores either from primary mixing water or external circumstances turns to ice and turns back to water periodically with the alternation of external temperature (referred to as freeze–thaw cycles). During this process, the fractures extend gradually and through the pores. This further worsens the frost resistance of concrete. Gradually, the damage accumulates and propagates to the deeper area of the concrete. This leads to a decrease of concrete strength and section reduction until the whole member is destroyed. Frost action is one of the most common reasons for deterioration of concrete structures in cold areas, especially those having frequent contact with water, such as marine structures, water pools, cooling towers of power plants, structural plinths, balconies, roads, and bridges that are suffering from wetting–drying cycles and freeze–thaw cycles.

In cold areas, deicing salt is commonly used on roads, overpasses, and open garages, which can accelerate freezing activity. Since the exposed surface of concrete is more susceptible to rapid cooling caused by dissolving of ice or snow, differential freezing may arise due to the salt concentration gradient from surface to the interior of the concrete especially when a concrete surface delaminates.

There are different pores such as gel pores and capillary pores which coexist with air bubbles in hardened concrete. These pores vary greatly in size. Capillary pores (0.01–10 μm) are much larger than gel pores (15–100Å) and normally connected to each other, while air bubbles are generally round and enclosed and introduced naturally during concrete casting and compacting or entrained deliberately by air-entraining agents. When concrete is immerged in water, saturation can easily be reached for capillary pores but not for air bubbles despite the water adsorbed on their internal surface under normal air pressure. Furthermore, the finer the pore is, the lower the freezing point becomes. That is because the liquid in fine pores is under high surface tension forces, which prevent the formation of ice nuclei. When the environmental temperature falls between −1 and −1.9 °C, freezing occurs first in relatively big pores and then spreads gradually to fine pores, at −12 °C at which point, where all the fluid in capillary pores should freeze completely. However, as discussed, water molecules in gel pores are adsorbed onto the surface of the cement matrix and very difficult to freeze. Therefore, only capillary pores are harmful to concrete frost resistance.

Separate and enclosed air bubbles with diameters ranging from 25 to 500 μm distribute in hardened cement paste when air-entraining agents are added to the concrete during casting, which resists saturation. The bubbles can improve the frost resistance of concrete by accommodating the water pushed into from surrounding capillaries and releasing the hydrostatic pressure. This is why air-entraining concrete has better frost resistance than normal concrete.

The frost resistance durability of water saturated concrete is defined as its capacity to work against freezing and thawing cycles. The capacity is evaluated quantitatively by frost resistance grade. It is regulated by the *Standard for test methods of long-term performance and durability of normal concrete* (GB/T 50082-2009), in which the frost resistance grade is defined as the maximum freeze–thaw cycles at 5 % mass loss and 25 % strength loss simultaneously for standard specimens under slow freezing, such as D25, …, D300. The frost resistance of concrete can be determined by Eq. (12.6) so that the designed service life under frost action can be guaranteed. Meanwhile, the maximum W/C ratio is also given by Chinese National Standards in terms of climate conditions as listed in Table 12.1.

$$D = \frac{t_{\text{lim}} m_{\text{s}}}{n_{\text{s}}} \tag{12.6}$$

where

D freeze resistance grade;

t_{lim} designed service life in years;

n_{s} equivalent natural freeze–thaw cycles for one freeze–thaw cycle indoor test; and

m_{s} estimated natural freeze–thaw cycles in one year and can be taken as 50 for marine structures in northern China.

12.2.3 Alkali–Aggregate Reaction

12.2.3.1 Mechanism

AAR is a chemical reaction between certain types of aggregates and hydroxyl ions (OH^-) associated with alkalis in cement. During the reaction, the volume can be increased. This may lead to destruction of concrete microstructure, swelling, and map-like cracking of concrete and eventually remarkable degradation of mechanical performance such as a reduction of compressive and flexural strength along with stiffness. AAR normally occurs several years after hardening of concrete; however, once the reaction starts, it is very difficult to stop it or make any repair or rehabilitation. For this reason, AAR is also called the *cancer of concrete*.

AAR normally can be divided into three classes: (1) the alkali–silica reaction (ASR), (2) alkali–carbonate reaction (ACR), and (3) alkali–silicate reaction in terms of the different active components in aggregates.

Alkali–silica reaction is the most common kind of AAR and has been the focus of much research. During the reaction, the active silica reacts with hydroxyl ions in concrete pore solution to form silicate gel, which takes up large amounts of water with a consequential increase in gel volume. The reaction can finally lead to swelling and cracking of concrete, as seen in Eq. (12.7). The active silica includes opal, chalcedony, tridymite, cristobalite, crypto-crystal, crystallite, vitreous crystal, and fine cracked or stressed coarse crystal. Rocks containing such active silica can be igneous rocks, metamorphic rocks, or sedimentary rocks including granite, rhyolite, latite, pearlite, basalt, quartzite, firestone, and diatomite.

$$2\mathrm{ROH} + n\mathrm{SiO_2} \rightarrow \mathrm{R_2O} \cdot n\mathrm{SiO_2}\cdot\mathrm{H_2O} \tag{12.7}$$

where R represents Na or K.

Alkali–carbonate reaction is a less common case, in which certain dolomitic limestone aggregates react with the hydroxyl ions in pore solution to form $Mg(OH)_2$, $CaCO_3$ and R_2CO_3. This reaction will continue until no $Ca(OH)_2$ or limestone is left anymore, because the produced R_2CO_3 can react with $Ca(OH)_2$ to form ROH. The latter also participates in the ACR.

$$\mathrm{CaCO_3}\cdot\mathrm{MgCO_3} + 2\mathrm{ROH} \rightarrow \mathrm{Mg(OH)_2} + \mathrm{CaCO_3} + \mathrm{R_2CO_3} \tag{12.8}$$

$$\mathrm{R_2CO_3} + \mathrm{Ca(OH)_2} \rightarrow 2\mathrm{ROH} + \mathrm{CaCO_3} \tag{12.9}$$

ACR will experience a volume decrease, so the reaction itself does not lead to any expansion. Nevertheless, the dry clay enclosed in limestone can be exposed through the destruction of dolomitic crystal, which can take up water and increase in volume.

Alkali–silicate reaction occurs between hydroxyl ions in concrete pore solution and stratified silicate aggregate from silicified rocks or phyllites. This reaction can enlarge the distance among the silicate layers of aggregate and leads to swelling and cracking of concrete.

12.2.3.2 Conditions

Three necessary conditions are required for AAR: (1) excessive alkali in cement, (2) alkali-active minerals in aggregate, and (3) a humid environment. Accordingly, the harmful reaction can be effectively prevented or alleviated by controlling the alkali content in cement and keeping the concrete away from water.

1. Alkali content in concrete
   Apart from cement, alkali in concrete can also be introduced from admixtures, aggregate, water, or even external environment. In cement, alkali comes from raw materials such as clay and fuel coal in the manufacturing process. Its content is referred to as the equivalent Na_2O, which can be calculated as Na_2O plus 0.658 K_2O. Normally, cement manufactured in Northern China contains about 1 % alkali, which is more than the critical value (0.6 %) for AAR to take place. Alkali concentration in concrete may be even higher considering the alkali from sodium water-reducing agents, accelerating agents, frost resistance agents, etc.
   Generally, the alkali content in concrete is decided by cement content and the "equivalent Na_2O" of specified cement types. The latter is normally evaluated based on the alkali content in unit volume concrete (kg/m^3). Because alkali content plays an important role in prevention of AAR, the maximum values in various environmental conditions are regulated in codes of different countries. The limit value is 3 kg/m^3 in GB 50010 unless nonalkali-active aggregate is involved or a Class I environment (seen in Table 12.1).
2. Alkali activity of aggregate
   Choosing appropriate aggregates and lowering the active mineral content is another major step to avoiding AAR. As discussed earlier, various kinds of rocks contain active silica, but only argillaceous limestone and dolomitic limestone are ACR-active. A distribution map is needed whereby alkali-active aggregates can be identified with warnings as to which precautions can be taken in large-scale projects.
3. Humid environment
   Water isolation by maintaining a dry environment or surface coating is also an effective way to avoid AAR. Since sufficient water is a necessary condition, a controlled amount of water should run very slowly even in the presence of excessive alkali unless concrete is in direct contact with water or in an environment where the relative humidity is greater than 80 %.

12.2.4 *Chemical Attacks*

Concrete deterioration may arise from chemical attacks when concrete is in contact with aggressive agents such as acid, alkali, sulfate, and flowing water. The chemical attacks can be divided into three classes: the first is leaching of hydrates under the action of flowing water; the second is dissolving of products caused by reaction of hydrates with chemical agents such as acid or alkali; and the third is expansion caused by reaction between hydrates and agents such as sulfate.

12.2.4.1 Leaching Erosion

When low-quality concrete is emerged in flowing water, $Ca(OH)_2$ will continuously dissolve into the water. This dissolution can lead to decomposition and dissolving of calcium silicate as well as aluminate hydrates because they are stable only in saturated $Ca(OH)_2$ solution. This process is accompanied by a strength decrease, according to the testing results from China Institute of Water Resources Research, from which it was reported that the compressive strength would suffer a decrease of 35.8 at 25 % mass loss of $Ca(OH)_2$ (calculated by equivalent CaO), and the decrease in the tensile strength could be even more rapid as high as 66.4 % at the same loss ratio as seen in Fig. 12.5.

12.2.4.2 Dissolving Erosion

Dissolving erosion can be caused by weak acid, strong acid, or alkali. Weak acid attacks can occur with carbonic acid and hydrosulfuric acid. The loss of $Ca(OH)_2$ through a reaction with carbonic acid (CO_2 > 15 mg/L) forms soluble calcium bicarbonate. The calcium bicarbonate can be washed away in direct contact with water. A loss of $Ca(OH)_2$ can also occur with the formation of soluble calcium hydrosulphide through a reaction between $Ca(OH)_2$ and hydrosulfuric acid which commonly exists in the water from polluted soil, drainage pipes, or runoff water from food industry manufacturing plants.

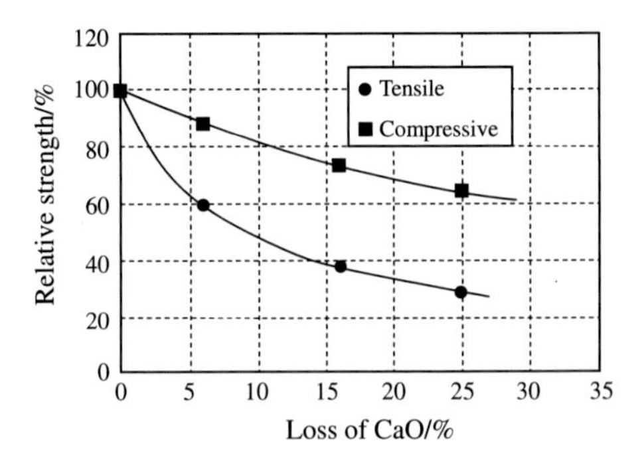

Fig. 12.5 Concrete strength decreasing curve in leaching process

Fig. 12.6 Strength curve of concrete under sulfate acid attack

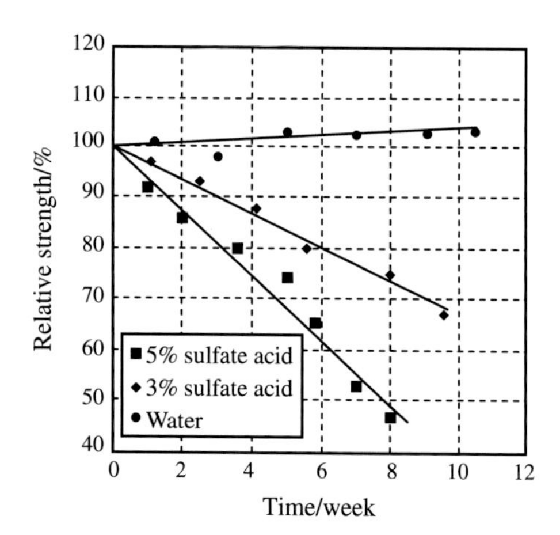

Mineral acid such as hydrochloric acid, sulfuric acid, and nitric acid can lead to strong acid attack as well as attacks from organic acids, including acetic acid, formic acid, and lactic acid. In the deterioration process, the acids first react with Ca $(OH)_2$; then, calcium silicate hydrates form soluble calcium salt. The reaction can take place continuously and result in decrease of concrete alkalinity and strength as shown in Eqs. (12.10) and (12.11). Furthermore, SO_4^{2-} can participate in the reaction by initiating a sulfate attack, and chloride can lead to corrosion of embedded steel. This further aggravates the damage effect. Figure 12.6 shows the experimental results of the concrete strength curve under sulfate acid attack. Phosphoric acid attack produces insoluble calcium salt, which clogs up the pores in hydrated cement paste and retards the subsequent erosion; however, the decrease of concrete strength will continue until the final destruction of members.

$$Ca(OH)_2 + 2H^+ \rightarrow Ca^{2+} + 2H_2O \tag{12.10}$$

$$3CaO{\cdot}2SiO_2{\cdot}2H_2O + 6H^+ \rightarrow 3Ca^{2+} + 2(SiO_2{\cdot}H_2O) + 3H_2O \tag{12.11}$$

Both high concentration alkaline solution and melted alkali can lead to deterioration of concrete. The erosive effect of caustic alkali (NaOH, KOH) includes combined chemical attack and crystallization. The reaction occurs between calcium silicate and aluminate hydrates to form weak bond products with good leachability.

$$3CaO{\cdot}2SiO_2{\cdot}3H_2O + 4NaOH \rightarrow 3Ca(OH)_2 + 2Na_2SiO_3 + 2H_2O \tag{12.12}$$

$$3CaO{\cdot}Al_2O_3{\cdot}12H_2O + 2NaOH \rightarrow 3Ca(OH)_2 + Na_2O{\cdot}Al_2O_3 + 10H_2O \tag{12.13}$$

12.2.4.3 Expansion Erosion

Sulfate attack is a typical expansion erosion and also the most common form of chemical attack, which can be caused by sodium sulfate, potassium sulfate, calcium sulfate, magnesium sulfate, etc. Structures that have endured sulfate attack can often be seen around sewage treatment facilities, chemical fiber production plants, saltworks, or soap manufacturing plants. In these areas, the underground water contains highly concentrated sulfate. Sulfate is also the major component of seawater.

The yielding of gypsum, ettringite, and wollastonite is the main course for sulfate-induced concrete deterioration.

Sodium sulfate, potassium sulfate, and magnesium sulfate in pore solution can react with $Ca(OH)_2$ to form gypsum. Take sodium sulfate as an example

$$Ca(OH)_2 + Na_2SO_4 \cdot 10H_2O \rightarrow CaSO_4 \cdot 2H_2O + 2NaOH + 8H_2O \quad (12.14)$$

In flowing water, the reaction goes on until $Ca(OH)_2$ runs out, while in static water, the chemical balance can be reached with the accumulation of NaOH, after which some reaction products become a part of the gypsum. Since the volume is almost doubled during the conversion from $Ca(OH)_2$ to gypsum, expansion damage will occur.

Gypsum can further react with calcium aluminate hydrates ($4CaO \cdot Al_2O_3 \cdot 19H_2O$) and calcium aluminosulfate hydrates ($3CaO \cdot Al_2O_3 \cdot CaSO_4 \cdot 18H_2O$) to form insoluble ettringite, which is likely to precipitate in water. In the end, the volume increases greatly and this aggravates the expansive deterioration.

$$\begin{aligned} &4CaO \cdot Al_2O_3 \cdot 19H_2O + 3CaSO_4 + 14H_2O \\ &\quad \rightarrow 3CaO \cdot Al_2O_3 \cdot 3CaSO_4 \cdot 32H_2O + Ca(OH)_2 \end{aligned} \quad (12.15)$$

$$\begin{aligned} &3CaO \cdot Al_2O_3 \cdot CaSO_4 \cdot 18H_2O + 2CaSO_4 + 14H_2O \\ &\quad \rightarrow 3CaO \cdot Al_2O_3 \cdot 3CaSO_4 \cdot 32H_2O \end{aligned} \quad (12.16)$$

Another expansive product generated in sulfate attack is wollastonite ($CaCO_3 \cdot CaSO_4 \cdot CaSiO_2 \cdot 15H_2O$), which is formed by $Ca(OH)_2$, $CaCO_3$, gypsum, and undefined SiO_2 in low temperature. It can lead to expansive cracking, swelling and weakening of the concrete surface, and subsequently to decreased concrete strength.

Besides the erosive reaction mentioned above, magnesium sulfate can also react with calcium silicate hydrate

$$\begin{aligned} &3CaO \cdot 2SiO_2 \cdot 3H_2O + 3MgSO_4 + 10H_2O \\ &\quad \rightarrow 3(CaSO_4 \cdot 2H_2O) + 3Mg(OH)_2 + 2SiO_2 \cdot 4H_2O \end{aligned} \quad (12.17)$$

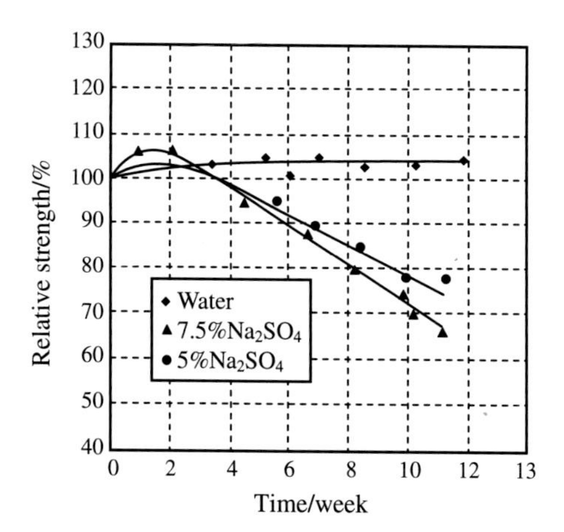

Fig. 12.7 Strength curve of concrete under Na_2SO_4 attack

The decomposition and gelatinization loss of the hydrates make it more harmful than other sulfate to the concrete.

Sulfate attack is characterized by the concrete surface turning white. The deterioration normally starts from corner; then, cracks extend until the surface peels off, after which concrete becomes fragile and loose. The concrete strength experiences a change of two stages: In the early stage, crystal grows in concrete pores and densifies the concrete; thus, concrete strength experiences a slight increase. Then, due to the large amounts of expansive products, high stress is generated, which can destroy the pore structure, expand the micro-cracks, and finally lead to a decrease in concrete strength. The time-dependent rule of concrete strength under sulfate attack is shown in Fig. 12.7, which has also been established through tests conducted at Tongji University.

12.3 Corrosion of Steel Embedded in Concrete

12.3.1 Mechanism

12.3.1.1 Destruction of Passive Oxide Film

The pH value of concrete pore solution could be as high as 13.0 with saturated Ca $(OH)_2$ and a small amount of NaOH/KOH. In this high alkaline environment, a $m\mathrm{Fe_2O_3}\cdot n\mathrm{H_2O}$ film with a thickness of only 20–60Å is formed covering the steel surface and passivates the steel, thereby protecting the steel from corrosion, even in the presence of water and oxygen. Without stray-current action, the protection film will not be destroyed unless a certain level of chloride ions accumulate on its surface or the alkaline environment is ruined by concrete neutralization.

The passivation effect diminishes during the concrete carbonization process, because the consumption of calcium hydroxide during this process will certainly lead to a decrease in pH value, which is responsible for the stable presence of passive oxide film. When the pH value drops below 11.5, the film is no longer stable and corrosion initiates. If the pH value drops to 9–10, the film can be destroyed completely.

The passive oxide film can also be destroyed by chloride ions in high alkaline environment with pH value as high as 11.5. Small and active chloride ions can be easily adsorbed on the film at flaw points like dislocation or grain boundary; then, it can penetrate through the oxide film and react with iron to form soluble $FeCl_2$, which may lead to local dissolving of the film and subsequent arising of pitting corrosion. If chloride ions distribute uniformly on the bar surface, the pitting corrosion will expand and merge to form a larger area of corrosion.

12.3.1.2 Electrochemical Mechanism of Steel Corrosion

Three necessary conditions are required for the electrochemical process of steel corrosion considering the electrochemical principle of metal corrosion and characteristics of passive oxide film protection: (1) Corrosion cells are formed due to potential difference on steel surface; (2) steel is active with the destruction of passive oxide film; and (3) water and oxygen are present on the steel surface.

The first condition can be automatically satisfied because of the isolation of carbon or other alloy compounds, nonuniform distribution of alkalinity and chloride, high concentration gradient of oxygen around fractures, internal residual stresses, etc. When the other two conditions are satisfied simultaneously, the electrochemical corrosion occurs in four basic steps:

1. Anode reaction: Iron atoms in the anode region break away from crystal lattice and change into surface adsorptive atoms before they can get across the electric double layer and release electrons.

$$Fe \rightarrow Fe^{2+} + 2e^- \tag{12.18}$$

2. Electron transformation: The electrons move into cathode region from anode region.
3. Cathode reaction: Oxygen in pore water takes in the electrons and initiates the reducing reaction.

$$O_2 + 2H_2O + 4e^- \rightarrow 4OH^- \tag{12.19}$$

4. Corrosion product formation: OH^- enters into anode region through migration and diffusion and then reacts with Fe^{2+} to form $Fe(OH)_2$, which can be further oxidized into $Fe(OH)_3$ and then converted into loose, porous, and incoherent red rust Fe_2O_3 by dewatering, or into black rust Fe_3O_4 partially with insufficient oxygen.

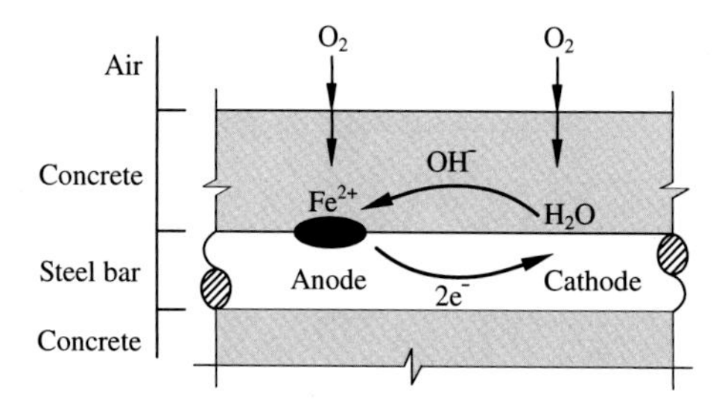

Fig. 12.8 Electrochemical mechanism of corrosion of steel embedded in concrete

$$Fe^{2+} + 2OH^{-} \rightarrow Fe(OH)_2 \quad (12.20)$$

$$4Fe(OH)_2 + O_2 + 2H_2O \rightarrow 4Fe(OH)_3 \quad (12.21)$$

$$2Fe(OH)_3 \rightarrow Fe_2O_3 + 3H_2O \quad (12.22)$$

$$6Fe(OH)_2 + O_2 \rightarrow 2Fe_3O_4 + 6H_2O \quad (12.23)$$

The final products of the electronic corrosion depend on the supply of oxygen. It can be seen from Eqs. (12.21) and (12.23) that oxygen around the steel is also included in the secondary reaction of the corrosion products. The electrochemical mechanism of above reaction is shown in Fig. 12.8.

12.3.2 *Corrosion Effect*

The decline in the bearing capacity is inevitable for RC members subjected to corrosion effect, as shown in Fig. 12.9. This can be explained by the combined reduction of the sectional area and mechanical performance degradation of reinforcing steel bars, including the decrease of strength, decrease of ductility, and increase of brittleness.

The serviceability of RC structures can also be severely impaired by corrosion. Once corrosion occurs, a loose layer of rust starts forming on the bar surface and simultaneously diffuses into the surrounding concrete pores. Meanwhile, an increase in volume by a magnitude of 2–6 times is also initiated, depending on the product forms and circumferential tensile stress engendered. When the stress overcomes the resistance of the concrete, radial cracks will initiate and extend to the concrete surface with corrosion aggravation, until finally the longitude cracks along the corroded steel bars become apparent or the whole cover spalls off.

The cracking or spalling off of the concrete cover can lead to deterioration or even complete loss of the bond and anchorage together with the lubrication of the

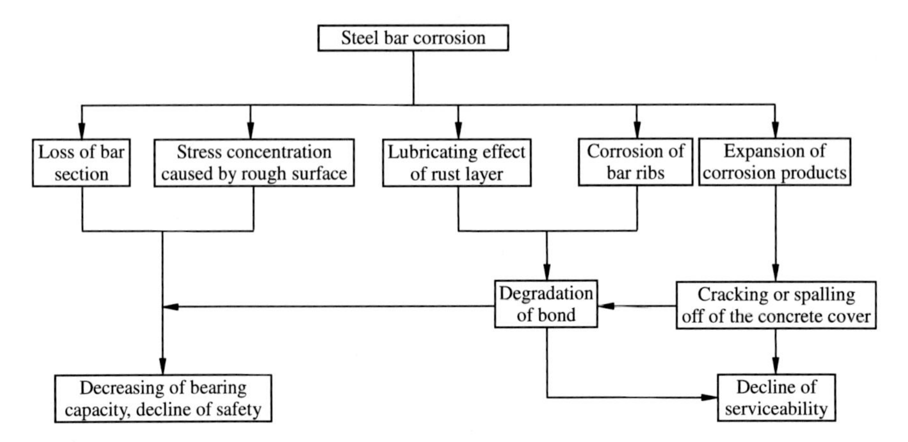

Fig. 12.9 Influence of steel bar corrosion on structural performance of reinforced concrete members

rust layer and loss of the deformed bar ribs, all of which can severely reduce the safety and serviceability of RC structures.

12.3.3 *Mechanical Properties of Corroded Steel Bars*

12.3.3.1 Elastic Modulus of Corroded Steel Bars

The relationship between the elastic modulus of corroded steel bars and average cross-sectional loss ratio or corrosion degree is shown in Fig. 12.10, which indicates that as the degree of corrosion increases, elastic modulus changes little, fluctuating up, and down near a certain value. Therefore, the effect of corrosion on the elastic modulus can be neglected.

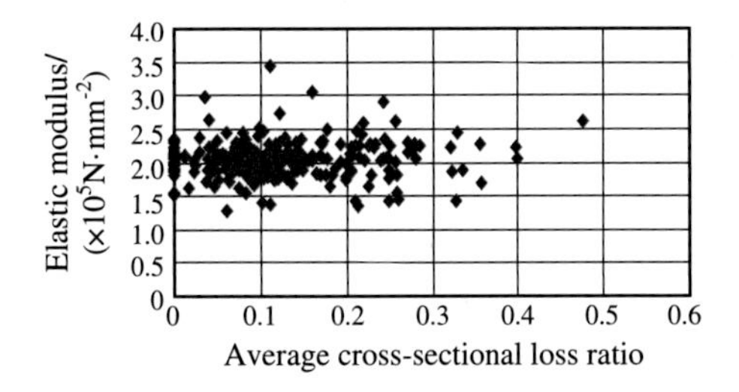

Fig. 12.10 Influence of corrosion on the elastic modulus

12.3.3.2 Strength of Corroded Steel Bars

The stretching resistance of a single corroded steel bar can be simply calculated by its residual sectional area A_{sc} multiplied by the ultimate tensile strength of the uncorroded steel f_u in the case of uniform corrosion; moreover, the weight has approximately the same loss ratio with the sectional area.

Uniform corrosion, however, can hardly occur to embedded reinforcing bars due to the nonhomogeneous property of concrete, diversity of working environments, and different stress conditions along the bars. Normally, the maximum section loss is greater than the weight loss. When corrosion further develops, the gap becomes bigger and bigger with descending corrosion uniformity. Therefore, stress concentration on the accidented steel surface, which can lower the yield strength as well as ultimate strength, is thought to be another factor besides section loss responsible for the decreased stretching resistance of corroded steel bars.

Take the nominal yield strength and nominal ultimate strength of corroded steel bars as the ratio of the yield load and ultimate load to the cross-sectional area before corrosion, respectively, and take the nominal relative strength as the ratio of the nominal strength of corroded steel bars to that of uncorroded ones. Coupon tests of corroded steel bars embedded in concrete show that, in normal air, under the aggravation of corrosion, the nominal yield strength and nominal ultimate strength degrade almost linearly, and the degradation is observed to be more severe for nominal ultimate strength compared to nominal yield strength (Fig. 12.11). The yield strength and ultimate strength of corroded steel bars can be obtained by Eqs. (12.24) and (12.25).

$$f_{yc} = \frac{P_{yc}}{A_{sc}} = \frac{P_{yc}}{P_{y0}} \cdot \frac{P_{y0}}{A_{s0}} \cdot \frac{A_{s0}}{A_{sc}} = \frac{\alpha_{yc}}{1 - \eta_s} f_{y0} \tag{12.24}$$

$$f_{uc} = \frac{P_{uc}}{A_{sc}} = \frac{P_{uc}}{P_{y0}} \cdot \frac{P_{u0}}{A_{s0}} \cdot \frac{A_{s0}}{A_{sc}} = \frac{\alpha_{uc}}{1 - \eta_s} f_{u0} \tag{12.25}$$

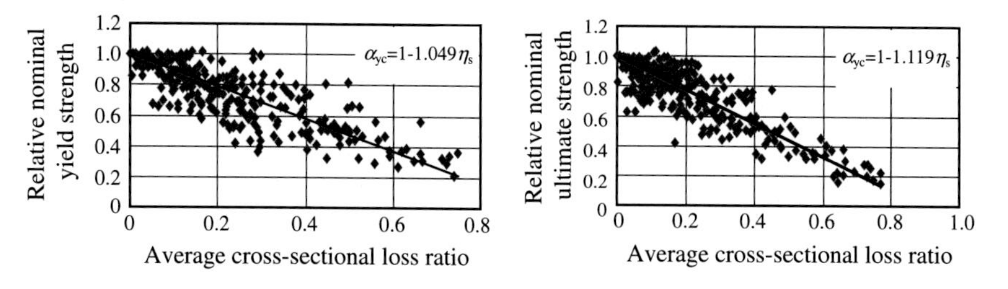

Fig. 12.11 Decline of yield strength and ultimate strength with bar corrosion

where

f_{yc} and f_{y0} yield strength of corroded bars and uncorroded bars, respectively;
f_{uc} and f_{u0} ultimate strength of corroded bars and uncorroded bars, respectively;
P_{yc} and P_{y0} yield tensile load of corroded bars and uncorroded bars, respectively;
P_{uc} and P_{u0} ultimate tensile load of corroded bars and uncorroded bars, respectively;
A_{sc} and A_{s0} effective sectional area of the corroded bars and uncorroded bars, respectively;
α_{yc}, α_{uc} relative values of the nominal yield strength and nominal ultimate strength of corroded steel bars, respectively; and
η_s average sectional loss ratio or weight loss ratio.

12.3.3.3 Deformation of Corroded Steel Bars

The ultimate elongation and ductility of corroded steel bars are much lower than those of uncorroded bars. The test results showing that the limited elongation in the code is likely to be satisfied when the sectional loss ratio is lower than 0.05; however, it will not be satisfied if the sectional loss ratio is higher than 0.1. Figure 12.12 presents the ultimate elongation of corroded steel bars versus average cross-sectional loss ratio in normal atmospheric conditions. This figure also shows that the ultimate elongation of corroded steel bars degrades almost exponentially with the increase of corrosion degree. The ultimate strain of corroded steel bars can be calculated with Eq. (12.26).

$$\varepsilon_{suc} = \alpha_{\delta c}\varepsilon_{su0} \tag{12.26}$$

where

ε_{suc}, ε_{su0} ultimate strains of corroded steel bars and uncorroded steel bars, respectively; and
$\alpha_{\delta c}$ relative ultimate elongation of corroded steel bars (Fig. 12.12).

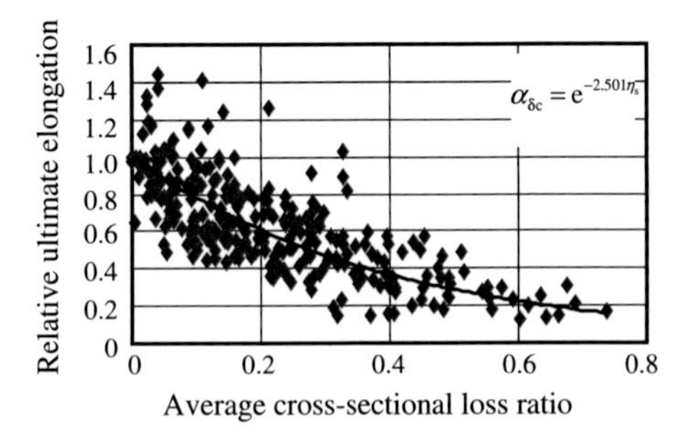

Fig. 12.12 Influence of corrosion degree on ultimate elongation

12.3.3.4 Stress–Strain Relationship of Corroded Steel Bars

Generally, hot rolled steel with common processing and the right chemical components is characterized by a clear yield point, yield plateau, and ratio of yield strength to ultimate strength being less than 0.67. When corrosion occurs, diminishing of yield plateau can be observed until it vanishes at a certain corrosion degree depending on bar types, diameters, and corrosion types, as shown in Fig. 12.13.

For corroded steel bars from concrete members in normal atmospheric conditions, statistic data analysis indicates that the yield plateau vanishes when the average cross-sectional loss ratio reaches 0.2 and 0.1 for deformed steel bars and plain steel bars, respectively. With the aggravation of corrosion, the ratio of yield strength to ultimate strength keeps increasing since the ultimate strength drops more rapidly. Meanwhile, abrupt failure of a member/structure is more likely to happen. Based on these experimental results, a model of the stress–strain relationship of corroded steel bars has been proposed by Tongji University, as shown in Fig. 12.14. This model reflects the influence of corrosion degree η_s on the stress–strain ($\sigma_{sc} - \varepsilon_{sc}$) relationship of corroded steel bars.

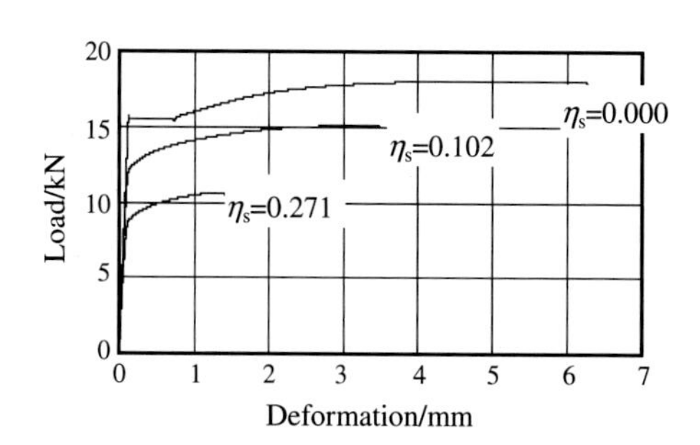

Fig. 12.13 Load–deformation relationship of corroded steel bars (ϕ6)

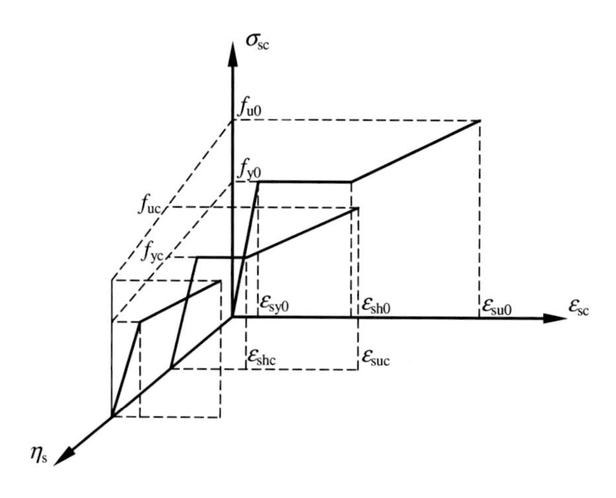

Fig. 12.14 Model of stress–strain relationship of corroded steel bars

12.3.4 Mechanical Properties of Corroded Prestressed Tendons

12.3.4.1 Elastic Modulus

Take β_{Ec} as the ratio of the elastic modulus of corroded prestressed tendons E_{pc} to average elastic modulus of uncorroded prestressed tendons E_p, referred to as relative elastic modulus of prestressed tendons. The relationship between β_{Ec} and corrosion degree is shown in Fig. 12.15a. As the corrosion degree increases, the elastic modulus of steel strands tends to decrease, while the elastic modulus of steel wires approximately fluctuates up and down near a certain value, which is consistent with the elastic modulus test results of corroded steel bars shown in Fig. 12.10. Therefore, it is concluded that corrosion has no effect on the elastic modulus of steel wires. For steel strands, steel wires become thinner after corrosion, and consequently, the squeezing effect among steel wires during the stretching process decreases, resulting in a decrease of the stiffness and elastic modulus.

12.3.4.2 Nominal Ultimate Strength

Define the nominal ultimate strength of corroded prestressed tendons as the ratio of ultimate tensile load to nominal cross-sectional areas of tendons before corrosion. Take relative nominal ultimate strength of corroded prestressed tendons α_{puc} as the ratio of nominal ultimate strength of corroded prestressed tendons to that of uncorroded prestressed tendons. The ultimate strength of corroded prestressed tendons f_{puc} can be calculated with α_{puc} by Eq. (12.27)

$$f_{puc} = \frac{F_{puc}}{A_{sc}} = \frac{F_{puc}/A_{s0}}{F_{pu}/A_{s0}} \cdot \frac{F_{pu} \cdot A_{s0}}{A_{s0} \cdot A_{sc}} = \frac{\alpha_{puc} \cdot f_{pu}}{1 - \eta_s} \tag{12.27}$$

where

F_{pu}, F_{puc} ultimate tensile load of prestressed tendons before and after corrosion, respectively; and

A_{s0}, A_{sc} cross-sectional areas of prestressed tendons before and after corrosion, respectively.

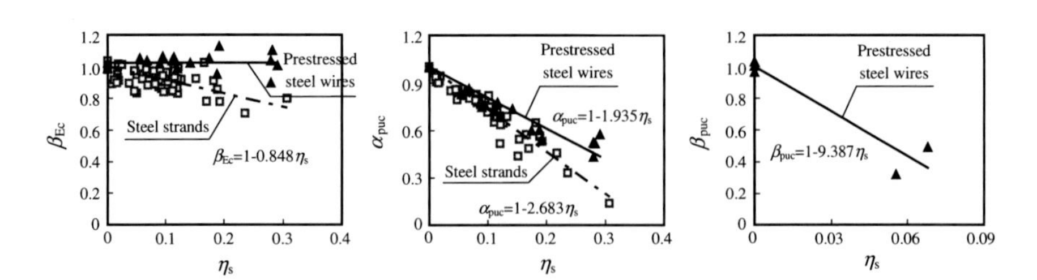

Fig. 12.15 Mechanical properties of corroded prestressed tendons

Statistic data of relative nominal ultimate strength of corroded steel strands and wires are given in Fig. 12.15b. It is revealed that the nominal ultimate strength of prestressed tendons decreases with the corrosion degree. Experimental phenomenon shows that prestressed tendons are prone to failure at corrosion pits when the corrosion degree is relatively large. Therefore, the nominal ultimate strength scatters under large corrosion degree. In comparison with steel wires, corrosion degree has more effect on the nominal ultimate strength of steel strands. The measured corrosion degree actually refers to the weight loss ratio, and the inside wire in steel strands generally has very little corrosion, which indicates that the real corrosion degree of outside wires is larger than the average corrosion degree. Fracture of corroded steel strands happens in the outside wires with the most serious corrosion. Therefore, steel strands are more sensitive to corrosion degree.

12.3.4.3 Ultimate Strain

We define relative ultimate strain of corroded prestressed tendons β_{puc} as the ratio of ultimate strain of corroded prestressed tendons ε_{puc} to that of uncorroded prestressed tendons ε_{pu}. It is observed from the load–deformation curve of corroded prestressed tendons that the relationship between load and deformation appears to be elastic when the corrosion degree exceeds 0.08. Therefore, the critical corrosion degree for prestressed tendons is defined to be 0.08. The ultimate strain can be obtained from the ultimate strength divided by the elastic modulus when the corrosion degree exceeds the critical value. The statistic data of ultimate strain versus corrosion degree (smaller than 0.08) of corroded steel wires are presented in Fig. 12.15c. It is found that the ultimate strain degrades with corrosion degree linearly. The ultimate strain of corroded steel strands versus corrosion degree is considered to be the same as corroded steel wires.

12.3.4.4 Stress–Strain Relationship of Corroded Prestressed Tendons

Figure 12.16 displays the stress–strain relationship curves of corroded prestressed tendons. Note that the deformability of prestressed tendons is very sensitive to corrosion. As the corrosion degree increases, the ultimate elongation decreases and the strain-hardening range diminishes until it vanishes. As aforementioned, the elastic modulus of steel strands decreases with corrosion degree, when the corrosion degree exceeds 0.08, corroded prestressed tendons rupture in the elastic state, and the ultimate strength and ultimate strain decrease linearly. Therefore, Tongji University proposes a stress–strain relationship of corroded prestressed tendons as shown in Fig. 12.17.

12.3.5 *Bond Between Concrete and Corroded Steel Bars*

12.3.5.1 Mechanism

The bond between concrete and steel bars is mainly formed by combined chemical adhesion, friction, and mechanical interlocking at the concrete–steel interface. The bond makes it possible for the two different materials to work together. In early stage before expansion crack occurs, the confinement of concrete to the embedded bars is increased due to the expansion of the corrosion products. The friction is also slightly increased especially for plain bars because their surface gets rough due to corrosion. As a result, the bond performance is improved slightly, as shown in Fig. 12.18. However, when corrosion further develops, a loose layer is formed on the interface and destroys the chemical adhesion and weakens the friction. Corrosion of the ribs of deformed bars reduces the mechanical interlocking. Besides, the concrete cover may crack or even peel off, which can reduce the confinement of the surrounding concrete. Consequently, the bond strength between embedded bars and concrete can be impaired or completely lost, as shown in Fig. 12.19.

12.3.5.2 Degradation Rule

A series of tests were conducted by the Yellow River Institute of Hydraulic Research in China to study the influencing mechanism of bar corrosion on the strength of bond between bars and concrete, as shown in Fig. 12.20a. In these tests, steel plates with the same corrosion degree were used instead of plain steel bars to investigate the influencing rule of corrosion degree and volume expansion force on bond strength. The normal force asserted by the confinement of the surrounding concrete was simulated by external force applied perpendicular to the steel plate. It was found from the test results that the bond strength increased linearly with the

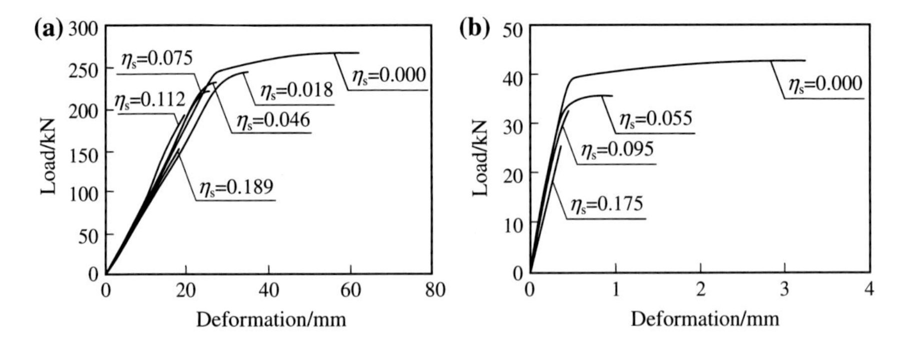

Fig. 12.16 Typical load–deformation curves of corroded prestressed tendons, **a** steel strands (grade 1860, $7\phi^S 15.2$). **b** prestressed steel wires (grade 1860, $7\phi^S 15.2$, inside wire)

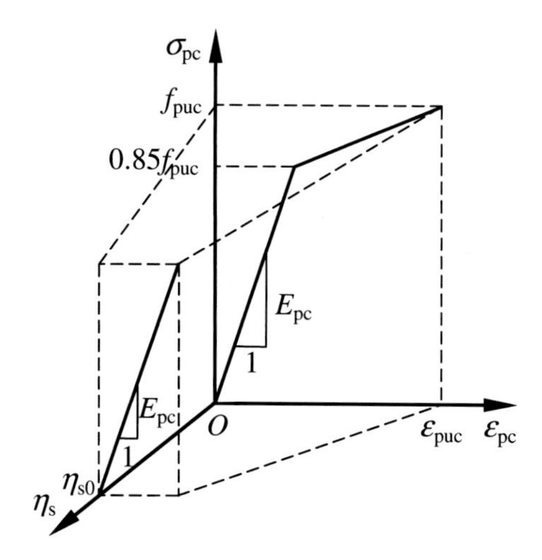

Fig. 12.17 Typical stress–strain relationship of corroded prestressed tendons

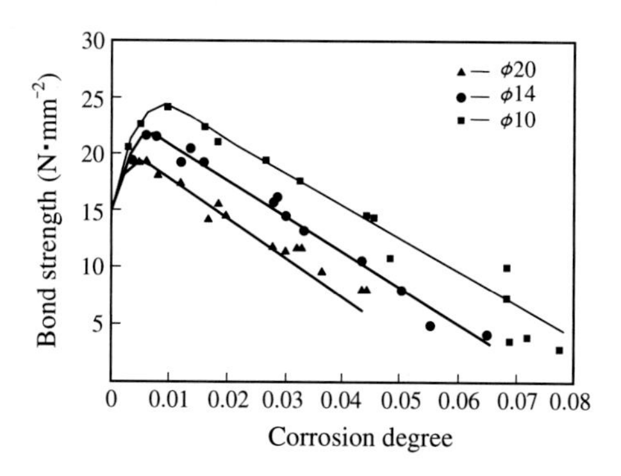

Fig. 12.18 Relationship between corrosion degree and average bond strength

external normal force and decreased with the corrosion aggravation, as shown in Fig. 12.20b.

Bond strength decreases remarkably with the aggravation of bar corrosion, especially after the expansive cracking of concrete cover. This is caused by the lubricating effect of corrosion products, loss of mechanical interlocking, and diminishing of concrete confinement. It was found from the half-beam tests (Fig. 12.21) conducted by Tongji University that the bond strength decreased by 16.2, 21.4, 27.2, and 51.6 % when compared to uncorroded members with expansive crack initiation and crack widths of 0.15, 0.3, and 0.6 mm, respectively. The comparison is shown in Fig. 12.22. The average bond strength after expansive cracking can be calculated by

$$\bar{\tau}_{u,w} = k_w \cdot \bar{\tau}_u \tag{12.28}$$

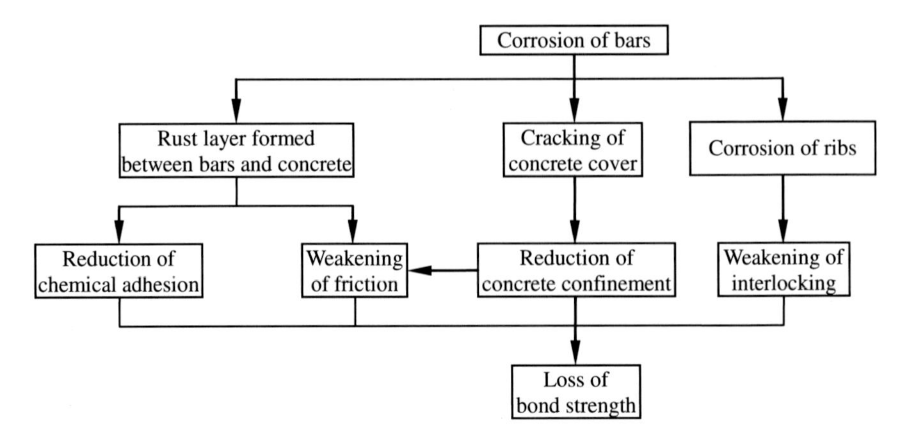

Fig. 12.19 Influence mechanism of bar corrosion on bond strength

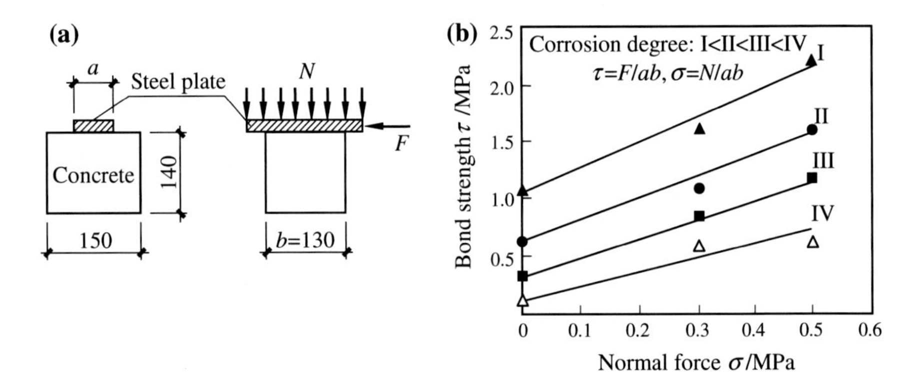

Fig. 12.20 Simulation tests of corrosion effect on bond strength before expansive cracking of concrete, **a** testing setup, **b** influence of corrosion and normal force

where

$\bar{\tau}_{\mathrm{u}}, \bar{\tau}_{\mathrm{u,w}}$ average bond strength before and after expansive cracking, respectively; and

k_{w} bond strength decreasing coefficient considering expansion cracks, which can be obtained by using Eq. (12.29).

$$k_{\mathrm{w}} = 0.9495\ \mathrm{e}^{-1.093\mathrm{w}} \tag{12.29}$$

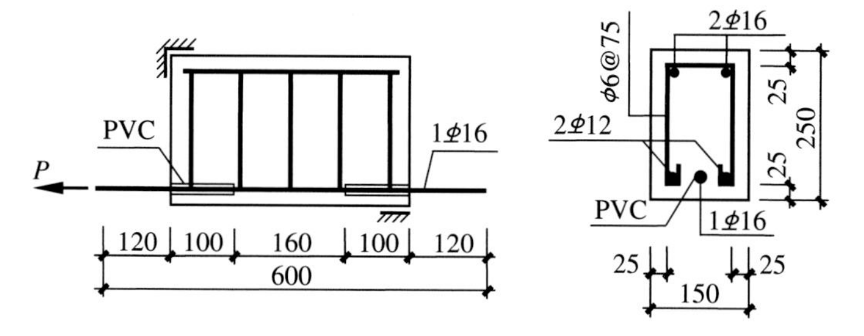

Fig. 12.21 Half-beam test setup

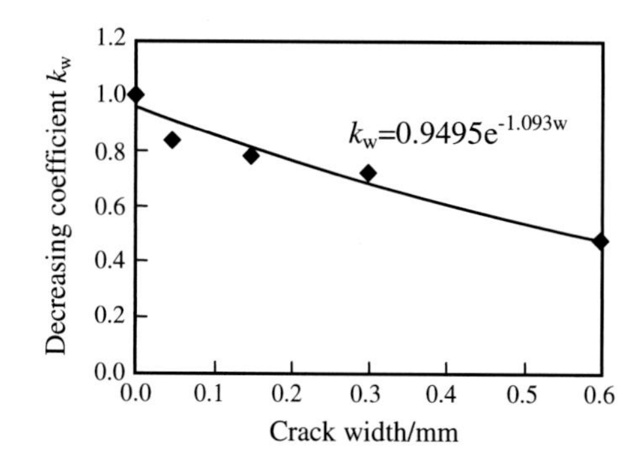

Fig. 12.22 Influence of expansive crack width on bond strength

12.3.5.3 Average Bond Stress–Slip Relationship

The average bond stress–slip relationship between corroded steel bars and concrete, which was established based on the half-beam tests conducted at Tongji University, can be approximately divided into five stages, as shown in Fig. 12.23.

The five stages of bond stress–slip relationship between steel bars and concrete are as follows:

1. Micro-slip stage: During this stage, the bond-slip exhibits a linear relationship under relatively small load. Very little slip is observed at the loading ends, while no slip is observed at the free end.
2. Slip stage: When the load reaches a certain level, slip will take place at the free end. This implies that the adhesion in the anchorage area has been totally destroyed. The bond-slip relationship is then no longer linear.
3. Cracking stage: With the load continuously increased, the expansion cracks on the concrete surface expand from the loading end to the free end. The ultimate load can only be reached by using a small load increment.
4. Descending stage: In this stage, the load decreases slowly after the peak load is reached, while the slip increases significantly.

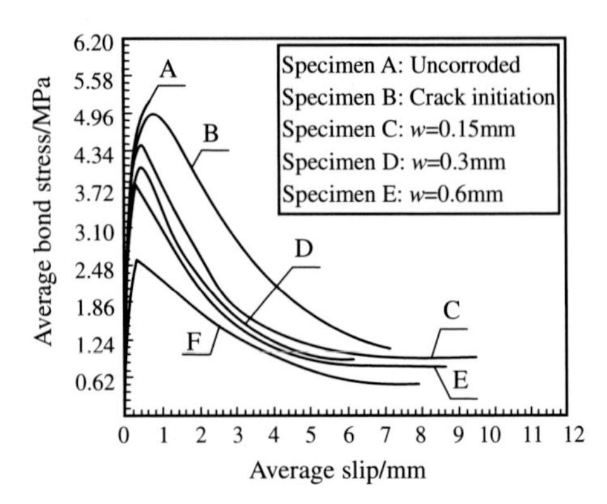

Fig. 12.23 $\bar{\tau}_w - \bar{s}_w$ relationship

5. Residual stage: The slip increases while the load stays constant at more or less than one-fifth of the ultimate load. Therefore, the bond-slip relationship can be expressed approximately as a straight line in most of the five stages except for the cracking and descending stages, where a parabolic curve should be used instead. It is concluded based on statistic analysis of the test results that

$$\bar{\tau}_{cr,w} = 0.58\bar{\tau}_{u,w}, \bar{\tau}_{r,w} = 0.2\bar{\tau}_{u,w} \tag{12.30}$$

by which the relationship between $\bar{\tau}_w$ and $\bar{s}_w$ can be established by considering different crack widths, as presented in the following Eqs. (12.31)–(12.34). A theoretical model for the $\bar{\tau}_w$ versus $\bar{s}_w$ relationship is shown schematically in Fig. 12.24.

$$\text{Slip stage}: \quad \bar{\tau}_w = k_1\bar{s}_w \quad (0 \leqslant \bar{s}_w \leqslant \bar{s}_{cr,w}) \tag{12.31}$$

$$\text{Cracking stage}: \quad \bar{\tau}_w = k_2\bar{s}_w^2 + k_3\bar{s}_w + k_4 \quad (\bar{s}_{cr,w} < \bar{s}_w \leqslant \bar{s}_{u,w}) \tag{12.32}$$

$$\text{Descending stage}: \quad \bar{\tau}_w = k_5\bar{s}_w^2 + k_6\bar{s}_w + k_7 \quad (\bar{s}_{u,w} < \bar{s}_w \leqslant \bar{s}_{r,w}) \tag{12.33}$$

$$\text{Residual stage}: \quad \bar{\tau}_w = \bar{\tau}_{r,w} \quad (\bar{s}_w > \bar{s}_{r,w}) \tag{12.34}$$

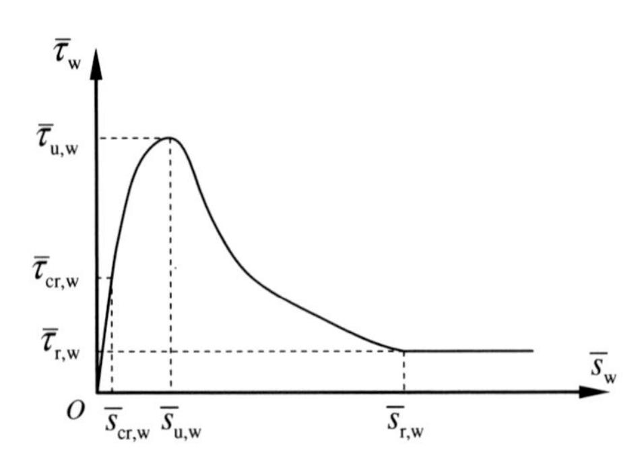

Fig. 12.24 Theoretic $\bar{\tau}_w - \bar{s}_w$ model

where

$\bar{\tau}_{\mathrm{w}}, \bar{s}_{\mathrm{w}}$	bond stress and relative slip, respectively, at an expansive crack width w;
$\bar{\tau}_{\mathrm{cr,w}}, \bar{s}_{\mathrm{cr,w}}$	bond stress and the relative slip, respectively, at slip initiation;
$\bar{\tau}_{\mathrm{u,w}}, \bar{s}_{\mathrm{u,w}}$	bond stress and the relative slip, respectively, at the ultimate load;
$\bar{\tau}_{\mathrm{r,w}}, \bar{s}_{\mathrm{r,w}}$	bond stress and the relative slip, respectively, at the beginning of the horizontal stage; and
k_1–k_7	constants related to the expansive crack width.

Provided $\bar{\tau}_{\mathrm{u,w}}$ is equal to $k_1(\bar{s}_{\mathrm{cr,w}} + \bar{s}_{\mathrm{u,w}})/2$, all of the constants can be calculated from the continuity of critical point and the condition of crest value at peak load: $k_1 = 35.372\,\mathrm{e}^{-1.683\mathrm{w}}$, $k_2 = \frac{k_1^2}{-1.68\bar{\tau}_{\mathrm{u,w}}}$, $k_3 = 1.69k_1$, $k_4 = 1.2\bar{\tau}_{\mathrm{u,w}}$, $k_5 = \frac{0.8\bar{\tau}_{\mathrm{u,w}}}{81\bar{s}_{\mathrm{u,w}}}$, $k_6 = -\frac{16\bar{\tau}_{\mathrm{u,w}}}{81\bar{s}_{\mathrm{u,w}}}$, and $k_7 = \frac{96.2}{81}\bar{\tau}_{\mathrm{u,w}}$.

12.4 Flexural Behavior of Corroded RC Members

12.4.1 Experimental Study

The flexural behavior of RC members is affected by bar corrosion mainly in three ways: (1) reduction in bar sectional area, (2) degradation of bar mechanical performance, and (3) deterioration of bond between embedded bars and surrounding concrete. Before expansive cracking, biaxial stress status arises in compressive concrete near to compressive corroded steel bars due to the radial tensile stress caused by corrosion. This lowers the compressive strength of concrete and thereby reduces the flexural bearing capacities of RC members.

The failure mode of corroded RC beams is similar to that of normal beams. However, as the corrosion degree increases, the bond strength between steel bars and concrete decreases gradually, and the crack spacing and crack width tend to increase. When the corrosion degree is relatively small, typical flexural failure occurs to the corroded members, i.e., steel bars yield and then concrete in compression crushes. Figure 12.25 presents typical load–deflection curves from corroded RC beam tests conducted at Tongji University. The cross-sectional dimensions of the beams were $b \times h$ = 150 mm × 200 mm. The length was 2000 mm and the span was 1800 mm. The bottom and top longitudinal reinforcement was 2ϕ16 and 2ϕ10, respectively. The stirrups were ϕ6@150. The concrete cover was 25 mm thick, and the beams were loaded symmetrically at one-third point. The measured yield strength and ultimate strength of longitudinal steel bars were 350 and 520 MPa, respectively, and the measured axial compressive strength of concrete was 28.6 MPa. L10 was an uncorroded beam, and the corrosion degrees of beams L11–L13 were 0.05, 0.13 and 0.24, respectively. When the corrosion degree is relatively large, the ultimate elongation of corroded steel bars

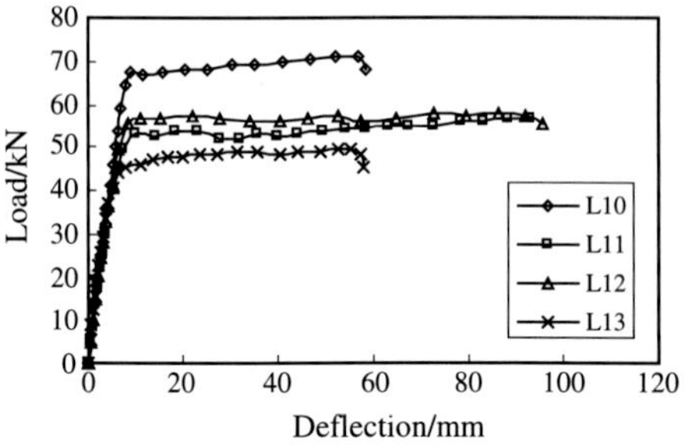

Fig. 12.25 Load–deflection curves of corroded reinforced RC beams

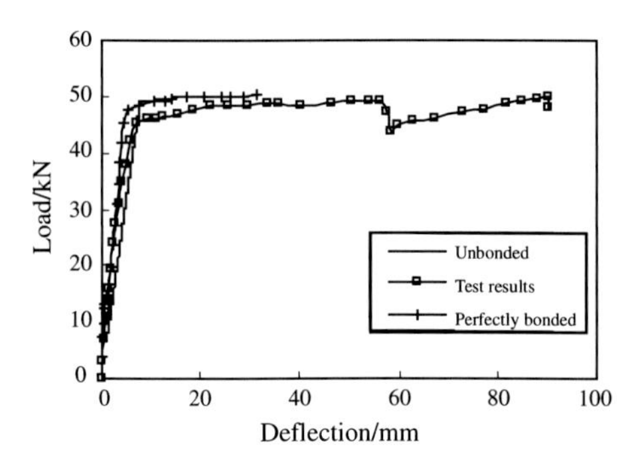

Fig. 12.26 Load–deflection curves of beam L13

decreases significantly due to the influence of nonuniform corrosion and corrosion pits. Therefore, corroded steel bars may break before crush of concrete in compression, which is similar to the failure mode of a lightly reinforced beam. In Fig. 12.25, Beam L13 failed as a result of corroded steel bars rupturing, while the other beams failed with the crush of concrete in compression.

Numerical simulation was performed to investigate the influence of degraded bond strength on the flexural behavior of corroded beams. In real practice, longitudinal steel bars extend into supports and consequently are seldom corroded. Therefore, assuming that the anchorage at the ends of longitudinal steel bars is good, just consider two extreme cases: (1) Perfect bond exists between steel bars and concrete; and (2) no bond exists between steel bars and concrete. In real construction, the bond behavior should be between these two cases. The numerical results of Beam L13 were compared with test results as shown in Fig. 12.26, where the flexural bearing capacities of beams with or without bond are both close to the test results, while the bending stiffness of the perfectly bonded beam is higher than that of the test one and that of the unbonded beam is the smallest. Therefore, it is concluded that, in the case of good anchorage, it is the reduction of cross-sectional area and degradation of mechanical properties of corroded steel bars that influence the flexural bearing capacity of a corroded RC beam, while it is the degradation of bond behavior between corroded steel bars and concrete that affect the bending stiffness of a corroded RC beam.

12.4.2 Flexural Bearing Capacities of Corroded RC Beams

Experimental results indicate that the aggravation of corrosion causes the bond behavior between steel bars and concrete to degrade to different extents. However, the plane section assumption is still valid according to the strain distribution on beams in experimental study. Meanwhile, the degradation of bond behavior between corroded steel bars and concrete has little effect on the flexural bearing capacity of a beam. Therefore, the plane section assumption is applicable to the calculation of flexural bearing capacity of corroded RC beams. Figure 12.27 shows the stress and strain distributions across the section at the ultimate state.

The equilibrium condition in the horizontal direction gives Eq. (12.35). Taking moment about the acting point of the resultant compressive force of the compression zone or the tensile force of the steel bars, the ultimate moment M_u can be calculated by (12.36)

$$\alpha_1 f_c b x + \sum_{i=1}^{n} \sigma_{sci} A_{sci} = 0 \tag{12.35}$$

$$M_u = \alpha_1 f_c b x \left(h_0 - \frac{x}{2} \right) = \sum_{i=1}^{n} \sigma_{sci} A_{sci} \left(h_0 - \frac{x}{2} \right) \tag{12.36}$$

where

x depth of equivalent rectangular stress block, $x = \beta_1 x_n$; and

α_1, β_1 graphic coefficients of concrete under compression.

For the concrete with a strength grade of no more than C50 (f_{cu} = 50 MPa), $\alpha_1 = 1.00$ and $\beta_1 = 0.80$. When f_{cu} = 80 MPa (C80), similar analysis gives $\alpha_1 = 0.94$ and $\beta_1 = 0.74$. For the concrete with the strength grade between C50 and C80, α_1 and β_1 can be obtained by using the linear interpolation method.

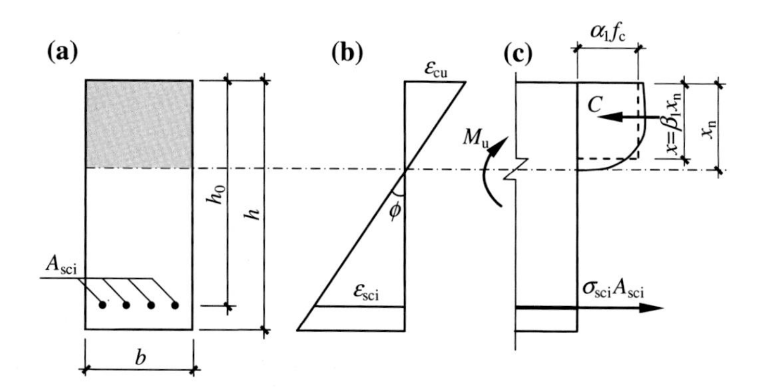

Fig. 12.27 Stress and strain distribution across the section of a corroded RC beam at the ultimate state. **a** cross section. **b** strain distribution. **c** stress distribution

A_{sci} cross-sectional area of the ith corroded steel bar;
σ_{sci} stress of the ith corroded steel bar, which is obtained from the strain based on the stress–strain relationship of corroded steel bars.

The strain in Fig. 12.27 can be expressed by Eq. (12.37) according to the plane section assumption.

$$\varepsilon_{sci} = \left(\frac{\beta_1 h_0}{x} - 1\right)\varepsilon_{cu} \tag{12.37}$$

12.4.3 Flexural Stiffness of Corroded RC Beams

Degradation of bond behavior between steel bars and concrete is one of the main factors which decrease the flexural stiffness of corroded RC beams.

First, with the degradation of bond behavior, lag of steel strain occurs to the cross section of maximum moment, which is considered by $m(\eta_s)$

$$m(\eta_s) = \bar{\varepsilon}_t / \bar{\varepsilon}_{sc} \tag{12.38}$$

where
$\bar{\varepsilon}_t$ average tensile strain of concrete at the position of tensile steel bars; and
$\bar{\varepsilon}_{sc}$ strain of tensile steel bars.

Second, the steel strain becomes even among cracks, while the coefficient which indicates the nonuniformity of steel strain tends to increase and can be obtained by the nonuniformity coefficient of uncorroded steel bars by multiplying a correction factor $n(\eta_s)$

$$\psi_c = n(\eta_s) \cdot \psi \tag{12.39}$$

where ψ, ψ_c = nonuniformity coefficients of steel strain before and after corrosion, respectively, and ψ is calculated by Eq. (11.40).

The average curvature of the cross section ϕ is expressed by Eq. (12.40) based on the geometrical relationship.

$$\phi = \frac{\bar{\varepsilon}_c + \bar{\varepsilon}_{sc}}{h_0} \tag{12.40}$$

where
$\bar{\varepsilon}_c$ average compressive strain of concrete in pure flexural section.

Since it varies little, it could be approximately taken as the compressive strain at the top of the cracked section ε_{cm}.

Substituting Eqs. (12.38) and (12.39) into (12.40) results in

$$\phi = \frac{\varepsilon_{cm} + n(\eta_s)\psi\varepsilon_{scm} \cdot m(\eta_s)}{h_0} \tag{12.41}$$

where

ε_{scm} steel strain of cracked section.

Therefore, the short-term flexural stiffness of a corroded RC beam can be calculated by

$$B_s = \frac{E_s A_{sc} h_0^2}{1.15k(\eta_s)\psi + 0.2 + 6\alpha_E \rho_{sc}} \tag{12.42}$$

where

$k(\eta_s)$ coefficient of comprehensive strain of corroded steel bars; hence,
$k(\eta_s)$ $m(\eta_s) \cdot n(\eta_s)$ and can be calculated by Eq. (12.43);
α_E ratio of elastic modulus of steel bars to that of concrete; and
ρ_{sc} reinforcement ratio based on the cross-sectional area of corroded steel bars.

$$\begin{cases} k(\eta_s) = 1 & \eta_s \leqslant 0.55/k_u \\ k(\eta_s) = 9k_u^2\eta_s^2 - 10.1k_u\eta_s + 3.83 & 0.55/k_u < \eta_s \leqslant 1/k_u \\ k(\eta_s) = 2.73 & \eta_s > 1/k_u \end{cases} \tag{12.43}$$

where

k_u an empirical coefficient which reflects the degradation rate of bond behavior between steel bars and concrete after corrosion;
k_u 10.544 − 1.5896 × (*c*/*d*);
c depth of concrete cover; and
d diameter of a steel bar.

12.5 Flexural Behavior of Corroded Prestressed Concrete Members

12.5.1 Experimental Study

Figure 12.28 presents load–deflection curves of corroded prestressed concrete beams symmetrically loaded at one-third points. The dimensions and reinforcement of the specimens are detailed in Fig. 12.29. The corrosion degrees of Beams L1–L4 are 0, 0.09. 0.14, and 1.0 (ruptured), respectively. Figure 12.28 indicates that as the corrosion degree of prestressed tendon increases, the characteristics of load–deflection curves tend to change from those of uncorroded prestressed beams to

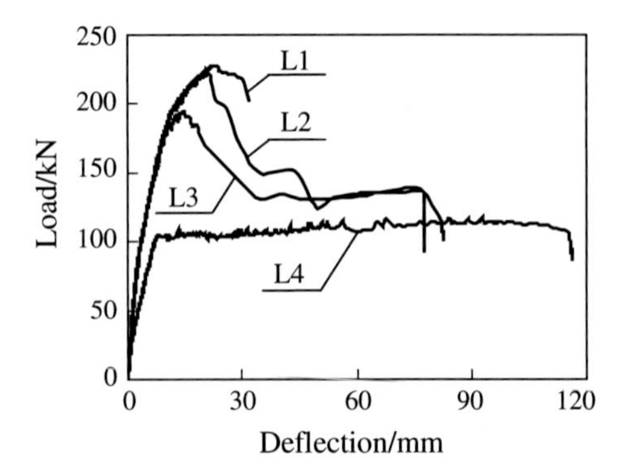

Fig. 12.28 Load–deflection curves of corroded prestressed concrete beams

those of normal reinforced beams and the flexural bearing capacity decreases significantly. When the corrosion degree is relatively small, the stiffness does not change much, but when the corrosion degree is large (even ruptured), the stiffness lowers considerably.

12.5.2 Flexural Bearing Capacities of Corroded Prestressed Concrete Beams

There are three possible failure modes of corroded prestressed RC beams: (1) crush of concrete at compression zone; (2) fracture of corroded normal steel bars; and (3) fracture of corroded prestressed tendons. Based on the statistical results showing the ultimate strains of corroded steel bars and corroded prestressed tendons in Sect. 12.3, take the test beams in Fig. 12.29 as example and assume that the corrosion degrees of prestressed tendons and normal steel bars are the same. It is found that, when the corrosion degree exceeds 0.05, the ratio of ultimate strain of corroded steel bars $\varepsilon_{\mathrm{suc}}$ to ultimate strain increment of corroded prestressed tendons ($\Delta\varepsilon_{\mathrm{puc}} = \varepsilon_{\mathrm{puc}} - \varepsilon_{\mathrm{poc}}$) is larger than 5, as shown in Fig. 12.30. Therefore, generally, corroded prestressed tendons break first and the failure mode with the fracture of corroded normal steel bars can be neglected.

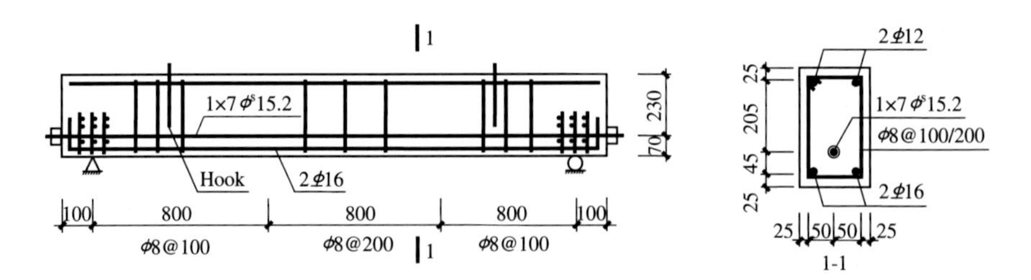

Fig. 12.29 Dimensions and reinforcement of test beams

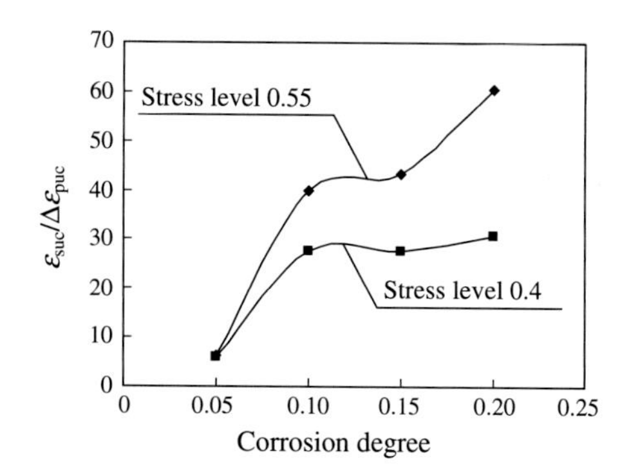

Fig. 12.30 Ratio of ultimate strain of corroded normal steel bars to ultimate strain increment of corroded tendons

Before calculating the flexural bearing capacities of corroded prestressed concrete beams, the following assumptions are made: (1) For the corroded prestressed concrete beams with good end anchorage, the degradation of the bond behavior between corroded prestressed tendons or corroded steel bars and concrete has little effect on the bearing capacities of corroded beams, based on the numerical results presented in Sect. 12.4; (2) the strain of concrete and prestressed tendons obeys the plane section assumption; and (3) the contribution of the concrete in tension to the bearing capacity is neglected.

The stress and strain distributions at the ultimate moment are shown in Fig. 12.31. The plane section assumption results in

$$\frac{\Delta\varepsilon_{pc}}{h_p - x_n} = \frac{\varepsilon_{sc}}{h_0 - x_n} = \frac{\varepsilon'_{sc}}{x_n - a'_s} = \frac{\varepsilon_c}{x_n} \quad (12.44)$$

where

$\Delta\varepsilon_{pc}$ strain increment of prestressed tendons under loading;
x_n actual compression zone depth of concrete;
h beam height;
h_0, h_p distances from the centroids of normal steel bars and prestressed tendons to the beam top, respectively;
ε_{sc}, ε'_{sc} strains of longitudinal tension and compression steel bars without prestressing, respectively;
ε_c compressive strain at the beam top; and
a'_s distance from the centroid of the compression steel bars without prestressing to the extreme compression fiber of concrete.

Assuming the equivalent compression zone depth $x = \beta_1 x_n$, the equilibrium equation in horizontal direction could be obtained as

$$\alpha_1 f_c bx + \sigma'_{sc}A'_{sc} = \sigma_{sc}A_{sc} + \sigma_{pc}A_{pc} \quad (12.45)$$

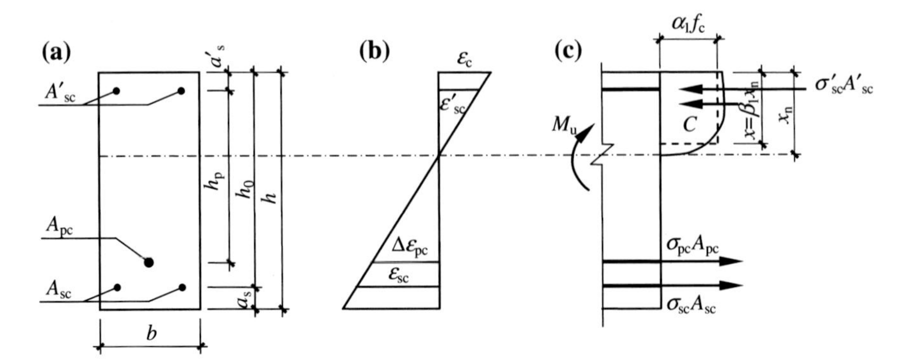

Fig. 12.31 Stress and strain distributions across the section of a corroded prestressed concrete beam at the ultimate state. **a** cross section. **b** strain distribution. **c** stress distribution

where

A_{sc}, A'_{sc} cross-sectional areas of corroded steel bars in tension and compression zones, respectively;

A_{pc} cross-sectional area of corroded prestressed tendons;

σ_{pc} stress of corroded prestressed tendons, which can be calculated based on the stress–strain relationship of corroded prestressed tendons; and

σ_{sc}, σ'_{sc} stresses of longitudinal tension and compression steel bars without prestressing, respectively, which can be calculated based on the stress–strain relationship of corroded steel bars;

When the concrete strain at the beam top reaches the ultimate compressive strain, or the strain of corroded prestressed tendons reaches the ultimate tensile strain, it indicates that the corroded prestressed RC beam has failed. In the calculation, the failure mode should be firstly determined. Therefore, the balanced failure mode is defined that the concrete strain reaches the limitation while the corroded prestressed tendons fracture. Define ρ_{pc}, ρ_{sc}, ρ'_{sc} as the residual reinforcement ratios of corroded prestressed tendons, steel bars in tension zone, and steel bars in compression zone, respectively, i.e., $\rho_{pc} = \frac{A_{pc}}{bh_0}$, $\rho_{sc} = \frac{A_{sc}}{bh_0}$, $\rho'_{sc} = \frac{A'_{sc}}{bh_0}$. The balanced residual reinforcement ratio of corroded prestressed tendons ρ_{pcb} could be calculated by Eq. (12.46)

$$\rho_{pcb} = \frac{\alpha_1 f_c \beta_1 x_b / h_0 + \sigma'_{sc}\rho'_{sc} - \sigma_{sc}\rho_{sc}}{f_{puc}} \tag{12.46}$$

where

x_b actual height of neutral axis when balanced failure happens, which is given in Eq. (12.47).

$$x_{\mathrm{b}} = \frac{\varepsilon_{\mathrm{cu}}}{\varepsilon_{\mathrm{cu}} + \Delta\varepsilon_{\mathrm{puc}}} h_{\mathrm{p}} \tag{12.47}$$

where

$\varepsilon_{\mathrm{cu}}$ ultimate strain of concrete; and
$\Delta\varepsilon_{\mathrm{puc}}$ ultimate strain increment of corroded prestressed tendons.

For corroded prestressed concrete beams, if $\rho_{\mathrm{pc}} \leqslant \rho_{\mathrm{pcb}}$, fracture of corroded prestressed tendons happens; if $\rho_{\mathrm{pc}} > \rho_{\mathrm{pcb}}$, crush of concrete happens.

After the failure mode is determined, the flexural bearing capacities of corroded prestressed beams can be evaluated according to the following steps:

1. Calculate x_{b} with Eq. (12.47); then, calculate σ'_{sc} and σ_{sc} with Eq. (12.44) and the stress–strain relationship of steel. Calculate ρ_{pcb} with Eq. (12.46) and determine the failure mode according to the actual reinforcement ratio.
2. If $\rho_{\mathrm{pc}} > \rho_{\mathrm{pcb}}$, crush of concrete will happen and then take $\varepsilon_{\mathrm{c}} = \varepsilon_{\mathrm{cu}}$. Based on Eq. (12.44), the geometric compatibility relationship results in

$$\varepsilon_{\mathrm{pc}} = \varepsilon_{\mathrm{sc}} \cdot \frac{h_{\mathrm{p}} - x_{\mathrm{n}}}{h_0 - x_{\mathrm{n}}} + \frac{f_{\mathrm{pt}} \cdot \varphi_{\mathrm{p}}}{E_{\mathrm{p}}}, \varepsilon_{\mathrm{sc}} = \varepsilon_{\mathrm{cu}} \cdot \frac{h_0 - x_{\mathrm{n}}}{x_{\mathrm{n}}}, \varepsilon'_{\mathrm{sc}} = \varepsilon_{\mathrm{cu}} \cdot \frac{x_{\mathrm{n}} - a'_{\mathrm{s}}}{x_{\mathrm{n}}} \tag{12.48}$$

where

φ_{p} coefficient of prestressing level; and
$f_{\mathrm{pt}}, E_{\mathrm{p}}$ ultimate strength and elastic modulus of uncorroded tendons, respectively.

Based on the stress–strain relationship, calculate the stress by strain and substitute it into Eq. (12.45). Thereafter, the equivalent compression zone depth $x = \beta_1 x_{\mathrm{n}}$ is obtained. Taking moment about the acting point of the resultant compressive force of the compression zone, the ultimate bending moment M_{u} can be evaluated

$$M_{\mathrm{u}} = f'_{\mathrm{yc}} A'_{\mathrm{sc}} \left(\frac{x}{2} - a'_{\mathrm{s}}\right) + \sigma_{\mathrm{sc}} A_{\mathrm{sc}} \left(h_0 - \frac{x}{2}\right) + \sigma_{\mathrm{pc}} A_{\mathrm{pc}} \left(h_{\mathrm{p}} - \frac{x}{2}\right) \tag{12.49}$$

where

f'_{yc} yield strength of corroded compressive steels; and
σ_{sc} stress of corroded tensile steels. If $\sigma_{\mathrm{sc}} > f_{\mathrm{yc}}$, take $\sigma_{\mathrm{sc}} = f_{\mathrm{yc}}$, where f_{yc} is the yield strength of corroded tensile steels.

When $x < 2a'_{\mathrm{s}}$, take $x = 2a'_{\mathrm{s}}$.

3. If $\rho_{\mathrm{puc}} \leqslant \rho_{\mathrm{pcb}}$, fracture of corroded prestressed tendons will happen; therefore, the geometric relationship at the ultimate state is

$$\frac{\Delta\varepsilon_{\rm puc}}{h_{\rm p}-x_{\rm n}}=\frac{\varepsilon_{\rm sc}}{h_0-x_{\rm n}}=\frac{\varepsilon'_{\rm sc}}{x_{\rm n}-a'_{\rm s}}=\frac{\varepsilon_{\rm c}}{x_{\rm n}} \tag{12.50}$$

The strains of concrete and corroded steel bars can be expressed as

$$\varepsilon_{\rm c}=\frac{\Delta\varepsilon_{\rm puc}x_{\rm n}}{h_{\rm p}-x_{\rm n}},\ \varepsilon_{\rm sc}=\Delta\varepsilon_{\rm puc}\cdot\frac{h_0-x_{\rm n}}{h_{\rm p}-x_{\rm n}},\ \varepsilon'_{\rm sc}=\Delta\varepsilon_{\rm puc}\cdot\frac{x_{\rm n}-a'_{\rm s}}{h_{\rm p}-x_{\rm n}} \tag{12.51}$$

The stresses of corroded steel bars $\sigma_{\rm sc}$ and $\sigma'_{\rm sc}$ can be obtained through the stress–strain relationship. Assume the equivalent compression zone depth $x=\beta'_1x_{\rm n}$ and the equivalent stress of concrete $\alpha'_1f_{\rm c}$. The equilibrium equation in horizontal direction indicated by Fig. 12.31 is

$$\alpha'_1f_{\rm c}bx+\sigma'_{\rm sc}A'_{\rm sc}=\sigma_{\rm sc}A_{\rm sc}+f_{\rm puc}A_{\rm pc} \tag{12.52}$$

The equivalent compression zone depth x can be evaluated by Eq. (12.52). Taking moment about the acting point of the resultant compressive force of the compression zone, we have the flexural bearing capacity $M_{\rm u}$

$$M_{\rm u}=\sigma'_{\rm sc}A'_{\rm sc}\left(\frac{x}{2}-a'_{\rm s}\right)+\sigma_{\rm sc}A_{\rm sc}\left(h_0-\frac{x}{2}\right)+f_{\rm puc}A_{\rm pc}\left(h_{\rm p}-\frac{x}{2}\right) \tag{12.53}$$

When the concrete strength $f_{\rm cu}$ is less than 50 MPa, α_1 and β_1 can be calculated by Eqs. (12.54) and (12.55), respectively.

$$\begin{cases}\alpha'_1=\frac{\varepsilon_{\rm c}}{\varepsilon_0}-\frac{\varepsilon_{\rm c}^2}{3\varepsilon_0^2} & 0\leqslant\varepsilon_{\rm c}\leqslant\varepsilon_0\\ \alpha'_1=1-\frac{\varepsilon_0}{3\varepsilon_{\rm c}} & \varepsilon_0\leqslant\varepsilon_{\rm c}\leqslant\varepsilon_{\rm cu}\end{cases} \tag{12.54}$$

$$\begin{cases}\beta'_1=2\cdot\frac{\frac{1}{3}-\frac{\varepsilon_{\rm c}}{12\varepsilon_0}}{1-\frac{\varepsilon_{\rm c}}{3\varepsilon_0}} & 0\leqslant\varepsilon_{\rm c}\leqslant\varepsilon_0\\ \beta'_1=2\cdot\frac{\frac{\varepsilon_{\rm c}}{2}-\frac{\varepsilon_0}{3}+\frac{\varepsilon_0^2}{12\varepsilon_{\rm c}}}{\varepsilon_{\rm c}-\frac{\varepsilon_0}{3}} & \varepsilon_0\leqslant\varepsilon_{\rm c}\leqslant\varepsilon_{\rm cu}\end{cases} \tag{12.55}$$

where

ε_0 peak strain of concrete under compression, taken as 0.002; and
$\varepsilon_{\rm cu}$ ultimate strain of concrete under compression, taken as 0.0033.

12.5.3 Flexural Stiffness of Corroded Prestressed Concrete Beams

The short-term flexural stiffness of concrete members prestressed with bonded tendons can be calculated by Eq. (11.65). *Technical specification for concrete structures prestressed with unbonded tendons* (JGJ 92) gives the calculation method of short-term flexural stiffness of concrete members prestressed with unbonded tendons, which is similar to Eq. (11.65). For concrete members where cracks are strictly prohibited, $B_s = 0.85E_cI_0$; for cracked prestressed concrete members, assuming that the moment–curvature relationship is still a bilinear curve, the intersection value is the cracking moment M_{cr} and the stiffness reduction factor at $\frac{M_{cr}}{M_k} = 0.6$ is applied; then, after cracking, the short-term stiffness of a prestressed concrete beam is

$$B_s = \frac{E_cI_0}{\frac{1}{\beta_{0.6}} + \frac{\frac{M_{cr}}{M_k} - 0.6}{1 - 0.6}\left(\frac{1}{\beta_{cr}} - \frac{1}{\beta_{0.6}}\right)} \tag{12.56}$$

where $\beta_{cr} = 0.85$; based on the experimental results, take $\frac{1}{\beta_{0.6}}$ as

$$\frac{1}{\beta_{0.6}} = \left(1.26 + 0.3\lambda' + \frac{0.07}{\alpha_E\rho}\right)(1 + 0.45\gamma_f) \tag{12.57}$$

where

λ' ratio of unbonded tendon reinforcement index to integrated reinforcement index, take $\lambda' = \frac{\sigma_{pe}A_p}{\sigma_{pe}A_p + f_yA_s}$;

σ_{pe}, A_p effective tensile stress and cross-sectional area of unbonded tendons after stress loss; and

f_y, A_s yield stress and cross-sectional area of tensile non-prestressed steel bars.

Substituting β_{cr} and $\frac{1}{\beta_{0.6}}$ into Eq. (12.56) results in

$$B_s = \frac{0.85E_cI_0}{\kappa_{cr} + (1 - \kappa_{cr})\omega_0} \tag{12.58}$$

where $\omega_0 = \left(1.0 + 0.8\lambda' + \frac{0.21}{\alpha_E\rho}\right)(1 + 0.45\gamma_f)$.

For a corroded prestressed concrete beam, the bond between tendons and concrete degrades gradually. When the corrosion ratio reaches a critical value η_{s1}, the beam cracks, and the bond between corroded tendons and concrete vanishes, which make it behave as a concrete beam prestressed with unbonded tendons. Consequently, the corroded tendons and concrete cannot work together well, and the strain of corroded tendons lags in comparison with that of concrete. Therefore, the flexural stiffness of the corroded prestressed concrete beam decreases, especially in the case with a large deformation after cracking. Assuming that the moment–curvature relationship is still a bilinear curve, the intersection value is the cracking moment M_{crc}, as shown in Fig. 12.32. The short-term flexural stiffness of a corroded prestressed concrete beam is

$$B_{sc} = \frac{E_cI_{0c}}{\frac{1}{\beta_\xi} + \frac{\frac{M_{crc}}{M_{kc}} - \xi}{1 - \xi} \cdot \left(\frac{1}{\beta_{cr}} - \frac{1}{\beta_\xi}\right)} \tag{12.59}$$

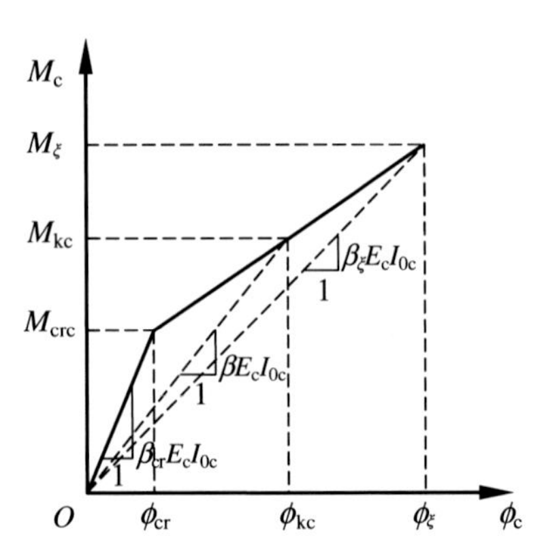

Fig. 12.32 Moment–curvature relationship of corroded prestressed concrete beams

where β_ζ, β_{cr} are stiffness reduction factors when $\frac{M_{crc}}{M_{kc}} = \xi$ and 1.0, respectively, $0.4 \leq \xi \leq 0.6$.

Assume that after cracking, the stiffness reduction factor of a corroded prestressed concrete beam, β_ξ, varies linearly between that of the corroded concrete beam prestressed with bonded tendons, $\beta_{0.4c}$, and that of the corroded concrete beam prestressed with unbonded tendons, $\beta_{0.6c}$, as expressed by Eq. (12.60)

$$\beta_\xi = \frac{\beta_{0.6c} - \beta_{0.4c}}{\eta_{s1}}\eta_s + \beta_{0.4c} \tag{12.60}$$

where

η_s corrosion degree of prestressed tendons; and

η_{s1} critical corrosion degree of prestressed tendons;

when $\eta_s > \eta_{s1}$, the bond strength between prestressed tendons and concrete decreases to 0, and we can then take $\xi = 0.6$ and $\beta_\xi = \beta_{0.6c}$

$\beta_{0.4c}$ and $\beta_{0.6c}$ can be calculated as follows

$$\beta_{0.4c} = \frac{1}{\left(0.8 + \frac{0.15}{\alpha_{Ec}\rho_c}\right)(1 + 0.45\gamma_f)} \tag{12.61}$$

$$\beta_{0.6c} = \frac{1}{\left(1.26 + 0.3\lambda'_c + \frac{0.07}{\alpha_{Ec}\rho_c}\right)(1 + 0.45\gamma_f)} \tag{12.62}$$

where

α_{Ec} ratio of elastic modulus of corroded normal steel bars to that of concrete;

ρ_c reinforcement ratio of longitudinal corroded steel bars and tendons;

γ_f ratio of cross-sectional area of tensile flange to effective area of web;

I_{0c} moment of inertia of transformed section; and

λ'_c ratio of corroded unbonded tendon reinforcement index to integrated reinforcement index; then, take $\lambda'_c = \frac{\sigma_{pec}A_{pc}}{\sigma_{pec}A_{pc} + f_{yc}A_{sc}}$;

M_{crc} cracking moment calculated according to *Code for design of concrete structures* (GB 50010), where the parameters of steel bars and tendons are taken as corroded ones.

M_{kc} maximum moment calculated under the characteristic combination of load effects.

Similar to β_{ξ}, assume that ξ varies linearly between 0.4 and 0.6 with the corrosion degree of tendons. $\xi = \frac{0.6-0.4}{\eta_{s1}}\eta_s + 0.4$ ($\eta_s \leqslant \eta_{s1}$); when $\eta_s > \eta_{s1}$, take $\xi = 0.6$, $\beta_{\xi} = \beta_{0.6c}$.

12.6 Durability Design and Assessment of Concrete Structures

12.6.1 Framework of Life Cycle Design Theory for Concrete Structures

It is pointed out in *Unified standard for reliability design of building structures* (GB 50068) that safety, serviceability, and durability should be always satisfied in the service life of a structure. Under the environmental action and loads, the structural safety and serviceability will decrease gradually with time. Therefore, time-dependent degradation should be taken into consideration in life cycle design, i.e., essentially, durability is time-dependent safety and serviceability. Taking time-dependent safety as an example, based on the probabilistic model of time-dependent resistances and effects, probability of failure within the design life should be less than the allowed value as shown in Fig. 12.33.

In order to realize life cycle design in concrete structures, the following steps are proposed to calculate the failure probability: (1) determine the environmental effects according to locations of the structures. (2) Evaluate the corrosion time of steel bars

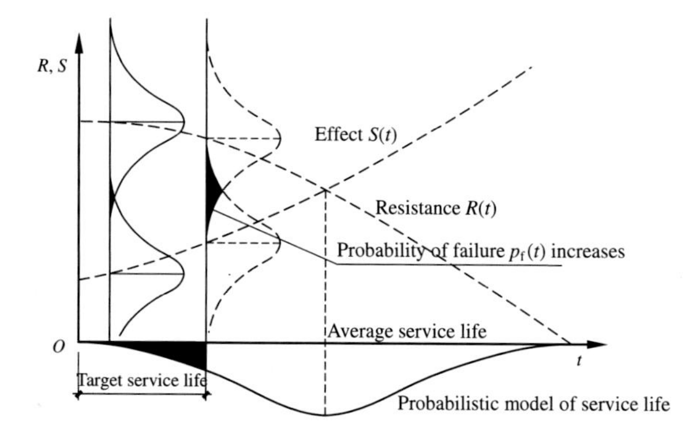

Fig. 12.33 Time-dependent probability of failure in a life cycle

t_0 based on the diffusion coefficients of CO_2 and chloride ions under load effects. (3) Determine the expansion cracking time of concrete cover t_1 by the corrosion rate model of steel bars before cracking and the prediction model of corrosion products to cause cracking. (4) Build the time-dependent resistance model of corroded RC members based on the prediction model of corrosion rate and degradation model of mechanical behavior of corroded members. (5) Divide the service life T into n equal times τ. At the same time, resistance $R(t)$ and load effect $S(t)$ are dispersed into n random variables $R(t_i)$, $S(t_i)$. Therefore, a time-dependent reliability calculation of a structure is transformed into a normal reliability calculation of a series system, as expressed by Eq. (12.63). To guarantee the service life of T years, $p_f(0, T)$ should be less than the failure threshold.

$$p_{\mathrm{f}}(0,T) = 1 - \Pr[G_{\tau 1}(X) > 0 \cap G_{\tau 2}(X) > 0 \cap \ldots \cap G_{\tau \mathrm{n}}(X) > 0] \qquad (12.63)$$

As an example, Fig. 12.34 gives the framework of life cycle design for reinforced concrete beams.

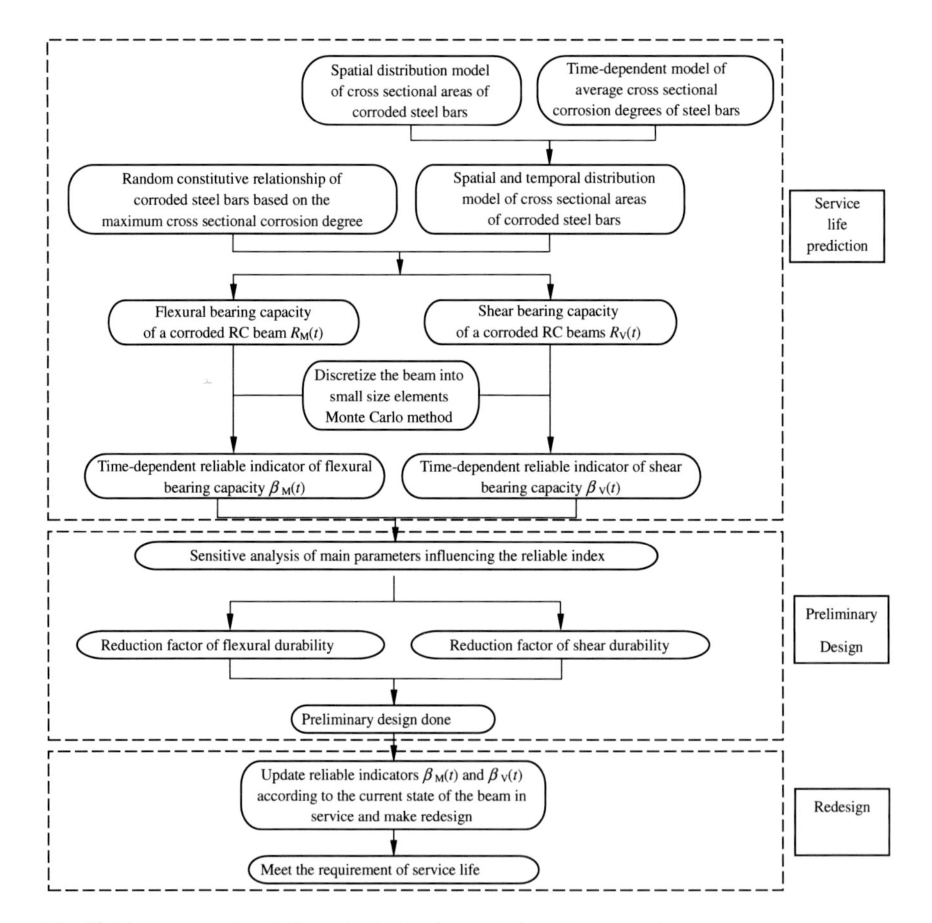

Fig. 12.34 Framework of life cycle design for a reinforced concrete beam

12.6.2 Durability Design

Various physical and numerical models have been developed for concrete carbonization, frost attack, chloride penetration, and bar corrosion, by which quantitative analysis can be carried out. Nevertheless, the quantitative durability design cannot be used in engineering practice, because the complex working environment and numerous factors, which include varying compositions of different concretes undergoing deterioration, result in big differences between the viewpoints and mechanism of the models. So far, no unified calculation method with sufficient accuracy has been developed to be used in real projects, and the durability of structures can only be guaranteed by some detailing measures. For instance, in GB 50010, structures are designed based on ultimate limit states and checked with serviceability limit states. For durability consideration, only some basic requirements are regulated in terms of the environment (See Table 11.1) and service life, including water/cement ratio, cement content, strength grade, alkali content, and chloride content (Table 12.1) together with cover depth and crack width mentioned in former chapters.

In severe conditions, some specific treatments can be applied to improve the durability of the concrete structures apart from the material/structural design and regulations mentioned above. These treatments include surface coating, adding corrosion resisting agents, employing electrochemical techniques, and using coated bars, corrosion-resistant bars, fiber-reinforced composites, etc.

Table 12.1 Regulations on durability of concrete structures

Environment	Maximum W/C ratio	Minimum strength grade	Maximum chloride content/%	Maximum alkali content/(kg·m^{-3})
I	0.60	C20	0.30	No limitation
IIa	0.55	C25	0.20	3.0
IIb	0.50 (0.55)	C30 (C25)	0.15	
IIIa	0.40 (0.50)	C35 (C30)	0.15	
IIIb	0.40	C40	0.10	

Notes (1) Chloride content refers to the percentage of its share in total cementitious material
(2) For pre-stressed concrete, the maximum chloride content is 0.05 %, the minimum concrete strength should be 2 grades higher than that in the table
(3) The water–cement ratio and minimum concrete strength for plain concrete members may be relaxed appropriately
(4) With reliable engineering experiences, the minimum concrete strength under second-class environment can be 1 grade lower than that in the table
(5) In cold areas, air entrainment should be adopted for the concrete members under IIb and IIIa environment. The parameters in parentheses can be used
(6) When using non-alkali reactive aggregate, the alkali content of the concrete may not be restricted

12.6.3 Durability Assessment for Existing Concrete Structures

Scientific durability assessment and service life prediction together with corresponding treatments to guarantee the safety and serviceability of existing structures have become another application field of durability study. Nowadays, more and more existing structures experience aging and simultaneously, numerous newly built structures in severe environment suffer shortening of service life. With the development of an accountable economy in the construction industry, requests for durability assessment and residual service life evaluation are arising from inspection of existing structures.

Durability assessment includes the following: (1) evaluation of durability grade and service life prediction of structures/members; (2) determining whether safety and serviceability can be satisfied in service life or the residual life can meet the demands of target service life under specified conditions and normal maintenance; and (3) measuring the minimum cost to extend service life and guarantee safety and serviceability in this period.

In durability assessment, a preliminary investigation should be carried out to determine whether detailed inspection is necessary, as shown in Fig. 12.35. This includes the following aspects of work.

1. Investigate the geological condition of the structure together with the design, construction, rehabilitation and maintenance history, retrofitting and extension history, accidents, treatments, and so on.
2. Investigate the structural function and history.

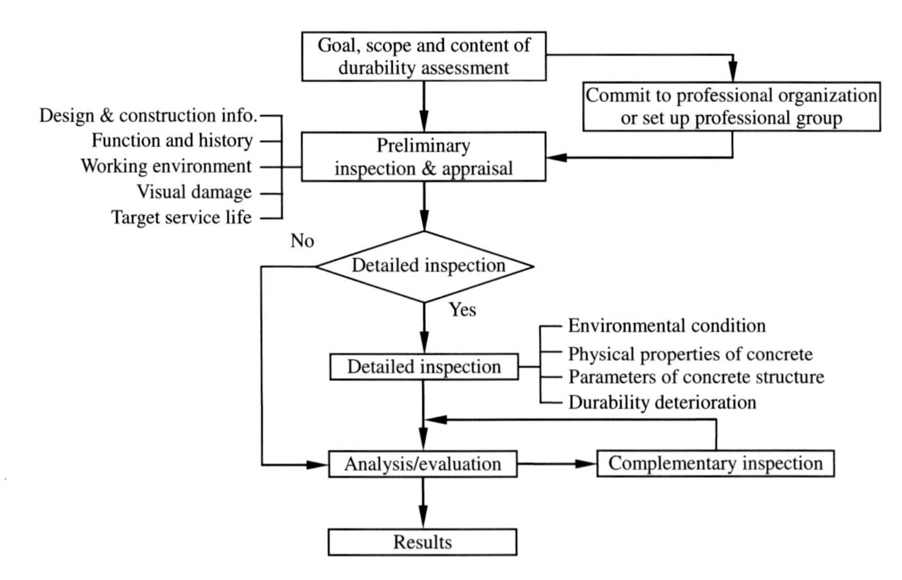

Fig. 12.35 Basic procedure of durability assessment

3. Investigate the working environment and protection measures of the structure.
4. Check for visual damage, examining structural systems, load transfer paths, and key joints; and run a preliminary analysis on important members or members in severe condition.
5. Investigate the target service life.
6. Make preliminary conclusion.

Durability assessment will be more complicated if time factor is included in the analysis. In order to simplify the work, assessment results can be given after checking the bearing capacity and serviceability of a structure. Approval can be given if the structures are (1) under stable mechanical loads and environmental actions, and if during its service life, the structure has worked well for a long time; and (2) in favorable environment with a service life of less than 10 years with no durability deterioration found in preliminary inspection. Otherwise, detailed inspection should be conducted with consideration of the durability deterioration with aging.

Durability assessment can be started from investigation of working environment, durability deterioration, and target service life. Then, if necessary, the assessment can proceed to a detailed inspection for some specific aspects. After all the basic information, including environmental conditions, physical properties of concrete, structural parameters, and durability deterioration, is obtained, the durability analysis and calculation can be done by the theories introduced in the previous sections with consideration of time, as shown in Fig. 12.35.

Questions

12.1 What are the mechanisms, influencing factors and consequences of concrete carbonization?

12.2 What are the mechanisms and consequences of concrete freeze–thaw activity?

12.3 What are the mechanisms and prevention methods of alkali–silica reaction?

12.4 List types of chemical agent attacks and their main forms.

12.5 What are the causes of steel bar corrosion and its effect on structural behavior?

12.6 Summarize the life prediction methods of concrete structures.

Appendix A
Basic Requirements of Experiments for Basic Principles of Concrete Structures

A.1 Objectives and Programs of Experiments

A.1.1 Objectives of Experiments

Understand the failure processes of concrete structural members through experimental studies, master the test methods for mechanical behavior of concrete members subjected to bending, compression and torsion loads, and deepen the understanding of basic principles of concrete structures. Train students' cognitive and practical abilities.

A.1.2 Experimental Programs

As an important part of experiments for civil engineering undergraduate students, experiments of *Basic Principles of Concrete Structures* are mainly composed of:

1. Experiments on normal sectional bending behavior of reinforced concrete flexural members;
2. Experiments on inclined sectional shear behavior of reinforced concrete flexural members;
3. Experiments on normal sectional compressive behavior of eccentrically loaded reinforced concrete members;
4. Experiments on torsional behavior of reinforced concrete members subjected to pure torsion.

X. Gu et al., *Basic Principles of Concrete Structures*,
DOI 10.1007/978-3-662-48565-1

A.2 Experiments on Normal Sectional Bending Behavior of Reinforced Concrete Flexural Members

A.2.1 Experimental Program

Test the normal sectional bending behavior of an under-reinforced beam, a lightly-reinforced beam and an over-reinforced beam under three-point concentrated loads, respectively (Fig. 5.8).

A.2.2 Basic Requirements

1. Design the experimental program and prepare the specimens;
2. Obtain the material properties of steel bars and concrete through coupon tests;
3. Measure and record geometrical parameters of the specimens;
4. Perform the beam tests, record the failure processes, and collect key data of the load and the deformation;
5. Summarize and analyze the test results, and propose your own understandings;
6. Calculate the relationships between moment and curvature ($M - \phi$) of the test beams by using the method presented in Sect. 5.5, and compare the results with the test data;
7. Calculate the normal sectional bearing capacities of the test beams by using the simplified method presented in Sect. 5.6, and compare the results with the test data;
8. Estimate the crack spaces and the widths of the test beams under different loads by using the method given in Chap. 11, and compare the results with the test data;
9. Estimate the deflections of the test beams under different loads by using the method given in Chap. 11, and compare the results with the test data;
10. Complete the experimental report.

A.3 Experiments on Inclined Sectional Shear Behavior of Reinforced Concrete Flexural Members

A.3.1 Experimental Program

Test the inclined sectional shear behavior of three beams with the shear span ratio of $1 \leqslant \lambda \leqslant 3$ and moderate reinforcement ratio of stirrups, shear span ratio of $\lambda < 1$, and shear span ratio of $\lambda > 3$ and very small reinforcement ratio of stirrups under symmetric concentrated loads, respectively (as shown in Fig. 7.2).

A.3.2 Basic Requirements

1. Design the experimental program and prepare the specimens;
2. Obtain the material properties of steel bars and concrete through coupon tests;
3. Measure and record geometrical parameters of the specimens;
4. Perform the beam tests, record the failure processes, and collect key data of the load and the deformation;
5. Summarize and analyze the test results, and propose your own understandings;
6. Calculate the inclined sectional bearing capacities of the test beams by using the method presented in Chap. 7, and compare the results with the test data;
7. Complete the experimental report.

A.4 Experiments on Normal Sectional Compressive Behavior of Eccentrically Loaded Reinforced Concrete Members

A.4.1 Experimental Program

Test the normal sectional compressive behavior of eccentrically loaded reinforced concrete members with a large eccentricity and a small eccentricity, as shown in Figs. 6.5 and 6.7, respectively.

A.4.2 Basic Requirements

1. Design the experimental program and prepare the specimens;
2. Obtain the material properties of steel bars and concrete through coupon tests;
3. Measure and record geometrical parameters of the specimens;
4. Perform the compressive tests, record the failure processes, and collect key data of the load and the deformation;
5. Summarize and analyze the test results, and propose your own understandings;
6. Calculate the normal sectional bearing capacities of the test members by using the method presented in Chap. 6, and compare the results with the test data;
7. Complete the experimental report.

A.5 Experiments on Torsional Behavior of Reinforced Concrete Members Subjected to Pure Torsion

A.5.1 Experimental Program

Test the torsional behavior of lightly-reinforced, under-reinforced and over-reinforced (or partially over-reinforced) concrete members, as shown in Fig. 8.3.

A.5.2 Basic Requirements

1. Design the experimental program and prepare the specimens;
2. Obtain the material properties of steel bars and concrete through coupon tests;
3. Measure and record geometrical parameters of the specimens;
4. Perform the torsion tests, record the failure processes, and collect key data of the load and the deformation;
5. Summarize and analyze the test results, and propose your own understandings;
6. Calculate the torsional bearing capacities by using the method presented in Chap. 8, and compare the results with the test data;
7. Complete the experimental report.

References

Beijing Shougang Co., Ltd, China Metallurgical Information and Standardization Institute, China Iron and Steel Research Institute Group (2010) GB/T 232-2010, Metallic materials-bent test. Standard Press of China, Beijing (in Chinese)
CCCC Highway Consultants Co., Ltd. (HPDI) (2004) JTG D62-2004, Code for design of highway reinforced concrete and prestressed concrete bridges and culverts. China Communications Press, Beijing (in Chinese)
Central Research Institute of Building and Construction, MCC Group Co., Ltd., Beijing Shougang Co., Ltd., Laiwu Iron and Steel Group Co., Ltd., et al (2008) GB1499.2-2007, Steel for the reinforcement of concrete-part 2: hot rolled ribbed bars. Standard Press of China, Beijing (in Chinese)
Changjiang Institute of Survey, Planning, Design and Research (2009) SL 191-2008, Design code for hydraulic concrete structures. China Water Power Press, Beijing (in Chinese)
Chen ZY, Zhu JQ, Wu PG (1992) High strength concrete and its applications. Tsinghua University Press, Beijing (in Chinese)
China Academy of Building Research (2001) GB50068-2001, Unified standard for reliability design of building structures. China architecture & Building Press, Beijing (in Chinese)
China Academy of Building Research (2002) GB50204-2002, Code for acceptance of constructional quality of concrete structures. China architecture & Building Press, Beijing (in Chinese)
China Academy of Building Research (2003) GB/T 50081-2002, Standard for test method of mechanical properties on ordinary concrete. China Architecture & Building Press, Beijing (in Chinese)
China Academy of Building Research (2006) JGJ 12-2006, Specification for design of lightweight aggregate concrete structures. China Architecture & Building Press, Beijing (in Chinese)
China Academy of Building Research (2009) GB/T50082-2009, Standard for test methods of long-term performance and durability ordinary concrete. China Architecture & Building Press, Beijing (in Chinese)
China Academy of Building Research (2009) JGJ92-2004, Technical specification for concrete structures prestressed with unbonded tendons. China Planning Press, Beijing (in Chinese)
China Academy of Building Research (2010) GB50010-2010, Code for design of concrete structures. China Architecture & Building Press, Beijing (in Chinese)
China Academy of Building Research (2012) GB50009-2012, Load code for the design of building structures. China Architecture & Building Press, Beijing (In Chinese)
China National Construction Steel Quality Supervision and Test Centre, Tianjin Tiantie Zhaer Steel Co., Ltd, Anshan Iron & Steel Group Corporation et al (2006) GB/T 20065-2006, Screw-thread steel bars for the prestressing of concrete. Standard Press of China, Beijing (in Chinese)

X. Gu et al., *Basic Principles of Concrete Structures*,
DOI 10.1007/978-3-662-48565-1

China National Construction Steel Quality Supervision and Test Centre, WuKun Steel Co. Ltd., China Metallurgical Information and Standardization Institute, et al (2008) GB1499.1-2008, Steel for the reinforcement of concrete-part 1: hot rolled plain bars. Standard Press of China, Beijing (in Chinese)

CI (2008) ACI 318-08 Building code requirements for structural concrete and commentary. American Concrete Institute, Detroit

Collins MP, Mitchell D (1991) Prestressed concrete structures. Prentice-Hall Inc., New Jersey

Cowan HJ (1977) A historical outline of architectural science. Applied Science Published Ltd., London

Ding DJ (1986) Crack resistance behavior, cracks and stiffness of reinforced concrete structures. Southeast University Press, Nanjing (in Chinese)

Ding DJ (1988) Concrete structures, vol 1. China Railway Publishing House, Beijing (in Chinese)

Editorial Group of Design of Partially Prestressed Concrete Structures (1985) Design of partially prestressed concrete structures (PPC suggestions for short). China Railway Publishing House, Beijing (in Chinese)

Fan JJ, Gao LD, Yu YY (1991) Reinforced concrete structures, vol 1. China Architecture & Building Press, Beijing (in Chinese)

Gu XL (1994) Shear capacities of reinforced concrete circular columns under seismic loadings. Struct Eng 10(4):6–10 (in Chinese)

Gu XL (1999) Prestressed FRP concrete structural systems. Eng Mech (Suppl):348–354 (in Chinese)

Gu XL, Sun FF (2002) Numerical simulation of concrete structures. Tongji University Press, Shanghai (in Chinese)

Gu XL, Zhang WP, Shang DF, et al (2010) Flexural behavior of corroded reinforced concrete beams. In: Proceedings of the 12th ASCE aerospace division international conference on engineering, science, construction, and operations in challenge environments and the 4th NASA/ASCE workshop on granular materials in lunar martin exploration, Honolulu, Hawaii, 14–17 Mar 2010, pp 3553–3558.

Hawkins NM (1968) The bearing strength of concretes loaded through rigid plates. Mag Concr Res 20(62):31–40

Lampert P, Collins MP (1972) Torsion, bending and confusion—an attempt to establish the facts. J ACI 69(8):500–504

Leonhardt F (1970) Shear and torsion in prestressed concrete. Eur Civil Eng 4:157–181

Leonhardt F, Monnlg E (1991) Design principles of reinforced concrete structures. China Communications Press, Beijing (in Chinese)

Li GP (2000) Design principles of prestressed concrete structures. China Communications Press, Beijing (in Chinese)

Li T, Liu XL (1999) Analysis and design of durability of concrete structures. Science Press, Beijing (in Chinese)

Lin TY, Burns NH (1981) Design of prestressed concrete structures, 3rd edn. Wiley, New York

Moe J (1961) Shearing strength of reinforced concrete slabs and footings under concentrated loads. Portland Cement Association, Development Department, Bulletin D47, PCA

Nilson AH, Winter G (1991) Design of concrete structures, 11th edn. Mc Graw-Hill Inc., New York

Niu DT (2003) Durability and life prediction of reinforced concrete structures. Science Press, Beijing (in Chinese)

Park R, Paulay T (1975) Reinforced concrete structures. Wiley, New York

Research group of torsional members (edited by Zheng ZQ, Wu YH) (1994) Experimental study on high strength concrete members under torsion. Selected research report on concrete structures (3). China Architecture & Building Press, Beijing (in Chinese)

Shanghai 3rd Steel Plant, Central Research Institute of Building and Construction, MCC Group Co., Ltd., China Metallurgical Information and Standardization Institute, et al (1991) GB13014-91, Remained heat treatment ribbed steel bars for the reinforcement of concrete. Standard Press of China, Beijing (in Chinese)

Shen PS, Liang XW (2007) Design principles of concrete structures, 3rd edn. Higher Education Press, Beijing (in Chinese)

Tarannath BS (1998) Steel, concrete, and composite design of tall buildings. McGraw-Hill Companies Inc, New York

Teng ZM (1987) Basic reinforced concrete members, 2nd edn. Tsinghua University Press, Beijing (in Chinese)

Tianjin 1st prestressed steel wire Co., Ltd., Tianjin Silvery Dragon Co., Ltd., Central Research Institute of Building and Construction, MCC Group Co., Ltd., Shanghai 3rd Steel Plant, et al (2002) GB/T 5223-2002, Steel wires for the prestressing of concrete. Standard Press of China, Beijing (in Chinese)

Tianjin 1st prestressed steel wire Co., Ltd., Xinhua Metal Products Co., Ltd., Zhuhai Hesheng Special Materials Co., Ltd., et al (2002) GB/T 5223.3-2005, Steel bars prestressed concrete. Standard Press of China, Beijing (in Chinese)

Tianjin 1st prestressed steel wire Co., Ltd., Xinhua Metal Products Co., Ltd., Tianjin Gaoli Yuyi Steel Strand Co, Ltd., et al (2003) GB/T 5224-2003, Steel strand for prestressed concrete. Standard Press of China, Beijing (in Chinese)

Tianjin University, Tongji University, Southeast University (1980) Reinforced concrete structures, vol 1. China Architecture & Building Press, Beijing (in Chinese)

Wang TC (2002) Theories of concrete structures. Tianjin University Press, Tianjin (in Chinese)

Wang CZ, Teng ZM (1985) Theories of reinforced concrete structures. China Architecture & Building Press, Beijing (in Chinese)

Wang YL, Yao Y (2001) Research and applications of concrete durability of key projects. China Building Materials Press, Beijing (in Chinese)

Water Transport Planning and Design Co. Ltd, China Communications Construction Company Limited (2011) JTS151-2011, Design code for concrete structures of port and waterway engineering. China Communications Press, Beijing (in Chinese)

Wight JK, MacGregor JG (2009) Reinforced concrete—mechanics and design, 5th edn. Pearson Prentice Hall, New Jersey

Wu J (1997) A history of Shanghai architecture (1840-1949). Tongji University Press, Shanghai (in Chinese)

Xu YL, Zhou D (2002) Understandings and applications of code for design of concrete structures. China Architecture & Building Press, Beijing (in Chinese)

Yam LCP (1991) Design of composite steel-concrete structures. Tongji University Press, Shanghai (in Chinese)

Yin ZL, Zhang Y, Wang ZD (1990) Torison. China Railway Publishing House, Beijing (in Chinese)

Yuan GG (1992) Design principles of reinforced concrete structures. Tongji University Press, Shanghai (in Chinese)

Zeng YH, Gu XL, Zhang WP et al (2010) Study on mechanical properties of corroded prestressed tendons. J Build Mater 13(2):169–174 (in Chinese)

Zeng YH, Huang QH, Gu XL et al (2010) Experimental study on bending behavior of corroded post-tensioned concrete beams. In: Proceedings of the 12th ASCE aerospace division international conference on engineering, science, construction, and operations in challenge environments and the 4th NASA/ASCE workshop on granular materials in lunar martin exploration, Honolulu, Hawaii, 14–17 Mar 2010, pp 3529–3536.

Zhang Y (2000) Basic principles of concrete structures. China Architecture & Building Press, Beijing (in Chinese)

Zhang Y, Jiang LX, Zhang WP et al (2003) Introduction to durability of concrete structures. Shanghai Science and Technology Press, Shanghai (in Chinese)

Zhang WP, Song XB, Gu XL et al (2012) Tensile and fatigue behavior of corroded rebars. Constr Build Mater 34:409–417

Zhang WP, Zhou BB, Gu XL et al (2014) Probability distribution model for cross-sectional area of corroded reinforcing steel bars. J Mater Civil Eng ASCE 26(5):822–832

Zhao GF (1991) Crack control of reinforced concrete structures. China Ocean Press, Beijing (in Chinese)

Zheng ZQ (1990) Punching shear capacities of reinforced concrete slabs with consideration of bend effects. In: Proceedings of 2nd symposium on theories and applications of concrete structures, Beijing (in Chinese)

Zhu BL (1992) Design principles of concrete structures, vol 1 & 2. Tongji University Press, Shanghai (in Chinese)

Printed in the United States
By Bookmasters